SURVEYING

(VOLUME I)

SURVEYING

(VOLUME I)

By

Dr. B.C. PUNMIA
Formerly,
Professor and Head, Deptt. of Civil Engineering, &
Dean, Faculty of Engineering
M.B.M. Engineering College,
Jodhpur

Er. ASHOK KUMAR JAIN
Director,
Arihant Consultants,
Jodhpur

Dr. ARUN KUMAR JAIN
Assistant Professor
M.B.M. Engineering College,
Jodhpur

EIGHTEENTH EDITION

(Thoroughly Revised and Enlarged)

LAXMI PUBLICATIONS (P) LTD
(An ISO 9001:2015 Company)
BENGALURU • CHENNAI • COCHIN • GUWAHATI • HYDERABAD
JALANDHAR • KOLKATA • LUCKNOW • MUMBAI • RANCHI • NEW DELHI

Printed and bound in India

First Edition: 1965; Seventeenth Edition : 2016, 2018

New Edition : 2025

ISBN : 978-81-7008-853-0

Branches

✆	Bengaluru	080-26 75 69 30
✆	Chennai	044-24 34 47 26
✆	Guwahati	0361-254 36 69
✆	Hyderabad	040-27 55 53 83
✆	Kochi	0484-405 13 03
✆	Kolkata	033-40 04 77 79
✆	Lucknow	0522-430 36 13
✆	Ranchi	0651-224 24 64

PUBLISHED IN INDIA BY

LAXMI PUBLICATIONS (P) LTD

An ISO 9001:2015 Company
113, GOLDEN HOUSE, GURUDWARA ROAD, DARYAGANJ,
NEW DELHI - 110002, INDIA

Corporate Office : B-33, Sector-5, Noida,
Gautambuddha Nagar (U.P.) - 201 301, INDIA
Telephone : +91-120-44 46 900
www.laxmipublications.com
info@laxmipublications.com

PREFACE

This volume is one of the two which offer a comprehensive course in those parts of theory and practice of Plane and Geodetic surveying that are most commonly used by civil engineers, and are required by the students taking examination in surveying for Degree, Diploma and A.M.I.E. The first volume covers in thirteen chapters the more common surveying operations.

Each topic introduced is thoroughly described, the theory is rigorously developed, and a large number of numerical examples are included to illustrate its application. General statements of important principles and methods are almost invariably given by practical illustrations. A large number of problems are available at the end of each chapter, to illustrate theory and practice and to enable the student to test his reading at different stages of his studies.

Apart from illustrations of old and conventional instruments, emphasis has been placed on new or improved instruments both for ordinary as well as precise work. A good deal of space has been given to instrumental adjustments with a thorough discussion of the geometrical principles in each case.

Metric system of units has been used throughout the text, and, wherever possible, the various formulae used in text have been derived in metric units. However, since the change over to metric system has still not been fully implemented in all the engineering institutions in our country, a few examples in F.P.S. system, have also been given.

I should like to express my thanks to M/s Vickers Instruments Ltd. (successors to M/s Cooke, Troughton & Simm's), M/s Wild Heerbrugg Ltd., M/s Hilger & Watts Ltd. and M/s W.F. Stanley & Co. Ltd. for permitting me to use certain illustrations from their catalogues or providing special photographs. My thanks are also due to various Universities and examining bodies of professional institution for permitting me to reproduce some of the questions from their examination papers.

Inspite of every care taken to check the numerical work, some errors may remain, and I shall be obliged for any intimation of theses readers may discover.

B.C. PUNMIA

JODHPUR

1st July, 1965

PREFACE TO THE THIRD EDITION

In this edition, the subject-matter has been revised thoroughly and the chapters have been rearranged. Two new chapters on 'Simple Circular Curves' and 'Trigonometrical Levelling (plane)' have been added. Latest Indian Standards on 'Scales', 'Chains' and 'Levelling Staff' have been included. A two-colour plate on the folding type 4 m Levelling Staff, conforming to IS 1779 : 1961 has been given. In order to make the book more useful to the students appearing at A.M.I.E. Examination in Elementary Surveying, questions from the examination papers of Section A, from May 1962 to Nov. 1970 have been given Appendix 2. Account has been taken throughout of the suggestions offered by the many users of the book, and grateful acknowledgement is made to them. Further suggestions will be greatly appreciated.

JODHPUR
1st Feb., 1972

B.C. PUNMIA

PREFACE TO THE FOURTH EDITION

In this edition, the subject-matter has been revised and updated. An appendix on 'Measurement of Distance by Electronic Methods' has been added.

JODHPUR
15-10-1975

B.C. PUNMIA

PREFACE TO THE FIFTH EDITION

In the Fifth Edition, the subject-matter has been thoroughly revised. An Appendix on SI units has been added.

JODHPUR
25-4-1978

B.C. PUNMIA

PREFACE TO THE SIXTH EDITION

In the Sixth Edition of the book, the subject-matter has been thoroughly revised and updated.

JODHPUR
2nd Jan., 1980

B.C. PUNMIA

PREFACE TO THE NINTH EDITION

In the Ninth Edition, the subject-matter has been revised and updated.

JODHPUR
1st Nov., 1984

B.C. PUNMIA

PREFACE TO THE TENTH EDITION

In the Tenth Edition, the book has been completely rewritten, and all the diagrams have been redrawn. Many new articles and diagrams/illustrations have been added. New instruments, such as precise levels, precise theodolites, precise plane table equipment, automatic levels, new types of compasses and clinometers etc. have been introduced. Two chapters on 'Setting Out Works' and 'Special Instruments' have been added at the end of the book. Knowledge about special instruments, such as site square, transit-level, Brunton's universal pocket transit, mountain compass-transit, automatic levels, etc. will be very much useful to the field engineers. Account has been taken throughout of the suggestions offered by the many users of the book, and grateful acknowledgement is made to them. Further suggestions will be greatly appreciated.

JODHPUR
10th July, 1987

B.C. PUNMIA
A.K. JAIN

PREFACE TO THE TWELFTH EDITION

In the Twelfth Edition, the subject matter has been revised and updated.

JODHPUR
30th March, 1990

B.C. PUNMIA
A.K. JAIN

PREFACE TO THE THIRTEENTH EDITION

In the Thirteenth Edition of the book, the subject matter has been thoroughly revised and updated. Many new articles and solved examples have been added. The entire book has been typeset using laser printer. The authors are thankful to Shri Mool Singh Gahlot for the fine laser typesetting done by him.

JODHPUR
15th Aug., 1994

B.C. PUNMIA
ASHOK K. JAIN
ARUN K. JAIN

PREFACE TO THE SIXTEENTH EDITION

In the Sixteenth Edition, the subject matter has been thoroughly revised, updated and rearranged. In each chapter, many new articles have been added. Three new Chapters have been added at the end of the book. Chapter 22 on 'Tacheometric Surveying', Chapter 23 on 'Electronic Theodolites' and Chapter 24 on 'Electromagnetic Distance Measurement (EDM)'. All the diagrams have been redrawn using computer graphics and the book has been computer type-set in bigger format keeping in pace with the modern trend. Account has been taken throughout of the suggestions offered by many users of the book and grateful acknowledgement is made to them. The authors are thankful to Shri M.S. Gahlot for the fine Laser type setting done by him. The Authors are also thankful Shri R.K. Gupta, Managing Director Laxmi Publications, for taking keen interest in publication of the book and bringing it out nicely and quickly.

JODHPUR
Mahaveer Jayanti
1st July, 2005

B.C. PUNMIA
ASHOK K. JAIN
ARUN K. JAIN

PREFACE TO THE SEVENTEENTH EDITION

In the Seventeenth Edition of this book, the subject matter has been revised and upgraded. This Edition has been resent in **Two-Colour** format. All the diagrams have been redrawn in double colour format so as to help the students in understanding the same.

It is hoped that this edition will be useful to the students, teachers and practising engineers.

All the suggestions offered by the users of the book have been seriously considered and the same have been gratefully acknowledged.

Any comments or suggestions from the users for the further improvement of this book will be gratefully acknowledged.

JODHPUR

ASHOK K. JAIN

PREFACE TO THE EIGHTEENTH EDITION

In the Eighteenth Edition, the subject-matter has been revised and updated.

JODHPUR

B.C. PUNMIA
A.K. JAIN

CONTENTS

CHAPTER 1

FUNDAMENTAL DEFINITIONS AND CONCEPTS

1.1 SURVEYING : OBJECT

Surveying is the art of *determining* the relative positions of points on, above or beneath the surface of the earth by means of direct or indirect measurements of distance, direction and elevation. It also includes the art of *establishing* points by predetermined angular and linear measurements. The application of surveying requires skill as well as the knowledge of mathematics, physics, and to some extent, astronomy.

Levelling is a branch of surveying the object of which is (*i*) to find the elevations of points with respect to a given or assumed datum, and (*ii*) to establish points at a given elevation or at different elevations with respect to a given or assumed datum. The first operation is required to enable the works to be designed while the second operation is required in the setting out of all kinds of engineering works. Levelling deals with measurements in a vertical plane.

The knowledge of surveying is advantageous in many phases of engineering. The earliest surveys were made in connection with land surveying. Practically, every engineering project such as water supply and irrigation schemes, railroads and transmission lines, mines, bridges and buildings etc. require surveys. Before plans and estimates are prepared, boundaries should be determined and the topography of the site should be ascertained. After the plans are made, the structures must be staked out on the ground. As the work progresses, lines and grades must be given.

In surveying, all measurements of lengths are horizontal, or else are subsequently reduced to horizontal distances. The object of a survey is to prepare plan or map so that it may represent the area on a horizontal plane. A plan or map is the horizontal projection of an area and shows only horizontal distances of the points. Vertical distances between the points are, however, shown by contour lines, hachures or some other methods. Vertical distances are usually represented by means of *vertical sections* drawn separately.

1.2 PRIMARY DIVISIONS OF SURVEY

The earth is an oblate spheroid of revolutions, the length of its polar axis (12,713,800 metres) being somewhat less than that of its equatorial axis (12,756,750 metres). Thus, the polar axis is shorter than the equatorial axis by 42.95 kilometres. Relative to the diameter of the earth this is less than 0.34 percent. If we neglect the irregularities of the earth, the surface of the imaginary spheroid is a curved surface, every element of which is normal to the plumb line. The intersection of such a surface with a plane passing through the centre of the earth will form a line continuous around the earth. The portion of such a line is known as *'level line'* and the circle defined by the intersection is known as *'great circle'*. Thus in Fig. 1.1, the distance between two points *P* and *Q* is the length of the arc of the great circle passing through these points and is evidently somewhat more than the chord intercepted by the arc.

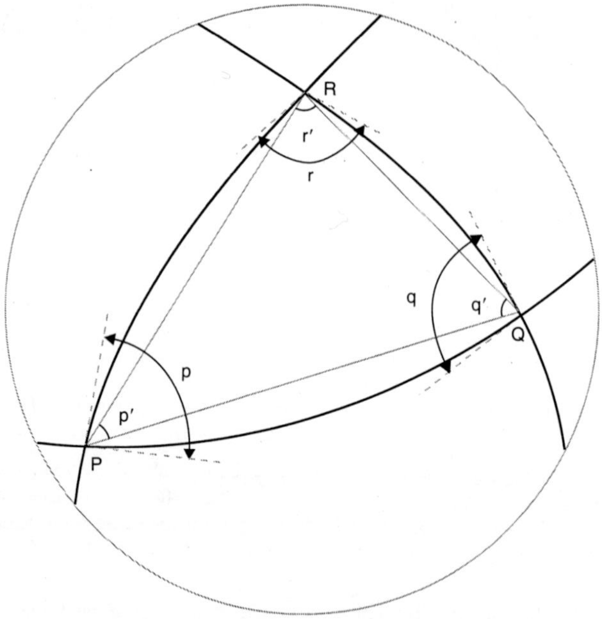

Fig. 1.1

Consider three points *P*, *Q* and *R* (Fig. 1. 1) and three level lines passing through these points. The surface within the triangle *PQR* so formed is a curved surface and the lines forming its sides are arcs of great circles. The figure is a *spherical triangle*. The angles *p*, *q* and *r* of the spherical triangle are somewhat more than corresponding angles *p′ q′* and *r′* of the plane triangle. If the points are far away, the difference will be considerable. If the points are nearer, the difference will be negligible.

As to whether the surveyor must regard the earth's surface as curved or may regard it is as plane depends upon the character and magnitude of the survey, and upon the precision required.

Thus, primarily, surveying can be divided into two classes:

1. *Plane Surveying* 2. *Geodetic Surveying*.

Plane surveying is that type of surveying in which the mean surface of the earth is considered as a plane and the spheroidal shape is neglected. All triangles formed by survey lines are considered as plane triangles. The level line is considered as straight and all plumb lines are considered parallel. In everyday life we are concerned with small portions of earth's surface and the above assumptions seem to be reasonable in light of the fact that the length of an arc 12 kilometres long lying in the earth's surface is only 1 cm greater than the subtended chord and further that the difference between the sum of the angles in a plane triangle and the sum of those in a spherical triangle is only one second for a triangle at the earth's surface having an area of 195 sq. km.

Geodetic surveying is that type of surveying in which the *shape* of the earth is taken into account. All lines lying in the surface are curved lines and the triangles are spherical triangles. It, therefore, involves spherical trigonometry. All geodetic surveys include work of larger magnitude and high degree of precision. *The object of geodetic survey is to determine the precise position on the surface of the earth, of a system of widely distant points which form control stations to which surveys of less precision may be referred.*

1.3 CLASSIFICATION

Surveys may be classified under headings which define the uses or purpose of the resulting maps.

(A) CLASSIFICATION BASED UPON THE NATURE OF THE FIELD SURVEY

1. Land Surveying

(i) ***Topographical Surveys:*** This consists of horizontal and vertical location of certain points by linear and angular measurements and is made to determine the natural features of a country such as rivers, streams, lakes, woods, hills, etc., and such artificial features as roads, railways, canals, towns and villages.

(ii) ***Cadastral Surveys:*** Cadastral surveys are made incident to the fixing of property lines, the calculation of land area, or the transfer of land property from one owner to another. They are also made to fix the boundaries of municipalities and of State and Federal jurisdictions.

(iii) ***City Surveying:*** They are made in connection with the construction of streets, water supply systems, sewers and other works.

2. Marine or Hydrographic Survey. Marine or hydrographic survey deals with bodies of water for purpose of navigation, water supply, harbour works or for the determination of mean sea level. The work consists in measurement of discharge of streams, making topographic survey of shores and banks, taking and locating soundings to determine the depth of water and observing the fluctuations of the ocean tide.

3. Astronomical Survey. The astronomical survey offers the surveyor means of determining the *absolute* location of any point or the absolute location and direction of any line on the surface of the earth. This consists in observations to the heavenly bodies such as the sun or any fixed star.

(B) CLASSIFICATION BASED ON THE OBJECT OF SURVEY

1. Engineering Survey. This is undertaken for the determination of quantities or to afford sufficient data for the designing of engineering works such as roads and reservoirs, or those connected with sewage disposal or water supply.

2. Military Survey. This is used for determining points of strategic importance.

3. Mine Survey. This is used for the exploring mineral wealth.

4. Geological Survey. This is used for determining different strata in the earth's crust.

5. Archaeological Survey. This is used for unearthing relics of antiquity.

(C) CLASSIFICATION BASED ON INSTRUMENTS USED

An alternative classification may be based upon the *instruments or methods* employed, the chief types being :

1. Chain survey
2. Theodolite survey
3. Traverse survey
4. Triangulation survey
5. Tacheometric survey
6. Plane table survey
7. Photogrammetric survey and
8. Aerial survey.

The book mainly deals with the principles and methods of the above types.

1.4 PRINCIPLES OF SURVEYING

The fundamental principles upon which the various methods of *plane surveying* are based are of very simple nature and can be stated under the following two aspects :

1. **Location of a point by measurement from two points of reference.** The relative positions of the points to be surveyed should be located by measurement from at least two points of reference, the positions of which have already been fixed. Let P and Q be the reference points on the ground. The distance PQ can be measured accurately and the relative positions of P and Q can be plotted on the sheet to some scale. The points P and Q will thus serve as reference points for fixing the relative positions of other points. *Any other point, such as R, can be located by any of the following direct methods* (Fig. 1.2)

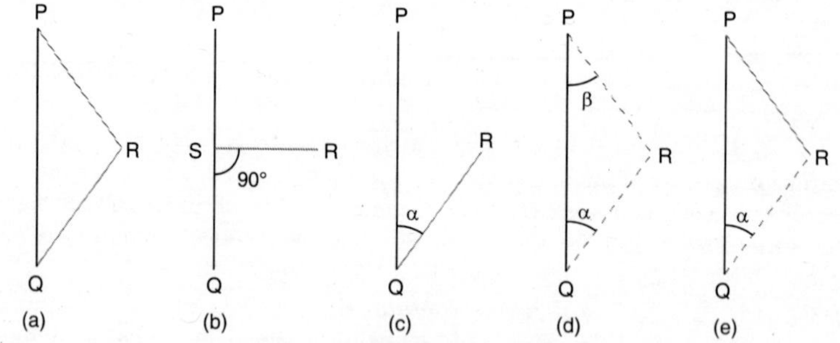

Fig. 1.2 Location of A point

(*a*) Distances PR and QR can be measured and point R can be plotted by swinging the two arcs to the same scale to which PQ has been plotted. *The principle is very much used in chain surveying.*

(*b*) A perpendicular RS can be dropped on the reference line PQ and lengths PS and SR are measured. The point R can then be plotted using set square. *This principle is used for defining details.*

(*c*) The distance QR and the angle PQR can be measured and point R is plotted either by means of a protractor or trigonometrically. *This principle is used in traversing.*

(*d*) In this method, the distances PR and QR are not measured but angle RPQ and angle RQP are measured with an angle-measuring instrument. Knowing the distance PQ, point R is plotted either by means of a protractor or by solution of triangle PQR. *This principle is very much used in triangulation and the method is used for very extensive work.*

(*e*) Angle RQP and distance PR are measured and point R is plotted either by protracting an angle and swinging an arc from P or plotted trigonometrically. *This principle, used in traversing, is of minor utility.*

Figs. 1.2 (b), (c) and (d) can also be used to illustrate the principles of determining relative elevations of points. Considering these diagrams to be in vertical plane, with PQ as horizontal. *Fig.1.2 (b) represents the principle of ordinary spirit levelling.* A horizontal line PQ is instrumentally established through P and the vertical height of R is measured by taking staff reading. *Similarly, Fig. 1.2 (c) and (d) represent the principles of trigonometrical levelling.*

2. **Working from whole to part.** The second ruling principle of surveying, whether plane or geodetic, is to work from whole to part. It is very essential to establish first a system of control points and to fix them with higher precision. Minor control points can then be established by less precise methods and the details can then be located using these minor control points by running minor traverses etc. The idea of working in this way is to prevent the accumulation of errors and to control and localise minor errors which, otherwise, would expand to greater magnitudes if the reverse process is followed, thus making the work uncontrollable at the end.

1.5 UNITS OF MEASUREMENTS

There are four kinds of measurements used in plane surveying :

1. Horizontal distance
2. Vertical distance
3. Horizontal angle, and
4. Vertical angle.

Linear measures. According to the Standards of Weights and Measures Act (India), 1956 the unit of measurement of distance is metres and centimetres. Prior to the introduction of metric units in India, feet, tenths and hundredths of a foot were used. Table 1.1 gives the basic linear measures, both in metric as well as in British system. while Tables 1 .2 and 1.3 give the conversion factors.

TABLE 1.1 *Basic Units of Length*

British Units	*Metric Units*
12 inches = 1 foot	10 millimetres = 1 centimetre
3 feet = 1 yard	10 centimetres = 1 decimetre
$5\frac{1}{2}$ yards = 1 rod, pole or perch	10 decimetres = 1 metre
4 poles = 1 chain (66 feet)	10 metres = 1 decametre
10 chains = 1 furlong	10 decametres = 1 hectometre
8 furlongs = 1 mile	10 hectometres = 1 kilometre
100 links = 1 chain = 66 feet	1852 metres = 1 nautical mile (International)
6 feet = 1 fathom	
120 fathoms = 1 cable length	
6080 feet = 1 nautical mile	

TABLE 1.2 *Conversion Factors*
(Metres, yards, feet and inches)

Metres	*Yards*	*Feet*	*Inches*
1	1.0936	3.2808	39.3701
0.9144	1	3	36
0.3048	0.3333	1	12
0.0254	0.0278	0.0833	1

TABLE 1.3 *Conversion Factors*
(Kilometres, Nautical miles and Miles)

Kilometres	*Nautical miles*	*Miles*
1	0.53996	0.6214
1.852	1	1.1508
1.6093	0.869	1

Basic units of area. The units of measurements of area are sq. metres, sq. decimetres, hectares and sq. kilometres. Table 1.4 gives the units of area both in metric as well as British systems. Tables 1.5 and 1.6 gives the conversion factors.

TABLE 1.4 *Basic Units of Area*

British Units		*Metric Units*	
144 sq. inches	= 1 sq. foot	10 sq. millimetres	= 1 sq. centimetre
9 sq. feet	= 1 sq. Yard	100 sq. centimetres	= 1 sq. decimetre
$30\frac{1}{4}$ sq. yards	= 1 sq. rod, pole or perch	100 sq. decimetres	= 1 sq. metre
40 sq. rods	= 1 rood	100 sq. metres	= 1 are or 1 sq. decametre
4 roods	= 1 acre	100 ares	= 1 hectare or 1 sq. hectometre
640 acres	= 1 sq. mile	100 hectares	= 1 sq. kilometre
484 sq. yards	= 1 sq. chain		
10 sq. chains	= 1 acre		

TABLE 1.5 *Conversion Factors*
(Sq. metres, Sq. yards, Sq. feet and Sq. inches)

Sq. metres	*Sq. yards*	*Sq. feet*	*Sq. inches*
1	1.196	10.7639	1550
0.8361	1	9	1296
0.0929	0.1111	1	144
0.00065	0.00077	0.0069	1

TABLE 1.6 *Conversion Factors*
(Ares, Acres and sq. yards)

Ares	*Acres*	*Sq. yards*
1	0.0247	119.6
40.469	1	4840
0.0084	0.00021	1

1 sq. mile = 640 acres = 258.999 hectares
1 acre = 10 sq. chains
1 are = 100 sq. metres

Basic units of volume. The units of measurements of volumes are cubic decimetres and cubic metres. Table 1.7 gives the basic units of measurement of volumes both in metric as well as British units. Tables 1.8 and 1.9 give the conversion factors.

TABLE 1.7 *Basic Units of Volume*

British Units		***Metric Units***	
1728 cu. inches	= 1 cu. foot	1000 cu. millimetres	= 1 cu. centimetre
27 cu. feet	= 1 cu. yard	1000 cu. centimetres	= 1 cu. decimetre
		1000 cu. decimetres	= 1 cu.metre

TABLE 1.8 *Conversion Factors*
(Cu. metres, Cu. yards and Imp. gallons)

Cu. metres	***Cu. yards***	***Gallons (Imp.)***
1	1.308	219.969
0.7645	1	168.178
0.00455	0.00595	1

TABLE 1.9 *Conversion Factors*
(Cubic metres, Acre feet, Imp. Gallons and Kilolitres)

Cu. metres	***Acre feet***	***Gallons (Imp.)***	***Kilolitres***
1	0.000811	219.969	0.99997
1233.48	1	271327	1233.45
0.00455	0.00000369	1	0.00455
1.000028	0.000811	219.976	1

Basic units of angular measure. An angle is the difference in direction of two intersecting lines. The *radian* is the unit of plane angle. The radian is the angle between two radii of a circle which cuts-off on the circumference of an arc equal in length to the radius. There are three popular systems of angular measurements:

(*a*) *Sexagesimal System*

1 circumference = 360° (degrees of arc)
1 degree = 60′ (minutes of arc)
1 minute = 60″ (seconds of arc)

(*b*) *Centesimal System*

1 circumference = 400^g (grads)
1 grad = 100^c (centigrads)
1 centigrad = 100^{cc} (centi-centigrads)

(*c*) *Hours System*

1 circumference = 24^h (hours).
1 hour = 60^m (minutes of time)
1 minute = 60^s (seconds of time).

The sexagesimal system is widely used in United States, Great Britain, India and other parts of the World. More complete tables are available in this system and most surveying instruments are graduated according to this system. However, due to facility in computation and interpolation, the centesimal system is gaining more favour in Europe. The hours system is mostly used in astronomy and navigation.

1.6 PLANS AND MAPS

A plan is the graphical representation, to some scale, of the features on, near or below the surface of the earth as projected on a horizontal plane which is represented by plane of the paper on which the plan is drawn. However, since the surface of the earth is curved and the paper of the plan or map is plane, no part of the surface can be represented on such maps without distortion. In plane surveying, the areas involved are small, the earth's surface may be regarded as plane and hence map is constructed by orthographic projection without measurable distortion.

The representation is called a *map* if the scale is small while it is called a *plan* if the scale is large. On a plan, generally, only horizontal distances and directions are shown. On a topographic map, however, the vertical distances are also represented by contour lines, hachures or other systems.

1.7 SCALES

The area that is surveyed is vast and, therefore, plans are made to some scale. *Scale is the fixed ratio that every distance on the plan bears with corresponding distance on the ground.* Scale can be represented by the following methods :

1. One cm on the plan represents some whole number of metres on the ground, such as 1 cm = 10 m etc. This type of scale is called *engineer's scale*.
2. One unit of length on the plan represents some number of same units of length on the ground, such as $\frac{1}{1000}$, etc. This ratio of map distance to the corresponding ground distance is independent of units of measurement and is called *representative fraction.* The representative fraction (abbreviated as R.F.) can be very easily found for a given engineer's scale. For example, if the scale is 1 cm = 50 m

$$\text{R.F.} = \frac{1}{50 \times 100} = \frac{1}{5000}$$

The above two types of scales are also known as *numerical scales.*

3. An alternative way of representing the scale is to draw on the plan a *graphical scale.* A graphical scale is a line sub-divided into plan distance corresponding to convenient units of length on the ground.

If the plan or map is to be used after a few years, the numerical scales may not give accurate results if the sheet or paper shrinks. However, if a graphical scale is also drawn, it will shrink proportionately and the distances can be found accurately. *That is why, scales are always drawn on all survey maps.*

Choice of Scale of a Map

The most common scales for ordinary maps are those in which the number of metres represented by one centimetre is some multiple of ten. The preliminary consideration in choosing the scale are : (1) the use to which the map will be put, and (2) the extent of territory to be represented. The following two general rules should be followed :

1. Choose a scale large enough so that in plotting or in scaling distance from the finished map, it will not be necessary to read the scale closer than 0.25 mm.
2. Choose as small a scale as is consistent with a clear delineation of the smallest details to be plotted. Table 1.10 gives the common scales generally used in various surveys.

TABLE 1.10

Type or purpose of survey	*Scale*	*R.F.*
(*a*) *Topographic Survey*		
1. Building sites	1 cm = 10 m or less	$\frac{1}{1000}$ or less
2. Town planning schemes, reservoirs etc.	1 cm = 50 m to 100 m	$\frac{1}{5000}$ to $\frac{1}{10000}$
3. Location surveys	1 cm = 50 m to 200 m	$\frac{1}{5000}$ to $\frac{1}{20000}$
4. Small scale topographic maps	1 cm = 0.25 km to 2.5 km	$\frac{1}{25000}$ to $\frac{1}{250000}$
(*b*) *Cadastral maps*	1 cm = 5 m to 0.5 km	$\frac{1}{500}$ to $\frac{1}{5000}$
(*c*) *Geographical maps*	1 cm = 5 km to 160 km	$\frac{1}{500000}$ to $\frac{1}{16000000}$
(*d*) *Longitudinal sections*		
1. Horizontal scale	1 cm = 10 m to 200 m	$\frac{1}{1000}$ to $\frac{1}{20000}$
2. Vertical scale	1 cm = 1 m to 2 m	$\frac{1}{100}$ to $\frac{1}{200}$
(*e*) *Cross-sections* (Both horizontal and vertical scales equal)	1 cm = 1 m to 2 m	$\frac{1}{100}$ to $\frac{1}{200}$

Types of Scales

Scales may be classified as follows :

1. Plain scale
2. Diagonal scale
3. Vernier scale
4. Scale of chords.

1.8 PLAIN SCALE

A plain scale is one on which it is possible to measure two dimensions only, such as units and lengths, metres and decimetres, miles and furlongs, etc.

Example 1.1: *Construct a plain scale 1 cm to 3 metres and show on it 47 metres.*

Solution: Take a 20 cm length and divide it into 6 parts, each representing 10 metres. Subdivide the first left hand division into 10 equal parts, each reading 1 metre. Place zero of the scale between the subdivided parts and the undivided part and mark the scale as shown in Fig. 1.3. To take 47 metres, place one leg of the divider at 40 and the other at 7, as shown in Fig. 1.3.

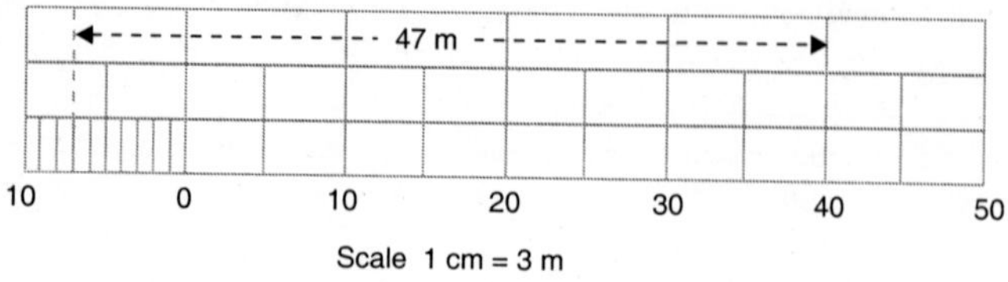

Fig. 1.3 Plain Scale

Indian Standard on plain scales

IS : 1491-1959 has recommended six different plain scales in metric units used by engineers, architects and surveyors. The scale designations along with their R.F. are given in the table below:

Designation	*Scale*	*R.F.*
A	1. Full size	$\frac{1}{1}$
	2. 50 cm to a metre	$\frac{1}{2}$
B	3. 40 cm to a metre	$\frac{1}{2.5}$
	4. 20 cm to a metre	$\frac{1}{5}$
C	5. 10 cm to a metre	$\frac{1}{10}$
	6. 5 cm to a metre	$\frac{1}{20}$
D	7. 2 cm to a metre	$\frac{1}{50}$
	8. 1 cm to a metre	$\frac{1}{100}$
E	9. 5 mm to a metre	$\frac{1}{200}$
	10. 2 mm to a metre	$\frac{1}{500}$
F	11. 1 mm to a metre	$\frac{1}{1000}$
	12. 0.5 mm to a metre	$\frac{1}{2000}$

1.9 DIAGONAL SCALE

On a diagonal scale, it is possible to measure three dimensions such as metres, decimetres and centimetres; units, tenths and hundredths; yards, feet and inches etc. A short length is divided into a number of parts by using the principle of similar triangles in which like sides are proportional. For example let a short length PQ be divided into 10 parts (Fig. 1.4). At Q draw a line QR perpendicular to PQ and of any convenient length. Divide it into ten equal parts. Join the diagonal PR. From each of the divisions, 1, 2, 3 etc., draw lines parallel to PQ to cut the diagonal in corresponding points 1, 2, 3 etc., thus dividing the diagonal into 10 equal parts.

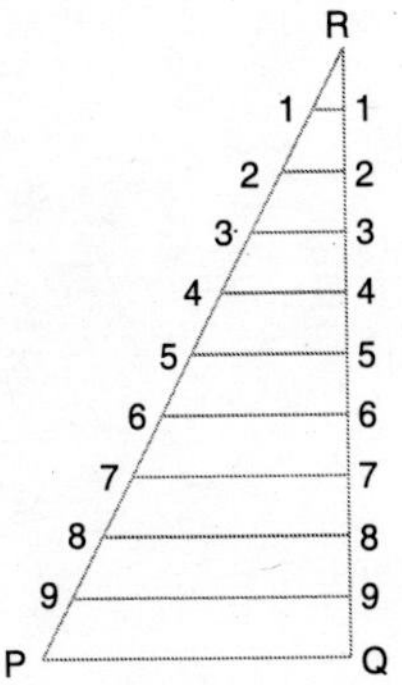

Fig. 1.4

Thus,

$$\text{1-1 represents } \frac{1}{10}\, PQ$$

$$\text{2-2 represents } \frac{2}{10}\, PQ$$

....

$$\text{9-9 represents } \frac{9}{10}\, PQ \text{ etc.}$$

Examples 1.2: *Construct a diagonal scale 1 cm = 3 metres to read metres and decimetres and show on that 33.3 metres.*

Solution: Take 20 cm length and divide it into 6 equal parts, each part representing 10 metres. Sub-divide the first left hand part into 10 divisions, each representing 1 metre. At the left of the first sub-division erect a perpendicular of any suitable length (say 5 cm) and divide it into 10 equal parts and draw through these parts lines parallel to the scale. Sub-divide the top parallel line into ten divisions (each representing 1 metre) and join these diagonally to the corresponding sub-divisions on the first parallel line as shown in Fig. 1.5 where a distance of 33.3 metres has been marked.

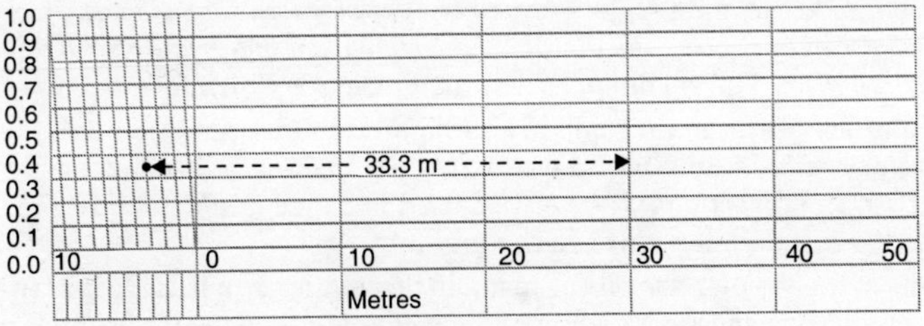

Scale 1 cm = 3 m

Fig. 1.5 Diagonal Scale

Indian Standard on diagonal scales

IS : 1562-1962 recommends four diagonal scales A, B, C and D, as shown in the table below :

Designation		*R.F.*	*Graduated length*
A		$\frac{1}{1}$	150 cm
B	1.	$\frac{1}{100000}$	100 cm
	2.	$\frac{1}{50000}$	
	3.	$\frac{1}{25000}$	
C	1.	$\frac{1}{100000}$	50 cm
	2.	$\frac{1}{50000}$	
	3.	$\frac{1}{25000}$	
D	1.	$\frac{1}{100000}$	150 cm
	2.	$\frac{1}{8000}$	
	3.	$\frac{1}{4000}$	

1.10 THE VERNIER

The vernier, invented in 1631 by Pierre Vernier, is a device for measuring the fractional part of one of the smallest divisions of a graduated scale. It usually consists of a small auxiliary scale which slides along side the main scale. *The principle of vernier is based on the fact that the eye can perceive without strain and with considerable precision when two graduations coincide to form one continuous straight line.* The vernier carries an index mark which forms the zero of the vernier.

If the graduations of the main scale are numbered in one direction only, the vernier used is called a *single vernier*, extending in one direction. If the graduations of the main scale are numbered in both the directions, the vernier used is called *double vernier*, extending in both the directions, having its index mark in the middle.

The divisions of the vernier are either just a little smaller or a little larger than the divisions of the main scale. The *fineness* of reading or *least count* of the vernier is equal to the difference between the smallest division on the main scale and smallest division on the vernier.

Whether single or double, a vernier can primarily be divided into the following two classes :

(*a*) Direct Vernier

(*b*) Retrograde Vernier.

(a) Direct Vernier. A direct vernier is the one which extends or increases in the same direction as that of the main scale and in which the smallest division on the vernier is shorter than the smallest division on the main scale. It is so constructed that $(n - 1)$ divisions of the main scale are equal in length of n divisions of the vernier.

Let s = Value of one smallest division on main scale.

v = Value of one smallest division on the vernier.

n = Number of divisions on the vernier.

Since a length of $(n - 1)$ divisions of main scale is equal to n divisions of vernier, we have

$$nv = (n - 1)\, s$$

$$\therefore \quad v = \left(\frac{n-1}{n}\right) s$$

$$\text{Least count} = s - v = s - \frac{n-1}{n} s = \frac{s}{n}.$$

Thus, the least count (L.C.) can be found by dividing the value of one main scale division by the total number of divisions on the vernier.

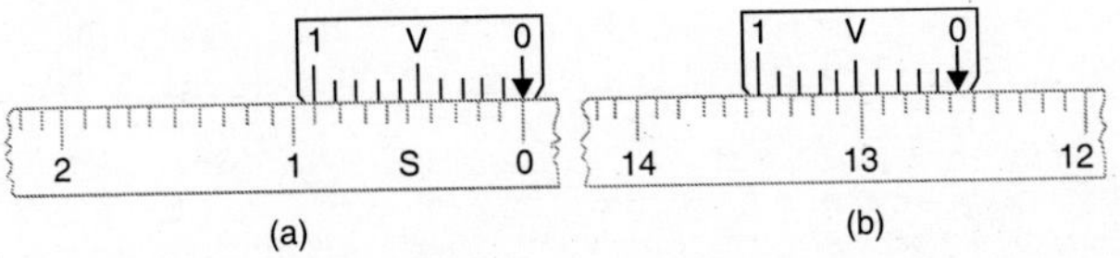

Fig. 1.6 Direct Vernier Reading to 0.01.

Fig. 1.6(a) shows a direct vernier in which 9 parts of the main scale divisions coincide with 10 parts of the vernier. The total number of the divisions on the vernier are 10 and the value of one main scale division is 0.1. The least count of the vernier is therefore, $\frac{0.1}{10} = 0.01$. The reading on the vernier [Fig. 1.6(b)] is 12.56.

Fig. 1.7 (a) shows a double vernier (direct type) in which the main scale is figured in both the directions and the vernier also extends to both the sides of the index mark.

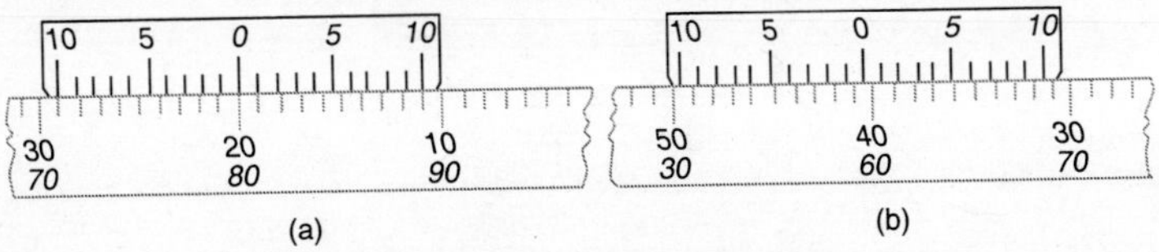

Fig. 1.7 Double Vernier (Direct).

The 10 spaces on either half of the vernier are equivalent to 9 scale divisions and hence least count is $\frac{s}{n} = \frac{1}{10} = 0.1$. The left-hand vernier is used in conjunction with the upper figures on the main scale (those sloping to the left) and the right-hand vernier is used in conjunction with the lower figures on the scale (those sloping to the right). Thus, in Fig. 1.7 (b), the reading on the left vernier is 40.6 and on the right vernier is 59.4.

(b) Retrograde Vernier. A retrograde vernier is the one which extends or increases in opposite direction as that of the main scale and in which the smallest division of the vernier is longer

than the smallest division on the main scale. It is so constructed that $(n + 1)$ divisions of the main scale are equal in length of n divisions of the vernier.

Thus, we have, for this case

$$nv = (n + 1)\, s \text{ ; or } v = \frac{n+1}{n} s$$

$$\text{The least count} = v - s = \left(\frac{n+1}{n}\right) s - s = \frac{s}{n}$$

which is the same as before.

Fig. 1.8 (*a*) illustrates a retrograde vernier in which 11 parts of the main scale divisions coincide with 10 divisions of the vernier. The value of one smallest division on the main scale is 0.1 and the number of division on the vernier are 10. Therefore, the least count is = $\frac{0.1}{10}$ = 0.01. The reading on the vernier [Fig.1.8 (*b*)] is 13.34.

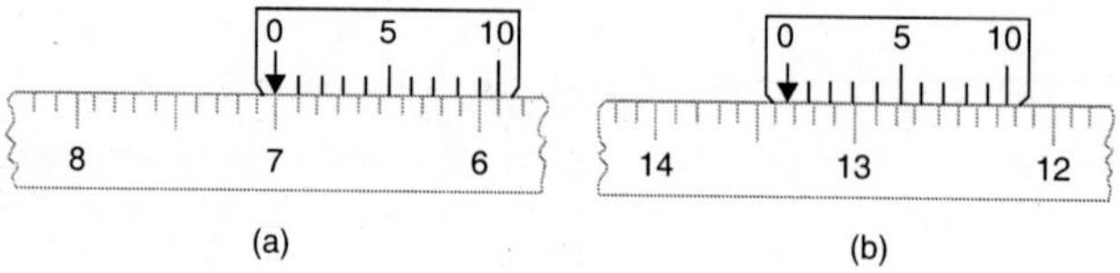

Fig. 1.8 Retrograde Vernier.

Special Forms of Verniers

The Extended Vernier. It may happen that the divisions on the main scale are very close and it would then be difficult, if the vernier were of normal length, to judge the exact graduation where coincidence occurred. In this case, an extended vernier may be used.

Here $(2n - 1)$ divisions on the main scale are equal to n divisions on the vernier, so that

$$nv = (2n - 1)\, s$$

or
$$v = \frac{2n-1}{n} s = \left(2 - \frac{1}{n}\right) s$$

The difference between two main scale spaces and one vernier space

$$= 2s - v = 2s - \frac{2n-1}{n} s = \frac{s}{n} = \text{least count.}$$

The extended vernier is, therefore, equivalent to a simple direct vernier to which only every second graduation is engraved. The extended vernier is regularly employed in the astronomical sextant. Fig. 1.9 shows an extended vernier. It has 6 spaces on the vernier equal to 11 spaces of the main scale each of 1°. The least count is therefore = $\frac{1}{6}$ degree = 10′.

The reading on the vernier illustrated in Fig. 1.9(*b*) is 3° 20′ and that in Fig. 1.9(*c*) is 2° 40′.

In the case of astronomical sextant, the vernier generally provided is of extended type having 60 spaces equal to 119 spaces of the main scale, each of 10′, the least count being $\frac{10}{60}$ minutes or 10 seconds.

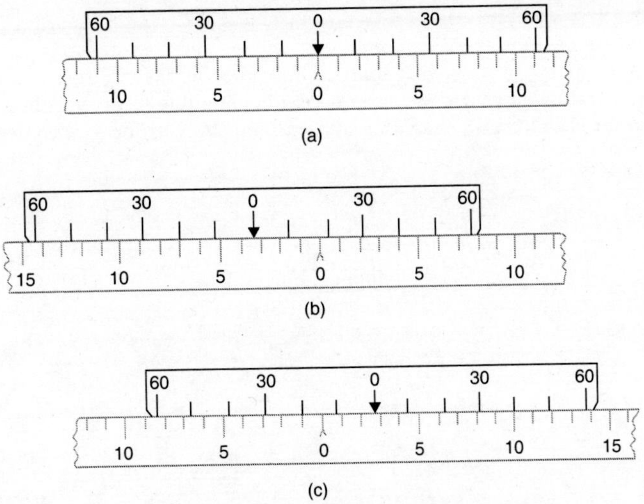

Fig. 1.9 Extended Vernier

The Double Folded Vernier. The double folded vernier is employed where the length of the corresponding double vernier would be so great as to make it impracticable. This type of vernier is sometimes used in compasses having the zero in the middle of the length. The full length of vernier is employed for reading angles in either direction. The vernier is read from the index towards either of the extreme divisions and then from the other extreme division in the same direction to the centre.

Fig. 1.10 shows double folded vernier in which 10 divisions of vernier are equal to $9\frac{1}{2}$ divisions of the main scale (or 20 vernier divisions = 19 main scale divisions). The least count of the vernier is equal to $\frac{s}{n} = \frac{1}{20}$ degrees = 3′. For motion to the right, the vernier is read from 0 to 30 at the right extremity and then from 30 at the left extremity to 60 (or zero) at the centre. Similarly,

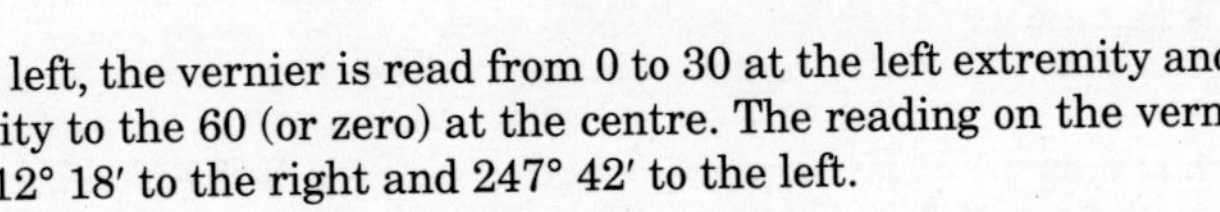

Fig. 1.10 Double-Folded Vernier.

for motion to the left, the vernier is read from 0 to 30 at the left extremity and then from 30 at the right extremity to the 60 (or zero) at the centre. The reading on the vernier illustrated in Fig. 1.10 (*b*) is 112° 18′ to the right and 247° 42′ to the left.

Verniers to Circular Scales

The above examples of verniers were for linear scales. Verniers are also extensively used to circular scales in a variety of surveying instruments such as theodolites, sextants, clinometers etc. Fig. 1.11 (*a*), (*b*) shows two typical arrangements of double direct verniers. In Fig. 1.11 (*a*), the scale is graduated to 30′ and the value of n = 30 on the vernier. Hence, least count = s/n = 30′/30 = 1′.

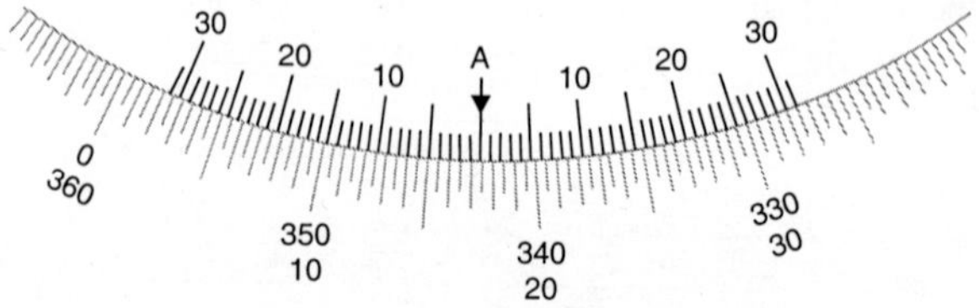

(A) Graduated to 30° : Reading to 1′

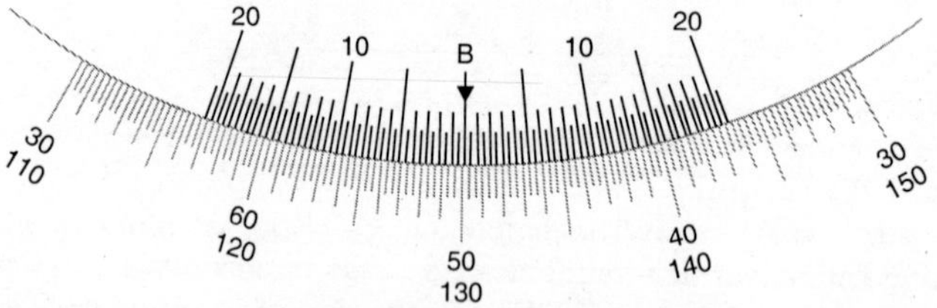

(B) Graduated to 20′ : Reading to 30″

Fig. 1.11 Verniers to Circular Scales

In Fig. 1.11 (*b*), the scale is graduated to 20 minutes, and the number of vernier divisions are 40.

Hence, least count = s/n = 20′/40 = 0.5′ = 30″.

Thus, in Fig. 1.11 (*a*), the clockwise angle reading (inner row) is 342° 30′ + 05′ = 342° 35′ and counter clockwise angle reading (outer row) is 17° 0′ + 26′ = 17° 26′. Similarly, in Fig. 1.11 (*b*), the clockwise angle reading (inner row) is 49° 40′ + 10′ 30″ = 49° 50′ 30″ and the counter clockwise angle (outer row) is 130° 00′ + 9′ 30″ = 130° 09′ 30″. *In both the cases, the vernier is always read in the same direction as the scale.*

Examples on Design of Verniers

Examples 1.3: Design a vernier for a theodolite circle divided into degrees and half degrees to read up to 30″.

Solution: Least Count = $\frac{s}{n}$; s = 30′, L.C. = 30″ = $\frac{30}{60}$ minutes

$$\therefore \quad \frac{30}{60} = \frac{30}{n} \text{ ; or } n = 60.$$

Fifty-nine such primary divisions should be taken for the length of the vernier scale and then divided into 60 parts for a direct vernier.

Example 1.4: *Design a vernier for a theodolite circle divided into degrees and one-third degrees to read to 20″.*

Solution:

$$\text{L.C.} = \frac{s}{n}\ ;\ s = \frac{1°}{3} = 20';\ \text{L.C.} = 20'' = \frac{20}{60}\ \text{minutes}$$

$$\therefore \quad \frac{20}{60} = \frac{20}{n} \ \text{ or } n = 60$$

Fifty-nine divisions should be taken for the length of the vernier scale and divided into 60 parts for a direct vernier.

Example 1.5: *The value of the smallest division of circle of a repeating theodolite is 10′. Design a suitable vernier to read up to 10″.*

Solution:

$$\text{L.C.} = \frac{s}{n}\ ;\ s = 10';\ \text{L.C.} = 10'' = \frac{10}{60}\ \text{minutes}$$

$$\therefore \quad \frac{10}{60} = \frac{10}{n}\ ;\ \text{or } n = 60$$

Take 59 such primary divisions from the main scale and divide it into 60 parts.

Example 1.6: *The circle of a theodolite is divided into degrees and 1/4 of a degree. Design a suitable decimal vernier to read up to 0.005°.*

Solution:

$$\text{L.C.} = \frac{s}{n};\ s = \frac{1°}{4};\ \text{L.C.} = 0.005°$$

$$0.005 = \frac{1}{4}\cdot\frac{1}{n}$$

or

$$n = \frac{1}{4 \times 0.005} = 50$$

Take 49 such primary divisions from the main scale and divide it into 50 parts for the vernier.

Example 1.7: *Design an extended vernier for an Abney level to read up to 10′. The main circle is divided into degrees.*

Solution:

$$\text{L.C.} = \frac{s}{n};\ s = 1°;\ \text{L.C.} = 10'$$

$$\therefore \quad \frac{10}{60} = \frac{1}{n};\quad \text{or} \quad n = 6$$

Take eleven spaces of the main scale and divide it into 6 spaces of the vernier.

1.11 MICROMETER MICROSCOPES

Generally, verniers are used when the finest reading to be taken is not less than 20″ or in some exceptional cases up to 10″. The micrometer microscope is a device which enables a measurement to be taken to a still finer degree of accuracy. Micrometer microscopes generally provided in geodetic theodolites can read to 1″ and estimate to 0.2″ or 0.1″.

The micrometer microscope consists of a small low-powered microscope with an object glass, an eye-piece and diaphragm which is capable of delicately controlled movement at right angles to the longitudinal axis of the tube. Fig. 1.12 shows a typical micrometer and one form of the field of view in taking a reading is shown in Fig. 1.13. The circle in Fig. 1.13 is divided into 10 minutes divisions. The micrometer has an objective lens close to the circle graduations. It forms an enlarged image of the circle near the micrometer eye-piece, which further enlarges the image. One pair of wires mounted on a movable frame is also in the image plane. The frame and the wires can be moved left and right by a micrometer screw drum. One complete revolution of the graduated drum moves the vertical wires across one division or 10′ of the circle. The graduated drum is divided into 10 large divisions (each of 1′) and each of the large divisions into 6 small ones of 10″ each. Fractional parts of a revolution of the drum, corresponding to fractional parts of a division on the horizontal circle, may be read on the graduated drum against an index mark fitted to the side.

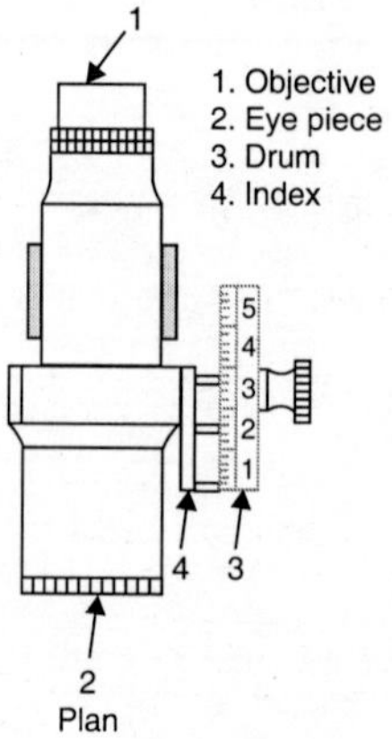

Fig. 1.12 Micrometer Microscope.

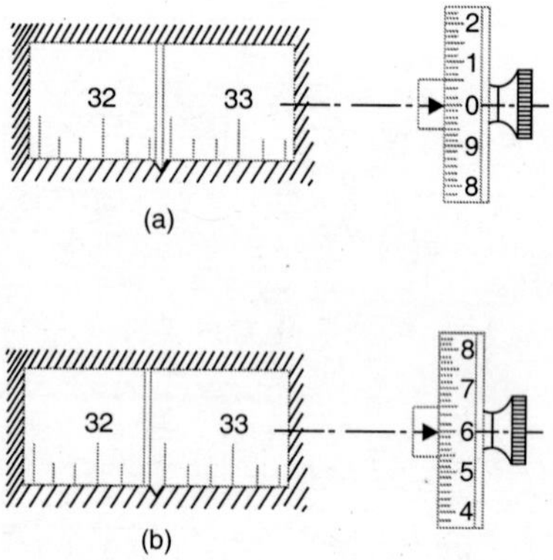

Fig. 1.13

The approximate reading is determined from the position of the specially marked V-notch. In the illustration of Fig. 1.13 (*a*), the circle reading is between 32° 20′ and 32° 30′ and the double wire index is on the notch.Turn the drum until the nearest division seems to be midway between the two vertical hairs and note the reading on the graduated drum, as shown in Fig. 1.13 (*b*) where the reading is 6′ 10″. The complete reading is 32° 26′ 10″. The object of using two closely spaced parallel wires instead of a single wire is to increase the precision of centering over graduations.

1.12 SCALE OF CHORDS

A scale of chords is used to measure an angle or to set-off an angle, and is marked either on a rectangular protractor or on an ordinary box wood scale.

1. Construction of a chord scale

1. Draw a quadrant *ABC*, making *AB* = *BC*. Prolong *AB* to *D*, making *AD* = *AC*.
2. Divide arc *AC* in nine equal parts, each part representing 10°.

3. With A as the centre, describe arc from each of the divisions, cutting ABD into points marked 10°, 20°, ... 90°.
4. Sub-divide each of these parts, if required, by first subdividing each division of arc AC, and then draw arcs with A as centre, as in step 3.
5. Complete the scale as shown in Fig. 1.14. *It should be noted that the arc through the 60° division will always pass through the point B* (since the chord of 60° is always equal to radius AB). The distance from A to any mark on the scale is equal to the chord of the angle of that mark. For example, the distance between A to 40° mark on the scale is equal to the chord of 40°.

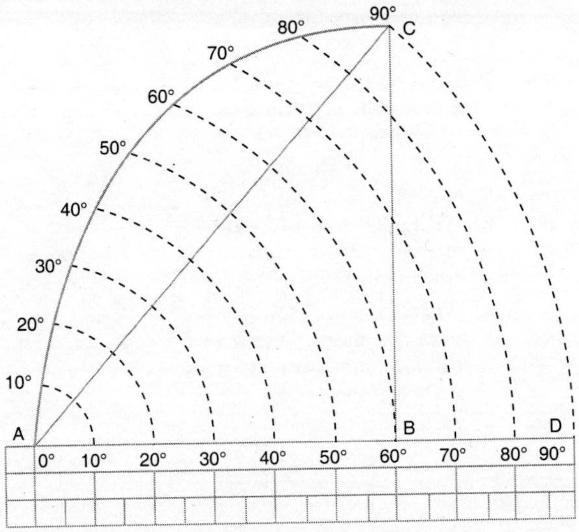

Fig. 1.14 Construction of A Chord Scale.

2. Construction of angles 30° and 80° with the scale of chords. (Fig. 1.15)

1. Draw a line AD, and on that mark AB = chord of 60° from the scale of chords.
2. With A as centre and AB as radius draw an arc.
3. With B as centre and radius equal to chord of 30° (*i.e.* distance from 0° to 30° on the scale of chords) draw an arc to cut the previous arc in E. Join AE. Then $\angle EAB = 30°$.
4. Similarly, with B as centre and radius equal to chord of 80° (i.e., distance from 0° to 80° on the scale of chords) draw an arc to cut previous arc in F. Join A and F. Then $\angle FAB = 80°$.

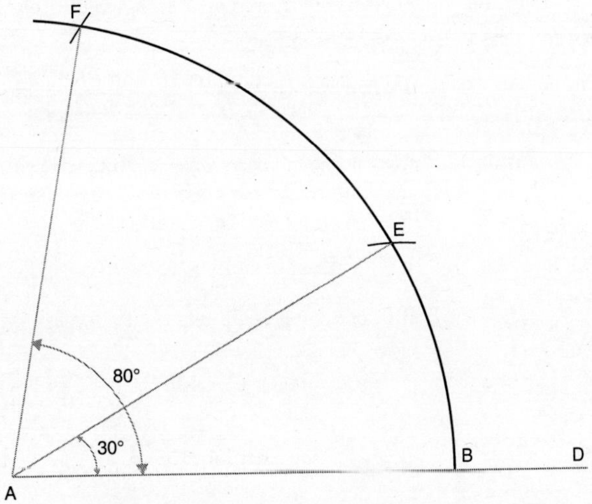

Fig. 1.15 Construction of an angle with the scale of chords.

3. Measurement of an angle with the scale of chords

1. Let the angle *EAD* be measured. On the line *AD*, measure *AB* = chord of 60° from the scale of chords.
2. With *A* as centre and *AB* as radius, draw an arc to cut line in *F*.
3. With the help of dividers take the chord distance *BF* and measure it on scale of chords to get the value of the angle θ.

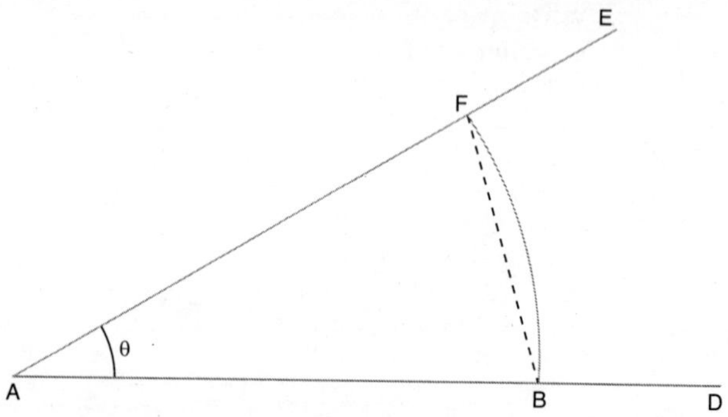

Fig. 1.16 Measurement of an angle with the scale of chords.

1.13 ERROR DUE TO USE OF WRONG SCALE

If the length of a line existing on a plan or a map is determined by means of measurement with a wrong scale, the length so obtained will be incorrect. The true or correct length of the line is given by the relation,

$$\textbf{\textit{Correct length}} = \frac{\textbf{\textit{R. F. of wrong scale}}}{\textbf{\textit{R. F. of correct scale}}} \times \textbf{\textit{measured length.}}$$

Similarly, if the area of a map or plan is calculated with the help of using a wrong scale, the correct area is given by

$$\textbf{\textit{Correct area}} = \left(\frac{\textbf{\textit{R. F. of wrong scale}}}{\textbf{\textit{R. F. of correct scale}}}\right)^2 \times \textbf{\textit{calculated area.}}$$

Example 1.8: *A surveyer measured the distance between two points on the plan drawn to a scale of 1 cm = 40 m and the result was 468 m. Later, however, he discovered that he used a scale of 1 cm = 20 cm. Find the true distance between the points.*

Solution:

$$\text{Measured length} = 468 \text{ m}$$

$$\text{R.F. of wrong scale used} = \frac{1}{20 \times 100} = \frac{1}{2000}$$

$$\text{R.F. of correct scale} = \frac{1}{40 \times 100} = \frac{1}{4000}$$

$$\therefore \quad \text{Correct length} = \left(\frac{1/2000}{1/4000}\right) \times 468 = \mathbf{936\ m.}$$

Alternative Solution : Map distance between two points measured with a scale of 1 cm to 20 m $= \frac{468}{20} = 23.4$ cm

Actual scale of the plan is 1 cm = 40 m

∴ True distance between the points = 23.4 × 40 = **936 m**

1.14 SHRUNK SCALE

If a *graphical scale* is not drawn on the plan and the sheet on which the plan is drawn shrinks due to variations in the atmospheric conditions, it becomes essential to find the shrunk scale of the plan. Let the original scale (*i.e.* 1 cm = x m) or its R.F. be known (stated on the sheet). The distance between any two known points on the plan can be measured with the help of the stated scale (*i.e.* 1 cm = x m) and this length can be compared with the actual distance between the two points. The *shrinkage ratio* or *shrinkage factor* is then equal to the ratio of the shrunk length to the actual length. The shrunk scale is then given by

"Shrunk scale = shrinkage factor × original scale."

For example, if the shrinkage factor is equal to $\frac{15}{16}$ and if the original scale is $\frac{1}{1500}$, the shrunk scale will have a R.F. = $\frac{15}{16} \times \frac{1}{1500} = \frac{1}{1600}$ (*i.e.* 1 cm = 16 m).

Example 1.9: *The area of the plan of an old survey plotted to a scale of 10 metres to 1 cm measures now as 100.2 sq. cm as found by a planimeter. The plan is found to have shrunk so that a line originally 10 cm long now measures 9.7 cm only. Find (i) the shrunk scale, (ii) true area of the survey.*

Solution:

(*i*) Present length of 9.7 cm is equivalent to 10 cm original length.

∴ Shrinkage factor $= \frac{9.7}{10} = 0.97$

True scale R.F. $= \frac{1}{10 \times 100} = \frac{1}{1000}$

∴ R.F. of shrunk scale $= 0.97 \times \frac{1}{1000} = \mathbf{\frac{1}{1030.93}}$

(*ii*) Present length of 9.7 cm is equivalent to 10 cm original length.

∴ Present area of 100.2 sq. cm is equivalent to

$\left(\frac{10}{9.7}\right)^2 \times 100.2$ sq. cm= 106.49 sq. cm = original area on plan.

Scale of plan is 1 cm=10 m

∴ Area of the survey =106.49 $(10)^2$ = **10649 sq. m.**

Example 1.10: *A rectangular plot of land measures 20 cm × 30 cm on a village map drawn to a scale of 100 m to 1 cm. Calculate its area in hectares. If the plot is re-drawn on a topo sheet to a scale of 1 km to 1 cm, what will be its area on the topo sheet? Also determine the R.F. of the scale of the village map as well as on the topo sheet.*

Solution:

(*i*) ***Village map :***

1 cm on map = 100 m on the ground

$\therefore$ 1 cm^2 on map = $(100)^2$ m^2 on the ground.

The plot measures 20 cm × 30 cm *i.e.*, 600 cm^2 on the map.

$\therefore$ Area of plot = $600 \times 10^4 = 6 \times 10^6$ m^2 = **600 hectares.**

(*ii*) ***Topo sheet***

1 km^2 is represented by 1 cm^2 or (1000 × 1000) m^2 is represented by 1 cm^2

$\therefore$ 6×10^6 m^2 is represented by $\frac{1}{1000 \times 1000} \times 6 \times 10^6 = \mathbf{6\ cm^2}$

(*iii*) R.F. of the scale of village map = $\frac{1}{100 \times 100} = \mathbf{\frac{1}{10000}}$

R.F. of the scale of topo sheet = $\frac{1}{1 \times 1000 \times 100} = \mathbf{\frac{1}{100000}}$

1.15 SURVEYING–CHARACTER OF WORK

The work of a surveyer may be divided into three distinct parts :

1. Field work
2. Office work
3. Care and adjustment of the instruments.

1. Field Work: The field work consists of the measurement of angles and distances and the keeping of a record of what has been done in the form of *field notes.* Some of the operations which a surveyor has to do in the field work are as follows :

1. Establishing stations and bench marks as points of reference and thus to extablish a system of horizontal and vertical control.
2. Measuring distance along the angles between the survey lines.
3. Locating details of the survey with respect to the stations and lines between stations, details such as boundary lines, streets, roads, buildings, streams, bridges and other natural or artificial features of the area surveyed.
4. Giving lines and elevations (or setting out lines and establishing grades) for a great variety of construction work such as that for building boundaries, roads, culverts, bridges, sewers and water supply schemes.
5. Determining elevations (or heights) of some existing points or establishing points at given elevations.
6. Surveying contours of land areas (topographic surveying) in which the field work involve both horizontal and vertical control.
7. Carrying out miscellaneous operations, such as :
   (*i*) Establishing parallel lines and perpendiculars.
   (*ii*) Taking measurements to inacessible points.
   (*iii*) Surveying past the obstacles, and carrying on a great veriety of similar field work that is based on geometric or trigonometric principles.
8. Making observations on the sun or a star to determine the meridian, latitude or longitude, or to determine the local time.

Field notes. *Field notes are written records of field work made at the time work is done.* It is obvious that, no matter how carefully the field measurements are made, the survey as a whole may be valueless if some of those measurements are not recorded or if any ambiguity exists as to the meaning of the records. The competency of the surveyor's planning and his knowledge of the work are reflected in the field record more than in any other element of surveying. The field notes should be legible, concise and comprehensive, written in clear, plain letters and figure. Following are some general important rules for note-keepers :

1. Record directly in the field book as observations are made.
2. Use a sharp 2H or 3H pencil. Never use soft pencil or ink.
3. Follow a consistant simple style of writing.
4. Use a liberal number of carefully executed sketches.
5. Make the notes for each day's work on the survey complete with a title of the survey, date, weather conditions, personnel of the crew, and list of equipment used.
6. Never erase. If a mistake is made, rule one line through the incorrect value and record the correction above the mistake.
7. Sign the notes daily.

The field notes may be divided into *three parts :*

1. **Numerical values.** These include the records of all measurements such as lengths of lines and offsets, staff readings (or levels) and angles or directions. All significant figures should be recorded. If a length is measured to the nearest 0.01 m, it should be so recorded; for example, 342.30 m and *not* 342.3 m. Record angles as 08° 06′ 20″, using at least two digits for each part of the angle.
2. **Sketches.** Sketches are made as records of outlines, relative locations and topographic features, Sketches are almost never made to scale. If measurements are put directly on the sketches, make it clear where they belong. Always make a sketch when it will help to settle beyond question any doubt which otherwise might arise in the interpretation of notes. Make sketches large, open and clear.
3. **Explanatory notes.** The object of the explanatory notes is to make clear that which is not perfectly evident from numerals and sketches, and to record such information concerning important features of the ground covered and the work done as might be of possible use later.

2. Office Work: The office work of a surveyor consist of

1. Drafting
2. Computing
3. Designing

The drafting mainly consists of preparations of the plans and sections (or plotting measurements to some scale) and to prepare topographic maps. The computing is of two kinds; (*i*) that done for purposes of plotting, and (*ii*) that done for determining areas and volumes. The surveyor may also be called upon to do some design work specially in the case of route surveying.

3. Care and Adjustments of Instruments: The practice of surveying requires experience in handling the equipment used in field and office work, a familiarity with the care and adjustment of the surveying instruments, and an understanding of their limitations. Many surveying instruments such as level, theodolite, compass etc. are very delicate and must be handled with great care *since there are many part of an instrument which if once impared cannot be restored to their original efficiency.* Before an instrument is taken out of the box, relative position of various parts should be carefully noted so that the instrument can be replaced in the box without undue strain on any of the parts. The beginner is advised to make a rough sketch showing the position of the instrument in the box. Following precautions must be taken :

1. While taking out the instrument from the box, do not lift it by the telescope or with hands under the horizontal circle plate. It should be lifted by placing the hands under the levelling base or the foot plate.
2. While carrying an instrument from one place to the other, it should be carried on the shoulder, setting all clamps tightly to prevent needless wear, yet loose enough so that if the parts bumped they will yield. If the head room available is less, such as carrying it through doors etc., it should be carried in the arms. If the distance is long, it is better to put it in box and then carried.
3. When the telescope is not in use, keep the cap over the lens. Do not rub lenses with silk or muslin. Avoid rubbing them altogether ; use a brush for removing dust.
4. Do not set an instrument on smooth floor without proper precautions. Otherwise the tripod legs are likely to open out and to let the instrument fall. If the instrument has been set up on a pavement or other smooth surface, the tripod legs should be inserted in the joints or cracks. The tripod legs should be spread well apart.
5. Keep the hands off the vertical circle and other exposed graduations to avoid tarnishing. Do not expose an instrument needlessly to dust, or to dampness, or to the bright rays of the sun. A water proof cover should be used to protect it.
6. To protect an instrument from the effects of salt water, when used near the sea coast, a fine film of watch oil rubbed over the exposed parts will often prevent the appearance of oxide. To remove such oxide-spots as well as possible, apply some watch-oil and allow it to remain for a few hours, then rub dry with a soft piece of linen. To preserve the outer apearance of an instrument, never use anything for dusting except a fine camel's hair brush. To remove water and dust spots, first use the camel's hair brush, and then rub-off with fine watch oil and wipe dry ; to let the oil remain would tend to accumulate dust on the instrument.
7. Do not leave the instrument unguarded when set on a road, street, foot-path or in pasture, or in high wind.
8. Do not force any screw or any part to move against strain. If they do not turn easily, the parts should be cleaned and lubricated.
9. The steel tape should be wiped clean and dry after using with the help of a dry cloth and then with a slightly oily one. Do not allow automobiles or other vehicles to run over a tape. Do not pull on a tape when there is kink in it, or jerk it unnecessarily.
10. In the case of a compass, do not let the compass needle swing needlessly. When not in use, it should be lifted off the pivot. Take every precaution to guard the point and to keep it straight and sharp.

PROBLEMS

1. Explain the following terms :
   (*i*) Representative fraction.
   (*ii*) Scale of plan
   (*iii*) Graphical scale

2. Give the designation and representative fraction of the following scales :
   (*i*) A line 135 metres long represented by 22.5 cm on plan.
   (*ii*) A plan 400 sq. metres in area represented by 4 sq. cm on plan.
3. Explain, with neat sketch, the construction of a plain scale. Construct a plain scale 1 cm = 6 m and show 26 metres on it.
4. Explain, with neat sketch, the construction of a diagonal scale. Construct a diagonal scale 1 cm = 5 m and show 18.70 metres on it.
5. Discuss in brief the principles of surveying.
6. Differentiate clearly between plane and geodetic surveying.
7. What is a vernier ? Explain the principle on which it is based.
8. Differentiate between :
   (*a*) Direct vernier and Retrograde vernier.
   (*b*) Double vernier and Extended vernier.
9. The circle of a theodolite is graduated to read to 10 minutes. Design a suitable vernier to read to 10 ″.
10. A limb of an instrument is divided to 15 minutes. Design a suitable vernier to read to 20 seconds.
11. Explain the principles used in the construction of vernier.
    Construct a vernier to read to 30 seconds to be used with a scale graduated to 20 minutes.
12. The arc of a sextant is divided to 10 minutes. If 119 of these divisions are taken for the length of the vernier, into how many divisions must the vernier be divided in order to read to (*a*) 5 seconds, and (*b*) 10 seconds ?
13. Show how to construct the following verniers :
    (*i*) To read to 10″ on a limb divided to 10 minutes.
    (*ii*) To read to 20″ on a limb divided to 15 minutes.
14. (*a*) Explain the function of a vernier.
    (*b*) Construct a vernier reading 114.25 mm on a main scale divided to 2.5 mm.
    (*c*) A theodolite is fitted with a vernier in which 30 vernier divisions are equal to 14°, 30′ on main scale divided to 30 minutes. Is the vernier direct or retrograde, and what is its least count ?

ANSWERS

2. (*i*) 6 m to 1 cm; $\frac{1}{600}$; (*ii*) 10 m to 1 cm; $\frac{1}{1000}$
9. $n = 60$
10. $n = 45$
11. $n = 40$
12. (*a*) $n = 120$ (direct vernier) (*b*) $n = 60$ (extended vernier)
13. (*i*) $n = 60$ (*ii*) $n = 45$
14. (*c*) Direct ; 1 minute.

CHAPTER 2

ACCURACY AND ERRORS

2.1 GENERAL

In dealing with measurements, it is important to distinguish between *accuracy* and *precision*. *Precision* is the degree of perfection *used* in the instruments, the methods and the observations. *Accuracy* is the degree of perfection *obtained*.

Accuracy depends on (1) *Precise instruments*, (2) *Precise methods* and (3) *Good planning*. The use of precise instruments simplify the work, save time and provide economy. The use of precise methods eliminate or try to reduce the effect of all types of errors. Good planning, which includes proper choice and arrangements of survey control and the proper choice of instruments and methods for each operation, saves time and reduces the possibility of errors.

The difference between a *measurement* and the true *value* of the quantity measured is the *true error* of the measurement, and is never known since the true value of the quantity is never known. However, the important function of a surveyor is to secure measurements which are correct within a certain *limit of error* prescribed by the nature and purpose of a particular survey.

A *discrepancy* is the difference between two measured values of the same quantity; it is not an error. A discrepancy may be small, yet the error may be great if each of the two measurements contains an error that may be large. It does not reveal the magnitude of systematic errors.

2.2 SOURCES OF ERRORS

Errors may arise from three sources :

1. **Instrumental.** Error may arise due to imperfection or faulty adjustment of the instrument with which measurement is being taken. For example, a tape may be too long or an angle measuring instrument may be out of adjustment. Such errors are known as *instrumental errors*.
2. **Personal.** Error may also arise due to want of perfection of human sight in observing and of touch in manipulating instruments. For example, an error may be there in taking the level reading or reading an angle on the circle of a theodolite. Such errors are known as *personal errors*.
3. **Natural.** Error may also be due to variations in natural phenomena such as temperature, humidity, gravity, wind, refraction and magnetic declination. If they are not properly

observed while taking measurements, the results will be incorrect. For example, a tape may be 20 metres at 20°C but its length will change if the field temeprature is different.

2.3 KINDS OF ERRORS

Ordinary errors met with in all classes of survey work may be classified as :

(*a*) Mistakes

(*b*) Systematic errors (Cumulative errors)

(*c*) Accidental errors (Compensating errors).

(*a*) **Mistakes.** *Mistakes are errors which arise from inattention, inexperience, carelessness and poor judgment or confusion in the mind of the observer.* If a mistake is undetected, it produces a serious effect upon the final result. Hence, every vaule to be recorded in the field must be checked by some independent field observation.

(*b*) **Systematic Errors (Cumulative Errors).** A systematic *error or cumulative error is an error that, under the same conditions, will always be of the same size and sign.* A systematic error always follows some definite mathematical or physical law, and a correction can be determined and applied. Such errors are of constant character and are regarded as *positive* or *negative* according as they make the *result too great* or *too small.* Their effect is, therefore, *cumulative.* For example, if a tape is P cm short and if it is stretched N times, the total error in the measurement of the length will be $P.N.$ cm.

If undetected, systematic errors are very serious. Therefore : (1) all surveying equipment must be designed and used so that whenever possible systematic errors will be *automatically* eliminated. (2) All systematic errors that cannot be surely eliminated by this means must be evaluated and their relationship to the conditions that cause them must be determined. For example, in ordinary levelling, the levelling instrument must first be adjusted so that the line of sight is as nearly horizontal as possible when bubble is centered. Also, the horizontal lengths for back-sight and fore-sight from each instrument position should be kept as nearly equal as possible. In precise levelling, every day the actual error of the instrument must be determined by careful peg test, the length of each sight is measured by stadia and a correction to the results is applied.

(*c*) **Accidental Errors (Compensating Errors).** *Accidental errors or compensating errors are those which remain after mistakes and systematic errors have been eliminated and are caused by a combination of reasons beyond the ability of the observer to control.* They tend sometimes in one direction and sometimes in the other, *i.e.* they are equally likely to make the apparent result too large or too small. An accidental error of a single determination is the difference between (1) the true value of the quantity, and (2) a determination that is free from mistakes and systematic errors. *Accidental errors represent the limit of precision in the determination of a value. They obey the laws of chance and therefore, must be handled according to the mathematical laws of probability.*

As stated above, accidental errors are of a *compensative nature* and tend to balance out in the final results. For example, an error of 2 cm in the tape may fluctuate on either side of the amount by reason of small variations in the pull to which it is subjected.

2.4 THEORY OF PROBABILITY

Investigations of observations of various types show that accidental errors follow a definite law, the *law of probability*. This law defines the occurrence of errors and can be expressed in the form of equation which is used to compute the probable value or the probable precision of a quantity. The most important features of accidental (or compensating) errors which usually occur are :

(*i*) Small errors tend to be more frequent than the large ones; that is, they are more *probable*.

(*ii*) Positive and negative errors of the same size happen with equal frequency ; that is they are *equally probable*.

(*iii*) Large errors occur infrequently and are *improbable*.

Probability Curve. The theory of probability describes these features by saying that the relative frequencies of errors of different extents can be represented by a curve as in Fig. 2.1. This curve, called the *curve of error* or *probability curve*, forms the basis for the mathematical derivation of theory of errors.

Principle of Least Square. According to the principle of least square, *the most probable value of an observed quantity available from a given set of observations is the one for which the sum of the squares of errors (residuals) is a minimum.*

Most Probable Value. The most probable value of a quantity is the one which has more chances of being correct than has any other. *The most probable error is defined as that quantity which when added to and subtracted from, the most probable value fixes the limits within which it is an even chance the true value of the measured quantity must lie.*

The probable error of a *single observation* is calculated from the equation,

$$E_s = \pm\, 0.6745 \sqrt{\frac{\Sigma v^2}{n-1}} \qquad \text{...(2.1)}$$

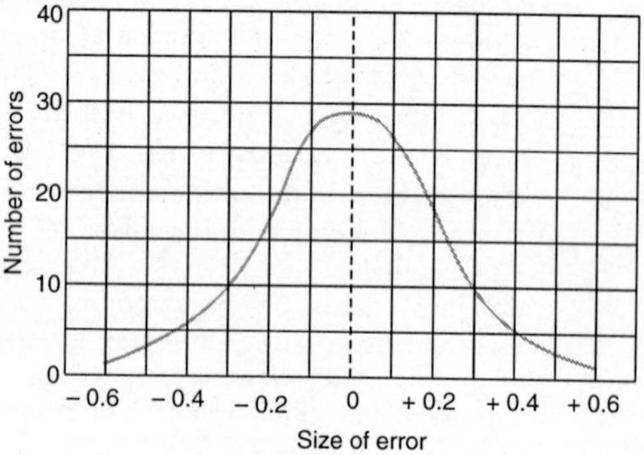

Fig. 2.1 Probability Curve.

The probable error of the *mean* of a number of observations of the same quantity is calculated from the equation :

$$E_m = \pm\, 0.6745 \sqrt{\frac{\Sigma v^2}{n(n-1)}} = \frac{E_s}{\sqrt{n}} \qquad \text{...(2.2)}$$

where E_s = Probable error of single observation

v = Difference between any single observation and the mean of the series

E_m = Probable error of the mean

n = Number of observations in the series.

Example 2.1: *In carrying a line of levels across a river, the following eight readings were taken with a level under identical conditions : 2.322, 2.346, 2.352, 2.306, 2.312, 2.300, 2.306, 2.326 metres.*

Calculate (*i*) *the probable error of single observation,* (*ii*) *probable error of the mean.*

Solution: The computations are arranged in the tabular form :

Rod reading (m)	*v (m)*	v^2
2.322	0.001	0.000001
2.346	0.025	0.000625
2.352	0.031	0.000961
2.306	0.015	0.000225
2.312	0.009	0.000081
2.300	0.021	0.000441
2.306	0.015	0.000225
2.326	0.005	0.000025
Mean : 2.321		$\Sigma v^2 = 0.002584$

From equation (1),

$$E_s = \pm\, 0.6745 \sqrt{\frac{0.002584}{8-1}} = \pm\, \mathbf{0.01295} \text{ metre}$$

and

$$E_m = \frac{E_s}{\sqrt{n}} = \pm \frac{0.01295}{\sqrt{8}} = \pm\, \mathbf{0.00458} \text{ metre.}$$

2.5 ACCURACY IN SURVEYING : PERMISSIBLE ERROR

The *permissible error* is the maximum allowable limit that a measurement may vary from the true value, or from a value previously adopted as correct. The value of the permissible error in any given case depends upon the scale, the purpose of the survey, the instruments available, class of work etc. The surveyor may be handicapped by rough country, too short a time, too small a party, poor instruments, bad weather and many other unfavourable conditions. The limit of error, therefore, cannot be given once for all. Examples of the permissible error for various classes of work have been mentioned throughout this book. However, *the best surveyor is not he who is extremely accurate in all his work, but he who does it just accurately enough for the purpose without waste of time or money.* A surveyor should make the precision of each step in the field work corresponding to the importance of that step.

Significant Figures in Measurement

In surveying, an indication of accuracy attained is shown by number of significant figures. Each such quantity, expressed in n number of digits in which $n - 1$ are the digits of definite value while the last digit is the least accurate digit which can be estimated and is subject to error. For example, a quantity 423.65 has five significant figures, with four *certain* and the last digit 5, *uncertain*. The error in the last digit may, in this case, be a maximum value of 0.005 or a probable value of ± 0.0025.

As a rule, the field measurements should be consistent, thus dictating the number of significant figures in desired or computed quantities. *The accuracy of angular and linear values should be compatible.* For small angles, arc = chord = $R\,\theta''/206265$, where θ is expressed in seconds of arc. Thus for 1″ of arc, the subtended value is 1 mm at 206.265 m while for 1′ of arc, the subtended value is 1 mm at 3.438 m or 1 cm at 34.38 m. In other words, the angular values measured to 1″ require distances to be measured to 1 mm, while the angular values measured to 1′ require distances to be measured to 1 cm.

Accumulation of Errors: In the accumulation of errors of known sign, the summation is algebraic while the summation of random errors of ± values can only be computed by the root mean square value:

$$e_t = \sqrt{\pm e_1^2 \pm e_2^2 \pm e_3^2 \pm \ldots\ldots \pm e_n^2} \qquad \ldots(2.3)$$

2.6 ERRORS IN COMPUTED RESULTS

The errors in computed results arise from (*i*) errors in *measured* or *derived* data, or (*ii*) errors in trigonometrical or logrithmical values used. During common arithmetical process (*i.e.* addition, subtraction, multiplication, division etc.), the resultant values are frequently given false accuracies as illustrated below.

(*a*) **Addition**

Let $s = x + y$, where x and y are measured quantities.

Then $s + \delta s = (x + \delta x) + (y + \delta y)$

where δs may be + or – .

Considering *probable errors* of indefinite values,

$$s \pm e_s = (x \pm e_x) + (y \pm e_y) \quad \text{or} \quad s \pm e_s = (x + y) \pm \sqrt{e_x^2 + e_y^2}$$

$$\therefore \text{ Probable Error } \pm e_s = \sqrt{e_x^2 + e_y^2} \qquad \ldots(2.4)$$

(*b*) **Subtraction**

Let $s = x - y$

$\therefore$ $s + \delta s = (x - y) + (\delta x + \delta y)$

The maximum error $= \delta s = (\delta x + \delta y)$

Considering probable errors of indefinite value, $s \pm e_s = (x - y) \pm \sqrt{e_x^2 + e_y^2}$

$$\therefore \text{ Probable error } \pm e_s = \sqrt{e_x^2 + e_y^2} \qquad \ldots(2.5)$$

which is the same as in addition.

(*c*) **Multiplication**

Let $s = x \,.\, y$

$\therefore$ $\delta s_x = y \,.\, \delta x$ and $\delta s_y = x \,.\, \delta y$

The maximum error $\delta s = y\delta x + x\delta y$

Considering probable errors of indefinite values,

$$e_s = \sqrt{y^2\, e_x^2 + x^2\, e_y^2} \qquad \ldots(2.6\ a)$$

or

$$e_s = xy\,\sqrt{(e_x / x)^2 + (e_y / y)^2} \qquad \ldots(2.6\ b)$$

and error ratio :

$$\frac{e_s}{s} = \sqrt{(e_x / x)^2 + (e_y / y)^2} \qquad \ldots(2.6)$$

(*d*) Division

Let $$s = \frac{x}{y}$$

$\therefore$ $$\delta s_x = \frac{\delta x}{y} \quad \text{and} \quad \delta s_y = \frac{x\,\delta y}{y^2}$$

The maximum error $$\delta s = \frac{\delta x}{y} + \frac{x\,.\,\delta y}{y^2}$$

Considering probable errors of indefinite values,

$$e_s = \sqrt{\left(\frac{e_x}{y}\right)^2 + \left(\frac{x e_y}{y^2}\right)^2} \qquad ...(2.7\ a)$$

or $$e_s = \frac{x}{y}\sqrt{\left(\frac{e_x}{x}\right)^2 + \left(\frac{e_y}{y}\right)^2} \qquad ...(2.7\ b)$$

and error ratio : $$\frac{e_s}{s} = \sqrt{(e_x / x)^2 + (e_y / y)^2} \qquad ...(2.7)$$

which is the same as for multiplication.

(*e*) Powers

Let $$s = x^n$$

$\therefore$ $$\delta s = n x^{n-1}\, \delta x$$

$\therefore$ Error ratio $$\frac{\delta s}{s} = \frac{n \delta x}{x} \qquad ...(2.8)$$

Example 2.2: *A quantity s is equal to the sum of two measured quantities x and y given by*

$$s = 4.88 + 5.637$$

Find the most probable error, the maximum limits and most probable limits of the quantity s.

Solution: The maximum errors (δx and δy) will be 0.005 and 0.0005 and the probable errors will be $\pm$ 0.0025 and $\pm$ 0.00025.

$$\therefore \quad s + \delta s = (x + y) \pm (\delta x + \delta y) = (4.88 + 5.637) \pm (0.005 + 0.0005)$$
$$= 10.517 \pm 0.0055 \qquad ...(1)$$
$$= 10.5225 \text{ and } 10.5115$$

Also $$s \pm e_s = (x + y) \pm \sqrt{e_x^2 + e_y^2} = (4.88 + 5.637) \pm \sqrt{(0.0025)^2 + (0.00025)^2}$$
$$= 10.517 \pm 0.00251 \qquad ...(2)$$
$$= 10.5195 \text{ and } 10.5145$$

Hence the most probable error = $\pm$ 0.00251 and most probable limits of the quantity (s) are 10.5195 and 10.5145. Similarly the maximum limits of quantity are 10.5225 and 10.5115.

From the above, it is clear that the quantity s, consisting of the sum of two quantities may be expressed either as 10.52 or as 10.51, and that the second decimal place is the most probable limit to which the derived quantity (s) may be quoted. Hence it is concluded that the *accuracy of the sum must not exceed the least accurate figure used.*

Example 2.3: *A quantity s is given by*

$$s = 5.367 - 4.88$$

Find the most probable error, and the most probable limits and maximum limits of the quantity.

Solution: The maximum errors will be 0.0005 and 0.005 and probable errors will be ± 0.0025.

$$\therefore \quad s + \delta s = (x - y) \pm (\delta x + \delta y) = (5.367 - 4.88) \pm (0.005 + 0.0005)$$
$$= 0.487 \pm 0.0055 = 0.4925 \quad \text{or} \quad 0.4815$$

Also
$$s \pm e_s = (x - y) \pm \sqrt{e_x^2 + e_y^2} = (5.367 - 4.88) \pm \sqrt{(0.00025)^2 + (0.0025)^2}$$
$$= 0.4925 \pm 0.00251 = 0.4950 \quad \text{or} \quad 0.4900$$

Hence the most probable error = ± **0.00251**

Most probable limits of s = 0.4950 and 0.4900 and maximum limits of s = 0.4925 and 0.4815.

Here again, the quantity s can only be 0.48 or 0.49, and the second decimal place is the most probable limit to which a *derived quantity* (s) can be given. Hence the accuracy of a subtraction must not exceed the least accurate figure used.

Example 2.4: *A derived quantity s is given by product of two measured quantities, as under :*

$$s = 2.86 \times 8.34$$

Find the maximum error and most probable error in the derived quantity.

Solution: The maximum errors in the individual measurements will be 0.005 and 0.005, while the most probable errors will be ± 0.0025 and ± 0.0025 respectively.

Now max. error
$$\delta s = y\,\delta x + x\delta y = (8.34 \times 0.005) + (2.86 \times 0.005)$$
$$= 0.0417 + 0.0143 = 0.056 \approx \mathbf{0.06}$$

The most probable error is

$$e_s = xy\sqrt{\left(\frac{e_x}{x}\right)^2 + \left(\frac{e_y}{y}\right)^2} = (2.86 \times 8.34)\sqrt{\left(\frac{0.0025}{2.86}\right)^2 + \left(\frac{0.0025}{8.34}\right)^2}$$
$$= \pm\, \mathbf{0.02}$$

Now
$$s = x \times y = 2.86 \times 8.34 = 23.85$$

Hence the most probable limits are thus 23.87 and 23.83, and by rounding off process, value may be given as 23.85, *i.e.* to the same accuracy as the least accurate figure used.

Example 2.5: *A derived quantity s is given by*

$$s = \frac{23.9}{8.34}$$

Find the maximum error and most probable error in the quantity.

Solution: The maximum error δs is given by

$$\delta s = \frac{\delta x}{y} + \frac{x.\delta y}{y^2}$$

where δx and δy (maximum errors in individual measurements) are 0.05 and 0.005 respectively

$$\therefore \quad \delta s = \frac{0.05}{8.34} + \frac{23.9 \times 0.005}{(8.34)^2} = 0.006 + 0.0017 = \mathbf{0.0077}$$

The probable errors in individual measurements are ± 0.025 and ± 0.0025. Hence the probable error in the derived quantity is

$$e_s = \frac{x}{y}\sqrt{\left(\frac{e_x}{x}\right)^2 + \left(\frac{e_y}{y}\right)^2} = \frac{23.9}{8.34}\sqrt{\left(\frac{0.025}{23.9}\right)^2 + \left(\frac{0.0025}{8.34}\right)^2}$$

$$= \pm\, \mathbf{0.003}$$

Now $$s = \frac{23.9}{8.34} = 2.8657 \approx 2.866$$

Hence the most probable limits of s are 2.869 and 2.863. For practical purposes, adopting rounding off, the value may be given as 2.87.

Example 2.6: *A derived quantity s is given by*

$$s = (4.86)^2$$

Find the maximum value of error and most probable value of error.

Solution:

$$s = (4.86)^2 = 23.6196$$

Now maximum error in the individual measurement is 0.005 and probable error in measurement is 0.0025.

Now, maximum error δs is given by

$$\delta s = n\,x^{n-1}\,\delta x = 2(4.86)^{2-1} \times 0.005$$

$$= \mathbf{0.0486}$$

The most probable value of error is

$$e_s = n\,x^{n-1}\,e_x = 2(4.86)^{2-1} \times 0.0025 = \pm\, \mathbf{0.0243.}$$

The most probable limits of s are thus 23.6439 and 23.5953, and rounding these off, we get s, practically, equal to 23.62.

Example 2.7: *The long and short sides of a rectangle measure 8.28 m and 4.36 m, with errors of ± 5 mm. Express the area to correct number of significant figures.*

Solution:

$$A = 8.28 \times 4.36 = 36.1008 \text{ m}^2$$

Maximum error in individual measurements = 0.005 m

$$\therefore \quad \text{Error ratios are : } = \frac{0.005}{8.25} \approx \frac{1}{1650} \quad \text{and} = \frac{0.005}{4.36} \approx \frac{1}{872}$$

$$\therefore \quad \delta A = 36.1008\left(\frac{1}{1650} + \frac{1}{872}\right) \approx \pm\, 0.06 \text{ m}$$

Hence area has limits of 36.16 and 36.04 m^2 and the answer can be quoted as **36.10 m^2** *correct to two significant figures compatible with the field measurements.*

Example 2.8: *A rectangle has sides approximately 380 metres and 260 metres. If the area is to be determined to the nearest 10 m^2 what will be maximum error permitted in each line and to what accuracy should the lines be measured. Assume equal precision ratio for each length.*

Solution:

$$A = 380 \times 260 = 98800 \text{ m}^2 \text{ ;}$$
$$\delta A = 10 \text{ m}^2$$

$$\therefore \quad \frac{\delta A}{A} = \frac{10}{98800} = \frac{1}{9880} = \frac{\delta x}{x} + \frac{\delta y}{y}$$

$$\text{But} \quad \frac{\delta x}{x} = \frac{\delta y}{y}$$

$$\therefore \quad \frac{\delta x}{x} + \frac{\delta y}{y} = \frac{2\,\delta x}{x} = \frac{1}{9880}$$

$$\therefore \quad \frac{\delta x}{x} = \frac{1}{2 \times 9880} = \frac{1}{19760}$$

Hence precision ratio of each line = $\mathbf{\frac{1}{19760}}$

$\therefore$ Max. error in 380 m length = $\frac{380}{19760}$ = **0.0192 m**

Max. error in 260 m length = $\frac{260}{19760}$ = **0.0132 m**

If the number of significant figures in area is 5 (*i.e.* nearest to 10 m^2), each line must be measured to at least 5 significant figures, *i.e.* 380.00 m and 260.00 m.

PROBLEMS

1. Explain the following terms :
   (*i*) Accuracy (*ii*) Precision (*iii*) Discrepancy (*iv*) True error.
2. Distinguish clearly between cumulative and compensating errors.
3. Discuss in brief the different sources of errors in surveying.
4. What are the characteristic features of accidental error ? Explain how will you find out the probable error in a quantity measured several times in the field.
5. An angle has been measured under different field conditions, with results as follows :

28° 24′ 20″	28° 24′ 00″
28° 24′ 40″	28° 23′ 40″
28° 24′ 40″	28° 24′ 20″
28° 25′ 00″	28° 24′ 40″
28° 24′ 20″	28° 25′ 20″

Find (*i*) the probable error of single observation (*ii*) probable error of the mean.

ANSWERS

5. (*i*) 19″.34 (*ii*) 6″.11.

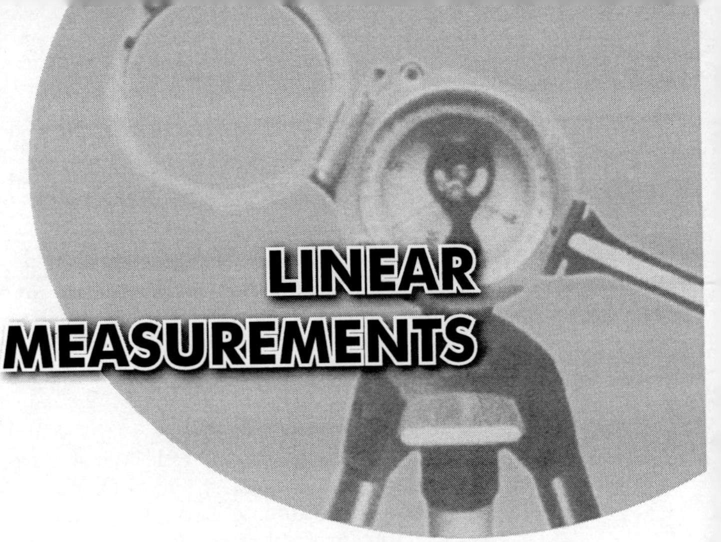

3 CHAPTER

LINEAR MEASUREMENTS

3.1 DIFFERENT METHODS

There are various methods of making linear measurements and their relative merit depends upon the degree of precision required.

They can be mainly divided into three heads :

1. Direct measurements.
2. Measurements by optical means.
3. Electro-magnetic methods.

In the case of direct measurements, distances are *actually measured* on the ground with help of a chain or a tape or any other instrument. In the optical methods, observations are taken through a telescope and calculations are done for the distances, such as in tacheometry or triangulation. In the electro-magnetic methods, distances are measured with instruments that rely on propagation, reflection and subsequent reception of either radio waves, light waves or infrared waves.

For measurement of distances by optical means, refer chapter 22 on 'Tacheometric Surveying'. For measurement of distances by electro-magnetic methods, refer chapter 24 on 'Electro-magnetic Distance Measurement (EDM)'.

3.2 DIRECT MEASUREMENTS

The various methods of measuring the distances directly are as follows :

1. Pacing
2. Measurement with passometer
3. Measurement with pedometer
4. Measurement by odometer and speedometer
5. Chaining.

1. **Pacing.** Measurements of distances by pacing is chiefly confined to the preliminary surveys and explorations where a surveyor is called upon to make a rough survey as quickly as possible. It may also be used to roughly check the distances measured by other means. The method consists in counting the number of paces between the two points of a line. The length of the line can then be computed by knowing the average

length of the pace. The length of the pace varies with the individual, and also with the nature of the ground, the slope of the country and the speed of pacing. A length of pace more nearly that of one's natural step is preferable. The length of one's natural step may be determined by walking on fairly level ground over various lines of known lengths. One can soon learn to pace distances along level, unobstructed ground with a degree of accuracy equivalent approximately to 1 in 100. However, pacing over rough ground or on slopes may be difficult.

2. **Passometer.** Passometer is an instrument shaped like a watch and is carried in pocket or attached to one leg. The mechanism of the instrument is operated by motion of the body and it automatically registers the number of paces, thus avoiding the monotony and strain of counting the paces, by the surveyor. The number of paces registered by the passometer can then be multiplied by the average length of the pace to get the distance.
3. **Pedometer.** Pedometer is a device similar to the passometer except that, adjusted to the length of the pace of the person carrying it, it registers the total distance covered by any number of paces.
4. **Odometer and Speedometer.** The odometer is an instrument for registering the number of revolutions of a wheel. The well-known speedometer works on this principle. The odometer is fitted to a wheel which is rolled along the line whose length is required. The number of revolutions registered by the odometer can then be multiplied by the circumference of the wheel to get the distance. Since the instrument registers the length of the surface actually passed over, its readings obtained on undulating ground are inaccurate. If the route is smooth, the speedometer of an automobile can be used to measure the distance approximately.
5. **Chaining.** Chaining is a term which is used to denote measuring distance either with the help of a chain or a tape and is the most accurate method of making direct measurements. For work of ordinary precision, a chain can be used, but for higher precision a tape or special *bar* can be used. The distances determined by chaining form the basis of all surveying. No matter how accurately angles may be measured, the survey can be no more precise than the chaining.

3.3 INSTRUMENTS FOR CHAINING

The various instruments used for the determination of the length of line by chaining are as follows :

1. Chain or tape
2. Arrows
3. Pegs
4. Ranging rods
5. Offset rods
6. Plasterer's laths and whites
7. Plumb bob.

1. Chain

Chains are formed of straight links of galvanised mild steel wire bent into rings at the ends and joined each other by three small circular or oval wire rings. These rings offer flexibility to the chain. The ends of the chain are provided with brass handle at each end with swivel joint, so that the chain can be turned without twisting. The length

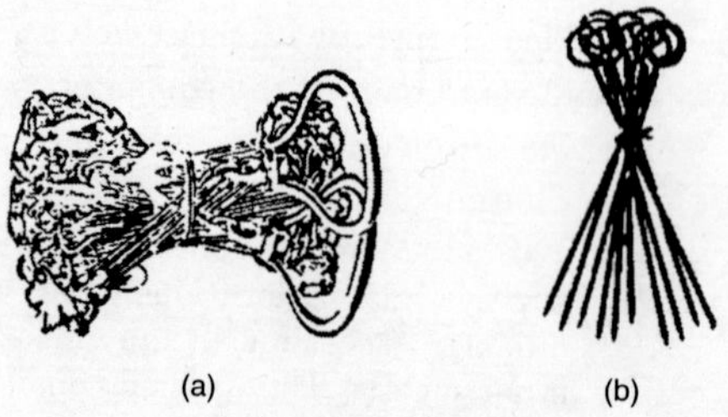

Fig. 3.1 Chain and Arrows.

of a link is the distance between the centres of two consecutive middle rings, while the length of the chain is measured from the outside of one handle to the outside of the other handle.

Following are various types of chains in common use :

(*i*) Metric chains (*ii*) Gunter's chain or Surveyor's chain

(*iii*) Engineer's chain (*iv*) Revenue chain

(*v*) Steel band or band chain.

Metric chains. After the introduction of metric units in India, the metric chains are widely used. Metric chains are generally available in lengths of 5, 10, 20 and 30 metres. IS : 1492-1970 covers the requirements of metric surveying chains. Figs. 3.2 and 3.3 show 5 m and 10 m chains respectively, while Figs. 3.4 and 3.5 show the 20 m and 30 m chains respectively. Fig. 3.6 shows the details of a metric chain.

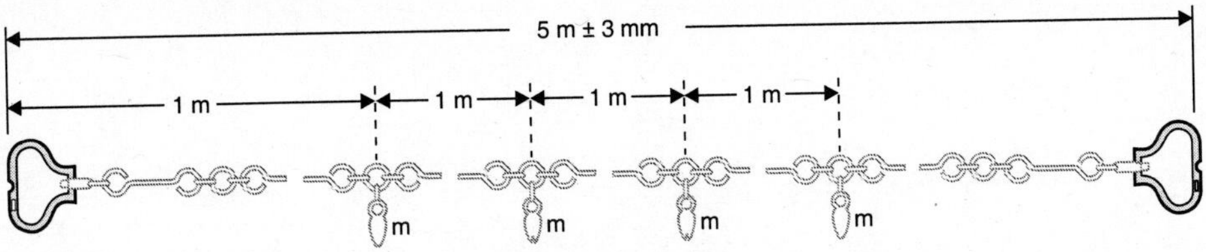

Fig. 3.2 5-Metre Chain

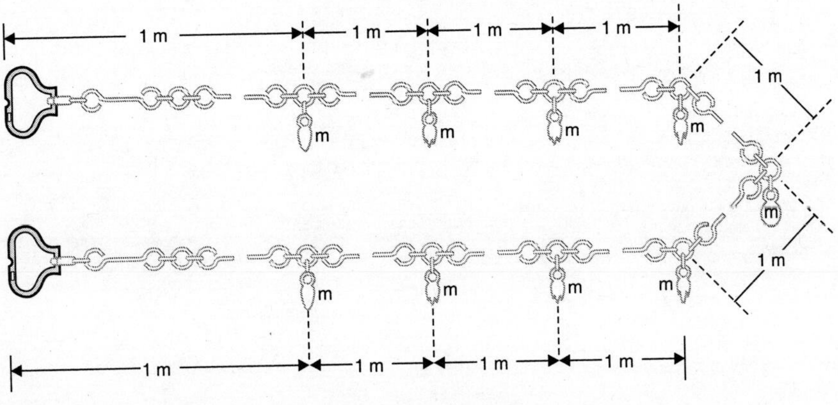

Fig. 3.3 10-Metre Chain

To enable the reading of fractions of a chain without much difficulty, tallies are fixed at every metre length for chains of 5 m and 10 m lengths (see Figs. 3.2 and 3.3) and at every five-metre length for chains of 20 m and 30 m lengths (see Figs. 3.4 and 3.5). In the case of 20 m and 30 m chains, small brass rings are provided at every metre length, except where tallies are attached. The shapes of tallies for chains of 5 m and 10 m lengths for different positions are shown in Fig. 3.7. To facilitate holding of arrows in position with the handle of the chain, a groove is cut on the outside surface of the handle, as shown in Fig. 3.6. The tallies used for marking distances in the metric chains are marked with the letters 'm' in the order to distinguish them from non-metric chains. The length of chain, 5 m, 10 m, 20 m or 30 m as the case may be, are engraved

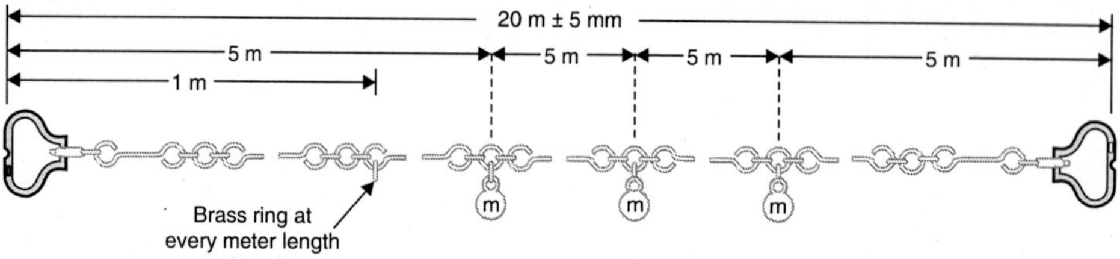

Fig. 3.4 20-Metre Chain

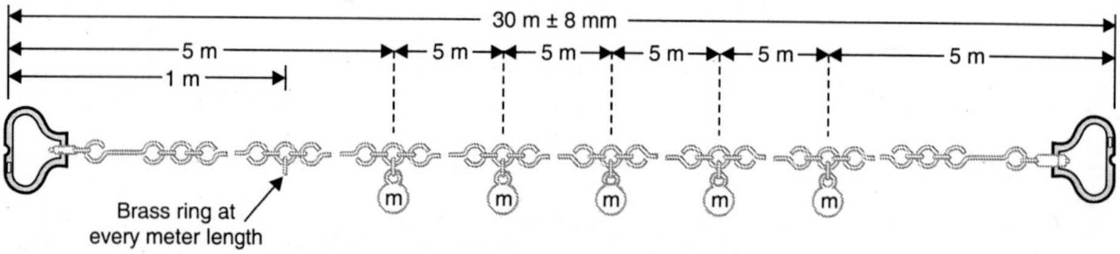

Fig. 3.5 30-Metre Chain

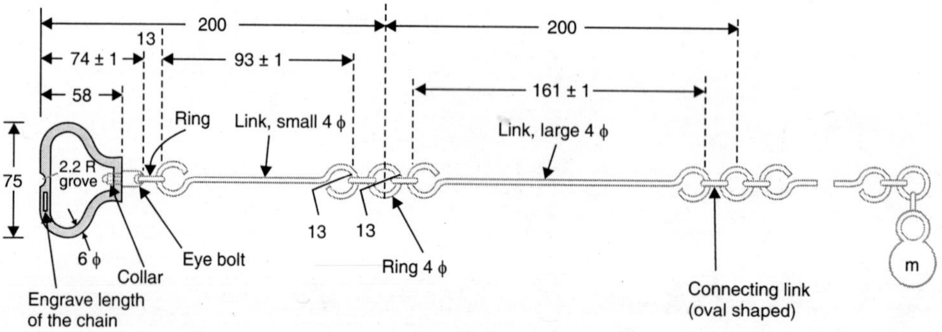

Fig. 3.6 Details of a Metric Chain

on both the handles to indicate the length and also to distinguish the chains from non-metric chains.

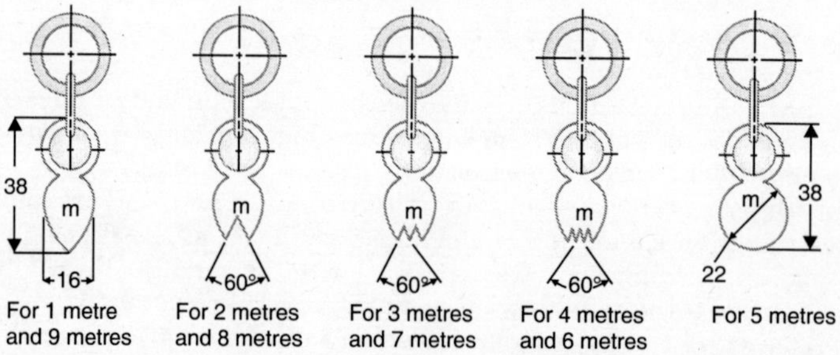

Fig. 3.7 Shapes of Tallies for 5 m and 10 m Chains.

Gunter's Chain or Surveyor's Chain: A Gunter's chain or surveyor's chain is 66 ft. long and consists of 100 links, each link being 0.6 ft. or 7.92 inches long. The length of 66 ft. was originally adopted for convenience in land measurement since 10 square chains are equal to 1 acre. Also, when linear measurements are required in furlongs and miles, it is more convenient since 10 Gunter's chains = 1 furlong and 80 Gunter's chains = 1 mile.

Engineer's Chain: The engineer's chain is 100 ft. long and consists of 100 links, each link being 1 ft. long. At every 10 links, brass tags are fastened, with notches on the tags indicating the number of 10 link segments between the tag and end of the chain. The distances measured are recorded in feet and decimals.

Revenue Chain: The revenue chain is 33 ft. long and consists of 16 links, each link being $2\frac{1}{16}$ ft. long. The chain is mainly used for measuring fields in cadastral survey.

Fig. 3.8 Steel Band.

Steel band or band chain (Fig. 3.8)

The steel band consists of a long narrow strip of blue steel, of uniform width of 12 to 16 mm and thickness of 0.3 to 0.6 mm. Metric steel bands are available in lengths of 20 or 30 m. It is divided by brass studs at every 20 cm and numbered at every metre. The first and last links (20 cm length) are subdivided into cm and mm. Alternatively, in the place of putting brass studs, a steel band may have graduations etched as metres, decimetres and centimetres on one side and 0.2 m links on the other. For convenience in handling and carrying, steel bands are almost invariably wound on special steel crosses or metal reels from which they can be easily unrolled.

For accurate work, the steel band should always be used in preference to the chain, but it should only be placed in the hands of careful chainmen. A steel band is lighter than the chain and is easier to handle. It is practically unalterable in length, and is not liable to kinks when in use. Its chief disadvantage is that it is easily broken and difficult to repair in the field.

Testing and Adjusting Chain: During continuous use, the length of a chain gets altered. Its length is shortened chiefly due to the bending of links. Its length is elongated either due to stretching of the links and joints and opening out of the small rings, or due to wear of wearing surface. For accurate work, it is necessary to test the length of the chain from time to time and make adjustments in the length.

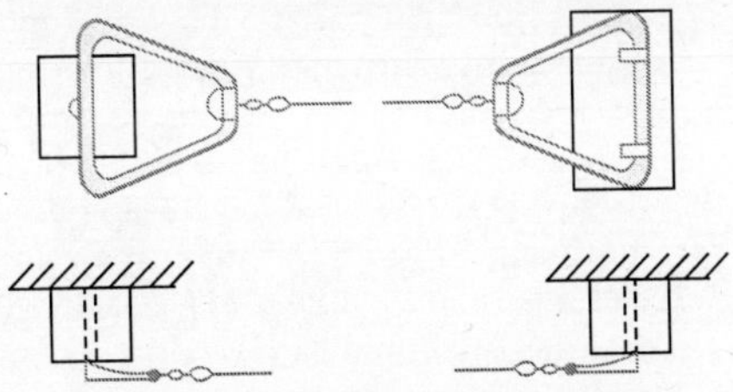

Fig. 3.9 Field Testing of Chain.

A chain may either be tested with reference to a steel tape. Sometimes, it is convenient to have a permanent *test gauge* established and the chain tested by comparing with the test gauge from time to time. In field, where no permanent test gauge exists, a test gauge is established by driving two pegs the requisite distance apart, and inserting

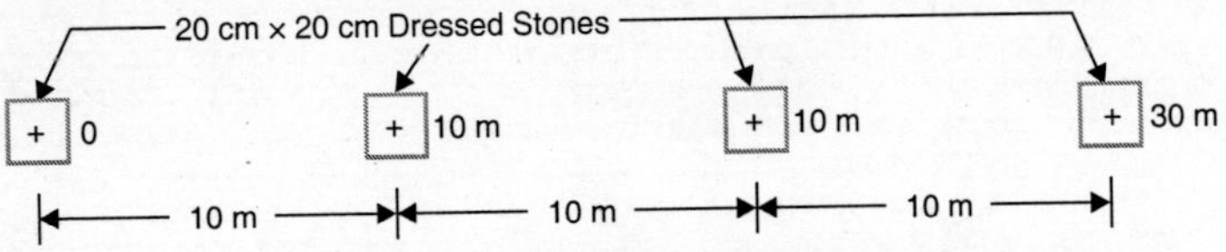

Fig. 3.10 Permanent Test Gauge.

nails into their tops to mark exact points, as shown in Fig. 3.9. Fig. 3.10 shows a permanent test gauge, made of dressed stones 20 cm × 20 cm.

The overall length of a chain, when measured at 8 kg pull and checked against a steel tape standardized at 20°C, shall be within the following limits :

20 metre chain: ± 5 mm and 30 metre chain: ± 8 mm

In addition to this, every metre length of the chain shall be accurate to within 2 mm.

On testing, if a chain is found to be long, it can adjusted by

(*i*) closing the joints of the rings if opened out
(*ii*) reshaping the elongated rings
(*iii*) removing one or more small circular rings
(*iv*) replacing worn out rings
(*v*) adjusting the links at the ends.

If, on the other hand, a chain is found to be short, it can be adjusted by :

(*i*) straightening the links
(*ii*) flattening the circular rings
(*iii*) replacing one or more small circular rings by bigger ones
(*iv*) inserting additional circular rings
(*v*) adjusting the links at the end.

However, in both the cases, adjustment must be done symmetrically so that the position of the central peg does not alter.

2. Tapes

Tapes are used for more accurate measurements and are classed according to the material of which they are made, such as follows :

(*i*) cloth or linen tape (*ii*) metallic tape
(*iii*) steel tape and (*iv*) invar tape.

Cloth or Linen Tape: Cloth tapes of closely woven linen, 12 to 15 mm wide varnished to resist moisture, are light and flexible and may be used for taking comparatively rough and subsidiary measurements such as offsets. A cloth tape is commonly available in lengths of 10 metres, 20 metres, 25 metres and 30 metres, and in 33 ft., 50 ft., 66 ft. and 100 ft. The end of the tape is provided with small brass ring whose length is included in the total length of the tape. A cloth tape is rarely used for making accurate measurements, because of the following reasons: (*i*) it is easily affected by moisture or dampness and thus shrinks; (*ii*) its length gets altered by stretching; (*iii*) it is likely to twist and tangle; (*iv*) it is not strong. Before winding up the tape in the case, it should be cleaned and dried.

Fig. 3.11 Metallic Tape

Fig. 3.12 Steel Tape

Metallic Tape: A metallic tape is made of varnished strip of waterproof linen interwoven with small brass, copper or bronze wires and does not stretch as easily as a cloth tape. Since metallic tapes are light and flexible and are not easily broken, they are particularly useful in cross-sectioning and in some methods of topography where small errors in length of the tape are of no consequence. Metallic tapes are made in lengths of 2, 5, 10, 20, 30 and 50 metres. In the case of tapes of 10, 20, 30 and 50 m lengths a metal ring is attached to the outer ends and fastened to it by a metal strip of the same width as the tape. This metal strip protects the tape, and at the same time inspector's stamp can be put on it. In addition to the brass ring, the outer ends of these tapes are reinforced by a strip of leather or suitable plastic material of the same width as the tape, for a length of at least 20 cm. Tapes of 10, 20, 30 and 50 metre lengths are supplied in a metal or leather case fitted with a winding device (Fig. 3.11).

Steel Tape: Steel tapes vary in quality and accuracy of graduation, but even a poor steel tape is generally superior to a cloth or metallic tape for most of the linear measurements that are made in surveying. A steel tape consists of a light strip of width 6 to 10 mm and is more accurately graduated. Steel tapes are available in lengths of 1, 2, 10, 20, 30 and 50 metres. The tapes of 10, 20, 30 and 50 metre lengths, are provided with a brass ring at the outer end, fastened to it by a metal strip of the same width as the tape. The length of the tape includes the metal ring. It is wound in a well-sewn leather case or a corrosion resisting metal case, having a suitable winding device (Fig. 3.12). Tapes of longer length (*i.e.*, more than 30) m are wound on metal reel (Fig. 3.13).

Fig. 3.13 Steel Tape on Reel.

A steel tape is a delicate instrument and is very light, and therefore, cannot withstand rough usage. The tape should be wiped clean and dry after using, and should be oiled with a little mineral oil, so that it does not get rusted.

Invar Tape: Invar tapes are used mainly for linear measurements of a very high degree of precision, such as measurements of base lines. The invar tape is made of alloy of nickel (36%) and steel, and has very low coefficient of thermal expansion—seldom more than about one-tenth of that of steel, and often very much less. The coefficient of thermal expansion varies a good deal with individual bands but an average value of 0.0000005 per 1° F may be taken. The other great advantage of invar is that bands and wires made of invar enable base lines to be measured very much more rapidly and conveniently. Invar tapes and bands are more expensive, much softer and are more easily deformed than steel tapes. Another great disadvantage of invar tape is that it is subjected to *creep* due to which it undergoes a small increase in length as time goes on. Its coefficient of thermal expansion also goes on changing. It is therefore, very essential to determine its length and coefficient of expansion from time to time. Invar tapes are normally 6 mm wide and are available in lengths of 20, 30 and 100 m.

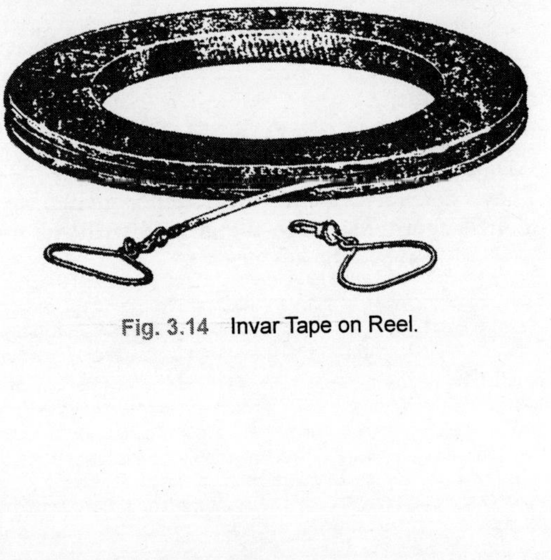

Fig. 3.14 Invar Tape on Reel.

The difficulty with invar tapes is that they are easily bent and damaged. They must, therefore, be kept on reels of large diameter, as shown in Fig. 3.14.

3. Arrows

Arrows or marking pins are made of stout steel wire, and generally, 10 arrows are supplied with a chain. An arrow is inserted into the ground after every chain length measured on the ground. Arrows are made of good quality hardened and tempered steel wire 4 mm (8 s.w.g.) in diameter, and are black enamelled. The length of arrow may vary from 25 cm to 50 cm, the most common length being 40 cm. One end of the arrow is made sharp and other end is bent into a loop or circle for facility of carrying. Fig. 3.15 shows the details of a 40 cm long arrow as recommended by the Indian Standard.

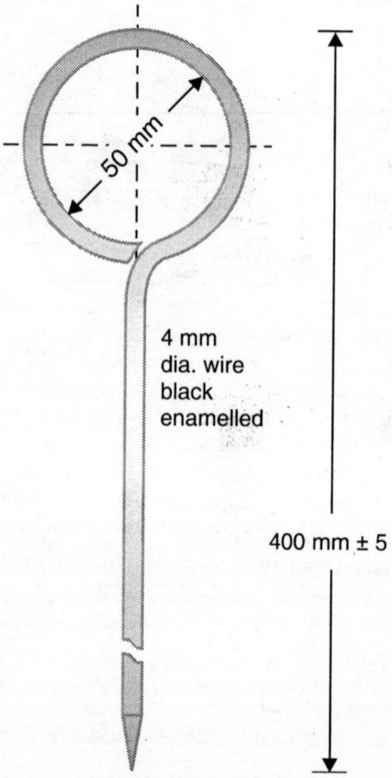

Fig. 3.15 Arrow.

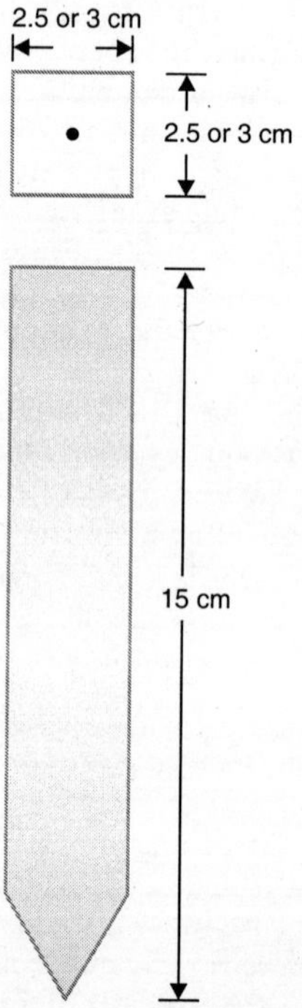

Fig. 3.16 Wooden Peg.

4. Pegs

Wooden pegs are used to mark the positions of the stations or terminal points of a survey line. They are made of stout timber, generally 2.5 cm or 3 cm square and 15 cm long, tapered at the end. They are driven in the ground with the help of a wooden hammer and kept about 4 cm projecting above the surface.

5. Ranging Rods

Ranging rods have a length of either 2 m or 3 m, the 2 metre length being more common. They are shod at the bottom with a heavy iron point, and are painted in alternative bands of either black and white or red and white or black, red and white in succession, each band being 20 cm deep so that on occasion the rod can be used for rough measurement of short lengths. Ranging rods are used to range some intermediate points in the survey line. They are circular or octagonal in cross-section of 3 cm nominal diameter, made of well-seasoned, straight grained timber. The rods are almost invisible at a distance of about 200 metres; hence when used on long lines each rod should have a red, white or yellow flag, about 30 to 50 cm square, tied on near its top [Fig. 3.17(*a*)].

Ranging poles: Ranging poles are similar to ranging rods except that they are longer and of greater diameter and are used in case of very long lines. Generally, they are not painted, but in all cases they are provided with a large flag. Their length may vary from 4 to 8 metres, and diameter from 6 to 10 cm. The foot of each pole is sunk about $\frac{1}{2}$ m into the ground, the pole being set quite vertical by aid of a plumb bob.

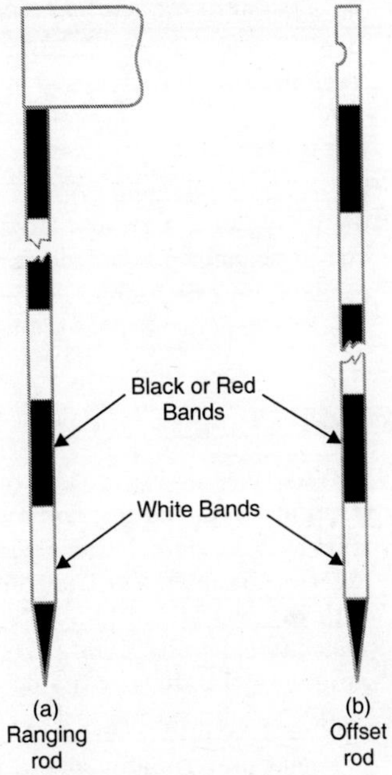

Fig. 3.17

6. Offset Rods

An offset rod is similar to a ranging rod and has a length of 3 m. They are round wooden rods, shod with pointed iron shoe at one end, and provided with a notch or a hook at the other. The hook facilitates pulling and pushing the chain through hedges and other obstructions. The rod is mainly used for measuring rough offsets nearby [Fig. 3.17 (*b*)]. It has also two narrow slots passing through the centre of the section, and set at right angles to one another, at the eye level, for aligning the offset line.

Butt rod: A butt rod is also used for measuring offsets, but it is often used by building surveyors or architects. It generally consists of two laths, each of 1 yard or 1 m in length loosely riveted together. The joint is also provided with a spring catch to keep the rod extended. The rod is painted black. The divisions of feet and inches are marked out with white and red paint.

7. Plasterer's Laths

In open level ground, intermediate points on a line may also be lined out with straight laths, 1/2 to 1 metre long, made of soft wood. They are light both in colour and weight, and can be easily carried about and sharpened with a knife when required. They are also very useful for ranging out a line when crossing a depression from which the forward rod is invisible, or when it is hidden by obstacles, such as hedges etc.

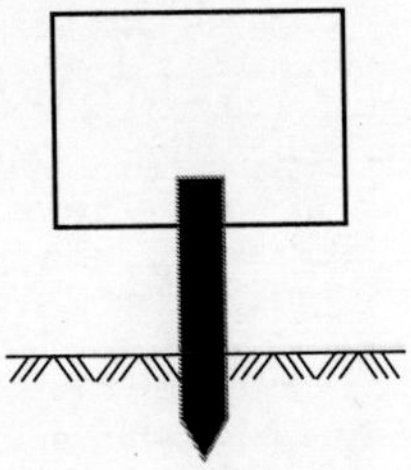

Fig. 3.18 Whites.

Whites: Whites are pieces of sharpened thin sticks cut from the nearest edge, and are used for the same purpose as the laths, though not so satisfactory in use. They are sharpened at one end and split with the knife at the top, and pieces of white paper are inserted in the clefts in order to make them more visible when stuck up in the grass. They are also useful in cross-sectioning or in temporary marking of contour points.

8. Plumb Bob

While chaining along sloping ground, a plumb-bob is required to transfer the points to the ground. It is also used to make ranging poles vertical and to transfer points from a line ranger to the ground. In addition, it is used as centering aid in theodolites, compass, plane table and a variety of other surveying instruments.

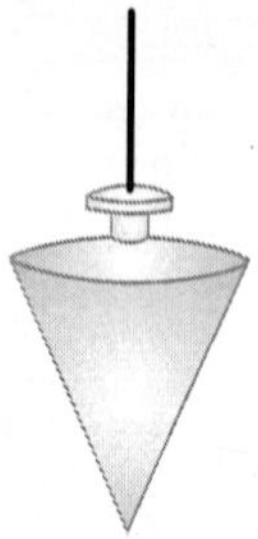

Fig. 3.19 Plumb bob.

3.4 RANGING OUT SURVEY LINES

While measuring the length of a survey line or 'chain line', the chain or tape must be stretched straight along the line joining its two terminal stations. If the length of line is less than the length of the chain, there will be no difficulty, in doing so. If however, the length of the line exceeds the length of the chain, some intermediate points will have to be established in line with the two terminal points before chaining is started. The process of fixing or establishing such intermediate points is known as *ranging*. There are two methods of ranging: (*i*) Direct ranging, (*ii*) Indirect ranging.

(*i*) **Direct Ranging:** Direct ranging is done when the two ends of the survey lines are *intervisible*. In such cases, ranging can either be done by eye or through some optical instrument such as a *line ranger* or a *theodolite*.

Ranging by eye: (Fig. 3.20). Let A and B be the two points at the ends of a survey line. One ranging rod is erected at the point B while the surveyor stands with another ranging rod at point A, holding the rod at about half metre length. The assistant then goes with another ranging rod and establishes the rod at a point approximately in the line with AB (by judgment) at a distance not greater than one chain length from A. The surveyor at A then signals the assistant to move transverse to the chain line, till he is in line with A and B. Similarly, other intermediate points can be established. The code of signals used for this purpose is given in the table below :

Fig. 3.20 Ranging by eye.

CODE OF SIGNALS FOR RANGING

S.No.	*Signal by the Surveyor*	*Action by the Assistant*
1	Rapid sweep with right hand	Move considerably to the right
2	Slow sweep with right hand	Move slowly to the right
3	Right arm extended	Continue to move to the right
4	Right arm up and moved to the right	Plumb the rod to the right
5	Rapid sweep with left hand	Move considerably to the left
6	Slow sweep with left hand	Move slowly to the left
7	Left arm extended	Continue to move to the left
8	Left arm up and moved to the left	Plumb the rod to the left
9	Both hands above head and then brought down	Correct
10	Both arms extended forward horizontally and the hands depressed briskly	Fix the rod

Ranging by Line Ranger

A line ranger consists of either two plane mirrors or two right angled isosceles prisms placed one above the other, as shown in Fig. 3.21. The diagonals of the two prisms are silvered so as to reflect the incidental rays. A handle with a hook is provided at the bottom to hold the instrument in hand to transfer the point on the ground with the help of plumb-bob.

To range a point *P*, two ranging rods are fixed at the ends *A* and *B*, and the surveyor at *P* holds the line ranger very near to the line *AB* (by eye judgment). The lower prism *abc* receives the rays from *A* which are reflected by the diagonal *ac* towards the observer. Similarly, the upper prism *dbc* receives the rays from *B* which are reflected by the diagonal *bd* towards the observer. Thus, the observer views the images of ranging rods at *A* and *B*, which may not be in the same vertical line as shown in Fig. 3.21 (*c*).

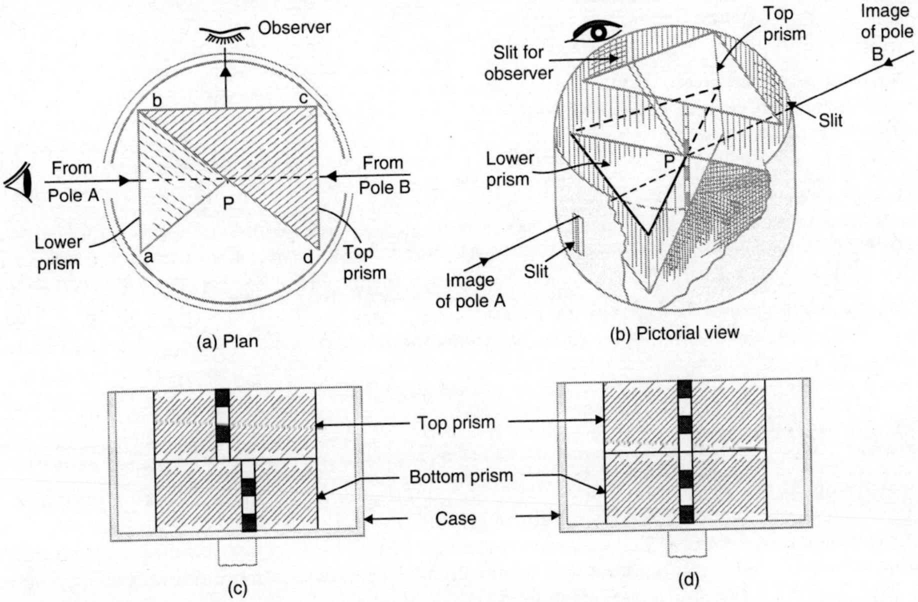

Fig. 3.21 Optical Line Ranger.

The surveyor then moves the instrument sideways till the two images are in the same vertical line as shown in Fig. 3.21 (*d*). The point *P* is then transferred to the ground with the help of a plumb bob. Thus, the instrument can be conveniently used for fixing intermediate points on a long line without going to either end. Also, only one person, holding the line ranger, is required in this case.

Fig. 3.18 shows a combined line ranger and a prism square.

Adjustment of Line Ranger: One of the mirrors or prisms is commonly made adjustable. To test the perpendicularity between the reflecting surfaces, three poles are ranged very accurately with the help of a theodolite. The line ranger is held over the middle pole. The instrument will be in perfect adjustment if the images of the two end poles appear in exact coincidence. If not, they are made to do so turning the movable prism by means of the adjusting screw.

(*ii*) Indirect or Reciprocal Ranging: *Indirect or Reciprocal ranging* is resorted to when both the ends of the survey line are *not intervisible* either due to high intervening ground or due to long distance between them. In such a case, ranging is done indirectly by selecting two intermediate points M_1 and N_1, very near to the chain line (by judgement) in such a way that from M_1, both N_1, and B are visible (Fig. 3.22) and from N_1, both M_1, and A are visible.

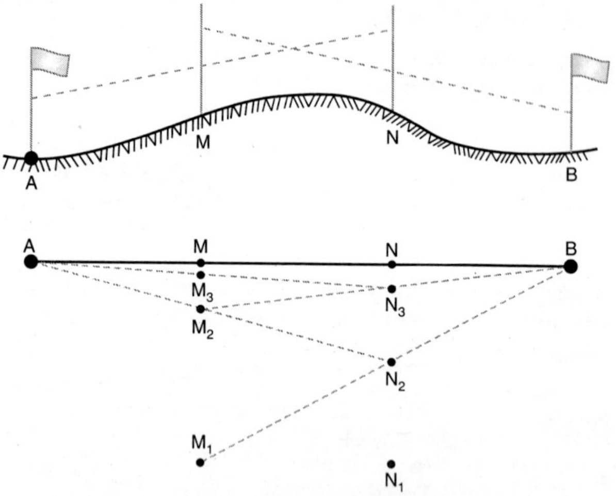

Fig. 3.22 Reciprocal Ranging.

Two surveyors station themselves at M_1 and N_1 with ranging rods. The person at M_1 then directs the person at N_1 to move to a new position N_2 in line with M_1 B. The person at N_2 then directs the person at M_1 to move to a new position M_2 in line with N_2A. Thus, the two persons are now at M_2 and N_2 which are nearer to the chain line than the positions M_1 and N_1. The process is repeated till the points M and N are located in such a way that the person at M finds the person at N in line with MB, and the person at N finds the person at M in line with NA. After having established M and N, other points can be fixed by direct ranging.

3.5 CHAINING

Two chainmen are required for measuring the length of a line which is greater than a chain length. The more experienced of the chainmen remains at the zero end or rear end of the chain and is called the *follower*. The other chainmen holding the forward handle is known as the *leader*. To start with, the leader takes a bundle of the arrows in one hand and a ranging rod, and the handle of the chain in the other hand.

Unfolding the chain: To unfold the chain, the chainmen keeps both the handles in the left hand and throws the rest of the portion of the chain in the forward direction with his right hand. The other chainman assists in removing the knots etc. and in making the chain straight.

Lining and marking: The follower holds the zero end of the chain at the terminal point while the leader proceeds forward with the other end in one hand and a set of 10 arrows and a ranging rod in the other hand. When he is approximately one chain length away, the follower directs him to fix his rod in line with the terminal poles. When the point is ranged, the leader makes a mark on the ground, holds the handle with both the hands and pulls the chain so that it becomes straight between the terminal point and the point fixed. Little jerks may be given for this purpose but the pull applied must be just sufficient to make the chain straight in line. The leader then puts an arrow at the end of the chain, swings the chain slightly out of the line and proceeds further with the handle in one hand and the rest of the arrows and ranging rod in the other hand. The follower also takes the end handle in one hand and a ranging rod in the other hand, follows the leader till the leader has approximately travelled one chain length. The follower puts the zero end of the chain at first arrow fixed by the leader, and ranges the leader who in turn, stretches the chain straight in the line and fixes the second arrow in the ground and proceeds further. The follower takes the first arrow and the ranging rod in one hand and the handle in the other and follows the leader. At the end of ten chains, the leader calls for the 'arrows'. The follower takes out the tenth arrow from the ground, puts a ranging rod there and hands over ten arrows to the leader. The transfer of ten arrows is recorded by the *surveyor*. To measure the fractional length at the end of a line, the leader drags the chain beyond the end station, stretches it straight and tight and reads the links.

3.6 MEASUREMENT OF LENGTH WITH THE HELP OF A TAPE

For accurate measurements and in all important surveys, the lengths are now measured with a tape, and not with a chain. However, the operation of measurement of the length of the line with the help of a tape is also conventionally called *chaining* and the two persons engaged in the measurement are called *'chainmen'*. The following procedure is adopted :

1. Let the length of a line *AB* be measured, point *A* being the starting point. Place a ranging rod behind the point *B* so that it is on the line with respect to the starting point *A*.
2. The follower stands at the point *A* holding one end of the tape while the leader moves ahead holding zero end of the tape in one hand and a bundle of arrows in the other. When he reaches approximately one tape length distant from *A*, the follower directs him for ranging in the line. The tape is then pulled out and whipped gently to make sure that its entire length lies along the line. The leader then pushes the arrow into the ground, opposite the zero. The pin is usually inclined from vertical about 20 or 30 degrees, starting at right angles to the line so that it slides under the tape, with its centre opposite the graduation point on the tape.
3. The follower then releases his end of the tape and the two move forward along the line, the leader dragging the tape. When the end of the tape reaches the arrow just placed, follower calls out "tape". He then picks up the end of tape and lines the leader in and the procedure is repeated as in step 2.
4. When the second arrow has been established by the leader, the follower picks up the first arrow, and both the persons move ahead as described in step 3. The procedure is repeated until ten tape lengths have been measured. At this stage, the leader will be out of arrows, while the follower will have nine arrows. The leader will then call "arrows" or "ten". When the leader moves further after the tape length has been measured, and reaches the tape length ahead, the follower takes out the tenth arrow, erects a ranging rod or a nail in its place and then transfers 10 arrows to the leader. The surveyor records the *transfer of arrows* in the field book.
5. At the end of the line, at *B*, the last measurement will generally be a partial tape length from the last arrow set to the end point of the line. The leader holds the end of the tape at *B* while the follower pulls the tape back till it becomes taut and then reads against the arrow.

3.7 ERROR DUE TO INCORRECT CHAIN

If the length of the chain used in measuring length of the line is not equal to the *true length* or the *designated length,* the measured length of the line will not be correct and suitable correction will have to be applied. If the chain is *too long,* the measured distance will be less. The error will, therefore, be *negative* and the correction is *positive.*

Similarly, if the chain is *too short,* the measured distance will be more, the error will *positive* and the correction will be *negative.*

Let L = True or designated length of the chain or tape.

L' = Incorrect (or actual) length of the chain or tape used.

(*i*) *Correction to measured length:*

Let l' = measured length of the line

l = actual or true length of the line.

Then, true length of line = measured length of line × $\frac{L'}{L}$

or $$l = l'\left(\frac{L'}{L}\right) \quad ...(3.1)$$

(*ii*) *Correction to area:*

Let A' = measured (or computed) area of the ground

A = actual or true area of the ground.

Then, true area = measured area × $\left(\frac{L'}{L}\right)^2$

or $$A = A'\left(\frac{L'}{L}\right)^2 \quad ...(3.2)$$

Alternatively, $$\frac{L'}{L} = \frac{L + \Delta L}{L} = 1 + \frac{\Delta L}{L}$$

Where ΔL = error in length of chain

Let $$\frac{\Delta L}{L} = e$$

∴ $$A = \left(\frac{L'}{L}\right)^2 \times A' = (1 + e)^2 \times A'$$

But $(1 + e)^2 = 1 + 2e + e^2 \approx 1 + 2e$, if e is small

∴ $$A = (1 + 2e)\,A' \quad ...(3.2\ a)$$

(*iii*) *Correction of volume:*

Let V' = measured or computed volume

V = actual or true volume.

Then, true volume = measured volume × $\left(\frac{L'}{L}\right)^3$

or $$V = V'\left(\frac{L'}{L}\right)^3 \quad ...(3.3)$$

Alternatively, $\frac{L'}{L} = \frac{L+\Delta L}{L} = 1 + \frac{\Delta L}{L}$

Let $\frac{\Delta L}{L} = e$

$\therefore$ $V = \left(\frac{L'}{L}\right)^3 V' = (1+e)^3 V'$

But $(1+e)^3 = 1 + e^3 + 3e^2 + 3e \approx (1+3e)$, if e is small

$\therefore$ $V = (1+3e)V'$ ...(3.3 *a*)

Example 3.1: *The length of a line measured with a 20 metre chain was found to be 250 metres. Calculate the true length of the line if the chain was 10 cm too long.*

Solution:

Incorrect length of the chain $= L' = 20 + \frac{10}{100} = 20.1$ m

Measured length $= l' = 250$ m

Hence true length of the line $= l'\left(\frac{L'}{L}\right) = 250\left(\frac{20.1}{20}\right)$ = **251.25 metres.**

Example 3.2: *The length of a survey line was measured with a 20 m chain and was found to be equal to 1200 metres. As a check, the length was again measured with a 25 m chain and was found to be 1212 m. On comparing the 20 m chain with the test gauge, it was found to be 1 decimetre too long. Find the actual length of the 25 m chain used.*

Solution: ***With 20 m chain :*** $L' = 20 + 0.10 = 20.10$ m

$\therefore$ $l = l'\left(\frac{L'}{L}\right) = 1200 \times \frac{20.10}{20} = 1206$ m = True length of line.

With 25 m chain : $l = \left(\frac{L'}{L}\right) l'$

or $1206 = \left(\frac{L'}{25}\right) 1212$

$\therefore$ $L' = \frac{1206 \times 25}{1212}$ = **24.88 m.**

Thus, the 25 m chain was 12 cm too short.

Example 3.3: *A 20 m chain was found to be 10 cm too long after chaining a distance of 1500 m. It was found to be 18 cm too long at the end of day's work after chaining a total distance of 2900 m. Find the true distance if the chain was correct before the commencement of the work.*

Solution: ***For first 1500 metres:***

Average error $= e = \frac{0+10}{2} = 5$ cm $= 0.05$ m

$\therefore \qquad L' = 20 + 0.05 = 20.05$ m

Hence $\qquad l_1 = \dfrac{20.05}{20} \times 1500 = 1503.75$ m

For next 1400 metres:

$$\text{Average error} = e = \frac{10+18}{2} = 14 \text{ cm} = 0.14 \text{ m}$$

$\therefore \qquad L' = 20 + 0.14 = 20.14$ m

Hence $\qquad l_2 = \dfrac{20.14}{20} \times 1400 = 1409.80$ m

$\therefore \qquad \text{Total length} = l = l_1 + l_2 = 1503.75 + 1409.80 = \mathbf{2913.55\ m.}$

Example 3.4: *A surveyor measured the distance between two points on the plan drawn to a scale of 1 cm = 40 m and the result was 468 m. Later, however, he discovered that he used a scale of 1 cm = 20 m. Find the true distance between the two points.*

Solution: Distance between two points measured with a scale of 1 cm to 20 m = $\dfrac{468}{20}$ = 23.4 cm

Actual scale of the plan is 1 cm = 40 m

True distance between the points = 23.4 × 40 = **936 m**

Example 3.5: *A 20 m chain used for a survey was found to be 20.10 m at the beginning and 20.30 m at the end of the work. The area of the plan drawn to a scale of 1 cm = 8 m was measured with the help of a planimeter and was found to be 32.56 sq. cm. Find the true area of the field.*

Solution:

$$L' = \text{Average length of the chain} = \frac{20.10 + 20.30}{2} = 20.20 \text{ m}$$

$$\text{Area of plan} = 32.56 \text{ sq. cm}$$

$$\text{Area of the ground} = 32.56(8)^2 = 2083.84 \text{ sq. m} = A' \text{ (say)}$$

$$\text{True area} = A = \left(\frac{L'}{L}\right)^2 A' = \left(\frac{20.20}{20}\right)^2 \times 2083.84 = \mathbf{2125.73\ sq.\ m.}$$

Alternatively, from Eq. 3.2 (*a*),

$$A \approx (1 + 2e)A'.$$

where $\qquad e = \dfrac{\Delta L}{L} = \dfrac{20.20 - 20}{20} = \dfrac{0.20}{20} = 0.01$

$\therefore \qquad A \approx (1 + 2 \times 0.01) \times 2083.84 = \mathbf{2125.52\ m^2}$

Example 3.6: *The area of the plan of an old survey plotted to a scale of 10 metres to 1 cm measures now as 100.2 sq. cm as found by planimeter. The plan is found to have shrunk so that a line originally 10 cm long now measures 9.7 cm only. There was also a note on the plan that the 20 m chain used was 8 cm too short. Find the true area of the survey.*

Solution: Present length of 9.7 cm is equivalent of 10 cm original length,

$\therefore$ Present area of 100.2 sq. cm is equivalent to $\left(\dfrac{10}{9.7}\right)^2 \times 100.2$ sq. cm = 106.49 sq. cm

= original area on plan

Scale of the plan is 1 cm = 10 m

∴ Original area of survey = (106.49) $(10)^2$ = 1.0649 × 10^4 sq. m

Faulty length of chain used = 20 – 0.08 = 19.92 m

$$\text{Correct area} = \left(\frac{19.92}{20}\right)^2 \times 1.0649 \times 10^4 \text{ sq. m.} = \mathbf{10564.7} \text{ sq. m}$$

3.8 CHAINING ON UNEVEN OR SLOPING GROUND

For all plotting works, horizontal distance between the points are required. It is therefore, necessary either to directly measure the horizontal distance between the points or to measure the sloping distance and *reduce* it to horizontal. Thus, there are two methods for getting the horizontal distance between two points: (1) Direct method, (2) Indirect method.

1. Direct Method

In the direct method or the *method of stepping,* as is sometimes called, the distance is measured in small horizontal stretches or steps. Fig. 3.23 (*a*) illustrates the procedure, where it is required to measure the horizontal distance between the two points *A* and *B*.

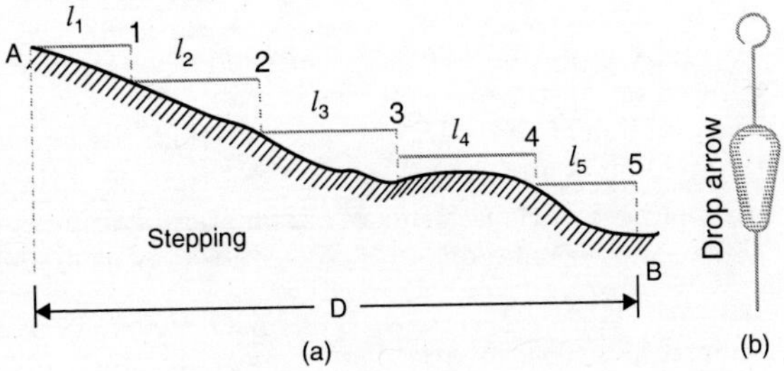

Fig. 3.23 Method of Stepping.

The follower holds the zero end of the tape at *A* while the leader selects any suitable length l_1 of the tape and moves forward. The follower directs the leader for ranging. The leader pulls the tape tight makes it horizontal and the point 1 is then transferred to the ground by a plumb bob. Sometimes, a special form of *drop arrow* is used to transfer the point to the surface, as shown in Fig. 3.23 (*b*). The procedure is then repeated. The total length *D* of the line is then equal to (l_1 + l_2 + l_3 ...). In the case of irregular slopes, this is the only suitable method.

It is more convenient to measure down-hill than to measure uphill, because in the latter case the follower end is off the ground and he is to plumb the point as well as to direct the leader. The tape must be kept horizontal either by eye judgment or by using a hand level. Sufficient amount of pull must be applied to avoid the *sag* otherwise the measured distance will be more. The lengths l_1, l_2 etc., to be selected depend on the steepness of the slope; steeper the slope, lesser the length and vice versa.

2. Indirect Method

In the case of a regular or even slope, the sloping distance can be measured and the horizontal distance can be calculated. In such cases, in addition to the sloping distance, the angle of the slope or the difference in elevation (height) between the two points is to be measured.

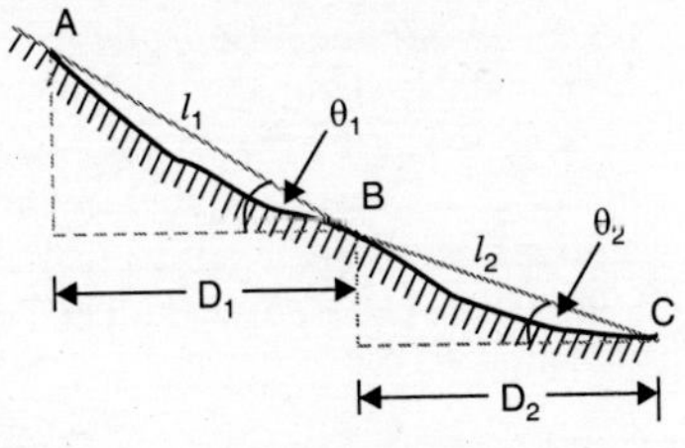

Fig. 3.24

Method 1. Angle measured: In Fig. 3.24, let l_1 = measured inclined distance between *AB* and θ_1 = slope of *AB* with horizontal. The horizontal distance D_1 is given by $D_1 = l_1 \cos \theta_1$.

Similarly, for *BC*, $D_2 = l_2 \cos \theta_2$

The required horizontal distance between any two points = $\Sigma l \cos \theta$.

The slopes of the lines can be measured with the help of a clinometer. A clinometer, in its simplest form, essentially consists of (*i*) A line of sight, (*ii*) a graduated arc, (*iii*) a light plumb bob with a long thread suspended at the centre.

Fig. 3.25 (*a*) shows a semicirular graduated arc with two pins at *A* and *B* forming the line of sight. A plumb bob is suspended from *C*, the central point. When the clinometer is horizontal, the thread touches the zero mark of the calibrated circle. To sight a point, the clinometer is tilted so that the line of sight *AB* may pass through the object. Since the thread still remains vertical, the reading against the thread gives the slope of the line of sight.

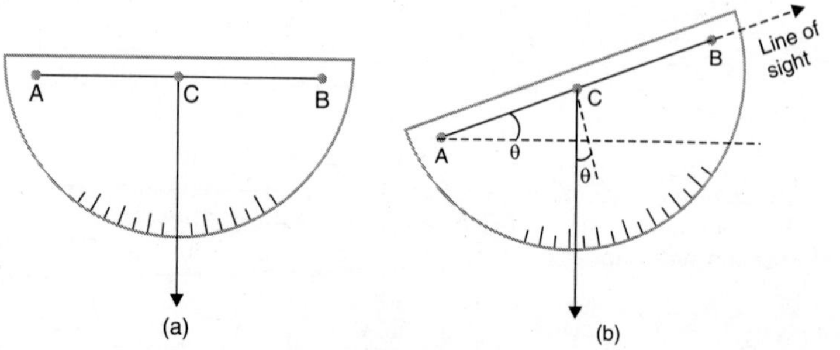

Fig. 3.25

There are various forms of clinometers available, using essentially the principle described above, and for detailed study, reference may be made to the Chapter 14 on *minor instruments.*

Method 2. Difference in level measured: Sometimes, in the place of measuring the angle θ, the difference in the level between the points is measured with the help of a levelling instrument and the horizontal distance is computed.

Thus, if *h* is the difference in level, we have

$$D = \sqrt{l^2 - h^2} \quad ...(3.4)$$

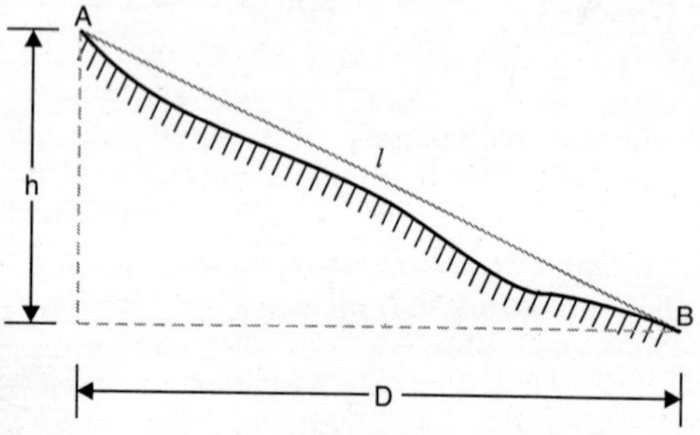

Fig. 3.26

Method 3. Hypotenusal allowance: In this method, a correction is applied in the field at every chain length and at every point where the slope changes. This facilitates in locating or surveying the intermediate points. When the chain is stretched on the slope, the arrow is not put at the end of the chain but is placed in *advance* of the end, by of an amount which allows for the slope correction. In Fig. 3.27, *BA′* is one chain length on slope. The arrow is not put at *A′* but is put at *A*, the distance *AA′* being of such magnitude that the horizontal equivalent of *BA* is equal to 1 chain. The distance *AA′* is sometimes called *hypotenusal allowance*.

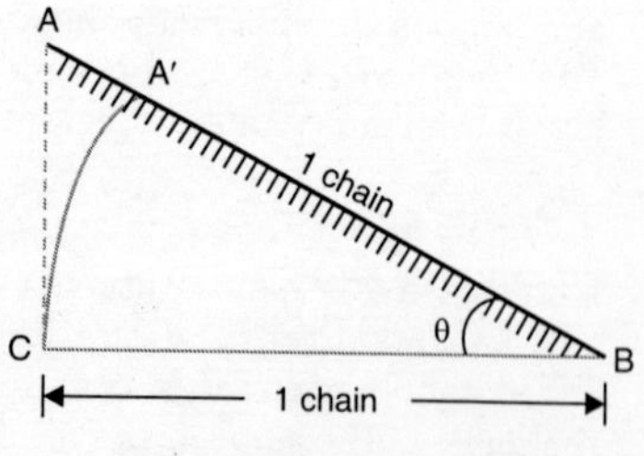

Fig. 3.27 Hypotenusal Allowance.

Thus, $BA = 100 \sec \theta$ links

$BA' = 100$ links

Hence $AA' = 100 \text{ see } \theta - 100 \text{ links} = 100 (\sec\theta - 1) \text{ links}$ ...(3.5)

Now $\sec\theta = 1 + \frac{\theta^2}{2} + \frac{5\theta^4}{24} + \ldots\ldots$ (where θ is in radian) $\approx \left(1 + \frac{\theta^2}{2}\right)$

∴ $AA' = 100\left(1 + \frac{\theta^2}{2} - 1\right)$ links

or $AA' = 50\,\theta^2$ links ...(3.5 *a*)

If, however, θ is in degrees, we have

$$\sec\theta \approx \left(1 + \frac{1.5}{10{,}000}\theta^2\right)$$

∴ $AA' = 100\left(1 + \frac{1.5}{10{,}000}\theta^2 - 1\right)$ links

or $AA' = \frac{1.5}{100}\theta^2$ links ...(3.5 *b*)

Thus, if $\theta = 10°$, $AA' = 1.5$ links.

If the slope is measured by levelling, it is generally expressed as 1 in *n*, meaning thereby a rise of 1 unit vertically for *n* units of horizontal distance.

Thus $\theta = \frac{1}{n}$ radians

Hence from Eq. 3.5 (*a*), $AA' = 50\,\theta^2 = \frac{50}{n^2}$ ...(3.5 *c*)

Thus, if the slope is 1 in 10,

$$AA' = \frac{50}{(10)^2} = 0.5 \text{ links.}$$

The distance AA' is an allowance which must be made for each chain length measured on the slope. As each chain length is measured on the slope, the arrow is *set forward* by this amount. In the record book, the horizontal distance between *B* and *A* is directly recorded as 1 chain. Thus, the slope is allowed for as the work proceeds.

Example 3.7: *The distance between the points measured along a slope is 428 m. Find the horizontal distance between them if (a) the angle of slope between the points is 8°, (b) the difference in level is 62 m (c) the slope is 1 in 4.*

Solution:

Let D = horizontal length; l = measured length = 428 m

(*a*) $D = l\cos\theta = 428\cos 8° = \mathbf{423.82\ m}$

(*b*) $D = \sqrt{l^2 - h^2} = \sqrt{(428)^2 - (62)^2} = \mathbf{423.48\ m}$

(*c*) For 1 unit vertically, horizontal distance is 4 units.

∴ $\tan\theta = \frac{1}{4} = 0.25$ or $\theta = 14°\,2'$

∴ $L = l\cos\theta = 428\cos 14°\,2' = \mathbf{415.23\ m.}$

Example 3.8: *Find the hypotenusal allowance per chain of 20 m length if (i) the angle of slope is 10° (ii) the ground rises by 4 m in one chain length.*

Solution:

(*i*) Hypotenusal allowance $= 100(\sec\theta - 1)$ links

$= 100\ (\sec 10° - 1) = 1.54$ links $=$ **0.31 m.**

(*ii*) $\tan\theta = \dfrac{4}{20} = \dfrac{1}{5} = 0.2$ or $\theta = 11°\ 19'$

Hypotenusal allowance $= 100\ (\sec 11°\ 19' - 1)$ links

$= 1.987$ links $=$ **0.4 m.**

Alternative approximate solution:

(*i*) From Eq. 3.5 (*b*),

Hypotenusal allowance $= \dfrac{1.5}{100}\ \theta^2$ links

Here $\theta = 10°$

Hypotenusal allowance $= \dfrac{1.5}{100}\ (10)^2 = 1.5$ links $=$ **0.3 m.**

(*ii*) Slope is 4 m in 20 m or 1 m in 5 m or 1 m in n m where $n = 5$.
Hence from Eq. 3.5 (*c*),

Hypotenusal allowance $= \dfrac{50}{n^2}$ links $= \dfrac{50}{(5)^2}$ links

$= 2$ links $\approx$ **0.4 m.**

Example 3.9: *In chaining a line, what is the maximum slope (a) in degrees, and (b) as 1 in n, which can be ignored if the error from this source is not to exceed 1 in 1000.*

Solution: While chaining on the sloping ground, the error is evidently equal to the hypotenusal allowance if this is not taken into account. The value of this error (*i.e.* hypotenusal allowance) is given by Eq. 3.5 (*a*), (*b*) and (*c*).

(*a*) Error per chain = 1 in 1000 = 0.1 link
Hence from Eq. 3.5 (*b*),

$$\frac{1.5}{100}\theta^2 = 0.1 \text{ link}$$

or

$$\theta^2 = \frac{0.1 \times 100}{1.5}$$

From which $\theta \approx$ **2.6°.**

(*b*) Error per chain = 0.1 link

Hence from Eq. 3.5 (*c*),

$$\frac{50}{n^2} = 0.1 \quad \text{or} \quad n^2 = \frac{50}{0.1} = 500$$

From which $n =$ **22.4.**

∴ Max. slope is **1 in 22.4.**

3.9 ERRORS IN CHAINING

A general classification of errors is given in Chapter 2 and it is necessary in studying this article to keep clearly in mind the difference between the cumulative and compensating errors, and between positive and negative errors.

A *cumulative error* is that which occurs in the same direction and tends to accumulate while a *compensating error* may occur in either direction and hence tends to compensate. Errors are regarded as *positive* or *negative* according as they make the result *too great* or *too small.*

Errors and mistakes may arise from :

1. Erroneous length of chain or tape
2. Bad ranging
3. Careless holding and marking
4. Bad straightening
5. Non-horizontality
6. Sag in chain
7. Variation in temperature
8. Variation in pull
9. Personal mistakes.

1. **Erroneous Length of Chain or Tape:** (Cumulative + or –). The error due to the wrong length of the chain is always cumulative and is the most serious source of error. If the length of the chain is more, the measured distance will be less and hence the error will be negative. Similarly, if the chain is too short, the measured distance will be more and error will be positive. However, it is possible to apply proper correction if the length is checked from time to time.
2. **Bad Ranging:** (Cumulative, +). If the chain is stretched out of the line, the measured distance will always be more and hence the error will be positive. For each and every stretch of the chain, the error due to bad ranging will be cumulative and the effect will be too great a result. The error is not very serious in ordinary work if only the length is required. But if offsetting is to be done, the error is very serious.
3. **Careless Holding and Marking:** (Compensating ±). The follower may sometimes hold the handle to one side of the arrow and sometimes to the other side. The leader may thrust the arrow vertically into the ground or exactly at the end of chain. This causes a variable systematic error. The error of marking due to an inexperienced chainman is often of a cumulative nature, but with ordinary care such errors tend to compensate.
4. **Bad Straightening:** (Cumulative, +). If the chain is not straight but is lying in an irregular horizontal curve, the measured distance will always be too great. The error is, therefore, of cumulative character and positive.
5. **Non-Horizontality:** (Cumulative, +). If the chain is not horizontal (specially in case of sloping or irregular ground), the measured distance will always be too great. The error is, therefore, of cumulative character and positive.
6. **Sag in Chain:** (Cumulative, +). When the distance is measured by 'stepping' or when the chain is stretched above the ground due to undulations or irregular ground, the chain sags and takes the form of a catenary. The measured distance is, therefore, too great and the error is cumulative and positive.
7. **Variation in Temperature:** (Cumulative, + or –). When a chain or tape is used at temperature different from that at which it was calibrated, its length changes. Due to the rise in the temperature, the length of the chain increases. The measured distance is thus less and the error becomes negative. Due to the fall in temperature, the length decreases. The measured distance is thus more and the error becomes negative. In either cases the error is cumulative.

8. **Variation in Pull:** (Compensating ±, or Cumulative + or –). If the pull applied in straightening the chain or tape is not equal to that of the standard pull at which it was calibrated, its length changes. If the pull applied is not measured but is irregular (sometimes more, sometimes less), the error tends to compensate. A chainman may, however, apply too great or too small a pull every time and the error becomes cumulative.
9. **Personal Mistakes:** Personal mistakes always produce quite irregular effects. The following are the most common mistakes :
   (*i*) *Displacement of arrows:* If an arrow is disturbed from its position either by knocking or by pulling the chain, it may be replaced wrongly. To avoid this, a cross must also be marked on the ground while inserting the arrows.
   (*ii*) *Miscounting chain length:* This is a serious blunder but may be avoided if a systematic procedure is adopted to count the number or arrows.
   (*iii*) *Misreading:* A confusion is likely between reading a 5 m tally for 15 m tally, since both are of similar shape. It can be avoided by seeing the central tag. Sometimes, a chainman may pay more attention on cm reading on the tape and read the metre reading wrong. A surveyor may sometimes read 6 in place of 9 or 28.26 in place of 28.62.
   (*iv*) *Erroneous booking.* The surveyor may enter 246 in place of 264 etc. To avoid such possibility, the chainman should first speak out the reading loudly and the surveyor should repeat the same while entering in the field book.

Summary of Errors in Chaining

1. Incorrect length of tape	Cumulative + or –
2. Bad ranging	Cumulative +
3. Tape not stretched horizontally	Cumulative +
4. Tape not stretched tight and straight, but both ends in line	Cumulative +
5. Error due to temperature	Cumulative + or –
6. Variation in pull	Compensating ±
7. Error due to sag	Cumulative +
8. Error in marking tape lengths	Compensating ±
9. Disturbing arrows after they are set	Blunder
10. Errors in reading the tape	Mistake
11. Incorrect counting of tape lengths	Blunder

Relative Importance of Errors

1. Cumulative errors are more important than compensating errors.
2. Not all the cumulative errors are equally important.
3. In a short line, a compensating error fails to compensate because such an error may occur only once or twice. The more tape lengths there are in a line, the more likely are such errors to be truly compensating.
4. The more times a line is measured, the more likely are accidental errors to disappear from the mean.
5. One cumulative error sometimes balances other cumulative error. For example, a greater pull may offset sag, or high temperature may offset a slight shortage in the length of the tape.
6. All things being equal it is most important to guard against those errors which are most likely to occur.

3.10 TAPE CORRECTIONS

We have seen the different sources of errors in linear measurements. In most of the errors, proper corrections can be applied. In ordinary chaining, however corrections are not necessary but in important and precise work, corrections must be applied. Since in most of the cases a tape is used for precise work, the corrections are sometimes called as 'tape corrections', though they can also be applied to the measurements taken with a chain or with a steel band.

A correction is positive when the erroneous or uncorrected length is to be increased and negative when it is to be decreased to get the true length.

After having measured the length, the correct length of the base is calculated by applying the following corrections :

1. Correction for absolute length
2. Correction for temperature
3. Correction for pull or tension
4. Correction for sag
5. Correction for slope
6. Correction for alignment
7. Reduction to sea level.
8. Correction to measurement in vertical plane

1. **Correction for Absolute Length:** If the *absolute length* (or actual length) of the tape or wire is not equal to its *nominal* or *designated length,* a correction will have to be applied to the measured length of the line. If the absolute length of the tape is greater than the nominal or the designated length, the measured distance will be too short and the correction will be additive. If the absolute length of the tape is lesser than the nominal or designated length, the measured distance will be too great and the correction will be subtractive.

Thus,
$$C_a = \frac{L.c}{l} \qquad ...(3.6)$$

where C_a = correction for absolute length
L = measured length of the line
c = correction per tape length
l = designated length of the tape
C_a will be of the same sign as that of c.

2. **Correction for Temperature:** If the temperature in the field is *more* than the temperature at which the tape was standardised, the length of the tape *increases,* measured distance becomes *less,* and the correction is therefore, *additive.* Similarly, if the temperature is *less,* the length of the tape *decreases,* measured distance becomes *more* and the correction is *negative.* The temperature correction is given by

$$C_t = \alpha(T_m - T_0)L \qquad ...(3.7)$$

where α = coefficient of thermal expansion
T_m = mean temperature in the field during measurement
T_0 = temperature during standardisation of the tape
L = measured length.

If, however, steel and brass wires are used simultaneously, as in Jaderin's Method, the corrections are given by

$$C_t \text{ (brass)} = \frac{\alpha_b (L_s - L_b)}{\alpha_b - \alpha_s} \quad \text{...[3.8 (a)]}$$

and

$$C_t \text{(steel)} = \frac{\alpha_s (L_s - L_b)}{\alpha_b - \alpha_s} \quad \text{...[3.8 (b)]}$$

To find the new standard temperature T_0' which will produce the nominal length of the tape or band

Some times, a tape is not of standard or designated length at a given standard temperature T_0. The tape/band will be of the designated length at a new standard temperature T_0'. Let the length at standard temperature T_0 be $l \pm \delta l$, where l is the designated length of the tape.

Let ΔT be the number of degrees of temperature change required to change the length of the tape by $= \delta l$

Then $$\delta l = (l \pm \delta l)\, \alpha\, \Delta T$$

$$\therefore \quad \Delta T = \frac{\delta l}{(l \pm \delta l)\alpha} \simeq \frac{\delta l}{l\alpha}$$

(Neglecting δl which will be very small in comparison to l)

If T_0' is the new standard temperature at which the length of the tape will be exactly equal to its designated length l, we have

$$T_0' = T_0 \pm \Delta T$$

or

$$T_0' = T_0 \pm \frac{\delta l}{l\alpha} \quad \text{...(3.9)}$$

See example 3.17 for illustration.

3. Correction for Pull or Tension: If the pull applied during measurement is *more* than the pull at which the tape was standardised, the length of the tape *increases,* measured distance becomes *less,* and the correction is *positive.* Similarly, if the pull is less, the length of the tape *decreases,* measured distance becomes more and the correction is negative. If C_p is the correction for pull, we have

$$C_p = \frac{(P - P_0)L}{AE} \quad \text{...(3.10)}$$

where P = Pull applied during measurement (N)

P_0 = Standard pull (N)

L = Measured length (m)

A = Cross-sectional area of the tape (cm^2)

E = Young's Modulus of Elasticity (N/cm^2)

The pull applied in the field should be less than 20 times the weight of the tape.

4. Correction for Sag: When the tape is stretched on supports between two points, it takes the form of a horizontal catenary. The horizontal distance will be less than the distance along the curve. The difference between horizontal distance and the measured length along catenary is called the *Sag Correction.* For the purpose of determining the correction, the curve may be assumed to be a parabola.

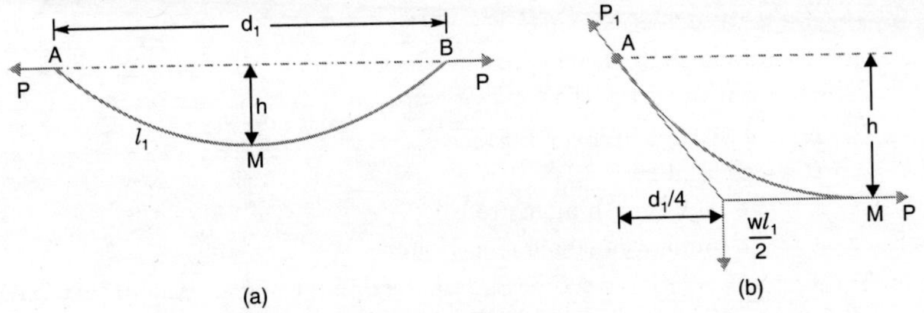

Fig. 3.28 Sag Correction

Let l_1 = length of the tape (in metres) suspended between A and B

M = centre of the tape

h = vertical sag of the tape at its centre

w = weight of the tape per unit length (N/m)

C_{s1} = Sag correction in metres for the length l_1

C_s = Sag correction in metres per tape length l

$W_1 = wl_1$ = weight of the tape suspended between A and B

d_1 = horizontal length or span between A and B.

The relation between the curved length (l_1) and the chord length (d_1) of a very flat parabola, $\left(i.e., \text{ when } \frac{h}{l_1} \text{ is small}\right)$ is given by

$$l_1 = d_1\left[1 + \frac{8}{3}\left(\frac{h}{d_1}\right)^2\right]$$

Hence $$C_{s1} = d_1 - l_1 = -\frac{8}{3}\frac{h^2}{d_1} \qquad ...(1)$$

The value of h can be found from statics [Fig. 3.28 (b)]. If the tape were cut at the centre (M), the exterior force at the point would be tension P. Considering the equilibrium of half the length, and taking moments about A, we get

$$Ph = \frac{wl_1}{2} \times \frac{d_1}{4} = \frac{wl_1 d_1}{8}$$

or $$h = \frac{wl_1 d_1}{8P} \qquad ...(2)$$

Substituting the value of h in (1), we get

$$C_{s1} = -\frac{8}{3}\frac{1}{a_1}\left(\frac{wl_1 d_1}{8P}\right)^2 = \frac{d_1}{24P^2}(wl_1)^2 \approx \frac{l_1(wl_1)^2}{24P^2} = \frac{l_1 W_1^2}{24P^2} \qquad ...(3.11)$$

If l is the total length of tape and it is suspended in n equal number of bays, the Sag Correction (C_s) per tape length is given by

$$C_s = n\,C_{sl} = \frac{nl_1(wl_1)^2}{24P^2} = \frac{l(wl_1)^2}{24P^2} = \frac{l\,(wl)^2}{24n^2P^2} = \frac{lW^2}{24n^2P^2} \qquad ...(3.12)$$

where C_s = tape correction per tape length
l = total length of the tape
W = total weight of the tape
n = number of equal spans
P = pull applied

If L = the total length measured
and N = the number of whole length tape

then, Total Sag Correction = NC_s + Sag Correction for any fractional tape length.

Note: Normally, the mass of the tape is given. In that case, the weight W (or wl) is equal to mass × g, where the value of g is taken as 9.81. For example, if the mass of tape is 0.8 kg, $W = 0.8 \times 9.81 = 7.848$ N.

It should be noted that the Sag Correction is always negative. If however, the tape was standardised on catenary, and used on flat, the correction will be equal to 'Sag Correction for standard pull – sag correction at the measured pull', and will be positive if the measured pull in the field is more than the standard pull.

For example, let the tape be standardised in catenary at 100 N pull.

If the pull applied in the field is 120 N, the Sag Correction will be = Sag Correction for 100 N pull – Sag Correction for 120 N pull

$$= \frac{l_1 W_1^2}{24(100)^2} - \frac{l_1 (W_1)^2}{24\,(120)^2}$$

$$= \frac{l_1 W_1^2}{24}\left[\frac{1}{(100)^2} - \frac{1}{(120)^2}\right]$$

and is evidently positive

If the pull applied in the field is 80 N, the Sag Correction will be

$$= \frac{l_1 W_1^2}{24(100)^2} - \frac{l_1 W_1^2}{24\,(80)^2} = \frac{l_1 W_1^2}{24}\left[\frac{1}{(100)^2} - \frac{1}{(80)^2}\right]$$

and is evidently negative.

If, however the pull applied in the field is equal to the standard pull, no Sag Correction is necessary. See Example 3.13.

Equation 3.12 gives the Sag Correction when the ends of the tape are at the same level. If, however, the ends of the tape are not at the same level, but are at an inclination θ with the horizontal, the Sag Correction given is by the formula,

$$C_s' = C_s \cos^2\theta \left(1 + \frac{wl}{P}\sin\theta\right) \qquad \text{...[3.13 (a)]}$$

when tension P is applied at the higher end ;

and $$C_s' = C_s \cos^2\theta \left(1 - \frac{wl}{P}\sin\theta\right) \qquad \text{...[3.13 (b)]}$$

when tension P is applied at the lower end.

If, however, θ is small, we can have

$$C_s' = C_s \cos^2\theta \qquad \text{...(3.14)}$$

irrespective of whether the pull is applied at the higher end or at the lower end. *It should be noted that equation 3.14 includes the corrections both for sag and slope, i.e. if equation 3.14 is used, separate correction for slope is not necessary.* See Example 3.15.

Normal Tension: Normal tension is the pull which, when applied to the tape, equalises the correction due to pull and the correction due to sag. Thus, at normal tension or pull, the effects of pull and sag are neutralised and no correction is necessary.

The correction for pull is $C_p = \dfrac{(P_n - P_0)l_1}{AE}$ (additive)

The correction for sag $C_{s1} = \dfrac{l_1(wl_1)^2}{24P_n^2} = \dfrac{l_1 W_1^2}{24P_n^2}$ (subtractive)

where P_n = the *normal pull* applied in the field.

Equating numerically the two, we get

$$\frac{(P_n - P_0)l_1}{AE} = \frac{l_1 W_1^2}{24P_n^2}$$

$$P_n = \frac{0.204\, W_1 \sqrt{AE}}{\sqrt{P_n - P_0}} \qquad ...(3.15)$$

The value of P_n is to be determined by trial and error with the help of the above equation.

5. **Correction for Slope or Vertical Alignment:** The distance measured along the slope is always greater than the horizontal distance and hence the correction is always *subtractive*.

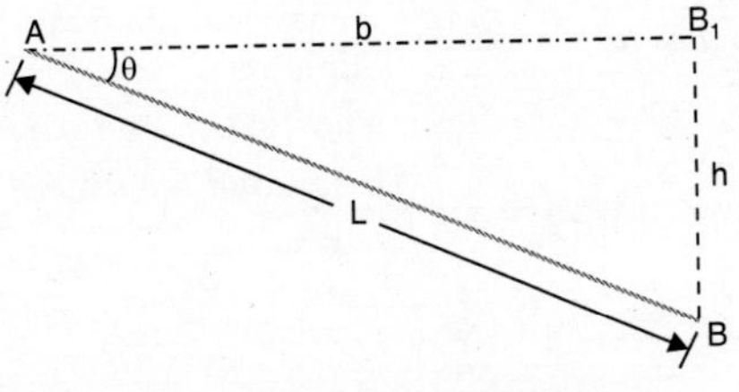

Fig. 3.29 Correction for Slope.

Let

$AB = L$ = inclined length measured

AB_1 = horizontal length

h = difference in elevation between the ends

C_v = slope correction, or correction due to vertical alignment

Then $C_v = AB - AB_1 = L - \sqrt{L^2 - h^2}$

$$= L - L\left(1 - \frac{h^2}{2L^2} - \frac{h^4}{8L^4}\right) = \frac{h^2}{2L} + \frac{h^4}{8L^3} + ...$$

The second term may safely be neglected for slopes flatter than about 1 in 25.

Hence, we get $C = \dfrac{h^2}{2L}$ (subtractive) ...(3.16)

Let L_1, L_2, ...etc. = length of successive uniform gradients

h_1, h_2, ...etc. = differences of elevation between the ends of each.

The total slope correction $= \dfrac{h_1^2}{2L_1} + \dfrac{h_2^2}{2L_2} + ... = \Sigma \dfrac{h^2}{2L}$

If the grades are of uniform length L, we get total slope correction = $\dfrac{\Sigma h^2}{2L}$

If the angle (θ) of slope is measured instead of h, the correction is given by

$$C_v = L - L\cos\theta = L(1 - \cos\theta) = 2L\sin^2\frac{\theta}{2} \qquad ...(3.17)$$

Effect of measured value of slope θ

Usually, the slope θ of the line is measured instrumentally, with a theodolite. In that case the following modification should be made to the measured value of the slope. See Fig. 3.30.

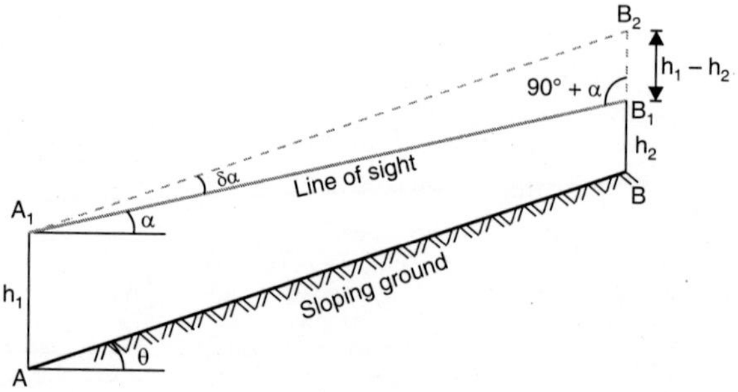

Fig. 3.30

Let h_1 = height of the instrument at A

h_2 = height of the target at B

α = measured vertical angle

θ = slope of the line AB

l = measured length of the line

Then $\theta = \alpha + \delta\alpha$.

From $\Delta\, A_1B_1B_2$, by sine rule, we get

$$\sin\delta\alpha = \frac{(h_1 - h_2)\sin(90° + \alpha)}{l} = \frac{(h_1 - h_2)\cos\alpha}{l}$$

$$\therefore \qquad \delta\alpha'' = \frac{206265(h_1 - h_2)\cos\alpha}{l} \qquad ...(3.18)$$

The sign of $\delta\alpha$ will be obtained by the above expression itself.

6. **Correction for horizontal alignment**

(*a*) **Bad ranging or misalignment:** If the tape is stretched out of line, measured distance will always be *more* and hence the *correction* will be *negative.* Fig. 3.31 shows the effect of wrong alignment, in which $AB = (L)$ is the measured length of the line, which is along the wrong alignment while the correct alignment is AC. Let d be the perpendicular deviation.

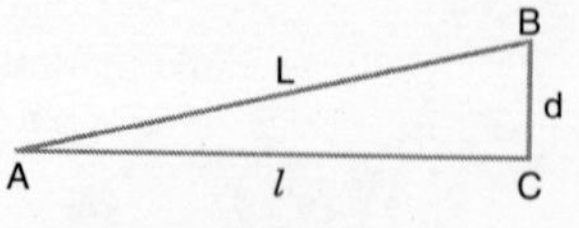

Fig. 3.31

Then $\qquad L^2 - l^2 = d^2$

or $\qquad (L + l)(L - l) = d^2$

Assuming $L = l$ and applying it to the first parenthesis only, we get

$$2L(L - l) \simeq d^2$$

or

$$L - l \simeq \frac{d^2}{2L}$$

Hence correction $$C_h = \frac{d^2}{2L} \qquad ...(3.19)$$

It is evident that smaller the value of d is in comparison to L, the more accurate will be the result.

(*b*) **Deformation of the tape in horizontal plane:** If the tape is not pulled straight and the length L_1 of the tape is out of the line by amount d, then

$$C_h = \frac{d^2}{2L_1} + \frac{d^2}{2L_2} \qquad ...(3.20)$$

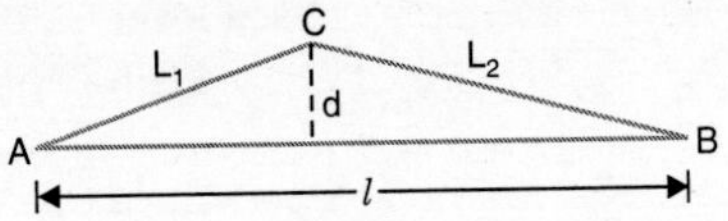

Fig. 3.32

(*c*) **Broken base:** Due to some obstructions etc., it may not be possible to set out the base in one continuous straight line. Such a base is then called a *broken base*. In Fig. 3.33,

let AC = straight base

AB and BC = two sections of the broken base

β = exterior angle measured at B.

$AB = c$; $BC = a$; and $AC = b$.

The correction (C_h) for horizontal alignment is given by

$C_h = (a+c) - b$ ....(subtractive)

The length b is given by the sine rule

$b^2 = a^2 + c^2 + 2\,ac \cos \beta$

or $a^2 + c^2 - b^2 = -2\,ac \cos \beta$

Adding $2ac$ to both the sides of the above equation, we get

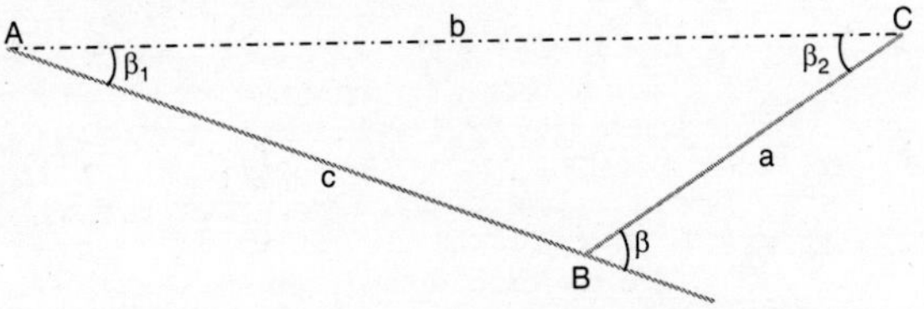

Fig. 3.33 Correction for Horizontal Alignment

$$a^2 + c^2 - b^2 + 2ac = 2ac - 2ac \cos \beta \quad \text{or} \quad (a+c)^2 - b^2 = 2ac\,(1 - \cos \beta)$$

$$\therefore \qquad (a+c) - b = \frac{2ac(1-\cos\beta)}{(a+c)+b} = \frac{4ac \sin^2 \frac{1}{2}\beta}{(a+c)+b}$$

$$\therefore \qquad C_h = (a+c) - b = \frac{4ac \sin^2 \frac{1}{2}\beta}{(a+c)+b} \qquad ...[3.21\ (a)]$$

Taking $\sin \frac{1}{2}\beta \approx \frac{1}{2}\beta$ and expressing β in minutes, we get

$$C_h = \frac{ac\,\beta^2 \sin^2 1'}{(a+c)+b} \qquad ...[3.21\ (b)]$$

Taking $b \approx (a+c)$ we get

$$C_h = \frac{ac\,\beta^2 \sin^2 1'}{2(a+c)} \qquad ...(3.21)$$

$$= \frac{ac\,\beta^2}{(a+c)} \times 4.2308 \times 10^{-8} \qquad ...[3.21\ (c)]$$

where $\frac{1}{2}\sin^2 1' = 4.2308 \times 10^{-8}$.

7. **Reduction to Mean Sea Level:** The measured horizontal distance should be reduced to the distance at the mean sea level, called the *Geodetic distance*. If the length of the base is reduced to mean sea level, the calculated length of all other triangulation lines will also be corresponding to that at mean sea level.

Let $AB = L$ = measured horizontal distance

$A'B' = D$ = equivalent length at M.S.L. = Geodetic M.S.L.

h = mean equivalent of the base line above M.S.L.

R = Radius of earth

θ = angle subtended at the centre of the earth, by AB.

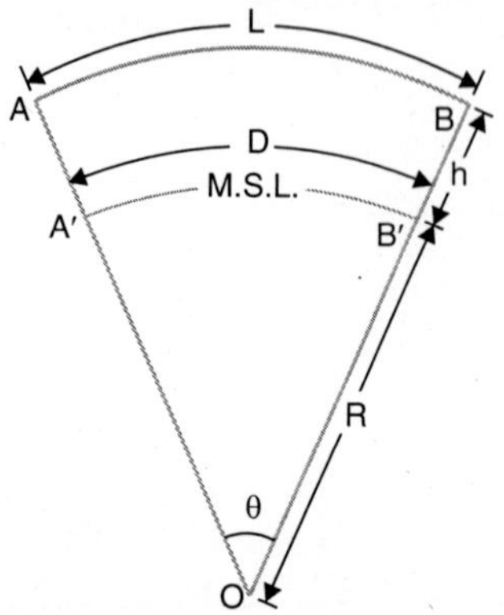

Fig. 3.34 Reduction to Mean Sea Level

Then
$$\theta = \frac{D}{R} = \frac{L}{R+h}$$

$\therefore$
$$D = L\frac{R}{R+h} = L\left(1+\frac{h}{R}\right)^{-1} = C\left(1-\frac{h}{R}\right) = L - \frac{Lh}{R}$$

$\therefore$
$$\text{Correction } (C_{ms\,l}) = L - D = \frac{Lh}{R} \text{ (subtractive)} \qquad ...(3.22)$$

8. **Correction to measurement in vertical plane:** Sometimes, as in case of measurements in mining shafts, it is required to make measurements in vertical plane, by suspending a metal tape vertically. When a metal tape AB, of length l, is *freely* suspended vertically, it will lengthen by value s due to gravitational pull on the mass ml of the tape. In other words, the tape will be subjected to a tensile force, the value of which will be zero at bottom point (B) of the tape, and maximum value of mgl at the fixed point A, where m is the mass of the tape per unit length.

Let a mass M be attached to the tape at its lower end B. Consider a section C, distant x from the fixed point A. It we consider a small length δx of the tape, its small increment δs_x in length is given by Hooke's law

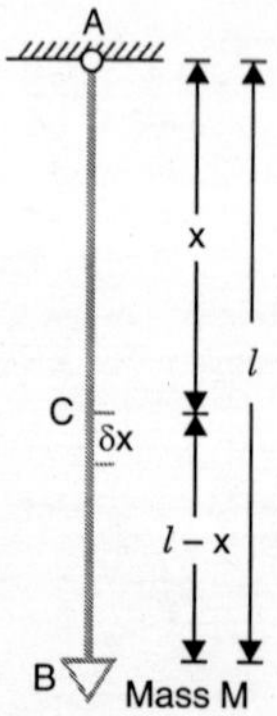

Fig. 3.35

$$\delta s_x = \frac{P(\delta x)}{AE},$$

where P = pull at point C, the value of which is given by,

$$P = Mg + mg(l - x)$$

Substituting this value, we get

or
$$AE\,\frac{\delta s_x}{\delta x} = Mg + mgl - mgx$$

Integrating,
$$AE\,s_x = Mg\,x + mglx - \frac{mgx^2}{2} + C$$

When $x = 0$ and $s_x = 0$, hence we have $C = 0$

$$\therefore \quad s_x = \frac{gx}{AE}\left[M + \frac{1}{2}m(2l - x)\right] \qquad ...(3.23\ a)$$

$$\text{If} \quad x = l, \quad s = \frac{gl}{AE}\left[M + \frac{ml}{2}\right] \qquad ...(3.23\ b)$$

$$\text{When} \quad M = 0, \quad s = \frac{mgl^2}{2\,AE} \qquad ...(3.23)$$

Taking into account the standardisation tension factor, a negative extension must be allowed initially as the tape is not tensioned up to standard tension or pull (P_0). Thus, the general equation for precise measurements is

$$s_x = \frac{gx}{AE}\left[M + \frac{1}{2}m(2l - x) - \frac{P_0}{g}\right] \qquad ...(3.24)$$

See example 3.19 for illustration.

Example 3.10: *A tape 20 m long of standard length at 84 °F was used to measure a line, the mean temperature during measurement being 65°. The measured distance was 882.10 metres, the following being the slopes:*

2° 10′	*for*	*100 m*
4° 12′	*for*	*150 m*
1° 6′	*for*	*50 m*
7° 48′	*for*	*200 m*
3° 0′	*for*	*300 m*
5° 10′	*for*	*82.10 m*

Find the true length of the line if the co-efficient of expansion is 65×10^{-7} per 1°F.

Solution: Correction for temperature of the whole length = C_t

$$= L\,\alpha\,(T_m - T_0) = 882.1 \times 65 \times 10^{-7}\,(65 - 84) = 0.109 \text{ m (Subtractive)}$$

Correction for slope = $\Sigma l(1 - \cos\theta)$

$$= 100\,(1 - \cos 2°\,10') + 150\,(1 - \cos 4°\,12') + 50\,(1 - \cos 1°\,6')$$
$$+ 200\,(1 - \cos 7°\,48') + 300\,(1 - \cos 3°) + 82.10\,(1 - \cos 5°\,10')$$
$$= 0.071 + 0.403 + 0.009 + 1.850 + 0.411 + 0.334$$
$$= 3.078 \text{ (m) (subtractive)}$$

∴ Total correction = 0.109 + 3.078 = 3.187 (subtractive)

∴ Corrected length = 882.1 – 3.187 = **878.913 m.**

Example 3.11: *(SI Units). Calculate the sag correction for a 30 m steel under a pull of 100 N in three equal spans of 10 m each. Weight of one cubic cm of steel = 0.078 N. Area of cross-section of tape = 0.08 sq. cm.*

Solution: Volume of tape per metre run = 0.08 × 100 = 8 cm^3

∵ Weight of the tape per metre run = 8 × 0.078 = 0.624 N

∴ Total weight of the tape suspended between two supports = $W = 8 \times 0.078 \times 10 = 6.24$ N

$$\text{Now correction of sag} = C_s = \frac{nl_1(wl_1)^2}{24P^2} = \frac{nl_1W^2}{24P^2} = \frac{3\times10\times(6.24)^2}{24\,(100)^2} = \mathbf{0.00487\ m.}$$

Example 3.12: A steel tape 20 m long standardised at 55° F with a pull of 10 kg was used for measuring a base line. Find the correction per tape length, if the temperature at the time of measurement was 80 °F and the pull exerted was 16 kg. Weight of 1 cubic cm of steel = 7.86 g, Wt. of tape = 0.8 kg and E = 2.109 × 10⁶ kg/cm². Coefficient of expansion of tape per 1°F = 6.2 × 10⁻⁶.

Solution: Correction for temperature = $20 \times 6.2 \times 10^{-6} (80 - 55) = 0.0031$ m (additive)

$$\text{Correction for pull} = \frac{(P - P_0)L}{AE}$$

Now, weight of tape = $A\,(20 \times 100)(7.86 \times 10^{-3})$ kg = 0.8 kg (given)

$$\therefore \quad A = \frac{0.8}{7.86\times2} = 0.051 \text{ sq. cm.}$$

$$\text{Hence,} \quad C_p = \frac{(16-10)\,20}{0.051\times2.109\times10^6} = 0.00112 \text{ (additive)}$$

$$\text{Correction for sag} = \frac{l_1(wl_1)^2}{24P^2} = \frac{20(0.8)^2}{24(16)^2} = 0.00208 \text{ m (subtractive)}$$

∴ **Total correction** = + 0.0031 + 0.00112 – 0.00208 = **+ 0.00214 m**

3.11 DEGREE OF ACCURACY IN CHAINING

Some conditions affecting the accuracy are (*i*) fineness of the graduations of the chain, (*ii*) nature of the ground, (*iii*) time and money available, (*iv*) weather etc. The error may be expressed as a ratio such as $1/n$ which means there is an error of 1 unit in the measured distance of n units. The value of n depends upon the purpose and extent of the different conditions:

1. For measurement with invar tape, spring balance, thermometers, etc. 1 in 10,000
2. For ordinary measurements with steel tape, plumb bob, chain pins etc. 1 in 1,000
3. For measurements made with tested chain, plumb bob, etc. 1 in 1,000
4. For measurements made with chain under average conditions 1 in 500
5. For measurements with chain on rough or hilly ground 1 in 250

3.12 PRECISE LINEAR MEASUREMENTS

In the linear measurements of high degree of precision, errors in measurements must be reduced to a far degree than in ordinary chaining. The method of linear measurements can be divided into three categories : (1) *Third order* (2) *Second order*, (3) *First order measurements. Third order* measurements, generally used in chain surveying and other minor surveys have been described in the previous articles. *Second order* measurements are made in the measurement of traverse lines in which theodolite is used for measuring directions. First order measurements are used in triangulation survey, for the determination of the length of base line.

1. Second Order Linear Measurements

The following specifications of second order chaining* are taken from *Manual 20,* entitled *Horizontal Control Surveys to supplement the Fundamental Net,* published by American Society of Civil Engineers.

1. **Method:** Length measurements should be made with 100 ft. tapes of invar or of steel, supported either at the 0 ft. and 100 ft. marks only, or throughout the entire tape. The *two point support method* can be adapted to all ground conditions and, therefore, is used almost exclusively. The *supported throughout method* should be used chiefly for measurements on rail road rails. It can be used on concrete road surfaces, but even when great care is taken, the wear on the tape is excessive. Reduction in cross-sections due to wear increases the length of the tape under tension because of the increased stretch and decreased sag.

   If possible, measurements should be made on hazy days, unless an invar tape is used. Measurement over bridges or other structures should always be made on cloudy days, or at night, and should be repeated several times to overcome errors due to the expansion of the structure.

2. **Equipment:** The equipment for one taping party should consist of the following: One tape; five to ten chaining tripods; one spring balance ; two standardized thermometers; two tape stretchers ; two rawhide thongs ; five to ten banker's pins for marking; two plumb bobs : adhesive tape, 1/2 in. and 1 in. widths ; one keel ; fifty stakes. 2 in. by 2 in. by 30 in; one transit, preferably with attached level ; one self-supporting target ; one level (if no transit level is available) ; one level rod, graduated to hundredths of a foot; two folding rules graduated to tenths of feet ; one brush hook, one hatchet ; one machete ; one 6 lb. or 8 lb. hammer to wooden maul ; one or two round-end shovels ; record books and pencils.

3. **Personnel:** The minimum taping party consists of the chief (who acts as marker), recorder, tension man, rear tapeman and instrument man. A level man must be added if the transit is not equipped with a level or if a hand level is used.

4. **Field Procedure : tape supported at two points:** A target is set at the point towards which measurement is to be made, and the tripods are distributed roughly in position. The transit is set up at the point of beginning and sighted on the target. Although alignment by transit is not necessary, it increases the speed of the party greatly. If the beginning point is not readily accessible to the tape, a taping tripod is placed under the instrument, carefully in order not to disturb it, and the starting point is transferred to the edge of the top of the taping tripod by the instrument plumb bob. The tripod is not removed until the taping of the section is completed.

   The tape is stretched out in the line of progress with the 100 ft. mark forward, and a thermometer is attached at the 2 ft. mark with adhesive tape so that the bulb is in contact with the measuring tape, but free from adhesive tape. A loop of rawhide or string is passed through the eye of the tape at the zero end, and tied 6 to 18 in. from the tape. The tape end is laid on the starting tripod. A rear tapeman passes his stretcher through the loop and places the lower end of the stretcher on the ground against the outside of his right foot. The upper end is under his right arm and behind his shoulder. In this position, he leans over the tape to see that the zero graduation is held exactly on the mark. This is readily controlled by adjusting his stance. However, he may find it helpful to grasp the

* "Surveying Theory and Practice" by John Clayton Tracy.

tape near its end and behind the mark, applying a slight kinking force, just sufficient to control the position of the zero graduation.

The tension man passes his stretcher through a 6-in. loop of rawhide attached to the spring balance, snaps the spring balance to the tape, and using the same position employed by the rear tapeman, applies a 200 lb. tension.

The chief of the party who acts as marker places a tripod in line (as directed by the instrument man) and under the 100 ft. graduation. The tension man slides his rawhide thong until the tape just clears the top of the tripod. The marker must see that the tape is dry, clean and free from all obstructions and may run a light sag along its entire length at this time to remove any moisture or dirt. The marker gently depresses the tape to touch the marking surface of the forward tripod and, on a signal from the tension man that he has exactly 20 lb., and from the rear tapeman that the mark is right, he marks the tripod at 100 ft. mark. When the tripod has a wooden top, the mark may be made with hard pencil or with a T-shaped banker's pin which is forced into the wood to mark the point, and is always left sticking in the tripod. Bristol board of the thickness of the tape may be secured to the top of the tripod with Scotch marking tape, so that the edge of tape butts against the edge of the Bristol board. The terminal mark of the tape can then be transferred to the board with a marking awl or a sharp, hard pencil. The Bristol board can be renewed at any time. On the heavier type of tripod, the mark may be made on the strip of white adhesive tape attached temporarily to the top of the tripod. Tension is released slowly, then re-applied for a check on the marking, signals from the tension man and rear tapeman being repeated. The recorder obtains the temperature from the rear tapemen, holds the rod on the tops of the chaining tripods for the instrument man and records the rod readings. A record is made for each individual tape length or partial tape length, which includes the length used, the temperature and the inclination.

The marker moves back to support the centre of the tape, and it is then carried forward, the tape being held clear of all contacts by the marker, the tension man and the rear tapeman. After the second tape length is measured, the recorder may begin picking up the tripods. He can carry about five of these, to be distributed later to the entire party. When it is necessary to bring the transit up, one of the tripods is placed accurately on line and the instrument is set up over it. For distance of less than a tape length, the tape is read independently by both the chief of party and the recorder. If the reverse side of the tape is graduated in metres, the metric reading should be recorded as well. The head of the tape is carried beyond the end point, the zero mark being at the back tripod as usual. If the set up is more than 50 ft., a 20 lb. tension is used; otherwise a pull of 10 lb. is used ; this affords a close approximation for proportional application of the standard tape correction.

5. **Field procedure : tape supported throughout:** When the tape is supported throughout, the procedure is much the same as in the foregoing description, except that no transit alignment is necessary on railroad rails. The rails themselves are sufficiently accurate. Stretchers are placed in from the foot, which is placed on the base of the fulcrum. The recorder must aid the rear tapeman in making contact. On railroad rails or asphalt roads, marks can be made with a sharp awl, but on concrete surfaces a piece of adhesive tape should be stuck to the pavement and marked with a hard pencil.

6. **Backward Measurement:** It is best to measure each section in two directions. Although this is not demanded by the accuracy required, it provides the only proper check against blunders. The results, reduced for temperature and inclination should agree within one part in 30,000.

7. **Levelling:** Levelling may be done with a surveyor's level, the attached level on a transit, or hand level or a clinometer. All have been used successfully, but the first two increase both speed and precision. When a surveyor's level or a transit level is used, readings are taken to hundredth's of a foot on the tops of the tripods. A reading is taken on the same tripod from each of the two instrument positions, when the instrument is moved, and care taken to denote which reading was obtained from each position.

   An extra man should be available when the hand level is used. He should carry a light notched stick to support the level, and standing near the 50 ft. mark, should take a reading to tenths of a foot on both tripods for each tape length, recording the difference in elevation. Collapsible foot rules, graduated to tenths of a foot, should be carried by the tension man and the rear tapeman for the levelman to sight on. The clinometer is most successfully employed when 4 ft. taping tripods are used. It is placed on one tripod and sighted on a small target on the next tripod. The angle of inclination or the percentage grade is recorded.

8. **Field Computations:** The measurements should be reduced as soon as possible, either in the field or in the field office. A form for computation is given below

FORM OF COMPUTATION FOR REDUCING MEASUREMENTS

Section	Date	Direction of Measurement	Tape No.	Tape Support	*Uncorrected length*		Temperature °F	*Correction*					Reduced length	Adopted Geodetic length
					Tape lengths	Metres		Temperature (*m*)	Inclination (*m*)	Tape and Catenary (*m*)	Set up (+) (= + Cor.) (*m*)	Sea Level (*m*)	(*m*)	(*m*)

2. First Order Measurements: Base Line Measurements

There are two forms of base measuring apparatus: (A) Rigid bars, and (B) Flexible apparatus.

(A) Rigid Bars: Before the introduction of invar tapes, rigid bars were used for work of highest precision. The rigid bars may be divided into two classes :

(*i*) *Contact apparatus:* in which the ends of the bars are brought into successive contacts. Example: The Eimbeck Duplex Apparatus.

(*ii*) *Optical apparatus:* in which the effective lengths of the bars are engraved on them and observed by microscopes. Example: The Colby apparatus and the Woodward Iced Bar Apparatus.

The rigid bars may also be divided into the following classes depending upon the way in which the uncertainties of temperature corrections are minimised :

(*i*) *Compensating base bars:* which are designed to maintain constant length under varying temperature by a combination of two or more metals. Example: The Colby Apparatus.

(*ii*) *Bimetallic non-compensating base bars:* in which two measuring bars act as a bimetallic thermometer. Example: The Eimbeck Duplex Apparatus (U.S. Coast and Geodetic Survey), Borda's Rod (French system) and Bessel's Apparatus (German system).

(*iii*) ***Monometallic base bars:*** in which the temperature is either kept constant at melting point of ice, or is otherwise ascertained. Example: The Woodward Iced Bar Apparatus and Struve's Bar (Russian system).

The Colby Apparatus (Fig. 3.36). This is compensating and optical type rigid bar apparatus designed by Maj-Gen. Colby to eliminate the effect of changes of temperature upon the measuring appliance. The apparatus was employed in the Ordinance Survey and the Indian Surveys. All the ten bases of G.T. Survey of India were measured with Colby Apparatus. The apparatus (Fig. 3.36) consists of two bars, one of steel and the other of brass, each 10 ft. long and riveted together at the centre of their length. The ratio of co-efficients of linear expansion of these metals having been determined as 3:5. Near each end of the compound bar, a metal tongue is supported by double conical pivots held in forked ends of the bars. The tongue projects on the side away from the brass rod. On the extremities of these tongues, two minute marks a and a' are put, the distance between them being exactly equal to 10′ 0″. The distance ab (or (a' b') to the junction with the steel is kept $\frac{3}{5}$ ths of distance ac (or a' c') to the brass junction. Due to change in temperature, if the distance bb' of steel change to b_1 b_1' by an amount x, the distance cc' of brass will change to c_1c_1' by an amount $\frac{3}{5}x$, *thus unaltering the positions of dots a and a'*. The brass is coated with a special preparation in order to render it equally susceptible to change of temperature as the steel. The compound bar is held in the box at the middle of its length. A spirit level is also placed on the bar. In India, five compound bars were simultaneously employed in the field. The gap between the forward mark of one bar and the rear bar of the next was kept constant equal to 6″ by means of a *framework* based on the same principles as that of the 10′ compound bar. The framework consists of two microscopes, the distance between the cross-wires of which was kept exactly equal to 6″. To start with, the cross-wires of the first microscope of the *framework* was brought into coincidence with the platinum dot, let into the centre of the one extremity of the base line. The platinum dot a of the first compound bar was brought into the coincidence with the cross-hairs of second microscope. The cross-hairs of the first microscope of the second framework (consisting two microscopes 6″ apart) is then set over the end a' of the first rod. The work is thus continued till a length of (10′ × 5 + 5 × 6″) = 52′ 6″ is measured at a time with the help of 5 bars and 2 frameworks. The work is thus continued till the end of the base is reached.

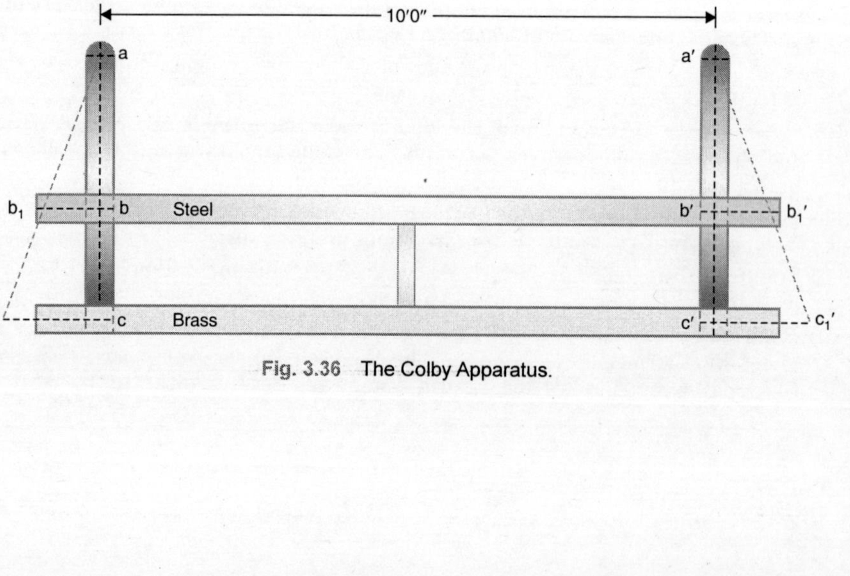

Fig. 3.36 The Colby Apparatus.

(B) Flexible Apparatus: In the recent years, the use of flexible instruments has increased due to the longer lengths that can be measured at a time without any loss in accuracy. The flexible apparatus consists of (*a*) steel or invar tapes, and (*b*) steel and brass wires. The flexible apparatus has the following *advantages* over the rigid bars :

(*i*) Due to the greater length of the flexible apparatus, a wider choice of base sites is available since rough ground with wider water gaps can be utilised.

(*ii*) The speed of measurement is quicker, and thus less expensive.

(*iii*) Longer bases can be used and more check bases can be introduced at closer intervals.

Equipment for base line measurement: The equipment for base line measurement by flexible apparatus consists of the following :

1. Three standardised tapes: out of the three tapes one is used for field measurement and the other two are used for standardising the field tape at suitable intervals.
2. Straining device, marking tripods or stakes and supporting tripods or staking.
3. A steel tape for spacing the tripods or stakes.
4. Six thermometers: four for measuring the temperature of the field and two for standardising the four thermometers.
5. A sensitive and accurate spring balance.

The Field Work: The field work for the measurement of base line is carried out by two parties :

1. *The setting out party,* consisting of two surveyors and a number of porters, have the duty to place the measuring tripods in alignment in advance of the measurement, and at correct intervals.
2. *The measuring party,* consisting of two observers, recorder, leveller and staffman, for actual measurements.

The base line is cleared of the obstacles and is divided into suitable sections of $\frac{1}{2}$ to 1 kilometre in length and is accurately aligned with a transit. Whenever the alignment changes, stout posts are driven firmly in the ground. The setting out party then places the measuring tripods in alignments in advance of the measurement which can be done by two methods :

(*i*) Measurement on Wheeler's method by Wheeler's base line apparatus.

(*ii*) Jaderin's method.

(i) Wheeler's base line apparatus (Fig. 3.37): The marking stakes are driven on the line with their tops about 50 cm above the surface of the ground, and at distance apart slightly less than the length of the tape. On the tops of the marking stakes, strips of zinc, 4 cm in width, are nailed for the purpose of scribing off the extremities of the tapes. Supporting stakes are also provided at interval of 5 to 15 metres, with their faces in the line. Nails are driven in the sides of the supporting stakes to carry hooks to support the tape. The points of supports are set either on a uniform grade between the marking stakes or at the same level. A weight is attached to the other end of the straining tripod to apply a uniform pull. To measure the length, the rear end of the tape is connected to the straining pole and the forward end to the spring balance to the other end of which a weight is attached. The rear end of the tape is adjusted to coincide with the mark on the zinc strip at the top of the rear marking stake by means of the adjusting screw of the side. The position of the forward end of the tape is marked on the zinc strip at the top of the forward marking stake after proper tension has been applied. The work is thus continued. The thermometers are also observed.

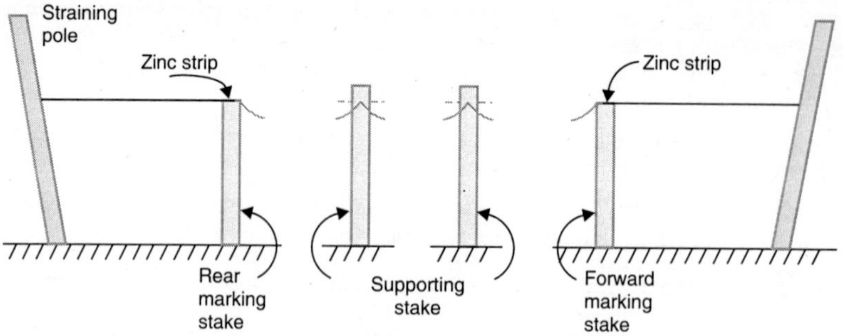

Fig. 3.37 Wheeler's base line Apparatus.

(ii) Jaderin's method (Fig. 3.38): In this method introduced by Jaderin, the measuring tripods are aligned and set at a distance approximately equal to the length of the tape. The ends of the tapes are attached to the straining tripods to which weights are attached. The spring balance is used to measure the tension. The rear mark of the tape is adjusted to coincide with the mark on rear measuring tripod. The mark on the forward measuring tripod is then set at the forward mark of the tape. The tape is thus suspended freely and is subjected to constant tension. An aligning and levelling telescope is also sometimes fitted to the measuring tripod. The levelling observations are made by a level and a light staff fitted with a rubber pad for contact with the tripod heads. The tension applied should not be less than 20 times the weight of the tape.

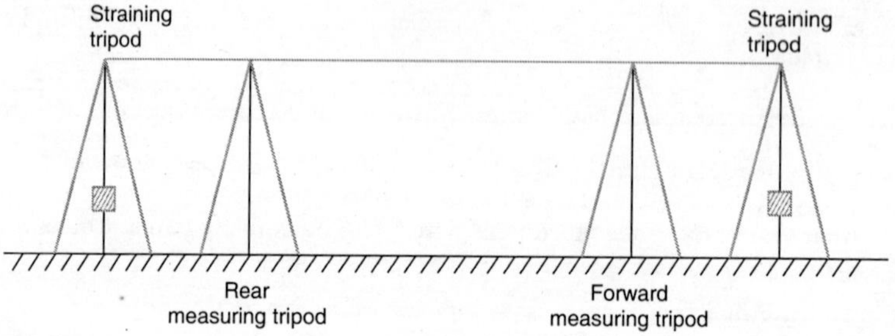

Fig. 3.38 Jaderin's Method.

Measurement by Steel and Brass Wires : Principle of Bimetallic Thermometer: The method of measurement by steel and brass wire is based on Jaderin's application of the principle of bimetallic thermometer to the flexible apparatus. The steel and brass wire are each 24 m long and 1.5 to 2.6 mm in diameter. The distance between the measuring tripods is measured first by the steel wire and then by the brass wire by Jaderin method as explained above (Fig. 3.38) with reference to invar tape or wire. Both the wires are nickel plated to ensure the same temperature conditions for both. From the measured lengths given by the steel and brass wires, the temperature effect is eliminated as given below :

Let L_s = distance as computed from the absolute length of the steel wire

L_b = distance as computed from the absolute length of the brass wire

α_s = co-efficient of expansion for steel

α_b = co-efficient of expansion for brass
D = corrected distance
T_m = mean temperature during measurement
T_s = temperature at standardisation
$T = T_m - T_s$ = temperature increase

Now $$D = L_s(1 + \alpha_s T) = L_b(1 + \alpha_b T) \quad ...(1)$$

or $$T(L_b \alpha_b - L_s \alpha_s) = L_s - L_b$$

$\therefore$ $$T = \frac{L_s - L_b}{L_b \alpha_b - L_s \alpha_s} \quad ...(2)$$

Substituting this value of T in (1) for steel wire, we get

$$D = L_s \left\{1 + \frac{\alpha_s(L_s - L_b)}{L_b \alpha_b - L_s \alpha_s}\right\}$$

$\therefore$ Correction for steel wire = $D - L_s$

$$= + \frac{L_s \alpha_s(L_s - L_b)}{L_b \alpha_b - L_s \alpha_s} \approx + \frac{\alpha_s(L_s - L_b)}{\alpha_b - \alpha_s} \quad ...(3.25)$$

with sufficient accuracy.

Similarly, correction for brass wire $= D - L_b \approx + \dfrac{\alpha_b(L_s - L_b)}{\alpha_b - \alpha_s}$ ...(3.26)

The corrections can thus be applied without measuring the temperature in the field. The method has however been superseded by the employment of invar tapes or wires.

Example 3.13: *A nominal distance of 30 metres was set out with a 30 m steel tape from a mark on the top of one peg to a mark on the top of another, the tape being in catenary under a pull of 10 kg and at a mean temperature of 70° F. The top of one peg was 0.25 metre below the top of the other. The top of the higher peg was 460 metres above mean sea level. Calculate the exact horizontal distance between the marks on the two pegs and reduce it to mean sea level, if the tape was standardised at a temperature of 60° F, in catenary, under a pull of (a) 8 kg, (b) 12 kg, (c) 10 kg.*

Take radius of earth = 6370 km
Density of tape = 7.86 g/cm³
Section of tape = 0.08 sq. cm
Co-efficient of expansion = 6×10^{-6} per 1° F
Young's modulus = 2×10^6 kg/cm²

Solution:

(*i*) Correction for standardisation = nil

(*ii*) Correction for slope $= \dfrac{h^2}{2L} = \dfrac{(0.25)^2}{2 \times 30} = 0.0010$ m (subtractive)

(*iii*) Temperature correction $= L.\,\alpha\,(T_m - T_0) = 30 \times 6 \times 10^{-6}\,(70 - 60)$
$= 0.0018$ m (additive)

(*iv*) Tension correction $= \dfrac{(P - P_0)L}{AE}$

(*a*) When $P_0 = 8$ kg

Tension correction $= \dfrac{(10-8)30}{0.08\times2\times10^6} = 0.0004$ m. (additive)

(*b*) When $P_0 = 12$ kg,

Tension correction $= \dfrac{(10-12)30}{0.08\times2\times10^6} = 0.0004$ m (subtractive)

(*c*) When $P_0 = 10$ kg, Tension correction = zero

(*v*) Sag correction $= \dfrac{LW^2}{24P^2}$

Now weight of tape per metre run = $(0.08 \times 1 \times 100) \times \dfrac{7.86}{1000}$ kg = 0.06288 kg/m

∴ Total weight of tape = 0.06288 × 30 = 1.886 kg

(*a*) When $P_0 = 8$ kg, sag correction $= \dfrac{30\times(1.886)^2}{24(8)^2} - \dfrac{30(1.886)^2}{24(10)^2}$

$= 0.0695 - 0.0445 = 0.0250$ (additive)

(*b*) When $P_0 = 12$ kg, sag correction $= \dfrac{30(1.886)^2}{24(12)^2} - \dfrac{33(1.886)^2}{24(10)^2}$

$= 0.0309 - 0.0445 = -0.0136$ m (subtractive)

(*c*) When $P_0 = 10$ kg $= P$, sag correction is zero.

Final correction

(*a*) Total correction = – 0.0010 + 0.0018 + 0.0004 + 0.0250 m = **+ 0.0262 m.**

(*b*) Total correction = – 0.0010 + 0.0018 – 0.0004 – 0.0136 = **– 0.0132 m**

(*c*) Total correction = – 0.0010 + 0.0018 + 0 + 0 = **+ 0.0008 m**

Example 3.14: *(SI units). It is desired to find the weight of the tape by measuring its sag when suspended in catenary with both ends level. If the tape is 20 metre long and the sag amounts to 20.35 cm at the mid-span under a tension of 100 N, what is the weight of the tape ?*

Solution: From expression for sag, we have

$$h = \frac{wl_1d_1}{8P}$$

But $h = 20.35$ cm (given)

Taking $l_1 = d_1$ (approximately), we get

$$h = \frac{wl_1^2}{8P}$$

or $w = \dfrac{8Ph}{l_1^2} = \dfrac{8\times100}{20\times20}\times\dfrac{20.35}{100}$ N/m = **0.407 N/m**

Example 3.15: *Derive an expression for correction to be made for the effects of sag and slope in base measurement, introducing the case where the tape or wire is supported at equidistant points between measuring pegs or tripods.*

Solution: (Fig. 3.39)

In Fig. 3.39, let tape be supported at A and B, and let C be the lowest point where the tension is horizontal having value equal to P. Let the horizontal length be l_1 and l_2 such that $l_1 + l_2 = l$. Let s_1 and s_2 be the lengths along the curve such that $s_1 + s_2 = s$ = total length along the curve. Let a = difference in elevation between A and C, and b = difference in elevation between B and C. Let $h = b - a$ = difference in level between B and A. Treating approximately the curve to be parabola, the equations are :

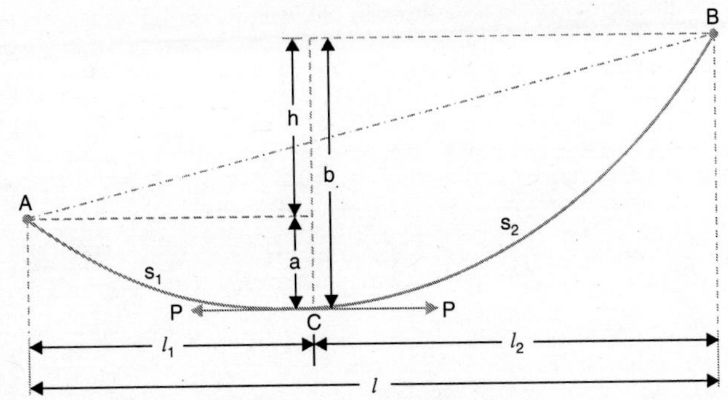

Fig. 3.39

$$y = k_1 x^2, \text{ for } CA \text{ and } y = k_2 x^2, \text{ for } CB$$

where the origin is at C in both the cases.

Now, when $x = l_1, y = a \; ; \quad \therefore \quad k_1 = \dfrac{a}{l_1^2}$

and, When $x = l_2, y = b \; ; \quad \therefore \quad k_2 = \dfrac{b}{l_2^2}$

Hence the equations are :

$$y = \frac{ax^2}{l_1^2} \text{ for } CA \quad \text{and} \quad y = \frac{bx^2}{l_2^2} \text{ for } CB$$

$$\therefore \quad \frac{dy}{dx} = \frac{2ax}{l_1^2} \text{ for } CA \quad \text{and} \quad \frac{dy}{dx} = \frac{2bx}{l_2^2} \text{ for } CB$$

Thus, the length of the curve

$$s = s_1 + s_2 = \int_0^{l_1} \left\{1 + \left(\frac{2ax}{l_1^2}\right)^2\right\} dx + \int_0^{l_2} \left\{1 + \left(\frac{2bx}{l_2^2}\right)^2\right\} dx$$

$$= \left[l_1 + l_2 + \frac{2}{3}\left(\frac{a^2}{l_1} + \frac{b^2}{l_2}\right)\right] = l + \frac{2}{3}\left(\frac{a^2}{l_1} + \frac{b^2}{l_2}\right) \qquad \text{...(1)}$$

Again, from the statics of the figure, we get

$$P \times a = \frac{wl_1^2}{2} \text{ for } CA, \text{ and } P \times b = \frac{wl_2^2}{2} \text{ for } CB$$

$$\therefore \quad P = \frac{wl_1^2}{2a} = \frac{wl_2^2}{2b} \qquad \text{...(2)}$$

and

$$\frac{a}{l_1^2} = \frac{b}{l_2^2} \qquad \text{...(3)}$$

Substituting these values in (1), we get

$$s-l=\frac{2}{3}\left\{\left(\frac{W}{2P}\right)^2.l_1^3+\left(\frac{W}{2P}\right)^2 l_2^3\right\}=\frac{1}{6}\frac{w^2}{P^2}(l_1^3+l_2^3)$$

Now, writing $l_1=\frac{1}{2}l-e$ and $l_2=\frac{1}{2}l+e$, we get

$$(s-l) = (\text{sag} + \text{level})\text{ correction}$$

$$=\frac{1w^2}{6P^2}\left[\left(\frac{1}{2}l-e\right)^3+\left(\frac{1}{2}l+e\right)^3\right]=\frac{w^2}{6P^2}\left\{\frac{l^3}{4}+\frac{3}{4}l(l_2-l_1)^2\right\}$$

$$=\frac{w^2l^3}{24P^2}+\frac{w^2}{8P^2}.\frac{l^2(l_2-l_1)^2}{l}=\frac{l(wl)^2}{24P^2}+\frac{w^2}{8P^2}.\frac{(l_2^2-l_1^2)^2}{l} \quad ...(4)$$

Now from (3), $\frac{b-a}{a}=\frac{l_2^2-l_1^2}{l_1^4}$ and from (2), $\frac{w^2}{4P^2}=\frac{a^2}{l_1^4}$

$$\therefore \quad \frac{w^2}{8P^2}.\frac{(l_2^2-l_1^2)^2}{l}=\frac{a^2}{2l_1^4}.\left\{\frac{b-a}{a}.l_1^2\right\}^2\frac{1}{l}=\frac{(b-a)^2}{2l}=\frac{h^2}{2l}$$

Substituting in (4), we get

$$(s-l)=\frac{l(wl)^2}{24P^2}+\frac{h^2}{2l}$$

Thus, the total correction is the sum of the separate corrections for sag and slope.

Example 3.16: *A flexible, uniform, inextensible tape of total weight 2W hangs freely between two supports at the same level under a tension T at each support. Show that horizontal distance between the supports is*

$$\frac{H}{w}log_e\frac{T+W}{T-W}$$

where H = horizontal tension at the centre of the tape and w = weight of tape per unit length.

Solution: Fig. 3.40 (*a*) shows the whole tape, being hung from two supports *A* and *B*. Let *O* be the lowest point, which is the origin of co-ordinates. Fig. 3.40 (*b*) shows a portion *OM* of the tape, of length *s*, such that the horizontal tension at *O* is *H*, and the tension *P* at point *M* makes an angle ψ with the *x*-axis. Resolving forces vertically and horizontally for this portion of tape,

$$P\sin\psi = w.s \quad ...(1)$$

$$P\cos\psi = H \quad ...(2)$$

$$\therefore \quad \tan\psi=\frac{w.s}{H} \quad \text{(From 1 and 2)}$$

Differentiating with respect to *x*,

$$\sec^2\psi\,\frac{d\psi}{dx}=\frac{w}{H}\frac{ds}{dx} \quad ...(3)$$

Now, from the elemental triangle [Fig. 3.40 (*c*)]

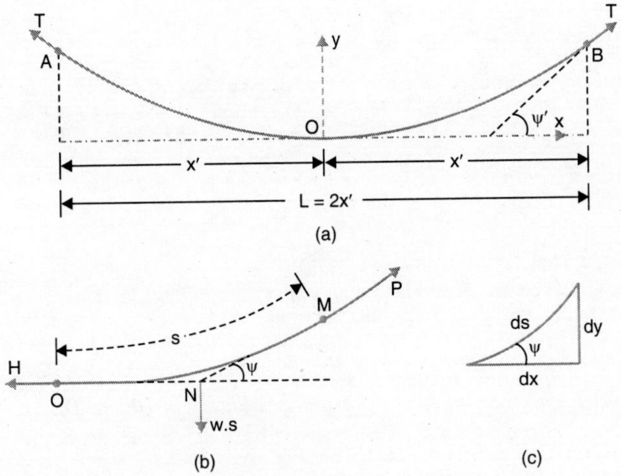

Fig. 3.40

$$\frac{ds}{dx} = \sec \psi$$

$\therefore$ $$\sec^2 \psi . \frac{d\psi}{dx} = \frac{w}{H} \sec \psi$$

or $$\sec \psi . \frac{d\psi}{dx} = \frac{w}{H} \qquad ...(4)$$

Let x' be half the length of tape, and ψ' be the inclination of tangent at the end. Integrating Eq. (4) from O to B, we get

$$\int_0^{\psi'} \sec \psi \, d\psi = \int_0^{x'} \frac{w}{H} dx$$

$\therefore$ $$[\log_e(\sec \psi + \tan \psi)]_0^{\psi'} = \frac{w}{H} x'$$

or $$x' = \frac{H}{w} \left(\log_e \frac{\sec \psi' + \tan \psi'}{1+0} \right)$$

or $$x' = \frac{H}{w} \log_e (\sec \psi' + \tan \psi') \qquad ...(5)$$

Again, resolving vertically for one-half of the tape,

$$T \sin \psi' = W \quad \text{or} \quad \sin \psi' = \frac{W}{T}.$$

$\therefore$ $$\cos \psi' = \sqrt{1 - \sin^2 \psi'} = \frac{\sqrt{T^2 - W^2}}{T}$$

Also, $\tan \psi' = \dfrac{W}{\sqrt{T^2 - W^2}}$

Substituting the values in Eq. (5), we get

$$x' = \frac{H}{w}\log_e\left[\frac{W}{\sqrt{T^2 - W^2}} + \frac{W}{\sqrt{T^2 - W^2}}\right] = \frac{H}{w}\log_e\left[\frac{T+W}{\sqrt{T^2 - W^2}}\right]$$

$$= \frac{H}{w}\log_e\sqrt{\frac{T+W}{T-W}} = \frac{1}{2}\frac{H}{w}\log_e\frac{T+W}{T-W}$$

The total horizontal distance = $2x'$

$$= \frac{H}{w}\log_e\frac{T+W}{T-W}$$ (Hence proved)

Example 3.17: *A field tape, standardised at 18°C measured 100.0056 m. Determine the temperature at which it will be exactly of the nominal length of 100 m. Take $\alpha = 11.2 \times 10^{-6}$ per °C.*

Solution: Given $\delta l = 0.0056$ m ; $T_0 = 18°C$

New standard temperature $T_0' = T_0 \pm \dfrac{\delta l}{l\alpha}$

$$= 18° - \frac{0.0056}{100 \times 11.2 \times 10^{-6}} = 18° - 5° = \mathbf{13°C}$$

Example 3.18: *A distance AB measures 96.245 m on a slope. From a theodolite set at A, with instrument height of 1.400 m, staff reading taken at B was 1.675 m with a vertical angle of 4° 30′ 40″. Determine the horizontal length of the line AB. What will be the error if the effect were neglected?*

Solution: Given h_1, = 1.400 m; h_2 = 1.675 m; α = 4° 30′ 20″ ; l = 96.245 m

$$\delta\alpha'' = \frac{206265(h_1 - h_2)\cos\alpha}{l} = \frac{206265(1.400 - 1.675)\cos 4°30'32''}{96.245}$$

$$= -588'' = -0°\ 09'\ 48''$$

∴ $\theta = \alpha + \delta\alpha$ = 4° 30′ 20″ – 0° 09′ 48″ = 4° 20′ 32″

∴ Horizontal length $L = l \cos\theta$ = 96.245 cos 4° 20′ 32″ = 95.966 m.

If the effect were neglected, L = 96.245 cos 4° 30′ 40″ = 95.947 m

∴ Error = **0.019 m**

Example 3.19: *(a) Calculate the elongation at 400 m of a 1000 m mine shaft measuring tape hanging vertically due to its own mass. The modulus of elasticity is 2×10^5 N/mm², the mass of the tape is 0.075 kg/m and the cross-sectional area of the tape is 10.2 mm².*

(b) If the same tape is standardised as 1000.00 m at 175 N tension, what is the true length of the shaft recorded as 999.126 m ?

Solution: (a) Taking $\boldsymbol{M} = 0$, we have

$$s_x = \frac{mgx}{2AE}(2l - x) = \frac{0.075 \times 9.81 \times 400(2000 - 400)}{2 \times 10.2 \times 2 \times 10^5} = \mathbf{0.115\ m}$$

(b) $$s = \frac{gx}{AE}\left[M + \frac{m}{2}(2l - x) - \frac{P_0}{g}\right]$$

Here $x = 99.126$, $M = 0$ and $P_0 = 175$

$\therefore$ $$s = \frac{9.81 \times 999.126}{10.2 \times 2 \times 10^5}\left[0 + \frac{0.075}{2}(2 \times 1000 - 999.126) - \frac{175}{9.81}\right]$$

$$= \mathbf{0.095\ m}$$

PROBLEMS

1. Describe different kinds of chains used for linear measurements.
   Explain the method of testing and adjusting a chain.
2. (a) How may a chain be standardized? How may adjustments be made to the chain if it is found to be too long?
   (b) A field was surveyed by a chain and the area was found to be 127.34 acres. If the chain used in the measurement was 0.8 per cent too long, what is the correct area of the field? *(A.M.I.E.)*
3. Explain, with neat diagram, the working of the line ranger.
   Describe how you would range a chain line between two points which are not intervisible.
4. Explain the different methods of chaining on sloping ground. What is hypotenusal allowance?
5. What are different sources of errors in chain surveying?
   Distinguish clearly between cumulative and compensating errors.
6. What are different tape corrections and how are they applied?
7. The length of a line measured with a chain having 100 links was found to be 2000 links. If the chain was 0.5 link too short, find the true length of line.
8. The true length of a line is known to be 500 metres. The line was again measured with a 20 m tape and found to be 502 m. What is the correct length of the 20 m tape?
9. The distance between two stations was measured with a 20 m chain and found to be 1500 metres. The same was measured with a 30 m chain and found to be 1476 metres. If the 20 m chain was 5 cm too short, what was the error in the 30 metre chain?
10. A 30 m chain was tested before the commencement of the day's work and found to be correct. After chaining 100 chains, the chain was found to be half decimetre too long. At the end of day's work, after chaining a total distance of 180 chains, the chain was found to be one decimetre too long. What was the true distance chained?
11. A chain was tested before starting the survey, and was found to be exactly 20 metres. At the end of the survey, it was tested again and was found to be 20.12 m. Area of the plan of the field drawn to a scale of 1 cm = 6 m was 50.4 sq. cm. Find the true area of the field in sq. metres.
12. The paper of an old map drawn to a scale of 100 m to 1 cm has shrunk, so that a line originally 10 cm has now become 9.6 cm. The survey was done with a 20 m chain 10 cm too short. If the area measured now is 71 sq. cm, find the correct area on the ground.
13. The surveyor measured the distance between two stations on a plan drawn to a scale of 10 m to 1 cm and the result was 1286 m. Later, however, it was discovered that he used a scale of 20 m to 1 cm. Find the true distance between the stations.

14. The distance between two points measured along a slope 126 m. Find the horizontal distance between them, if (*a*) the angle of slope between the points is 6° 30′, (*b*) the difference in level is 30 m, (*c*) the slope is 1 in 4.

15. Find the hypotenusal allowance per chain of 30 m length if the angle of slope is 12°30′.

16. Find the sag correction for a 30 m steel tape under a pull of 8 kg in three equal spans of 10 m each. Weight of 1 cubic cm of steel = 7.86 g. Area of cross-section of the tape = 0.10 sq. cm.

17. A steel tape is 30 m long at a temperature of 65°F when lying horizontally on the ground. Its sectional area is 0.082 sq. cm, its weight 2 kg and the co-efficient of expansion 65×10^{-7} per 1°F. The tape is stretched over three equal spans. Calculate the actual length between the end graduations under the following conditions: temp. 85°F, pull 18 kg. Take $E = 2.109 \times 10^6$ kg/cm^2.

18. A 30 m steel tape was standardized on the flat and was found to be exactly 30 m under no pull at 66°F. It was used in catenary to measure a base of 5 bays. The temperature during the measurement was 92°F and the pull exerted during the measurement was 10 kg. The area of cross-section of the tape was 0.08 sq. cm. The specific weight of steel is 7.86 g/cm^3.

$$\alpha = 0.0000063 \text{ per } 1°\text{F and } E = 2.109 \times 10^6 \text{ kg/cm}^2.$$

Find the true length of the line.

19. (*a*) What are the sources of cumulative errors in long chain line?
    (*b*) What is the limit of accuracy obtainable in chain surveying?
    (*c*) An engineer's chain was found to be 0′6″ too long after chaining 5,000 ft. The same chain was found to be 1′0″ too long after chaining a total distance of 10,000 ft. Find the correct length at the commencement of chaining. (*A.M.I.E. May, 1966*)

20. Derive an expression for correction per chain length to be applied when chaining on a regular slope in terms of (*a*) the slope angle and (*b*) the gradient expressed as 1 in *n*.
    What is the greatest slope you would ignore if the error from this source is not to exceed 1 in 1500? Give you answer (*a*) as an angle (*b*) as a gradient.

ANSWERS

2. (*b*) 129.34 acres
7. 1990 links
8. 19.92 m
9. 41 cm too long
10. 180.28 chains or 5408.4 m
11. 1825 sq. m.
12. 0.763 sq. km.
13. 643 m.
14. (*a*) 125.19 m (*b*) 122.37 m (*c*) 122.24 m
15. 0.71 m
16. 0.01206 m
17. 30.005 m
18. 30.005 m
19. 10,050 ft.
20. (*a*) 2.11° (*b*) 1 in 27.4.

CHAPTER 4

CHAIN SURVEYING

4.1 CHAIN TRIANGULATION

Chain surveying is that type of surveying in which only linear measurements are made in the field. This type of surveying is suitable for surveys of small extent on open ground to secure data for exact description of the boundaries of a piece of land or to take simple details.

The principle of chain survey or *Chain Triangulation,* as is sometimes called, is to provide a skeleton or framework consisting of a number of connected triangles, as triangle is the only simple figure that can be plotted from the lengths of its sides measured in the field. To get good results in plotting, the framework should consist of triangles which are as nearly equilateral as possible.

4.2 SURVEY STATIONS

A survey station is a prominent point on the chain line and can be either at the beginning of the chain line or at the end. Such station is known as *main station.*However, *subsidiary* or *tie station* can also be selected anywhere on the chain line and subsidiary or tie lines may be run through them.

A survey station may be marked on the ground by driving pegs if the ground is soft. However, on roads and streets etc., the survey station can be marked or located by making two or preferably three tie measurements with respect to some permanent reference objects near the station. The more nearly the lines joining the peg to the reference points intersect at right angles, the more definitely will the station be fixed. A diagram of the survey lines with main stations numbered should be inserted in the beginning of the field note book.

4.3 SURVEY LINES

The lines joining the main survey stations are called *main survey lines*. The biggest of the main survey line is called the *base line* and the various survey stations are plotted with reference to this. If the area to be surveyed has more than three straight boundaries, the field measurements must be so arranged that they can be plotted by laying down the triangles as shown in Fig. 4.1 (*a*) or (*b*).

Check lines: *Check lines or proof lines* are the lines which are run in the field to check the accuracy of the work. The length of the check line measured in the field must agree with its length on the plan. A check line may be laid by joining the apex of the triangle to any point on the opposite side or by joining two points on any two sides of a triangle. Each triangle must have a check line. For the framework shown in Fig. 4.1 (*a*), the various arrangements of the check lines are shown in Fig. 4.2 (*a*), (*b*), (*c*) and (*d*) by dotted lines. In Fig. 4.1 (*b*), the dotted lines show the arrangements of check lines for the framework.

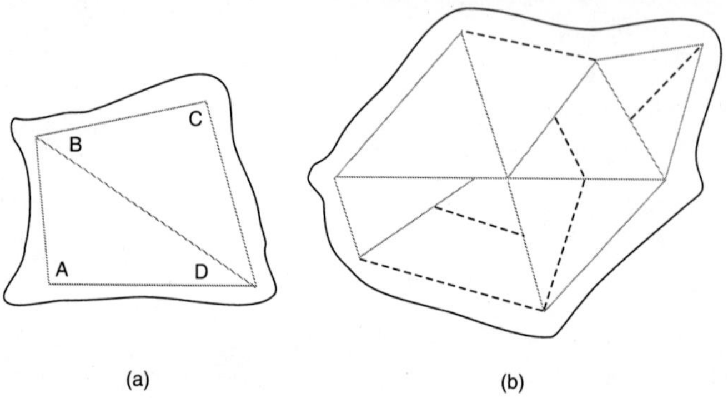

Fig. 4.1

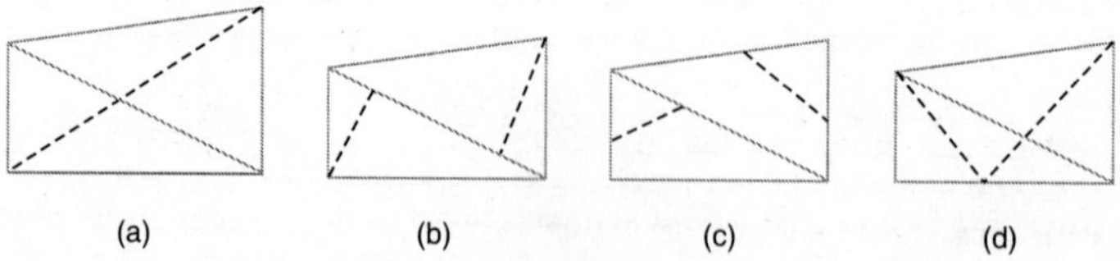

Fig. 4.2

Tie lines: A tie line is a line which joins subsidiary or tie stations on the main line. The main object of running a tie line is to take the details of nearby objects but it also serves the purpose of a check line. The accuracy in the location of the objects depends upon the accuracy in laying the tie line. A framework may have one or more tie lines depending upon the circumstances (Fig. 4.3).

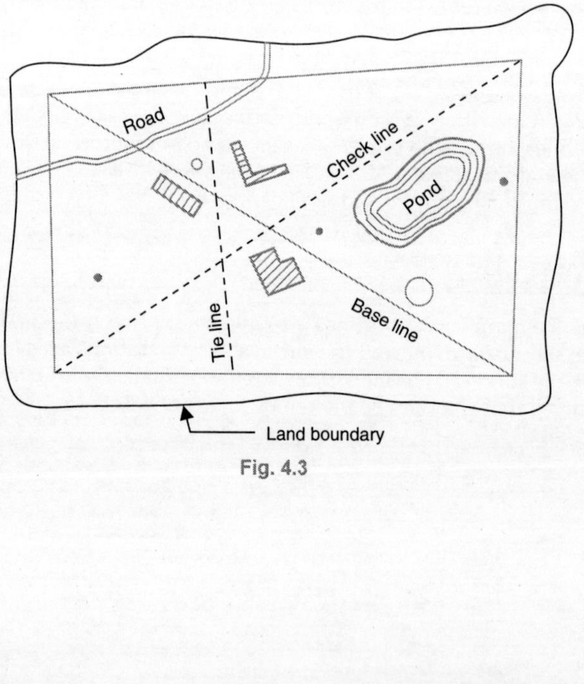

Fig. 4.3

Arrangements of Survey Lines

Let us take the case of plotting a simple triangle *ABC*. Let *a* and *b* represent two points *A* and *B* correctly plotted with respect to each other and *c* be the correct position of point *C* to be plotted. Let there be some error in the measurement of side *AC* so that *c'* is the wrong position. The corresponding displacement of the plotted position of *C* will depend upon the angle *ACB*. Fig. 4.4. (*a*) shows the case when *acb* is a right angle ; in this case the displacement of *C* will be nearly equal to the error in the side *AC*. Fig. 4.4. (*b*) shows the case when *ACB* is 60° ; in this case the displacement of *C* will be nearly 1.15 times the error. In Fig. 4.4 (c), the angle *ACB* is 30° : the displacement of C will be nearly twice the error. Hence, to get more accurate result, angle *C* must be a right angle. If, however, there is equal liability of error in all the three sides of a triangle the best form is equilateral triangle. In any case, to get a *well-proportioned* or *well-shaped* triangle, no angle should be less than 30°.

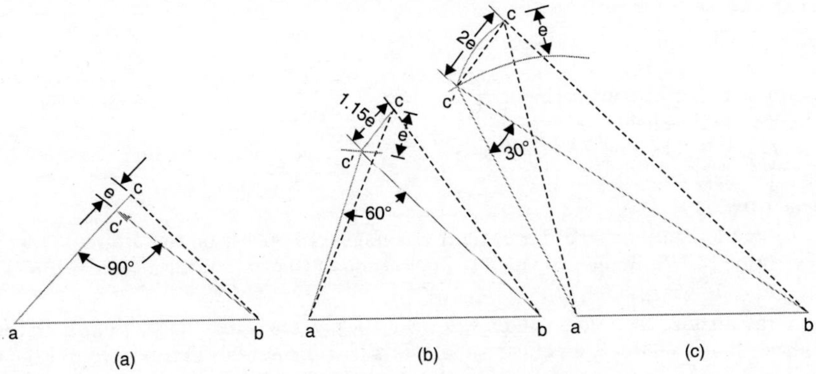

Fig. 4.4 Well Conditioned Triangles.

Conditions to be Fulfilled by Survey Lines or Survey Stations

The survey stations should be so selected that a good system of lines is obtained fulfilling the following conditions :

1. Survey stations must be mutually visible.
2. Survey lines must be as few as possible so that the framework can be plotted conveniently.
3. The framework must have one or two base lines. If one base line is used, it must run along the length and through the middle of the area. If two base lines are used, they must intersect in the form of letter *X*.
4. The lines must run through level ground as possible.
5. The main lines should form well-conditioned triangles.
6. Each triangle or portion of skeleton must be provided with sufficient check lines.
7. All the lines from which offsets are taken should be placed close to the corresponding surface features so as to get short offsets.
8. As far as possible, the main survey lines should not pass through obstacles.
9. To avoid trespassing, the main survey lines Should fall within the boundaries of the property to be surveyed.

4.4 LOCATING GROUND FEATURES : OFFSETS

An offset is the lateral distance of an object or ground feature measured from a survey line. By method of offsets, the point or object is located by measurement of a distance and angle (usually 90°) from a point on the chain line. When the angle of offset is 90°, it is called *perpendicular offset* [Fig. 4.5 (*a*), (*c*)] or sometimes, simply, *offset* and when the angle is other than 90°, it is called an *oblique offset* [Fig. 4.5 (*b*)]. Another method of locating a point is called the method of 'ties' in which the distance of the point is measured from two separate points on the chain line such that the three points form, as nearly as possible an equilateral triangle [Fig. 4.5 (*d*)]. The method of perpendicular offsets involves less measuring on the ground.

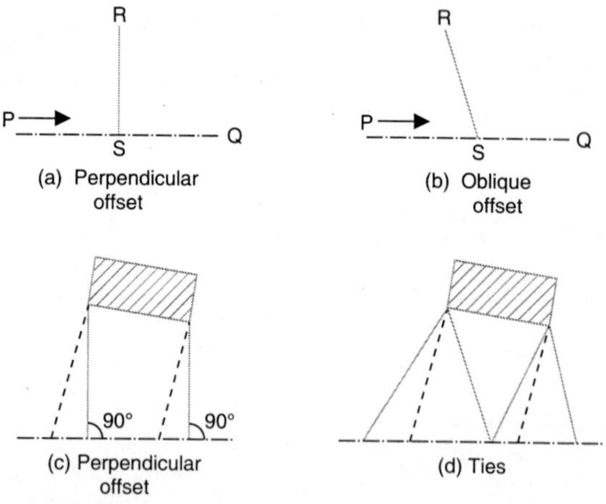

Fig. 4.5

Offsets should be taken in order of their chainages. In offsetting to buildings, check can be made by noting the chainages at which the directions of the walls cut the survey line, as shown by dotted lines in Fig. 4.5 (*c*), (*d*).

In general, an offset should be taken wherever the outline of an object changes. In the case of a straight wall or boundary, an offset at each end is sufficient. To locate irregular boundaries, sufficient number of offsets are taken at suitable interval and at such point where the direction suddenly changes, as shown in Fig. 4.6 (*a*). In the case of a nallah, offsets should be taken to both the sides of its width, as shown in Fig. 4.6 (*b*). However, in the case of regular curves with constant width, the offsets should be taken to the centre line only and the width should also be measured.

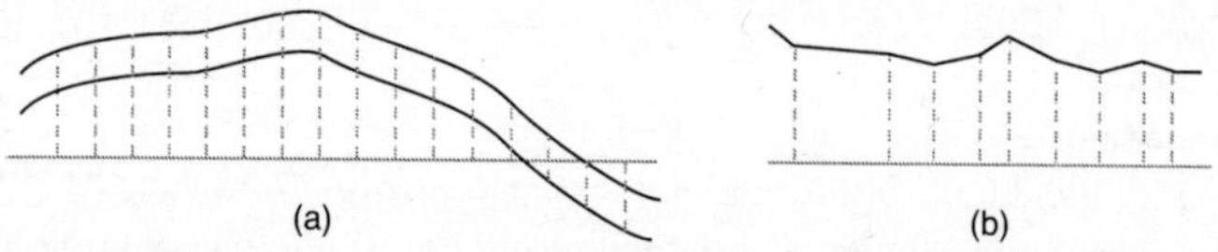

Fig. 4.6

Taking Perpendicular Offsets: Fig. 4.7 illustrates the procedure for finding the length and position of the perpendicular offset. The leader holds the zero end of the tape at the point *P* to be located and the follower carries the tape box and swings the tape along the chain. The length of the offset is the shortest distance from the object to the chain obtained by swinging the tape about the object as centre. Such an offset is called *swing offset.* The position of the offset on the chain is located by the point where the arc is tangential to the chain.

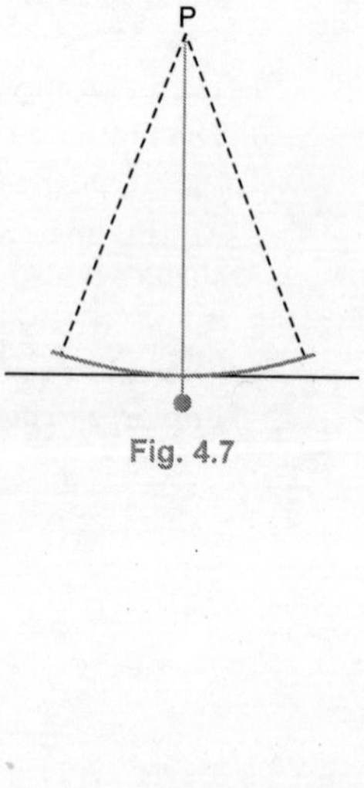

Fig. 4.7

Degree of Precision in Measuring the Offsets: Before commencing the field measurements, one should know the degree of precision to be maintained in measuring the length of the offset. This mainly depends on the scale of survey. Normally, the limit of precision in plotting is 0.25 mm. If the scale of plotting is 1 cm = 2 m, 0.25 mm on paper will correspond to $\frac{2 \times 0.25}{10}$ = 0.05 m on the ground. Hence, in such a case, the offsets should be measured to the nearest 5 cm. On the other hand, if the scale of plotting is 1 cm = 10 m. 0.25 mm on paper will correspond to $\frac{10 \times 0.25}{10}$ = 0.25 m on the ground. Hence the offset should be measured to the nearest 25 cm. However, if there is likelihood of changing the scale of plotting at a later stage, it is better always to be over-accurate than to be under-accurate.

Long Offsets: The survey work can be accurately and expeditiously accomplished if the objects and features that are to be surveyed are near to the survey lines. The aim should always be to make the offset as small as possible. Long offset may be largely obviated by judiciously placing the main lines of the survey near the object or by running subsidiary lines from the main lines. Fig. 4.8 shows a well-proportioned subsidiary triangle *abc* run to locate the deep bend of the outline of the fence. The base of the triangle is on the main line and *bd* is the check line.

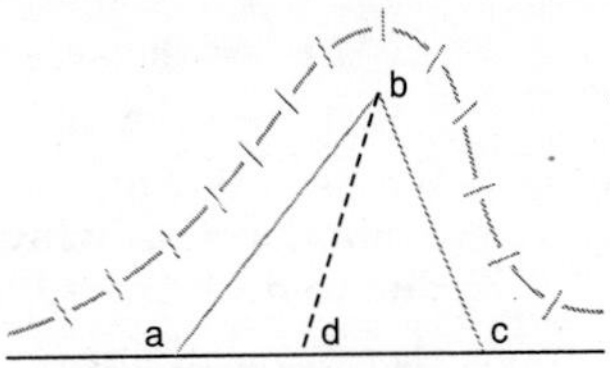

Fig. 4.8

Limiting Length of Offset

The allowable length of offset depends upon the degree of accuracy required, scale, method of setting out the perpendicular and nature of ground. The only object is that the error produced by taking longer lengths of offsets should not be appreciable on the paper.

(*i*) **Effect of error in laying out the direction:** Let us first consider the effect of error in laying out the perpendicular.

Let the offset *CP* be laid out from a point *C* on the chain line to the object *P*, and let the angle *BCP* be (90° – α) where α is the error in laying the perpendicular. Let the length *CP* be *l*. While plotting, the point *P* will be plotted at P_1, CP_1 being perpendicular to *AB* and of length *l*. Thus, the displacement of the point *P* along the chain line is given by

$$PP_2 = \frac{l \sin \alpha}{s} \text{ cm}$$

where

l = length of offset in metres

s = scale (*i.e.* 1 cm = *s* metres)

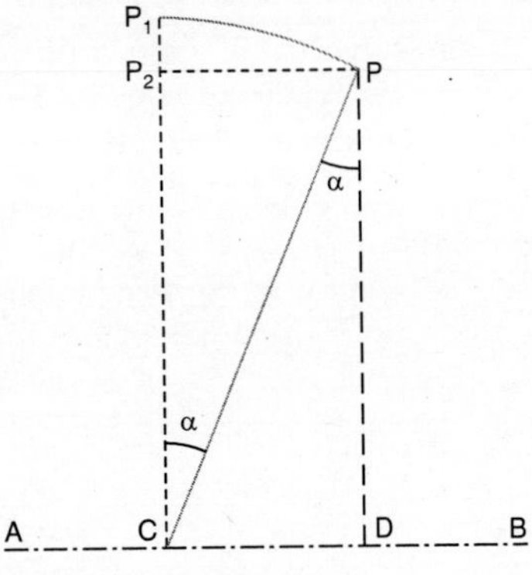

Fig. 4.9

Taking 0.25 mm as the limit of accuracy in plotting, we have

$$\frac{l \sin \alpha}{s} = \frac{0.25}{10} \quad \text{or} \quad l = 0.025\, s \operatorname{cosec} \alpha \qquad ...(4.1)$$

Also, displacement of the point perpendicular to the chain line is

$$P_1 P_2 = CP_1 - CP_2 = \frac{l - l\cos\alpha}{s} \text{ cm (on the paper)} \qquad ...(4.2)$$

(ii) **Combined effect of error due to length and direction** (Fig 4.10).

Let P = actual position of the point
CP = true length of the offset
$CP_1 = l$ = measured length of the offset
$CP_2 = l$ = plotted length of the offset
α = angular error
PP_2 = total displacement of P
1 in r = the accuracy in measurement of the offset
1 cm = s metres (scale).

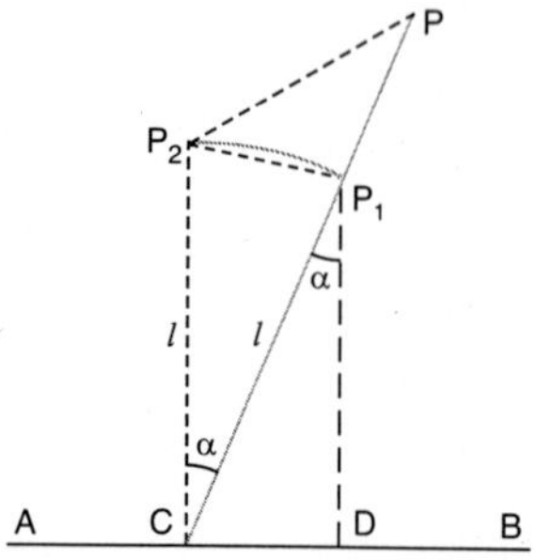

Fig. 4.10

(a) **Given the angular error, to find the degree of accuracy with which the length of offset should be measured so that the error due to both the sources may be equal.**

Displacement due to angular error = $P_1 P_2 = l \sin\alpha$ (nearly)

Displacement due to linear error = $\frac{l}{r}$

Assuming both the errors equal, we get $l \sin\alpha = \frac{l}{r}$

or $$r = \text{cosec}\,\alpha \qquad ...(4.3)$$

If $\alpha = 3°$, cosec $3° = 19 = r$. Hence the degree of accuracy in linear measurement should be 1 in 19.

Similarly, if $r = 100$, $\alpha = \text{cosec}^{-1}\,100 = 34'$ *i.e.*, the offset should be laid out with an accuracy of nearly $\frac{1}{2}°$.

(b) **Given the scale, to find the limiting length of the offset so that error due to both the sources may not exceed 0.25 mm on the paper.**

Taking $P_1 P_2 = P_1 P$ and $\angle PP_1 P_2 \approx 90°$ we have

$$PP_2 = \sqrt{2}\, PP_1 = \sqrt{2}\frac{l}{r}, \text{ on the ground.}$$

Hence the corresponding displacement on the paper will be equal to $\sqrt{2}\,\frac{l}{r}.\frac{1}{s}$. If this error is not to be appreciable on the paper, we have

$$\sqrt{2}\,\frac{l}{rs} = 0.025$$

or $$l = \frac{0.025}{\sqrt{2}}\, rs \text{ metres} \qquad ...(4.4)$$

(c) **Given the maximum error in the length of the offset, the maximum length of the offset and the scale, to find the maximum value of α so that maximum displacement on the paper may not exceed 0.25 mm.**

Let e = maximum error in measurement of offset (metres)

$$PP_1 = e \text{ metres (given)}$$
$$P_1 P_2 = l \sin \alpha \text{ (approx)}$$
$$PP_2 = \sqrt{e^2 + l^2 \sin^2 \alpha} \text{ Approximately (on ground)}$$

$\therefore$ $$PP_2 \text{ on paper} = \frac{1}{s}\sqrt{e^2 + l^2 \sin^2 \alpha} = 0.025$$

Hence $$\sin^2 \alpha = \left(\frac{6.25 s^2}{100^2} - e^2\right)\frac{1}{l^2} \quad ...(4.5)$$

From which α can be calculated.

Example 4.1: *An offset is laid out 5° from its true direction on the field. Find the resulting displacement of the plotted point on the paper (a) in a direction parallel to the chain line, (b) in a direction perpendicular to the chain line, given that the length of the offset is 20 m and the scale is 10 m to 1 cm.*

Solution:

(*a*) Displacement parallel to the chain $= \dfrac{l \sin \alpha}{s}\text{cm} = \dfrac{20 \sin 5°}{10} = \mathbf{0.174\ cm}$

(*b*) Displacement perpendicular to the chain $= \dfrac{l(1 - \cos \alpha)}{s}\text{ cm} = \dfrac{20}{10}(1 - \cos 5°)$

$= \mathbf{0.0076\ cm}$ (inappreciable).

Example 4.2: *An offset is laid out 2° from its true direction on the field. If the scale of plotting is 10 m to 1 cm, find the maximum length of the offset so that the displacement of the point on the paper may not exceed 0.25 mm.*

Solution:

Displacement of the point on the paper $= \dfrac{l \sin \alpha}{s} = \dfrac{l \sin 2°}{10}\text{cm}$

This should not exceed 0.025 cm.

Hence $$\frac{l \sin 2°}{10} = 0.025.$$

or $$l = \frac{0.025 \times 10}{\sin 2°} = \mathbf{7.16\ m}$$

Example 4.3: *An offset is laid out 1° 30′ from its true direction on the field. Find the degree of accuracy with which the offset should be measured so that the maximum displacement of the point on the paper from both the sources may be equal.*

Solution: Displacement due to angular error $= l \sin \alpha$

Displacement due to linear error $= \dfrac{l}{r}$

Taking both these equal, $l \sin \alpha = \dfrac{l}{r}$

or $$r = \text{cosec}\ \alpha = \text{cosec}\ 1°\ 30' = \mathbf{38.20}$$

Hence, the offset should be measured with an accuracy of 1 in 39.

Example 4.4: *An offset is measured with an accuracy of 1 in 40. If the scale of plotting is 1 cm = 20 m, find the limiting length of the offset so that the displacement of the point on the paper from both sources of error may not exceed 0.25 mm.*

Solution:

The total displacement of the paper = $\dfrac{\sqrt{2}\, l}{rs}$ cm

But this is not to exceed 0.025 cm.

Hence $$\frac{\sqrt{2}\, l}{rs} = 0.025$$

or $$l = \frac{0.025}{\sqrt{2}} \times 40 \times 20 = \mathbf{14.14\ m}$$

Example 4.5: *The length of an offset is 16 m and is measured with a maximum error of 0.2 m. Find the maximum permissible error in laying off the direction of the offset so that the maximum displacement may not exceed 0.25 mm on the plan drawn to a scale of 1 cm = 40 m.*

Solution:

$$\sin^2 \alpha = \frac{1}{l^2}\left(\frac{6.25}{100^2} s^2 - e^2\right)$$

∴ $$\sin \alpha = \frac{1}{16}\sqrt{\frac{6.25}{(100)^2}(40)^2 - (0.2)^2} = \frac{1}{16}\sqrt{0.96} = 0.0612$$

or $$\alpha = \mathbf{3°\ 30'}.$$

4.5 FIELD BOOK

The book in which the chain or tape measurements are entered is called the *field book*. It is an oblong book of size about 20 cm × 12 cm and opens lengthwise. The main requirements of the field book are that it should contain good quality stout opaque paper, it should be well-bound and of a size convenient for the pocket. The chain line may be represented either by a single line or by two lines spaced about $1\frac{1}{2}$ to 2 cm apart, ruled down the middle of each page. The *double line* field book (Fig 4.12) is most commonly used for ordinary work, the distance along the chain being entered between the two lines of the page. *Single line* field book (Fig. 4.11) is used for a comparatively large scale and most detailed dimension work. A chain line is started from the bottom of the page and works upwards. All distances along the chain line are entered in the space between the two ruled lines while the offsets are entered either to the left or to the right of the chain line, as the case may be. Offsets are entered in the order they appear at the chain line. As the various details within offsetting distances are reached, they are sketched and entered as shown in Fig. 4.11 and Fig. 4.12. Every chain line must be started from a fresh page. All the pages must be machine numbered.

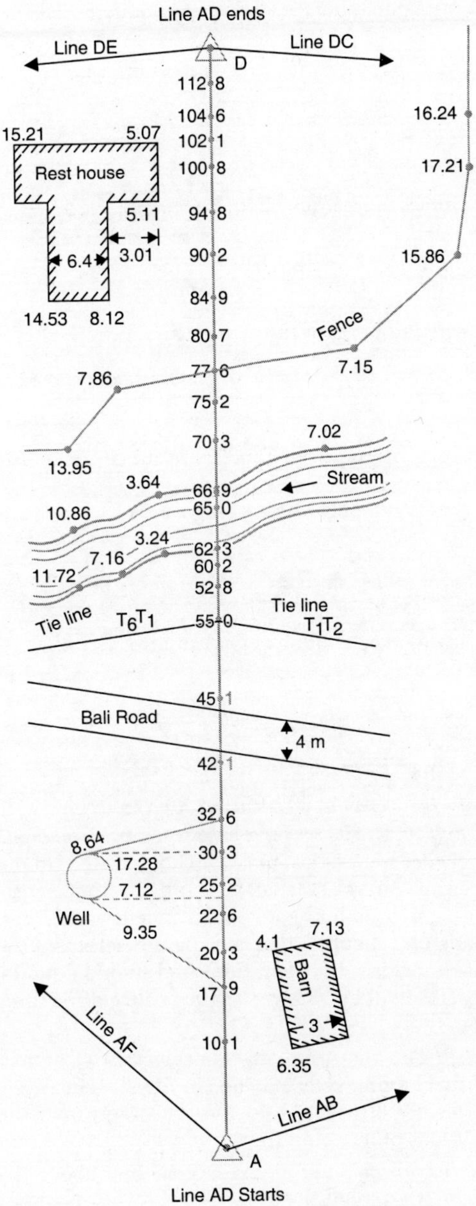

Fig. 4.11 Single Line Booking

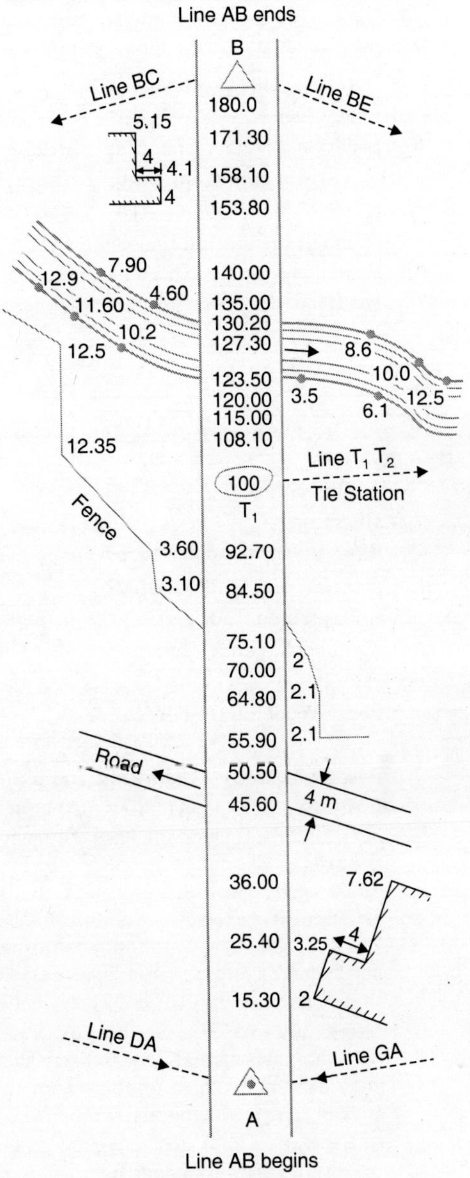

Fig. 4.12 Double Line Booking

At the beginning of a particular chain survey, the following details must be given :

(*i*) Date of survey and names of surveyors
(*ii*) General sketch of the layout of survey lines
(*iii*) Details of survey lines
(*iv*) Page index of survey lines
(*v*) Location sketches of survey stations.

At the starting of a chain or survey lines, the following details should be given :

(*i*) Name of the line (say, *AB*)
(*ii*) Name of the station marked either by an oval or by a triangle.
(*iii*) Bearing of the line (it measured)
(*iv*) Details of any other line meeting at the starting point of the survey line.

4.6 FIELD WORK

Equipment: The following is the list of equipment required for chain survey or chain triangulation :

(*i*) A 20 m chain
(*ii*) 10 arrows
(*iii*) Ranging rods and offset rods
(*iv*) A tape (10 m or 20 m length)
(*v*) An instrument for setting right angles : say a cross staff or optical square
(*vi*) Field book, pencil etc., for note-keeping
(*vii*) Plumb bob
(*viii*) Pegs, wooden hammer, chalks, etc.

A chain survey may be done in the following steps :

(*a*) Reconnaissance (*b*) Marking and fixing survey stations (*c*) Running survey lines.

(*a*) **Reconnaissance:** The first principle of any type of surveying is to work from *whole to part*. Before starting the actual survey measurements, the surveyor should walk around the area to fix best positions of survey lines and survey stations. During reconnaissance, a *reference sketch* of the ground should be prepared and general arrangement of lines, principal features such as buildings, roads etc. should be shown. Before selecting the stations, the surveyor should examine the intervisibility of stations and should note the positions of buildings, roads, streams etc. He should also investigate various difficulties that may arise and think of their solution.

(*b*) **Marking and Fixing Survey Stations:** The requirements for selection of survey stations have already been discussed. After having selected the survey stations, they should be marked to enable them to be easily discovered during the progress of the survey. The following are some of the methods of marking the stations :

(*i*) In soft ground, wooden pegs may be driven, leaving a small projection above the ground. The name of the station may be written on the top.
(*ii*) Nails or spikes may be used in the case of roads or streets. They should be flush with the pavement.
(*iii*) In hard ground, a portion may be dug and filled with cement mortar etc.

(*iv*) For a station to be used for a very long time, a stone of any *standard shape* may be embedded in the ground and fixed with mortar etc. On the top of the stone, description of the station etc. may be written.

Whenever possible, a survey station must be *fixed* with reference to two or three permanent objects and a reference or location sketch should be drawn in the field book. Fig. 4.13 shows a typical *location sketch* for a survey station.

(*c*) **Running Survey Lines:** After having completed the preliminary work, the chaining may be started from the base line. The work in running a survey line is two-fold : (*i*) to chain the line, and (*ii*) to locate the adjacent details. Offsets should be taken in order of their chainages. To do this, the chain is stretched along the line on the ground. Offsets are then measured. After having assured that no offset has been omitted, the chain must be pulled forward. The process of chaining and offsetting is repeated until the end of the line is reached. The distances along the survey line at which fences, streams, roads, etc., and intersected by it must also be recorded.

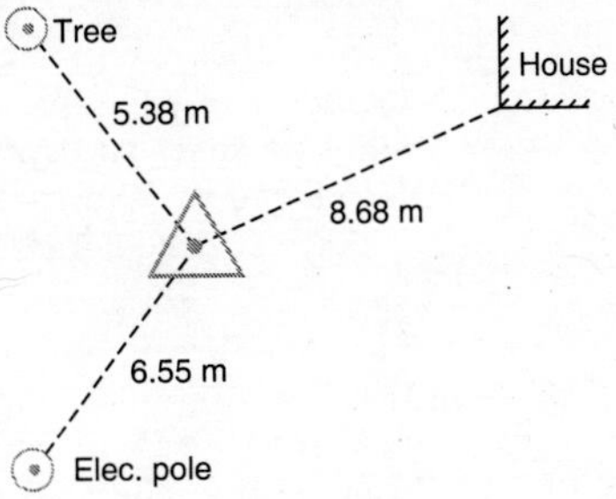

Fig. 4.13

4.7 INSTRUMENTS FOR SETTING OUT RIGHT ANGLES

There are several types of instruments used to set out a right angle to a chain line, the most common being (*i*) cross staff (*ii*) optical square (*iii*) prism square (*iv*) site square.

(*i*) **Cross Staff:** The simplest instrument used for setting out right angles is a *cross staff*. It consists of either a frame or box with two pairs of vertical slits and is mounted on a pole shod for fixing in the ground. The common forms of cross staff are (*a*) open cross staff (*b*) French cross staff (*c*) adjustable cross staff.

(*a*) **Open Cross Staff:** Fig. 4.14. (*a*) shows an open cross staff. It is provided with two pairs of vertical slits giving two lines of sights at right angles to each other. The cross staff is set up at a point on the line from which the right angle is to run, and is then turned until one line of sight passes through the ranging pole at the end of the survey line. The line of sight through the other two vanes will be a line at right angles to the survey line and a ranging rod may be established in that direction. If, however, it is to be used to take offsets, it is held vertically on the chain line at a point where the foot of the offsets is likely to occur. It is then turned so that one line of sight passes through the ranging rod fixed at the end

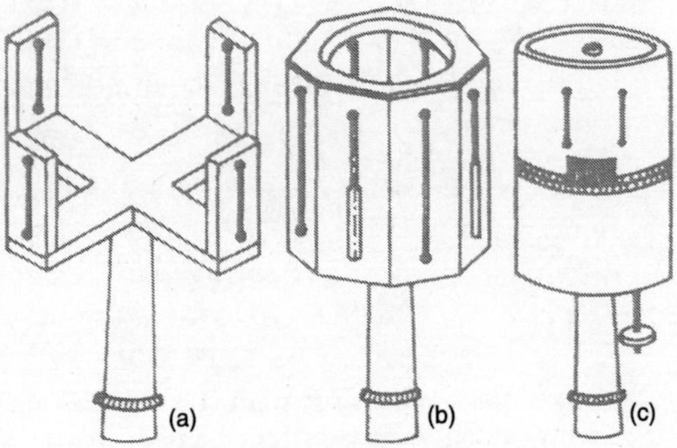

Fig. 4.14 Various Forms of Cross Staff.

of the survey line. Looking through the other pair of slits, it is seen if the point to which the offset is to be taken is bisected. If not, the cross staff is moved backward or forward till the line of sight also passes through the point.

(*b*) **French Cross Staff:** Fig. 4.14 (*b*) shows a French cross staff. If consists of a hollow octagonal box. Vertical sighting slits are cut in the middle of each face, such that the lines between the centres of opposite slits make angles of 45° with each other. It is possible, therefore, to set out angles of either 45° or 90° with this instrument.

(*c*) **Adjustable Cross Staff:** The adjustable cross staff [Fig. 4.14 (*c*)] consists of two cylinders of equal diameter placed one on top of the other Both are provided with sighting slits. The upper box carries a vernier and can be rotated relatively to the lower by a circular rack and pinion arrangement actuated by a milled headed screw. The lower box is graduated to degrees and sub-divisions. It is, therefore, possible to set out any angle with the help of this instrument.

(*ii*) **Optical Square:** Optical square is somewhat more convenient and accurate instrument than the cross staff for setting out a line at right angles to another line. Fig. 4.15 (*a*) illustrates the principle on which it works.

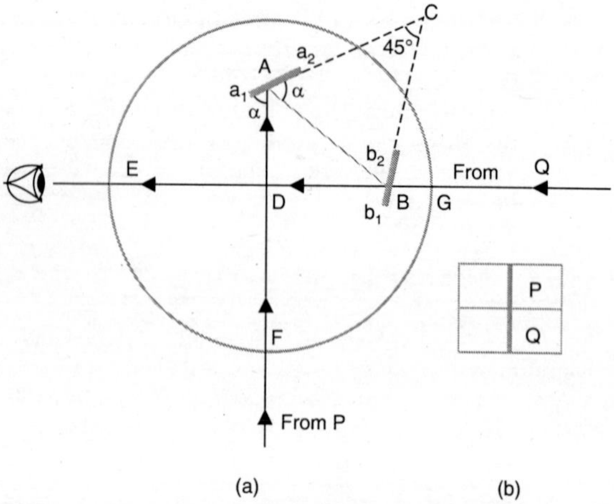

Fig. 4.15 Optical Square.

It consists of a circular box with three slits at *E*, *F* and *G*. In line with the openings *E* and *G*, a glass silvered at the top and unsilvered at the bottom, is fixed facing the opening *E*. Opposite to the opening *F*, a silvered glass is fixed at *A* making an angle of 45° to the previous glass. A ray from the ranging rod at *Q* passes through the lower unsilvered portion of the mirror at *B*, and is seen directly by eye at the slit *E*. Another ray from the object at *P* is received by the mirror at *A* and is reflected towards the mirror at *B* which reflects it towards the eye. Thus, the images of *P* and *Q* are visible at *B*. If both the images are in the same vertical line as shown in Fig. 4.15 (*b*), the line *PD* and *QD* will be at right angles to each other.

Let the ray *PA* make an angle α with the mirror at *A*,

$$\angle ACB = 45° \quad \text{or} \quad \angle ABC = 180° - (45° + \alpha) = 135° - \alpha$$

By law of reflection $\angle EBb_1 = \angle ABC = 135° - \alpha$

Hence $\angle ABE = 180° - 2(135° - \alpha) = 2\alpha - 90°$ ...(*i*)

Also $\angle DAB = 180° - 2\alpha$ ...(*ii*)

From $\Delta ABD, \angle ADB = 180° - (2\alpha - 90°) - (180° - 2\alpha)$

$$= 180° - 2\alpha + 90° - 180° + 2\alpha = 90°$$

Thus, if the images of *P* and *Q* lie in the same vertical line, as shown in Fig. 4.15 (*b*), the line *PD* and *QD* will be at right angles to each other.

To set a right angle: To set a right angle on a survey line, the instrument is held on the line with its centre on the point at which perpendicular is erected. The slits F and G are directed towards the ranging rod fixed at the end of the line. The surveyor (holding the instrument) then directs person, holding a ranging rod and stationing in a direction roughly perpendicular to the chain line, to move till the two images described above coincide.

Testing the Optical Square (Fig. 4.16)

(*i*) Hold the instrument in hand at any intermediate point C on AB, sight a pole held at A and direct an assistant to fix a ranging rod at a, such that the images of the ranging rods at a and A coincide in the instrument.

(*ii*) Turn round to face B and sight the ranging rod at a. If the image of the ranging rod at B coincides with the image of ranging rod at a, the instrument is in adjustment.

(*iii*) If not, direct the assistant to move to a new position b so that both the images coincide. Mark a point d on the ground mid-way between a and b. Fix a ranging rod at d.

(*iv*) Turn the *adjustable mirror* till the image of the ranging rod at d coincide with the image of the ranging rod at B. Repeat the test till correct.

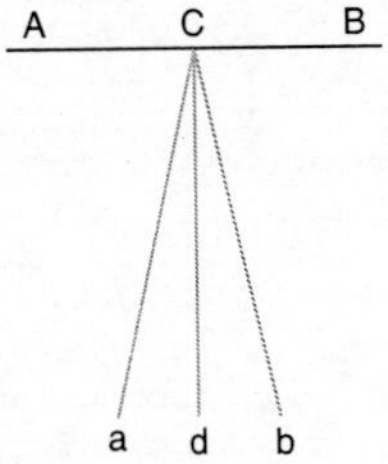

Fig. 4.16

(*iii*) **Prism Square:** The prism square shown in Fig. 4.17 works on the same principle as that of optical square. It is a more modern and precise instrument and is used in a similar manner. It has the merit that no adjustment is required since the angle between the reflecting surfaces (*i.e.* 45°) cannot vary. Fig. 4.18 shows a combined prism square as well as line ranger.

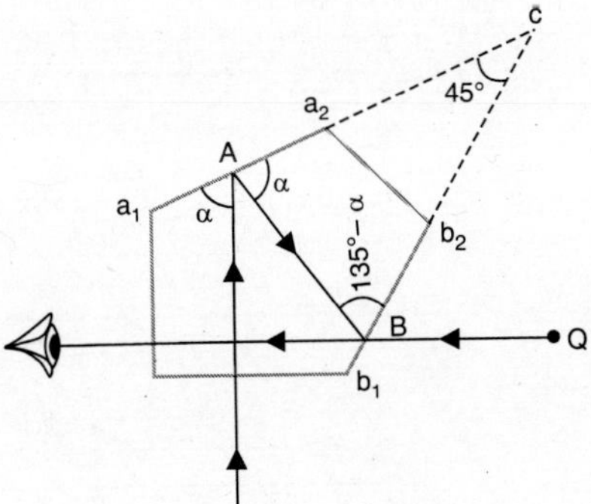

Fig. 4.17 Prism Square

Fig. 4.18 Combined Prism Square and Line Ranger.

(*iv*) **Site Square (Fig. 4.19):** A site square, designed for setting out straight lines and offset lines at 90 °, consists of a cylindrical metal case containing two telescopes set at 90° to each other, a fine setting screw near the base, a circular spirit level at the top and a knurled ring at the base. It is used in conjunction with a datum rod screwed into the base of the instrument.

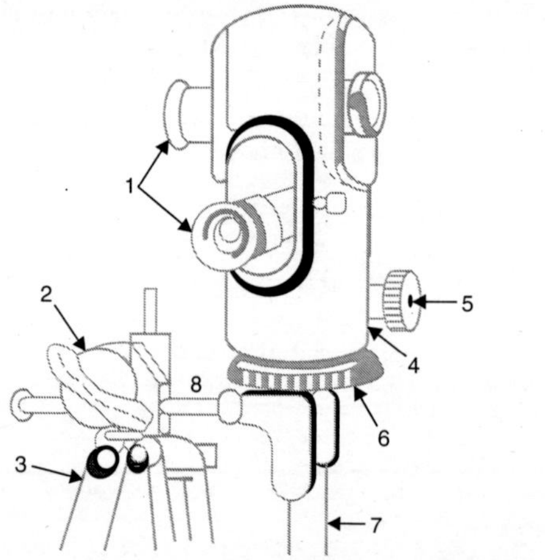

Fig. 4.19 The Site Square.

4.8 BASIC PROBLEMS IN CHAINING

(A) **To Erect a Perpendicular to a Chain Line from a Point on it:** The method of establishing perpendiculars with the tape are based on familiar geometric constructions. The following are some of the methods most commonly used. The illustrations given are for a 10 m tape. However, a 20 m tape may also be used.

(i) ***The 3-4-5 method:*** Let it be required to erect a perpendicular to the chain line at a point C in it [Fig. 4.20 (*a*)]. Establish a point *E* at a distance of 3 m from *C*. Put the 0 end of the tape (10 m long) at *E* and the 10 m end at *C*. The 5 m and 4 m marks are brought together to form a loop of 1 m. The tape is now stretched tight by fastening the ends *E* and *C*. The point *D* is thus established. Angle *DCE* will be 90°. One person can set out a right angle by this method.

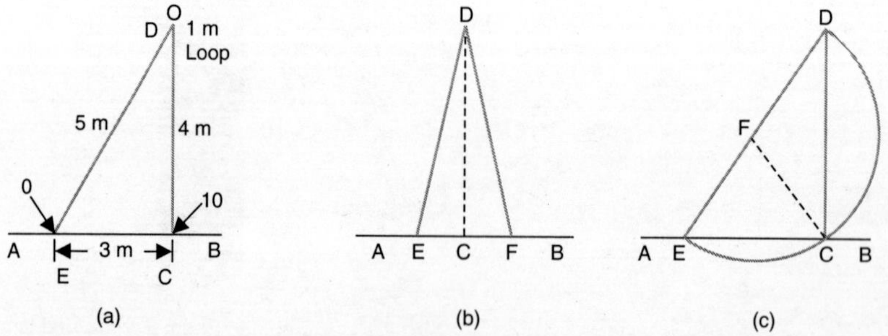

Fig. 4.20

(ii) ***Second method*** **[Fig. 4.20 (b)]:** Select *E* and *F* equidistant from *C*. Hold the zero end of the tape at *E,* and 10 m end at *F.* Pick up 5 m mark, stretch the tape tight and establish *D.* Join *DC.*

(*iii*) *Third method* [Fig. 4.20 (c)]: Select any point F outside the chain, preferably at 5 m distance from C. Hold the 5 m mark at F and zero mark at C, and with F as centre draw an arc to cut the line at E. Join EF and produce it to D such that $EF = FD = 5$ m.

Thus, point D will lie at the 10 m mark of tape laid along EF with its zero end at E. Join DC.

(B) **To Drop a Perpendicular to a Chain Line from a Point outside it:** Let it be required to drop a perpendicular to a chain line AB from a point D outside it.

(*i*) *First method* [Fig. 4.21 (a)]: Select any point E on the line. With D as centre and DE as radius, draw an arc to cut the chain line in F. Bisect EF at C. CD will be perpendicular to AB.

(*ii*) *Second method* [Fig. 4.21 (b)]: Select any point E *on* the line. Join ED and bisect it at F. With F as centre and EF or FD as radius, draw an arc to cut the chain line in C. CD will be perpendicular to the chain line.

(*iii*) *Third method* [Fig. 4.21 (c)]: Select any point E on the line. With E as centre and ED as radius, draw an arc to cut the chain line in F. Measure FD and FE. Obtain the point C on the line by making $FC = \frac{FD^2}{2EF}$. Join C and D. CD will be perpendicular to the chain line.

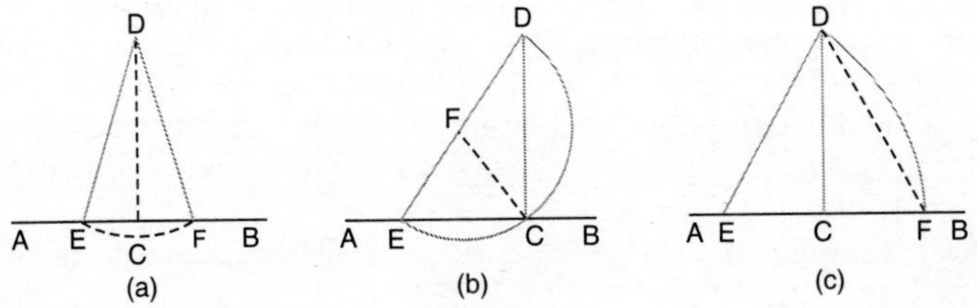

Fig. 4.21

(C) **To run a Parallel to Chain Line through a given Point:** Let it be required to run a parallel to a chain line AB through a given point C.

(*i*) *First method* [Fig. 4.22 (a)]: Through C, drop a perpendicular CE to the chain line. Measure CE. Select any other point F on line and erect a perpendicular FD. Make $FD = EC$. Join C and D.

(*ii*) *Second method* [Fig. 4.22 (b)]: Select any point F on the chain line. Join CF and bisect at G. Select any other point E on the chain line. Join EG and prolong it to D such that $EG = GD$. Join C and D.

(*iii*) *Third method* [Fig. 4.22 (c)]: Select any point G outside the chain line and away from C (but to the same side of it). Join GC and prolong it to meet the chain

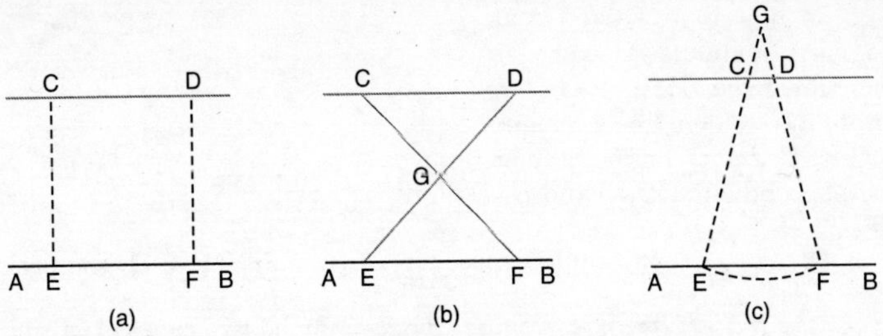

Fig. 4.22

line in E. With G as centre and GE as radius, draw an arc to cut AB in F. Join GF and make $GD = GC$. Join C and D.

(D) **To run a Parallel to a given Inaccessible Line through a Given Point:** Let AB be the given inaccessible line and C be the given point through which the parallel is to be drawn (Fig. 4.23).

Select any point E in line with A and C. Similarly, select any other convenient point F. Join E and F. Through C, draw a line CG parallel to AF. Through G, draw a line GD parallel to BF, cutting BE in D. CD will then be the required line.

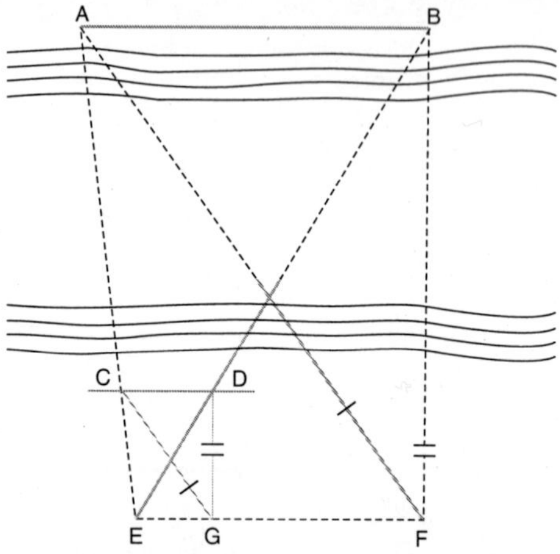

Fig. 4.23

4.9 OBSTACLES IN CHAINING

Obstacles to chaining prevent chainman from measuring directly between two points and give rise to a set of problems in which distances are found by indirect measurements. Obstacles to chaining are of three kinds :

(*a*) Obstacles to ranging

(*b*) Obstacles to chaining

(*c*) Obstacles to both chaining and ranging.

(a) Obstacle to Ranging But Not Chaining: This type of obstacle, in which the ends are not intervisible, is quite common except in flat country. There may be two cases of this obstacle.

(*i*) Both ends of the line may be visible from intermediate points on the line.

(*ii*) Both ends of the line may not be visible from intermediate points on the line (Fig. 4.24).

Case (i): Method of reciprocal ranging of § 3.3 may be used.

Case (ii): In Fig. 4.24, let AB be the line in which A and B are not visible from intermediate point on it. Through A, draw a random line AB_1 in any convenient direction but as nearly towards B as possible. The point B_1 should be so chosen that (*i*) B_1 is visible from B and (*ii*) BB_1 is perpendicular to the random line. Measure BB_1. Select points C_1 and D_1 on the random line and erect perpendicular C_1C and D_1D on it.

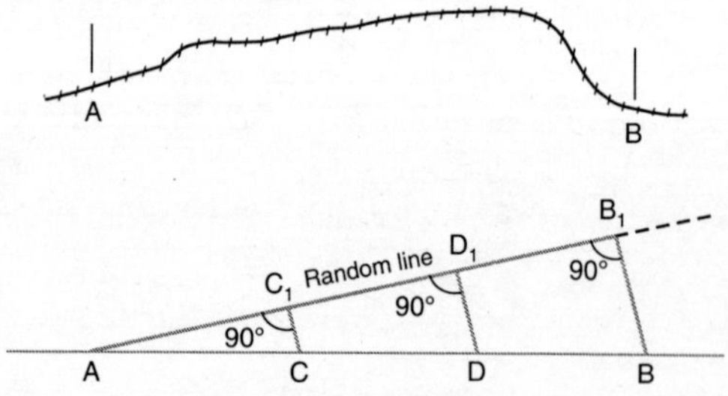

Fig. 4.24

Make $$CC_1 = \frac{AC_1}{AB_1} \cdot BB_1 \text{ and } DD_1 = \frac{AD_1}{AB_1} \cdot BB_1.$$ Join C and D, and prolong.

(b) Obstacle to Chaining But Not Ranging: There may be two cases of this obstacle :

(*i*) When it is possible to chain round the obstacle, *i.e.* a pond, hedge etc.

(*ii*) When it is not possible to chain round the obstacle, *e.g.* a river.

Case (I): Following are the chief methods (Fig. 4.25).

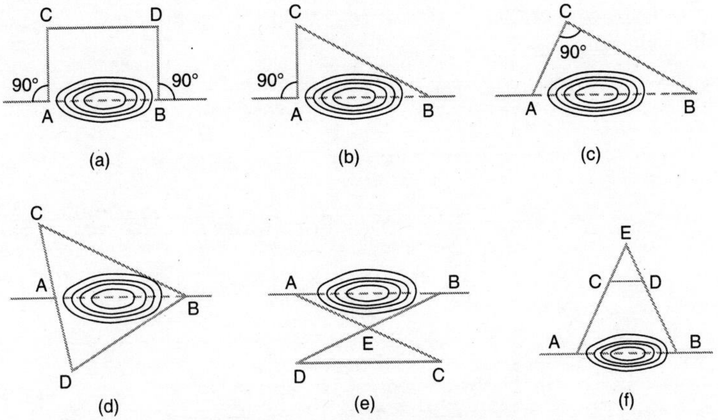

Fig. 4.25 Obstacles to Chaining

Method (a): Select two points A and B on either side. Set out equal perpendiculars AC and BD. Measure CD; then $CD = AB$ [Fig. 4.25 (*a*)].

Method (b): Set out AC perpendicular to the chain line. Measure AC and BC [Fig. 4.25 (*b*)]. The length AB is calculated from the relation $AB = \sqrt{BC^2 - AC^2}$.

Method (c): By optical square or cross staff, find a point C which subtends 90° with A and B. Measure AC and BC [Fig. 4.25 (*c*)]. The length AB is calculated from the relation :

$$AB = \sqrt{AC^2 + BC^2} .$$

Method (d): Select two points C and D to both sides of A and in the same line. Measure AC, AD, BC and BD [Fig. 4.25 (*d*)]. Let angle BCD be equal to θ.

From ΔBCD, $BD^2 = BC^2 + CD^2 - 2BC \times CD \cos\theta$

or
$$\cos\theta = \frac{BC^2 + CD^2 - BD^2}{2\,BC \times CD} \qquad ...(i)$$

Similarly from ΔBCA,
$$\cos\theta = \frac{BC^2 + AC^2 - AB^2}{2\,BC \times AC} \qquad ...(ii)$$

Equating (*i*) and (*ii*) and solving for AB we get

$$AB = \sqrt{\frac{(BC^2 \times AD) + (BD^2 \times AC)}{CD} - (AC \times AD)}$$

Method (e): Select any point E and range C in line with AE, making $AE = EC$. Range D in line with BE and make $BE = ED$. Measure CD ; then $AB = CD$ [Fig. 4.25 (*e*)].

Method (f): Select any suitable point E and measure AE and BE. Mark C and D on AE and BE such that $CE = \frac{AE}{n}$ and $DE = \frac{BE}{n}$ Measure CD ; then

$$AB = n \,.\, CD. \qquad \text{[Fig. 4.25 }(f)].$$

Case (II): (Fig. 4.26)

Method (a): Select point B on one side and A and C on the other side. Erect AD and CE as perpendiculars to AB and range B, D and E in one line. Measure AC, AD and CE [Fig. 4.26 (*a*)]. If a line DF is drawn parallel to AB, cutting CE in F perpendicularly, then triangles ABD and FDE will be similar.

$$\therefore \qquad \frac{AB}{AD} = \frac{DF}{FE}$$

But $\qquad FE = CE - CF = CE - AD$, and $DF = AC$.

$$\therefore \qquad \frac{AB}{AD} = \frac{AC}{CE - AD} \qquad \text{From which} \qquad AB = \frac{AC \times AD}{CE - AD}$$

Method (b) : Erect a perpendicular AC and bisect it at D. Erect perpendicular CE at C and range E in line with BD. Measure CE [Fig. 4.26 (*b*)]. Then $AB = CE$.

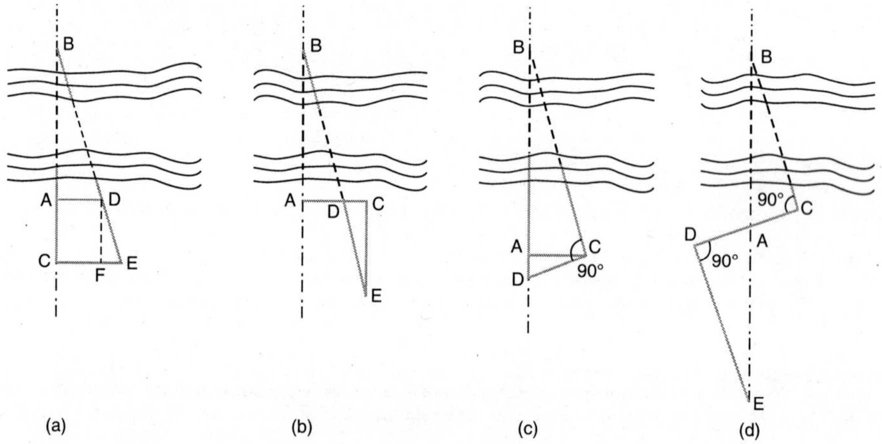

Fig. 4.26 Obstacles to Chaining.

Method (c): Erect a perpendicular AC *at* A and choose any convenient point C. With the help of an optical square, fix a point D on the chain line in such a way that BCD is a right angle [Fig. 4.26 (c)]. Measure AC and AD. Triangles ABC and DAC are similar. Hence $\frac{AB}{AC} = \frac{AC}{AD}$.

Therefore, $AB = \frac{AC^2}{AD}$

Method (d): Fix point C in such a way that it subtends 90° with AB. Range D in line with AC and make $AD = AC$. At D, erect a perpendicular DE to cut the line in E [Fig. 4.26 (*d*)]. Then $AB = AE$.

(c) Obstacles to Both Chaining and Ranging: A building is the typical example of this type of obstacle. The problem lies in prolonging the line beyond the obstacle and determining the distance across it. The following are some of the methods (Fig. 4.27).

Method (a): Choose two points A and B to one side and erect perpendiculars AC and BD of equal length. Join CD and prolong it past the obstacle. Choose two points E and F on CD and erect perpendiculars EG and FH equal to that of AC (or BD). Join GH and prolong it. Measure DE. Evidently, $BG = DE$ [Fig. 4.27 (*a*)].

Method (b): Select a point A and erect a perpendicular AC of any convenient length. Select another point B on the chain line such that $AB = AC$. Join B and C and prolong it to any convenient point D. At D, set a right angle DE such that $DE = DB$. Choose another point F on DE such that $DF = DC$. With F as centre and AB as radius, draw an arc. With E as centre, draw another arc of the same radius to cut the previous arc in G. Join GE which will be in range with the chain line. Measure CF [Fig. 4.27 (*b*)]. Then $AG = CF$.

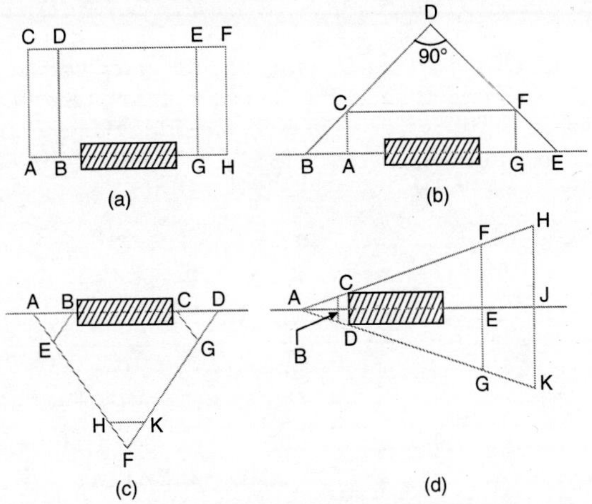

Fig. 4.27 Obstacles to Chaining

Method (c): Select two points A and B on the chain line and construct an equilateral triangle ABE by swinging arcs. Join AE and produce it to any point F. On AF, choose any point H and construct an equilateral triangle FHK. Join F and K and produce it to D such that $FD=FA$. Choose a point G on FD and construct an equilateral triangle CDG. The direction CD is in range with the chain line [Fig. 4.27 (*c*)]. The length BC is given by

$$BC = AD - AB - CD = AF - AB - CD$$

Method (d): Select two points A and B on the chain line and set a line CBD at any angle. Join A and C and produce it to F such that $AF = n \,.\, AC$. Similarly join A and D and produce it to G such that $AG = n \cdot AD$. Join F and G and mark point E on it such that $FE = n \,.\, BC$. Similarly, produce AF and AG to H and K respectively such that $AH = n' \,.\, AC$ and $AK = n' \,.\, AD$. Join H and K and mark J on it in such a way that $HJ = n' \,.\, CB$. Join EJ, which will be in range with chain line. The obstructed distance BE is given by [Fig. 4.27 (*d*)] :

$$BE = AE - AB. \quad \text{But } AE = n \cdot AB$$

$$\therefore \quad BE = n \cdot AB - AB = (n - 1)\, AB.$$

Example 4.6: *To continue a survey line AB past an obstacle, a line BC 200 metres long was set out perpendicular to AB, and from C angles BCD and BCE were set out at 60° and 45° respectively. Determine the lengths which must be chained off along CD and CE in order that ED may be in AB produced. Also, determine the obstructed length BE.*

Solution: (Fig. 4.28).

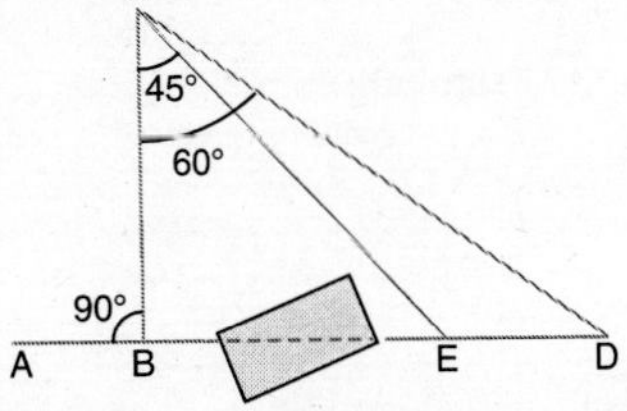

Fig. 4.28

$$\angle ABC \text{ is } 90°$$

From ΔBCD, $CD = BC \sec 60° = 200 \times 2 =$ **400 m.**

From ΔBCE, and $CE = BC \sec 45° = 200 \times 1.4142$

$= $ **282.84 m.**

$BE = BC \tan 45° = 200 \times 1 =$ **200 m.**

Example 4.7: *In passing an obstacle in the form of a pond, stations A and D, on the main line, were taken on the opposite sides of the pond. On the left of AD, a line AB, 200 m long was laid down and a second line AC. 250 m long, was ranged on the right of AD, the points B, D and C being in the same straight line. BD and DC were then chained and found to be 125 m and 150 m respectively. Find the length of AD.*

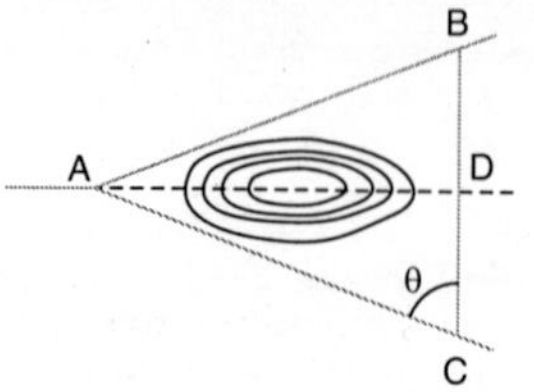

Fig. 4.29

Solution: (Fig. 4.29). In ΔABC, Let $\angle ACD = \theta$

$$AC = 250 \text{ m} ;\ AB = 200 \text{ m} ;\ BC = BD + DC = 125 + 150 = 275 \text{ m}$$

Now, $$\cos\theta = \frac{AC^2 + CB^2 - AB^2}{2\,AC \times CB} = \frac{(250)^2 + (275)^2 - (200)^2}{2 \times 250 \times 275} = \frac{9.813}{13.75} = 0.7137$$

From $$\Delta ADC,\ AD^2 = AC^2 + CD^2 - 2\,AC\,.\,CD\cos\theta$$

$$= (250)^2 + (150)^2 - 2(250)\,(150) \times 0.7137 = 31474.5$$

Hence $$AD = \mathbf{177.41\ m.}$$

Example 4.8: *A survey line BAC crosses a river, A and C being on the near and distant banks respectively. Standing at D, a point 50 metres measured perpendicularly to AB from A, the bearings of C and B are 320° and 230° respectively, AB being 25 metres. Find the width of the river.*

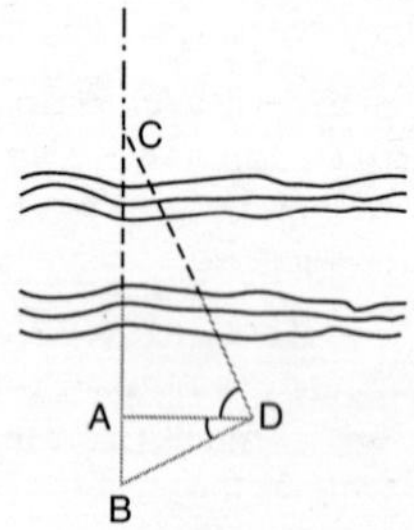

Fig. 4.30

Solution: (Fig. 4.30).

In $$\Delta ABD,\ AB = 25 \text{ m} ;\ AD = 50 \text{ m}$$

$\therefore$ $$\tan\angle BDA = \frac{25}{50} = 0.5 \quad \text{or} \quad \angle BDC = 26°\ 34'$$

$$\angle BDC = -\ 320° - 230° = 90° \quad \text{and} \quad \angle ADC = 90° - 26°\ 34' = 63°\ 26'$$

Again, from ΔADC, $CA = AD \tan ADC = 50 \tan 63°\ 26' = \mathbf{100\ m}$

Example 4.9: *A survey line ABC cuts the banks of a river at B and C, and to determine the distance BC, a line BE, 60 m long was set out roughly parallel to the river. A point D was then found in CE produced and middle point F of DB determined. EF was then produced to G, making FG equal to EF, and DG produced to cut the survey line in H. GH and HB were found to be 40 and 80 metres long respectively. Find the distance from B to C .*

Solution: (Fig. 4.31)

In $BEDG$, $\quad BF = FD$ and $GF = FE$

Hence $BEDG$ is a parallelogram.

Hence $$GD = BE = 60 \text{ m}$$

$\therefore$ $$HD = HG + GD = 40 + 60 = 100 \text{ m}$$

From similar triangles CHD and CBE, we get

$$\frac{CB}{CH} = \frac{BE}{HD}$$

or $$\frac{CB}{CB + BH} = \frac{BE}{HG + GD} \quad \text{or} \quad \frac{CB}{CB + 80} = \frac{60}{40 + 60} = 0.6$$

$\therefore$ $$CB = 0.6\,CB + 48 \quad \text{or} \quad CB = \mathbf{120\ m}$$

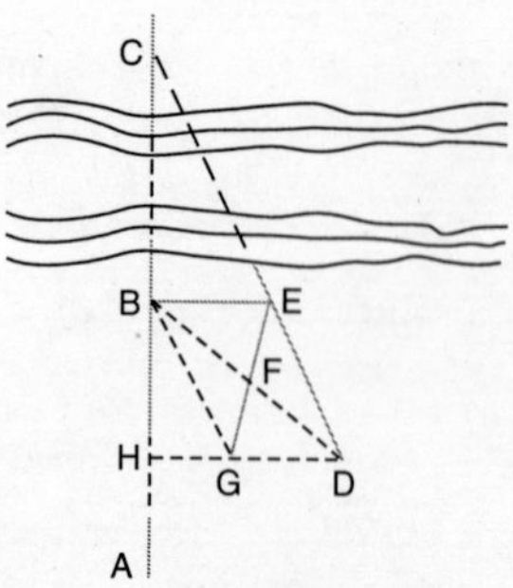

Fig. 4.31

4.10 CROSS STAFF SURVEY

Cross staff survey is done to locate the boundaries of a field and to determine its area. A chain line is run through the centre of the area which is divided into a number of triangles and trapezoids. The offsets to the boundary are taken in order of their chainages. The instruments required for cross staff survey are chain, tape, arrows and a cross staff. After the field work is over, the survey is plotted to a suitable scale.

Example 4.10: *Plot the following cross staff survey of a field ABCDEFG and calculate its area* [Fig. 4.32 (*a*)].

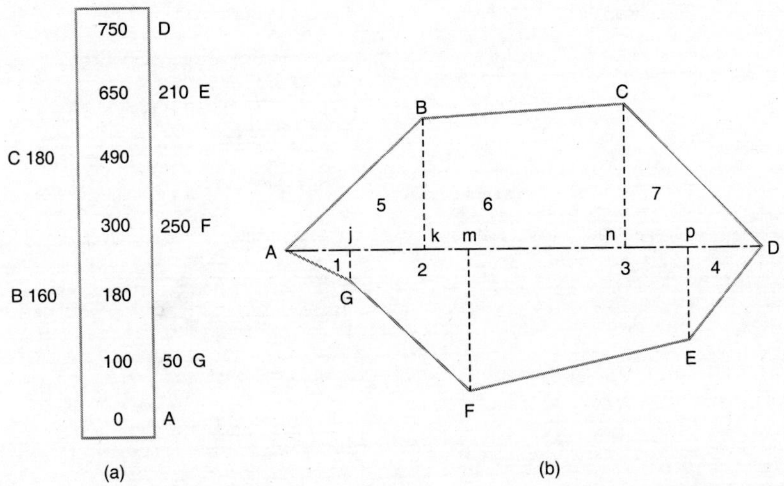

Fig. 4.32

Solution: Fig. 4.32 (*b*) shows the field *ABCDEFG*. The calculations for the area are given in the table below :

S.No.	*Figure*	*Chainage (m)*	*Base (m)*	*Offsets (m)*	*Mean (m)*	*Area (m^2)*
1.	*AjG*	0 & 100	100	0 & 50	25	2,500
2.	*jGFm*	100 & 300	200	50 & 250	150	30,000
3.	*mFEp*	300 & 650	350	250 & 210	230	80,500
4.	*pED*	650 & 750	100	210 & 0	105	10,500
5.	*ABk*	0 & 180	180	0 & 160	80	14,400
6.	*BknC*	180 & 490	310	160 & 180	170	52,700
7.	*CnD*	490 &750	260	180 & 0	90	23,400
					Total	214,000

∴ Area of Field = 214000 m^2 = **21.4 hectares.**

4.11 PLOTTING A CHAIN SURVEY

Generally, the scale of plotting a survey is decided before the survey is started. In general, the scale depends on the purpose of survey, the extent of survey and the finances available.

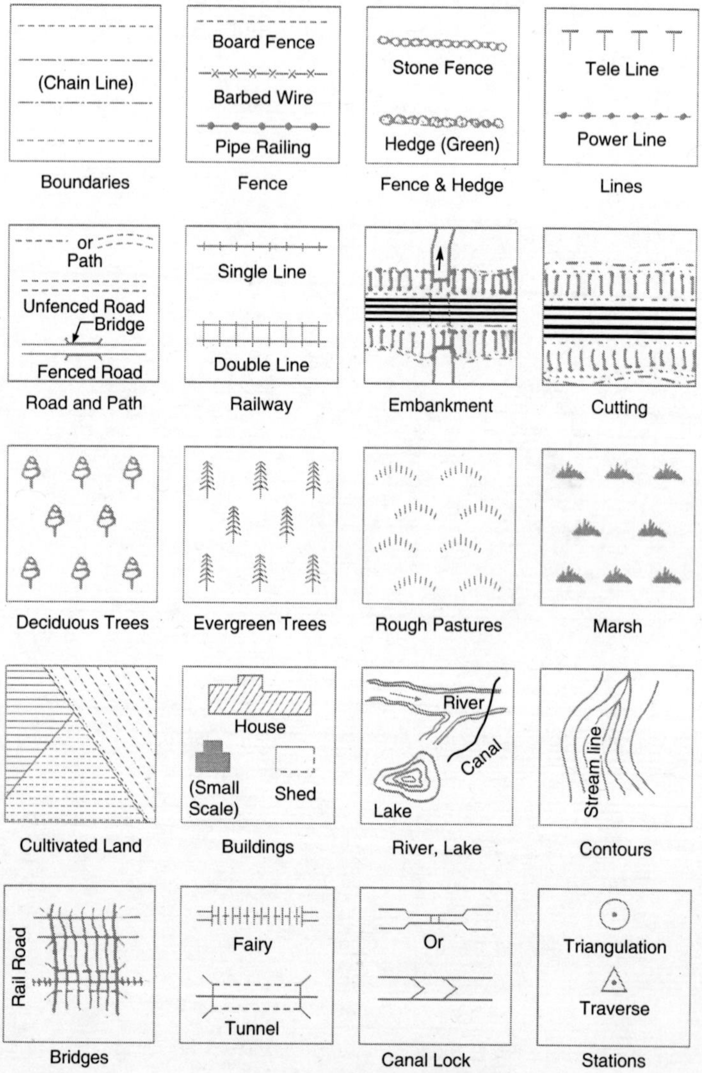

Fig. 4.33 Conventional Symbols

The plan must be so oriented on the sheet that the north side of the survey lies towards the top of the sheet and it is centrally placed. The way to achieve this, is to first plot the skeleton on a tracing paper and rotate it on the drawing paper. After having oriented it suitably, the points may be pricked through. To begin with, base line is first plotted. The other triangles are then laid by intersection of arcs. Each triangle must be verified by measuring the check line on the plan and comparing it with its measured length in the field. If the discrepancy is not within the limits, measurements may be taken again. If it is less, the error may be adjusted suitably.

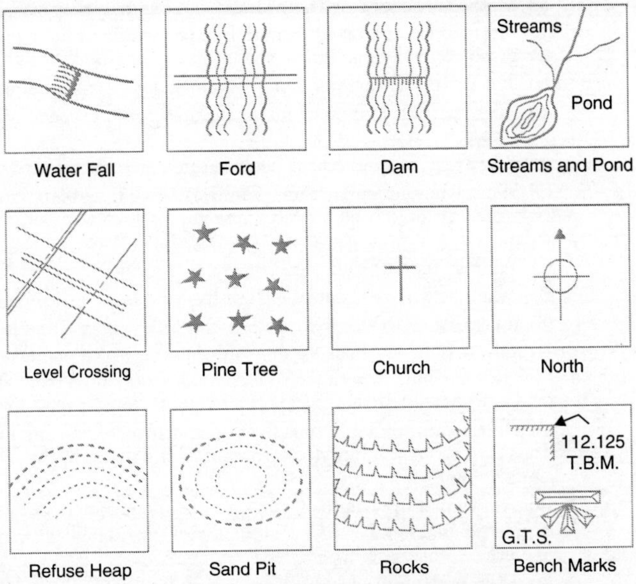

Fig. 4.34 Conventional Symbols.

After having drawn the skeleton consisting a number of triangles, offsets may be plotted. There are two methods of plotting the offsets. In the first method, the chainages of the offsets are marked on the chain line and perpendicular to the chain line are erected with the help of a set-square. In the other method, the plotting is done with the help of an offset scale. A long scale is kept parallel to the chain line and a distance equal to half the length of the offset scale. The offset scale consists of a small scale having zero mark in the middle. The zero of the long scale is kept in line with the zero of the chain line. Chainages are then marked against the *working edge* of the offset scale and the offsets are measured along its edge. Thus, the offsets can be plotted to both the sides of the line.

Different features on the ground are represented by different symbols. Figs. 4.33 and 4.34 shows some *conventional symbols* commonly used.

PROBLEMS

1. Explain the principle on which chain survey is based.
2. Explain, with neat diagrams the construction and working of the following :
   (*a*) Optical square (*b*) Prism square (*c*) Cross staff.
3. What are the instruments used in chain surveying ? How is a chain survey executed in the field ?
4. What is a well conditional triangle ? Why is it necessary to use well-conditioned triangles ?
5. (*a*) Explain clearly the principle of chain surveying.
   (*b*) How would you orient in direction a chain survey plot on the drawing sheet ?
   (*c*) Set out clearly the precautions a surveyor should observe in booking the field work of a chain survey. (*A.M.I.E.*)
6. Illustrate any four of the following by neat line diagrams (explanation and description not required) :
   (*a*) Permanent reference of a survey station.
   (*b*) Construction and working of either an optical square or a prism square.

(*c*) Methods of checking the triangle of a chain survey.

(*d*) Methods of setting out a chain line perpendicular to a given chain line and passing through a given point laying outside the latter.

(*e*) The prismatic reading arrangement in a prismatic compass.

(*f*) Map conventional signs for a metalled road, a hedge with fence, a tram line, a house and a rivulet. (*A.M.I.E.*)

7. (*a*) What factors should be considered in deciding the stations of a chain survey ?

(*b*) What detailed instructions would you give to a fresh trainee surveyor regarding the care and use of his field book for recording survey measurements?

8. Explain the following terms : (*a*) Base line (*b*) Check line (*c*) Tie line (*d*) Swing offset (*e*) Oblique offset (*f*) Random line.

9. Explain how will you continue chaining past the following obstacles :

(*a*) a pond (*b*) a river (*c*) a hill (*d*) a tall building.

10. Explain various methods for determining the width of a river.

11. Find the maximum length of an offset so that the displacement of a point on the paper should not exceed 0.25 mm, given that the offset was laid out 3° from its true direction and the scale was 20 m to 1 cm.

12. To what accuracy should the offset be measured if the angular error in laying out the direction is 4° so that the maximum displacement of the point on the paper from one source of error may be same as that from the other source.

13. Find the maximum length of offset so that displacement of the point on the paper from both sources of error should not exceed 0.25 mm, given that the offset is measured with an accuracy of 1 in 50 and scale is 1 cm = 8 m.

14. Find the maximum permissible error in laying off the direction of offset so that the maximum displacement may not exceed 0.25 mm on the paper, given that the length of the offset is 10 metres, the scale is 20 m to 1 cm and the maximum error in the length of the offset is 0.3 m.

15. A main line of a survey crosses a river about 25 m wide. To find the gap in the line, stations *A* and *B* are established on the opposite banks of the river and a perpendicular *AC*, 60 m long is set out at *A*. If the bearings of *AC* and *CB* are 30° and 270° respectively, and the chainage at *A* is 285.1 m, find the chainage at *B*.

16. A chain line *ABC* crosses a river. *B* and *C* being on the near and distant banks respectively. The respective bearings of *C* and *A* taken at *D*, a point 60 m measured at right angles to *AB* from *B* are 280° and 190°, *AB* being 32 m. Find the width of the river.

17. In passing an obstacle in the form of a pond, stations *A* and *D*, on the main line, were taken on the opposite sides of the pond. On the left of *AD*, a line *AB*, 225 m long was laid down, and a second line *AC*, 275 m long, was ranged on the right of *AD*, the points *B*, *D* and *C* being in the same straight line. *BD* and *DC* were then chained and found to be 125 m and 137.5 m respectively. Find the length of *AD*.

18. (*a*) What are the conventional signs used to denote the following : (*i*) road, (*ii*) railway double line, (*iii*) cemetery, (*iv*) railway bridge, and (*v*) canal with lock ?

(*b*) Differentiate between a Gunter's chain and an Engineer's chain. State relative advantages of each. (*A.M.I.E. May 1966*)

19. *B* and *C* are two points on the opposite banks of a river along a chain line *ABC* which crosses the river at right angles to the bank. From a point *P* which is 150 ft. from *B* along the bank, the bearing of *A* is 215° 30′ and the bearing of *C* is 305° 30′. If the length *AB* is 200 ft., find the width of the river. (*A.M.I.E. May 1966*)

ANSWERS

11. 9.5 m	12. 1 in 14.3	13. 7.07 m	14. 2° 18′	15. 386 m
16. 112.5 m	17. 212.9 m	19. 112.5 ft.		

CHAPTER 5

THE COMPASS

5.1 INTRODUCTION

Chain surveying can be used when the area to be surveyed is comparatively small and is fairly flat. However, when large areas are involved, methods of chain surveying *alone* are not sufficient and convenient. In such cases, it becomes essential to use some sort of instrument which enables angles or directions of the survey lines to be observed. In engineering practice, following are the instruments used for such measurements :

(*a*) ***Instruments for the direct measurement of directions :***

(*i*) Surveyor's Compass

(*ii*) Prismatic Compass

(*b*) ***Instruments for measurements of angles***

(*i*) Sextant

(*ii*) Theodolite

Traverse Survey. *Traversing* is that type of survey in which a number of connected survey lines form the framework and the directions and lengths of the survey line are measured with the help of an angle (or direction) measuring instrument and a tape (or chain) respectively. When the lines form a circuit which ends at the starting point, it is known as a *closed traverse.* If the circuit ends elsewhere, it is said to be an *open traverse.* The various methods of traversing have been dealt with in detail in Chapter 7.

Units of Angle Measurement. An angle is the difference in directions of two intersecting lines. There are three popular systems of angular measurement :

(*a*) ***Sexagesimal System:***	1 circumference	= 360° (degrees of arc)
	1 degree	= 60′ (minutes of arc)
	1 minute	= 60″ (seconds of arc)
(*b*) ***Centesimal System:***	1 circumference	= 400^g (grads)
	1 grad	= 100^c (centigrads)
	1 centigrad	= 100^{cc} (centicentigrads)
(*c*) ***Hours System:***	1 circumference	= 24^h (hours of time)
	1 hour	= 60^m (minutes of time)
	1 minute	= 60^s (seconds of time)

The sexagesimal system is widely used in United States, Great Britain. India and other parts of the world. More complete tables are available in this system and most surveying instruments are graduated according to this system. However, due to facility in computation and interpolation, the centesimal system is gaining more favour in Europe. The Hours system is mostly used in astronomy and navigation.

5.2 BEARINGS AND ANGLES

The direction of a survey line can either be established (*a*) with relation to each other, or (*b*) with relation to any meridian. The first will give the angle between two lines while the second will give the *bearing* of the line.

Bearing. Bearing of a line is its direction relative to a given meridian. A meridian is any direction such as (1) True Meridian (2) Magnetic Meridian (3) Arbitrary Meridian.

1. **True Meridian.** True meridian through a point is the line in which a plane, passing that point and the north and south poles, intersects with surface of the earth. It, thus, passes through the true north and south. The direction of true meridian through a point can be established by astronomical observations.

   True Bearing. True bearing of a line is the horizontal angle which it makes with the true meridian through one of the extremities of the line. Since the direction of true meridian through a point remains fixed, the true bearing of a line is a constant quantity.

2. **Magnetic Meridian.** Magnetic meridian through a point is the direction shown by a freely floating and balanced magnetic needle free from all other attractive forces. The direction of magnetic meridian can be established with the help of a magnetic compass.

   Magnetic Bearing. The magnetic bearing of a line is the horizontal angle which it makes with the magnetic meridian passing through one of the extremities of the line. A magnetic compass is used to measure it.

3. **Arbitrary Meridian.** Arbitrary meridian is any convenient direction towards a permanent and prominent mark or signal, such as a church spire or top of a chimney. Such meridians are used to determine the relative positions of lines in a small area.

   Arbitrary Bearing. Arbitrary bearing of a line is the horizontal angle which it makes with any arbitrary meridian passing through one of the extremities. A theodolite or sextant is used to measure it.

Designation of Bearings

The common systems of *notation* of bearings are :

(*a*) The whole circle bearing system (W.C.B.) or Azimuthal system.

(*b*) The Quadrantal bearing (Q.B.) system.

(*a*) **The Whole Circle Bearing System.** (Azimuthal system).

In this system, the bearing of a line is measured with magnetic north (or with south) in clockwise direction. The value of the bearing thus varies from 0° to 360°. Prismatic compass is graduated on this system. In India and U.K., the W.C.B. is measured clockwise with magnetic north.

Referring to Fig. 5.1, the W.C.B. of *AB* is θ_1, of *AC* is θ_2, of *AD* is θ_3 and of *AF* is θ_4.

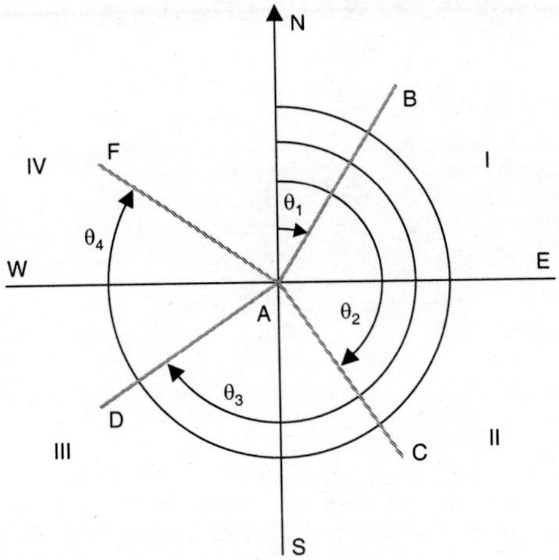

Fig. 5.1 W.C.B. system

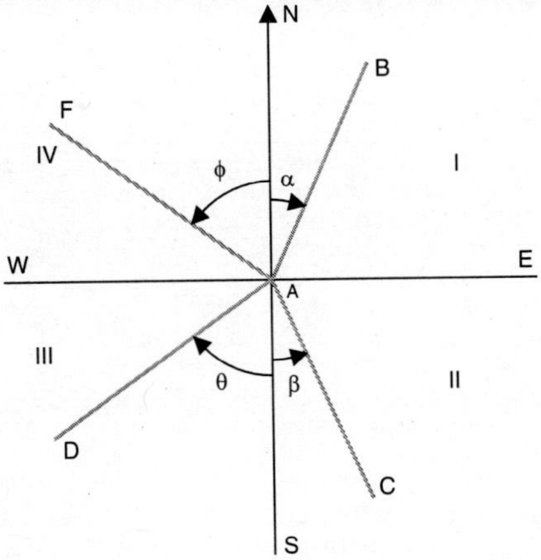

Fig. 5.2 Q.B. system.

(b) The Quadrantal Bearing System: (Reduced bearing)

In this system, the bearing of a line is measured eastward or westward from north or south, whichever is nearer. Thus, both North and South are used as reference meridians and the directions can be either clockwise or anti-clockwise depending upon the position of the line. In this system, therefore, the quadrant, in which the line lies, will have to be mentioned. These bearings are observed by Surveyor's compass.

Referring Fig. 5.2, the Q.B. of the line *AB* is α and is written as N α E, the bearing being measured with reference to North meridian (since it is nearer), towards East. The bearing of *AC* is β and is written as S β E, it being measured with reference of South and in anti-clockwise direction towards East. Similarly, the bearings of *AD* and *AF* are respectively S θ W and N ϕ W.

Thus, in the quadrantal system, the reference meridian is prefixed and the direction of measurement (Eastward or Westward) is affixed to the numerical value of the bearing. The Q.B. of a line varies from 0° to 90°. The bearings of this system are known as *Reduced Bearings* (R.B.)

Conversion of Bearings from One System to the Other

The bearing of a line can be very easily converted from one system to the other, with the aid of a diagram. Referring to Fig. 5.1, the conversion of W.C.B. into R.B. can be expressed in the following Table :

TABLE 5.1 Conversion of W.C.B. into R.B.

Line	*W.C.B. between*	*Rule for R.B.*	*Quadrant*
AB	0° and 90°	R.B.= W.C.B.	NE
AC	90° and 180°	R.B.= 180° – W.C.B.	SE
AD	180° and 270°	R.B.= W.C.B.– 180°.	SW
AF	270° and 360°	R.B.= 360° – W.C.B.	NW

Similarly, referring to Fig. 5.2, the conversion of R.B. into W.C.B. can be expressed in the following Table :

TABLE 5.2 Conversion of R.B. into W.C.B.

Line	*R.B.*	*Rule for W.C.B.*	*W.C.B. between*
AB	N α E	W.C.B. = R.B.	0° and 90°
AC	S β E	W.C.B. = 180° – R.B.	90° and 180°
AD	S θ W	W.C.B. = 180° + R.B.	180° and 270°
AF	N ϕ W	W.C.B. = 360° – R.B.	270° and 360°

Fore and Back Bearing

The bearing of line, whether expressed in W.C.B. system or in Q.B. system, differs according as the observation is made from one end of the line or from the other. If the bearing of a line *AB* is measured from *A* towards *B*, it is known as forward bearing or *Fore Bearing* (F.B). If the bearing of the line *AB* is measured from *B* towards *A*, it is known as backward bearing or *Back Bearing* (B.B.), since it is measured in backward direction.

Considering first the W.C.B. system and referring to Fig. 5.3 (*a*), the back bearing of line *AB* is ϕ and fore bearing of *AB* is θ . Evidently $\phi = 180° + \theta$. Similarly, from Fig. 5.3 (*b*), the back bearing of *CD* is ϕ and fore bearing θ, hence, $\phi = \theta - 180°$. *Thus, in general, it can be stated that B.B.= F.B. ± 180°, using plus sign when F.B is less than 180° and minus sign when F.B. is greater than 180°.*

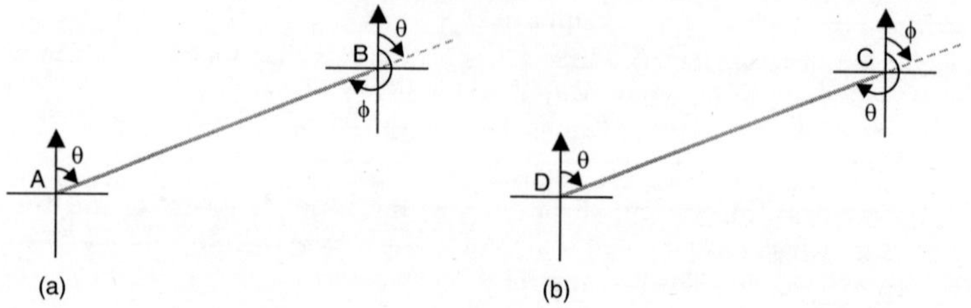

Fig. 5.3 Fore and back bearings.

Again, considering the Q.B. system and referring to Fig. 5.4 (*a*), the fore bearing of line *AB* is NθE and, therefore, the back bearing is equal to SθW. Similarly, from Fig. 5.4 (*b*), the fore bearing of the line *CD* is SθW and back bearing is equal to NθE. *Thus, it can be stated that to convert the fore bearing to back bearing, it is only necessary to change the cardinal points by substituting* N *for* S, *and* E *for* W *and vice versa, the numerical value of the bearing remaining the same.*

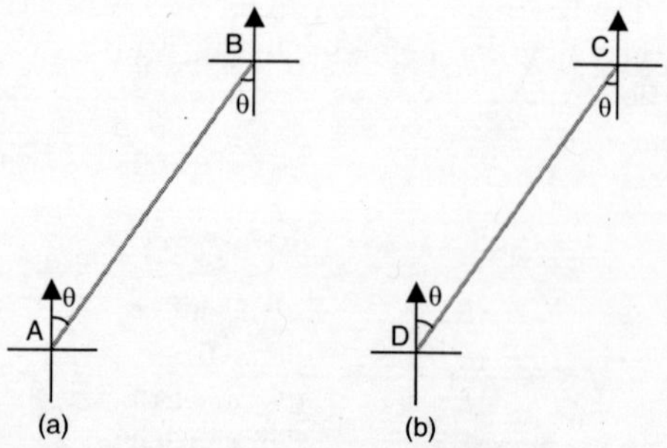

Fig. 5.4 Fore and back bearings.

Calculation of Angles from Bearings

Knowing the bearing of two lines, the angle between the two can very easily be calculated with the help of a diagram,

Ref. to Fig. 5.5 (*a*), the included angle α between the lines *AC* and $AB = \theta_2 - \theta_1$ = F.B. of one line – F.B. of the other line, both bearings being measured from a common point *A*. Ref. to Fig. 5.5 (*b*), the angle $\alpha = (180° + \theta_1) - \theta_2$ = B. B. of previous line – F.B. of next line.

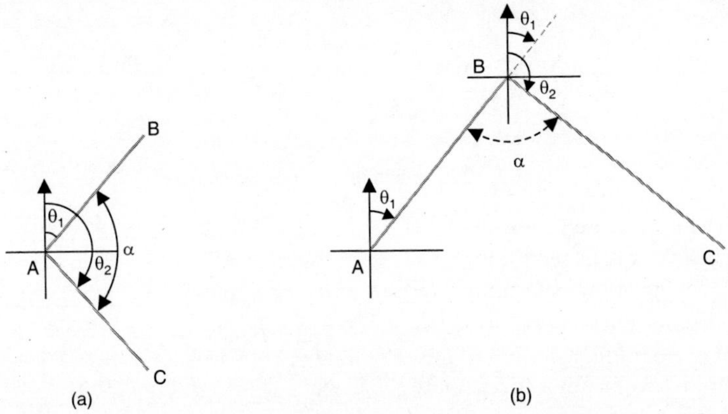

Fig. 5.5 Calculation of angles from bearings.

Let us consider the quadrantal bearing. Referring to Fig. 5.6 (*a*) in which both the bearings have been measured to the *same side* of common meridian, the included angle $\alpha = \theta_2 - \theta_1$. In Fig. 5.6 (*b*), both the bearings have been measured to the *opposite sides* of the common meridian, and included angle $\alpha = \theta_1 + \theta_2$. In Fig. 5.6 (*c*) both the bearings have been measured to the *same side of different meridians* and the included angle $\alpha = 180° - (\theta_2 + \theta_1)$. In Fig. 5.6 (*d*), both the bearings have been measured to the *opposite sides of different meridians,* and angle $\alpha = 180° - (\theta_1 - \theta_2)$.

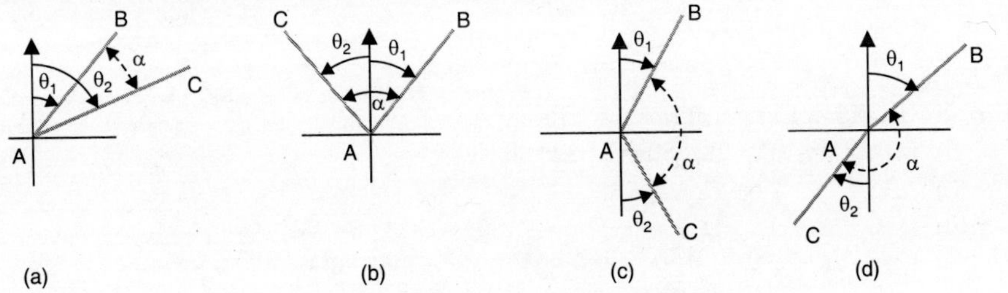

Fig. 5.6 Calculation of angles from bearings

Calculation of Bearings from Angles

In the case of a traverse in which included angles between successive lines have been measured, the bearings of the lines can be calculated provided the bearing of any one line is also measured.

Referring to Fig. 5.7, let α, β, γ, δ, be the included angles measured *clockwise* from back stations and θ_1 be the measured bearing of the line *AB*.

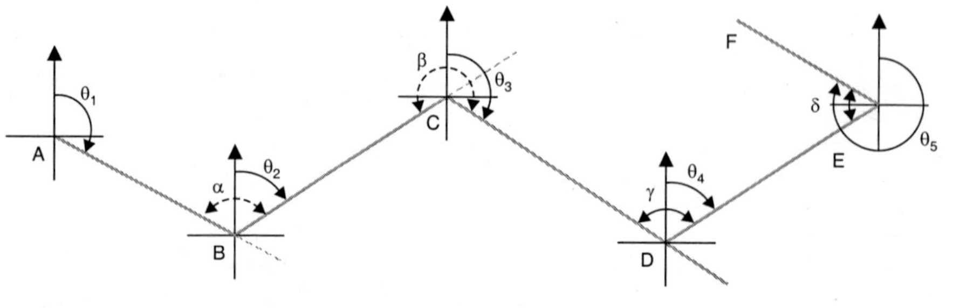

Fig. 5.7 Calculation of bearings from angles.

$\therefore$ The bearing of the next line $BC = \theta_2 = \theta_1 + \alpha - 180°$ ...(1)

The bearing of the next line $CD = \theta_3 = \theta_2 + \beta - 180°$ ...(2)

The bearing of the next line $DE = \theta_4 = \theta_3 + \gamma - 180°$ ...(3)

The bearing of the next line $EF = \theta_5 = \theta_4 + \delta + 180°$ ...(4)

As is evident from Fig. 5.7, $(\theta_1 + \alpha)$, $(\theta_2 + \beta)$, and $(\theta_3 + \gamma)$ are more than 180° while $(\theta_4 + \delta)$ is less than 180°. Hence in order to calculate the bearing of the next line, the following statement can be made :

"Add the measured clockwise angles to the bearing of the previous line. If the sum is more than 180°, deduct 180°. If the sum is less than 180°, add 180° ".

In a closed traverse, clockwise angles will be obtained if we proceed round the traverse in the anti-clockwise direction.

Examples on Angles and Bearings

Example 5.1: *(a) Convert the following whole circle bearings to quadrantal bearings: (i) 22° 30′ (ii) 170° 12′ (iii) 211° 54′ (iv) 327° 24′.*

(b) Convert the following quadrantal bearing to whole circle bearings : (i) N 12° 24′E (ii) S 31° 36′E (iii) S 68° 6′W (iv) N 5° 42′W.

Solution.

(*a*) Ref. to Fig. 5.1 and Table 5.1 we have

(*i*) R.B. = W.C.B. = 22° 30′ = N **22° 30′ E.**

(*ii*) R.B. = 180° – W.C.B. = 180° – 170° 12′ = **S 9° 48′ E.**

(*iii*) R.B. = W.C.B. – 180° = 211° 54 – 180° = **S 31° 54′ W.**

(*iv*) R.B.= 360° – W.C.B. = 360° – 327° 24′ = **N 32° 36′ W.**

(*b*) Ref. to Fig. 5.2 and Table 5.5 we have

(*i*) W.C.B. = R.B.= **12° 24′**

(*ii*) W.C.B. = 180° – R.B. = 180° – 31° 36′ = **148° 24′**

(*iii*) W.C.B. = 180° + R.B. = 180° + 68° 6′ = **248° 6′**

(*iv*) W.C.B. = 360° – R.B. = 360° – 5° 42′ = **354° 18′**

Example 5.2: *The following are observed fore-bearings of the lines (i) AB 12° 24′ (ii) BC 119° 48′ (iii) CD 266° 30′ (iv) DE 354° 18′ (v) PQ N 18° 0′ E (vi) QR S12° 24′ E (vii) RS S59° 18′ W (viii) ST N 86° 12′W. Find their back bearings.*

Solution: B.B. = F.B. ± 180°, using + sign when F.B. is less than 180° and – sign when it is more than 180°.

(*i*) B.B. of AB = 12° 24′ + 180° = **192° 24′.**

(*ii*) B.B. of BC = 119° 48′ + 180° = **299° 48′**

(*iii*) B.B. of CD = 266° 30′ – 180° = **86° 30′**

(*iv*) B.B. of DE = 354° 18′ – 180° = **174° 18′**

(*v*) B.B. of PQ = **S 18° 0′ W**

(*vi*) B.B. of QR = **N 12° 24′ W**

(*vii*) B.B. of RS = **N 59° 18′ E**

(*viii*) B.B. of ST = **S 86° 12′ E**

Example 5.3: *The following bearings were observed with a compass. Calculate the interior angles.*

Line	*Fore Bearing*
AB	*60° 30′*
BC	*122° 0′*
CD	*46° 0′*
DE	*205° 30′*
EA	*300° 0′.*

Solution: Fig. 5.8 shows the plotted traverse.

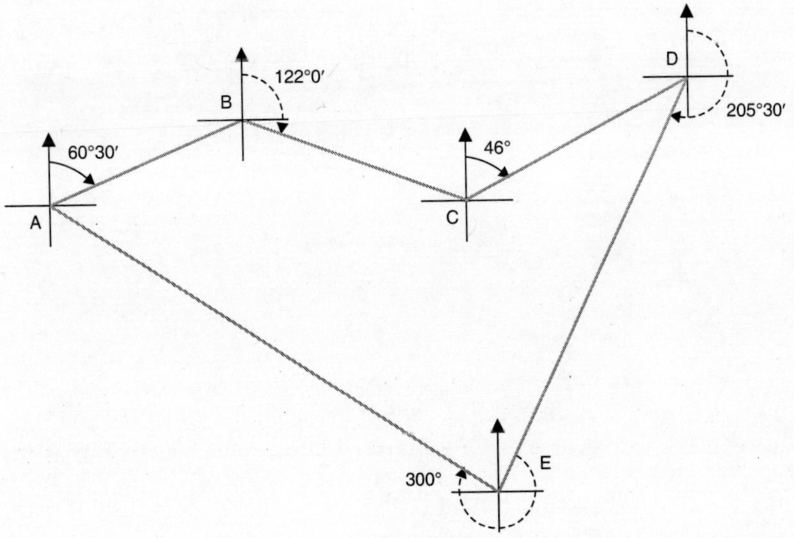

Fig. 5.8

Included angle = Bearing of previous line – Bearing of next line

$\angle A$ = Bearing of AE – Bearing of AB

$= (300° - 180°) - 60° 30' =$ **59° 30′**.

$\angle B$ = Bearing of BA – Bearing of BC

$= (60° 30' + 180°) - 122° =$ **118° 30′**.

$\angle C$ = Bearing of CB – Bearing of CD

$= (122° + 180°) - 46° =$ **256°**

$\angle D$ = Bearing of DC – Bearing of DE

$= (46° + 180°) - 205° 30' =$ **20° 30′**.

$\angle E$ = Bearing of ED – Bearing of EA

$= (205° 30' - 180°) - 300° + 360° =$ **85° 30′**

Sum = 540° 00′.

Check : $(2n - 4)\, 90° = (10 - 4)\, 90° = 540°$.

Example 5.4: *The following interior angles were measured with a sextant in a closed traverse. The bearing of the line AB was measured as 60° 00′ with prismatic compass. Calculate the bearings of all other line if* $\angle A = 140°\ 10'$; $\angle B = 90°8'$; $\angle C = 60°\ 22'$; $\angle D = 69°\ 20'$.

Solution: Fig. 5.9 shows the plotted traverse.

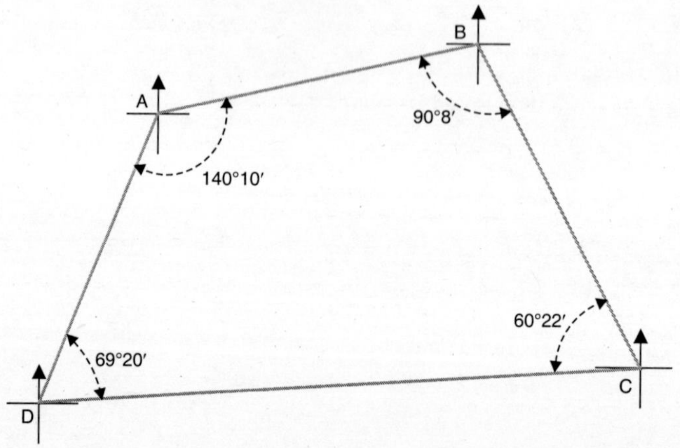

Fig. 5.9

To find the bearing of a line, add the measured clockwise angle to the bearing of the previous line. If the sum is more than 180°, deduct 180°. If the sum is less than 180°, add 180°.

Clockwise angles will be obtained if we proceed in the anticlockwise direction round the traverse.

Starting with A and proceeding toward D, C, B etc., we have

Bearing of AD = Bearing of BA + 140° 10′ – 180°
= (180° + 60°) + 140° 10′ – 180° = 200° 10′
∴ Bearing of DA = **20° 10′**

Bearing of DC = Bearing of AD + 69° 20′ – 180°
= 200° 10′ + 69° 20′ – 180° = 89° 30′
∴ Bearing of CD = **269° 30′**.

Bearing of CB = Bearing of DC + 60° 22′ + 180°
= 89° 30′ + 60° 22′ + 180° = 329° 52′
∴ Bearing of BC = **149° 52′**

Bearing of BC = Bearing of CB + 90° 8′ – 180°
= 329° 52′ + 90° 8′ – 180° = 240°
∴ Bearing of AB = **60°** (check).

5.3 THE THEORY OF MAGNETIC COMPASS

Magnetic compass gives directly the magnetic bearings of lines. The bearings may either be measured in the W.C.B. system or in Q.B. system depending upon the form of the compass used. The bearings so measured are entirely independent on any other measurement.

The general principle of all magnetic compass depends upon the fact that if a long, narrow strip of steel or iron is magnetised, and is suitably suspended or pivoted about a point near its centre so that it can oscillate freely about the vertical axis, it will tend to establish itself in the magnetic meridian at the place of observation.

The most essential features of a magnetic compass are :

(*a*) *Magnetic needle,* to establish the magnetic meridian.

(*b*) *A line of sight,* to sight the other end of the line.

(*c*) *A graduated circle,* either attached to the box or to the needle, to read the directions of the lines.

(*d*) *A compass box* to house the above parts.

In addition, a tripod or suitable stand can be used to support the box.

The various compasses exhibiting the above features are :

(1) Surveyor's compass

(2) Prismatic compass

(3) Transit or Level Compass.

Earth's Magnetic Field and Dip: The earth acts as a powerful magnet and like any magnet, forms a field of magnetic force which exerts a directive influence on a magnetised bar of steel or iron. If any slender symmetrical bar magnet is freely suspended at its centre of gravity so that it is free to turn in azimuth, it will align itself in a position parallel to the lines of magnetic force of the earth at that point.

The lines of force of earth's magnetic field run generally from South to North (Fig. 5.10). Near the equator, they are parallel to the earth's surface. The horizontal projections of the lines of force define the magnetic meridian. The angle which these lines of force make with the surface of the earth is called the *angle of dip* or simply the *dip* of the needle. In elevation, these lines of force (*i.e.* the North end of the needle), are inclined downward towards the north in the

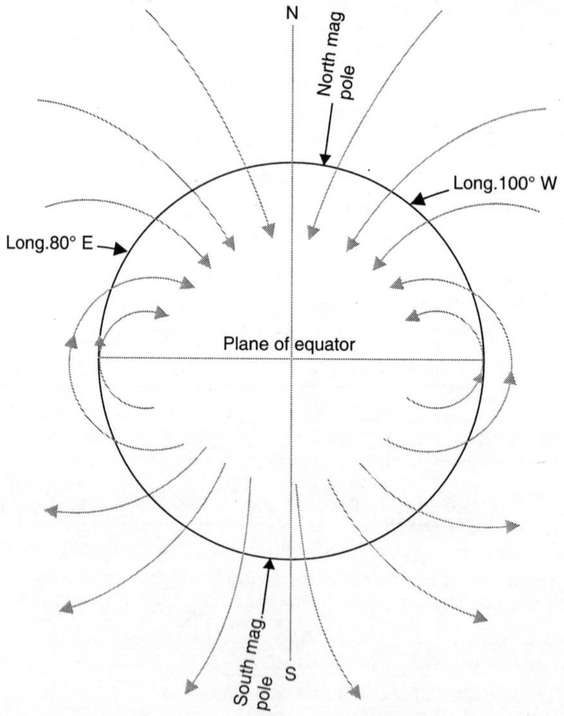

Fig. 5.10 Cross-section of earth's magnetic field.

Northern hemisphere and downward towards South in Southern hemisphere. At a place near 70° North latitude and 96° West longitude, it will dip 90°. This area is called *North magnetic pole.* A similar area in Southern hemisphere is called the *South magnetic pole.* At any other place, the magnetic needle will not point towards the North magnetic pole, but it will take a direction and dip in accordance with the lines of force at the point. Since the lines of force are parallel to the surface of the earth only at equator, the dip of the needle will be zero at equator and the needle will remain horizontal. At any other place, one end of the needle will dip downwards. By suitably weighting the high end of the needle may be brought to a horizontal position.

The Magnetic Needle: The compass needle is made of a slender symmetrical bar of magnetised steel or iron. It is hung from a conical jewel bearing supported on a sharp, hardened steel pivot. Before magnetisation, the needle is free to rotate both vertically and horizontally and does not tend to move away from any direction in which it is originally pointed. When it is magnetised, it will dip downwards and take a definite direction of magnetic meridian, small coil of brass wire is wrapped around it to balance the force tending to make the needle dip. The position of the coil is adjustable for the dip in the locality where the compass is to be used.

Fig. 5.11 shows a typical needle in section, which can either be a "broad needle" or "edge bar" needle type.

The pivot is a sharp and hard point and the slightest jar will break its tip or make it blunt. A lever arrangement is usually provided for lifting the needle off its bearing when not in use, so as to prevent unnecessary wear of the bearing with consequent increase in friction.

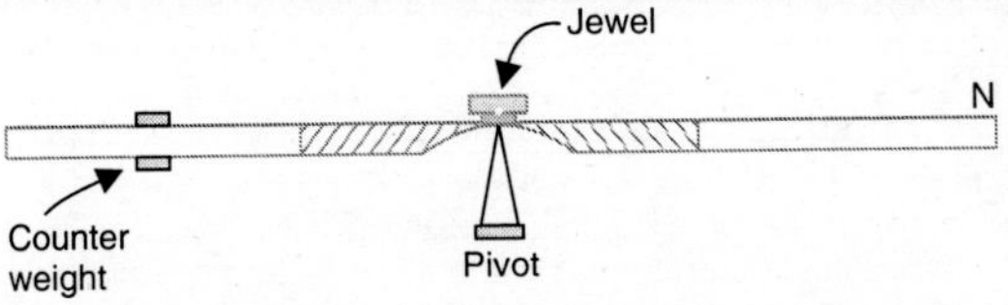

Fig. 5.11 The magnetic needle.

Requirements of a Magnetic Needle: The following are the principal requirements of a magnetic needle :

(1) The needle should be straight and symmetrical and the magnetic axis of the needle should coincide with the geometrical axis. If not, the bearing reading will not be with reference to the magnetic axis, and, therefore, will be wrong. However, the included angles calculated from the observed bearings will be correct.

(2) The needle should be sensitive. It may loose its sensitivity due to (*a*) loss of polarity, (*b*) wear of the pivot. If the polarity has been lost, the needle should be remagnetised. The pivot can either be sharpened with the help of very fine oil stone or it may be completely replaced. Suitable arrangement should be provided to lift the needle off the pivot when not in use.

(3) The ends of the needle should lie in the same horizontal and vertical planes as those of the pivot point. If the ends are not in the same horizontal plane as that of the pivot point, they will be found to quiver when the needle swings, thus causing inconvenience in reading.

(4) For stability, the centre of the gravity of the needle should be as far below the pivot as possible.

In addition to the above requirements of the needle, the compass box along with other accessories should be of non-magnetic substance so that needle is uninfluenced by all other attractive forces except that of the earth's.

5.4 THE PRISMATIC COMPASS

Prismatic compass is the most convenient and portable form of magnetic compass which can either be used as a hand instrument or can be fitted on a tripod. The main parts of the prismatic compass are shown in Fig. 5.12.

As illustrated in the diagram, the magnetic needle is attached to the circular ring or compass card made up of aluminium, a non-magnetic substance. When the needle is on the pivot, it will orient itself in the magnetic meridian and, therefore, the N and S ends of the ring will be in this direction. The line of sight is defined by the object vane and the eye slit, both attached to the compass box. The object vane consists of a vertical hair attached to a suitable frame while the eye slit consists of a vertical slit cut into the upper assembly of the prism unit, both being hinged to the box. When an object is sighted, the sight vanes will rotate with respect to the NS end of ring through an angle which the line makes with the magnetic meridian. A triangular prism is fitted below the eye slit, having suitable arrangement for focusing to suit different eye sights. The prism has both horizontal and vertical faces convex, so that a magnified image of the ring graduation is formed. When the line of sight is also in the magnetic meridian, the South end of the ring comes vertically below the horizontal face of the prism. The 0° or 360° reading

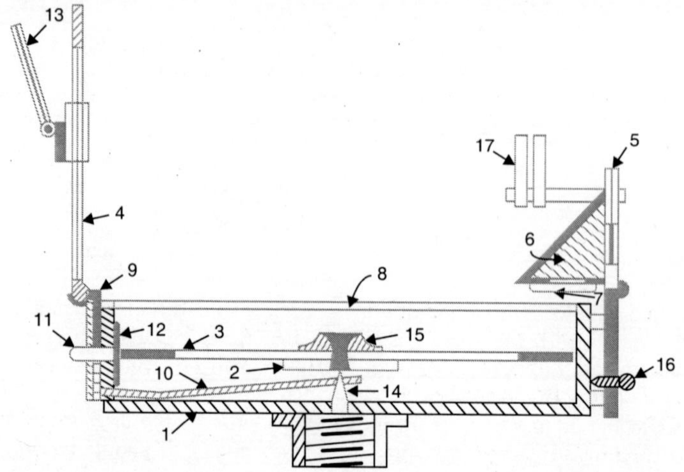

1. Box
2. Needle
3. Graduated ring
4. Object vane
5. Eye vane
6. Prism
7. Prism cap
8. Glass cover
9. Lifting pin
10. Lifting lever
11. Brake pin
12. Spring brake
13. Mirror
14. Pivot
15. Agate cap
16. Focusing stud
17. Sun glass

Fig. 5.12 The prismatic compass.

is, therefore, engraved on the South end of the ring, so that bearing of the magnetic meridian is read as 0° with the help of the prism which is vertically above South end in this particular position. The readings increase in clockwise direction from 0° at South end to 90° at West end, 180° at North end and 270° at East end. This has been clearly illustrated in Fig. 5.13 (*a*) and (*b*).

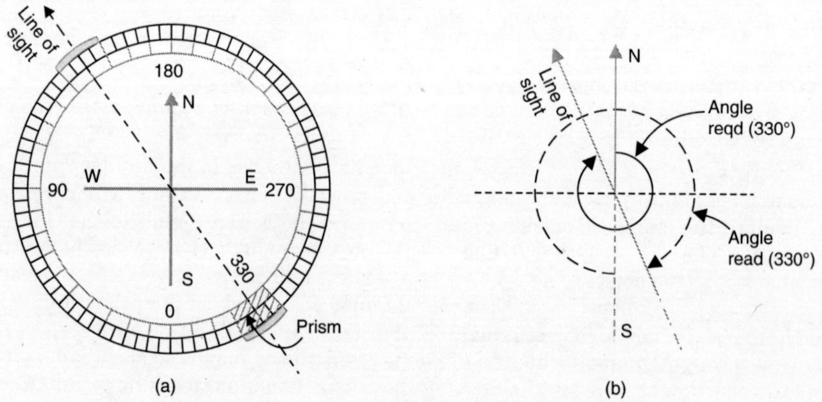

Fig. 5.13 System of graduation in prismatic compass.

When not in use, the object vane frame can be folded on the glass lid which covers the top of the box. The object vane, thus presses against a bent lever which lifts the needle off the pivot and holds it against the glass lid. By pressing knob or brake-pin placed at the base of the object vane, a light spring fitted inside the box can be brought into the contact with the edge of the graduated ring to damp the oscillations of the needle when about to take the reading. The prism can be folded over the edge of the box. A metal cover fits over the circular box, when not in use. To sight the objects which are too high or too low to be sighted directly, a hinged mirror capable of sliding over the object vane is provided and the objects sighted by reflection. When bright objects are sighted, dark glasses may be interposed into the line of sight.

The greatest advantage of prismatic compass is that both sighting the object as well as reading circle can be done simultaneously without changing the position of the eye. The circle is read at the reading at which the hair line appears to cut the graduated ring.

Adjustment of Prismatic Compass: The following are the adjustments usually necessary in the prismatic compass.

(*a*) ***Station or Temporary Adjustments:***

(*i*) Centring (*ii*) Levelling (*iii*) Focusing the prism.

(*b*) ***Permanent Adjustments.*** The permanent adjustments of prismatic compass are almost the same as that of the surveyor's compass except that there are no bubble tubes to be adjusted and the needle cannot be straightened. The sight vanes are generally not adjustable. (See the permanent adjustments of Surveyor's compass).

Temporary Adjustments: Temporary adjustments are those adjustments which have to be made at every set up of the instrument. They comprise the following:

(*i*) ***Centring.*** Centring is the process of keeping the instrument exactly over the station. Ordinary prismatic compass is not provided with fine centring device as is generally fitted to engineer's theodolite. The centring is invariably done by adjusting or manipulating the legs of the tripod. A plumb-bob may be used to judge the centring and if it is not available, it may be judged by dropping a pebble from the centre of the bottom of the instrument.

(*ii*) ***Levelling.*** If the instrument is a hand instrument, it must be held in hand in such a way that graduated disc is swinging freely and appears to be level as judged from the top edge of the case. Generally, a tripod is provided with ball and socket arrangement with the help of which the top of the box can be levelled.

(*iii*) ***Focusing the Prism.*** The prism attachment is slided up or down for focusing till the readings are seen to be sharp and clear.

5.5 THE SURVEYOR'S COMPASS

Fig. 5.14 shows the essential parts of a surveyor's compass. As illustrated in the figure, the graduated ring is directly attached to the box, and not with needle. The edge bar needle freely floats over the pivot. Thus, the graduated card or ring is not oriented in the magnetic meridian, as was the case in the prismatic compass. The object vane is similar to that of prismatic compass. The eye vane consists of a simple metal vane with a fine slit. Since no prism is provided, the object is to be sighted first with the object and eye vanes and the reading is then taken against the North end of the needle, by looking vertically through the top glass. Fig. 5.15 shows the plan view of a surveyor's compass.

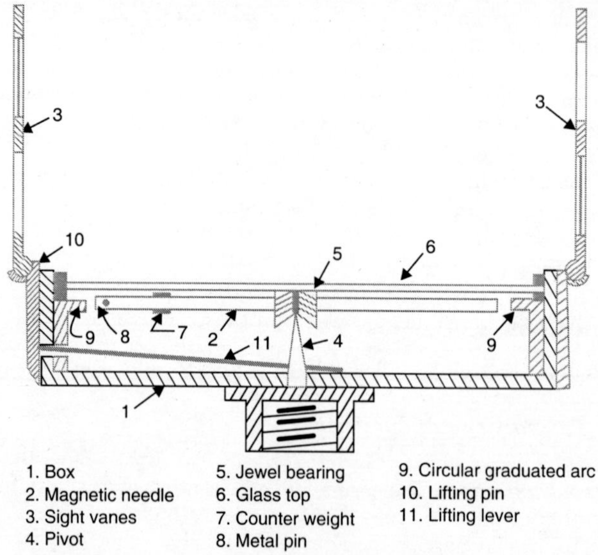

Fig. 5.14 The surveyor's compass

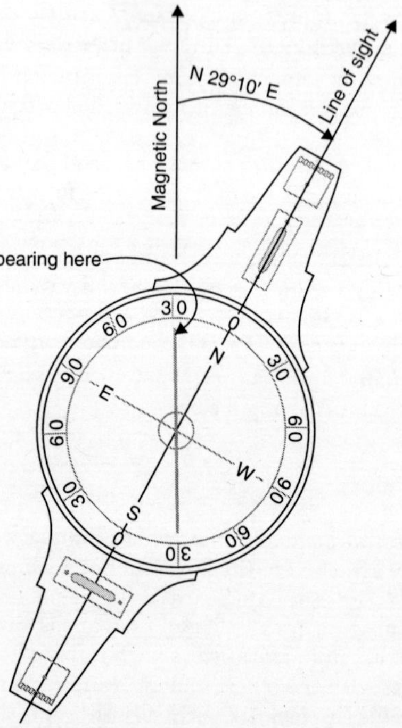

Fig. 5.15 Surveyor's compass (plan).

When the line of sight is in magnetic meridian, the North and South ends of the needle will be over the 0° N and 0° S graduations of the graduated card. The card is graduated in quadrantal system having 0° at N and S ends and 90° at East and West ends. Let us take the case of a line *AB* which is in North-East quadrant. In order to sight the point *B*, the box will have to be rotated about the vertical axis. In doing so, the pointer of the needle remains fixed in position (pointing always to the magnetic meridian) while the 0° N graduation of the card moves in a clockwise direction. In other words, the North end of the needle moves in the anti-clockwise direction with relation to the 0° N graduation of the card. Taking the extreme case when the line has a bearing of 90° in East direction, the pointer appears to move by 90° from the 0° N graduation in anti-clockwise direction ; in this position, therefore, the pointer must read the reading 90° E. Thus, on the graduated card, the East and West are interchanged. See Fig. 5.16 (*a*) and (*b*).

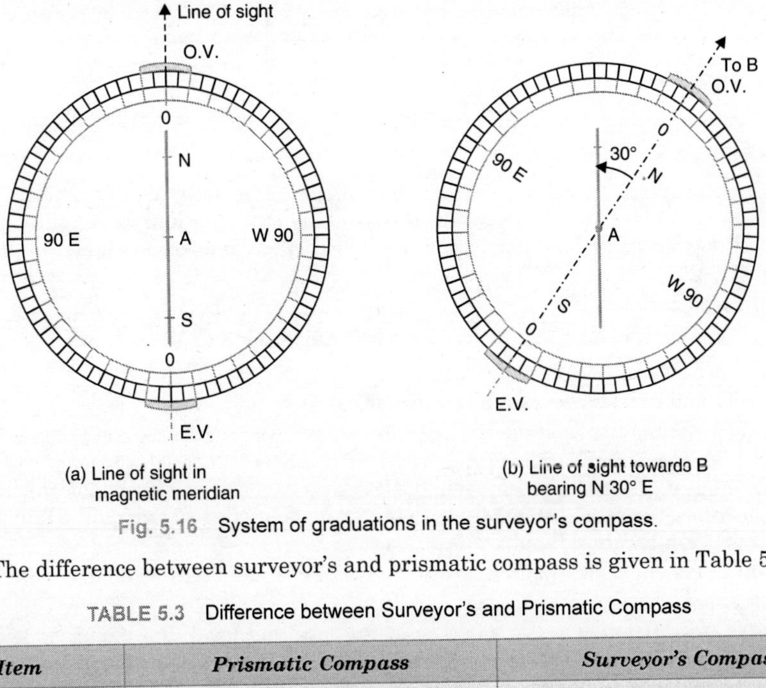

(a) Line of sight in magnetic meridian

(b) Line of sight towards B bearing N 30° E

Fig. 5.16 System of graduations in the surveyor's compass.

The difference between surveyor's and prismatic compass is given in Table 5.3.

TABLE 5.3 Difference between Surveyor's and Prismatic Compass

Item	*Prismatic Compass*	*Surveyor's Compass*
1. ***Magnetic Needle***	The needle is of 'broad needle' type. The needle does not act as index.	The needle is of 'edge bar' type. The needle acts as the index also.
2. ***Graduated Card***	(*i*) The graduated card ring is attached with the needle. The ring does not rotate along with the line of sight. (*ii*) The graduations are in W.C.B. system, having 0° at South end, 90° at West, 180° at North and 270° at East. (*iii*) The graduations are engraved inverted.	(*i*) The graduated card is attached to the box and not to the needle. The card rotates along with the line of sight. (*ii*) The graduations are in Q.B. system, having 0° at N and S and 90° at East and West. East and West are interchanged. (*iii*) The graduations are engraved erect.

3. ***Sighting Vanes***	(*i*) The object vane consists of metal vane with a vertical hair. (*ii*) The eye vane consists of a small metal vane with slit.	(*i*) The object vane consists of a metal vane with a vertical hair. (*ii*) The eye vane consists of a metal vane with a fine slit.
4. ***Reading***	(*i*) The reading is taken with the help of a prism provided at the eye slit. (*ii*) Sighting and reading taking can be done simultaneously from one position of the observer.	(*i*) The reading is taken by directly seeing through the top of the glass. (*ii*) Sighting and reading taking cannot be done simultaneously from one position of the observer.
5. ***Tripod***	Tripod may or may not be provided. The instrument can be used even by holding suitably in hand.	The instrument cannot be used without a tripod.

Temporary Adjustments: Same as for prismatic compass, except for the focusing of the prism.

Permanent Adjustments of Surveyor's Compass: Permanent adjustments are those adjustments which are done only when the fundamental relations between the parts are disturbed. They are, therefore, not required to be repeated at every set up of the instrument. These consist of :

(*i*) Adjustment of levels.
(*ii*) Adjustment of sight vanes.
(*iii*) Adjustment of needle.
(*iv*) Adjustment of pivot point.

(*i*) Adjustment of levels

Object: To make the levels, when they are fitted, perpendicular to the vertical axis.

Test: Keep the bubble tube parallel to two foot screws and centre the bubble. Rotate the instrument through 90° about the vertical axis, till it comes over the third foot screw and centre the bubble. Repeat till it remain central in these two positions. When the bubble is central in any of these positions, turn the instrument through 180° about vertical axis. If the bubble remains central, it is in adjustment. If not,

Adjustment: Bring the bubble half way by foot screws and half by adjusting the screws of the bubble tube.

Note: If the instrument is not fitted with the levelling head, the bubble is levelled with the help of ball and socket arrangement, turned through 180° and tested. In case it needs adjustment, it is adjusted half way by the adjusting screw of the bubble tube and half by the ball and the socket. Generally, this adjustment is an unnecessary refinement and the levels are not provided on the instrument.

(*ii*) Adjustment of Sight Vanes

Object: To bring the sight vanes into a vertical plane when the instrument is levelled.

Test: Level the instrument properly. Suspend a plumb line at some distance and look at it. first through one of the sight vanes and then through the other.

Adjustment: If the vertical hair in the object vane or the slit in the eye vane is not seen parallel to the plumb line, remove the affected vane and either file the higher side of the bed or insert a suitable packing under the lower side.

(iii) Adjustment of Needle

The needle is adjusted for : (*a*) Sensitivity, (*b*) Balancing the needle, (*c*) Straightening vertically, and (*d*) Straightening horizontally.

(*a*) *Sensitivity:* The needle may loose its sensitivity either by the loss of its magnetism or *by* the pivot becoming blunt. To *test* it, level the instrument and lower the needle on its pivot. If it comes to rest quickly, it shows the sign of sluggishness. To adjust it find the reason, whether it is due to loss of magnetism or due to the blunt pivot. Remagnetise the needle, if necessary. The pivot point can be sharpened with the help of fine oil stone or can be completely replaced.

(*b*) *Balancing the needle:* Due to the effect of the dip, the needle may not the balanced on its pivot. To *test* it, level the instrument and lower the needle on its pivot. Note the higher end, remove the compass glass and slide the counter weight towards the higher end, till it balances.

(*c*) *Straightening the needle vertically:* If the needle is bent vertically, a vertical seesaw motion of the ends will take place with its horizontal swing when the needle is lowered on the pivot. In such a case, the needle may be taken off the pivot and may be suitably bent in the vertical direction so that the seesaw motion ceases.

(d) Straightening Horizontally

Object: To straighten the needle so that its two ends shall lie in the same vertical plane as that of its centre.

Test: Note the reading of both ends of the needle in different positions of the graduated arc.

If the difference between both end readings is always some constant quantity other than 180°, the needle is bent horizontally but the pivot coincides the centre of the graduations. On the other hand, if the difference varies, the error may be both in the needle as well as in the pivot. In order to know, in such a case, whether the needle is straight or not, level the instrument and read both ends of the needle in any position. Revolve the compass until the South end of the needle comes against the previous reading of the North end; read the North end now. If the reading at the North end is the same as that of the South end in the previous position, the needle is not bent. Otherwise, it is bent and needs adjustment.

Adjustment: If not, note the difference. Remove the needle from the pivot and bend the North end halfway towards the new position of the original reading at the South end. Replace and repeat till correct.

(iv) Adjustment of the Pivot

Object: To bring the pivot point exactly in the centre of the graduated circle.

Test and Adjustment

1. Bring the North end of the needle against the North 0° mark of the graduated circle. Note the reading of the South end of the needle. If it does not read 0°, correct the error by bending the pivot pin slightly in a direction at right angles to the line between the North and South zeros.
2. Bring the North end of the needle exactly against 90° mark, and note the reading against the South end. If it does not read 90°, correct the error by bending the pivot pin in a direction at right angles to the line between the two 90° marks. Repeat (1) and (2) until the readings for the opposite ends of the needle agree for any position of the needle.

5.6 WILD B3 PRECISION COMPASS

Fig. 5.18 shows the photograph of Wild B3 tripod compass. It is a precision compass for simple, rapid surveys. It is particularly valuable whenever a small, light weight survey-instrument is required. It derives its precision from the *fine pivot system,* the *balanced circle* and the *strong magnet.*

The B3 is set up on a tripod and levelled with foot screws and circular bubble like other surveying instruments. On *pulling out* the circular clamp, the magnet brings the zero graduation of the circle to magnetic north, and the magnetic bearing to the target can be read to 0.1°. On *releasing* the clamp, after the reading has been taken, the circle is lifted automatically off the pivot and is held again in a fixed position so that the damage to the pivot cannot occur during transport.

With the circle in the clamped position, the B3 can be used as a simple angle measuring instrument. The small vertical arc alongside the telescope allows slopes to be measured within a range of ± 70%.

The circle has a spring mounted sappire bearing. The pivot is sharp and made of extremely hard metal. The instrument can be adjusted for earth's magnetic field (*i.e.* for dip) by moving tiny adjustment weights, thus balancing the circle so that it will swing horizontally in any part of the world.

The small sighting telescope has 2 X magnification and stadia hairs for approximate distance measurement from a staff.

5.7 MAGNETIC DECLINATION

Magnetic declination at a place is the horizontal angle between the true meridian and the magnetic meridian shown by the needle at the *time* of observation. If the magnetic meridian is to the right side (or eastern side) of the true meridian, declination is said to be *eastern or positive* [see Fig. 5.19 (*a*)]; if it to be the left side (or western side), the declination is said to be *western* or *negative* [see Fig. 5.19 (*b*)].

Mariners call declination by the name *variation.*

The declination at any particular location can be obtained by establishing a true meridian from astronomical observations and then reading the compass while sighting along the true meridian.

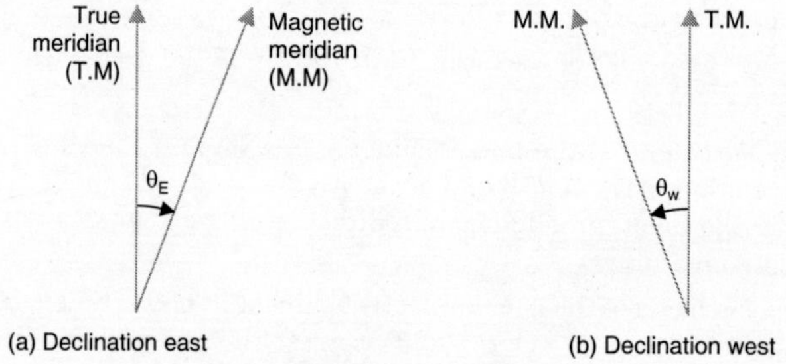

Fig. 5.19 Magnetic declination.

Isogonic line is the line drawn through the points of same declination. The distribution of earth's magnetism is not regular and consequently, the isogonic lines do not form complete great circles, but radiating from the North and South magnetic regions they follow irregular paths. *Agonic line* is the line made up of points having a zero declination.

Variations in Declination: The value of declination at a place never remains constant but changes from time to time. There are four types of variations in declination

(*a*) Diurnal variation (*b*) Annual variation (*c*) Secular variation (*d*) Irregular variation.

(*a*) **Diurnal Variation:** The diurnal variation or daily variation is the systematic departure of the declination from its mean value during a period of 24 hours. It generally varies with the phase of the sunspot period. The difference in declination between morning and afternoon is often as much as 10° of arc. The extent of daily variations depend upon the following factors:

(*i*) The Locality : More at magnetic poles and less at equator.

(*ii*) Season of the year : Considerably more in summer than in winter.

(*iii*) Time: More in day and less in night. The rate of variation during 24 hours is variable.

(*iv*) The amount of daily variation changes from year to year.

(*b*) **Annual Variation:** The variation which has a yearly period is known as annual variation. The declination has a yearly swing of about 1′ or 2′ in amplitude. It varies from place to place.

(*c*) **Secular Variation:** Due to its magnitude, secular variation is the most important in the work of surveyor. It appears to be of periodic character and follows a roller-coaster (sine-curve) pattern. It swings like a pendulum. For a given place, the compass needle after moving continuously for a period of years in one direction with respect to the true North, gradually comes to a stand still and then begins to move in opposite direction. Secular change from year to year is not uniform for any given locality and is different for different places. Its period is approximately 250 years. In Paris, the records show a range from 11° E in 1680 to 22° W in 1820. This magnitude of secular variation is very great, it is very important in the work of the surveyor, and unless otherwise specified, it is the change commonly referred to.

(*d*) **Irregular Variation:** The irregular variations are due to what are known as 'magnetic storms', earthquakes and other solar influences. They may occur at any time and cannot be predicted. Change of this kind amounting to more than a degree have been observed.

Determination of True Bearing: All important surveys are plotted with reference to true meridian, since the direction of magnetic meridian at a place changes with time. If however, the magnetic declination *at a place, at the time of observation* is known, the true bearing can be calculated from the observed magnetic bearing by the following relation (Fig. 5.19):

True bearing = magnetic bearing ± declination.

Use plus sign if the declination is to the East and minus sign if it is to the West.

The above rule is valid for whole circle bearings only. If however, a reduced bearing has been observed, it is always advisable to draw the diagram and calculate bearing.

Example 5.5: *The magnetic bearing of a line is 48° 24′ . Calculate the true bearing if the magnetic declination is 5° 38′ East.*

Solution: Declination = + 5° 38′

∴ True bearing = 48° 24′ + 5° 38′ = **54° 02′.**

Example 5.6: *The magnetic bearing of a line AB is S 28° 30′ E. Calculate the true bearing if the declination is 7° 30′ West.*

Solution: The positions of true meridian, magnetic meridian and the line have been shown in Fig. 5.20. Since the declination is to be West, the magnetic meridian will be to the West of true meridian.

Hence, true bearing = S 28° 30′ E + 7° 30′.

= **S 36° 00′ E.**

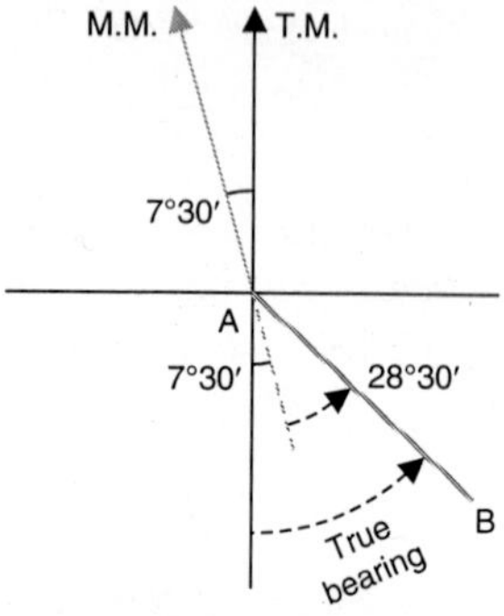

Fig. 5.20

Example 5.7: *In an old map, a line AB was drawn to a magnetic bearing of 5° 30′ the magnetic declination at the time being 1° East. To what magnetic bearing should the line be set now if the present magnetic declination is 8° 30′ East.*

Solution:

True bearing of the line = 5° 30′ + 1° = 6° 30′

Present declination = + 8° 30′ (East)

Now, True bearing = Magnetic bearing + 8° 30′

∴ Magnetic bearing = True bearing – 8° 30′

= 6° 30′ – 8° 30′ = – 2° (*i.e.* 2° in the anti-clockwise direction) = **358°.**

Example 5.8: *Find the magnetic declination at a place if the magnetic bearing of the sun at noon is (a) 184° (b) 350° 20′.*

Solution: (*a*) At noon, the sun is exactly on the geographical meridian. Hence, the true bearing of the sun at noon is zero or 180° depending upon whether it is to the North of the place or to the South of the place. Since the magnetic bearing of the sun is 184°, the true bearing will be 180°.

Now, True bearing = Magnetic bearing + Declination

∴ 180° = 184°+ Declination

or Declination = **– 4° = 4° W**

(*b*) Since the magnetic bearing of the sun is 350° 20′, it is at the North of the place and hence the true bearing of the sun, which is on the meridian, will be 360°.

Now, True bearing = Magnetic bearing + Declination

360° = 350° 20′ + Declination

or Declination = 360° – 350° 20′ = 9° 40′ = **9° 40′ E.**

5.8 LOCAL ATTRACTION

A magnetic meridian at a place is established by a magnetic needle which is uninfluenced *by other* attracting forces. However, sometimes, the magnetic needle may be attracted and prevented from indicating the true magnetic meridian when it is in proximity to certain magnetic substances. *Local attraction is a term used to denote any influence, such as the above, which prevents the needle from pointing to the magnetic North in a given locality.* Some of the sources of local attraction are: magnetite in the ground, wire carrying electric current, steel structures, railroad rails, underground iron pipes, keys, steel-bowed spectacles, metal buttons, axes, chains, steel tapes etc., which may be lying on the ground nearby.

Detection of Local Attraction: The local attraction at a particular place can be detected by observing the fore and back bearings of each line and finding its difference. If the difference between fore and back bearing is 180°, it may be taken that both the stations are free from local attraction, provided there are no observational and instrumental errors. If the difference is other than 180°, the fore bearing should be measured again to find out whether the discrepancy is due to *avoidable attraction* from the articles on person, chains, tapes etc. It the difference still remains, the local attraction exists at one or both the stations.

Strictly speaking, the term local attraction does not include avoidable attraction due to things about the person or to other sources not connected with the place where the needle is read.

Elimination of Local Attraction: If there is local attraction at a station, all the bearings measured at that place will be incorrect and the *amount of error* will be equal in all the bearings. There are two methods for liminating the effects of local attraction.

First Method: In this method, the bearings of the lines are calculated on the basis of the bearing of that line which has a difference of 180° in its fore and back bearings. It is, however, assumed that there are no observational and other instrumental errors. The amount and direction of error due to local attraction at each of the affected station is found. If, however, there is no such line in which the two bearings differ by 180°, the corrections should be made from the mean value of the bearing of that line in which there is least discrepancy between the back sight and fore sight readings.

If the bearings are expressed in quadrantal system, the corrections must be applied in *proper direction.* In 1st and 3rd quadrants, the numerical value of bearings increase in clockwise direction while they increase in anti-clockwise direction in 2nd and 4th quadrants. *Positive corrections are applied clockwise and negative corrections counter-clockwise.*

Examples 5.9, 5.10 and 5.11 completely illustrate the procedure for applying the corrections by the first method.

Second Method: This is more a general method and is based on the fact that though the bearings measured at a station may be incorrect due to local attraction, the included angle calculated from the bearings will be correct since the amount of error is the same for all the bearings measured at the station. The included angles between the lines are calculated at all the stations. If the traverse is a closed one, the sum of the internal included angles must be $(2n - 4)$ right angles. If there is any discrepancy in this, observational and instrumental errors also exist. Such error is distributed equally to all the angles. Proceeding now with the line, the bearings of which differ by 180°, the bearings of all other lines are calculated, as illustrated in example 5.12.

Special case: Special case of local attraction may arise when we find no line which has a difference of 180° in its fore and back bearings. In that case select the line in which the difference in its fore and back bearings is closest to 180°. The mean value of the bearing of that line is found by applying half the correction to both the fore and back bearings of that line, thus obtaining the *modified* fore and back bearings of that line differing exactly by 180°. Proceeding with the modified bearings of that line, corrected bearings of other lines are found. See example 5.13 for illustration.

Example 5.9: *The following bearings were observed while traversing with a compass.*

Line	*F.B.*	*B.B.*	*Line*	*F.B.*	*B.B.*
AB	*45° 45′*	*226° 10′*	*CD*	*29° 45′*	*209° 10′*
BC	*96° 55′*	*277° 5′*	*DE*	*324° 48′*	*144° 48′*

Mention which stations were affected by local attraction and determine the corrected bearings. (U.B.)

Solution: On examining the observed bearings of the lines, it will be noticed that difference between back and fore bearings of the line *DE* is exactly 180°. Hence both stations *D* and *E* are free from local attraction and all other bearings measured at these stations are also correct. Thus, the observed bearing of *DC* (*i.e.* 209° 10′) is correct. The correct bearing of *CD* will, therefore, be 209° 10′ – 180° = 29° 10′ while the observed bearing is 29° 45′. The error at C is therefore + 35′ and a correction – 35′ must be applied to all the bearings measured at C. The correct bearings of *CB* thus becomes 277° 5′ – 35′ = 276° 30′ and that of *BC* as 276° 30′ – 180° = 96° 30′. The observed bearing of *BC* is 96° 55′. Hence the error at *B* is + 25′ and a correction of – 25′ must be applied to all the bearings measured at *B*. The correct bearing of ***BA*** thus becomes 226° 10′ – 25′ = 225° 45′, and that of *AB* as 225° 45′ – 180° = 45° 45′ which is the same as the observed one. Station *A* is, therefore, free from local attraction.

The results may be tabulated as under:

Line	*Observed bearing*	*Correction*	*Corrected bearing*	*Remarks*
AB	45° 45′	0 at *A*	45° 45′	Stations *B* and *C* are affected by local attraction
BA	226° 10′	– 25′ at *B*	225° 45′	
BC	96° 55′	– 25′ at *B*	96° 30′	
CB	277° 5′	– 35′ at *C*	276° 30′	
CD	29° 45′	– 35′ at *C*	29° 10′	
DC	209° 10′	0 at *D*	209° 10′	
DE	324° 48′	0 at *D*	324° 48′	
ED	144° 48′	0 at *E*	144° 48′	

Example 5.10: *Apply the corrections if the bearings of the previous example are measured in the quadrantal system as under:*

Line	*F.B.*	*B.B.*	*Line*	*F.B.*	*B.B.*
AB	*N 45° 45′ E*	*S 46° 10′ W*	*CD*	*N 29° 45′ E*	*S 29° 10′ W*
BC	*S 83° 05′ E*	*N 82° 55′ W*	*DE*	*N 35°12′ W*	*S 35° 12′ E*

Solution: By inspection of the observed bearings, stations *D* and *E* are free from local attraction and hence bearings of *ED, DE* and *DC* are correct. The correct bearing of *CD* will, therefore, be N 29° 10′ E. Since the observed bearing of *CD* is N 29° 45′ E, *the magnetic needle* at C is deflected by 35′ towards West. The corrected bearings of *CB* will, therefore, be N 82° 55′ W + 35′ = N 83° 30′ W.

The corrected bearing of *BC* will be S 83° 30′ E. Since the observed bearing of *BC* is S 83° 05′ E, the needle at *B* is deflected by 25′ towards East. Hence the corrected bearing of *BA* will be S 46° 10′ W – 25′ = S 45° 45′ W. The bearing of line *AB* will be N 45° 45′ E, which is the same as the observed one. Station *A* is, therefore, not affected by local attraction.

Example 5.11: *The following bearings were observed in running a closed traverse:*

Line		*F.B.*	*B.B.*
AB	...	*75° 5′*	*254° 20′*
BC	...	*115° 20′*	*296° 35′*
CD	...	*165° 35′*	*345° 35′*
DE	...	*224° 50′*	*44° 5′*
EA	...	*304° 50′*	*125° 5′*

At what stations do you suspect the local attraction ? Determine the correct magnetic bearings. If declination was 5° 10′ E, what are the true bearings ?

Solution: By inspection of the observed bearings it will be noticed that stations *C* and *D* are free from local attractions since the B.B. and F.B. of *CD* differ by 180°. All the bearings measured at *C* and *D* are, therefore, correct. Thus, the observed bearing of *CB* (i.e. 296° 35′) is correct. The correct bearing of *BC* will be 296°35′ – 180° = 116°35′. Since the observed bearing of *BC* is 115° 20′, a correction of + 1° 15′ will have to be applied to the bearing of *BA* measured at *B*. Thus, the correct bearing of *BA* becomes 254° 20′ + 1°15 = 255°35′. The correct bearing of *AB* will, therefore, be 255°35′ – 180° = 75° 35′. Since the observed bearing of *AB* is 75° 5′ a correction of + 30′ will be have to be applied to the bearing of *AE* measured at *A*. Thus, correct bearing of *AE* becomes 125° 5′ + 30′ = 125°35′. The corrected bearing of *EA* will be 125° 35′ + 180° = 305° 35′. Since the observed bearing of *EA* is 304° 50′, a correction of + 45′ will have to be applied to the bearing of *ED* measured at *E*. The correct bearing of *ED* will thus be 44° 5′ + 45′ = 44° 50′. The correct bearing of *DE* will be 44° 50′ + 180° = 224° 50′, which is the same as the observed one, since the station *D* is not affected by local attraction.

Thus, results may be tabulated as given below. Since the magnetic declination is + 5° 10′ E, the true bearings of the lines will be obtanied by adding 5° 10′ to corrected magnetic bearings.

Line	*Observed bearing*	*Correction*	*Corrected bearing*	*True bearing*	*Remarks*
AB	75° 5′	+ 30′ at *A*	75° 35′	80° 45′	
BA	254° 20′	+ 1° 15′ at *B*	255° 35′	260° 45′	
BC	115° 20′	+ 1° 15′ at *B*	116° 35′	121° 45′	
CB	296° 35′	0 at *C*	296° 35′	301° 45′	
CD	165° 35′	0 at *C*	165° 35′	170° 45′	Stations *A* , *B* and *E* are affected by local attraction
DC	345° 35′	0 at *D*	345° 35′	350° 45′	
DE	224° 50′	0 at *D*	224° 50′	230° 0′	
ED	44° 5′	+ 45′ at *E*	44° 50′	50° 0′	
EA	304° 50′	+ 45′ at *E*	305° 35′	310° 45′	
AE	125° 5′	+ 30′ at *A*	125° 35′	130° 45′	

Example 5.12: *The following are bearings taken on a closed compass traverse :*

Line	*F.B.*	*B.B.*
AB	*80° 10′*	*259° 0′*
BC	*120° 20′*	*301° 50′*
CD	*170° 50′*	*350° 50′*
DE	*230° 10′*	*49° 30′*
EA	*310° 20′*	*130° 15′*

Compute the interior angles and correct them for observational errors. Assuming the observed bearing of the line CD to be correct adjust the bearing of the remaining sides.

Solution: $\angle A$ = Bearing of AE – Bearing of AB = 130° 15′ – 80° 10′ = 50° 5′

$\angle B$ = Bearing of BA – Bearing of BC = 259° – 120° 20′ = 138° 40′

$\angle C$ = Bearing of CB – Bearing of CD = 301° 50′ – 170° 50′ = 131° 0′

$\angle D$ = Bearing of DC – Bearing of DE = 350° 50′ – 230° 10′ = 120° 40′

$\angle E$ = Bearing of ED – Bearing of EA = 49° 30′ – 310° 20′ + 360° = 99° 10′

$\angle A + \angle B + \angle C + \angle D + \angle E$ = 50° 5′ + 138° 40′ + 131° 0′ + 120° 40′ + 99° 10′ = 539° 35′

Theoretical sum = $(2n - 4)$ 90° = 540°

∴ Error = – 25′

Hence a correction of + 5′ is applied to all the angles. The corrected angles are:

$\angle A$ = 50° 10′; $\angle B$ = 138° 45′; $\angle C$ = 131° 5′; $\angle D$ = 120° 45′ and $\angle E$ = 99° 15′

Starting with the corrected bearing of CD, all other bearings can be calculated as under:

Bearing of DE = Bearing of DC – $\angle D$ = 350° 50′ – 120° 45 = **230° 5′**

∴ Bearing of ED = 230° 5′ – 180° = **50° 5′**

∴ Bearing of EA = Bearing of ED – $\angle E$ = 50° 5′ – 99° 15′ + 360° = **310° 50′**

∴ Bearing of AE = 310° 50′ – 180° = **130° 50′**

Bearing of AB = Bearing of AE – $\angle A$ = 130° 50′ – 50° 10′ = **80° 40′**

∴ Bearing of BA = 80° 40′ + 180° = **260° 40′**

Bearing of BC = Bearing of BA – $\angle B$ = 260° 40′ – 138° 45′ = **121° 55′**

∴ Bearing of CB =121° 55 + 180° = **301° 55′**

Bearing of CD = Bearing of CB – $\angle C$ = 301° 55′ – 131° 5′ = **170° 50′**

∴ Bearing of DC = 170° 50′ + 180° = **350° 50′**. (Check)

Example 5.13: *The following bearings were observed in running a closed traverse.*

Line	*F.B.*	*B.B.*
AB	*71° 05′*	*250°20′*
BC	*110° 20′*	*292° 35′*
CD	*161°35′*	*341°45′*
DE	*220° 50′*	*40° 05′*
EA	*300°50′*	*121°10′*

Determine the correct magnetic bearings of the lines.

Solution: By inspection, we find that there is no line whose F.B. and B.B. differ exactly by 180°. However, the F.B. and B.B. of line CD differ by 180°10′, the difference being only + 10′. Hence the correct F.B. of CD is obtained by adding half the difference.

Hence corrected F.B. of CD = 161°35′ + 5′ = 161°40′

and corrected B.B. of CD = 341°45′ – 5′ = 341°40′

Difference = 180°0′

$\angle ABC$ = 250° 20′ – 110° 20′ = 140° 0′

$\angle BCD$ = 292°35′ – 161°35′ = 131°0′

$\angle CDE$ = 341°45′ – 220°50′ = 120°55′

$\angle DEA = 300°50' - 40°05' = 260°45'$ (Exterior)

$= 99°\ 15'$ (Interior)

$\angle EAB = 121°10' - 71°5' = 50°5'$

Sum $= 541°15'$

Theoretical sum $= (2N - 4)\ 90° = 540°$

$\therefore$ Error $= 541°15' - 540° = 1°15'$

$\therefore$ Correction for each angle $= -\ 15'$

Hence the corrected angles are

$\angle ABC = 140°0' - 15' = 139°45'$

$\angle BCD = 131°0' - 15' = 130°45'$

$\angle CDE = 120°55' - 15' = 120°40'$

$\angle DEA = 99°15' - 15' = 99°00'$

$\angle EAB = 50°05' - 15' = 49°50'$

sum $= 540°00'$

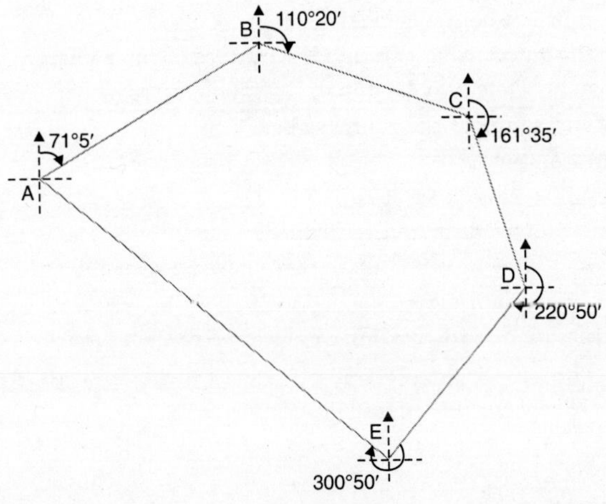

Fig. 5.21

The corrected bearings of all the lines are obtained from the included angles and the corrected bearing of *CD*.

Corrected F.B. of $DE = 341°40' - 120°40' = 221°00'$

B.B. of $DE = 221°00' - 180° = 41°00'$

F.B. of $EA = 41°00' + 261° = 302°00'$

B.B. of $EA = 302°00' - 180° = 122°00'$

F.B. of $AB = 122°00' - 49°50' = 72°10'$

B.B. of $AB = 72°10' + 180° = 252°10'$

F.B. of $BC = 252°10' - 139°45' = 112°25'$

B.B. of $BC = 112°25' + 180° = 292°25'$

F.B. of $CD = 292°25' - 130°45' = 161°40'$ (check)

5.9 ERRORS IN COMPASS SURVEY

The errors may be classified as:

(*a*) Instrumental errors:

(*b*) Personal errors

(*c*) Errors due to natural causes.

(*a*) **Instrumental errors.** They are those which arise due to the faulty adjustments of the instruments. They may be due to the following reasons:

1. The needle not being perfectly straight.
2. Pivot being bent.
3. Sluggish needle.
4. Blunt pivot point.
5. Improper balancing weight.
6. Plane of sight not being vertical.
7. Line of sight not passing through the centres of the right.

(*b*) **Personal errors.** They may be due to the following reasons:

1. Inaccurate levelling of the compass box.
2. Inaccurate centring.
3. Inaccurate bisection of signals.
4. Carelessness in reading and recording.

(*c*) **Natural errors.** They may be due to the following reasons:

1. Variation in declination.
2. Local attraction due to proximity of local attraction forces.
3. Magnetic changes in the atmosphere due to clouds and storms.
4. Irregular variations due to magnetic storms etc.

PROBLEMS

1. Explain, with the help of neat sketch, the graduations of a prismatic compass and a surveyor's compass.
2. Give, in a tabular form, the difference between prismatic compass and surveyor's compass.
3. What are the sources of errors in compass survey and what precautions will you take to eliminate them?
4. What is local attraction ? How is it detected and eliminated?
5. Define the terms: True and magnetic bearing, local attraction, back bearings and magnetic declination. (*A.M.I.E.*)
6. Determine the values of included angles in the closed compass traverse *ABCD* conducted in the clockwise direction, given the following fore bearings of their respective lines:

Line	F.B.
AB	40°
BC	70°
CD	210°
DA	280°

Apply the check. (U.B.)

7. The following angles were observed in clockwise direction in an open traverse : $\angle ABC = 124° 15'$; $\angle BCD = 156° 30'$; $\angle CDE = 102° 0'$; $\angle DEF = 95° 15'$; $\angle EFG = 215° 45'$.
Magnetic bearing of the line AB was 241° 30′. What would be the bearing of line *FG*? (G.U.)

8. In an old survey made when the declination was 4° W, the magnetic bearing of a given line was 210°. The declination in the same locality is now 10° E. What are the true and present magnetic bearings of the line? (U.B.)

9. The magnetic bearing of line as observed by the prismatic compass at a survey station is found to be 272°. If the local attraction at this station is known to be 5° E and the declination is 15° West, what is the true bearing of the line ? (P.U.)

10. (*a*) What is back bearing and what are the advantages of observing it in a traverse ?
(*b*) At a place the bearing of sun is measured at local noon and found to be 175° 15′. What is the magnitude and direction of magnetic declination of the place ?
(*c*) Show by a neat diagram the graduations on the circle of a prismatic compass.

11. The following bearings were taken in running a compass traverse

Line	F.B.	B.B.	Line	F.B.	B.B.
AB	124° 30′	304° 30′	CD	310° 30′	135° 15′
BC	68° 15′	246° 0′	DA	200° 15′	17° 45′

At what stations do you suspect local attraction ? Find the correct bearings of the lines and also compute the included angles.

12. The following fore and back bearings were observed in traversing with a compass in place where local attraction was suspected.

Line	F.B.	B.B.	Line	F.B.	B.B
AB	38° 30′	219° 15′	CD	25° 45′	207° 15′
BC	100° 45′	278° 30′	DE	325° 15′	145° 15′

Find the corrected fore and back bearings and the true bearing of each of the lines given that the magnetic declination was 10° W.

13. The following are the bearings taken on a closed compass traverse:

Line	F.B.	B.B.	Line	F.B.	B.B
AB	S 37° 30′ E	N 37° 30′ W	DE	N 12° 45′ E	S 13° 15′ W
BC	S 43° 15′ W	S 44° 15′ E	EA	N 60° 00′ E	S 59° 00′ W
CD	N 73° 00′ W	S 72° 15′ E			

Compute the interior angles and correct them for observational errors. Assuming the observed bearing of the line *AB* to be correct, adjust the bearing of the remaining sides.

14. (*a*) Derive rules to calculate reduced bearing from whole circle bearing for all the quadrants.
(*b*) The following bearings were observed with a compass:

AB	74° 0′	BA	254° 0′
BC	91° 0′	CB	271° 0′
CD	166° 0′	DC	343° 0′

DE	177° 0′	*ED*	0° 0′
EA	189° 0′	*AE*	9° 0′

Where do you suspect the local attraction ? Find the correct bearings.

ANSWERS

6. ∠*A* = 60° ; ∠*B* = 150° ; ∠*C* = 40° ; ∠*D* = 110° ; sum = 360°
7. 35° 35′
8. T.B. = 206° 0′; M.B. = 196° 0′.
9. 262°
10. (*b*) 4°45′ *E*
11. Stations *C* and *D*.

Line	*F.B.*	*B.B.*	*Line*	*F.B.*	*B.B.*
AB	124° 30′	304° 30′	*CD*	312° 45′	132° 45′
BC	68° 15′	248° 15′	*DA*	197°45′	17° 45′

∠A = 106° 45′ ; ∠B = 123°45′ ; ∠C = 64°30′ ; ∠D = 65°.

(**Note:** Take F.B. of CD = 310° 30′)

12.

Line	*F.B.*	*B.B.*	*True F.B.*
AB	38° 30′	218° 30′	28° 30′
BC	100° 0′	280° 0′	90° 0′
CD	27° 15′	207°15′	17°15′
DE	325° 15′	145°15′	315°15′

13. Summation error = + 1° 15′

Line	*F.B.*	*B.B.*	*Line*	*F.B.*	*B.B.*
BC	S 43° 30′ W	N 43° 30′ E	*DE*	N 11° 45′ E	S 11° 45′ W
CD	N 73° 30′ W	S 73° 30′ E	*EA*	N 58° 45′ E	S 58° 45′ W

14. (*b*)

AB	74° 0′	*BA*	254° 0′
BC	91° 0′	*CB*	271° 0′
CD	166° 0′	*DC*	346° 0′
DE	180° 0′	*ED*	0° 0′
EA	189° 0′	*AE*	9° 0′

CHAPTER 6

THE THEODOLITE

6.1 GENERAL

The Theodolite is the most precise instrument designed for the measurement of horizontal and vertical angles and has wide applicability in surveying such as laying off horizontal angles, locating points on line, prolonging survey lines, establishing grades, determining difference in elevation, setting out curves etc.

Theodolites may be classified as :

(*i*) Transit theodolite.

(*ii*) Non-transit theodolite.

A *transit theodolite* (or simply *'transit'*) is one is which the line of sight can be reversed by revolving the telescope through 180° in the vertical plane. The *non-transit* theodolites are either *plain theodolites* or *Y-theodolites* in which the telescope cannot be transited. The *transit* is mainly used and non-transit theodolites have now become obsolete.

6.2 THE ESSENTIALS OF THE TRANSIT THEODOLITE

Fig. 6.1. and 6.2 show diagrammatic sections of a vernier theodolite while Fig. 6.3 shows the photograph of a vernier theodolite. A transit consists of the following essential parts (Ref. Figs. 6.1 and 6.2) :

(*i*) **The Telescope.** The telescope (1) is an integral part of the theodolite and is mounted on a spindle known as horizontal axis or *trunnion axis* (2). The telescope may be internal focusing type or external focusing type. In most of the transits, and internal focusing telescope is used.

(*ii*) **The Vertical Circle.** The vertical circle is a circular graduated arc attached to the trunnion axis of the telescope. Consequently the graduated arc rotates with the telescope when the latter is turned about the horizontal axis. By means of *vertical circle clamp* (24) and its corresponding slow motion or *tangent screw* (25), the telescope can be set accurately at any desired position in vertical plane. The circle is either graduated continuously from 0° to 360° in clockwise direction or it is divided into four quadrants (Fig. 6.11).

(*iii*) **The Index Frame (or T-Frame or Vernier Frame).** The index frame (3) is a T-shaped frame consisting of a vertical leg known as *clipping arm* (28) and a horizontal bar known as vernier arm or *index arm* (29). At the two extremities of the index arm are fitted two

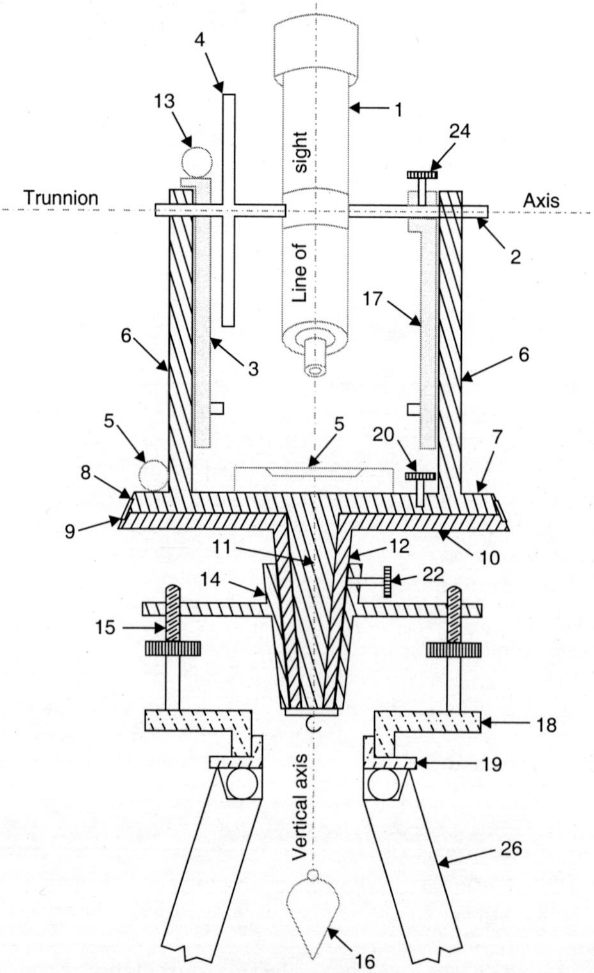

Fig. 6.1 The essentials of a transit.

1. Telescope
2. Trunnion Axis
3. Vernier Frame
4. Vertical Circle
5. Plate Levels
6. Standards (A-Frame)
7. Upper Plate
8. Horizontal Plate Vernier
9. Horizontal Circle
10. Lower Plate
11. Inner Axis
12. Outer Axis
13. Altitude Level
14. Levelling Head
15. Levelling Screw
16. Plumb Bob
17. Arm of Vertical Circle Clamp
18. Foot Plate
19. Tripod Head
20. Upper Clamp
22. Lower Clamp
24. Vertical Circle Clamp
26. Tripod

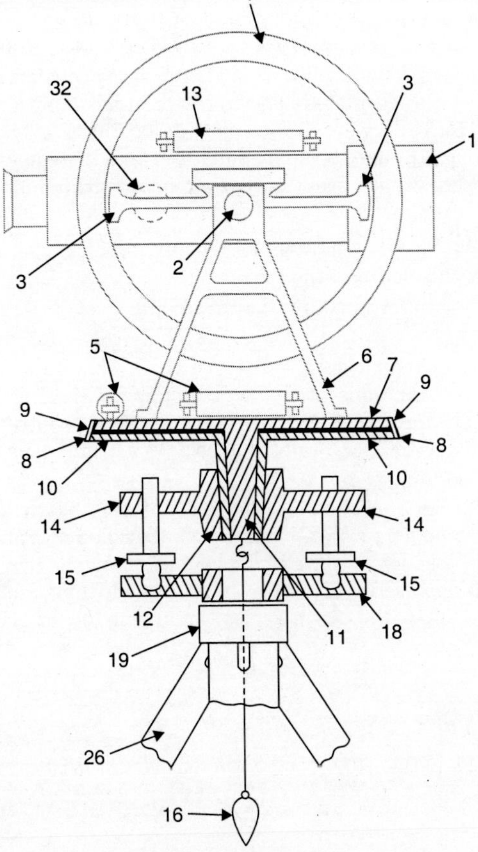

Fig. 6.2 The essentials of a transit.

1. Telescope
2. Trunnion Axis
3. Vernier Frame
4. Vernier Circle
5. Plate Levels
6. Standards (A-Frame)
7. Upper Plate
8. Horizontal Plate Vernier
9. Horizontal Circle
10. Lower Plate
11. Inner Axis
12. Outer Axis
13. Altitude Level
14. Levelling Head
15. Levelling Screw
16. Plumb Bob
18. Foot Plate
19. Tripod Head
26. Tripod
32. Focusing Screw

verniers to read the vertical circle. The index arm is centered on the trunnion axis in front of the vertical circle and remains *fixed*. When the telescope is moved in the vertical plane, the vertical circle moves relative to the verniers with the help of which reading can be taken. For adjustment purposes, however, the index arm can be rotated slightly with the help of a *clip screw* (27) fitted to the clipping arm at its lower end. Glass magnifiers (30) are placed in front of each vernier to magnify the reading. A long sensitive bubble tube, sometimes known as the *altitude bubble* (13) is placed on the top of the index frame.

(*iv*) **The Standards (or A-Frame).** Two standards (6) resembling the letter A are mounted on the upper plates (7). The trunnion axis of the telescope is supported on these. The T-frame and the *arm of vertical circle clamp* (17) are also attached to the A-frame.

(*v*) **The Levelling Head.** The levelling head (14) usually consists of two parallel triangular plates known as tribrach plates. The upper tribrach has three arms each carrying a levelling screw (15). The lower tribrach plate or foot plate (18) has a circular hole through which a plumb bob (16) may be suspended. In some instruments, four levelling screws (also called foot screws) are provided between two parallel plates. A levelling head has three distinctive functions:

(*a*) To support the main part of the instrument.

(*b*) To attach the theodolite to the tripod.

(*c*) To provide a mean for levelling the theodolite.

(*vi*) **The Two Spindles (or Axes or Centres).** The *inner spindle or axis* (11) is soild and conical and fits into the outer spindle (12) which is hollow and ground conical in the interior. The inner spindle is also called the upper axis since it carries the vernier or upper plate (7). The outer spindle carries the scale or lower plate (10) and is, therefore, also, known as the lower axis. Both the axes have a common axis which form the vertical axis of the instrument.

(*vii*) **The Lower Plate (or Scale Plate).** The lower plate (10) is attached to the outer spindle. The lower plate carries a horizontal circle (9) at its bevelled edge and is. therefore, also known as the scale plate. The lower plate carries a *lower clamp screw* (22) and a corresponding slow motion or *tangent screw* (23) with the help of which it can be fixed accurately in any desired position. Fig. 6.4 shows a typical arrangement for clamp and tangent screws.

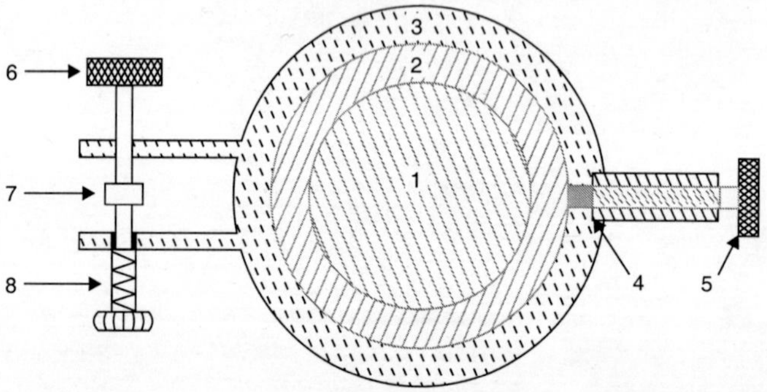

Fig. 6.4 Clamp and tangent screw for lower plate.

1. Inner axis
2. Outer axis
3. Casing
4. Pad
5. Lower clamp screw
6. Tangent screw
7. Lug on levelling head
8. Antagonising spring

When the clamp is tightened, the lower plate is fixed to the upper tribrach of the levelling head. On turning the tangent screw, the lower plate can be rotated slightly. Usually, the size of a Theodolite is represented by the size of the scale plate, *i.e.*, a 10 cm theodolite or 12 cm theodolite etc.

(*viii*) **The Upper Plate (or Vernier Plate).** The upper plate (7) or vernier plate is attached to the inner axis and carries two verniers (8) with magnifiers (3) at two extremities diametrically opposite. The upper plate supports the standards (6). It carries an *upper clamp screw* (2) and a corresponding *tangent screw* (21) for purpose of *accurately* fixing it to the lower plate. On clamping the upper and unclamping the lower clamp, the instrument can rotate on its outer axis without any relative motion between the two plates. If, however, the lower clamp is clamped and upper clamp unclamped, the upper plate and the instrument can rotate on the inner axis with a relative motion between the vernier and the scale. For using any tangent screw, its corresponding clamp screw must be tightened.

(*ix*) **The Plate Levels.** The upper plate carries two plate levels (5) placed at right angles to each other. One of the plate level is kept parallel to the trunnion axis. In some theodolites only one plate level is provided. The plate level can be centred with the help of foot screws (15).

(*x*) **Tripod.** When in use, the theodolite is supported on a tripod (26) which consists of three solid or framed legs. At the lower ends, the legs are provided with pointed steel shoes. The tripod head carries at its upper surface an external screw to which the foot plate (18) of the levelling head can be screwed.

(*xi*) **The Plumb Bob.** A plumb bob is suspended from the hook fitted to the bottom of the inner axis to centre the instrument exactly over the station mark.

(*xii*) **The Compass.** Some theodolites are provided with a compass which can be either tubular type or trough type.

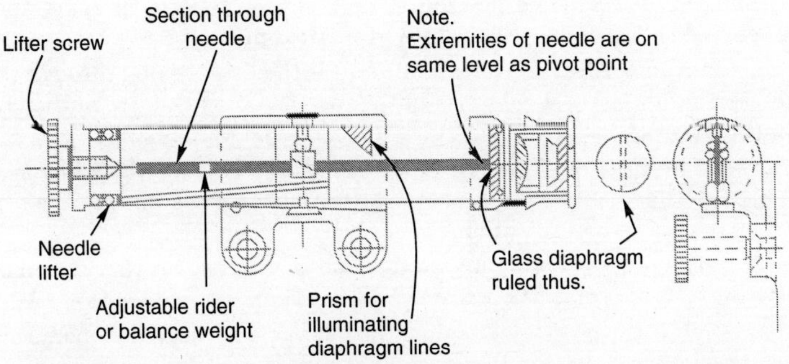

Fig. 6.5 Tubular compass.
(By courtesy of Messrs Vickers Instruments Ltd.)

Fig. 6.5 shows a tubular compass for use on a vernier theodolite. The compass is fitted to the standards.

A trough compass consists of a long narrow rec-tangular box along the longitudinal axis of which is provided a needle balanced upon a steel pivot. Small flat curve scales of only a few degrees are provided on each side of the trough.

(*xiii*) **Striding Level.** Some theodolites are fitted with a striding level. Fig. 6.6 shows a striding level in position. It is used to test the horizontality of the transit axis or trunnion axis.

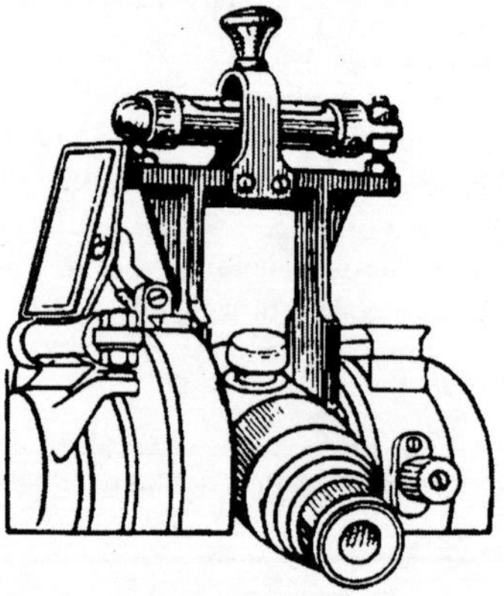

Fig. 6.6 Striding level in position.

6.3 DEFINITIONS AND TERMS

1. **The vertical axis:** The vertical axis is the axis about which the instrument can be rotated in a horizontal plane. This is the axis about which the lower and upper plates rotate.
2. **The horizontal axis:** The horizontal or trunnion axis is the axis about which the telescope and the vertical circle rotate in vertical plane.
3. **The line of sight or line of collimation:** It is the line passing through the intersection of the horizontal and vertical cross-hairs and the optical centre of the object glass and its continuation.
4. **The axis of level tube:** The axis of the level tube or the *bubble line* is a straight line tangential to the longitudinal curve of the level tube at its centre. The axis of the level-tube is horizontal when the bubble is central.
5. **Centring:** The process of setting the theodolite exactly over the station mark is known as centring.
6. **Transiting:** It is the process of turning the telescope in vertical plane through 180° about the trunnion axis. Since the line of sight is reversed in this operation, it is also known as *plunging* or *reversing.*
7. **Swinging the telescope:** It is the process of turning the telescope in horizontal plane. If the telescope is rotated in clock-wise direction, it is known as *right swing.* If telescope is rotated in the anti-clockwise direction, it is known as the *left swing.*
8. **Face left observation:** If the face of the vertical circle is to the left of the observer, the observation of the angle (horizontal or vertical) is known as face left observation.
9. **Face right observation:** If the face of the vertical circle is to the right of the observer, the observation is known as face right observation.
10. **Telescope normal:** A telescope is said to be *normal* or *direct* when the face of the vertical circle is to the left and the "bubble (of the telescope) up".

11. **Telescope inverted:** A telescope is said to *inverted* or *reversed* when of the vertical circle is to the right and the "bubble down".
12. **Changing face:** It is an operation of bringing the face of the telescope from left to right and *vice versa*.

6.4 TEMPORARY ADJUSTMENTS

Temporary adjustments or *station adjustments* are those which are made at every instrument setting and preparatory to taking observations with the instrument. The temporary adjustments are :

1. Setting over the station.
2. Levelling up
3. Elimination parallax.

1. **Setting up:** The operation of setting up includes :
   (*i*) *Centring* of the instrument over the station mark by a plumb bob or by optical plummet, and
   (*ii*) *approximate levelling* with the help of tripod legs. Some instruments are provided with *shifting head* with the help of which accurate centring can be done easily. *By moving the leg radially, the plumb bob is shifted in the direction of the leg while by moving the leg circumferentially or side ways considerable change in the inclination is effected without disturbing the plumb bob.* The second movement is, therefore, effective in the approximate levelling of the instrument. The approximate levelling is done either with reference to a small circular bubble provided on tribrach or is done by eye judgment.
2. **Levelling up:** After having centred and approximately levelled the instrument, accurate levelling is done with the help of foot screws and with reference to the plate levels. The purpose of the levelling is to make the vertical axis truly vertical. The manner of levelling the instrument by the plate levels depends upon whether there are three levelling screws or four levelling screws.

   Three Screw Head

   1. Turn the upper plate untill the longitudinal axis of the plate level is roughly parallel to a line joining any two (such as *A* and *B*) of the levelling screws [Fig. 6.7 (*a*)].

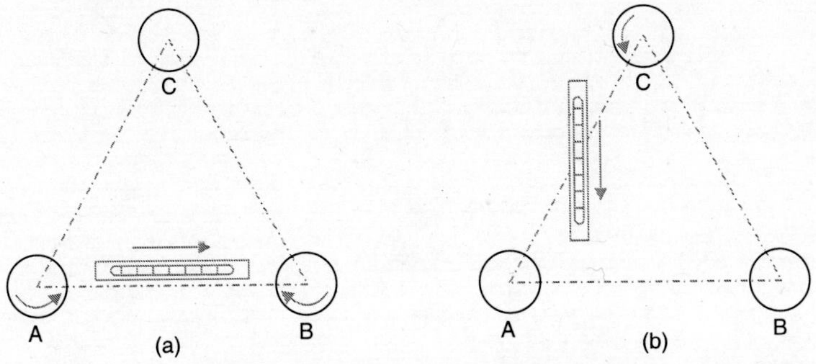

Fig. 6.7 Levelling up with three foot screws.

2. Hold these two levelling screws between the thumb and first finger of each hand and turn them uniformly so that the thumbs move either towards each other or away from each other until the bubble is central. *It should be noted that the bubble will move in the direction of movement of the left thumb* [Fig. 6.7 (*a*)].
3. Turn the upper plate through 90°, *i.e.*, until the axis of the level passes over the position of the third levelling screw *C* [Fig. 6.7 (*b*)].
4. Turn this levelling screw until the bubble is central.
5. Return the upper plate through 90° to its original position [Fig. 6.7 (*a*)] and repeat step (2) till the bubble is central.
6. Turn back again through 90° and repeat step (4).
7. Repeat steps (2) and (4) till the bubble is central in both the positions.
8. Now rotate the instrument through 180°. The bubble should remain in the centre of its run, provided it is in correct adjustment. The vertical axis will then be truly vertical. If not, it needs permanent adjustment.

Note: *It is essential to keep to the same quarter circle for the changes in direction and not to swing through the remaining three quarters of a circle to the original position.*

If two plate levels are provided in the place of one, the upper plate is *not* turned through 90° as is done in step (2) above. In such a case, the longer plate level is kept parallel to any two foot screws, the other plate level will automatically be over the third screw. Turn the two foot screws till the longer bubble is central. Turn now the third foot screw till the other bubble is central. The process is repeated till both the bubbles are central. The instrument is now rotated about the vertical axis through a complete revolution. Each bubble will now traverse, *i.e.*, remain in the centre of its run, if they are in adjustment.

Four Screw Head

1. Turn the upper plate until the longitudinal axis of the plate level is roughly parallel to the line joining two diagonally opposite screws (such as *D* and *B*) [Fig. 6.8 (*a*)].
2. Bring the bubble central exactly in the same manner as described in step (2) above.
3. Turn the upper plate through 90° until the spirit level axis is parallel to the other two diagonally opposite screws (such as *A* and *C*) [Fig. 6.8 (*b*)].
4. Centre the bubble as before.
5. Repeat the above steps till the bubble is central in both the positions.

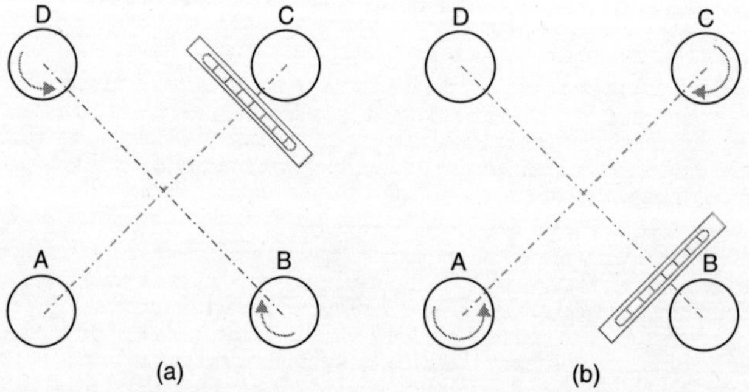

Fig. 6.8 Levelling up with four foot screws.

6. Turn through 180° to check the permanent adjustment, as for the three screw instrument.

3. Elimination of Parallax: Parallax is a condition arising when the image formed by the objective is not in the plane of the cross-hairs. Unless parallax is eliminated, accurate sighting is impossible. Parallax can be eliminated in two steps : (*i*) by focusing the eye-piece for distinct vision of the cross-hairs, and (*ii*) by focusing the objective to bring the image of the object in the plane of cross-hairs.

(*i*) *Focusing the eye-piece:* To focus the eye-piece for distinct vision of the cross-hairs, point the telescope towards the sky (or hold a sheet of white paper in front of the objective) and move eye-piece in or out till the cross-hairs are seen sharp and distinct. In some telescopes, graduations are provided at the eye-piece end so that one can always remember the particular graduation position to suit his eyes. This may save much of time.

(*ii*) *Focusing the objective:* The telescope is now directed towards the object to be sighted and the focusing screw is turned till the image appears clear and sharp. The image so formed is in the plane of cross-hairs.

6.5 MEASUREMENT OF HORIZONTAL ANGLES: GENERAL PROCEDURE

To measure the horizontal angle *PQR* (Fig. 6.9):

1. Set up the instrument at *Q* and level it accurately.
2. Release all clamps. Turn the upper and lower plates in opposite directions till the zero of one of the vernier (say *A*) is against the zero of the scale and the vertical circle is to the left. Clamp both the plates together by upper clamp and lower clamp and bring the two zeros into exact coincidence by turning the upper tangent screw. Take both vernier readings. The reading on vernier *B* will be 180°, if there is no instrumental error.

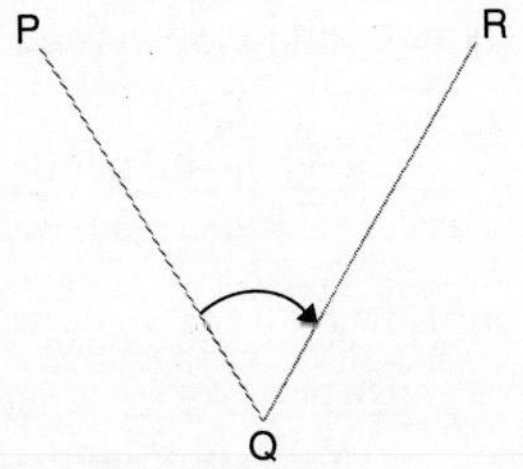

Fig. 6.9

3. Loose the lower clamp and turn the instrument towards the signal at *P*. Since both the plates are clamped together, the instrument will rotate about the outer axis. Bisect point *P* accurately by using lower tangent screw. Check the readings of verniers *A* and *B*. There should be no change in the previous reading.
4. Unclamp the upper clamp and rotate the instrument clockwise about the inner axis to 'bisect the point *R*. Clamp the upper clamp and bisect *R* accurately by using upper tangent screw. (The point of intersection of the horizontal and vertical cross-hairs should be brought into exact coincidence with the station mark by means of vertical circle clamp and tangent screw).
5. Read both verniers. The reading of vernier *A* gives the angle *PQR* directly while the vernier *B* gives by deducting 180°. While entering the reading, the full reading of vernier *A* (*i.e.*, degrees, minutes and seconds) should be entered, while only minutes and seconds of the vernier *B* are entered. The mean of the two such vernier readings gives angle with one face.
6. Change the face by transiting the telescope and repeat the whole process. The mean of the two vernier readings gives the angle with other face.

The average horizontal angle is then obtained by taking the mean of the two readings with different faces. Table 6.1 gives the specimen page for recording the observations.

TABLE 6.1

Instrument At	Sighted to	*Face : Left*								*Swing : Right*				*Face : Right*								*Swing : Right*				*Average Horizontal Angle*		
		A			B		Mean			*No. of Repetitions*	*Horizontal Angle*			A			B		Mean			*No. of Repetitions*	*Horizontal Angle*					
		°	′	″	′	″	°	′	″		°	′	″	°	′	″	′	″	°	′	″		°	′	″	°	′	″
Q	*P*	0	0	0	0	0	0	0	0					0	0	0	0	0	0	0	0							
	R	52	41	20	41	20	52	41	20	1	52	41	20	52	41	40	41	40	52	41	40	1	52	41	40	52	41	30

To Measure a Horizontal Angle by Repetition Method

The method of repetition is used to measure a horizontal angle to a finer degree of accuracy than that obtainable with the least count of the vernier. By this method, an angle is measured two or more times by allowing the vernier to remain clamped each time at the end of each measurement instead of setting it back at zero when sighting at the previous station. Thus an angle reading is mechanically added several times depending upon the number of repetitions. The average horizontal angle is then obtained by dividing the final reading by the number of repetitions.

To measure the angle *PQR* (Fig. 6.9) :

1. Set the instrument at *Q* and level it. With the help of upper clamp and tangent screw, set 0° reading on vernier *A*. Note the reading of vernier *B*.
2. Loose the lower clamp and direct the telescope towards the point *P*. Clamp the lower clamp and bisect point *P* accurately by lower tangent screw.
3. Unclamp the upper clamp and turn the instrument *clockwise* about the inner axis towards *R*. Clamp the upper clamp and bisect *R* accurately with the upper tangent screw. Note the reading of verniers *A* and *B* to get the approximate value of the angle *PQR*.
4. Unclamp the lower clamp and turn the telescope clockwise to sight *P* again. Bisect *P* accurately by using the lower tangent screw. *It should be noted that the vernier readings will not be changed in this operation since the upper plate is clamped to the lower.*
5. Unclamp the upper clamp, turn the telescope clockwise and sight *R*. Bisect *R* accurately by upper tangent screw.
6. Repeat the process until the angle is repeated the required number of times (usually 3). The average angle with face left will be equal to final reading divided by three.
7. Change face and make three more repetitions as described above. Find the average angle with face right, by dividing the final reading by three.
8. The average horizontal angle is then obtained by taking the average of the two angles obtained with face left and face right.

Any number of repetitions may be made. However, three repetitions with the telescope normal and three with the telescope inverted are quite sufficient for any thing except very precise work. Table 6.2 gives the method of recording observations by method of repetition for ordinary work.

TABLE 6.2 Repetition method

Q	*P*	0	0	0	0	0	0	0	0					0	0	0	0	0	0	0	0							
	R	52	41	20	41	20	52	41	20	1				52	41	40	41	40	52	41	40	1						
	R	158	04	40	04	40	158	04	40	3	52	41	33	158	04	40	04	40	158	04	40	3	52	41	33	52	41	33

'Sets' by Method of Repetition for High Precision: For measuring an angle to the highest degree of precision, several *sets* of repetitions are usually taken. *There are two methods of taking a single set.*

First Method :

1. Keeping the telescope normal throughout, measure the angle *clockwise* by 6 repetitions. Obtain the *first value* of the angle by dividing the final reading by 6.
2. Invert the telescope and measure the angle *counter-clockwise* by 6 repetitions. Obtain the *second value* of the angle by dividing the final reading by 6.
3. Take the mean of the first and second values to get the average value of the angle by *first set.*

Take as many sets in this way as may be desired. For first order work, five or six sets are usually required. The final value of the angle will be obtained by taking the mean of the values obtained by different sets.

Second Method :

1. Measure the angle clockwise by six repetitions, the first three with the telescope normal and the last three with the telescope inverted. Find the *first value* of the angle by dividing the final by six.
2. Without altering the reading obtained in the sixth repetition, measure the explement of the angle (*i.e.*, 360° – *PQR*) *clockwise* by six repetitions, the first three with telescope inverted and the last three with telescope normal. Take the reading which should theoretically by equal to zero (or the initial value). If not, note the error and distribute half the error to the *first value* of the angle. *The result is the corrected value of the angle by the first set.* Take as many sets as are desired and find the average angle. For more accurate work, the initial reading at the beginning of each set may not be set to zero but to two different values.

   Note. *During an entire set of observations, the transit should not be relevelled.*

Elimination of Errors by Method of Repetition: The following errors are eliminated by method of repetition:

1. Errors due to eccentricity of verniers and centres are eliminated by taking both vernier readings.
2. Errors due to inadjustments of line of collimation and the trunnion axis are eliminated by taking both face readings.
3. The error due to inaccurate graduations are eliminated by taking the readings at different parts of the circle.
4. Errors due to inaccurate bisection of the object, eccentric centring etc., may be to some extent counter-balanced in different observations.

It should be noted, however, that in repeating angles, operations such as sighting and clamping are multiplied and hence opportunities for error are multiplied. The limit of precision in the measurement of an angle is ordinarily reached after the fifth or sixth repetition.

Errors due to slip, displacement of station signals, and want of verticality of the vertical axis etc., are not eliminated since they are all cumulative.

To Measure a Horizontal Angle by Direction Method (or Reiteration Method)

The method known as *'direction method'* or *reiteration method* or *method of series* is suitable for the measurements of the angles of a group having a common vertex point. Several angles are measured successively and finally the *horizon is closed.* (Closing the horizon is the process of measuring the angles around a point to obtain a check on their sum, which should equal 360°).

To measure the angles *AOB, BOC, COD* etc., by reiteration, proceed as follows (Fig. 6.10).

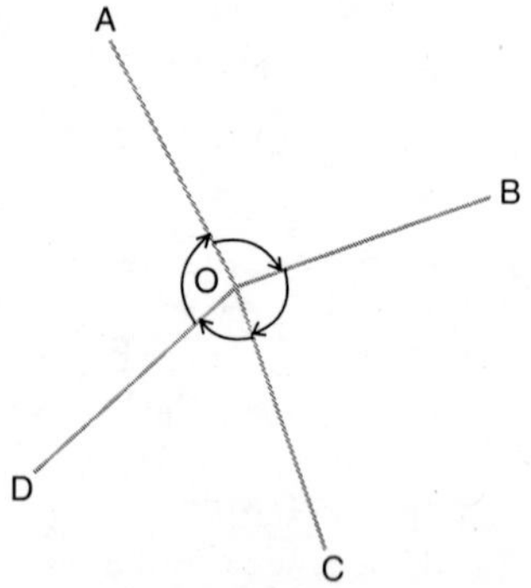

Fig. 6.10

1. Set the instrument over *O* and level it. Set one vernier to zero and bisect point *A* (or any other reference object) accurately.
2. Loose the upper clamp and turn the telescope clockwise to point *B*. Bisect *B* accurately using the upper tangent screw. Read both the verniers. The mean of the vernier readings will give the angles *AOB*.
3. Similarly, bisect successively, C, *D*, etc., thus closing the circle. Read both the verniers at each bisection. Since the graduated circle remains in a fixed position throughout the entire process, each included angle is obtained by taking the difference between two consecutive readings.
4. On final sight to *A*, the reading of the vernier should be the same as the original setting. It not, note the reading and find the error due to slips etc., and if the error is small, distribute it *equally* to all angles. If large, repeat the procedure and take a fresh set of readings.
5. Repeat steps 2 to 4 with the other face.

Table 6.3 illustrates the method of recording the observations.

Sets by the Direction Method. For precise work, several sets of readings are taken. The procedure for each set is as follows:

1. Set zero reading on one vernier and take a back sight on *A*. Measure *clockwise* the angles *AOB, BOC, COD, DOA,* etc., exactly in the same manner as explained above and close the horizon. Do not distribute the error.
2. Reverse the telescope, unclamp the lower clamp and back sigh on *A*. Take reading and foresight on *D, C, B* and *A*, in *counter-clockwise* direction and measure angles *AOD, DOC, COB* and *BOA*.

From the two steps, two values of each of the angles are obtained. The mean of the two is taken as the average value of each of the uncorrected angles. The sum of all the average angles so found should be 360°. In the case of discrepancy, the error (if small) may be distributed equally to all the angles. The values so obtained are the corrected values for the *first set*. Several such sets may be taken by setting the initial angle on the vernier to different values.

TABLE 6.3 Reiteration method

Instrument At	*Sighted to*	*Face : Left*								*Swing : Right*			*Face : Right*									*Swing : Left*			*Average Horizontal Angle*		
		A			*B*		*Mean*			*Horizontal Angle*			*A*			*B*		*Mean*			*Horizontal Angle*					
		°	′	″	′	″	°	′	″	°	′	″	°	′	″	′	″	°	′	″	°	′	″	°	′	″
O	*A*	0	0	0	0	0	0	0	0				0	0	0	0	0	0	0	0						
	B	54	31	20	31	20	54	31	20	AOB 54	31	20	54	31	40	31	40	54	31	40	54	31	40	54	31	30
	C	102	25	40	25	40	102	25	40	BOC 47	54	20	102	26	00	26	00	102	26	00	47	54	20	47	54	20
	D	239	49	40	49	40	239	40	40	COD 137	24	00	239	49	49	49	40	239	49	40	137	23	40	137	23	50
	A	360	0	0	0	0	360	0	0	DOA 120	10	20	360	0	0	0	0	360	0	0	120	10	20	120	10	20

The number of *sets* (or positions, as is sometimes called) depends on the accuracy required. For first order triangulation, *sixteen* such sets are required with a 1″ direction theodolite, while for second order triangulation, *four* and for third order triangulation *two*. For ordinary work, however, one set is sufficient.

6.6 MEASUREMENT OF VERTICAL ANGLES

Vertical angle is the angle which the inclined line of sight to an object makes with the horizontal. It may be an angle of elevation or angle of depression depending upon whether the object is above or below the horizontal plane passing through the trunnion axis of the instrument. To measure a vertical angle, the instrument should be levelled with reference to the altitude bubble. *When the altitude bubble is on the index frame, proceed as follows:*

1. Level the instrument with reference to the plate level, as already explained.
2. Keep the altitude level parallel to any two foot screws and bring the bubble central. Rotate the telescope through 90° till the altitude bubble is on the third screw. Bring the bubble to the centre with the third food screw. Repeat the procedure till the bubble is central in both the positions. If the bubble is in adjustment it will remain central for all pointings of the telescope.
3. Loose the vertical circle clamp and rotate the telescope in vertical plane to sight the object. Use vertical circle tangent screw for accurate bisection.
4. Read both verniers (*i.e. C* and *D)* of vertical circle. The mean of the two gives the vertical circle. Similar observation may be made with another face. The average of the two will give the required angle.

Note. *It is assumed that the altitude level is in adjustment and that index error has been eliminated by permanent adjustments. The clip screw should not be touched during these operations.*

In some instruments, the altitude bubble is provided both on index frame as well as on the telescope. In such cases, the instrument is levelled with reference to the altitude bubble on the index frame and *not* which reference to the altitude bubble on the telescope. Index error will be then equal to the reading on the vertical circle when the bubble on the telescope is central. If, however, the theodolite is to be used as a level, it is to be levelled with reference to the altitude bubble placed on the telescope.

If it is required to measure the vertical angle between two points A and B as subtended at the trunnion axis, sight first the higher point and take the reading of the vertical circle. Then sight the lower point and take the reading. The required vertical angle will be equal to the algebraic difference between the two readings taking angle of elevation as positive and angle of depression as negative. Table 6.4 illustrates the method of recording the observations.

Graduations on Vertical Circle: Fig. 6.11 shows two examples of vertical circle graduations. In Fig. 6.11(*a*), the circle has been divided into four quadrants. Remembering that the vernier is fixed while-circle is moved with telescope, it is easy to see how the readings are taken.

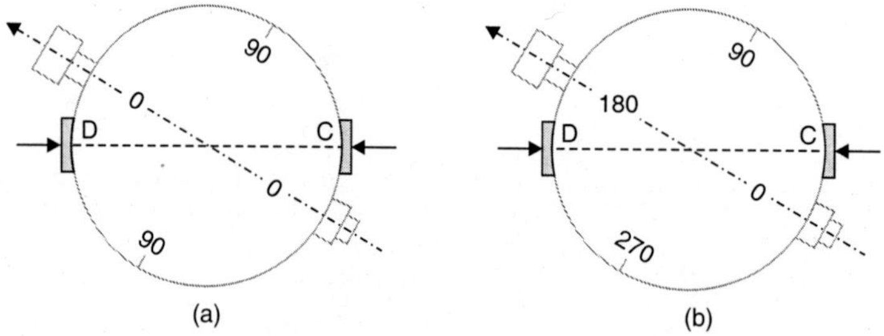

Fig. 6.11 Examples of vertical circle graduation.

For an elevated line of sight with face left, verniers C and D read 30° (say) as angle of elevation. In Fig. 6.11 (*b*), the circle is divided form 0° to 360° with zero at vernier C. For angle of elevation with face left, vernier C reads 30° while D reads 210°. In this system, therefore, 180° are to be deducted from vernier D to get the correct reading. However, it is always advisable to take full reading (*i.e.*, degrees, minutes and seconds) on one vernier and part reading (*i.e.*, minutes and seconds) of the other.

TABLE 6.4 Vertical angles

Instrument at	*Sighted to*	*Face : Left*													*Face : Right*													*Average Vertical Angle*		
		C			*D*		*Mean*			*Vertical Angle*			*C*			*D*		*Mean*			*Vertical Angle*					
		°	′	″	′	″	°	′	″	°	′	″	°	′	″	′	″	°	′	″	°	′	″	°	′	″
O	*A*	–5	12	20	12	00	–5	12	10				–5	12	40	12	20	–5	12	30						
	B	+2	25	40	25	20	+2	25	30	7	37	40	+2	26	00	25	40	+2	25	50	7	38	20	7	38	00

6.7 MISCELLANEOUS OPERATIONS WITH THEODOLITE

1. To Measure Magnetic Bearing of a Line

In order to measure the magnetic bearing of a line, the theodolite should be provided with either a tubular compass or trough compass. The following are the steps (Fig. 6.12):

1. Set the instrument at P and level it accurately.
2. Set accurately the vernier A to zero.
3. Loose the lower clamp. Release the needle of the compass. Rotate the instrument about its outer axis till the magnetic needle roughly points to north. Clamp the lower clamp. Using the lower tangent screw, bring the needle exactly against the mark so that it is in magnetic meridian. The line of sight will also be in the magnetic meridian.
4. Loose the upper clamp and point the telescope towards Q. Bisect Q accurately using the upper tangent screw. Read verniers A and B.
5. Change the face and repeat steps 2, 3 and 4. The average of the two will give the correct bearing of the line PQ.

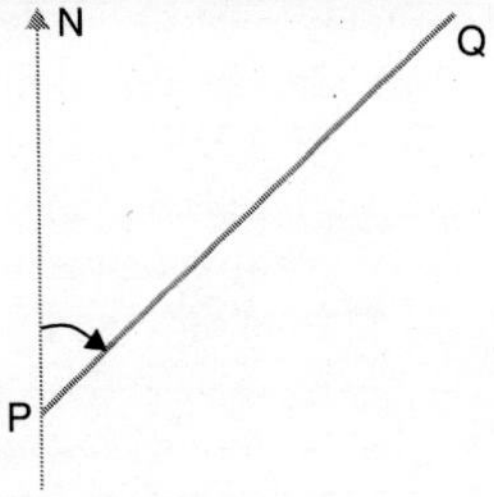

Fig. 6.12

2. To Measure Direct Angles

Direct angles are the angles measured *clockwise* from the preceding (previous) line to the following (*i.e.*, next) line. They are also known as *angles to the right* or *azimuths from the back line* and may vary from 0° to 360°.To measure the angle PQR (Fig. 6.13):

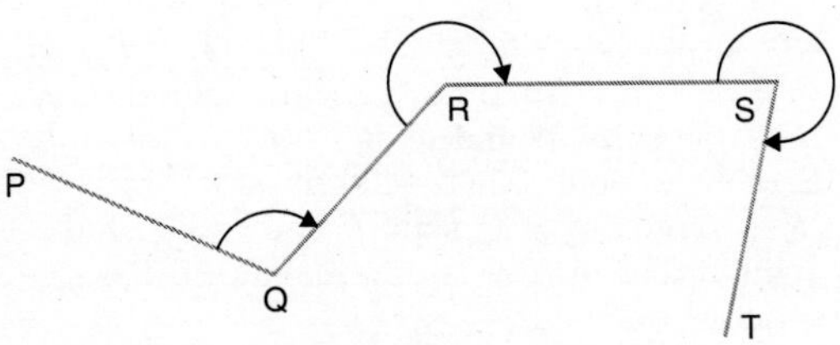

Fig. 6.13

1. Set the theodolite at Q and level it accurately. With face left, set the reading on vernier A to zero.
2. Unclamp the lower clamp and direct the telescope to P. Bisect it accurately using the lower tangent screw.
3. Unclamp the upper clamp and swing telescope *clockwise* and sight R. Bisect R accurately using the upper tangent screw. Read both verniers.
4. Plunge the telescope, unclamp the lower clamp and take backsight on P. Reading on the vernier will be the same as in step (3).
5. Unclamp the upper clamp and bisect R again. Read the verniers. The reading will be equal to twice the angle. $\angle PQR$ will then be obtained by dividing the final reading by two.

Similarly, angles at other stations may also be measured.

3. To Measure Deflection Angles

A deflection angle is the angle which a survey line makes with the *prolongation* of the preceeding line. It is designated as Right (R) or Left (L) according as it is measured to the clockwise or to anti-clockwise from the prolongation of the previous line. Its value may vary from 0° to 180°. The deflection angle at Q is $\alpha°$ R and that at R is $\theta°$ L (Fig. 6.14).

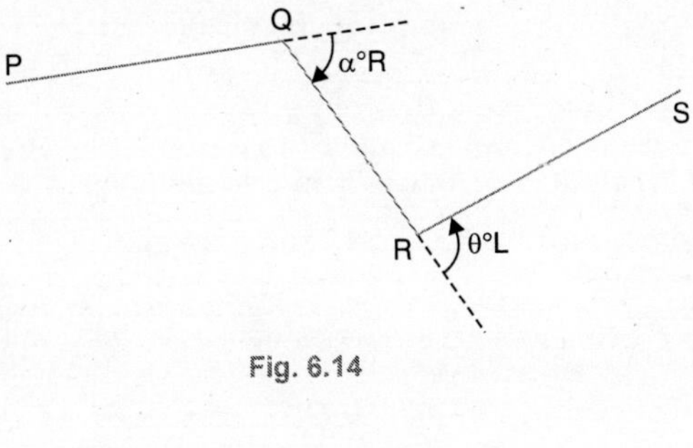

Fig. 6.14

To measure the deflection angles at Q:

1. Set the instrument at Q and level it.
2. With both plates clamped at 0°, take back sight on P.
3. Plunge the telescope. Thus the line of sight is in the direction PQ produced when the reading on vernier A is 0°.
4. Unclamp the upper clamp and turn the telescope clockwise to take the foresight on R. Read both the verniers.
5. Unclamp the *lower* clamp and turn the telescope to sight P again. The verniers still read the same reading as in (4). Plunge the telescope.
6. Unclamp the upper clamp and turn the telescope to sight R. Read both verniers. Since the deflection angle is doubled by taking both face readings, one-half of the final reading gives the deflection angle at Q.

4. To Prolong a Straight Line

There are three methods of prolonging a straight line such as AB to a point P which is not already defined upon the ground and is invisible from A and B (Fig. 6.15).

First method [Fig. 6.15 (*a*)]. Set the instrument at A and sight B accurately. Establish a point C in the line of sight. Shift the instrument at B, sight C and establish point D. The process is continued until P is established.

Second Method [Fig. 6.15 (*b*)]. Set the instrument at B and take a back sight on A. With both the motions clamped, plunge the telescope and establish C in the line of sight. Similarly, shift the instrument to C, back sight on B, plunge the telescope and establish D. The process is continued until P is established. If the instrument is in adjustment, B, C, D etc. will be in one straight line. If however, the line of sight is not perpendicular to the horizontal axis, points C', D', P' established will not be in a straight line.

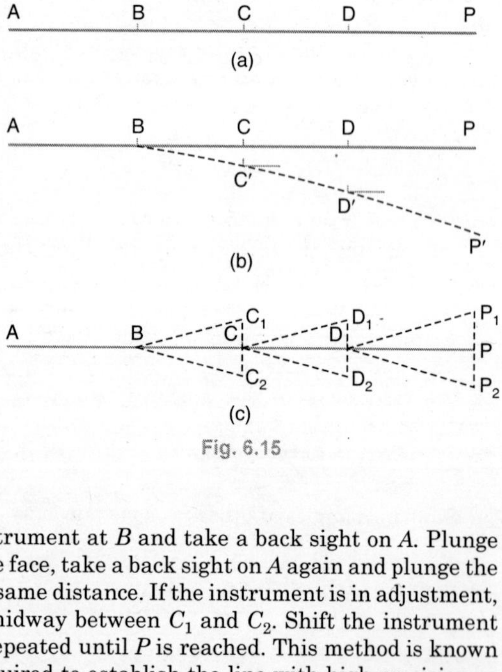

Fig. 6.15

Third Method [Fig. 6.15 (*c*)]. Set the instrument at B and take a back sight on A. Plunge the telescope and establish a point C_1. Change face, take a back sight on A again and plunge the telescope to establish another point C_2 at the same distance. If the instrument is in adjustment, C_1 and C_2 will coincide. If not, establish C midway between C_1 and C_2. Shift the instrument to C and repeat the process. The process is repeated until P is reached. This method is known as *double sighting* and is used when it is required to establish the line with high precision or when the instrument is in poor adjustment.

5. To Run a Straight Line between Two Points

Case 1. Both ends intervisible (Fig. 6.16). Set the instrument at A and take sight on B. Establish intermediate points C, D, E etc., in the line of sight.

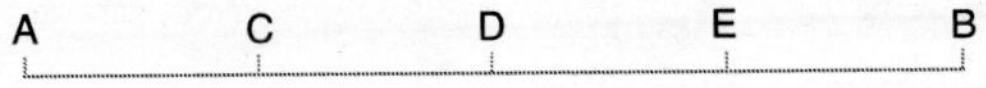

Fig. 6.16

Case 2. Both ends not intervisible, but visible from an intervening point (Fig. 6.17).

Set the instrument at *C* as nearly in line *AB* as possible (by judgment). Take backsight on *A* and plunge the telescope to sight *B*. The line of sight will not pass exactly through *B*. The amount by which the transit must be shifted laterally is estimated. The process is repeated till, on plunging the telescope, the line of sight passess through *B*. The location of the point C so obtained may then be checked by double sighting. The process is also known as *balancing in*.

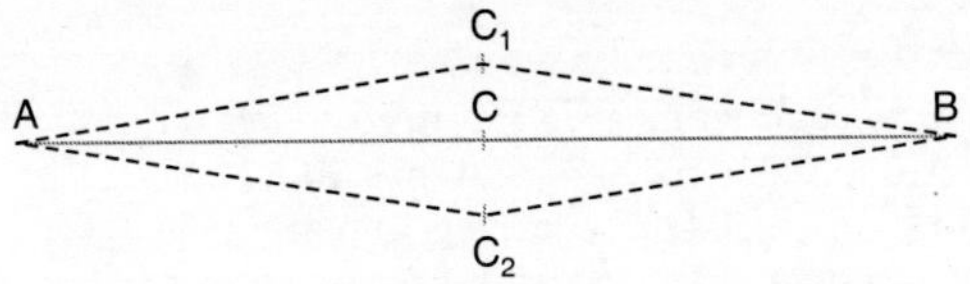

Fig. 6.17

Case 3. Both ends not visible from any intermediate point (Fig. 6.18). Let *A* and *B* be the required points which are not visible from intermediate points and it is required to establish intermediate points as *D, E,* etc.

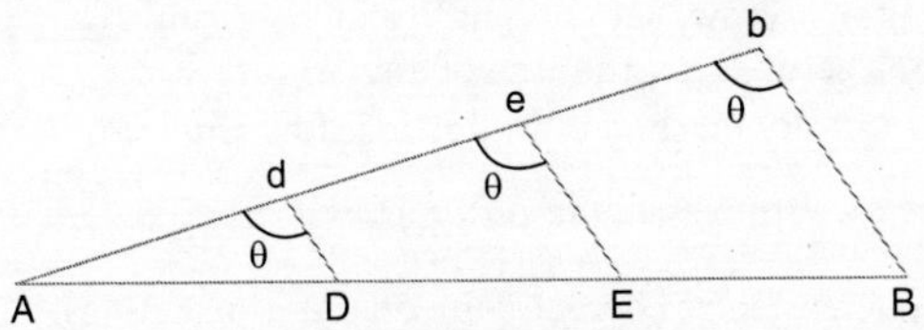

Fig. 6.18

Run a *random line Ab* by double sighting to a point *b* which is as near to *AB* as possible. Set the transit at *b* and measure angle *BbA*. Measure *Ab* and *Bb*. To locate *D* on *AB,* set the instrument at *d* on *Ab,* lay off angle $AdD = \theta$ and measure $dD = Bb \,.\, Ad/Ab$. The point *D* is then on the line *AB*. Other points can similarly be located.

6. To Locate the Point of Intersection of Two Straight Lines

Let it be required to locate the point of intersection *P* of the two lines *AB* and *CD* (Fig. 6.19). Set the instrument at *A,* sight *B* and set two stakes *a* and *b* (with wire nails) a short distance apart on either side of the estimated position of point *P*. Set the instrument at *C* and sight *D*. Stretch a thread or string between *ab* and locate *P*, where the line of sight cuts the string.

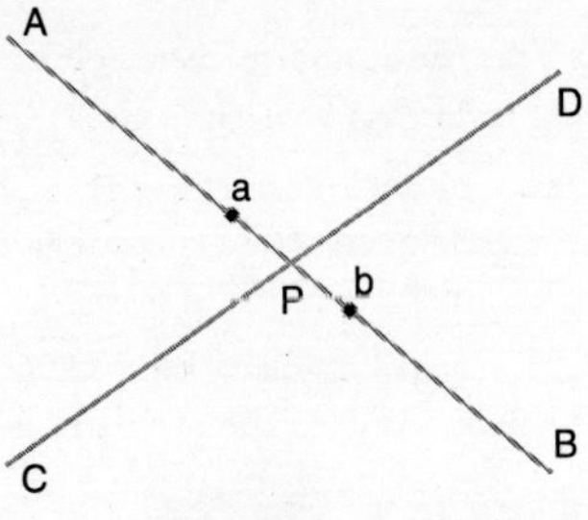

Fig. 6.19

7. To Lay Off a Horizontal Angle

Let it be required to lay off the angle *PQR*, say 42° 12′ 20″ (Fig. 6.20).

1. Set the instrument at Q and level it.
2. Using upper clamp and upper tangent screw, set the reading on vernier A to 0°.
3. Loose the lower clamp and sight P. Using lower tangent screw, bisect P accurately.
4. Loose upper clamp and turn the telescope till the reading is approximately equal to the angle PQR. Using upper tangent screw, set the reading exactly equal to 42° 12′ 20″
5. Depress the telescope and establish R in the line of sight.

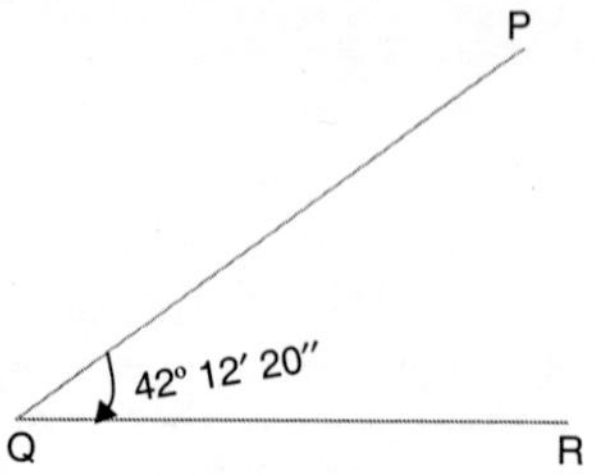

Fig. 6.20

8. To Lay Off an Angle by Repetition

The method of repetition is used when it is required to lay off an angle with the greater precision than that possible by a single observation. In Fig. 6.21. let QP be a fixed line and it is required to lay off QR at angle 45° 40′ 16″ with an instrument having a least count of 20″.

1. Set the instrument to Q and level it accurately.
2. Fix the vernier A at 0° and bisect P accurately.
3. Loose the upper clamp and rotate the telescope till the reading is approximately equal to the required angle. Using upper tangent screw, set the angle exactly equal to 45° 40′ 20″. Set point R_1 in the line of sight.
4. Measure angle PQR_1 by method of repetition. Let angle PQR_1 (by six repetition) be 274° 3′ 20″. The average value of the angle PQR_1 will be $\dfrac{274°3'20''}{6} = 45°\ 40'\ 33''$.
5. The angle PQR_1 is now to be corrected by an angular amount R_1QR to establish the true angle PQR. Since the correction (*i.e.*, 45° 40′ 33″ – 45° 40′ 16″ = 17″) is very small, it is applied linearly by making offset $R_1R = QR_1 \tan R_1QR$. Measure QR_1. Let it be 200 m. Then, $R_1R = 200 \tan 17'' = 0.017$ m (taking $\tan 1' = 0.0003$ nearly). Thus, point R is established by making $R_1 R = 0.017$ m
6. As a check, measure $\angle PQR$ again by repetition.

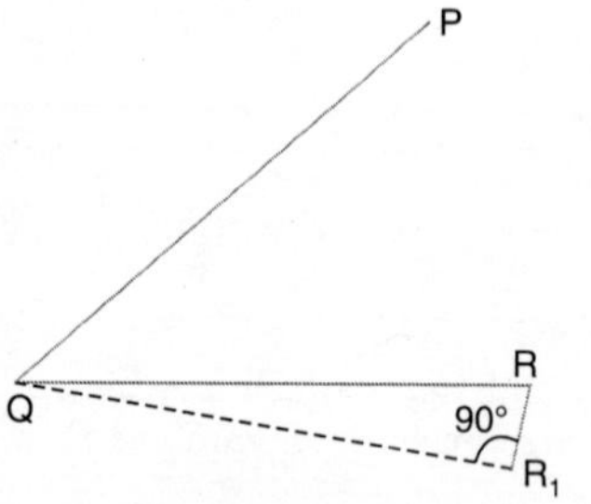

Fig. 6.21

6.8 FUNDAMENTAL LINES AND DESIRED RELATIONS

The *fundamental lines* of a transit are:

1. The vertical axis.
2. The horizontal axis (or trunnion axis or transit axis).
3. The line of collimation (or line of sight).
4. Axis of plate level.
5. Axis of altitude level.
6. Axis of the striding level, if provided.

Desired Relations: Fig. 6.22 shows the relationship between the line of sight, the axes and the circles of the theodolite. The following relationship should exist:

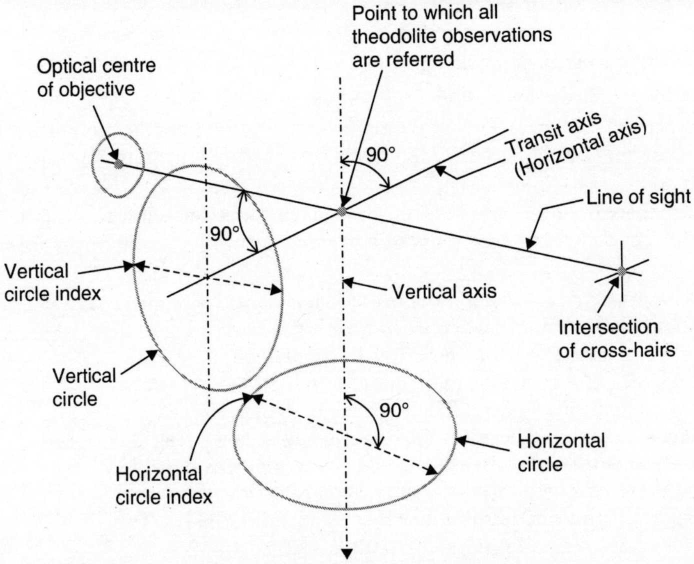

Fig. 6.22 Line of sight, axes and circles of the theodolite.

1. *The axis of the plate level must lie in a plane perpendicular to the vertical axis.*
   If this condition exists, the vertical axis will be truly vertical when the bubble is in the centre of its run.
2. *The line of collimation must be perpendicular to the horizontal axis at its intersection with the vertical axis. Also, if the telescope is external focusing type, the optical axis, the axis of the objective slide and the line of collimation must coincide.*
   If this condition exists, the line of sight will generate a vertical plane when the telescope is rotated about the horizontal axis.
3. *The horizontal axis must be perpendicular to the vertical axis.*
   If this condition exists, the line of sight will generate a vertical plane when the telescope is plunged.
4. *The axis of the altitude level (or telescope level) must be parallel to line of collimation.*
   If this condition exists, the vertical angles will be free from index error due to lack of parallelism.
5. *The vertical circle vernier must read zero when the line of collimation is horizontal.*
   If this condition exists, the vertical angles will be free from index error due to displacement of the vernier.
6. *The axis of the striding level (if provided) must be parallel to the horizontal axis.*
   If this condition exists, the line of sight (if in adjustment) will generate a vertical plane when the telescope is plunged, the bubble of striding level being in the centre of its run.

6.9 SOURCES OF ERROR IN THEODOLITE WORK

The sources of error in transit work are:

1. Instrumental 2. Personal, and 3. Natural.

1. **Instrumental Errors.** The instrumental errors are due to (*a*) imperfect adjustment of the instrument, (*b*) structural defects in the instrument, and (*c*) imperfections due to wear.

The total instrumental error to an observation may be due solely to one or to a combination of these. The following are errors due to imperfect adjustment of the instrument.

(*i*) **Error due to imperfect adjustment of plate levels.** If the upper and lower plates are not horizontal when the bubbles in the plate levels are centred, the vertical axis of the instrument will not be truly vertical (Fig. 6.23). The horizontal angles will be measured in an inclined plane and not in a horizontal plane. The vertical angles measured will also be incorrect. The error may be serious in obserying the points the difference in elevation of which is considerable. The error can be eliminated only by careful levelling with respect to the altitude bubble if it is in adjustment. The errors cannot be eliminated by double sighting.

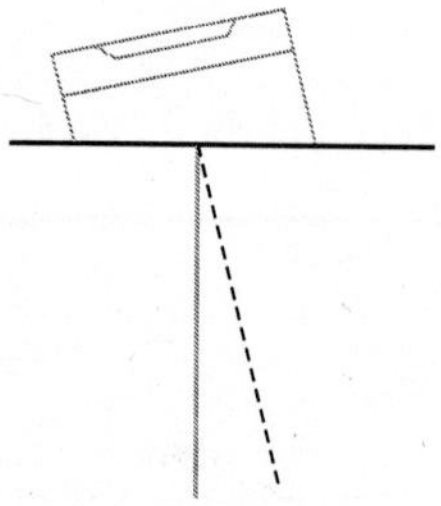

Fig. 6.23

(*ii*) **Error due to line of collimation not being perpendicular to the horizontal axis.** If the line of sight is not perpendicular to the trunnion axis of the telescope, it will not revolve in a plane when the telescope is raised or lowered but instead, it will trace out the surface of a cone. The trace of the intersection of the conical surface with the vertical plane containing the point will be hyperbolic. This will cause error in the measurement of horizontal angle between the points which are at considerable difference in elevation. Thus, in Fig. 6.24, let P and Q be two points at different elevation and let P_1 and Q_1 be their projections on a horizontal trace. Let the line AP be inclined at an angle α_1 to horizontal line AP_1. When the telescope is lowered

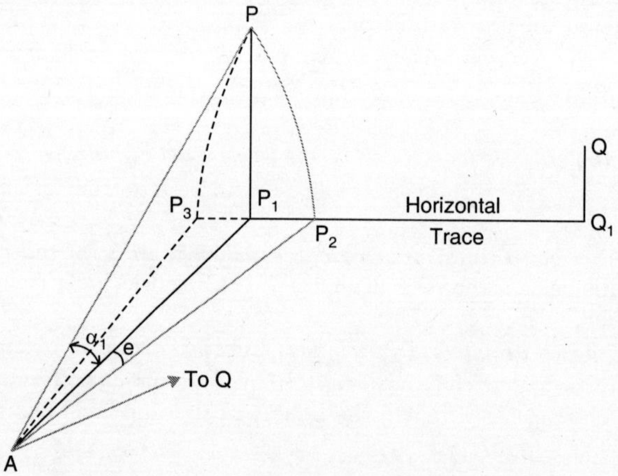

Fig. 6.24

after sighting P the hyperbolic trace will cut the horizontal trace P_1Q_1 in P_2 if the intersection of the cross-hairs is to the left of the optical axis. The horizontal angle thus measured will be with respect of AP_2 and not with respect to AP_1. The error e introduced will thus be $e = \beta \sec \alpha_1$, where β is the error in the collimation. On changing the face, however, the intersection of the cross-hairs will be to the right of the optical axis and the hyperbolic trace will intersect the line P_1Q_1 in P_3. The horizontal angle thus measured will be with respect to AP_3, the error being $e = \beta \sec \alpha_1$ to the other side. It is evident, therefore, that by taking both face observations the error can be eliminated. At Q also, the error will be $e' = \beta \sec \alpha_2$, where α_2 is the inclinations of AQ with horizontal, and the error can be eliminated by taking both face observations. If, however, only one face observations are taken to P and Q, the residual error will be equal to $\beta (\sec \alpha_1 - \sec \alpha_2)$ and will be zero when both the points are at the same elevation.

(*iii*) **Error due to horizontal axis not being perpendicular to the vertical axis.** If the horizontal axis is not perpendicular to the vertical axis, the line of sight will move in an inclined plane when the telescope is raised or lowered. Thus, the horizontal and vertical angles measured will be incorrect. The error will be of serious nature if the points sighted are at very different levels. Let P and Q be the two points to be observed, P_1 and Q_1 being their projection on a horizontal trace (Fig. 6.25). Let the line of sight AP make an angle α_1 with horizontal. When the telescope is lowered after sighting P, it will move in an inclined plane APP_2 and not in the vertical plane APP_1. The horizontal angle measured will now be with reference to AP_2 and not with AP_1. If β is the instrumental error and e is the resulting error, we get

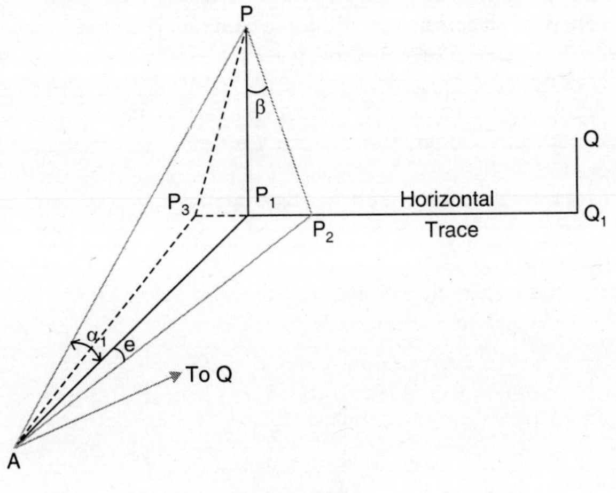

Fig. 6.25

$$\tan e = \frac{P_1P_2}{AP_1} = \frac{PP_1 \tan \beta}{AP_1} = \tan \alpha_1 \tan \beta$$

Since e and β will be usually small, we get

$$e = \beta \tan \alpha_1.$$

On changing the face and lowering the telescope after observing P, the line of sight will evidently move in the inclined plane AP_3. The angle measured will be with reference to AP_3 and not with AP_1, the error being $e = \beta \tan \alpha_1$ on the other side. It is quite evident, therefore, that the error can be eliminated by taking both face observations. At Q also, the error will be $e' = \beta \tan \alpha_2$, where α_2 is inclination of AQ with horizontal and the error can be eliminated by taking both face observations. If however, only one face observation is taken to both P and Q the residual error will be equal to $\beta (\tan \alpha_1 - \tan \alpha_2)$ and will be zero when both the points are at the same elevation.

(*iv*) **Error due to non-parallelism of the axis of telescope level and line of collimation.** If the line of sight is not parallel to the axis of telescope level, the measured vertical angles will be incorrect since the zero line of the vertical verniers will not be a true line of reference. It will also be a source of error when the transit is used as a level. The error can be eliminated by taking both face observations.

(*v*) **Error due to imperfact adjustment of the vertical circle vernier.** If the vertical circle verniers do not read zero when the line of sight is horizontal, the vertical angles measured will be incorrect. The error is known as the *index error* and can be eliminated either by applying index correction or by taking both face observations.

(*vi*) **Error due to eccentricity of inner and outer axes.** If the centre of the graduated horizontal circle does not coincide with the centre of the vernier plate, the reading against either vernier will be incorrect. In Fig. 6.26, let o be the centre of the circle and o_1 be the centre of the vernier plate. Let a be the position of vernier A while taking a back sight and a_1 be its corresponding position when a foresight is taken on another object. The positions of the vernier B are represented by b and b_1 respectively. The telescope is thus turned through an angle $ao_1 a_1$ while the arc aa_1 measures an angle aoa_1 and not the true angle ao_1a_1.

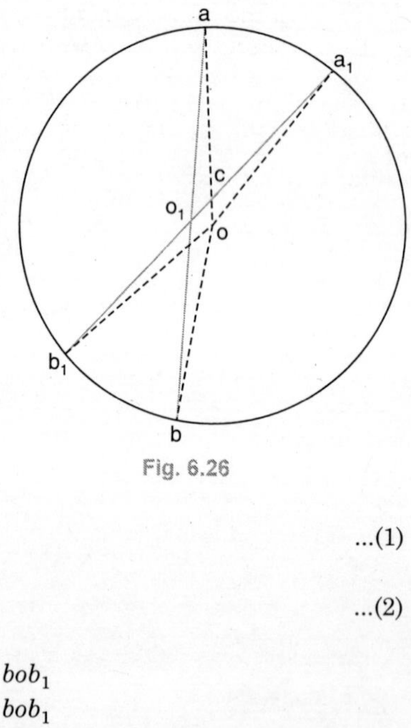

Fig. 6.26

Now $\quad ao_1a_1 = aca_1 - o_1ao$

or $\quad ao_1a_1 = (aoa_1 + o_1a_1o) - o_1ao \quad$ ...(1)

Similarly, $\quad bo_1b_1 = (bob_1 + o_1bo) - o_1b_1o$

or $\quad bo_1b_1 = bob_1 + o_1ao - o_1a_1o \quad$ ...(2)

Adding (1) and (2), we get

$$ao_1a_1 + bo_1b_1 = aoa_1 + bob_1$$

or
$$2ao_1a_1 = aoa_1 + bob_1$$

or
$$ao_1a_1 = \frac{aoa_1 + bob_1}{2}$$

Thus, the true angle is obtained by taking the mean of the two vernier readings.

(*vii*) **Error due to imperfect graduations.** The error due to defective graduations in the measurement of an angle may be eliminated by taking the mean of the several readings distributed over different portions of the graduated circle.

(*viii*) **Error due to eccentricity of verniers.** The error is introduced when the zeros of the vernier are not at the ends of the same diameter. Thus, the difference between the two vernier readings will not be 180°, but there will be a constant difference of other than 180°. The error can be eliminated by reading both the verniers and taking the mean of the two.

2. **Personal Errors.** The personal errors may be due to *(a)* Errors in manipulation, *(b)* Errors in sighting and reading.

(*a*) **Errors in manipulation.** They include:

(*i*) **Inaccurate centring:** If the vertical axis of the instrument is not exactly over the station mark, the observed angles will either be greater or smaller than the true angle. Thus in Fig. 6.27, C is the station mark while instrument is centred over C_1. The correct angle ACB will be given by

$$\angle ACB = \angle AC_1B - \alpha - \beta = \angle AC_1B - (\alpha + \beta)$$

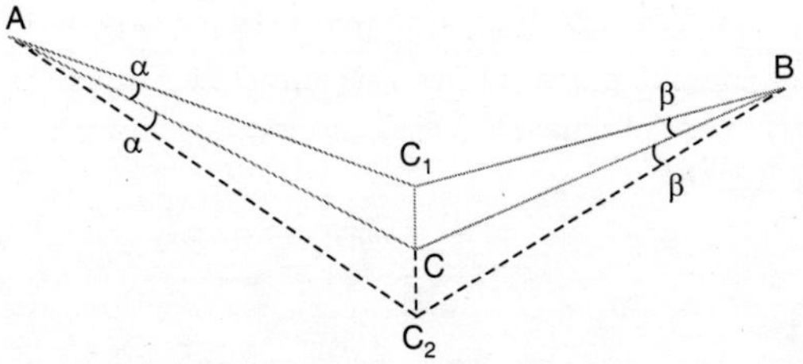

Fig. 6.27

If, however, the instrument is centred over C_2

$$\angle ACB = \angle AC_2B + (\alpha + \beta)$$

The error, *i.e.* $\pm(\alpha + \beta)$ depends on (*i*) the length of lines of sight, and (*ii*) the error in centring. The angular error due to defective centring varies inversely as the lengths of sights. The error is, therefore, of a very serious nature if the sights are short. It should be remembered that the error in sight is about 1′ when the error of centring is 1 cm and the length of sight is 35 m.

(*ii*) **Inaccurate levelling:** The error due to inaccurate levelling is similar to that due to non-adjustment of the plate levels. The error will be of serious nature when the points observed are at considerable difference in elevation. The error can be minimised by levelling the instrument carefully.

(*iii*) **Slip:** The error is introduced if the lower clamp is not properly clamped, or the shifting head is loose, or the instrument is not firmly tightened on the tripod head. The error is of a serious nature since the direction of the line of sight will change when such slip occurs, thus making the observation incorrect.

(*iv*) **Manipulating wrong tangent screw:** The error is introduced by using the upper tangent screw while taking the backsight or by using the lower tangent screw while taking a foresight. The error due to the former can be easily detected by checking the vernier reading after the backsight point is sighted, but the error due to the latter cannot be detected.

It should always be remembered to use lower tangent screw while taking a backsight and to use upper tangent screw while taking the foresight reading.

(*b*) **Errors in sighting and reading:** They include:

(*i*) **inaccurate bisection of points observed.** The observed angles will be incorrect if the station mark is not bisected accurately due to some obstacles etc. Care should be always be taken to intersect the lowest point of a ranging rod or an arrow placed at the station mark if the latter is not distinctly visible. The error varies inversely as the length of the line of sight.

If the ranging rod put at the station mark is not held vertical, the error e is given by

$$\tan e = \frac{\text{Error in verticality}}{\text{Length of sight}}$$

(*ii*) **Parallax:** Due to parallax, accurate bisection is not possible. The error can be eliminated by focusing the eye-piece and objective.

(*iii*) **Mistakes:** in setting the vernier, taking the reading and wrong booking of the readings.

3. **Natural Errors.** Sources of natural errors are :

(*i*) Unequal atmospheric refraction due to high temperature.

(*ii*) Unequal expansion of parts of telescope and circles due to temperature changes.

(*iii*) Unequal settlement of tripod.

(*iv*) Wind producing vibrations.

PROBLEMS

1. Define the terms: face right and face left observations; swinging the telescope ; transiting the telescope ; telescope normal.
2. (*a*) What are 'face left' and 'face right' observations? Why is it necessary to take both face observations? (*b*) Why both Verniers are read?
3. Explain how you would take field observations with a theodolite so as to eliminate the following verniers.
   (*i*) Error due to eccentricity of verniers.
   (*ii*) Error due to non-adjustment of line of sight.
   (*iii*) Error due to non-uniform graduations.
   (*iv*) Index error of vertical circle.
   (*v*) Error due to slip etc.
4. Explain the temporary adjustments of a transit.
5. Explain how you would measure with a theodolite:
   (*a*) Horizontal angle by repetition, (*b*) Vertical angle, (*c*) Magnetic bearing of line.
6. What are the different errors in theodolite work? How are they eliminated?
7. State what errors are eliminated by repetiton method. How will you set out a horizontal angle by method of repetition?

CHAPTER 7

TRAVERSE SURVEYING

7.1 INTRODUCTION

Traversing is that type of survey in which a number of connected survey lines form the framework and the directions and lengths of the survey lines are measured with the help of an angle (or direction) measuring instrument and a tape (or chain) respectively. When the lines form a circuit which ends at the starting point, it is known as a *closed traverse*. If the circuit ends elsewhere, it is said to be an *open traverse*. The closed traverse is suitable for locating the boundaries of lakes, woods etc., and for the survey of large areas. The open traverse is suitable for surveying a long narrow strip of land as required for a road or canal or the coast line.

Methods of Traversing: There are several methods of traversing, depending on the instruments used in determining the relative directions of the traverse lines. The following are the principal methods:

(*i*) Chain traversing.

(*ii*) Chain and compass traversing (loose needle method).

(*iii*) Transit tape traversing:

  (*a*) By fast needle method.

  (*b*) By measurement of angles between the lines.

(*iv*) Plane-table traversing (see Chapter 11).

Traverse survey differs from chain surveying in that the arrangement of the survey lines is not limited to any particular geometrical figure as in chain surveying, where a system of connected triangles forms the fundamental basis of the skeleton. Also, check lines etc. are not necessary in traversing as the traverse lines may be arranged near the details. The details etc. are directly located with respect to the survey lines either by offsetting (as in chain survey) or by any other method.

7.2 CHAIN TRAVERSING

In this method, the whole of the work is done with the chain and tape. No angle measuring instrument is used and the directions of the lines are fixed entirely by liner measurements. Angles fixed by linear or tie measurements are known as *chain angles*.

Fig. 7.1 (*a*) shows a closed chain traverse. At *A*, the directions *AB* and *AD* are fixed by internal measurements Aa_1, Ad_1, and a_1d_1. However, the direction may also be fixed by external measurements such as at station B [Fig. 7.1 (*a*) and 7.1 (*b*)]. Fig. 7.1 (*b*) shows an open chain traverse.

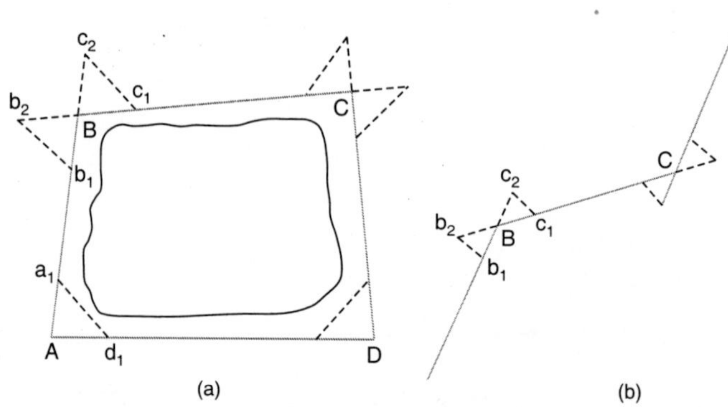

Fig. 7.1

The method is unsuitable for accurate work and is generally not used if an angle measuring instrument such as a compass, sextant, or theodolite is available.

7.3 CHAIN AND COMPASS TRAVERSING: FREE OR LOOSE NEEDLE METHOD

In chain and compass traversing, the magnetic bearings of the survey lines are measured by a compass and the lengths of the lines are measured either with a chain or with a tape. The direction of magnetic meridian is established at each traverse station independently. The method is also known as *free or loose needle method.* A theodolite fitted with a compass may also be used for measuring the magnetic bearings of the traverse line (see § 6.7). However, the method is not so accurate as that of transit tape traversing. The methods of taking the details are almost the same as for chain surveying.

7.4 TRAVERSING BY FAST NEEDLE METHOD

In this method also, the magnetic bearings of traverse lines are measured by a theodolite fitted with a compass. However, the direction of the magnetic meridian is *not* established at each station but instead, the magnetic bearings of the lines are measured with reference so the direction of magnetic meridian established at the *first station.* The method is, therefore, more accurate than the loose needle method. The lengths of the lines are measured with a 20 m or 30 m steel tape. There are three methods of observing the bearings of lines by fast needle method.

(*i*) Direct method with transiting.

(*ii*) Direct method without transiting.

(*iii*) Back bearing method.

(*i*) Direct Method with Transiting

Procedure (Fig. 7.2):

1. Set the theodolite at P and level it. Set the vernier A exactly to zero reading. Loose the clamp of the magnetic needle. Using lower clamp and tangent screw, point the telescope to magnetic meridian.
2. Loose the upper clamp and rotate the telescope clockwise to sight Q. Bisect Q accurately by using upper tangent screw. Read vernier A which gives the magnetic bearing of the line PQ.

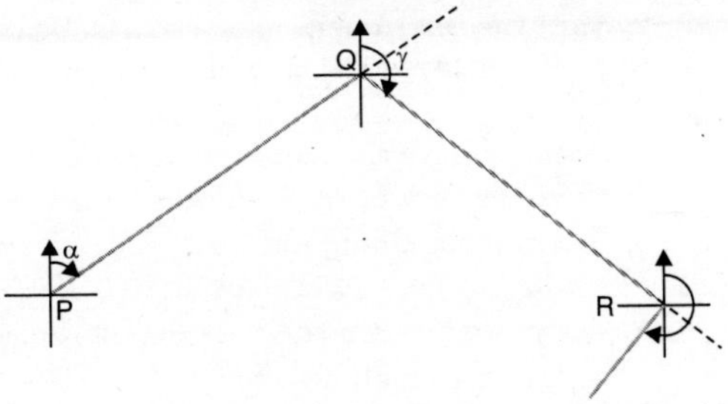

Fig. 7.2

3. With both the clamps clamped, move the instrument and set up at Q. Using lower clamp and tangent screw, take a back sight on P. See that the reading on the vernier A is still the same as the bearing of PQ.
4. Transit the telescope. The line of sight will now be in the direction of PQ while the instrument reads the bearing of PQ. The instrument is, therefore, oriented.
5. Using the upper clamp and tangent screw, take a foresight on R. Read vernier A which gives the magnetic bearing of QR.
6. Continue the process at other stations.

It is to be noted here that the telescope will be normal at one station and inverted at the next station. The method is, therefore, suitable only if the instrument is in adjustment.

(*ii*) Direct Method Without Transiting

Procedure (Fig. 7.2):

1. Set the instrument at P and orient the line of sight in the magnetic meridian.
2. Using upper clamp and tangent screw take a foresight on Q. The reading on vernier A gives the magnetic bearing of PQ.
3. With both plates clamped, move the instrument and set it at Q. Take a backsight on P. Check the reading on vernier A which should be the same as before. The line of sight is out of orientation by 180°.
4. Loosen the upper clamp and rotate the instrument *clockwise* to take a foresight on R. Read the vernier. Since the orientation at Q is 180° out, a correction of 180° is to be applied to the vernier reading to get the correct bearing of QR. Add 180° if the reading on the vernier is less than 180° and subtract 180° if it is more than 180°.
5. Shift the instrument of R and take backsight on Q. The orientation at R will be out by 180° with respect to that at Q and 360° with respect to that at P. Thus, after taking a foresight on the next station, the vernier reading will *directly* give magnetic bearing of the next line, without applying any correction of 180°.

The application of 180° correction is, therefore, necessary only at 2nd, 4th, 6th station occupied. Instead of applying correction at *even* station, opposite vernier may be read alternatively, *i.e.*, vernier A at P, vernier B at Q, verniers A at R, etc. However, it is always convenient to read one vernier throughout and apply the correction at alternate stations.

(*iii*) Back Bearing Method

Procedure (Fig. 7.2):

1. Set the instrument at P and measure the magnetic bearing of PQ as before.

2. Shift the instrument and set at *Q*. Before taking backsight on *P*, set vernier *A* to read back bearing of *PQ*, and fix the upper clamp.
3. Using lower clamp and tangent screw, take a backsight on *P*. The instrument is now oriented since the line of sight is along *QP* when the instrument is reading the bearing of *QP* (or back bearing of *PQ*).
4. Loose upper clamp and rotate the instrument clockwise to take a foresight on *R*. The reading on vernier *A* gives directly the bearing on *QR*.
5. The process is repeated at other stations.

Of the three methods of fast needle, the second method is the most satisfactory.

7.5 TRAVERSING BY DIRECT OBSERVATION OF ANGLES

In this method, the angles between the lines are *directly* measured by a theodolite. The method is, therefore, most accurate in comparison to the previous three methods. The magnetic bearing of any one line can also be measured (if required) and the magnetic bearing of other lines can be calculated as described in § 5.2 . The angles measured at different stations may be either (*a*) included angles or (*b*) deflection angles.

Traversing by Included Angles: An included angle at a station is either of the two angles formed by the two survey lines meeting there. The method consists simply in measuring each angle directly from a backsight on the preceding station. The angles may also be measured by repetition, if so desired. Both face observations must be taken and both the verniers should be read. Included angles can be measured either clockwise or counter-clockwise but it is better to measure all angles clockwise, since the graduations of the theodolite circle increase in this direction. The angles measured clockwise from the back station may be interior or exterior depending upon the direction of progress round the survey. Thus, in Fig. 7.3. (*a*), direction of progress is counter-clockwise and hence the angles measured clockwise are directly the interior angles. In Fig. 7.3 (*b*), the direction of progress around the survey is clockwise and hence the angles measured clockwise are exterior angles.

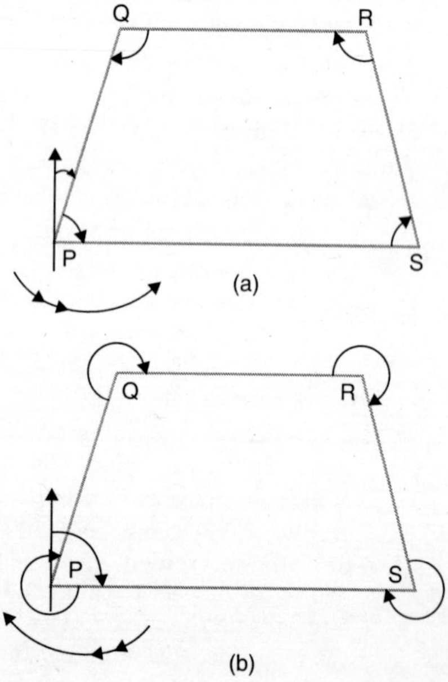

Fig. 7.3

Traversing by Deflection Angles: A deflection angle is the angle which a survey line makes with the prolongation of the preceding line. It is designated as right (*R*) or left (*L*) according as it is measured clockwise or anti-clockwise from the prolongation of the previous line. The procedure for measuring a deflection angle has been described in § 6.7.

This method of traversing is more suitable for survey of roads, railways, pipe-lines etc., where the survey lines make small deflection angles. Great care must be taken in recording and plotting whether it is right deflection angle or left deflection angle. However, except for specialised work in which deflection angles are required, it is preferable to read the included angles by reading clockwise from the back station. The lengths of lines are measured precisely using a steel tape. Table 7.1 shows the general method of recording the observation of transit tape traverse by observations of included angles.

7.6 LOCATING DETAILS WITH TRANSIT AND TAPE

Following are some of the methods of locating the details in theodolite traversing:

1. **Locating by angle and distance from one transit station:** A point can be located from a transit station by taking an angle to the point and measuring the corresponding distance from the station to the point. Any number of points can thus be located. The angles are usually taken from the same backsight, as shown in Fig. 7.4. The method is suitable specially when the details are near the transit station.

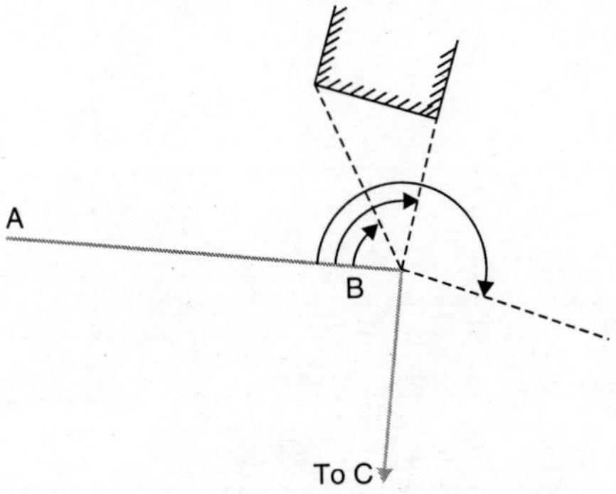

Fig. 7.4

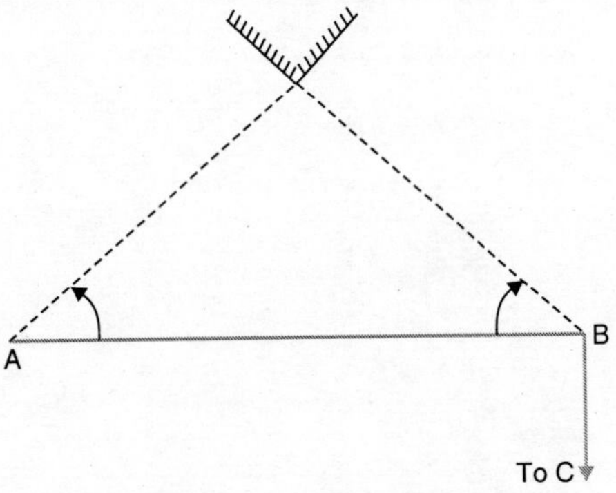

Fig. 7.5

2. **Locating by angles from two transit stations:** If the point or points are away from the transit stations or if linear measurements cannot be made, the point can be located by measuring angles to the point from at least two stations. This method is also known as method of intersection. For good intersection, the angle to the point should not be less than 20° (Fig. 7.5.).
3. **Locating by distances from two stations:** Fig. 7.6 illustrates the method of locating a point by measuring angle at one station and distance from the other. The method is suitable when the point is inaccessible from the station at which angle is measured.
4. **Location by distances from two points on traverse line:** If the point is near a transit line but is away from the transit station, it can located by measuring its distance from two points on the traverse line. The method is more suitable if such reference points (such as x and y in Fig. 7.7) are full chain points so that they can be staked when the traverse line is being chained.
5. **Locating by offsets from the traverse line:** If the points to be detailed are more and are near to traverse line, they can be located by taking offsets to the points as explained in chain surveying. The offsets may be oblique or may be perpendicular.

TABLE 7.1

Instrument at	*Sighted to*	*Face: Left*								*Swing: Right*				*Face: Right*								*Swing: Right*				*Average Horizontal Angle*				*Bearing*
		A			*B*		*Mean*			*No. of Repetitions*	*Horizontal Angle*			*A*			*B*		*Mean*			*No. of Repetitions*	*Horizontal Angle*							
		°	‴	″	″	″	°	″	″		°	″	″	°	″	″	″	″	°	″	″		°	″	″	°	″	″		
A	***F***	0	0	0	0	0	0	0	0					0	0	0	0	0	0	0	0									
	B	84	40	20	41	20	84	40	20	1	84	40	20	84	40	30	40	30	84	40	30	1	84	40	30	84	40	25	***AB***	
																													51.5	130°20′00″
B	***A***	0	0	0	0	0	0	0	0					0	0	0	0	0	0	0	0									
	C	56	15	20	15	20	56	15	20	1	56	15	20	56	15	30	15	30	56	15	30	1	56	15	30	56	15	25	***BC***	
																													120	6°35′25″
C	***B***	0	0	0	0	0	0	0	0					0	0	0	0	0	0	0	0									
	C	123	26	40	26	40	123	26	40	1	123	26	40	123	27	00	27	00	123	27	00	1	1123	27	00	123	26	00	***CD***	
																													84.2 etc.	310°20′15″

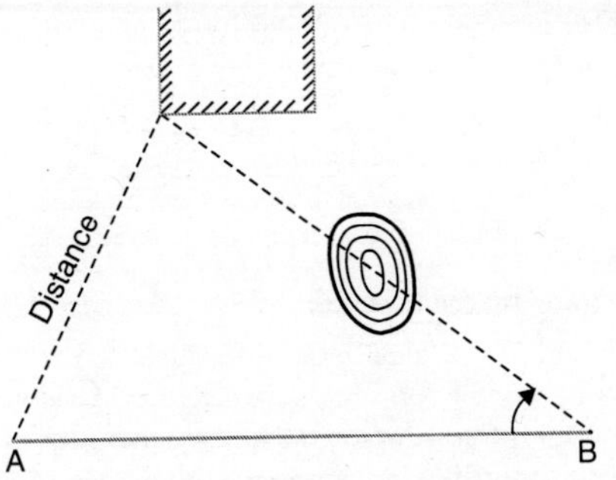

Fig. 7.6

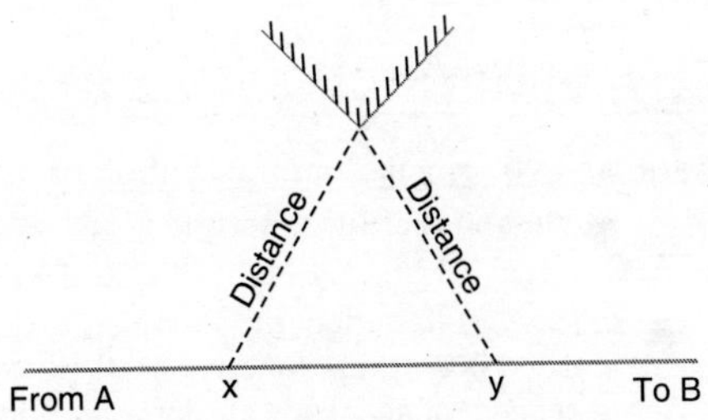

Fig. 7.7

7.7 CHECKS IN CLOSED TRAVERSE

The errors involved in traversing are two kinds: linear and angular. For important work the most satisfactory method of checking the linear measurements consists in chaining each survey line a second time, preferably in the reverse direction on different dates and by different parties. The following are the checks for the angular work:

1. *Traverse by included angles*
   (*a*) The sum of measured interior angles should be equal to $(2N - 4)$ right angles, where N = number of sides of the traverse.
   (*b*) If the exterior angles are measured, their sum should be equal to $(2N + 4)$ right angles.
2. *Traverse by deflection angles:* The algebraic sum of the deflection angles should be equal to 360°, taking the right-hand deflection angles as positive and left-hand angles as negative.
3. *Traverse by direct observation of bearings:* The fore bearing of the last line should be equal to its back bearing ± 180° measured at the initial station.

Checks in Open Traverse: No direct check of angular measurement is available. However, indirect checks can be made, as illustrated in Fig. 7.8.

As illustrated in Fig. 7.8 (*a*), in addition to the observation of bearing of *AB* at station *A*, bearing of *AD* can also be measured, if possible. Similarly, at *D*, bearing of *DA* can be measured and check applied. If the two bearings differ by 180°, the work (upto *D*) may be accepted as correct. If there is small discrepancy, it can be adjusted before proceeding further.

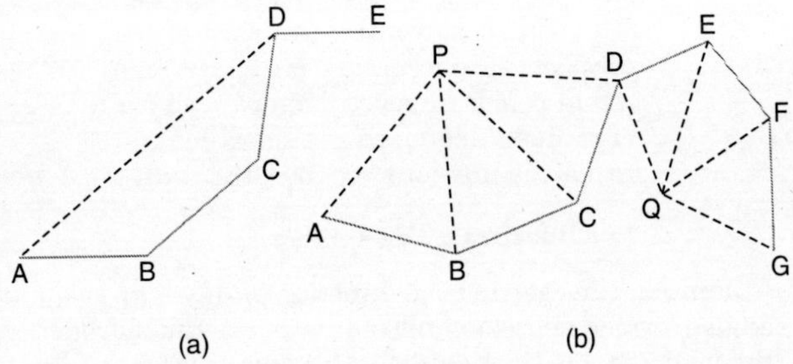

Fig. 7.8

Another method, which furnishes a check when the work is plotted is as shown in Fig. 7.8 (*b*), and consists in reading the bearings to any prominent point *P* from each of the consecutive stations. The check in plotting consists in laying off the lines *AP*, *BP*, *CP* etc. and noting whether the lines pass through one point.

In the case of long and precise traverse, the angular errors can be determined by astronomical observations for bearing at regular intervals during the progress of the traverse.

7.8 PLOTTING A TRAVERSE SURVEY

There are two principal methods of plotting a traverse survey:

1. Angle and distance method, and
2. Co-ordinate method.

1. **Angle and Distance Method:** In this method, distances between stations are laid off to scale and angles (or bearings) are plotted by one of the methods outlined below. This method is suitable for the small surveys, and is much inferior to the co-ordinate method in respect of accuracy of plotting. The more commonly used angle and distance methods of plotting an angle (or bearing) are:

   (*a*) By Protractor.
   (*b*) By the tangent of the angle.
   (*c*) By the chord of the angle.

   (*a*) ***The Protractor Method:*** The use of the protractor in plotting direct angles, deflection angles, bearings and azimuths requires no explanation. The ordinary protractor is seldom divided more finely than 10′ or 15′ which accords with the accuracy of compass traversing but not of theodolite traversing. A good form of protractor for plotting survey lines is the large circular cardboard type, 40 to 60 cm in diameter.

   (*b*) ***The Tangent Method:*** The tangent method is a trigonometric method based upon the fact that in right angled triangle, the *perpendicular* = *base* × *tan* θ where θ is the angle. From the end of the base, a perpendicular is set off, the length of the perpendicular being equal to *base* × *tan* θ. The station point is joined to the point so obtained; the line so obtained includes θ with the given side. The values of tan θ are taken from the table of natural tangents. If the angle is little over 90°, 90° of it is plotted by erecting a perpendicular and the remainder by the tangent method, using the *perpendicular as a base*.

   (*c*) ***The Chord Method:*** This is also a geometrical method of laying off an angle. Let it be required to draw line *AD* at an angle θ to the line *AB* in Fig. 7.9. With *A* as centre, draw an arc of any convenient radius (*r*) to cut line *AB* in *b*. With *b* as centre draw an arc of radius *r*′ (equal to the chord length) to cut the previous arc in *d*, the radius *r*′ being given $r' = 2\,r \sin\dfrac{\theta}{2}$.

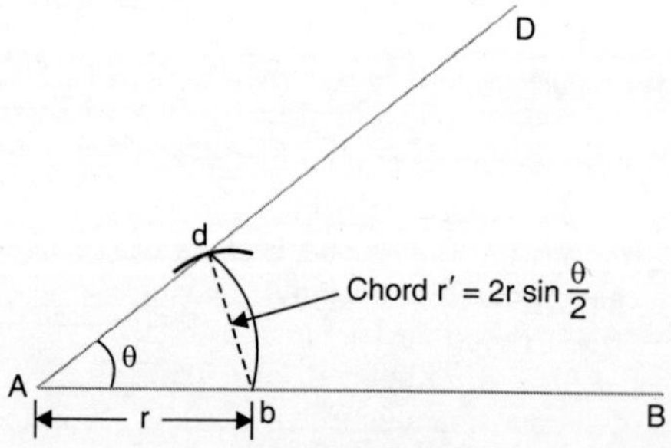

Fig. 7.9

Join *Ad*, thus getting the direction of *AD* at an inclination θ to *AB*. The lengths of chords of angles corresponding to unit radius can be taken from the *table of chords*. If an angle is greater than 90°, the construction should be done only for the part less than 90° because the intersections for greater angles become unsatisfactory.

2. **Co-ordinate Method:** In this method, survey stations are plotted by calculating their co-ordinates. This method is by far the most practical and accurate one for plotting traverses or any other extensive system of horizontal control. The biggest advantage in this method of plotting is that the *closing error* can be eliminated by *balancing*, prior

to plotting. The methods of calculating the co-ordinates and of balancing a traverse are discussed in the next article.

TRAVERSE COMPUTATIONS

7.9 CONSECUTIVE CO-ORDINATES: LATITUDE AND DEPARTURE

The *latitude* of a survey line may be defined as its co-ordinate length measured parallel to an assumed meridian direction (*i.e.* true north or magnetic north or any other reference direction). The *departure* of survey line may be defined as its co-ordinate length measured at right angles to the meridian direction. The latitude (L) of the line is *positive* when measured northward (or upward) and is termed as *northing*; the latitude is *negative* when measured southward (or downward) and is termed as *southing*. Similarly, the *departure* (D) of the line is *positive* when measured eastward and is termed as *easting;* the *departure* is negative when measured westward and is termed as *westing*.

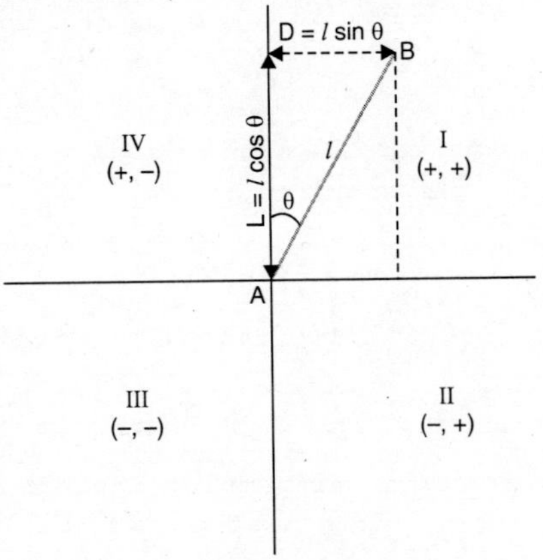

Fig. 7.10

Thus, in Fig. 7.10, the latitude and departure of the line AB of length l and reduced bearing θ are given by

$$L = + l \cos \theta \quad \text{and} \quad D = + l \sin \theta \qquad ...(7.1)$$

To calculate the latitudes and departure of the traverse lines, therefore, it is first essential to reduce the bearing in the quadrantal system. The sign of latitudes and departures will depend upon the reduced bearing of a line. The following table (Table 7.2) gives signs of latitudes and departures:

TABLE 7.2

W.C.B.	*R.B. and Quadrant*	*Sign of*	
		Latitude	*Departure*
0° to 90°	N θ E ; I	+	+
90° to 180°	S θ E ; II	–	+
180° to 270°	S θ W ; III	–	–
270° to 360°	N θ W ; IV	+	–

Thus, latitude and departure co-ordinates of any point with reference to the preceding point are equal to the latitude and departure of the line joining the preceding point to the point under consideration. Such co-ordinates are also known as *consecutive co-ordinates* or *dependent co-ordinates*.

Table 7.3 illustrates *systematic method* of calculating the latitudes and departures of a traverse.

TABLE 7.3 Calculations of Latitudes and Departures

Line	*Length (m)*	*W.C.B.*	*R.B.*	*Latitude*		*Departure*	
				Log length and Log cosine	*Latitude*	*Log length and Log sine*	*Departure*
AB	232	32° 12′	N 32° 12′E	2.36549 $\bar{1}$.92747 2.29296	+ 196.32	2.36549 $\bar{1}$.72663 2.09212	+ 123.63
BC	148	138° 36′	S 41° 24′E	2.17026 $\bar{1}$.87513 2.04539	– 111.02	2.17026 $\bar{1}$.82041 1.99067	+ 97.88
CD	417	202° 24′	S 22° 24′ W	2.62014 $\bar{1}$.96593 2.58607	– 385.54	2.62014 $\bar{1}$.58101 2.20115	– 158.90
DE	372	292° 0′	N 68° 0′ W	2.57054 $\bar{1}$.57358 2.14412	+ 139.36	2.57054 $\bar{1}$.96717 2.53771	– 329.39

Independent Co-ordinates

The co-ordinates of traverse stations can be calculated with respect to a common origin. The *total latitude and departure* of any point with respect to a common origin are known as *independent co-ordinates* or *total co-ordinates* of the point. The two reference axes in this case may be chosen to pass through any of the traverse station but generally a most westerly station is chosen for this purpose. The independent co-ordinates of any point may be obtained by adding algebracially the latitudes and the departure of the lines between that point and the origin.

Thus, *total latitude (or departure) of end point of a traverse = total latitudes* (*or departures*) *of first point of traverse plus the algebraic sum of all the latitudes* (*or departures*).

Table 7.4 shows the calculations of total co-ordinates of the traverse of Table 7.3. The axes are so chosen that the whole of the survey lines lie in the north east quadrant with respect to the origin so that the co-ordinates of all the points are positive. To achieve this, arbitrary values of co-ordinates are assigned to the starting point and co-ordinates of other points are calculated.

TABLE 7.4

Line	Latitude		Departure		Station	Total Co-ordinates	
	N	S	E	W		N	E
					A	400 assumed	400 assumed
AB	196.32		123.63				
					B	596.32	523.63
BC		111.02	97.88				
					C	485.30	621.51
CD		385.54		158.90			
					D	99.76	462.61
DE	139.36			329.39			
					E	239.12	133.22

7.10 CLOSING ERROR

If a closed traverse is plotted according to the field measurements, the end point of the traverse will not coincide exactly with the starting point, owing to the errors in the field measurements of angles and distances. Such error is known as *closing error* (Fig. 7.11). In a closed traverse, the algebraic sum of the latitudes (*i.e.* ΣL) should be zero and the algebraic sum of the departures (*i.e.* ΣD) should be zero. The *error of closure* for such traverse may be ascertained by finding ΣL and ΣD, both of these being the components of error e parallel and perpendicular to the meridian.

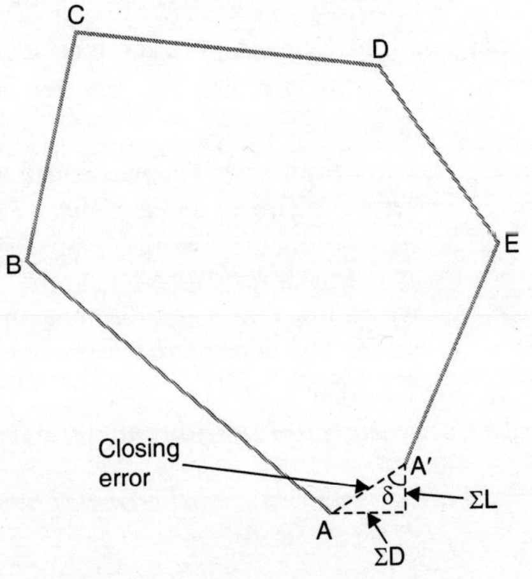

Fig. 7.11

Thus, in Fig. 7.11,

$$\text{Closing error } e = AA' = \sqrt{(\Sigma L)^2 + (\Sigma D)^2} \quad \text{...(7.2 a)}$$

The direction of closing error is given by

$$\tan \delta = \frac{\Sigma D}{\Sigma L} \quad \text{...(7.2 b)}$$

The sign of ΣD and ΣL will thus define the quadrant in which the closing error lies. The *relative error of closure,* the term sometimes used, is $= \dfrac{\text{Error of closure}}{\text{Perimeter of traverse}} = \dfrac{e}{p} = \dfrac{1}{p/e}$ ...(7.3)

Adjustment of the Angular Error: Before calculating latitudes and departures, the traverse angles should be adjusted to satisfy geometric conditions. In a closed traverse, the sum of interior angles should be equal to $(2N - 4)$ right angles (or the algebraic sum of deflection angles should be 360°). If the angles are measured with the same degree of precision, the error in the sum of angles may be distributed *equally* to each angle of the traverse. If the angular error is small, it may be arbitrarily distributed among two or three angles.

Adjustment of Bearings: In a closed traverse in which bearings are observed, the closing error in bearing may be determined by comparing the two bearings of the last line as observed at the first and last stations of traverse. Let e be the closing error in bearing of last line of a closed traverse having N sides. We get

$$\text{Correction for first line} = \frac{e}{N}$$

$$\text{Correction for second line} = \frac{2e}{N}$$

$$\text{Correction for third line} = \frac{3e}{N}$$

$$\text{Correction for last line} = \frac{Ne}{N} = e.$$

7.11 BALANCING THE TRAVERSE

The term '*balancing*' is generally applied to the operation of applying corrections to latitudes and departures so that $\Sigma L = 0$ and $\Sigma D = 0$. This applies only when the survey forms a closed polygon. The following are common methods of adjusting a traverse:

1. Bowditch's method
2. Transit method
3. Graphical method
4. Axis method.

1. **Bowditch's Method.** The basis of this method is on the assumptions that the errors in linear measurements are proportional to $\sqrt{l}$ and that the errors in angular measurements are inversely proportional to $\sqrt{l}$ where l is the length of a line. The *Bowditch's rule,* also termed as the *compass rule,* is mostly used to balance a traverse where linear and angular measurements are of equal precision. The total error in latitude, and in the departure is distributed in proportion to the lengths of the sides.

The Bowditich Rule is:

Correction to latitude (or departure) of any side =

$$\textit{Total error in latitude (or departure)} \times \frac{\textit{Length of that side}}{\textit{Perimeter of traverse}}$$

Thus, if
C_L = correction to latitude of any side
C_D = correction to departure of any side
ΣL = total error in latitude
ΣD = total error in departure
Σl = length of the perimeter
l = length of any side

We have
$$C_L = \Sigma L \, . \, \frac{l}{\Sigma l} \quad \text{and} \quad C_D = \Sigma D \, . \, \frac{l}{\Sigma l} \qquad ...(7.4)$$

2. **Transit Method:** The *transit rule* may be employed where angular measurements are more precise that the linear measurements. According to this rule, the total error in latitudes and in departures is distributed in proportion to the latitudes and departures of the sides. It is claimed that the angles are less affected by corrections applied by transit method than by those by Bowditch's method.

The transit rule is:

Correction to latitude (or departure) of any side =

$$\textit{Total error in latitude (or departure)} \times \frac{\textit{Latitude (or departure) of that line}}{\textit{Arithmetic sum of latitudes (or departures)}}$$

Thus, if L = latitude of any line

D = departure of any line

L_T = arithmetic sum of latitudes

D_T = arithmetic sum of departure

We have,

$$C_L = \Sigma L \,.\, \frac{L}{L_T} \quad \text{and} \quad C_D = \Sigma D \,.\, \frac{D}{D_T}. \qquad ...(7.5)$$

3. **Graphical Method:** For rough survey, such as a compass traverse, the Bowditch rule may be applied graphically without doing theoretical calculations. Thus, according to the graphical method, it is not necessary to calculate latitudes and departures etc. However, before plotting the traverse directly from the field notes, the angles or bearings may be adjusted to satisfy the geometric conditions of the traverse.

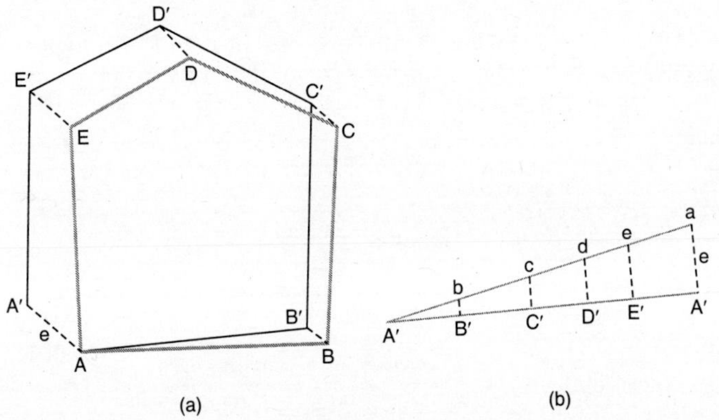

Fig. 7.12

Thus, in Fig. 7.12 (*a*), polygon *AB'C'D'E'A* represents an unbalanced traverse having a closing error equal to *A'A* since the first point *A* and the last point *A'* are not coinciding. The total closing error *AA'* is distributed linearly to all the sides in proportion to their length by a graphical construction shown in Fig. 7.12 (*b*). In Fig. 7.12 (*b*), *AB'*, *B'C'*, *C'D'* etc. represent the length of the sides of the traverse either to the same scale as that of Fig. 7.12 (*a*) or to a reduced scale. The ordinate *aA'* is made equal to the closing error *A'A* [of Fig. 7.12 (*a*)]. By constructing similar triangles, the corresponding errors *bB'*, *cC'*, *dD'*, *eE'* are found. In Fig. 7.12 (*a*), lines *E'E, D'D, C'C, B'B* are drawn parallel to the closing error *A'A* and made equal to *eE'*, *dD'*, *cC'*, *bB'* respectively. The polygon *ABCDE* so obtained represents the adjusted traverse. It should be remembered that the ordinates bB', cC', dD', eE', aA', of Fig. 7.12(*b*) represent the corresponding errors in *magnitude* only but not in *direction*.

4. **The Axis Method:** This method is adopted when the angles are measured very accurately, the corrections being applied to lengths only. Thus, only directions of the line are unchanged and the general shape of the diagram is preserved. To adjust the closing error aa_1 of a traverse $abcdefa_1$ (Fig. 7.13) following procedure is adopted:

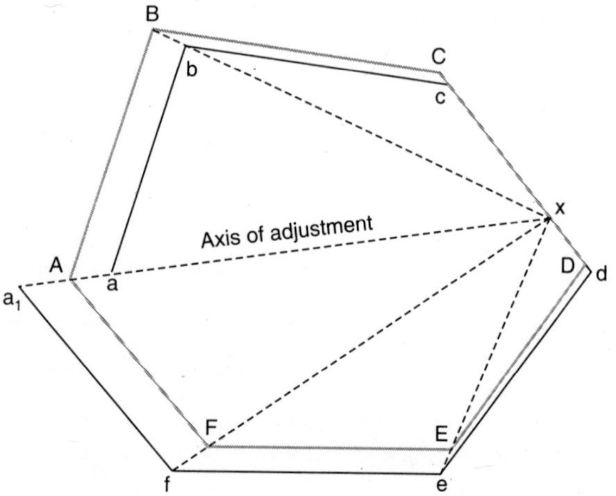

Fig. 7.13 Axis Method of Balancing Traverse.

1. Join a_1a and produce it to cut the side cd in x. The line a_1x is known as the *axis of adjustment*. The axis divides the traverse in two parts *i.e.* $a\ b\ c\ x$ and $a_1\ f\ e\ d\ x$
2. Bisect a_1a in A.
3. Join xb, xe and xf.
4. Through A, draw a line AB parallel to ab cutting $x\ b$ produced in B. Through B, draw a line BC parallel to $b\ c$ cutting $x\ c$ produced in C.
5. Similarly, through A, draw AF parallel to a_1f to cut $x\ f$ in F. Through F, draw FE parallel to $f\ e$ to cut $x\ e$ in E. Through E, draw ED parallel to $e\ d$ to cut $x\ d$ in D.

ABCDEF (thick lines) is the adjusted traverse.

Now, $$AB = \frac{Ax}{ax} . ab$$

$$\text{Correction to } ab = AB - ab = \frac{Ax}{ax} . ab - ab = \frac{Aa}{ax} .ab$$

$$= \frac{1}{2} \cdot \frac{a_1\,a}{a\,x} \cdot ab. = \frac{\frac{1}{2}\text{ closing error}}{a\,x} \cdot ab \qquad \text{...(1) ...(7.6 } a)$$

$$\text{Similarly, correction to } a_1\,f = \frac{1}{2} \cdot \frac{a_1\,a}{a_1\,x} . a_1 f = \frac{\frac{1}{2}\text{ closing error}}{a_1\,x} \cdot a_1\,f \qquad \text{...(2) ...(7.6 } b)$$

Taking $ax \approx a_1\,x$ = length of axis, we get the general rule:

$$\textit{Correction to any length = that length} \times \frac{\frac{1}{2}\textbf{ closing error}}{\textbf{Length of axis}} \qquad \text{...(7.6)}$$

The axis $a_1\,x$ should be so chosen that it divides the figure approximately into two equal parts. However, in some cases the closing error aa_1 may not cut the traverse or may cut it in very unequal parts. In such cases, the closing error is transferred to some other point. Thus, in Fig. 7.14, aa_1 when produced does not cut the traverse in two parts. Through a, a line ae' is drawn *parallel* and *equal* to $a_1\,e$. Through e', a line $e'\,d'$ is drawn *parallel* and equal to ed. A new unadjusted traverse $dcbae'd'$ is thus obtained in which the closing error dd' cuts the opposite side in x, thus dividing the traverse in two approximately equal parts. The adjustment is made with reference to the axis $d\,x$. The figure $ABCDE$ shown by thick lines represents the adjusted figure.

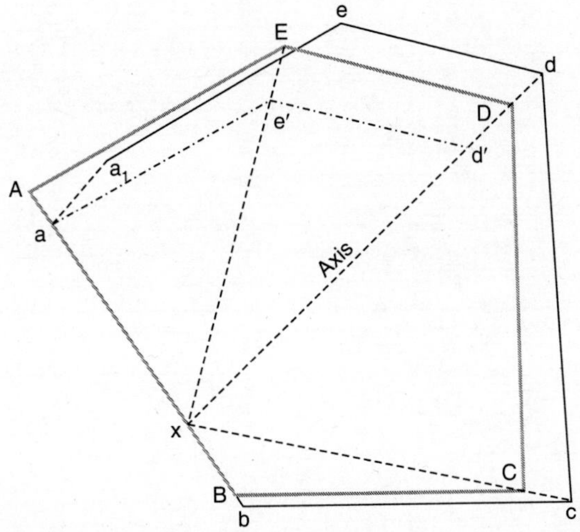

Fig. 7.14

Gales Traverse Table

Traverse computations are usually done in a tabular form, a more common form being *Gales Traverse Table* (Table 7.5). For complete traverse computations, the following steps are usually necessary:

(*i*) Adjust the interior angles to satisfy the geometrical conditions, *i.e.* sum of interior angles to be equal to $(2N - 4)$ right angles and exterior angles $(2N + 4)$ right angles.

In the case of a compass traverse, the bearings are adjusted for local attraction, if any.

(*ii*) Starting with observed bearings of one line, calculate the bearings of all other lines. Reduce all bearings to quadrantal system.

(*iii*) Calculate the consecutive co-ordinates (*i.e.* latitudes and departures).

(*iv*) Calculate ΣL and ΣD.

(*v*) Apply necessary corrections to the latitudes and departures of the lines so that $\Sigma L = 0$ and $\Sigma D = 0$. The corrections may be applied either by transit rule or by compass rule depending upon the type of traverse.

(*vi*) Using the corrected consecutive co-ordinates, calculate the independent co-ordinates to the points so that they are all positive, the whole of the traverse thus lying in the North East quadrant.

Table 7.5 illustrates completely the procedure.

Computation of Area of a Closed Traverse: (See Chapter 12).

TABLE 7.5 Gales Traverse Table

(1)	(2)	(3)	(4)	(5)	(6)	(7)	(8) Consecutive Co-ordinates				(9) Correction				(10) Corrected Consecutive Co-ordinates				(11) Independent Co-ordinates	
Line and length (m)	*Point*	*Angle*	*Correction*	*Corrected Angle*	*W.C.B.*	*R.B*	*N*	*S*	*E*	*W*	*N*	*S*	*E*	*W*	*N*	*S*	*E*	*W*	*N*	*E*
	A	95°24′	– 6′	95°18′			107.97		3.77		+0.26		–0.01		108.23		3.76		200*	100*
AB 250					86°42′	N86°42′E														
	B	88°42′	–6′	88°36′			14.39		249.57		+0.03		–0.71		14.42		248.86		214.42	348.85
BC 123					178°06′	S1°54′E														
	C	88°12′	–6′	88°06′				122.94	4.12			–0.29	–0.01			122.65	4.11		91.77	352.96
CD 256					270°00′	N90°W														
	D	88°06′	–6′	88°00′			0			256.00	+ 0			+0.73	0			256.73	91.77	96.23
DA 108					2°00′	N2°E														
SUM		360°24′	–24′	360°00′			122.36	122.94	257.46	256.00	+0.29	–0.29	–0.73	+0.73	122.65	122.65	256.73	256.73		
							–0.58		+1.46		+0.58		–1.46		0		0			

Remarks

1. All interior angles were measured. Bearing of *AB* was also observed.
2. The corrections to latitudes and departures have been applied by transit rule.

Note: *assumed, #observed

7.12 DEGREE OF ACCURACY IN TRAVERSING

Since both linear and angular measurements are made in traversing, the degree of accuracy depends upon the types of instruments used for linear and angular measurements and also upon the purpose and extent of survey. The degree of precision used in angular measurements must be consistent with the degree of precision used in linear measurements so that the effect of error in angular measurement will be the same as that of error in linear measurements. To get a relation between precision of angular and linear measurements consider Fig. 7.15.

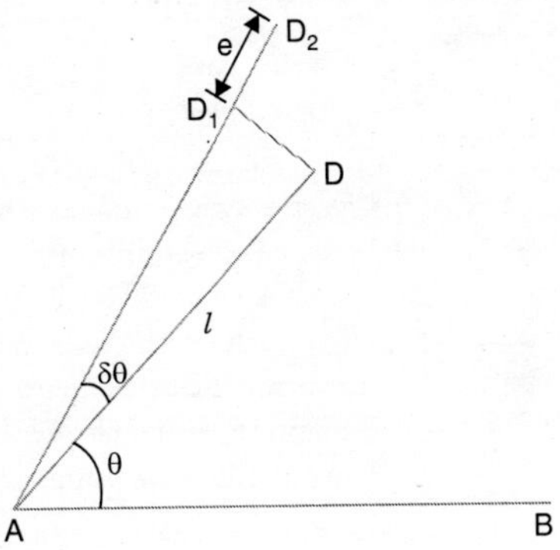

Fig. 7.15

Let D be the correct position of point with respect to a point A such that $AD = l$ and $\angle BAD = \theta$. In the field measurement, let $\delta\theta$ be the error in the angular measurement and e be the error in the linear measurement so that D_2 is the faulty location of the point D as obtained from the field measurements.

Now, displacement of D due to angular error $(\delta\theta) = DD_1 = l \tan \delta\theta$.

Displacement of D due to linear error $= D_1 D_2 = e$.

In order to have same degree of precision in the two measurements

$$l \tan \delta\theta = e \quad \text{or} \quad \delta\theta = \tan^{-1} \frac{e}{l}. \qquad ...(7.7)$$

In the above expression, $\frac{e}{l}$ is the *linear error* expressed as a ratio. If the precision of linear measurements is $\frac{1}{5000}$, the allowable angular error $= \delta\theta = \tan^{-1} \frac{1}{5000} = 41''$. Thus, the angle should be measured to the nearest $40''$. Similarly, if the allowable angular error is $20''$, the corresponding precision of linear measurement will be $= \tan 20'' = \frac{1}{10,300}$ (or about 1 metre in 1 kilometre).

The angular error of closure in theodolite traversing is generally expressed as equal to $C\sqrt{N}$, where the value of C may vary from $15''$ to $1'$ and N is the number of angles measured. The degree of precision in angular and linear measurement in theodolite traverse under different circumstances are given in Table 7.6 below:

TABLE 7.6 Errors of Closure

Type of Traverse	*Angular error of closure*	*Total linear error of closure*
(1) First order traverse for horizontal control	$6''\sqrt{N}$	1 in 25,000
(2) Second order traverse for horizontal control and for important and accurate surveys	$15''\sqrt{N}$	1 in 10,000
(3) Third order traverse for surveys of important boundaries etc.	$30''\sqrt{N}$	1 in 5,000
(4) Minor theodolite traverse for detailing	$1'\sqrt{N}$	1 in 300
(5) Compass traverse	$15'\sqrt{N}$	1 in 300 to 1 in 600

PROBLEMS

1. Distinguish clearly between:
   (*a*) Chain surveying and traverse surveying.
   (*b*) Closed traverse and open traverse.
   (*c*) Loose needle method and fast needle method
2. Discuss various methods of theodolite traversing.
3. Explain clearly, with the help of illustrations, how a traverse is balanced.
4. What is error of closure? How is it balanced graphically?
5. (*a*) Explain the principle of surveying (traversing) with the compass.
   (*b*) Plot the following compass traverse and adjust it for closing error if any:

Line	***Length (m)***	***Bearing***
AB	130	S 88° E
BC	158	S 6° E
CD	145	S 40° W
DE	308	N 81° W
EA	337	N 48° E

   Scale of plotting 1 cm = 20 m.
6. Describe 'Fast needle method' of theodolite traversing.

CHAPTER 8

OMITTED MEASUREMENTS

8.1 CONSECUTIVE CO-ORDINATES: LATITUDE AND DEPARTURE

There are two principal methods of plotting a traverse survey: (1) the angle and distance method, and (2) the co-ordinate method. If the length and bearing of a survey line are known, it can be represented on plan by two rectangular co-ordinates. The axes of the co-ordinates are the North and South line, and the East and West line. The *latitude* of survey line may be defined as its co-ordinate length measured parallel to the meridian direction. The *departure* of the survey line may be defined as its co-ordinate length measured at right angles to the meridian direction. The latitude (L) of the line is *positive* when measured northward (or upward) and is termed as *northing*. The latitude is *negative* when measured southward (or downward) and is termed as *southing*. Similarly, the departure (D) of the line is *positive* when measured eastward and is termed as *easting*. The departure is negative when measured westward and is termed as *westing*.

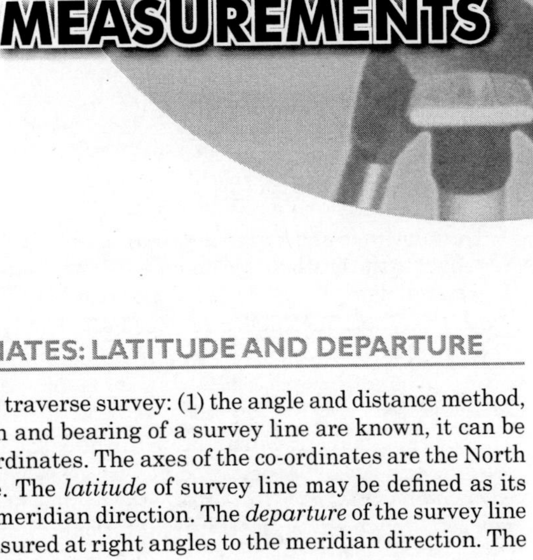

Fig. 8.1 Latitude and Departure

Thus, in Fig. 8.1, the latitude and departure of the line OA of length l_1 and reduced bearing θ_1 is given by

$$L_1 = + l_1 \cos \theta_1$$

and $$D_1 = + l_1 \sin \theta_1 \qquad ...(8.1)$$

To calculate the latitudes and departures of the traverse lines, therefore, it is first essential to reduce the bearing in the quadrantal system. The sign of latitude and departures will depend upon the reduced bearing of line.

The following table gives the signs of latitudes and departures.

TABLE 8.1

W.C.B.	*R.B. and Quadrant*	*Sign of*	
		Latitude	*Departure*
0° to 90°	N θ E ; I	+	+
90° to 180°	S θ E ; II	–	+
180° to 270°	S θ W ; III	–	–
270° to 360°	N θ W ; IV	+	–

Thus, latitude and departure co-ordinates of any point with reference to the preceding point are equal to the latitude and departure of the line joining the preceding point to the point under consideration. Such co-ordinates are also known as *consecutive co-ordinates* or *dependent co-ordinates. Table 7.3 illustrates systematic method of calculating the latitudes and departures of a traverse.*

Independent Co-ordinates

The co-ordinates of traverse station can be calculated with respect to a common origin. The *total latitude and departure* of any point with respect to a common origin are known as *independent co-ordinates* or *total co-ordinates* of the point. The two reference axes in this case may be chosen to pass through any of the traverse stations but generally a most westerly station is chosen for this purpose. The independent co-ordinates of any point may be obtained by adding algebraically the latitudes and the departure of the lines between the point and the origin.

Thus, total latitude (or departure) of end point of a traverse = total latitudes (or departures) of first point of traverse plus the algebraic sum of all the latitudes (or departures).

8.2 OMITTED MEASUREMENTS

In order to have a check on field work and in order to balance a traverse, the length and direction of each line is generally measured in the field. There are times, however, when it is not possible to take all measurements due to obstacles or because of some over-sight. Such *omitted measurements* or missing quantities can be calculated by latitudes and departures provided the quantities required are not more than two. In such cases, there can be no check on the field work nor can the survey be balanced. All errors propagated throughout the survey are thrown into the computed values of the missing quantities.

Since for a closed traverse, ΣL and ΣD are zero, we have

$$\Sigma L = l_1 \cos \theta_1 + l_2 \cos \theta_2 + l_3 \cos \theta_3 + ... = 0 \qquad ...(1)...(8.2\ a)$$

and

$$\Sigma D = l_1 \sin \theta_1 + l_2 \sin \theta_2 + l_3 \sin \theta_3 + ... = 0 \qquad ...(2)...(8.2\ b)$$

where l_1, l_2, l_3, ...etc, are the lengths of the lines and $\theta_1, \theta_2, \theta_3$,...etc. their reduced bearings. With the help of the above two equations, the two missing quantities can be calculated. Table 8.2 below gives the trigonometric relations of a line with its latitude and departure, and may be used for the computation of omitted measurements.

TABLE 8.2

Given	*Required*	*Formula*
l, θ	L	$L = l \cos \theta$
l, θ	D	$D = l \sin \theta$
L, D	$\tan \theta$	$\tan \theta = D/L$
L, θ	l	$l = L \sec \theta$
D, θ	l	$l = D \operatorname{cosec} \theta$
L, l	$\cos \theta$	$\cos \theta = L/l$
D, l	$\sin \theta$	$\sin \theta = D/l$
L, D	l	$l = \sqrt{L^2 + D^2}$

There are four general cases of omitted measurements:

I. (*a*) When the *bearing* of one side is omitted.

(*b*) When the *length* of one side is omitted.

(*c*) When the *bearing* and *length* of one side is omitted.

II. When the *length* of one side and the *bearing* of another side are omitted.

III. When the *lengths* of two sides are omitted.

IV. When the *bearings* of two sides are omitted.

In case (I), only one side is affected. In case II, III and IV two sides are affected both of which may either be adjacent or may be away.

8.3 CASE I: BEARING, OR LENGTH, OR BEARING AND LENGTH OF ONE SIDE OMITTED

In Fig. 8.2, let it be required to calculate either bearing or length or both bearing and length of the line *EA*. Calculate $\Sigma L'$ and $\Sigma D'$ of the four known sides *AB, BC, CD* and *DE*. Then

$$\Sigma L = \text{Latitude of } EA + \Sigma L' = 0$$

or $$\text{Latitude of } EA = -\Sigma L'$$

Similarly, $$\Sigma D = \text{Departure of } EA + \Sigma D' = 0$$

or $$\text{Departure } EA = -\Sigma D'$$

Knowing latitude and departure of *EA*, its length and bearing can be calculated by proper trigonometrical relations.

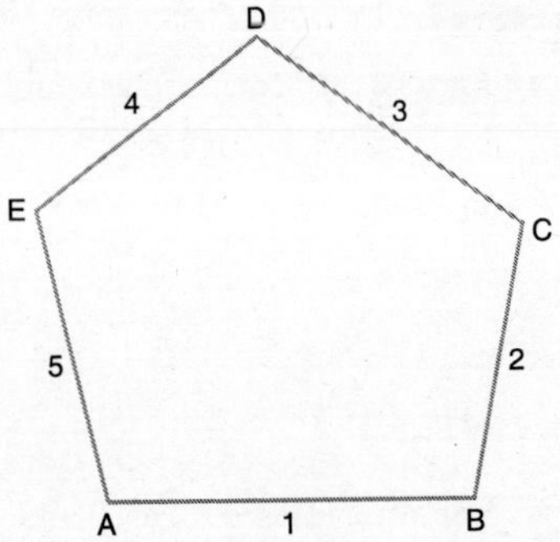

Fig. 8.2

8.4 CASE II: LENGTH OF ONE SIDE AND BEARING OF ANOTHER SIDE OMITTED

In Fig. 8.3, let the length of *DE* and bearing of *EA* be omitted. Join *DA* which becomes the closing line of the traverse *ABCD* in which all the quantites are known. Thus the length and bearing of *DA* can be calculated as in case *I*.

In ΔADE, the length of sides DA and EA are known, and angle ADE (α) is known. The angle β and the length DE can be calculated as under:

$$\sin \beta = \frac{DA}{EA} \sin \alpha \quad \text{...(8.3 a)}$$

$$\gamma = 180° - (\beta + \alpha) \quad \text{...(8.3 b)}$$

$$DE = EA \frac{\sin \gamma}{\sin \alpha} = DA \frac{\sin \gamma}{\sin \beta} \quad \text{...(8.3 c)}$$

Knowing γ, the bearing of EA can be calculated.

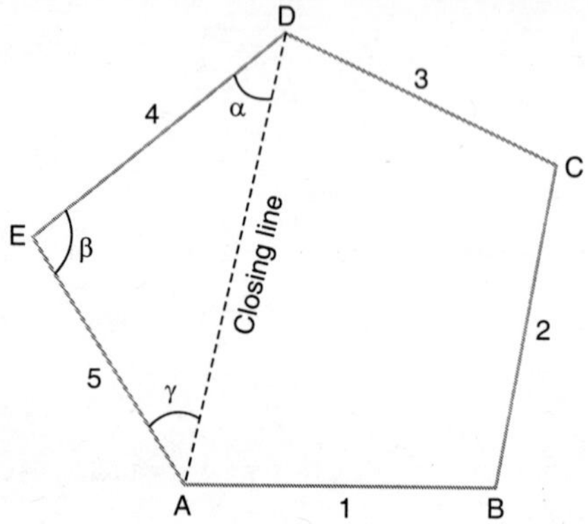

Fig. 8.3

8.5 CASE III: LENGTHS OF TWO SIDES OMITTED

In Fig. 8.3, let the length of DE and EA be omitted. The length and bearing of the closing line DA can be calculated as in the previous case. The angles α, β and γ can then be computed by the known bearing. The lengths of DE and EA can be computed by the solution of the triangle DEA.

Thus, $$DE = \frac{\sin \gamma}{\sin \beta} DA \quad \text{...(8.4)}$$

and $$EA = \frac{\sin \alpha}{\sin \beta} DA \quad \text{...(8.5)}$$

8.6 CASE IV: BEARING OF TWO SIDES OMITTED

In Fig. 8.3 let bearing of DE and EA be omitted. The length and bearing of the closing line DA can be calculated. The angles can be computed as under:

The area $$\Delta = \sqrt{s(s-a)(s-d)(s-e)} \quad \text{...(1)}$$

where s = half the perimeter $= \frac{1}{2}(a + d + e)$; $a = ED$, $e = AD$ and $d = AE$

Also, $$\Delta = \frac{1}{2} ad \sin \beta = \frac{1}{2} de \sin \gamma = \frac{1}{2} ae \sin \alpha \quad \text{...(2)}$$

Equating (1) and (2), α, β and γ can be calculated. Knowing the bearing of DA and the angles α, β, γ, the bearings of DE and EA can be calculated.

Alternatively, the angles can be found by the following expressions, specially helpful when an angle is an obtuse angle:

$$\tan \frac{\beta}{2} = \sqrt{\frac{(s-a)(s-d)}{s(s-e)}}; \quad \tan \frac{\gamma}{2} = \sqrt{\frac{(s-d)(s-e)}{s(s-a)}}; \quad \tan \frac{\alpha}{2} = \sqrt{\frac{(s-a)(s-e)}{s(s-d)}}$$

8.7 CASE II, III, IV: WHEN THE AFFECTED SIDES ARE NOT ADJACENT

If the affected sides are not adjacent, one of these can be shifted and brought adjacent to the other by drawing lines parallel to the given lines. Thus in Fig. 8.4 let BC and EF be the affected sides. In order to bring them adjacent, choose the starting point (say B) of any one affected side (say BC) and draw line BD' parallel and equal to CD. Through D', draw line $D'E'$ parallel and equal to ED. Thus evidently, $EE' = BC$ and FE and BC are brought adjacent. The line $E'F$ becomes the *closing line* of the traverse $ABD'E'F$. The length and bearing of $E'F$ can be calculated. Rest of the procedure for calculating the omitted measurements is the same as explained earlier.

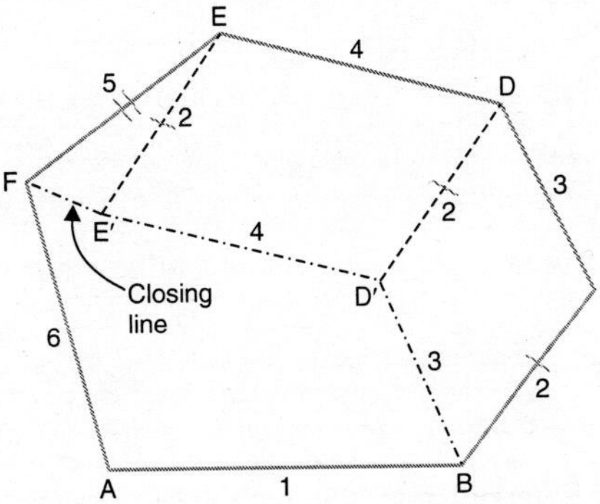

Fig. 8.4

Analytical Solution

(*a*) Case II: When the length of one line and bearing of another line missing

Let θ_1 and l_3 be missing.

Then $\quad l_1 \sin \theta_1 + l_2 \sin \theta_2 + l_3 \sin \theta_3 + \ldots\ldots l_n \sin \theta_n = 0$

or $\quad l_1 \sin \theta_1 + l_3 \sin \theta_3 = - l_2 \sin \theta_2 - \ldots\ldots - l_n \sin \theta_n = P$ (say) ...(1)

and $\quad l_1 \cos \theta_1 + l_2 \cos \theta_2 + l_3 \cos \theta_3 + \ldots\ldots l_n \cos \theta_n = 0$

or $\quad l_1 \cos \theta_1 + l_3 \cos \theta_3 = - l_2 \cos \theta_2 - \ldots\ldots - l_n \cos \theta_n = Q$ (say) ...(2)

Squaring and adding (1) and (2), we get

$$l_1^2 = P^2 + Q^2 + l_3^2 - 2\, l_3\, (P \sin \theta_3 + Q \cos \theta_3)$$

or
$$l_3^2 - 2\, l_3\, (P \sin \theta_3 + Q \cos \theta_3) + (P^2 + Q^2 - l_1^2) = 0 \quad \ldots(8.6)$$

This is a quadratic equation in terms of l_3 from which l_3 can be obtained.

Knowing l_3, θ_1 may be obtained from (1). Thus

$$\theta_1 = \sin^{-1} \left[\frac{P - l_3 \sin \theta_3}{l_1} \right] \quad \ldots(8.7)$$

(*b*) Case III: When the lengths of two lines are missing

Let l_1 and l_3 be missing.

Then $\quad l_1 \sin \theta_1 + l_2 \sin \theta_2 + l_3 \sin \theta_3 + \ldots\ldots l_n \sin \theta_n = 0$

or $\quad l_1 \sin \theta_1 + l_3 \sin \theta_3 = - l_2 \sin \theta_2 - \ldots\ldots - l_n \sin \theta_n = P$ (say) ...(1)

and $\quad l_1 \cos \theta_1 + l_2 \cos \theta_2 + l_3 \cos \theta_3 + \ldots\ldots l_n \cos \theta_n = 0$

or $\quad l_1 \cos \theta_1 + l_3 \cos \theta_3 = - l_2 \cos \theta_2 - \ldots\ldots - l_n \cos \theta_n = Q$ (say) ...(2)

In equations (1) and (2), only l_1 and l_3 are unknowns. Hence these can be found by solution of the two simultaneous equations.

(c) **Case IV: When the bearings of the two sides are missing**

Let θ_1 and θ_3 be missing.

Then, as before, $l_1 \sin \theta_1 + l_3 \sin \theta_3 = P$ ...(1)

and $l_1 \cos \theta_1 + l_3 \cos \theta_3 = Q$ ...(2)

From (1), $l_1 \sin \theta_1 = P - l_3 \sin \theta_3$ ...(3)

and from (2), $l_1 \cos \theta_1 = Q - l_3 \cos \theta_3$ ...(4)

Squaring (3) and (4) and adding $l_1^2 = P^2 + Q^2 + l_3^2 - 2\, l_3 (P \sin \theta_3 + Q \cos \theta_3)$

or $$\frac{P}{\sqrt{P^2+Q^2}} \sin \theta_3 + \frac{Q}{\sqrt{P^2+Q^2}} \cos \theta_3 = \frac{P^2+Q^2+l_3^2-l_1^2}{2l_3\sqrt{P^2+Q^2}} = k \text{ (say)}$$

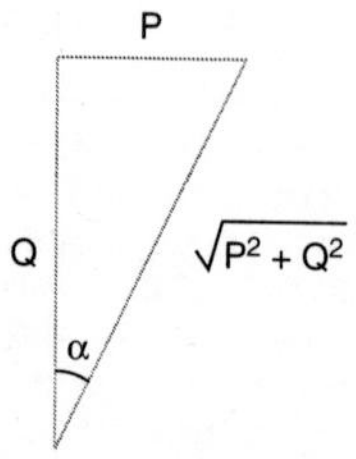

Fig. 8.5

Referring to Fig. 8.5 and taking $\tan \alpha = \dfrac{P}{Q}$, we have

$$\frac{P}{\sqrt{P^2+Q^2}} = \sin \alpha \quad \text{and} \quad \frac{Q}{\sqrt{P^2+Q^2}} = \cos \alpha$$

$\therefore$ $\sin \alpha \sin \theta_3 + \cos \alpha \cos \theta_3 = k$ ...(8.8)

or $\cos (\theta_3 - \alpha) = k$

From which $$\theta_3 = \alpha + \cos^{-1} k = \tan^{-1} \frac{P}{Q} + \cos^{-1} k \quad \text{...(8.9)}$$

Knowing θ_3, θ_1 is computed from Eq. (3):

Thus $$\theta_1 = \sin^{-1} \left[\frac{P - l_3 \sin \theta_3}{l} \right] \quad \text{...(8.10)}$$

See example 8.8 for illustration.

Example 8.1: *The Table below gives the lengths and bearings of the lines of a traverse ABCDE, the length and bearing of EA having been omitted. Calculate the length and bearing of the line EA.*

Line	*Length (m)*	*Bearing*
AB	*204.0*	*87° 30′*
BC	*226.0*	*20° 20′*
CD	*187.0*	*280° 0′*
DE	*192.0*	*210° 3′*
EA	?	?

Solution: Fig. 8.2 shows the traverse *ABCDE* in which *EA* is the closing line of the polygon. Knowing the length and bearing of the lines *AB*, *BC*, *CD* and *DE*, their latitudes and departures can be calculated and tabulated as under:

Line	*Latitude*		*Departure*	
	+	–	+	–
AB	8.90		203.80	–
BC	211.92		78.52	
CD	32.48			184.16
DE		165.44		97.44
Sum	253.30	165.44	282.32	281.60
	$\Sigma L' = +87.86$		$\Sigma D' = +0.72$	

∴ Latitude of $EA = -\Sigma L' = -87.86$ m and Departure $EA = -\Sigma D' = -0.72$ m.

Since the latitude of *EA* is negative and departure is also negative it lies in the SW quadrant. The reduced bearing (θ) of *EA* is given by

$$\tan\theta = \frac{\text{Departure}}{\text{Latitude}} = \frac{0.72}{87.86} \quad \text{or} \quad \theta = 0°\ 28'$$

∴ Bearing of EA = S 0° 28′ W = **180° 28′**

Also
$$\text{Length of } EA = \frac{\text{Latitude}}{\cos\theta} = \frac{87.86}{\cos 0°\ 28'} = \mathbf{87.85}$$

Example 8.2: *A closed traverse was conducted round an obstacle and the following observations were made. Work out the missing quantities:*

Side	*Length (m)*	*Azimuth*
AB	*500*	*98° 30′*
BC	*620*	*30° 20′*
CD	*468*	*298° 30′*
DE	*?*	*230° 0′*
EA	*?*	*150° 10′*

Solution: The affected sides are adjacent. Fig. 8.3 shows the traverse *ABCDE* in which *DA* is closing line of the polygon *ABCD*. The latitude and departure of the closing line *DA* can be calculated. The calculations are shown in the tabular form below:

	Latitude		*Departure*	
Line	+	–	+	–
AB		73.91	494.50	
BC	535.11		313.11	
CD	223.45			411.29
Sum	758.56	73.91	807.61	411.29
	$\Sigma L' = +684.55$		$\Sigma D' = +396.32$	

∴ Latitude of $DA = -\Sigma L' = -684.55$ and Departure $DA = -\Sigma D' = -396.32$

Since both latitude and departure are negative, the line DA is in third (*i.e.* SW) quadrant, the reduced bearing (θ) of DA is given by

$$\tan\theta = \frac{D}{L} = \frac{396.32}{684.55} \qquad \therefore \theta = 30^\circ\ 4'$$

∴ Bearing of DA = S 30° 4′ W = 210° 4′

Length of $DA = l = L \sec\theta = 684.55 \sec 30^\circ\ 4' = 791.01$ m

From Fig. 8.3, $\angle ADE = \alpha = 230^\circ\ 0' - 210^\circ\ 4' = 19^\circ\ 56'$

$\angle DEA = \beta = 150^\circ 10' - (230^\circ - 180^\circ) = 100^\circ\ 10'$

$\angle DAE = \gamma = 210^\circ\ 4' - 150^\circ\ 10' = 59^\circ\ 54'$

(*Check:* $\alpha + \beta + \gamma = 19^\circ\ 56' + 100^\circ\ 10' + 59^\circ\ 54' = 180^\circ$)

From triangle ADE, using the sine rule, we get

$$DE = DA\,\frac{\sin\gamma}{\sin\beta} = 791.01\,\frac{\sin 59^\circ\ 54'}{\sin 100^\circ\ 10'} = \mathbf{695.27\ m}$$

$$EA = DA\,\frac{\sin\alpha}{\sin\beta} = 791.01\,\frac{\sin 19^\circ\ 56'}{\sin 100^\circ\ 10'} = \mathbf{273.99\ m}$$

Example 8.3: *A four sided traverse ABCD, has the following lengths and bearings:*

Side	*Length (m)*	*Bearing*
AB	500	Roughly East
BC	245	178°
CD	Not obtained	270°
DA	216	10°

Find the exact bearing of the side AB.

Solution: Fig. 8.6 shows the traverse $ABCD$ in continous lines. The affected sides AB and CD are not adjacent. To bring AB adjacent to CD. draw a line AC' equal and parallel to BC. The line CC' is thus equal and parallel to BA and is adjacent to CD. The line $C'D$ becomes the closing line of the traverse $AC'D$. The latitude and departure of $C'D$ can be calculated as usual. The calculations are shown below:

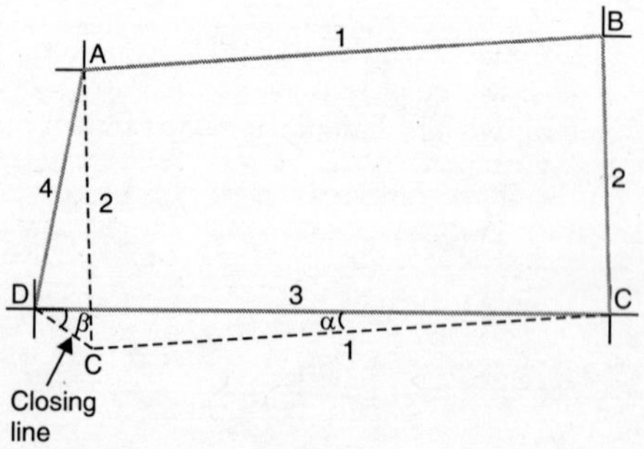

Fig. 8.6

Line	*Latitude*	*Departure*
DA	+ 212.71	+ 37.51
AC' (BC)	– 239.64	+ 50.94
Sum	– 26.93	+ 88.45

∴ Latitude of $C'D = +26.93$ and Departure of $C'D = -88.45$

The bearing (θ) of $C'D$ is given by

$$\tan\theta = \frac{D}{L} = \frac{88.45}{26.93} \quad \text{or} \quad \theta = 73° 4'$$

∴ Bearing of $C'D$ = N 73° 4′ W = 286° 56′

Angle $\beta = (270° - 180°) - 73° 4' = 16° 56'$

Length of $C'D = L \sec\theta = 26.93 \sec 73° 4' = 92.47$ m

From triangle CDC', we get

$$\frac{CC'}{\sin\beta} = \frac{DC'}{\sin\alpha} \text{ or } \sin\alpha = \frac{DC'}{CC'}\sin\beta = \frac{92.47}{500}\sin 16° 56'$$

∴ $\alpha = 3° 55'$.

∴ Bearing of BA = bearing of CC' = 270° – 3° 55′ – 266° 5′

∴ Bearing of AB = 266° 5′ – 180° = **86° 5′.**

Example 8.4: *A straight tunnel is to be run between two points A and B, whose co-ordinates are given below:*

Point	*Co-ordinates*	
	N	*E*
A	*0*	*0*
B	*3014*	*256*
C	*1764*	*1398*

It is desired to sink a shaft at D, the middle point of AB, but it is impossible to measure along AB directly, so D is to be fixed from C, a third known point.

Calculate: (*a*) *The co-ordinates of D.*

(*b*) *The length and bearing of CD.*

(*c*) *The angle ACD, given that the bearing of AC is 38° 24′ E of N.*

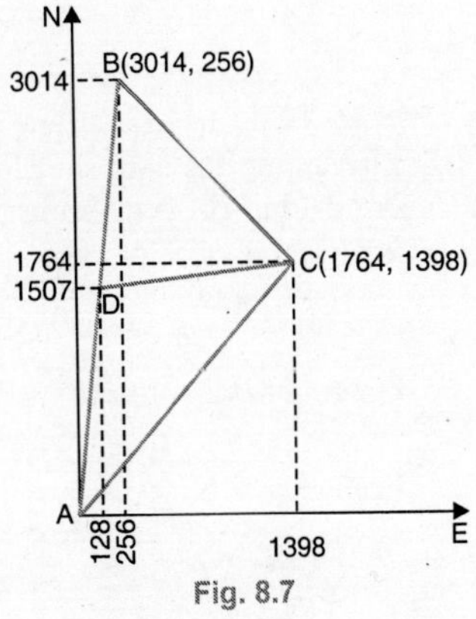

Fig. 8.7

Solution: Fig. 8.7 shows the points *A, B, C* and *D*. The co-ordinate axes have been chosen to pass through point *A*.

(*a*) Since *D* is midway between *A* and *B*, its co-ordinates will be **1507** and **128**.

(*b*) From Fig. 8.7.

Latitude of $AD = 1507$

Departure of $AD = 128$

Latitude of $AC = 1764$

Departure of $AC = 1398$

$\therefore$ Latitude of $DC = 1764 - 1507 = 257$

and Departure of $DC = 1398 - 128 = 1270$

Hence, Latitude of $CD = -\,257$

Departure of $CD = -\,1270$

Since both latitude and departure are negative, line *CD* is in the third quadrant with respect to the co-ordinate axes passing through *C*.

The bearing (θ) of *CD* is given by

$$\tan\theta = \frac{D}{L} = \frac{1270}{257} \qquad \therefore\ \theta = 78°\ 34'$$

$\therefore$ Bearing of CD = S 78° 34′ W = **258° 34′**

$$\text{Length of } CD = \sqrt{(1270)^2 + (257)^2} = \mathbf{1295.7.}$$

(*c*) $\angle ACD$ = Bearing of CD – Bearing of CA

$= 258°\ 34' - 38°\ 24' - 180° = \mathbf{40°\ 10'}.$

Example 8.5: *A and B are two stations of a location traverse, their total co-ordinates in metres being:*

	Total latitude	*Total Departure*
A	*34,321*	*7,509*
B	*33,670*	*9,652*

A straight reach of railway is to run from C, roughly south of A, to D, invisible from C and roughly north of B, the offsets perpendicular to the railway being AC=130 m and BD = 72 m. Calculate the bearing of CD.

Solution: (Fig. 8.8)

Co-ordinates of *A*, referred to *B*:

Latitude $= 34321 - 33670 = +\,651$

Departure $= 7509 - 9652 = -\,2143.$

Since latitude is positive and departure is negative, line *BA* is in the NW quadrant. The bearing (θ) of *BA* is given by

$$\tan\theta = \frac{D}{L} = \frac{2143}{651} \text{ or } \theta = 73°\ 6'.$$

$$\text{Length of } BA = \sqrt{(651)^2 + (2143)^2} = 2238 \text{ m}$$

From Fig. 8.8, $\dfrac{OB}{OA} = \dfrac{BD}{AC}$

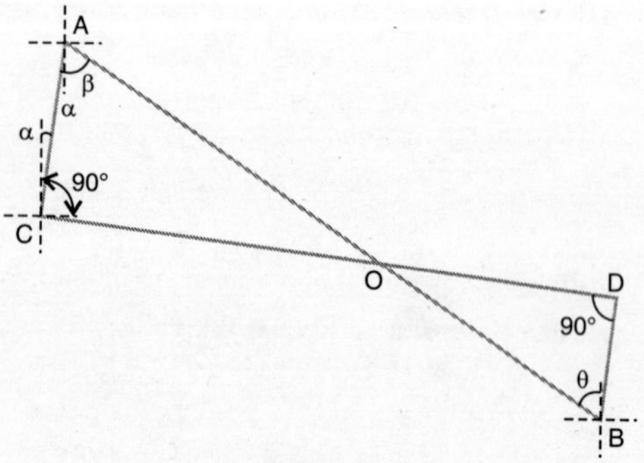

Fig. 8.8

or
$$\frac{OB+OA}{OA} = \frac{BD+AC}{AC}$$

$$\frac{AB}{OA} = \frac{130+72}{130} = \frac{202}{130}$$

$\therefore$
$$OA = \frac{130}{202} \times 2238 = 1440 \text{ m}$$

$$\cos\beta = \frac{AC}{AO} = \frac{130}{1440}$$

$\therefore$ $\beta = 84° \; 49'.5$

$\therefore$ $\alpha = 84° \; 49'.5 - 73° \; 6' = 11° \; 43'.5$

$\therefore$ Bearing of $CD = 90° + \alpha = 90° + 11° \; 43'.5 =$ **101° 43′.5.**

Example 8.6: *A and B are two of the stations used in setting out construction lines of harbour works. The total latitude and departure of A, referred to the origin of the system, are respectively + 542.7 and – 331.2, and those of B are + 713.0 and + 587.8 m (north latitude and east departure being reckoned as positive). A point C is fixed by measuring from A a distance of 432 m on a bearing of 346° 14', and from it a line CD, 1152 m in length is set out parallel to AB. It is required to check the position of D by a sight from B. Calculate the bearing of D from B.*

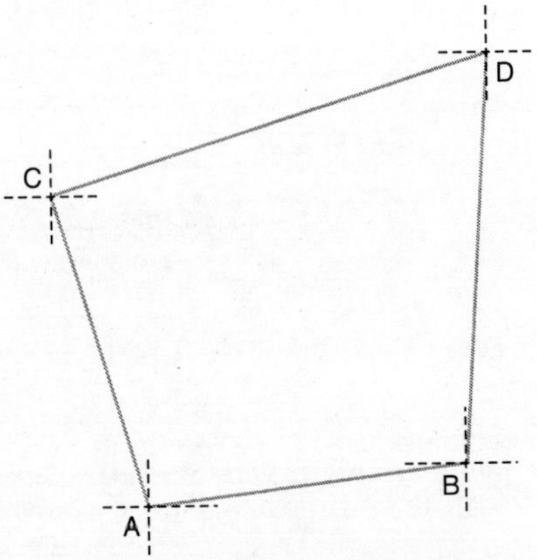

Fig. 8.9

Solution: (Fig. 8.9)

Latitude of $A = +\,542.7$

Departure of $A = -\,331.2$

Latitude of $B = +\,713$

Departure of $B = +\,587.8$

$\therefore$ Latitude of B referred to A

$= 713 - 542.7 = 170.3$

Departure of B referred to A

$= 587.8 - (-\,331.2) = 919.0$

Since both latitude and departure of B are positive, line AB lies in NE quadrant. The bearing (θ) of AB is given by

$$\tan\theta = \frac{D}{L} = \frac{919.0}{170.3}$$

$\therefore$ $\theta = 79° \; 30' =$ bearing of AB

$\therefore$ Bearing of AD = Bearing of AB = N 79° 30′ E

Fig. 8.9 shows the traverse $ABCD$ in which length and bearing of the line BD are not known. Table below shows the calculations for the latitudes and departures of the lines:

Line	*Length*	*Bearing*	*Latitude*		*Departure*	
			+	–	+	–
AB			170.3		919.0	
DC	1152	S 79° 30′ W		211.8		1132.0
CA	432	S 13° 46′ E		419.9	102.8	
		Sum	170.3	631.7	1021.8	1132.0
			$\Sigma L' = -461.4$		$\Sigma D' = -110.2$	

∴ Latitude of $BD = -\Sigma L' = +461.4$ and Departure of $BD = -\Sigma D' = +110.2$

Since both latitude and departure of *BD* are positive, it lies in NE quadrant, its bearing (α) being given by

$$\tan \alpha = \frac{D}{L} = \frac{110.2}{461.4} \quad \text{or} \quad \alpha = 13° 26'$$

∴ Bearing of *BD* = **N 13° 26′ E.**

Example 8.7: *For the following traverse, compute the length CD, so that A, D and E may be in one straight line.*

Line	*Length in metres*	*Bearing*
AB	*110*	*83° 12′*
BC	*165*	*30° 42′*
CD	–	*346° 6′*
DE	*212*	*16° 18′*

Solution: Fig. 8.10 shows the traverse *ABCD* in which *A, D* and *E* are in the same line. Treating *CA* as the closing line of the traverse *ABC,* its length and bearing can be calculated as under:

Line	*Latitude*	*Departure*
AB	+ 13.03	+ 109.23
BC	+ 141.88	+ 84.24
	$\Sigma L' = +154.91$	$\Sigma D' = +193.47$

∴ Latitude of $CA = -\Sigma L' = -154.81$

Departure of $CA = -\Sigma D' = -193.47$.

Since both latitude and departure are negative, it lies in *SW* quadrant, its bearing being given by

$$\tan \theta = \frac{D}{L} = \frac{193.47}{154.91}$$

$$\theta = \text{S } 51° 18' \text{ W}$$

$$\text{Length of } CA = L \sec \theta$$

$$= 154.91 \sec 51°18' = 247.92 \text{ m.}$$

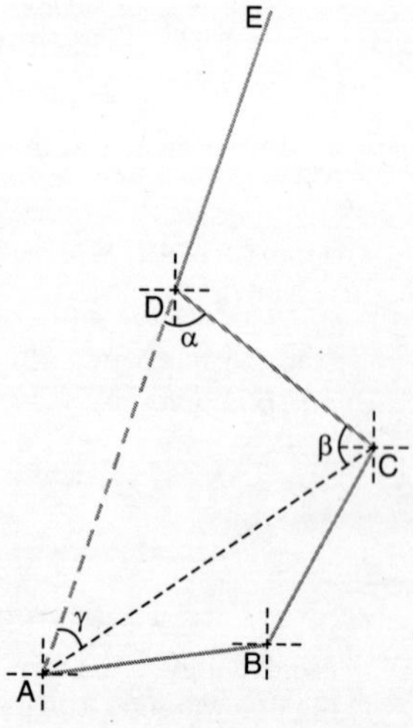

Fig. 8.10

Now, since *A, D* and *E* are in the same line, bearing of *AD* = bearing of *DE* = 16° 18′.

From ΔACD, $\alpha = (16° 18' + 180°) - (346° 6' - 180°) = 30° 12'$

$$\beta = 346° 6' - (51° 18' + 180°) = 114° 48'$$

$$\gamma = 51° 18' - 16° 18' = 35°$$

Check: $\alpha + \beta + \gamma = 30° 12' + 114° 48° + 35° = 180°$

Again, $$\frac{CD}{\sin\gamma} = \frac{CA}{\sin\alpha}$$

$\therefore$ $$CD = CA\,\frac{\sin\gamma}{\sin\alpha} = 247.92\,\frac{\sin 35°}{\sin 30°12'} = \mathbf{282.70\ m.}$$

Example 8.8: *In a closed traverse ABCD, with the data given below, the bearings of lines AB and CD are missing.*

Line	*Length (m)*	*Bearing*
AB	*160.00 m*	–
BC	*280.00 m*	*102°36′*
CD	*120.00 m*	–
DA	*320.00 m*	*270°00′*

Determine the missing bearings.

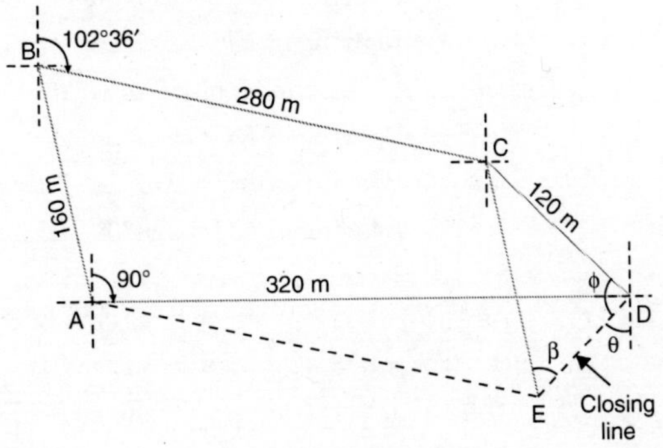

Fig. 8.11

Solution: Here is a case in which the bearings of two lines, not adjacent, are missing.

Method 1: Semi-analytical solution

Refer Fig. 8.11. The lines *AB* and *CD*, whose bearings θ_1 and θ_3 are missing, are not adjacent to each other. From *C*, draw *CE* parallel and equal to *BA*. Naturally, *AE* will be parallel and equal to *BC*. In the triangle *ADE*, the lengths and bearings of *AD* and *AE* are known. Hence the length and bearing of the closing line *DE* can be found, with the computations in the tabular form below.

Line	*Length (m)*	*Bearing*	*Latitude*	*Departure*
EA	280.0	282°36′	+ 61.08	– 273.26
AD	320.0	90°00′	0.00	+ 320.00
		Σ	+ 61.08	+ 46.74
DE			– 61.08	–46.74

Hence line *DE* falls in the third quadrant.

$$\theta - \tan^{-1}\frac{D}{L} = \tan^{-1}\left(\frac{-46.74}{-61.08}\right) = 37°.424$$

$\therefore$ Bearing of $DE = 180° + 37°.424 = 217°.424 = 217°26'$

$$\text{Length } DE = \sqrt{D^2 + L^2} = \sqrt{(46.74)^2 + (61.08)^2} = 76.91\text{ m}$$

Now for triangle *CDE*, $s = \frac{1}{2}(160 + 120 + 76.91) = 178.455$

Now $$\tan \varphi/2 = \sqrt{\frac{(178.455-76.91)(178.455-120)}{178.455(178.455-160)}}$$

$\therefore$ $\varphi/2 = 53°.319$ or $\varphi = 106°.64 = 106°38'$

Also $$\beta = \sin^{-1}\left(\frac{DC \sin \varphi}{EC}\right) = \sin^{-1}\left(\frac{120 \sin 106°.64}{160}\right) = 45°.94 = 45°56'$$

Let us now calculate the bearings of *AB* and *CD*.

Bearing of *DE* = 217°26′

$\therefore$ Bearing of *ED* = 217°26′ – 180° = 37° 26′

$\therefore$ Bearing of *AB* = Bearing of *EC* = 37°26′ + (360° – 45°56′) = **351°30′**

Again, Bearing of *DE* = 217°26′

$\therefore$ Bearing of *DC* = 217°26′ + 106°38′ = 324°04′

$\therefore$ Bearing of *CD* = 324°04′ – 180° = **144°04′**

Method 2: Analytical solution

Let us use suffixes 1, 2, 3, 4, for lines *AB*, *BC*, *CD* and *DA* respectively.

Since $\Sigma D = 0$, $l_1 \sin \theta_1 + l_3 \sin \theta_3 = - l_2 \sin \theta_2 - l_4 \sin \theta_4$

or $160 \sin \theta_1 + 120 \sin \theta_3 = - 280 \sin 102°36' - 320 \sin 270°00'$

or $160 \sin \theta_1 + 120 \sin \theta_3 = - 273.26 + 320 = 46.74 = P$ ...(1)

Also, $\Sigma L = 0$, $\therefore$ $l_1 \cos \theta_1 + l_3 \cos \theta_3 = - l_2 \cos \theta_2 - l_4 \cos \theta_4$

or $160 \cos \theta_1 + 120 \cos \theta_3 = - 280 \cos 102° 36' - 320 \cos 270° 00'$

or $160 \cos \theta_1 + 120 \cos \theta_3 = 61.08 + 0.0 = 61.08 = Q$ ...(2)

From (1) $160 \sin \theta_1 = 46.74 - 120 \sin \theta_3$ ...(3)

From (2) $160 \cos \theta_1 = 61.08 - 120 \cos \theta_3$ ...(4)

Squaring and adding

$$(160)^2 = (46.74)^2 + (61.08)^2 + (120)^2 - 2 \times 120 (46.74 \sin \theta_3 + 61.08 \cos \theta_3)$$

or $$46.74 \sin \theta_3 + 61.08 \cos \theta_3 = \frac{(46.74)^2 + (61.08)^2 + (120)^2 - (160)^2}{240} = -22.019$$

$\therefore$ $$\cos(\theta_3 - \alpha) = \frac{-22.019}{\sqrt{(46.74)^2 + (61.08)^2}} = -0.28629$$

$\therefore$ $\theta_3 - \alpha = 106°.64 = 106°38'$

But $$\tan \alpha = \frac{46.74}{61.08} \text{ or } \alpha = 37°.424 = 37°26'$$

$\therefore$ $\theta_3 = 106° 38' + 37° 26' = \mathbf{144°04'}$

Also, from (3), $$\theta_1 = \sin^{-1}\left[\frac{46.74 - 120 \sin 144°04'}{160}\right] = -8° 30'$$

or $\theta_1 = 360° - 8° 30' = \mathbf{351° 30'}$

Example 8.9: *The following measurements were made in a closed traverse ABCD:*
$AB = 97.54$ m; $CD = 170.69$ m; $AD = 248.47$ m; $\angle DAB = 70°45'$; $\angle ADC = 39°15'$
Calculate the missing measurements.

Solution: Taking the W.C.B. of line $AB = 90°$, the traverse is shown in Fig. 8.12.

Let the angle $ABC = \theta$ and the length BC be l.

Sum of interior angles = 360°.

$$\therefore \quad \angle BCD = 360° - (70°45' + 39°15' + \theta)$$
$$= 250° - \theta$$

Bearing of $AB = 90°$

Bearing of $AD = 90° - 70°45' = 19°\ 15'$

Bearing of $DA = 19°15' + 180° = 199°15'$

Bearing of $DC = 199°15' - 39°15' = 160°0'$

Bearing of $CD = 160° + 180° = 340°$

Bearing of $CB = 340° - (250° - \theta) = 90° + \theta$

Bearing of $BC = (90° + \theta) - 180° = \theta - 90°$

Bearing of $BA = [(\theta - 90°) - \theta] + 360° = 270°$

Bearing of $AB = 270° - 180° = 90°$ (check)

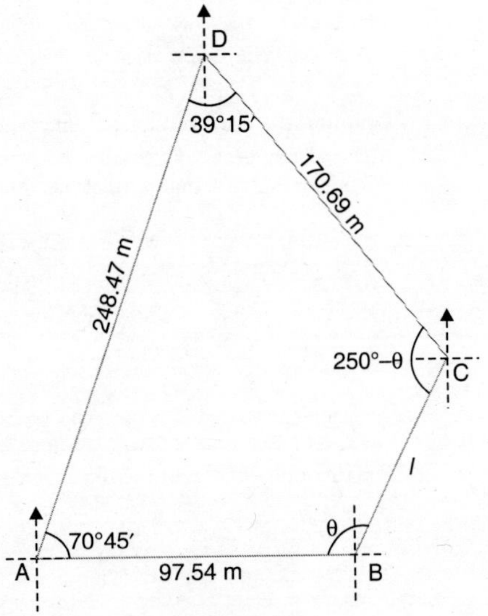

Fig. 8.12

Now, for the whole traverse, $\Sigma L = 0$ and $\Sigma D = 0$

$\therefore \quad 97.54 \cos 90° + l \cos (\theta - 90°) + 170.69 \cos 340° + 248.47 \cos 199°\ 15' = 0$

or $\quad 0 + l \sin \theta + 160.40 - 234.58 = 0$

or $\quad l \sin \theta = 74.18$ ...(1)

and $\quad 97.54 \sin 90° + l \sin (\theta - 90°) + 170.69 \sin 340° + 248.47 \sin 199°\ 15' = 0$

or $\quad 97.54 - l \cos \theta - 58.38 - 81.92 = 0$

or $\quad l \cos \theta = -42.76$ ...(2)

From (1) and (2) $\quad l = \sqrt{(74.18)^2 + (42.76)^2} = \mathbf{85.62\ m}$

Also, $\quad \theta = \tan^{-1} \dfrac{-74.18}{42.76} = 119°58'$

$\therefore \quad \angle ABC = 119°58'$

$\therefore \quad \angle BCD = 250° - \theta = 250° - (119°\ 58') = \mathbf{130°02'}$

PROBLEMS

1. From a point C, it is required to set out a line CD parallel to a given line AB, such that ABD is a right angle. C and D are not visible from A and B, and traversing is performed as follows:

Line	Length in m	Bearing
BA	–	360°0′
BE	51.7	290°57′
EF	61.4	352°6′
FC	39.3	263°57′

Compute the required length and bearing of *CD*.

2. A closed traverse was conducted round an obstacle and the following observations were made. Work out the missing quantities:

Side	Length in m	Azimuth
AB	–	33° 45′
BC	300	86° 23′
CD	–	169° 23′
DE	450	243° 54′
EA	268	317° 30′

3. For the following traverse, find the length of *DE* so that *A*, *E* and *F* may be in the same straight line:

Line	Length in metres	R.B.
AB	200	S 84° 30′ E
BC	100	N 75° 18′ E
CD	80	N 18° 45′ E
DE	–	N 29° 45′ E
EF	150	N 64° 10′ E

4. Two points *A* and *D* are connected by a traverse survey *ABCD* and the following records are obtained.

$$AB = 219 \text{ m} ; \quad BC = 170.5 \text{ m} ; CD = 245.75 \text{ m}$$

$$\text{Angle } ABC = 118° \ 15' ; \quad \text{Angle } BCD = 180° \ 40'$$

Assuming that *AB* is in meridian, determine:

1. The latitude and departure of *D* relative to *A*.
2. The length *AD*.
3. The angle *BAD*.

5. Find the co-ordinates of the point at which a line run from *A* on a bearing of N 10° E will cut the given traverse, and find the length of this line.

Line	Latitude		Departure	
	N	S	E	W
AB	1650		440	
BC	2875		120	
CD	3643			326
DE		1450		376
EA	0		0	

6. Surface and underground traverses have been run between two mine shafts *A* and *B*. The co-ordinates of *A* and *B* given by the underground traverse are 8560 N, 24860 W and 10451 N, 30624 W respectively. The surface traverse gave the co-ordinates of *B* as 10320 N and 30415 W, those of *A* being as before. Assuming the surface traverse to be correct, find the error in both bearing and distance of the line *AB*, as given by the underground traverse.
7. The following lengths and bearings were recorded in running a theodolite traverse in the counter clockwise direction, the length of *CD* and bearing of *DE* having been omitted.

Line	*Length in m*	*R.B.*
AB	281.4	S 69° 11′ E
BC	129.4	N 21° 49′ E
CD	?	N 19° 34′ W
DE	144.5	?
EA	168.7	S 74° 24′ W

Determine the length of *CD* and the bearing of *DE*.

ANSWERS

1. 74.82 m ; 180°
2. *AB* = 322.5 m ; *CD* = 305.7 m.
3. 66.5 m.
4. (*i*) 378.25 and 383.0 (*ii*) 538.5 m (*iii*) 45° 21′
5. 1991 N ; 351 E ; 2021.
6. 0° 34′.5 ; 238.
7. 131 m ; S 46° 9′ W.

CHAPTER 9

LEVELLING

9.1 DEFINITIONS (Ref. Fig. 9.1)

Levelling: Levelling is a branch of surveying the object of which is: (1) to find the elevations of given points with respect to a given or assumed datum, and (2) to establish points at a given elevation or at different elevations with respect to a given or assumed datum. The first operation is required to enable the works to be designed while the second operation is required in the setting out of all kinds of engineering works. Levelling deals with measurements in a vertical plane.

Level Surface: A level surface is defined as a curved surface which at each point is perpendicular to the direction of gravity at the point. The surface of a still water is a truly level surface. Any surface parallel to the mean spheroidal surface of the earth is, therefore, a level surface.

Level Line: A level line is a line lying in a level surface. It is, therefore, normal to the plumb line at all points.

Horizontal Plane: Horizontal plane through a point is a plane tangential to the level surface at that point. It is, therefore, perpendicular to the plumb line through the point.

Horizontal Line: It is straight line tangential to the level line at a point. It is also perpendicular to the plumb line.

Vertical Line: It is a line normal to the level line at a point. It is commonly considered to be the line defined by a plumb line.

Datum: Datum is any surface to which elevations are referred. The *mean sea level* affords a convenient datum world over, and elevations are commonly given as so much above or below sea level. It is often more convenient, however, to assume some other datum, specially if only the relative elevations of points are required.

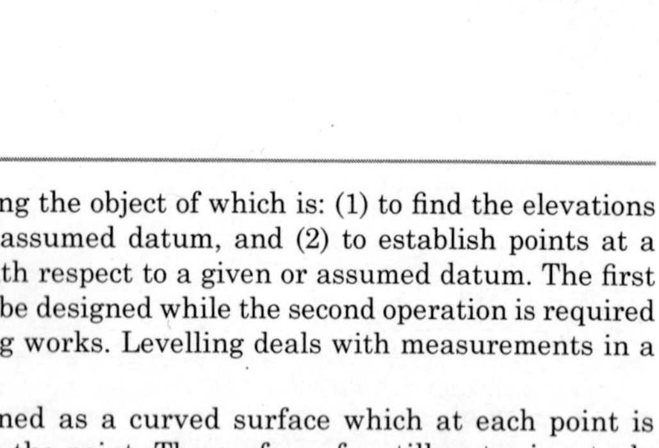

Fig. 9.1

Elevation: The elevation of a point on or near the surface of the earth is its vertical distance above or below an arbitrarily assumed level surface or datum. The *difference in elevation* between two points is the vertical distance between the two level surfaces in which the two points lie.

Vertical Angle: Vertical angle is an angle between two intersecting lines in a vertical plane. Generally, one of these lines is horizontal.

Mean Sea Level: Mean sea level is the average height of the sea for all stages of the tides. At any particular place it is derived by averaging the hourly tide heights over a long period of 19 years.

Bench Mark: Bench Mark is a relatively permanent point of reference whose elevation with respect to some assumed datum is known. It is used either as a starting point for levelling or as a point upon which to close as a check.

9.2 METHODS OF LEVELLING

Three principal methods are used for determining difference in elevation, namely, *barometric* levelling, *trigonometric* levelling and *spirit* levelling.

Barometric levelling: Barometric levelling makes use of the phenomenon that difference in elevation between two points is proportional to the difference in atmospheric pressures at these points. A barometer, therefore, may be used and the readings observed at different points would yield a measure of the relative elevations of those points.

At a given point, the atmospheric pressure does not remain constant in the course of the day, even in the course of an hour. The method is, therefore, relatively inaccurate and is little used in surveying work except on reconnaissance or exploratory surveys.

Trigonometric Levelling (Indirect levelling): Trigonometric or indirect levelling is the process of levelling in which the elevations of points are computed from the vertical angles and horizontal distances measured in the field, just as the length of any side in any triangle can be computed from proper trigonometric relations. In a modified form called *stadia levelling,* commonly used in mapping, both the difference in elevation and the horizontal distance between the points are directly computed from the measured vertical angles and staff readings.

Spirit Levelling (Direct Levelling): It is that branch of levelling in which the vertical distances with respect to a horizontal line (perpendicular to the direction of gravity) may be used to determine the relative difference in elevation between two adjacent points. A horizontal plane of sight tangent to level surface at any point is readily established by means of a spirit level or a level vial. In spirit levelling, a spirit level and a sighting device (telescope) are combined and vertical distances are measured by observing on graduated rods placed on the points. The method is also known as *direct levelling.* It is the most precise method of determining elevations and the one most commonly used by engineers.

9.3 LEVELLING INSTRUMENTS

The instruments commonly used in *direct levelling* are:

- A level
- A levelling staff.

Level

The purpose of a level is to provide a horizontal line of sight. Essentially, a level consists of the following four parts:

(*a*) A *telescope* to provide line of sight

(*b*) A *level tube* to make the line of sight horizontal

(*c*) A *levelling head* (tribrach and trivet stage) to bring the bubble in its centre of run

(*d*) A *tripod* to support the instrument.

There are the following chief types of levels:

(*i*) Dumpy level
(*ii*) Wye (or *Y*) level
(*iii*) Reversible level
(*iv*) Tilting level.

(i) Dumpy Level: The dumpy level originally designed by Gravatt, consists of a telescope tube firmly secured in two collars fixed by adjusting screws to the stage carried by the vertical spindle.

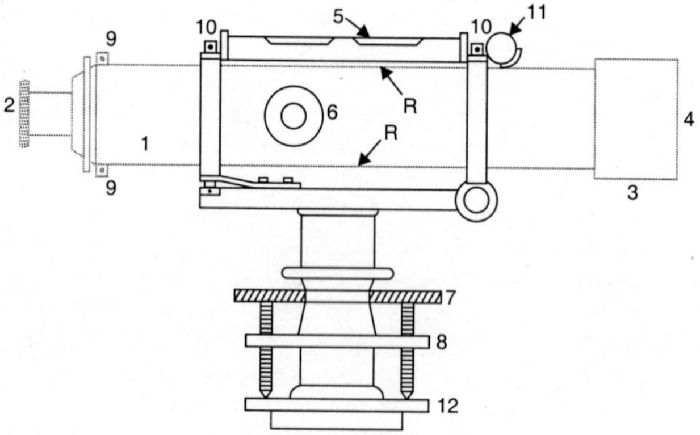

Fig. 9.2 Dumpy Level.

1. Telescope
2. Eye-Piece
3. Ray Shade
4. Objective End
5. Longitudinal Bubble
6. Focusing Screws
7. Foot Screws
8. Upper Parallel Plate (Tribrach)
9. Diaphragm Adjusting Screws
10. Bubble Tube Adjusting Screws
11. Transverse Bubble Tube
12. Foot Plate (Trivet Stage)

The modern form of dumpy level has the telescope tube and the vertical spindle cast in one piece and a long bubble tube is attached to the top of the telescope. This form is known as solid dumpy.

Fig. 9.2 shows the diagrammetic sketch of a dumpy level. Fig. 9.3 shows the section of a dumpy level. Figs. 9.4 and 9.5 show the photographs of dumpy levels manufactured by M/s Wild Heerbrugg and M/s Fennel Kessel respectively. Fig. 9.6 shows a dumpy level by M/s W.F. Stanley & Co. The name 'dumpy level' originated from the fact that formerly this level was equipped with an inverting eye-piece and hence was shorter than Wye level of the same magnifying power. However, modern forms generally have erecting eye-piece so that inverted image of the staff is visible in the field of view.

In some of the instruments, a clamp screw is provided to control the movement of the spindle about the vertical axis. For small or precise movement, a slow motion screw (or tangent screw) is also provided.

The levelling head generally consists of two parallel plates with either three-foot screws or four-foot screws. The upper plate is known as *tribrach* and the lower plate is known as *trivet* which can be screwed on to a tripod.

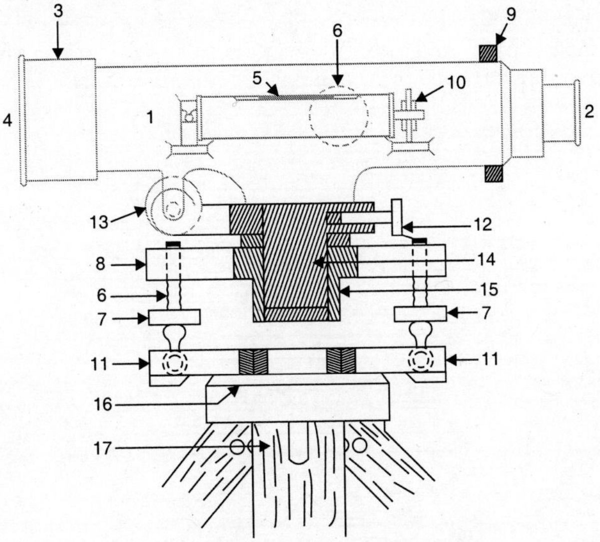

Fig. 9.3 Sectional View of a Dumpy Level.

1. Telescope
2. Eye-Piece
3. Ray Shade
4. Objective End
5. Longitudinal Bubble
6. Focusing Screw
7. Foot Screws
8. Upper Parallel Plate (Tribrach)
9. Diaphragm Adjusting Screws
10. Bubble Tube Adjusting Screws
11. Foot Plate (Trivet Stage)
12. Clamp Screw
13. Slow Motion Screw
14. Inner Cone
15. Outer Cone
16. Tripod Head
17. Tripod

The advantages of the dumpy level over the Wye level are:

(*i*) Simpler construction with fewer movable parts.

(*ii*) Fewer adjustments to be made.

(*iii*) Longer life of the adjustments.

(ii) Wye Level: The essential difference between the dumpy level and the Wye level is that in the former case the telescope is fixed to the spindle while in the Wye level, the telescope is carried in two vertical 'Wye' supports. The Wye support consists of curved clips. If the clips are raised, the telescope can be rotated in the Wyes, or removed and turned end for end. When the

clips are fastened, the telescope is held from turning about its axis by a lug on one of the clips. The bubble tube may be attached either to the telescope or to the stage carrying the wyes. In the former case, the bubble tube must be of reversible type.

Fig. 9.7 shows the essential features of Y-level. The levelling head may be similar to that of a dumpy level. In some cases, the instrument is fitted with a clamp and fine motion tangent screw for controlled movement in the horizontal plane. Fig. 9.8 shows the photograph of a Wye level by Fennel Kessel.

The Wye level has an advantage over the dumpy level in the fact that the adjustments can be tested with greater rapidity and ease. However, the adjustments do not have longer life and are disturbed more frequently due to large number of movable parts.

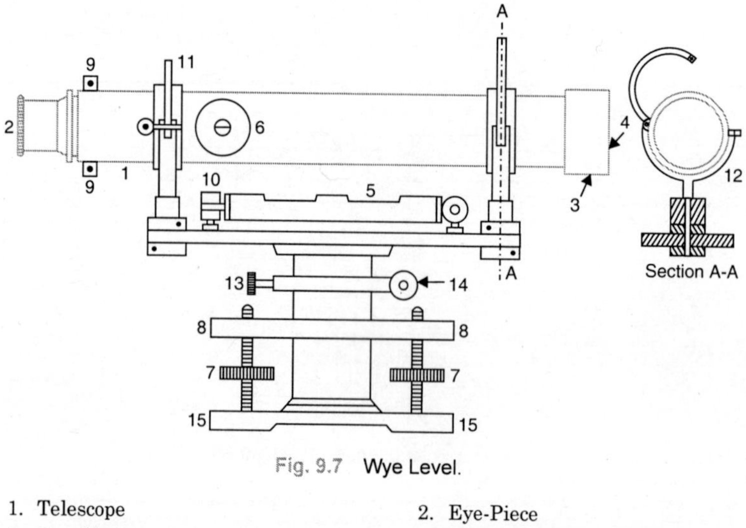

Fig. 9.7 Wye Level.

1. Telescope
2. Eye-Piece
3. Ray Shade
4. Objective End
5. Bubble Tube
6. Focusing Screw
7. Foot Screw
8. Tribrach
9. Diaphragm Adjusting Screws
10. Bubble Tube Adjusting
11. Wye Clip
12. Clip Half Open
13. Clamp Screws
14. Tangent Screw
15. Trivet Stage

(iii) Reversible Level: A reversible level combines the features of both the dumpy level and the Wye level. The telescope is supported by two rigid sockets into which the telescope can be introduced from either end and then fixed in position by a screw. The sockets are rigidly connected to the spindle through a stage. Once the telescope is pushed into the sockets and the screw is tightened, the level acts as a dumpy level. For testing and making the adjustments, the screw is slackened and the telescope can be taken out and reversed end for end. The telescope can also be turned within the socket about the longitudinal axis.

(iv) Tilting Level: In the case of a dumpy level and a Wye level, the line of sight is perpendicular to the vertical axis. Once the instrument is levelled, the line of sight becomes horizontal and

the vertical axis becomes truly vertical, provided the instrument is in adjustment. In the case of tilting level, however, the line of sight can be tilted slightly without tilting the vertical axis. Thus, the line of sight and the vertical axis need not be exactly perpendicular to each other. This feature, therefore, helps in quick levelling. The instrument is levelled roughly by the three-foot screws with respect either to the bubble tube or to a small circular bubble, thus making the vertical axis *approximately vertical*. While taking the sight to a staff, the line of sight is made exactly horizontal by centring the bubble by means of a fine pitched tilting screw which tilts the telescope with respect to the vertical axis. It is, however, essential that the observer should have the view of the bubble tube while sighting the staff.

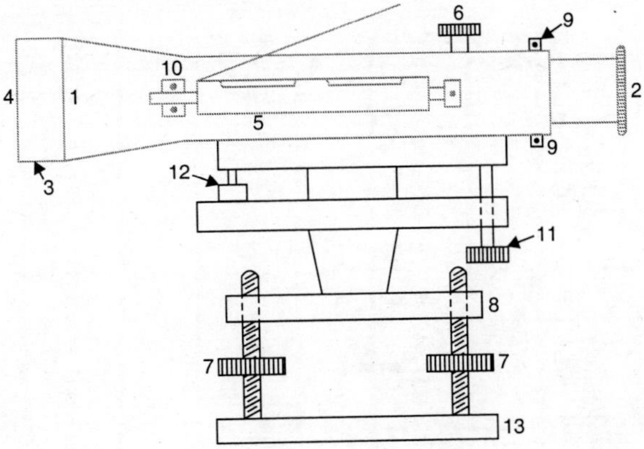

Fig. 9.9 Tilting Level.

1. Telescope
2. Eye-Piece
3. Ray Shade
4. Objective End
5. Level Tube
6. Focusing Screws
7. Foot Screws
8. Tribrach
9. Diaphragm Adjusting Screws
10. Bubble Tube Fixing Screws
11. Tilting Screws
12. Spring Loaded Plunger
13. Trivet Stage

Fig. 9.9 shows the essential features of a tilting level. A tilting level is mainly designed for precise levelling work. It has the advantage that due to the tilting screw, levelling can be done much quicker. However, this advantage is not so apparent when it is required to take so many readings from one instrument setting. Fig. 9.10 shows the photograph of a tilting level by M/s Vickers Instruments Ltd.

9.4 LEVELLING STAFF

A levelling staff is a straight rectangular rod having graduations, the foot of the staff representing zero reading. The purpose of a level is to establish a horizontal line of sight. The purpose of the levelling staff is to determine the amount by which the station (*i.e.*, foot of the staff) is above or below the line of sight. Levelling staves may be divided into two classes: (*i*) Self-reading staff, and (*ii*) Target staff. A *Self-Reading Staff* is the one which can be read directly by the instrument man through the telescope. A *Target Staff*, on the other hand, contains a moving target against which the reading is taken by staff man.

(i) Self-Reading Staff: There are usually three forms of self-reading staff:

(*a*) Solid staff; (*b*) Folding staff; (*c*) Telescopic staff (Sopwith pattern). Figs. 9.11 (*a*) and (*b*) show the patterns of a solid staff in English units while (*c*) and (*d*) show that in metric unit. In the most common forms, the smallest division is of 0.01 ft. or 5 mm. However, some staves may have fine graduations upto 2 mm. The staff is generally made of well seasond wood having a length of 10 feet or 3 metres.

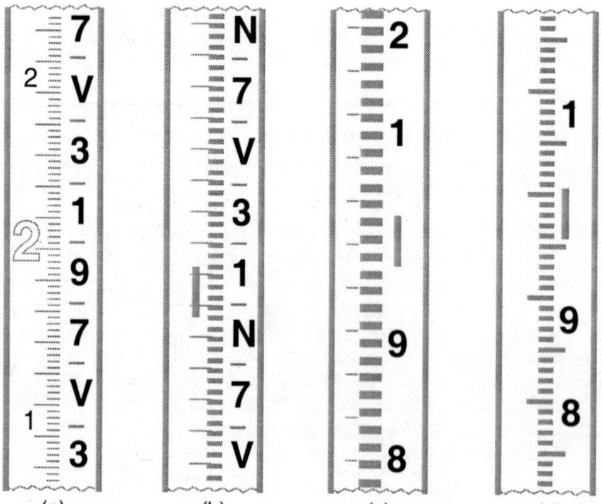

(a) (b) (c) (d)

English | Metric

Hundredths. Fiftieths. Centimetres. Half-Centimetres.

Fig. 9.11 (By Courtesy of M/s Vickers Instruments Ltd.)

Fig. 9.12 shows a sopwith pattern staff arranged in three telescopic lengths. When fully extended, it is usually of 14 ft (or 5 m) length. The 14 ft. staff has solid top length of 4′ 6″ sliding into the central box of 4′ 6″ length. The central box, in turn, slides into lower box of 5′ length. In the 5 m staff, the three corresponding lengths are usually 1.5 m, 1.5 m and 2 m.

Fig. 9.13 shows a folding staff usually 10 ft long having a hinge at the middle of its length. When not in use, the rod can be folded about the hinge so that it becomes convenient to carry it from one place to the other.

Since a self-reading staff is always seen through the telescope, all readings appear to be inverted. The readings are, therefore, taken from above downwards.

The levelling staves graduated in English units generally have whole number of feet marked in red to the left side of the staff (shown by hatched lines in Fig. 9.12). The odd lengths of the feet are marked in black to the right hand side. The top of these black graduations indicates the odd tenth while the bottom shows the even tenth. The hundredths of feet are indicated by alternate white and black spaces, the top of a black space indicating odd hundredths and top of a white space indicating even hundredths. Sometimes when the staff is near the instrument, the red mark of whole foot may not appear in the field of view. In that case, the staff is raised slowly until the red figure appears in the field of view, the red figure thus indicating the whole feet.

Fig. 9.12 Telescopic Staff

Fig. 9.13 Folding Staff

Fig. 9.14 Target Staff

(By Courtesy of M/s Vickers Instruments Ltd.)

Folding Levelling Staff in Metric Units

Fig. 9.15 (*a*) shows a 4 m folding type levelling staff (IS: 1779-1961). The staff comprises two 2 m thoroughly seasoned wooden pieces with the joint assembly. Each piece of the staff is made of one longitudinal strip without any joint. The width and thickness of staff is kept 75 mm and 18 mm respectively. The folding joint of the staff is made of the detachable type with a locking device at the back. The staff is jointed together in such a way that:

(*a*) the staff may be folded to 2 m length.

(*b*) the two pieces may be detached from one another, when required, to facilitate easy handling and manipulation with one piece, and

(*c*) when the two portions are locked together, the two pieces become rigid and straight.

A circular bubble, suitably cased, of 25 minute sensitivity is fitted at the back. The staff has fittings for a plummet to test and correct the back bubble. A brass is screwed on to the bottom brass cap. The staff has two folding handles with spring acting locking device or an ordinary locking device.

Each metre is subdivided into 200 divisions, the thickness of graduations being 5 mm. Fig. 9.15 (*b*) shows the details of graduations. Every decimetre length is figured with the corresponding numerals (the metre numeral is made in red and the decimetre numeral in black). The decimetre numeral is made continuous throughout the staff.

(ii) Target Staff: Fig. 9.14 shows a target staff having a sliding target equipped with vernier. The rod consists of two sliding lengths, the lower one of approx. 7 ft and the upper one of 6 ft. The rod is graduated in feet, tenths and hundredths, and the vernier of the target enables the readings to be taken upto a thousandth part of a foot. For readings below 7 ft the target is slided to the lower part while for readings above that, the target is fixed to the 7 ft mark of the upper length. For taking the reading, the level man directs the staff man to raise or lower the target till it is bisected by the line of sight. The staff holder then clamps the target and takes the reading. The upper part of the staff is graduated from top downwards. When higher readings have to be taken, the target is set at top (*i.e.* 7 ft mark) of the sliding length and the sliding length carrying the target is raised until the target is bisected by the line of sight. The reading is then on the back of the staff where a second vernier enables readings to be taken to a thousandth of a foot.

Relative Merits of Self-Reading and Target Staffs

(*i*) With the self-reading staff, readings can be taken quicker than with the target staff.

(*ii*) In the case of target staff, the duties of a target staff-man are as important as those of the leveller and demand the services of a trained man. In the case of a self-reading staff, on the other hand, ordinary man can hold the staff concentrating more on keeping the staff in plumb.

(*iii*) The reading with target staff can be taken with greater fineness. However, the refinement is usually more apparent than real as the target man may not be directed accurately to make the line of sight bisect the target.

9.5 THE SURVEYING TELESCOPE

The optical principles of the surveying telescope are based on the fact that all parallel rays of light reaching a convex lens are bent when they leave it in such a manner that they intersect at a common point, called the focus and that all the rays passing through another point called the optical centre pass through the geometrical centre of lens without bending.

The surveyor's telescope is an adaptation of Kepler's telescope which employs two convex lenses; the one nearest to the object is called the *objective* and the other near the eye is called the *eye-piece.*

The object glass provides a real inverted image in front of the eye-piece which, in turn, magnifies the image to produce an inverted virtual image. Fig. 9.16 shows the optical diagram of such a telescope.

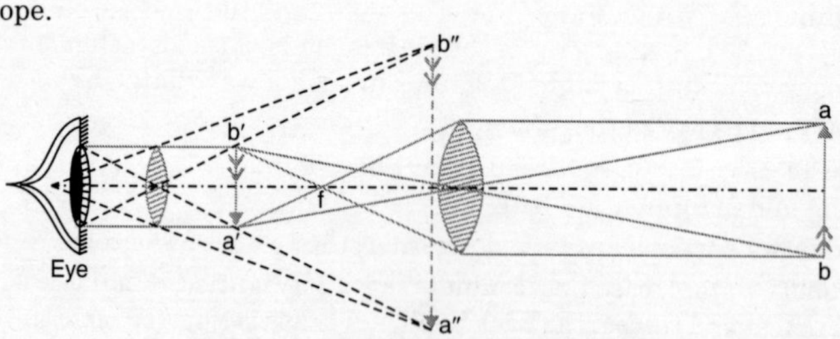

Fig. 9.16 Optical Diagram of Telescope.

The *line of sight* or *line of collimation* is a line which passes through the optical centre of the objective and the intersection of *cross hairs.* The axis` of the telescope is the line which passes through the optical centres of objective and eye-piece. The cross-hairs are placed in front of eye-piece and in the plane where the real inverted image is produced by the objective. Thus, the eye-piece magnifies the cross-hairs also.

The distance from the objective of the image formed by it is connected with the distance of the object by the relation:

$$\frac{1}{v} + \frac{1}{u} = \frac{1}{f}$$

where u = distance of object from optical centre of objective

v = distance of image from the optical centre of objective

f = focal length of the objective.

The focal length of an objective is constant. The establishment of a telescopic line of sight, therefore, involves the following two essential conditions:

1. The real image must be formed in front of the eye-piece.
2. The plane of the image must coincide with that of the cross-hairs.

Focusing: For quantitative measurements, it is essential that the image should always be formed in the fixed plane in the telescope where the cross-hairs are situated. *The operation of forming* or *bringing the clear image of the object in the plane of cross-hairs is known as focusing.* Complete focusing involves two steps:

(*i*) ***Focusing the eye-piece:*** The eye-piece unit is moved in or out with respect to the cross-hairs so that the latter are clearly visible. By doing so, the cross-hairs are brought in the plane of *distinct vision* which depends on eye-sight of a particular person.

(*ii*) ***Focusing the objective:*** The purpose of focusing the objective is to bring the image of object in the plane of cross-hairs which are clearly visible. The focusing can be done *externally* or *internally.*

The telescope in which the focusing is done by the external movement of either objective or eye-piece is known as an *external focusing telescope* (Fig. 9.24) and the one in which the focusing is done internally with a negative lens is known as *internal focusing telescope* (Fig. 9.25).

Parallax: If the image formed by objective lens is not in the same plane with cross-hairs, any movement of the eye is likely to cause an apparent movement of the image with respect to the cross-hairs. This is called *parallax.* The parallax can be eliminated by focusing as described above.

Whether internal focusing or external focusing, a telescope consists of the following essential parts:

(*i*) Objective (*ii*) Eye-piece (*iii*) Diaphragm (*iv*) Body and focusing device.

(i) Objective: If simple (single) lenses are used, the telescope would have various optical defects, known as *aberrations,* which would result in curvature, distortion, unwanted colours and indistinctness of the image. In order to eliminate these defects as much as possible, the objective and eye-piece lens are made up of two or more simple lenses The objective (Fig. 9.17) is invariably a compound lens consisting of (*a*) the front double convex lens made of crown glass and (*b*) the back concavo-convex lens made of flint glass, the two being cemented together with balsm at their common surface. Such compound lens is known as *achromatic lens,* and two serious optical defects *viz.,* spherical aberration and chromatic aberration are nearly eliminated.

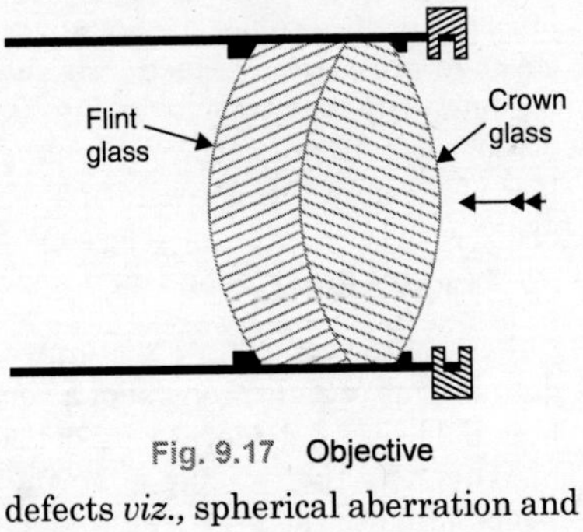

Fig. 9.17 Objective

(ii) Eye-Piece: In most of surveying telescopes, *Ramsden eye-piece* is used. It is composed of *plano-convex* lenses of equal focal length (Fig. 9.18), the distance between them being two-thirds of the focal length of either. This eye-piece gives a flat field of view. It is also known as positive or non-erecting eye-piece for the inverted image, which is formed by the object glass, appears still inverted to the observer.

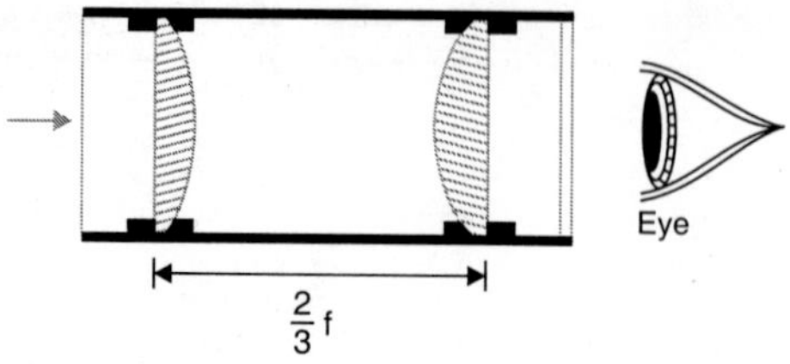

Fig. 9.18 Ramsden's Eye-piece

Another type of eye-piece, though not commonly used, is *Huygen's eye-piece.* It is composed of two plano-convex lenses (Fig. 9.19), the distance between them being two-third of the focal length of the larger and twice the focal length of the smaller. The chromatic aberration of this combination is slightly less and spherical aberration is more than that of Ramsden's. This is also a non-erecting eye-piece.

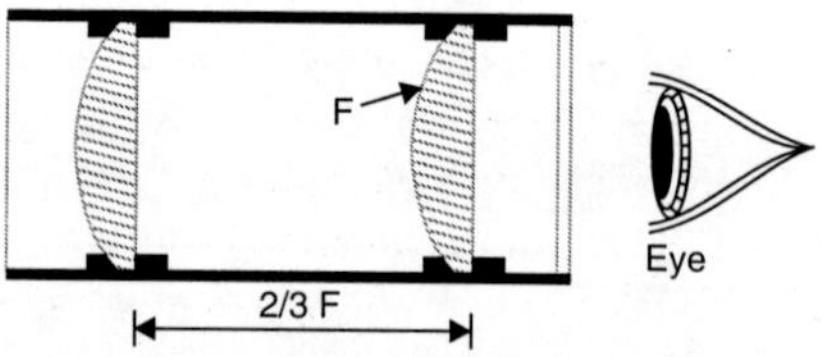

Fig. 9.19 Huygen's Eye-piece

Some telescopes are fitted with special *erecting eye-pieces* which give a magnified but inverted image of the image formed by the objective and hence, the latter itself forms an inverted image. The result is a magnified, but erect image of the original object. The eye-piece consists of four lenses (Fig. 9.20). The eye-piece involves the use of extra lenses. This results in loss of brilliancy of the image, which is a decided disadvantage. Additional advantages of a non-erecting telescope are (*i*) for any desired magnifying power, the length of non-erecting telescope is shorter than the erecting telescope, (*ii*) the definition is certain to be better because the image does not have to be erected but instead is formed by an achromatic eye-piece. For all precise work, the Ramsden eye-piece is to be preferred, as inverted images are not a great disadvantage and a surveyor very soon gets used to them.

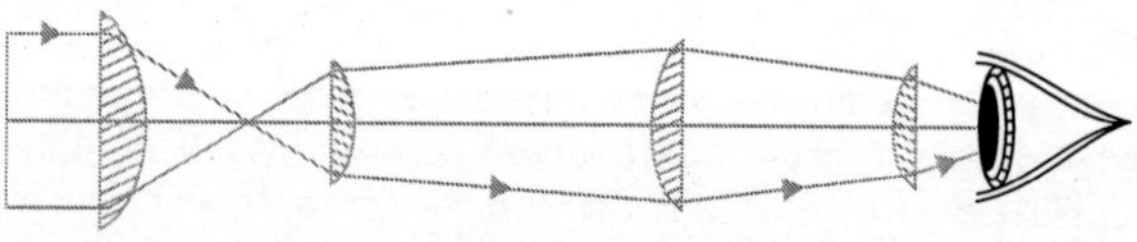

Fig. 9.20 Erecting Eye-piece.

When the line of sight is very much inclined to the horizontal, it becomes inconvenient for the eye to view through ordinary eye-piece. In such a case, a *diagonal eye-piece,* such as shown in Fig. 9.21 is used. Diagonal eye-piece, generally of Ramsden type, consists of the two lenses and a reflecting prism or a mirror fitted at an angle of 45° with the axis of the telescope. Such eye-piece is very much useful in astronomical observations.

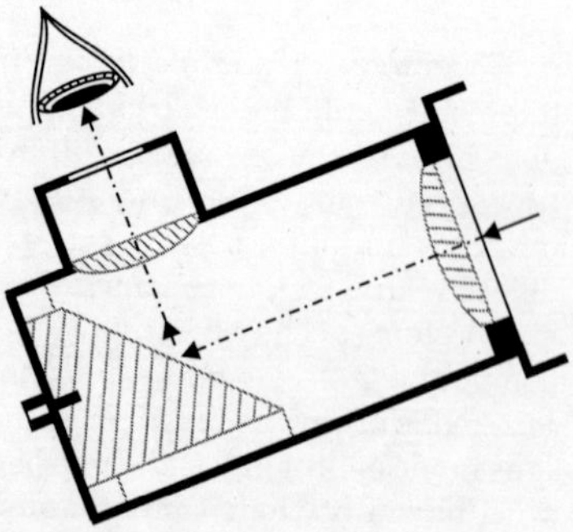

Fig. 9.21 Diagonal Eye-piece.

(iii) Diaphragm: The cross-hairs, designed to give a definite line of sight, consist of a vertical and a horizontal hair held in a flat metal ring called *reticule.* In modern instruments, the reticule is an interchangeable capsule which fits into the diaphragm, a flanged metal ring held

in telescope barrel by four capstan-headed screws (Fig. 9.22). With the help of these screws, the position of the cross-hairs inside of the tube can be adjusted slightly, both horizontally and vertically, and a slight rotational movement is also possible.

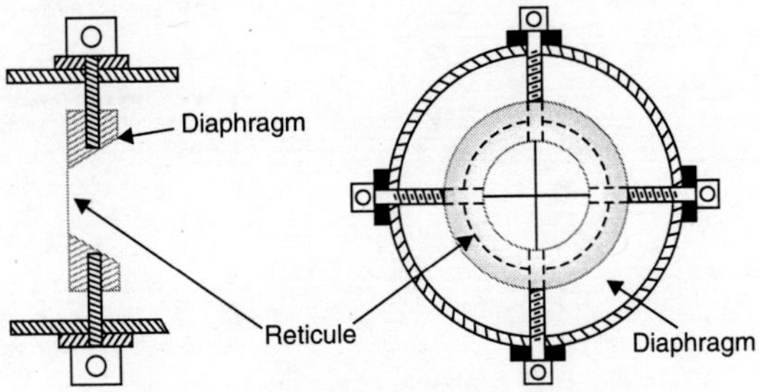

Fig. 9.22 Diaphragm and Reticule.

The hairs are usually made of threads from cocoon of the brown spider, but may be of very fine platinum wire of filaments of silk. In some instruments, the reticule consists of a glass plate on which are etched fine vertical and horizontal lines which serve as cross-hairs. A few typical arrangements of the lines and points are illustrated in Fig. 9.23 of which (*a*), (*b*) and (*c*) are used in levels.

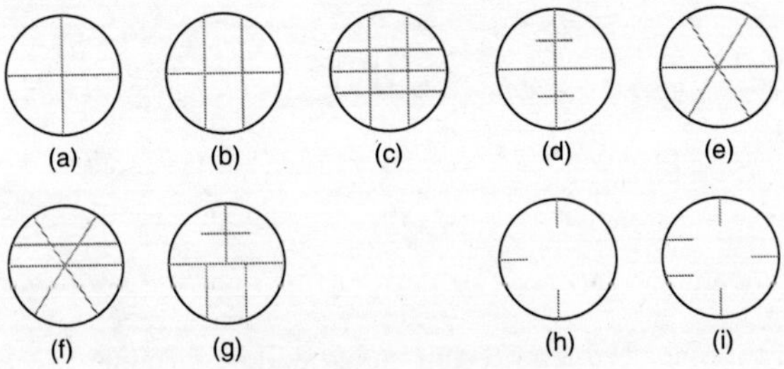

Fig. 9.23 Cross-Hairs.

The horizontal hair is used to read the staff and the two vertical hairs enable the surveyor to see if the staff is vertical laterally. Most telescopes are also equipped with two more horizontal hairs called *stadia hairs,* one above and other on equal distances below the horizontal cross-hair, for use in computing distances by stadia tacheometry.

(iv) Body and Focusing Device: The focusing device depends upon whether it is an external focusing telescope (Fig. 9.24) or an internal focusing telescope (Fig. 9.25). In the external focusing telescope, the body is formed of two tubes one capable of sliding axially within the other by means

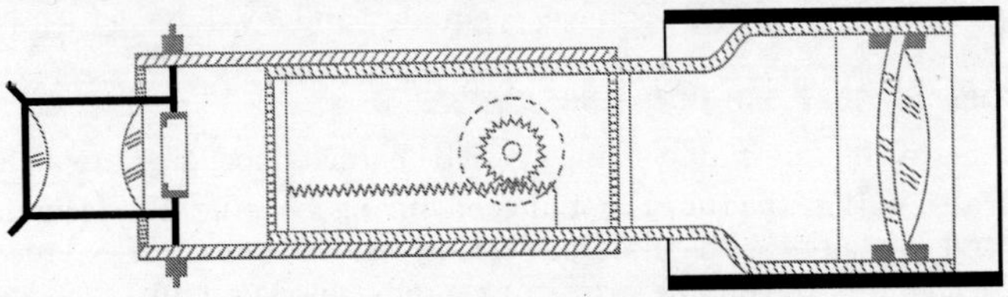

Fig. 9.24 External Focusing Telescope.

of rack gearing with a pinion attached to the milled focusing screw. In some cases, the objective cell is screwed to the inner tube, so that the focusing movement is effected by movement of the objective relatively to the outer fixed rube carrying the cross-hairs and eye-piece. In other cases, the objective is mounted on the outer tube and the eye-piece end, carrying the eye-piece and the diaphragm, moves in focusing.

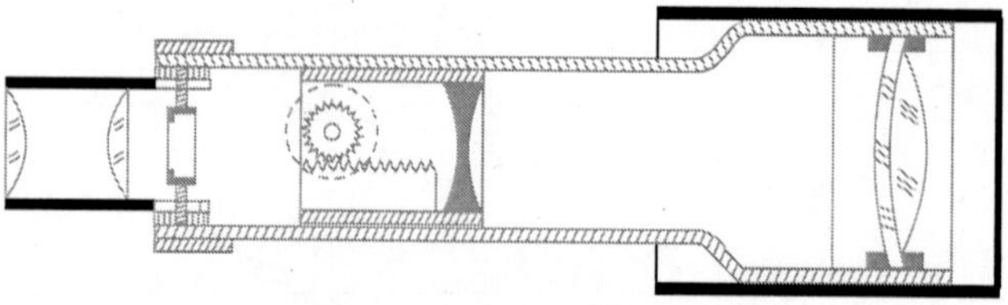

Fig. 9.25 Internal Focusing Telescope.

In internal focusing telescope, the objective and eye-piece are kept fixed and the focusing is done with help of a supplementary double concave lens mounted in a short tube which can be moved to and for between the diaphragm and the objective. This short tube holding the lens is moved along and inside the tube carrying the objective by means of rack and pinion and an external milled head. Fig. 9.26 illustrates the principle underlying the focusing with a negative lens.

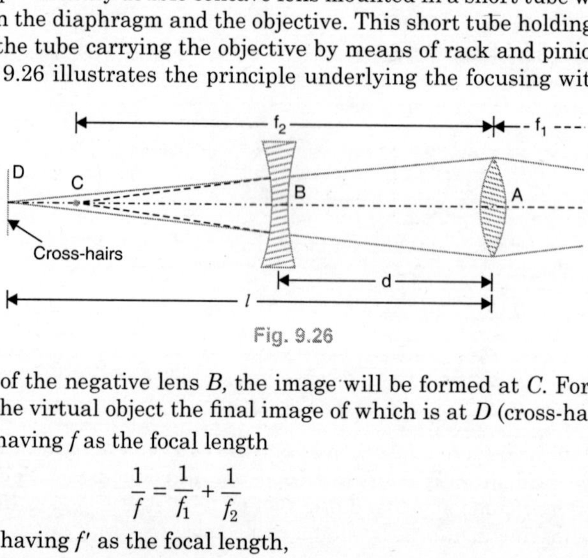

Fig. 9.26

In the absence of the negative lens *B*, the image will be formed at *C*. For the negative lens *B*, point *C* forms the virtual object the final image of which is at *D* (cross-hairs)

For the lens *A* having *f* as the focal length

$$\frac{1}{f} = \frac{1}{f_1} + \frac{1}{f_2} \qquad ...(1)$$

For the lens *B*, having f' as the focal length,

$$\frac{1}{f'} = \frac{1}{(f_2 - d)} + \frac{1}{(l - d)} \qquad ...(2)$$

From the above two conjugate focal equations, the distance *d* can be known for a given value of f_1. When the object is at infinite distance, f_2 equals *f* and *d* will have its minimum value.

Advantages of Internal Focusing Telescope

The advantages of the internal focusing over the external focusing are as follows:

1. The overall length of the tube is not altered during focusing. The focusing slide is light in weight and located near the middle of the telescope. Hence, the balance of the telescope is not affected and the bubble is less liable to be displaced during focusing.

2. Risk of breaking the parallel plate bubble tube or glass cover of compass box, when transiting a theodolite telescope, is eliminated.
3. Wear on the rack and pinion is less due to lesser movement of negative lens.
4. Line of collimation is less likely to be affected by focusing.
5. As the draw tube is not exposed to weather, oxidation is less likely to occur and the telescope can be made practically dust and water proof.
6. In making measurements by the stadia method, an instrument constant is almost eliminated and the computations are thus simplified.
7. The negative focusing lens serves a useful optical purpose because the focal length of the combined objective and negative lens is greater than the distance between the objective lens and focal plane. This extra equivalent focal length can be utilised to increase the power.
8. Internal focusing also gives the optical designer an extra lens to work with, in order to reduce the aberrations of the system and to increase the diameter of the objective lens.

Disadvantages of Internal Focusing Telescope

The principal disadvantages of an internal focusing telescope are as follows:

1. The internal lens reduces the brilliancy of the image.
2. The interior of the telescope is not so easily accessible for field cleaning and repairs.

Optical Defects of a Single Lens

The optical defects of a single lens are:

1. Spherical aberration
2. Chromatic aberration
3. Coma
4. Astigmation
5. Curvature of field
6. Distortion

Aberrations: *Aberration* is the deviation of the rays of light when unequally refracted by a lens so that they do not converge and meet at a focus but separate, forming an indistinct image of the object or an indistinct image with prismatically coloured images.

1. Spherical Aberration: In a single lens having truly spherical surfaces, the rays from a given point are not all collected exactly at one point. The rays through the edges converge slightly nearer the lens than those through the centre (Fig. 9.27). This defect or imperfection, arising from the form of curvature of the lens is known as *spherical aberration* or sometimes as *axial spherical aberration.* The spherical aberration of a negative lens tends to neutralise that of a positive lens. Hence the positive and negative elements of an achromatic lens can be so shaped as practically to eliminate spherical aberration also. (See Fig. 9.17 also).

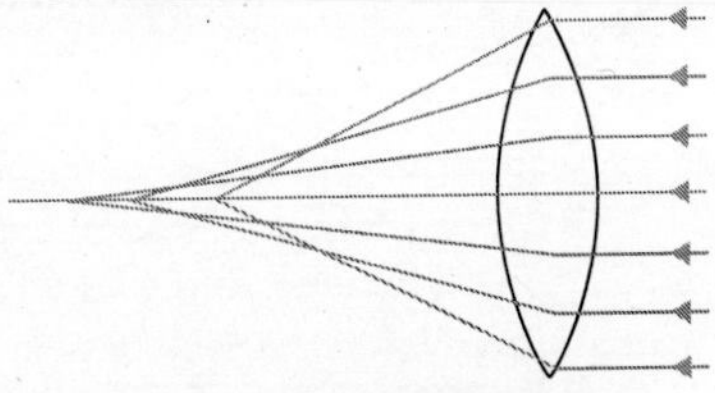

Fig. 9.27 Spherical Aberration.

2. Chromatic Aberration: A beam of white light is made up of seven colours—red, orange, yellow, green, blue, indigo and violet. Since the focal length of any *single* lens is different for each different colour of light, a beam of white light instead of converging at a focus after passing through a *single* lens, is distributed along the axis in a series

of focal points. The violet ray is refracted most and the red is refracted least (Fig. 9.28). This defect is known as *chromatic aberration* due to which a blurred and coloured image is formed. The chromatic aberrations are of opposite signs in positive and negative lenses. It is, therefore, possible to make a combination lens in which the chromatic aberration of a negative lens of relatively low power is sufficient practically to neutralize the chromatic aberration of positive lens of relative high power. Fig 9.17 shows such an *achromatic lens* in which the outer double convex lens is made of crown glass and the inner concavo-convex lens of flint glass.

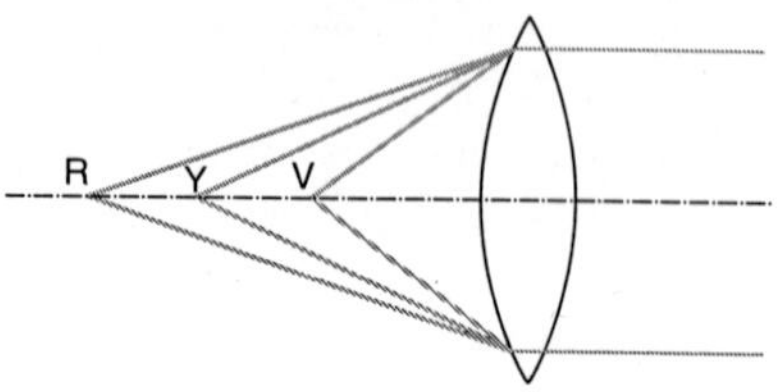

Fig. 9.28 Chromatic Aberration.

The *elimination of aberrations* is only one of the requirements in the design of a telescope. The extent to which this aim is achieved determines to a considerable degree the quality of telescope.

The other possible defects in a single lens, *i.e.* coma, astigmation, curvature, distortion etc., are of little importance to the majority of surveyors. If further information be required, reference should be made to any of the elementary standard textbooks on physics or optics.

Optical Characteristics of the Telescope

The desirable optical characteristics of surveying telescope are as follows:

1. **Aplanation:** Aplanation is the absence of spherical aberration. A compound lens, free from spherical aberration, is known as an *aplanatic* combination.
2. **Achromatism:** Achromation is the absence of chromatic aberration. A compound lens, free from chromatic aberration, is known as an *achromatic* combination.
3. **Definition:** *Definition,* as applied to a telescope, is its capability of producing a sharp image. This *resolving power* of a telescope is the power to form distinguishable images of objects separated by small angular distances and it wholly depends upon *definition.* The definition depends upon the extent to which the defects of a single lens have been eliminated and also upon the accuracy in centring the lenses on one axis.
4. **Illumination or Brightness:** The illumination or brightness of the image of telescope depends upon the magnifying power and the number and quality of the lenses. Illumination is inversely proportional to magnification and number of lenses.
5. **Magnification:** Magnification is the ratio between the angle subtended at the eye by the virtual image, and that subtended by the object and depends upon the ratio of the focal length of the objective lens to the focal length of the eye-piece. The magnification should be proportional to the aperture (*i.e.* to the amount of light which enters the telescope), because if the magnification is too high for the aperture, the ordinary objects will appear faint and if magnification is too low the objects will appear too small for accurate sighting. Provision of higher magnification reduces brilliancy of image, reduces the field of view and wastes more time in focusing.
6. **Size of Field:** By *field of view* is meant the whole circular area seen at one time through the telescope. The field of view is not merely dependent upon the size of the hole in the cross-hair reticule, but it also increases as the magnification of the telescope decreases.

9.6 TEMPORARY ADJUSTMENTS OF A LEVEL

Each surveying instrument needs two types of adjustments: 1. temporary adjustments, and 2. permanent adjustments. *Temporary adjustments or Station adjustments* are those which are made at every instrument setting and preparatory to taking observations with the instrument. *Permanent adjustments* need be made only when the fundamental relations between some parts or lines are disturbed (See Chapter 16).

The temporary adjustments for a level consist of the following:

1. Setting up the level 2. Levelling up 3. Elimination of parallax.

1. **Setting up the Level:** The operation of setting up includes (*a*) fixing the instrument on the stand, and (*b*) levelling the instrument approximately by leg adjustment. To fix the level to the tripod, the clamp is released, instrument is held in the right-hand and is fixed on the tripod by turning round the lower part with the left hand. The tripod legs are so adjusted that the instrument is at the convenient height and the tribrach is approximately horizontal. Some instruments are also provided with a small circular bubble on the tribrach.
2. **Levelling up:** After having levelled the instrument approximately, accurate levelling is done with the help of foot screws and with reference to the plate levels. The purpose of levelling is to make the vertical axis truly vertical. The manner of levelling the instrument by the plate levels depends upon whether there are three levelling screws or four levelling screws.

(a) Three Screw Head

1. Loose the clamp. Turn the instrument until the longitudinal axis of the plate level is roughly parallel to a line joining any two (such as *A* and *B*) of the levelling screws [Fig. 9.29 (*a*)].
2. Hold these two levelling screws between the thumb and first finger of each hand and turn them uniformly so that the thumbs move either towards each other or away from each other until the bubble is central. *It should be noted that the bubble will move in the direction of movement of the left thumb* [Fig. 9.29 (*a*)].

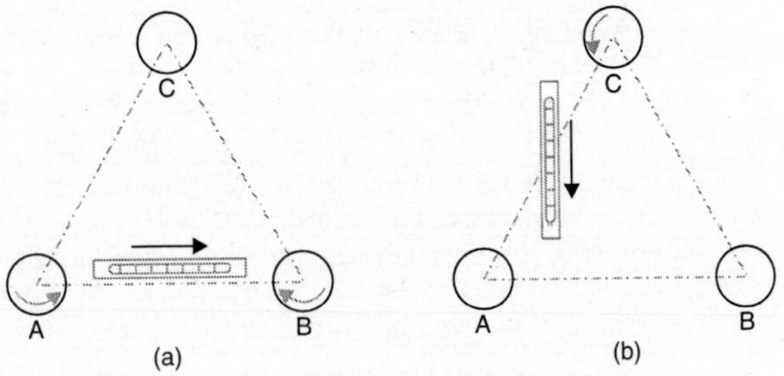

Fig. 9.29 Levelling-Up with Three Foot Screws.

3. Turn the upper plate through 90°, *i.e.,* until the axis on the level passes over the position of the third levelling screw *C* [Fig. 9.29 (*b*)].
4. Turn this levelling screw until the bubble is central.
5. Return the upper part through 90° to its original position [Fig. 9.29 (*a*)] and repeat step (2) till the bubble is central.

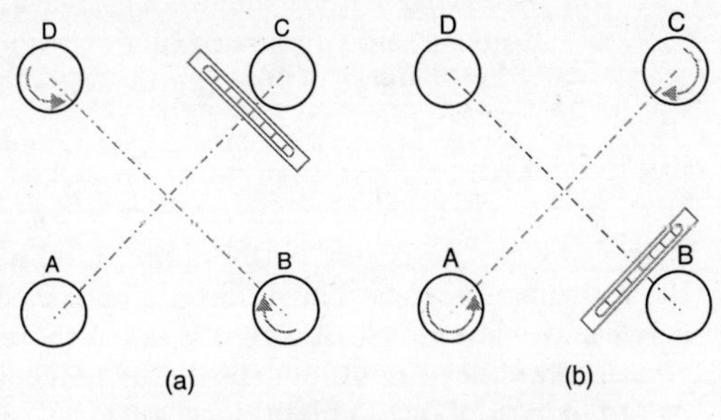

Fig. 9.30 Levelling-Up with Four-Foot

6. Turn back again through 90° and repeat step (4).
7. Repeat steps (2) and (4) till the bubble is central in both the positions.
8. Now rotate the instrument through 180°. The bubble should remain in the centre of its run, provided it is in correct adjustment. The vertical axis will then be truly vertical.

If not, it needs *permanent adjustment.*

Note: *It is essential to keep the same quarter circle for the changes in direction and not to swing through the remaining three quarters of a circle to the original position.*

(b) Four Screw Head

1. Turn the upper plate until the longitudinal axis of the plate level is roughly parallel to the line joining two diagonally opposite screws such as *D* and *B* [Fig. 9.30 (*a*)].
2. Bring the bubble central exactly in the same manner as described in step (2) above.
3. Turn the upper part through 90° until the spirit level axis is parallel to the other two diagonaily opposite screws such as *A* and *C* [Fig. 9.30 (*b*)].
4. Centre the bubble as before.
5. Repeat the above steps till the bubble is central in both the positions.
6. Turn through 180° to check the permanent adjustment as for three screw instrument.

   In modern instruments, three-foot screw levelling head is used in preference to a four foot screw levelling head. The three-screw arrangement is the better one, as three points of support are sufficient for stability and the introduction of an extra point of support leads to uneven wear on the screws. On the other hand, a four-screw levelling head is simpler and lighter as a three-screw head requires special casting called a *tribrach.* A three-screw instrument has also the important advantage of being more rapidly levelled.

3. **Elimination of Parallax:** Parallax is a condition arising when the image formed by the objective is not in the plane of the cross-hairs. Unless parallax is eliminated, accurate sighting is impossible. Parallax can be eliminated in two steps: (*i*) by focusing the eye-piece for distinct vision of the cross-hairs, and (*ii*) by focusing the objective to bring the image of the object in the plane of cross-hairs.

   (*i*) *Focusing the eye-piece:* To focus the eye-piece for distinct vision of the cross-hairs, point the telescope towards the sky (or hold a sheet of white paper in front of the objective) and move eye-piece in or out till the cross-haris are seen sharp and distinct. In some telescopes, graduations are provided at the eye-piece so that one can always remember the particular graduation position to suit his eyes. This may save much of time.

   (*ii*) *Focusing the objective:* The telescope is now directed towards the staff and the focusing screw in turned till the image appears clear and sharp. The image so formed is in the plane of cross-hairs.

9.7 THEORY OF DIRECT LEVELLING (SPIRIT LEVELLING)

A level provides horizontal line of sight, *i.e.*, a line tangential to a level surface at the point where the instrument stands. The difference in elevation between two points is the vertical distance between two level lines. Strictly speaking, therefore, we must have a level line of sight and not a horizontal line of sight; but the distinction between a level surface and a horizontal plane is not an important one in plane surveying.

Neglecting the curvature of earth and refraction, therefore, the theory of direct levelling is very simple. With a level set up at any place, the difference in elevation between any two points within proper lengths of sight is given by the difference between the rod readings taken on these points. By a succession of instrument stations and related readings, the difference in elevation between widely separated points is thus obtained.

Special Methods of Spirit Levelling

(*a*) **Differential Levelling:** It is the method of direct levelling the object of which is solely to determine the difference in elevation of two points regardless of the horizontal positions of the points with respect of each other. When the points are apart, it may be necessary to set up the instruments serveral times. This type of levelling is also known as *fly levelling.*

(*b*) **Profile Levelling:** It is the method of direct levelling the object of which is to determine the elevations of points at measured intervals along a given line in order to obtain a profile of the surface along that line.

(*c*) **Cross-Sectioning:** Cross-sectioning or cross-levelling is the process of taking levels on each side of a main line at right angles to that line, in order to determine a vertical cross-section of the surface of the ground, or of underlying strata, or of both.

(*d*) **Reciprocal Levelling:** It is the method of levelling in which the difference in elevation between two points is accurately determined by two sets of reciprocal observations when it is not possible to set up the level between the two points.

(*e*) **Precise Levelling:** It is the levelling in which the degree of precision required is too great to be attained by ordinary methods, and in which, therefore, special, equipment or special precautions or both are necessary to eliminate, as far as possible, all sources of error.

Terms and Abbreviations

(*i*) **Station:** In levelling, a station is that point where the level rod is held and not where level is set up. It is the point whose elevation is to be ascertained or the point that is to be established at a given elevation.

(*ii*) **Height of Instrument (H.I.):** For any set up of the level, the height of instrument is the elevation of plane of sight (line of sight) with respect to the assumed datum. It does not mean the height of the telescope above the ground where the level stands.

(*iii*) **Back Sight (B.S.):** Back sight is the sight taken on a rod held at a point of *known elevation,* to ascertain the amount by which the line of sight is *above* that point and thus to obtain the height of the instrument. *Back sighting* is equivalent to measuring *up* from the point of known elevation to the line of sight. It is also known as a *plus sight* as the back sight reading is always added to the level of the datum to get the height of the instrument. *The object of back sighting is, therefore, to ascertain the height of the plane of sight.*

(*iv*) **Fore Sight (F.S.):** Fore sight is a sight taken on a rod held at a point of unknown elevation, to ascertain the amount by which the point is *below* the line of sight and thus to obtain the elevation of the station. *Fore sighting* is equivalent to measuring *down* from the line of sight. It is also known as a *minus* sight as the fore sight reading is always subtracted (except in speical cases of tunnel survey) from the height of the instrument to get the elevation of the point. *The object of fore sighting is, therfore, to ascertain the elevation of the point.*

(*v*) **Turning Point (T.P.):** Turning point or *change point* is a point on which both minus sight and plus sight are taken on a line of direct levels. The minus sight (fore sight) is taken on the point in one set of instrument to ascertain the elevation of the point while the plus sight (back sight) is taken on the same point in other set of the instrument to establish the new height of the instrument.

(*vi*) **Intermediate Station (I.S.):** Intermediate station is a point, intermediate between two turning points, on which only one sight (minus sight) is taken to determine the elevation of the station.

Steps in Levelling (Fig. 9.31)

There are two steps in levelling: (*a*) to find by how much amount the line of sight is above the bench mark, and (*b*) to ascertain by how much amount the next point is below or above the line of sight.

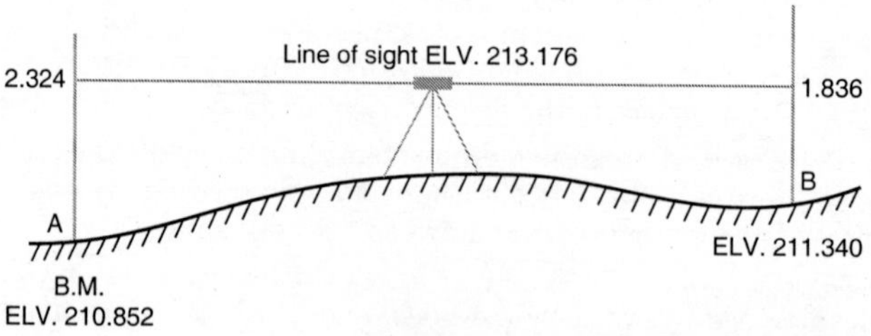

Fig. 9.31

A level is set up approximately midway between the bench mark (or a point of known elevation) and the point, the elevation of which is to be ascertained by direct levelling. A back sight is taken on the rod held at the bench mark. Then

$$\textbf{H.I. = Elv. of B.M. + B.S.} \qquad ...(1)$$

Turning the telescope to bring into view the rod held on point *B*, a foresight (minus sight) is taken. Then

$$\textbf{Elv. = H.I. – F.S.} \qquad ...(2)$$

For example, if elevation of *B.M.* = 210.852 m, *B.S.* = 2.324 m and *F.S.* = 1.836 m.

Then $H.I. = 210.852 + 2.324 = 213.176$ m

and Elv. of $B = 213.176 - 1.836 = 211.340$ m.

It is to be noted that if a back sight is taken on a bench mark located on the roof of a tunnel or on the ceiling of a room with the instrument at a lower elevation, the back sight must be subtracted from the elevation to get the height of the instrument. Similarly, if a foresight is taken on a point higher than the instrument, the foresight must be added to the height of the instrument, to get the elevation of the point.

9.8 DIFFERENTIAL LEVELLING

The operation of levelling to determine the elevation of points at some distance apart is called *differential levelling* and is usually accomplished by direct levelling. When two points are at such a distance from each other that they cannot both be within range of the level at the same time, the difference in elevation is not found by single setting but the distance between the points is divided in two stages by turning points on which the staff is held and the difference of elevation of each of succeeding pair of such turning points is found by separate setting up of the level.

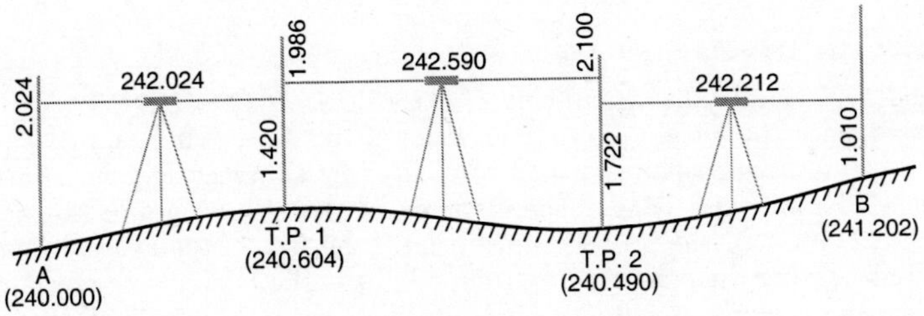

Fig. 9.32

Referring to Fig. 9.32, *A* and *B* are the two points. The distance *AB* has been divided into three parts by choosing two additional points on which staff readings (both plus sight and minus sight) have been taken. Points 1 and 2 thus serve as *turning points.*

The *R.L.* of point *A* is 240.00 m. The height of the first setting of the instrument is therefore = 240.00 + 2.024 = 242.024. If the following *F.S.* is 1.420, the *R.L.* of *T.P.* 1 = 242.024 – 1.420 = 240.604 m. By a similar process of calculations, *R.L.* of *T.P.* 2 = 240.490 m and of *B* = 241.202 m.

9.9 HAND SIGNALS DURING OBSERVATIONS

When levelling is done at construction site located in busy, noisy areas, it becomes difficult for the instrument man to give instructions to the man holding the staff at the other end, through vocal sounds. In that case, the following hand signals are found to be useful (Table 9.1 and Fig. 9.33)

TABLE 9.1 Hand Signals

Refer Fig. 9.33	*Signal*	*Message*
(*a*)	Movement of left arm over 90°	Move to my left
(*b*)	Movement of right arm over 90°	Move to my right
(*c*)	Movement of left arm over 30°	Move top of staff to my left
(*d*)	Movement of right arm over 30°	Move top of staff to my right
(*e*)	Extension of arm horizontally and moving hand upwards	Raise height peg or staff
(*f*)	Extension of arm horizontally and moving hand downwards	Lower height peg or staff
(*g*)	Extension of both arms and slightly thrusting downwards	Establish the position
(*h*)	Extension of arms and placement of hand on top of head.	Return to me

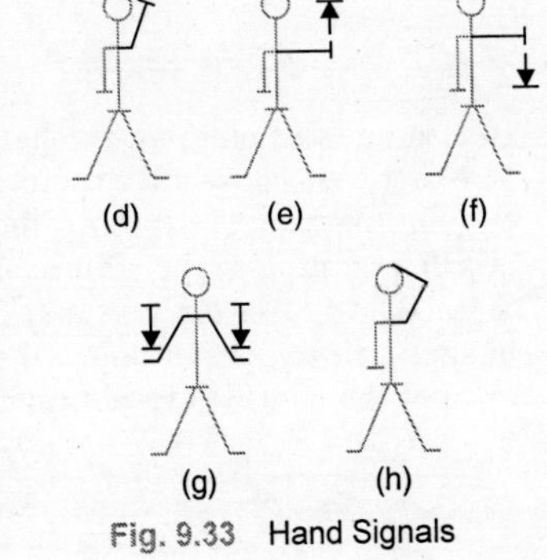

Fig. 9.33 Hand Signals

9.10 BOOKING AND REDUCING LEVELS

There are two methods of booking and reducing the elevation of points from the observed staff readings: (1) *Collimation or Height of Instrument* method ; (2) *Rise and Fall* method.

1. Height of Instrument Method

In this method, the height of the instrument (*H.I.*) is calculated for each setting of the instrument by adding back sight (plus sight) to the elevation of the *B.M.* (First point). The elevation of reduced level of the turning point is then calculated by subtracting from *H.I.* the fore sight (minus sight). For the next setting of the instrument, the *H.I.* is obtained by adding the *B.S.* taken on *T.P.* 1 to its *R.L.* The process continues till the *R.L.* of the last point (a fore sight) is obtained by subtracting the staff reading from height of the last setting of the instrument. If there are some intermediate points, the *R.L.* of those points is calculated by subtracting the intermediate sight (minus sight) from the height of the instrument for that setting.

The following is the specimen page of a level field book illustrating the method of booking staff readings and calculating reduced levels by height of instrument method.

Station	*B.S.*	*I.S.*	*F.S.*	*H.I.*	*R.L.*	*Remarks*
A	0.865			561.365	560.500	B.M. on Gate
B	1.025		2.105	560.285	559.260	
C		1.580			558.705	Platform
D	2.230		1.865	560.650	558.420	
E	2.355		2.835	560.270	557.815	
F			1.760		558.410	
Check	6.475		8.565		558.410	Checked
			6.475		560.500	
			2.090	Fall	2.090	

Arithmetic Check: The difference between the sum of back sights and the sum of fore sights should be equal to the difference between the last and the first *R.L.* Thus

$$\Sigma B.S. - \Sigma F.S. = \textbf{Last } R.L. - \textbf{First } R.L.$$

The method affords a check for the *H.I.* and *R.L.* of turning points but not for the intermediate points.

2. Rise and Fall Method

In rise and fall method, the height of instrument is not at all calculated but the difference of level between consecutive points is found by comparing the staff readings on the two points for the same setting of the instrument. The difference between their staff readings indicates a *rise* or *fall* according as the staff reading at the point is *smaller* or *greater* than that at the preceding point. The figures for 'rise' and 'fall' worked out thus for all the points give the vertical distance of each point above or below the preceding one, and if the level of any one point is known the level of the next will be obtained by adding its rise or subtracting its fall, as the case may be.

The following is the specimen page of a level field book illustrating the method of booking staff readings and calculating reduced levels by rise and fall method:

Station	*B.S.*	*I.S.*	*F.S.*	*Rise*	*Fall*	*R.L.*	*Remarks*
A	0.865					560.500	B.M. on Gate
B	1.025		2.105		1.240	559.260	
C		1.580			0.555	558.705	Platform
D	2.230		1.865		0.285	558.420	
E	2.355		2.835		0.605	557.815	
F			1.760	0.595		558.410	
Check	6.475		8.565	0.595	2.685	558.410	
			6.475		0.595	560.500	Checked
		Fall	2.090	Fall	2.090	2.090	

Arithmetic Check: The difference between the sum of back sights and sum of fore sights should be equal to the difference between the sum of rise and the sum of fall and should also be equal to the difference between the *R.L.* of last and first point. Thus,

$$\Sigma \boldsymbol{B.S.} - \Sigma \boldsymbol{F.S.} = \Sigma \textbf{\textit{Rise}} - \Sigma \textbf{\textit{Fall}} = \textbf{\textit{Last R.L.}} - \textbf{\textit{First R.L.}}$$

This provides a complete check on the intermediate sights also. The arithmetic check would only fail in the unlikely, but possible, case of two more errors occurring in such a manner as to balance each other.

It is advisable that on each page the rise and fall calculations shall be completed and checked by comparing with the difference of the back and fore sight column summations, before the reduced level calculations are commenced.

Comparison of the Two Methods: The height of the instrument (or collimation level) method is more rapid, less tedious and simple. However, since the check on the calculations for intermediate sights is not available, the mistakes in their levels pass unnoticed. The rise and fall method though more tedious, provides a full check in calculations for all sights. However, the height of instrument method is more suitable in case, where it is required to take a number of readings from the same instrument setting, such as for constructional work, profile levelling etc.

Example 9.1: *The following staff readings were observed successively with a level, the instrument having been moved after third, sixth and eighth readings: 2.228; 1.606; 0.988; 2,090; 2.864; 1.262; 0.602; 1.982; 1.044; 2.684 metres.*

Enter the above readings in a page of a level book and calculate the R.L. of points if the first reading was taken with a staff held on a bench mark of 432.384 m.

Solution: Since the instrument was shifted after third, sixth and eighth readings, these readings will be entered in the *F.S.* column and therefore, the fourth, seventh and ninth readings will be entered on the *B.S.* column. Also, the first reading will be entered in the *B.S.* column and the last reading in the *F.S.* column. All other readings will be entered in the *I.S.* column.

The reduced levels of the points may be calculated by rise and fall method as tabulated below:

Station	*B.S.*	*I.S.*	*F.S.*	*Rise*	*Fall*	*R.L.*	*Remarks*
1	2.228					432.384	B.M.
2		1.606		0.622		433.006	
3	2.090		0.988	0.618		433.624	T.P. 1
4		2.864			0.774	432.850	
5	0.602		1.262	1.602		434.452	T.P. 2
6	1.044		1.982		1.380	433.072	T.P. 3
7			2.684		1.640	431.432	
			6.916	2.842	3.794	432.384	
Check	5.964		5.964		2.842	431.432	
		Fall	0.952	Fall	0.952	0.952	Checked

Example 9.2: *It was required to ascertain the elevation of two points P and Q and a line of levels was run from P to Q. The levelling was then continued to a bench mark of 83.500, the readings obtained being as shown below. Obtain the R.L. of P and Q.*

B.S.	*I.S.*	*F.S.*	*R.L.*	*Remarks*
1.622				*P*
1.874		*0.354*		
2.032		*1.780*		
	2.362			*Q*
0.984		*1.122*		
1.906		*2.824*		
		2.036	*83.500*	*B.M.*

Solution: To find the *R.Ls.* of *P* and *Q*, we will have to proceed from bottom to the top. To find the *H.I.*, therefore, *F.S.* readings will have to be added to the *R.L.* of the known point and to find the *R.L.* of the previous point, the *B.S.* will have to be subtracted from the so obtained *H.I.* as clearly shown in the table below:

Station	*B.S.*	*I.S.*	*F.S.*	*H.I.*	*R.L.*	*Remarks*
P	1.622			84.820	83.198	
	1.874		0.354	86.340	84.466	
	2.032		1.780	86.592	84.560	
Q		2.362			84.230	
	0.984		1.122	86.454	85.470	
	1.906		2.824	85.539	83.630	
			2.036		83.500	B.M.
Check	8.418		8.116		83.500	
	8.116				83.198	Checked
Rise	0.302				0.302	

Example 9.3: *The following consecutive readings were taken with a level and 5 metre levelling staff on continuously sloping ground at a common interval of 20 metres: 0.385; 1.030; 1.925; 2.825; 3.730; 4.685; 0.625; 2.005; 3.110; 4.485. The reduced level of the first point was 208.125 m. Rule out a page of a level field book and enter the above readings. Calculate the reduced levels of the points by rise and fall method and also the gradient of the line joining the first and the last point.*

Solution: Since the readings were taken on a continuously sloping ground, the maximum staff reading can be 5 metres only, and therefore, sixth reading will be a fore sight taken on a turning point and the seventh reading will be a back sight. Also, the first reading will be a back sight and the last reading will be a fore sight. The levels can be readily calculated as shown in the tabular form below:

Station	*B.S.*	*I.S.*	*F.S.*	*Rise*	*Fall*	*R.L.*	*Remarks*
1	0.385					208.125	
2		1.030			0.645	207.480	
3		1.925			0.895	206.585	
4		2.825			0.900	205.685	
5		3.730			0.905	204.780	
6	0.625		4.685		0.955	203.325	
7		2.005			1.380	202.445	
8		3.110			1.105	201.340	
9			4.485		1.375	199.965	
Check	1.010		9.170	0.000	8.160	208.125	
			1.010		0.000	199.965	
		Fall	8.160	Fall	8.160	8.160	

$$\text{Gradient of the line} = \frac{8.160}{20 \times 8} = \frac{1}{19.61} = 1 \text{ in } 19.61 \text{ (falling)}.$$

Example 9.4: *The following figures were extracted from a level field book, some of the entries being illegible owing to exposure to rain. Insert the missing figures and check your results. Rebook all the figures by the 'rise' and 'fall' method.*

Station	*B.S.*	*I.S.*	*F.S.*	*Rise*	*Fall*	*R.L.*	*Remarks*
1	*2.285*					*232.460*	B.M. 1
2	*1.650*		×	*0.020*			
3		*2.105*			×		
4	×		*1.960*	×			
5	*2.050*		*1.925*		*0.300*		
6		×		×		*232.255*	B.M. 2
7	*1.690*		×	*0.340*			
8	*2.865*		*2.100*		×		
9			×	×		*233.425*	B.M. 3

Solution:

(*i*) The *F.S.* of station 2 is missing, but it can be calculated from the known rise. Since station 2 is higher than station 1, its *F.S.* will be lesser than the *B.S.* of station 1 (higher the point, lesser the reading). Hence,

$F.S.$ of station 2 = 2.285 – 0.020 = 2.265 m

and $R.L.$ of station 2 = 232.460 + 0.02 = 232.480 m

(*ii*) Fall of station 3 = 2.105 – 1.650 = 0.455 m

∴ $R.L.$ of station 3 = 232.480 – 0.455 = 232.025 m

(*iii*) *B.S.* of station 4 can be calculated from the fact that the *F.S.* of station 5, having a fall of 0.300 m, is 1.925 m

Thus, $B.S.$ of station 4 = 1.925 – 0.300 = 1.625 m

Also, Rise of station 4 = 2.105 – 1.960 = 0.145 m

and $R.L.$ of station 4 = 232.025 + 0.145 = 232.170 m

(*iv*) $R.L.$ of station 5 = 232.170 – 0.300 = 231.870 m

(*v*) From the known *R.L.* of stations 6 and 5, the rise of station 6 can be calculated

Thus, Rise of station 6 = 232.255 – 231.870 = 0.385

$I.S.$ of station 6 = 2.050 – 0.385 = 1.665

(*vi*) $F.S.$ of station 7 = 1.665 – 0.340 = 1.325

and $R.L.$ of station 7 = 232.255 + 0.340 = 232.595

(*vii*) Fall of station 8 = 2.100 – 1.690 = 0.410

$R.L.$ of station 8 = 232.595 – 0.410 = 232.185

(*viii*) Since the elevation of station 9 is 233.425 m, it has a rise of (233.425 – 232.185) = 1.240 m.

∴ $F.S.$ of station 8 = 2.865 – 1.240 = 1.625 m.

The above results and calculations are shown in the tabular form below:

Station	*B.S.*	*I.S.*	*F.S.*	*Rise*	*Fall*	*R.L.*	*Remarks*
1	2.285					232.460	B.M. 1
2	1.650		2.265	0.020		232.480	
3		2.105			0.455	232.025	
4	1.625		1.960	0.145		232.170	
5	2.050		1.925		0.300	231.870	
6		1.665		0.385		232.255	B.M. 2
7	1.690		1.325	0.340		232.595	
8	2.865		2.100		0.410	232.185	
9			1.625	1.240		233.425	B.M. 3
Check	12.165		11.200	2.130	1.165	233.425	
	11.200			1.165		232.460	
	0.965	Rise		0.965	Rise	0.965	Checked

Example 9.5: *During a construction work, the bottom of a R.C. Chhajja A was taken as a temporary B.M. (R.L. 63.120). The following notes were recorded.*

Reading on inverted staff on B.M. No. A.	*2.232*
Reading on peg P on ground:	*1.034*
Change of instrument	
Reading on peg P on ground:	*1.328*
Reading on inverted staff on bottom of cornice B:	*4.124*

Enter the readings in a level book page and calculate the R.L. of cornice B.

Solution: The first reading was taken on an inverted staff and therefore it will have to be subtracted from the *R.L.* to get the *H.I.* Similarly, the last reading was taken on an inverted staff, and the *R.L.* of the cornice *B* will be obtained by adding the *F.S.* reading to the *H.I.* Use (–) sign for the *B.S.* of *A* and *F.S.* of *B* since both of these have been taken in reverse directions than the normal ones. The calculations are shown in table the below:

Point	*B.S.*	*I.S.*	*F.S.*	*H.I.*	*R.L.*	*Remarks*
A	– 2.232			60.888	63.120	
P	1.328		1.034	61.182	59.854	
B			– 4.124		65.306	
Check	– 0.940		– 3.090		65.306	
	– 3.090				63.120	
	+					
Rise	+ 2.186			Rise	2.186	Checked

9.11 BALANCING BACKSIGHTS AND FORESIGHTS

When the difference in elevation between any two points is determined from a single set-up by backsighting on one point and foresighting on the other, the error due to non-parallelism of line of collimation and axis of the bubble tube (when the bubble is in the centre of the run) and also the error due to curvature and refraction may be eliminated if the lengths of two sights can be made equal.

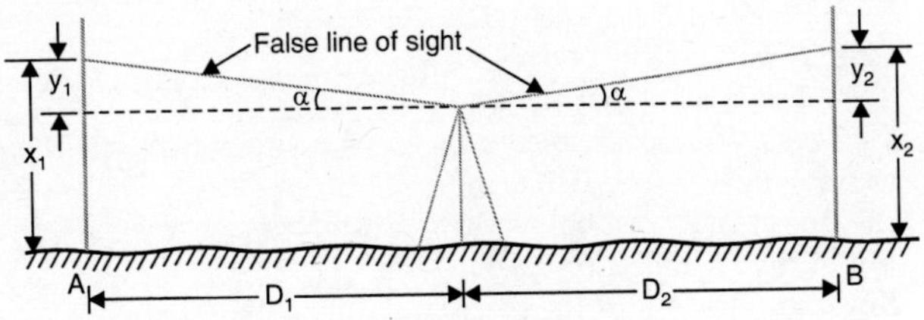

Fig. 9.34 Balancing B.S. and F.S.

In Fig. 9.34, let observations be made with a level in which the line of collimation is inclined upwards by an amount a from horizontal, when the bubble is in the centre of its run, the level being kept exactly midway between the two points *A* and *B*. The observed backsight and foresight are x_1 and x_2. The correct backsight on *A* will be equal to $x_1 - y_1$, where $y_1 = D_1 \tan \alpha$. The correct foresight on *B* will be equal to $x_2 - y_2$ where $y_2 = D_2 \tan \alpha$. Hence the correct difference in level between *A* and *B*

$$= (x_1 - y_1) - (x_2 - y_2) = (x_1 - x_2) + (y_2 - y_1)$$
$$= (x_1 - x_2) + (D_2 \tan \alpha - D_1 \tan \alpha) = (x_1 - x_2) \text{ if } D_1 = D_2$$

Thus, if backsight and foresight distances are balanced, the difference in elevation between two points can be directly calculated by taking difference of the two readings and no correction for the inclination of the line of sight is necessary.

Fig. 9.35 illustrates how the error due to curvature can be eliminated by equalising backsight and foresight distances. Since the level provides horizontal line of sight (and not a level line), the staff reading at point $A = h_a$ and at a point $B = h_b$. The correct staff readings should have been H_a and H_b so that

$$H_a = h_a - h_a' \quad \text{and} \quad H_b = h_b - h_b'$$

The correct difference in elevation between A and B, therefore is given by

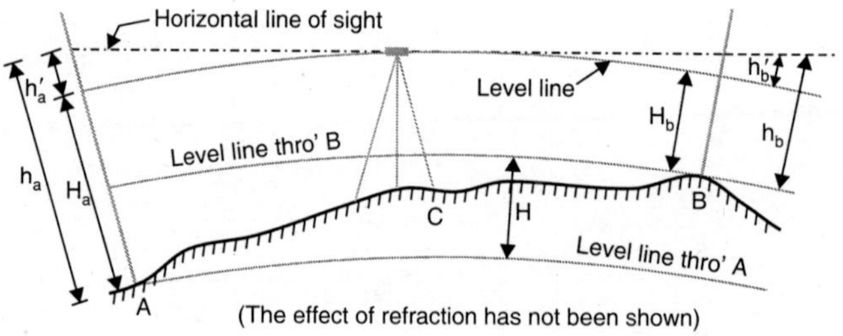

Fig. 9.35

$$H = H_a - H_b = (h_a - h_a') - (h_b - h_b') = (h_a - h_b) - (h_a' - h_b')$$

If the horizontal distance AC and BC are not equal, true difference in elevation H cannot be found unless h_a' and h_b' are numerically found (see Art. 9.7). But if the distances AC and BC are balanced (*i.e.*, made equal), h_a' and h_b' would be equal and H will equal to $(h_a - h_b)$.

Thus, if the backsight and foresight distances are balanced, the elevation between two points is equal to the difference between the rod readings taken to the two points and no correction for curvature and refraction is necessary.

Balancing Sight on a Slope

When the points lie on a sloping ground, the level should be set off to one side far enough to equalise, as nearly as practicable the uphill and downhill sights.

In Fig. 9.36, it is required to set the level between two points *T.P.* 1 (turning point) and *T.P.* 2. Let the level be set up at *A* far enough uphill to bring the line of sight just below the top of an extended rod when held on the turning point *T.P.* 1. A turning point *T.P.* 2 can then be establised far enough uphill to bring the line of sight *just* above the bottom of the rod when held on the turning point *T.P.* 2. The level at *A* is nearly on the line between *T.P.* 1 and *T.P.* 2, the corresponding distance being 20 m and 12 m (say). On the contrary, if the level is set up at *B*, instead of *A*, off to one side but at nearly the

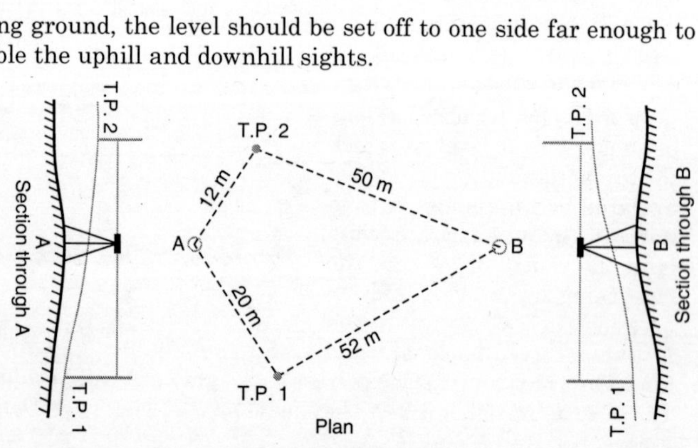

Fig. 9.36

same elevation as at *A*, so that sights on *T.P.* 1 and *T.P.* 2 can still be taken and distances of *T.P.* 2 and *T.P.* 1 from *B* can be equal, the error due to non-adjustment of collimation will be eliminated.

To take a numerical example, let the level have line of sight inclined upwards by an amount 0.008 metres in every 100 metres. When the level is at *A*, the error in the rod readings will be:

For *T.P.* 2: $0.12 \times 0.008 = 0.00096$ m

For *T.P.* 1: $0.20 \times 0.008 = \underline{0.00160}$ m

Error in the levelling = 0.00064 m

Again, if the level were at *B*, the errors in rod readings would be

at *T.P.* 2: $0.50 \times 0.008 = 0.0040$ m

T.P. 1: $0.52 \times 0.008 = \underline{0.00416}$ m

Error in levelling = 0.00016 m

Thus, when the level is at *B*, the error in levelling is about $\frac{1}{4}$th of the error if the level is set at *A*. By moving *B* farther away, the error may be reduced until it approaches zero, as the lengths of the two sights from *B* become nearly equal.

Example 9.6: *A level set up an extended line BA in a position 70 metres from A and 100 metres from B reads 1.684 on a staff held at A and 2.122 on a staff held at B, the bubble having been carefully brought to the centre of its run before each reading. It is known that the reduced levels of the tops of the pegs A and B are 89.620 and 89.222 respectively. Find (a) the collimation error, and (b) the readings that would have been obtained had there been no collimation error.*

Solution: Exact difference in elevation in *B* and *A*

$= 89.620 - 89.222 = 0.398$ m, *B* being lower.

As per observations, difference in elevation

$= 2.122 - 1.684 = 0.438$ m, *B* lower.

This shows *B* to be lower than what it is. We know that lower is the point, greater is the staff reading. Hence, the staff reading at *B* is greater than what it should be and thus, the line of sight is inclined upwards, as shown in Fig. 9.37, by an amount $0.438 - 0.398 = 0.040$ m in a distance of 30 m.

Therefore $\tan \alpha = \dfrac{0.04}{30} = 0.0013333$

We know that $\tan 60'' = 0.0002909$

Hence by proportion, $\alpha = \dfrac{13333 \times 60}{2909}$ seconds $= 4'\ 34''$ upwards.

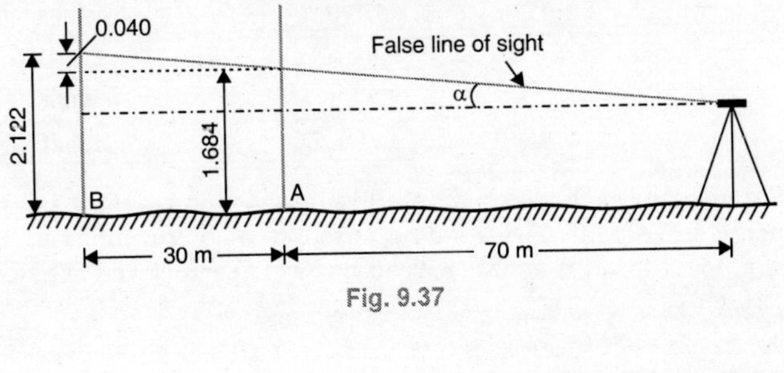

Fig. 9.37

Exact reading if there were no collimation error, would be

$$\text{at } A: \quad 1.684 - \left(\frac{0.040}{30} \times 70\right) = 1.684 - 0.093 = \mathbf{1.591} \text{ m}$$

$$\text{at } B: \quad 2.122 - \left(\frac{0.040}{30} \times 100\right) = 2.122 - 0.133 = \mathbf{1.989} \text{ m.}$$

So that the true difference in elevation = 1.989 – 1.591 = 0.398 m as given in the question.

Example 9.7: *A page of a level book was defaced so that the only legible figures were (a) consecutive entries in the column of reduced levels: 55.565 (B.M.); 54.985 (T.P.); 55.170; 56.265; 53.670; 53.940; (TP.); 52.180; 52.015; 51.480 (T.P.); 53.145; 54.065 (T.B.M.); (b) entries in the backsight column: 1.545; 2.310; 0.105; 3.360 in order from the top of the page.*

Reconstruct the page as booked and check your work. Calculate the corrected level of the T.B.M. if the instrument is known to have an elevated collimation error of 60″ and back and foresight distance averaged 80 and 30 metres respectively.

Solution: There are three turning points on which both back and foresights have been taken. The first sight is a backsight. The four backsight readings will, therefore, be entered in order, one against the *B.M.* point and other three against the three turning points. The last *R.L.* corresponds to *T.B.M.* on which a foresight is missing. All other sights will be *I.S.* and *F.S.* which are to be found. Knowing *R.L.* and *B.S.* of any point, the *F.S.* of the point can very easily be calculated. The readings having (×) mark are missing quantities which have been computed as shown in the tabular form.

Station	*B.S.*	*I.S.*	*F.S.*	*H.I.*	*R.L.*	*Remarks*
1	1.545			57.110	55.565	B.M.
2	2.310		× 2.125	57.295	54.985	T.P.
3		× 2.125			55.170	
4		× 1.030			56.265	
5		× 3.625			53.670	
6	0.105		× 3.355	54.045	53.940	T.P.
7		× 1.865			52.180	
8		× 2.030			52.015	
9	3.360		× 2.565	54.840	51.480	T.P.
10		× 1.695			53.145	
11			× 0.775		54.065	T.B.M.
Check	7.320		8.820		55.565	
			7.320		54.065	
		Fall	1.500	Fall	1.500	Checked

Due to collimation error each backsight staff reading is too great by an amount (80 tan 60″) metres. Also each change point *F.S.* reading is too great by an amount (30 tan 60″) metre. Taking both errors together, it is as if *F.S.* readings were correct and *B.S.* too great by amount (50 tan 60″) metres.

As there are four set-ups, the total *B.S.* reading are great by an amount 4 × 50 tan 60″ = 200 × 0.0002909 = 0.05818 ≈ 0.058 metres. Now greater the *B.S.* readings, higher will be the *H.I.* and, therefore, greater will be reduced levels calculated. The actual level of the *T.B.M.* will therefore, be = 54.065 – 0.058 = **54.007 m.**

9.12 CURVATURE AND REFRACTION

From the definition of a level surface and a horizontal line it is evident that a horizontal line departs from a level surface because of the curvature of the earth. Again, in the long sights, the horizontal line of sight does not remain straight but it slightly bends downwards having concavity towards earth surface due to refraction.

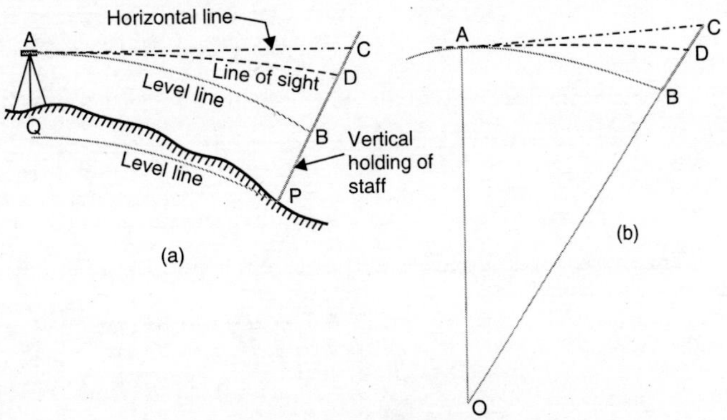

Fig. 9.38 Curvature and Refraction.

In Fig. 9.38 (*a*), *AC* is the horizontal line which deflects upwards from the level line *AB* by an amount *BC*. *AD* is the actual line of sight.

Curvature: *BC* is the departure from the level line. Actually the staff reading should have been taken at *B* where the level line cuts the staff, but since the level provides only the horizontal line of sight (in the absence of refraction), the staff reading is taken at the point *C*. Thus, the apparent staff reading is *more* and, therefore, the object appears to be *lower* than it really is. *The correction for curvature is, therefore, negative as applied to the staff reading,* its numerical value being equal to the amount *BC*. In order to find the value *BC*, we have, from Fig. 9.38 (*b*).

$$OC^2 = OA^2 + AC^2,\ \angle CAO \text{ being } 90°$$

Let

$BC = C_c$ = correction for curvature

$AB = d$ = horizontal distance between *A* and *B*

$AO = R$ = radius of earth in the same unit as that of d

$$\therefore \quad (R + C_c)^2 = R^2 + d^2$$

$$\text{or} \quad R^2 + 2RC_c + C_c^2 = R^2 + d^2$$

$$\therefore \quad C_c(2R + C_c) = d^2$$

$$\text{or} \quad C_c = \frac{d^2}{2R + C_c} \simeq \frac{d^2}{2R}, \text{ (Neglecting } C_c \text{ in comparison to } 2R)$$

That is, to find the curvature correction, divide the square of the length of sight by earth's diameter. Both d and R may be taken in the same units, when the answers will also be in terms of that unit. The radius of the earth can be taken equal to 6370 km. If d is to be in km, and R = 6370 km, $C_c = 0.07849\ d^2$ metres. In the above expression, d is to be substituted in km, while C_c will be in metres.

Refraction: The effect of refraction is the same as if the line of sight was curved downward, or concave towards the earth's surface and hence the rod reading is decreased. Therefore, the effect of refraction is to make the objects appear higher than they really are. *The correction, as applied to staff readings, is positive.* The refraction curve is irregular because of varying atmospheric conditions, but for average conditions it is assumed to have a diameter about seven times that of the earth.

The correction of refraction, C_r is therefore, given by

$$C_r = \frac{1}{7}\frac{d^2}{2R} \text{ (+ ve)} = 0.01121\, d^2 \text{ metres, when } d \text{ is in km.}$$

The combined correction due to curvature and refraction will be given by

$$C = \frac{d^2}{2R} - \frac{1}{7}\frac{d^2}{2R} = \frac{6}{7}\frac{d^2}{2R} \text{ (subtractive)}$$

$$= 0.06728\, d^2 \text{ metres, } d \text{ being in km.}$$

The corresponding values of the corrections in English units are:

$$\left.\begin{aligned} C_c &= \frac{2}{3} d^2 = 0.667\, d^2 \text{ feet} \\ C_r &= \frac{2}{21} d^2 = 0.095\, d^2 \text{ feet} \\ C &= \frac{4}{7} d^2 = 0.572\, d^2 \text{ feet} \end{aligned}\right\} \begin{array}{l} d \text{ is in miles and} \\ \text{radius of earth} = 3958 \text{ miles.} \end{array}$$

Distance to the Visible Horizon

In Fig. 9.39, let P be the point of observation, its height being equal to C and let A be the point on the *horizon i.e.,* a point where the tangent from P meets the level line. If d is the distance to visible horizon, it is given by

$$d = \sqrt{\frac{C}{0.06728}} \text{ km}$$

$$= 3.8553\sqrt{C} \text{ km, } C \text{ being in metres.}$$

(Taking both curvature and refraction into account).

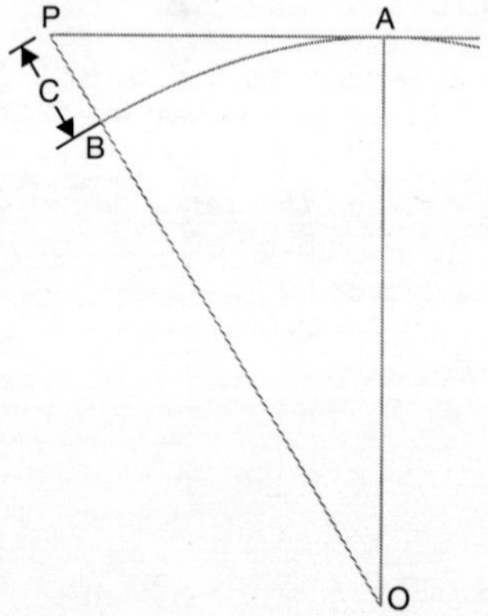

Fig. 9.39

Example 9.8: *Find the correction for curvature and for refraction for a distance of (a) 1200 metres (b) 2.48 km.*

Solution:

(*a*) Correction for curvature = $0.07849\, d^2$ metres (where d is in km)

$= 0.07849\, (1.2)^2 =$ **0.113 m**

Correction for refraction = $\frac{1}{7}C_c =$ **0.016 m**

(*b*) Correction for curvature = $0.07849\, (2.48)^2 =$ **0.483 m**

Correction for refraction = $\frac{1}{7}C_c =$ **0.069 m.**

Example 9.9: Find the combined correction for curvature and refraction for distance of (a) 3400 metres (b) 1.29 km.

Solution:

(*a*) Combined correction for curvature and refraction

$= 0.06728\ d^2\ \text{m} = 0.06728\ (3.40)^2 = \mathbf{0.778\ m.}$

(*b*) Combined correction $= 0.06728\ (1.29)^2 = \mathbf{0.112\ m.}$

Example 9.10: In order to find the difference in elevation between two points P and Q, a level was set upon the line PQ, 60 metres from P and 1280 metres from Q. The readings obtained on staff kept at P and Q were respectively 0.545 metre and 3.920 m. Find the true difference in elevation between P and Q.

Solution: Since the distance of P from instrument is small, the correction for curvature etc. is negligible.

Combined correction for Q $= 0.06728\ (1.280)^2 = 0.110$ m (Subtractive)

$\therefore$ Correct staff reading at Q $= 3.920 - 0.110 = 3.810$ m

$\therefore$ Difference in elevation between P and Q $= 3.810 - 0.545 = \mathbf{3.265\ m,}$ Q being lower.

Example 9.11: A light-house is visible just above the horizon at a certain station at the sea level. The distance between the station and the light-house is 50 km. Find the height of the light-house.

Solution: The height of the light-house is given by

$$C = 0.06728\ d^2 \text{ metres} = 0.06728\ (50)^2 \text{ metres} = \mathbf{168.20\ m}$$

Example 9.12: An observer standing on the deck of a ship just sees a light-house. The top of the light-house is 42 metres above the sea level and the height of the observer's eye is 6 metres above the sea level. Find the distance of the observer from the light-house.

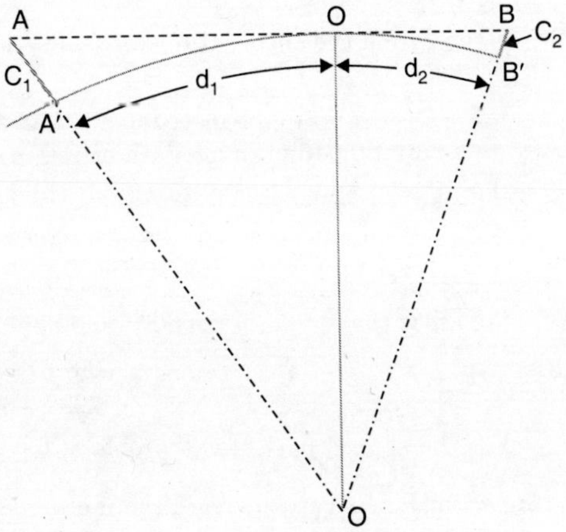

Fig. 9.40

Solution: (Fig. 9.40)

Let A be the position of the top of light-house and B be the position of observer's eye. Let AB be tangential to water surface at O.

The distances d_1 and d_2 are given by

$$d_1 = 3.8553\ \sqrt{C_1} \text{ km}$$

$$= 3.8553\ \sqrt{42} = 24.985 \text{ km}$$

and $$d_2 = 3.8553\ \sqrt{6} = 9.444 \text{ km}$$

$\therefore$ Distance between A and $B = d_1 + d_2$

$= 24.985 + 9.444 = \mathbf{34.429\ km.}$

Example 9.13: The observation ray between two triangulation stations A and B just grazes the sea. If the heights of A and B are 9,000 metres and 3,000 metres respectively, determine approximately the distance AB (Diameter of earth 12,880 km).

Solution: In Fig. 9.40, let A and B be the two triangulation stations and let O be the point of tangency on the horizon.

Let
$$A'A = C_1 = 9000 \text{ metres} = 9 \text{ km}$$
$$B'B = C_2 = 3000 \text{ metres} = 3 \text{ km}$$

The distance d_1 is given by $C_1 = \dfrac{d_1^2}{2R}$

or
$$d_1 = \sqrt{2RC_1} \text{ in which } d_1, R \text{ and } C_1 \text{ are in same units}$$

$\therefore$
$$d_1 = \sqrt{2 \times 6440 \times 9.0} = 340.48 \text{ km}$$

Similarly
$$d_2 = \sqrt{2RC_2} = \sqrt{2 \times 6440 \times 3.0} = 196.58 \text{ km}$$

$\therefore$
$$\text{Distance } AB = d_1 + d_2 = 340.48 + 196.58 = \mathbf{537.06 \text{ km.}}$$

Example 9.14: *Two pegs A and B are 150 metres apart. A level was set up in the line AB produced and sights were taken to a staff held in turn on the pegs, the reading being 1.962 (A) and 1.276(B), after the bubble has been carefully brought to the centre of its run in each case. The reduced level of the tops of the pegs A and B are known to be 120.684 and 121.324 m respectively.*

Determine (a) the angular error of the collimation line in seconds, and (b) the length of sight for which the error due to curvature and refraction would be the same as collimation error. Assume the radius of the earth to be 6370 km.

Solution: Observed difference in elevation between A and B = 1.962 – 1.276 = 0.686 m (A being lower)

The difference in elevation = 121.324 – 120.684 = 0.640 m, A being lower.

Hence, from the observations, A seems to be lower by an additional amount = 0.686 – 0.640 = 0.046 m.

Since B is nearer to the instruments than A, it is clear that the line of sight is inclined upwards by an amount 0.046 m in a length of 150 m.

If α is the angular inclination (upwards) of the line of sight with horizontal,

$$\tan \alpha = \frac{0.046}{150} = 0.0003067$$

We know that
$$\tan 60'' = 0.0002909$$

$\therefore$
$$\alpha = \frac{3067 \times 60}{2909 \times 60} \text{ minutes} = \mathbf{1'\ 3''} \text{ (upwards).}$$

For the second part of the problem, let the required line of sight be L km. The combined correction for curvature and refraction would be $\dfrac{6}{7}\dfrac{L^2}{2R}$ (negative). The correction for collimation error in a length L will be $L \tan \alpha$. Equating the two,

$$\frac{6}{7}\frac{L^2}{2R} = L \tan \alpha = L\ (0.0003067)$$

$\therefore$
$$L = \frac{0.0003067 \times 7 \times 2}{6} \times 6370 = \mathbf{4.557 \text{ km.}}$$

9.13 RECIPROCAL LEVELLING

When it is necessary to carry levelling across a river, ravine or any obstacle requiring a long sight between two points so situated that no place for the level can be found from which the lengths of foresight and backsight will be even approximately equal, special method *i.e., reciprocal levelling* must be used to obtain accuracy and to eliminate the following: (1) error in instrument adjustment; (2) combined effect of earth's curvature and the refraction of the atmosphere, and (3) variations in the average refraction.

Let A and B be the points and observations be made with a level, the line of sight of which is inclined upwards when the bubble is in the centre of its run. The level is set at a point near A and staff readings are taken on A and B with the bubble in the centre of its run. Since B.M. A is very near to instrument, no error due to curvature, refraction and collimation will be introduced in the staff readings at A, but there will be an error e in the staff reading on B. The level is then shifted to the other bank, on a point very near B.M. B, and the readings are taken on staff held at B and A. Since B is very near, there will be no error due to the three factors in reading the staff, but the staff reading on A will have an error e. Let h_a, and h_b be the corresponding staff readings on A and B for the first set of the level and h_a' and h_b' be the readings for the second set.

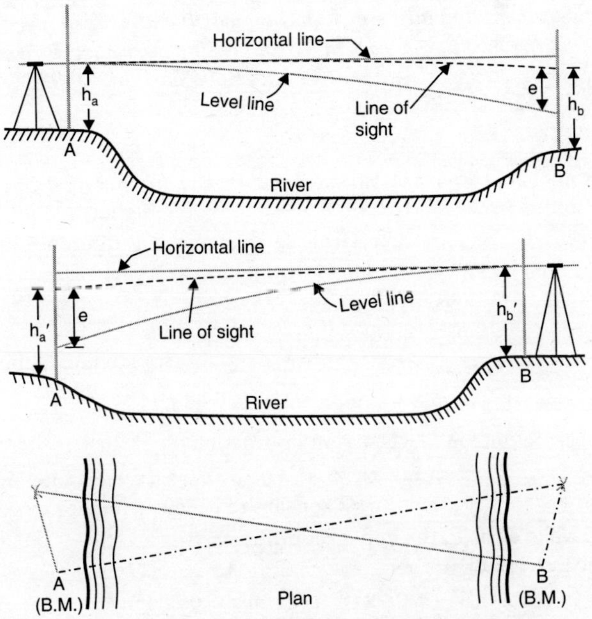

Fig. 9.41 Reciprocal Levelling.

From Fig. 9.41, it is evident that for the first set of the level, the correct staff readings will be

$$\text{On } A : h_a \text{ ; On } B : h_b - e$$

$\therefore$ True difference in elevation $= H = h_a - (h_b - e)$

Similarly for second set, the correct staff reading will be:

$$\text{On } A : h_a' - e \text{ ; On } B : h_b'$$

∴ True difference in elevation = $H = (h_a' - e) - h_b'$

Taking the average of the two true differences in elevations, we get

$$2H = [h_a - (h_b - e) + (h_a' - e) - h_b'] = (h_a - h_b) + (h_a' - h_b')$$

$$\therefore \quad H = \tfrac{1}{2}[(h_a - h_b) + (h_a' - h_b')].$$

The true difference in elevation, therefore, is equal to the mean of the two appearent differences in elevations, obtained by reciprocal observations.

Example 9.15: *The following notes refer to reciprocal levels taken with one level:*

Inst. at	*Staff readings on*		*Remarks*
	P	*Q*	
P	*1.824*	*2.748*	*Distance between P and Q = 1010 m*
Q	*0.928*	*1.606*	*R.L. of P = 126.386.*

Find (a) true R.L. of Q, (b) the combined correction for curvature and refraction, and (c) the angular error in the collimation adjustment of the instrument.

What will be the difference in answers of (a) and (c) if observed staff readings were 2.748 on P and 1.824 on Q, the instrument being at P; and 1.606 on P and 0.928 on Q, the instrument being at Q.

Solution:

(*a*) When the observations are taken from *P*, the apparent difference in elevation between *P* and *Q* = 2.748 – 1.824 = 0.924 m, *P* being higher

When the observations are taken from *Q*, the apparent difference in elevation between *P* and *Q* = 1.606 – 0.928 = 0.678, *P* being higher.

Hence, the true difference in elevation

$$= \frac{0.924 + 0.678}{2} = 0.801 \text{ m, } P \text{ being higher}$$

and true elevation of *Q* = 126.386 – 0.801 = **125.585** m.

(*b*) Combined correction for curvature and refraction

$$= 0.06728\, d^2 = 0.06728\,(1.010)^2 = \mathbf{0.069\ m}$$

(*Q* appears to be lower further by 0.069 m due to this).

(*c*) When the level was at *P*, the apparent difference in elevation = 0.924 m. The difference in elevation = 0.801 m

Error in observation = 0.924 – 0.801 m = + 0.123 m

This error consists of (*i*) error due to curvature and refraction (*ii*) collimation error. Error due to curvature and refraction = + 0.069 m

Error due to collimation = 0.123 – 0.069 = + 0.054 m.

Collimation error is said to be positive when the line of sight is so inclined that it increases the staff reading at the farther point thereby making that point appear lower than what it is. Hence, the line of sight is inclined upward by an amount 0.054 m in a distance of 1010 metres.

$$\therefore \qquad \tan \alpha = \frac{0.054}{1010} = 0.0000535$$

But $$\tan 60'' = 0.0002909$$

$$\therefore \qquad \alpha = \frac{536 \times 60}{2909} = \mathbf{11''} \text{ (upwards)}$$

If the staff readings are interchanged, then

(*a*) True difference in R.L. between *P* and *Q* will be 0.801 m (*Q* being higher)

R.L. of $Q = 126.386 + 0.801 = \mathbf{127.187\ m.}$

(*b*) When the instrument is at *P*, the apparent difference in elevation between *P* and *Q* = 0.924 m, *Q* being higher.

Hence, *Q* appears to be *higher by a further amount of* 0.924 – 0.801 = **0.123 m.**

This error is due to (*i*) curvature and refraction, and (*ii*) faulty adjustment of line of collimation.

Considering (*i*), the curvature and refraction tends to increase the staff reading at *Q*, thereby making *Q* appear *lower* than what it is by an amount 0.069 m (as already found out), but by actual observations, the point *Q* has been made to appear *higher* than what it is by an amount 0.123 m. Hence, it is clear that the line of sight is inclined *downwards* by an amount 0.123 + 0.069 = 0.192 m in a distance of 1010 m.

If α is the inclination of line of sight, we have

$$\tan \alpha = \frac{0.192}{1010} = 0.000190$$

But $$\tan 60'' = 0.0002909$$

$$\therefore \qquad \alpha = \frac{1900 \times 60}{2909} = \mathbf{39''} \text{ (downwards).}$$

Example 9.16: *In levelling between two points A and B on opposite banks of a river, the level was set up near A, and the staff readings on A and B were 1.285 and 2.860 m respectively. The level was then moved and set up near B and the respective readings on A and B were 0.860 and 2.220. Find the true difference of level between A and B.*

Solution: *When the instrument is at A,*

Appearent difference in elevation between *A* and *B*

$$= 2.860 - 1.285 = 1.575 \text{ m } (A \text{ higher})$$

When the instrument is at B,

Apparent difference in elevation between *A* and *B*

$$= 2.220 - 0.860 = 1.360 \text{ m } (A \text{ higher})$$

$$\text{True difference in elevation} = \frac{1.575 + 1.360}{2} = \mathbf{1.468\ m} \text{ } (A \text{ higher}).$$

Example 9.17: *Two points A and B are 1530 m apart across a wide river. The following reciprocal levels are taken with one level:*

Level at	*Readings on A*	*Readings on B*
A	*2.165*	*3.810*
B	*0.910*	*2.355*

The error in the collimation adjustments of the level is – 0.004 m in 100 m. Calculate the true difference of level between A and B and the refraction.

Solution:

(*i*) True difference in level between A and B

$$= \frac{(3.810-2.165)+(2.355-0.910)}{2} = \mathbf{1.545\ m.}$$

(*ii*) Error due to curvature = 0.07849 d^2 metres = 0.07849 $(1.53)^2$ = 0.184 m

∴ When the level is at A, corrected staff reading on $B = 3.810 - (C_c - C_r) + C_1$

where C_c = correction due to curvature = 0.184 m

C_r = correction due to refraction

$$C_1 = \text{correction due to collimation} = \frac{0.004}{100} \times 1530 = 0.0612 \text{ m}$$

∴ Corrected staff reading on $B = 3.810 - (0.184 - C_r) + 0.0612 = 3.6872 + C_r$

∴ True difference in level between A and $B = (3.6872 + C_r - 2.165) = (1.5222 + C_r)$

But it is equal to 1.545 m.

∴ $1.5222 + C_r = 1.545$

or $C_r = 1.545 - 1.5222 = 0.0228 \simeq \mathbf{0.023\ m.}$

9.14 PROFILE LEVELLING (LONGITUDINAL SECTIONING)

Profile levelling is the process of determining the elevations of points at short measured intervals along a fixed line such as the centre line of a railway, highway, canal or sewer. The fixed line may be a single straight line or it may be composed of a succession of straight lines or of a series of straight lines connected by curves. It is also known as *longitudinal sectioning*. By means of such sections the engineer is enabled to study the relationship between the existing ground surface and the levels of the proposed construction in the direction of its length. The profile is usually plotted on specially prepared profile paper, on which the vertical scale is much larger than the horizontal, and on this profile, various studies relating to the fixing of grades and the estimating of costs are made.

Field Procedure: Profile levelling, like differential levelling, requires the establishment of turning points on which both back and foresights are taken. In addition, any number of intermediate sights may be obtained on points along the line from each set up of the instrument (Fig. 9.42). In fact, points on the profile line are merely intermediate stations. It is generally best to set up the level to one side of the profile line to avoid too short sights on the points near the instrument. *For each set up, intermediate sights should be taken after the foresight on the next turning station has been taken.* The level is then set up in an advanced position and a backsight is taken on that turning point. The position of the intermediate points on the profile are simultaneously located by chaining along the profile and noting their distances from the point of commencement. When the vertical profile of the ground is regular or gradually curving, levels are taken on points at equal distances apart and generally at intervals of a chain length. On irregular ground where abrupt changes of slope occur, the points should be chosen nearer. For purose of checking and future reference, temporary bench marks should be established along the section.

Field notes for profile levelling are commonly kept in the standard form shown in the table on next page. The method is almost the same as given for 'collimation height' method as computations are easier by that method. The distances of the points on the profile are also recorded. The values shown in the table are same as those illustrated in Fig. 9.42.

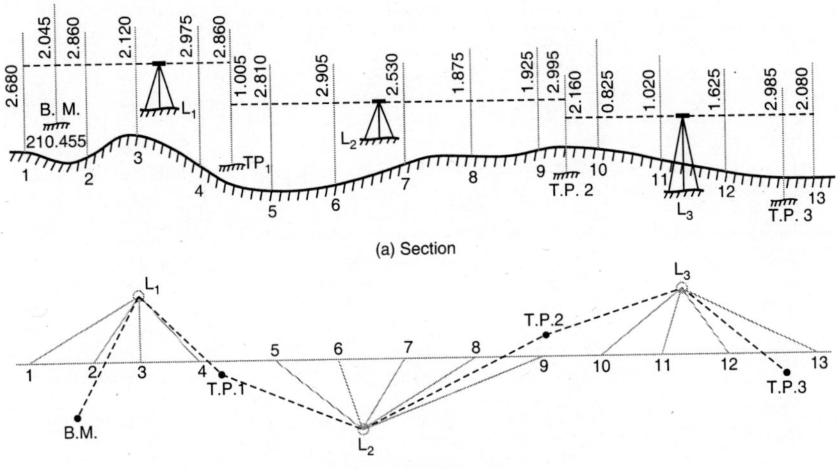

Fig. 9.42 Profile Levelling.

Plotting the Profile (Fig. 9.43)

The horizontal distances are plotted along the horizontal axis to some convenient scale and the distances are also marked. The elevations are plotted along the vertical axis. Each ground point is thus plotted by the two co-ordinates (*i.e.*, horizontal distance and vertical elevation). The various points so obtained are joined by straight lines, as shown in Fig. 9.43, where the readings of the above table are plotted.

Generally, the horizontal scale is adopted as 1 cm = 10 m (or 1″ = 100 ft). The vertical scale is not kept the same but is exaggerated so that the inequalities of the ground appear more apparent. The vertical scale is kept 10 times the horizontal scale (*i.e.*, 1 cm = 1 m). The reduced levels of the points are also written along with the horizontal distances.

LEVEL FIELD NOTES FOR PROFILE LEVELLING

Station	*Distance*	*B.S.*	*I.S.*	*F.S.*	*H.I.*	*R.L.*	*Remarks*
B.M.		2.045			212.500	210.455	
1	0		2.680			209.820	
2	10		2.860			209.640	
3	20		2.120			210.380	
4	35		2.975			209.525	
T.P. 1		1.005		2.860	210.645	209.640	
5	45		2.810			207.835	
6	63		2.905			207.740	

(Contd.)

(Contd.)

Station	*Distance*	*B.S.*	*I.S.*	*F.S.*	*H.I.*	*R.L.*	*Remarks*
7	80		2.530			208.115	
8	98		1.875			208.770	
9	115		1.925			208.720	
T.P. 2		2.160		2.995	209.810	207.650	
10	125		0.825			208.985	
11	145		1.020			208.790	
12	162		1.625			208.185	
13	180		2.080			207.730	
T.P. 3				2.985		206.825	
		5.210		8.840		210.455	
				5.210		206.825	
Check			Fall	3.630	Fall	3.630	

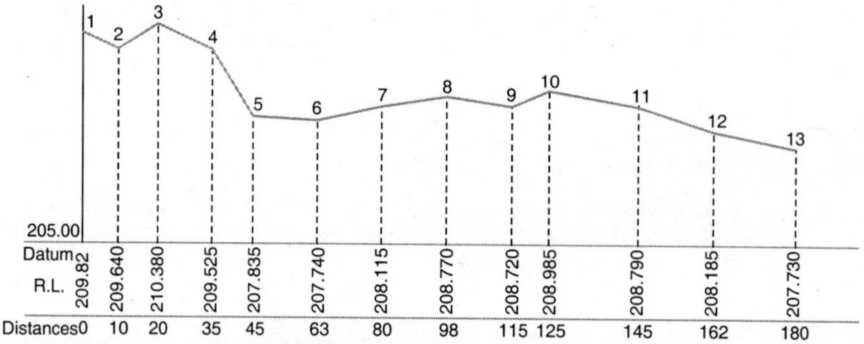

Longitudinal Section

Scale { Hor. 1 cm = 10 m; Ver. 1 cm = 1 m

Fig. 9.43

Levelling to Establish Grade Points: This kind of levelling, often referred to as *giving elevations* is used in all kinds of engineering construction. The operation of establishing grade points is similar to profile levelling and follows the latter. After the profile has been plotted and the grade line has been established on the profile map, the grade elevation for each station is known. The amounts of cut or fill at each point are thus determined before going into the field. The levelling operation starts from the bench mark and is carried forward by turning points. The grade point is established by *measuring down* from the height of the instruments a distance equal to the *grade rod reading,* using the following relation.

Grade point elevation = H.L. – Grade rod reading.

A grade stake is driven in the ground and grade rod is kept on the top of it and read with the help of level. The stake is driven in or *out* till the grade rod reading is the same as calculated above. Before proceeding the work in the field, a table is generally prepared giving the rod readings at each point to set it on a given gradient. Example 9.18 makes the procedure clear.

Example 9.18: *In running fly levels from a bench mark of R.L. 183.215, the following readings were obtained:*

B.S.	*1.215*	*2.035*	*1.980*	*2.625*
F.S.	*0.965*	*3.830*	*0.980*	

From the last position of the instrument, five pegs at 20 metres intervals are to be set out on a uniform rising gradient of 1 in 40; the first peg is to have a R.L. of 181.580. Work out the staff readings required for setting the tops of the pegs on the given gradient.

Solution: In the first part of the question, fly levelling was done, the computations for which can be done as usual. For the last setting of the instrument, when a backsight is taken on station No. 4, the height of collimation comes out to be 185.205. The R.L. of the first peg is to be 181.580. Hence Grade rod reading = H.I. – Grade point elevation = 185.205 – 181.580 = 3.625. The reading is entered in the I.S. column. The R.L. of the next peg at the rising gradient of 1 in 40 will be $181.580 + 1 \times \frac{20}{40} = 182.080$ and its grade rod reading will be 185.205 – 182.080 = 3.125. Similarly, the rod readings for other pegs are calculated as entered in the table given below:

S. No.	*Dist.*	*B.S.*	*I.S.*	*F.S.*	*H.I.*	*R.L.*	*Remarks*
1		1.215			184.340	183.125	
2		2.035		0.965	185.410	183.375	
3		1.980		3.830	183.560	181.580	
4		2.625		0.980	185.205	182.580	
5	0		3.625			181.580	Peg 1
6	20		3.125			182.080	Peg 2
7	40		2.625			182.580	Peg 3
8	60		2.125			183.080	Peg 4
9	80			1.625		183.580	Peg 5
Check		7.855		7.400		183.580	
		7.400				183.125	
	Rise	0.455			Rise	0.455	Checked

9.15 CROSS-SECTIONING

Cross-sections are run at right angles to the longitudinal profile and on either side of it for the purpose of lateral outline of the ground surface. They provide the data for estimating quantities of earth work and for other purposes. The cross-sections are numbered consecutively from the commencement of the centre line and are set out at right angles to the main line of section with the chain and tape, the cross-staff or the optical square and the distances are measured left and right from the centre peg (Fig. 9.44). Cross-section may be taken at each chain. The length of cross-section depends upon the nature of work.

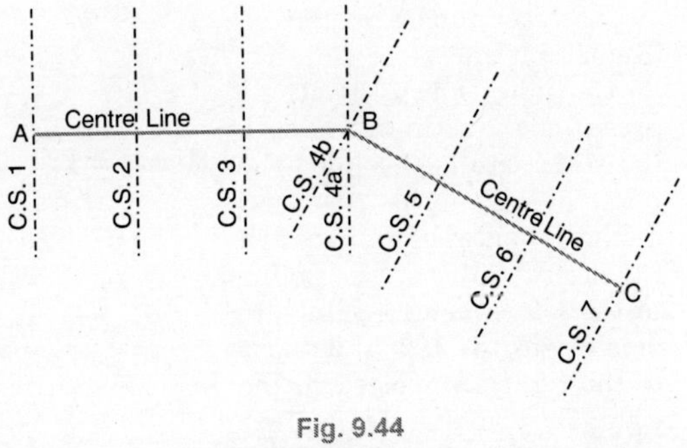

Fig. 9.44

The longitudinal and cross-sections may be worked together or separately. In the former case, two additional columns are required in the level field book to give the distances, left and right of the centre line, as illustrated in table below. To avoid confusion, the bookings of each cross-section should be entered separately and clearly and full information as to the number of the cross-section, whether on the left or right of the centre line, with any other matter which may be useful, should be recorded.

Station	*Distance (m)*			*B.S.*	*I.S.*	*F.S.*	*H.I.*	*R.L.*	*Remarks*
	L	*C*	*R*						
B.M.				1.325			101.325	100.000	
0		0			1.865			99.460	Cross-
L_1	3				1.905			99.420	section
L_2	6				2.120			99.205	at 0 m
L_3	9				2.825			98.500	chainage
R_1			3		1.705			99.620	
R_2			7.5		1.520			99.805	
R_3			10		1.955			99.370	
I		20			1.265			100.060	Cross-
L_1	3				1.365			99.960	section at
L_2	6				0.725			100.600	20 m
L_3	9				2.125			99.200	chainage
R_1			3		1.925			99.400	
R_2			7		2.250			99.075	
R_3			10		0.890			100.435	
T.P.						2.120		99.205	
Check				1.325		2.120		100.000	
						1.325		99.205	
					Fall	0.795	Fall	0.795	

Plotting the Cross-section (Fig. 9.45)

Cross-sections are plotted almost in the same manner as the longitudinal sections except that in this case both the scales are kept equal. The point along the longitudinal section is plotted at the centre of the horizontal axis. The points to the left of centre point are plot-ted to the left and those to the right are plotted to the right. The points so obtained are joined by straight lines.

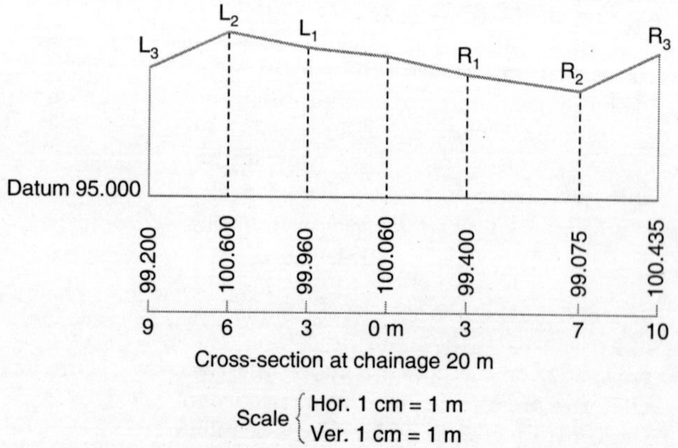

Fig. 9.45

9.16 LEVELLING PROBLEMS

The following are some of the difficulties commonly encountered in levelling:

1. **Levelling on Steep Slope.** See § 9.11.
2. **Levelling on Summits and Hollows:** In levelling over summit, level should be set up sufficiently high so that the summit can be sighted without extra setting (Fig. 9.46). Similarly, in levelling across a hollow, level should be set *only sufficiently low* to enable the levels of all the required points to be observed (Fig. 9.47).

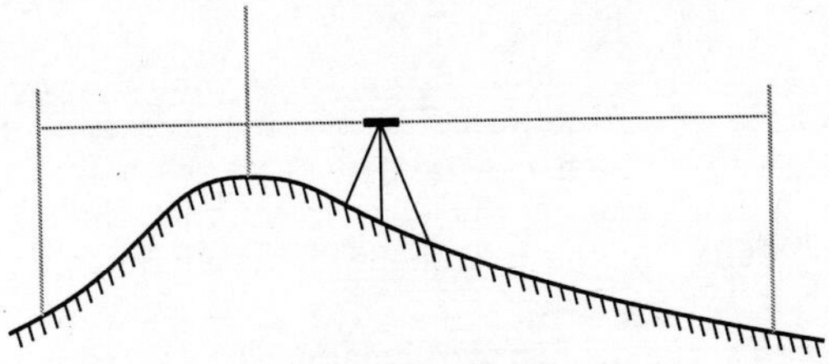

Levelling over summit

Fig. 9.46

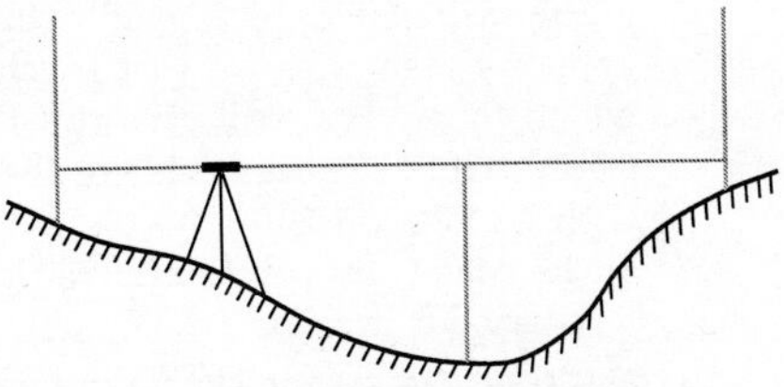

Levelling across hollow

Fig. 9.47

3. **Taking Level of an Overhead Point:** When the point under observation is higher than the line of sight, staff should be kept inverted on the overhead point keeping the foot of the staff touching the point, and reading should be taken. Such reading will show height of that point above the line of sight and should be added to the H.I. to get the R.L. of the point (Fig. 9.48). On the contrary, if such point happens to be temporary bench mark, the backsight reading on the point should be subtracted from the reduced level to get the H.I. (See example 9.5). So that, there may be no opportunity for mistake, it is well also to make a note on the description page that the staff has been held inverted.

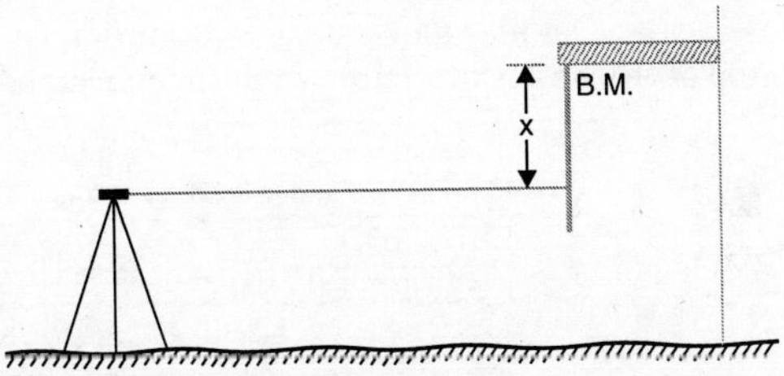

Fig. 9.48

4. **Levelling Ponds and Lakes too Wide to be Sighted Across** (Fig. 9.49): When the ponds and lakes are too wide to be sighted across, advantage may be taken of the fact that the surface of still water is a level surface. A peg may be driven at one end of the pond, keeping its top flush with the water surface. A similar peg may be driven to the other side. Level may be first set to one side the staff kept on the peg and reading taken. The R.L. of the top of the peg and of water surface is thus known. The instrument is then set on other side near the bank

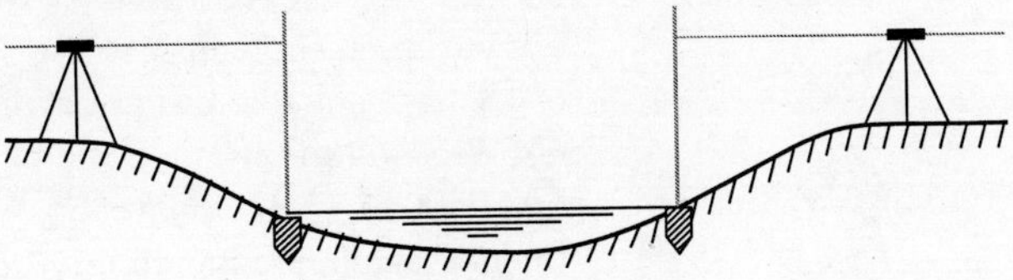

Levelling across ponds

Fig. 9.49

and reading is taken by keeping the staff on the top of the second peg. Adding the staff reading to the R.L. of the peg, the R.L. of instrument axis is known and the levelling operations can be carried further.

5. **Levelling Across River:** If the width of the river is less, the method of reciprocal levelling is to be used. If the river is too wide to be sighted *across,* levelling may be continued from one side to the other in the manner shown in (4) with little error, provided care is taken to choose a comparatively still stretch and to see that water levels are taken at points directly opposite each other.

6. **Levelling Past High Wall:** Two cases may arise. In the first case, when the height of the wall *above* the line of sight is lesser than the length of the staff, the staff can be kept inverted with its foot touching the top and reading taken. Such reading, when added to the H.I. will give the R.L. of the top of the wall. The instrument may then be shifted to the other side of the wall and reading may be taken on the inverted staff with its foot touching the top of the wall. Such reading when subtracted from the R.L. of the top of the wall will give the H.I. Knowning the H.I., the levelling operation can be carried forward.

In the second case, when the height of the wall above the line of collimation is more than the length of the staff, a suitable mark is made at the height, where the line of sight intersects the face of the wall. The vertical distance between the mark and the top of the wall is measured. The R.L. of the top of the wall is thus known. The instrument is then set to the other side of the wall and a similar mark at the collimation level is made on the wall. The vertical height of the top of the wall is measured from the mark and the height of the instrument is then calculated.

9.17 ERRORS IN LEVELLING

All levelling measurements are subject to three principal sources of errors:

1. **Instrumental**
   (*a*) Error due to imperfect adjustment.
   (*b*) Error due to sluggish bubble.
   (*c*) Error due to movement of objective slide.
   (*d*) Rod not of standard length.
   (*e*) Error due to defective joint.
2. **Natural**
   (*a*) Earth's curvature.
   (*b*) Atmospheric refraction.
   (*c*) Variations in temperature.
   (*d*) Settlement of tripod or turning points.
   (*e*) Wind vibrations.
3. **Personal**
   (*a*) Mistakes in manipulation.
   (*b*) Mistake in rod handling.
   (*c*) Errors in sighting.
   (*d*) Mistakes in reading the rod.
   (*e*) Mistakes in recording and computing.

Instrumental Errors

(*a*) **Error due to Imperfect Adjustment:** The essential adjustment of a level is that the line of sight shall be parallel to axis of the bubble tube. If the instrument is not in this adjustment, the line of sight will either be inclined upwards or downwards when the bubble is centred and the rod readings will be incorrect. The error in the rod reading will be proportional to the distance and can be eliminated by balancing the backsight and foresight distances. The error is likely to be cumulative, particularly in going up or down a steep hill, where all backsights are longer or shorter than all foresights unless care is taken to run a zigzag line.

(b) **Error due to Sluggish Bubble:** If the bubble is sluggish, it will come to rest in wrong position, even though it may creep back to correct position while the sight is being taken. Such a bubble is a constant source of annoyance and delay. However, the error may be partially avoided by observing the bubble *after* the target has been sighted. The error is compensating.

(c) **Error due to Movement of Objective Slide:** In the case of external focusing instruments, if the objective slide is slightly worn out, it may not move in truly horizontal direction. In the short sights, the objective slide is moved out nearly its entire length and the error is, therefore, more. Due to this reason, extremely short sights are to be avoided. The error is *compensating* and can be eliminated by balancing backsight and foresight, since in that case, focus is not changed and hence, the slide is not moved.

(d) **Rod not of Standard Length:** Incorrect lengths of divisions on a rod cause errors similar to those resulting from incorrect marking on a tape. The error is systematic and is directly proportional to the difference in elevation. If the rod is too long, the correction is added to a measured difference in elevation; if the rod is too short, the correction is subtracted. Uniform wearing of the shoe at the bottom of the rod makes H.I. values incorrect, but the effect is cancelled when included in both back and foresight readings. For accurate levelling, the rod graduation should be tested and compared with any standard tape.

(e) **Error due to Defective Joint:** The joint of the extendable rods may be worn out from setting the rod down *'on the run'* and from other sources. The failure to test the rod at frequent interval may result in a large *cumulative* error.

Natural Errors

(a) **Earth's Curvature:** The effect of curvature is to increase the rod readings. When the distances are small the error is negligible, but for greater distances when the back and foresights are not balanced, a systematic error of considerable magnitude is produced.

(b) **Refraction:** Due to refraction, the ray of light bends downwards in the form of curve with its concavity towards the earth surface, thus decreasing the staff readings. Since the atmospheric refraction often changes rapidly and greatly in short distance, it is impossible to eliminate entirely the effect of refraction even though the backsight and foresight distances are balanced. It is particularly uncertain when the line of sight passes close to the ground. Errors due to refraction tend to be compensating over a long period of time but may be cumulative on a full day's run.

(c) **Variation in Temperature:** The effect of variation in temperature on the adjustment of the instrument is not of much consequence in levelling of ordinary precision, but it may produce an appreciable error in precise work. The adjustment of the instrument is temporarily disturbed by unequal heating and the consequent warping and distortion. The heating of the level vial will cause the liquid to expand and bubble to shorten. If one end of the vial is warmed more than the other, the bubble will move towards the heated end and appreciable errors will be produced. In precise levelling, it is quite possible that errors from change of length of levelling rod from variations in temperature may exceed the errors arising from the levelling itself. Heat waves near the ground surface or adjacent to heated objects make the rod appear to wave and prevent accurate sighting. The heating effect is practically eliminated by shielding the instrument from the rays of the sun. The error is usually accidental, but under certain conditions it may become systematic.

(*d*) **Settlement of Tripod on Turning Point:** If the tripod settles in the interval that elapses between taking a backsight and the following foresight, the observed foresight will be too small and the elevation of the turning point will be too great. Similarly, if a turning point settles in the interval that elapses between taking a foresight and the following backsight in the next set up, the observed backsight will be too great and H.I. calculated will be too great. Thus, whether the tripod settles or the turning point settles, the error is always systematic and the resulting elevation will always be too *high*.

(*e*) **Wind Vibrations:** High wind shakes the instrument and thus disturbs the bubble and the rod. Precise levelling work should never be done in high wind.

Personal Errors

(*a*) **Mistakes in Manipulation:** These include mistakes in setting up the level, imperfect focusing of eye-piece and of objective, errors in centring the bubble and failure to watch it after each sight, and errors due to resting the hands on tripods or telescope. In the long sights, the error due to the bubble not being centred at the time of sighting are more important. Habit should be developed of checking the bubble before and after each sight. Parallax caused by improper focusing result in incorrect rod readings; it produces an accidental error and can be eliminated by carefully focusing.

(*b*) **Mistake in Rod Handling:** If the rod is not in plumb, the reading taken will be too great. The error varies directly with the magnitude of the rod reading and directly as the square of the inclination. In running a line of levels uphill, backsight readings are likely to be increased more than foresight from this source and the elevation of a bench mark on top will be too great. Similarly, the elevation of a bench mark at the bottom, while levelling downhill, will be too small. Thus, a positive systematic error results. Over level ground, the resultant error is accidental since the backsights are about equal to the foresights. The error can be minimised by carefully plumbing the rod either by eye estimation or by using a rod level, a special attachment devised for plumbing the rod or by waving the level rod slowly towards or away from the level thereby taking the minimum rod reading. Vertical cross-hair may be used to plumb the rod in the direction transverse to the line of sight.

(*c*) **Errors in Sighting:** The error is caused when it is difficult to tell when the cross-hair coincides with the centre of the target in a target rod and to determine the exact reading which the cross-hair appears to cover in the case of self-reading rod. This is an accidental error the magnitude of which depends upon the coarseness of the cross-hair, the type of rod, the form of target, atmospheric conditions, length of sight and the observer.

(*d*) **Mistakes in Reading the Rod**

The common mistakes in reading the rod are:

(*i*) Reading upwards, instead of downwards.

(*ii*) Reading downwards, instead of upwards when the staff is inverted.

(*iii*) Reading wrong metre mark when the staff is near the level and only *one* metre mark is visible through the telescope.

(*iv*) To omit a zero or even two zeros from a reading. For example, 1.28 instead of 1.028 or 1.06 instead of 1.006.

(*v*) Reading against a stadia hair.

(*vi*) Concentrating more attention on decimal part of the reading and noting whole metre reading wrongly.

(*e*) **Mistakes in Recording and Computing**

The common mistakes are:

(*i*) Entering the reading with digits interchanged *i.e.*, 1.242 instead of 1.422.

(*ii*) Entering backsights and foresights in a wrong column.

(*iii*) Mistaking the numerical value of reading called out by the level man.

(*iv*) Omitting the entry.

(*v*) Entering wrong remark against a reading.

(*vi*) Adding a foresight instead of subtracting it and/or subtracting a backsight reading instead of adding it.

(*vii*) Ordinary arithmetical mistakes.

Example 9.19: *Find the error of reading of a level staff if the observed reading is 3.845 m at the point sighted, the staff being 15 cm off the vertical through the bottom.*

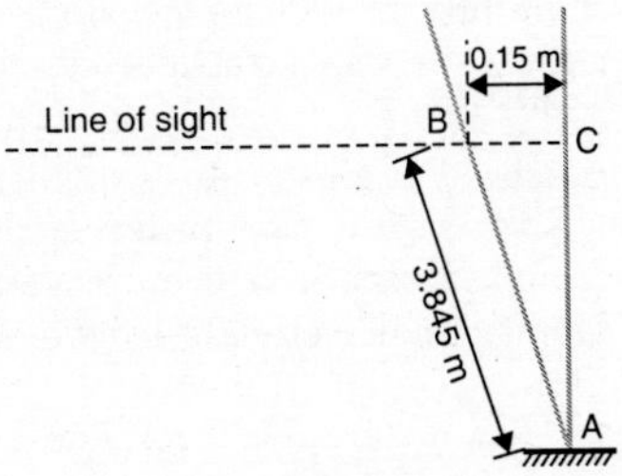

Fig. 9.50

Solution: In Fig. 9.50, let *AB* be the observed staff reading and let *AC* be the correct staff reading.

Evidently,

$$AC = \sqrt{AB^2 - BC^2}$$

$$= \sqrt{(3.845)^2 - (0.15)^2}$$

$$= \mathbf{3.842.}$$

9.18 DEGREE OF PRECISION

The degree of precision depends upon (*i*) the type of instrument, (*ii*) skill of observer, (*iii*) character of country, and (*iv*) atmospheric conditions. For a given instrument and atmospheric conditions, the precision depends upon the number of set-ups and also upon the length of sights. Thus, the precision on plains will be more than that on hills. No hard and fast rules can be laid down by means of which a desired precision can be maintained. However, the permissible closing error can be expressed as

$$E = C\sqrt{M} \text{ (in English units)} \quad \text{or} \quad E' = C'\sqrt{K} \text{ (in metric units)}$$

where E = permissible closing error in feet; C = constant; M = distance in miles

E' = permissible closing error in mm; C' = constant; K = distance in km.

The following table gives the different values:

Type of survey and purpose	Error in feet (E)	Error in mm (E')
1. *Rough levelling* for reconnaissance or preliminary surveys.	$\pm 0.4\sqrt{M}$	$\pm 100\sqrt{K}$
2. *Ordinary* levelling for location and construction surveys.	$\pm 0.1\sqrt{M}$	$\pm 24\sqrt{K}$
3. *Accurate levelling* for principal bench marks or for extensive surveys.	$\pm 0.05\sqrt{M}$	$\pm 12.0\sqrt{K}$
4. *Precise levelling* for bench marks of widely distributed points.	$\pm 0.017\sqrt{M}$	$\pm 4\sqrt{K}$

9.19 THE LEVEL TUBE

The level tube or bubble tube gives the direction of horizontal plane because the surface of a still liquid at all points is at right angles to the direction of gravity, and the liquid alone will, therefore, provide a *level surface*. For ordinary surveys the radius of the earth is so large that a level surface is considered to be the same thing as a *horizontal plane*.

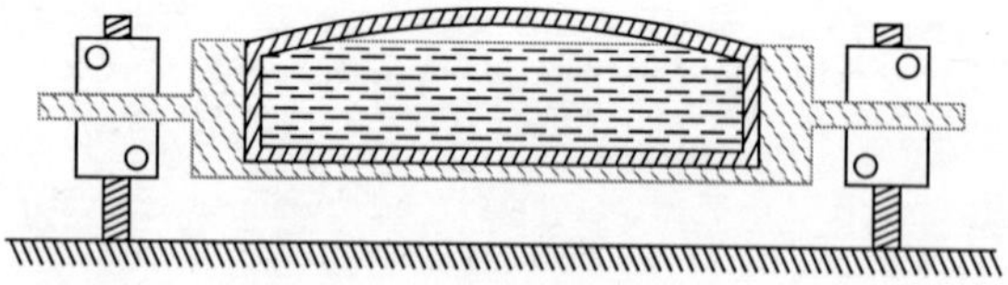

Fig. 9.51 Bubble Tube.

The spirit level or bubble tube consist of a glass tube partially filled with a liquid, the inner surface of which is carefully ground so that a longitudinal section of it by a vertical plane through the axis of the tube is part of circular arc. The tube is graduated on its upper surface and is enclosed for safety in a metal casing. At the ends of the casing are capstan headed screws for securing it to the telescope or any other part (Fig. 9.51).

Before it is sealed, the tube is partially filled with a liquid of low viscosity, such as alcohol, chloroform or sulphuric ether, leaving a small space which forms a bubble of mixed air and vapour. Spirituous liquids are used because they are less viscous, *i.e.,* flow more freely than water. Also, these liquids have a relatively low freezing point but a greater expansion than water. To minimize the effect of expansion, the proportion of liquid and vapour must be carefully regulated. Under the action of gravity, the bubble will always rise to the highest point of the tube, and thus comes to rest so that a tangent plane to the inner surface of the tube at the highest point of the bubble defines a horizontal plane.

The sensitiveness of a level tube is defined as the angular value of one division marked on the tube. It is the amount the horizontal axis has to be tilted to cause the bubble to move from one graduation to another. For example, if the tilting is 20″ of arc when the bubble moves 2 mm (one division), the sensitiveness of the level tube is expressed as 20″ per 2 mm. A tube is said to be more sensitive if the bubble moves by more divisions for a given change in the angle. The sensitiveness of a bubble tube can be increased by:

(*i*) increasing the internal radius of the tube,
(*ii*) increasing the diameter of the tube,
(*iii*) increasing the length of the bubble,
(*iv*) decreasing the roughness of the walls, and
(*v*) decreasing the viscosity of the liquid.

The sensitiveness of a bubble tube should never be greater than is compatible with accuracy achieved with the remainder of the accessories.

9.20 SENSITIVENESS OF BUBBLE TUBE

The sensitiveness of the bubble tube is defined as the *angular value* of one division of the bubble tube. Generally, the linear value of one division is kept as 2 mm. There are two methods of determining the sensitivity.

First Method (Fig. 9.52)

1. Set the instrument at O and level it accurately.
2. Sight a staff kept at C, distant D from O. Let the reading be CF.
3. Using a foot screw, deviate the bubble over n number of divisions and again sight the staff. Let the reading be CE.
4. Find the difference between the two staff readings. Thus,

$$s = CE - CF.$$

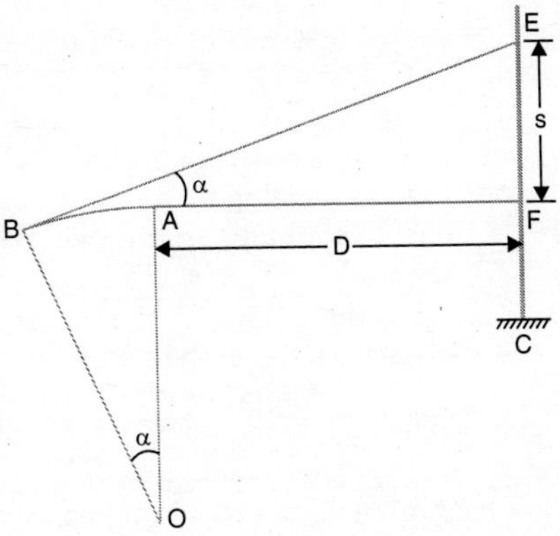

Fig. 9.52

From ΔBEF (approximately), we have

$$\tan \alpha \approx \alpha = \frac{s}{D} \qquad ...(i)$$

Similarly, from ΔAOB, $\alpha = \frac{AB}{R} = \frac{nl}{R}$ ...(*ii*)

where R = radius of curvature of the bubble tube

l = length of one division on the bubble tube (usually 2 mm or 0.1 in.)

Equating (*i*) and (*ii*), we get

$$\frac{s}{D} = \frac{nl}{R} \quad \text{or} \quad R = \frac{nlD}{s} \qquad ...(1)$$

Equation (1) above gives an expression for the radius of curvature of the bubble tube. It is to be noted that l, D and s are expressed in the same units.

Again, from (*ii*) we have $\alpha = \frac{nl}{R}$ ...(2)

$\therefore$ α' = sensitivity of the bubble tube = angular value of one division is given by

$$\alpha' = \frac{l}{R} \text{ by putting } n = l \qquad ...(3)$$

But $$R = \frac{nlD}{s} \text{ (from 1)}$$

$$\therefore \quad \alpha' = \frac{l}{\frac{nlD}{s}} = \frac{s}{nD} \text{ radians} = \frac{s}{nD} \times 206265 \text{ seconds} \qquad ...(4)$$

$$\left(\text{Since 1 radian} = 206265 \text{ seconds} = \frac{1}{\sin 1''}\right)$$

or $$\alpha' = \frac{s}{nD \sin 1''} \text{ seconds} \quad ...(5)$$

Equations (3), (4) and (5) give the expression for the sensitivity of the bubble tube.

Second Method (Fig. 9.53)

1. Set the instrument at O and keep a staff at C.
2. Move the bubble to the extreme left division. Read both ends of the bubble. Let the reading on the left end of the bubble be l_1 and on the right be r_1. Let the staff reading be CE.
3. Move the bubble to the extreme right division. Read both ends of the bubble. Let the reading on the left end of the bubble l_2 be and on the right end be r_2. Let the staff reading be CF.

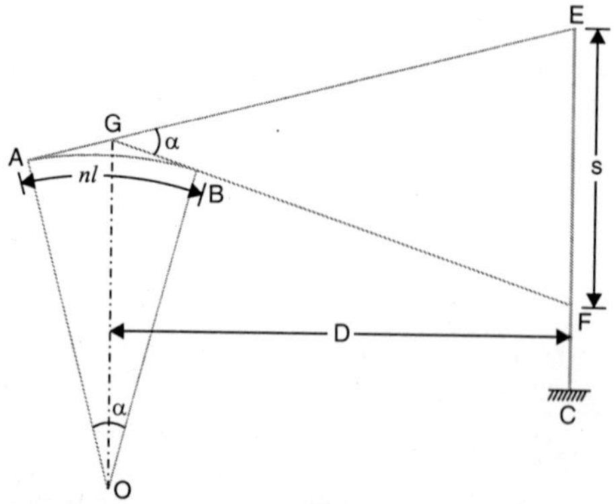

Fig. 9.53

4. Find the difference between the two staff readings
$$s = CE - CF$$
5. Let A and B represent the centres of the bubble in the two positions
The net travel of the bubble will be equal be
$$\frac{(l_1 - r_1) - (l_2 - r_2)}{2} \text{ divisions.}$$
Consider the left divisions as positive and right divisions as negative. Let n = total number of divisions through which the bubble has been moved. Then
$$n = \frac{(l_1 - r_1) - (l_2 - r_2)}{2}$$
6. Considering similar triangles GEF and ABO, we get, as before,

$$R = \frac{nlD}{s} \quad ...(1)$$

$$\alpha = \frac{s}{D} = \frac{nl}{R} \quad ...(2)$$

$$\alpha' = \frac{l}{R} \text{ radians} \quad ...(3)$$

$$= \frac{s}{nD} \times 206265 \text{ seconds} \quad ...(4)$$

$$= \frac{s}{nD \sin 1''} \text{ seconds} \quad ...(5)$$

The sensitivity of a bubble tube depends mainly on the radius of curvature of the tube (the larger the radius, the greater the sensitiveness). However, sensitiveness also depends upon (*i*) the diameter of the tube (the larger the diameter, the greater the sensitivity), (*ii*) length of the vapour bubble, (*iii*) viscosity and surface tension of the liquid (the lesser the viscosity and surface tension, the greater the sensitivity). A very smooth internal surface also increases sensitivity.

Example 9.20: *The reading taken on a staff 100 m from the instrument with the bubble central was 1.872 m. The bubble is then moved 5 divisions out of the centre, and the staff reading is observed to be 1.906 m. Find the angular value of one division of the bubble, and the radius of curvature of the bubble tube. The length of one division of the bubble is 2 mm.*

Solution: Staff intercept for 5-division deviation of the bubble = 1.906 – 1.872 = 0.034 m.

(*i*) The radius of curvature (R) is given by

$$R = \frac{nlD}{s}$$

Here $n = 5,\ l = 2 \text{ mm} = \frac{2}{1000} \text{ m};\ D = 100 \text{ m},\ s = 0.034 \text{ m}$

$\therefore$ $$R = \frac{5 \times 2 \times 100}{1000 \times 0.034} \text{ metres} = \mathbf{29.41\ m.}$$

(*ii*) The sensitivity of the bubble tube (α') is given by

$$\alpha' = \frac{s}{nD} \times 206265 \text{ seconds} = \frac{0.034}{5 \times 100} \times 206265 = \mathbf{14.03\ seconds.}$$

Example 9.21: *Find the radius of curvature of the bubble tube if the length of one division is 2 mm and if the angular value of one division is (a) 20 seconds, (b) 1 minute.*

Solution:

$$\alpha' = \frac{l}{R} \text{ radians} = \frac{l}{R} \times 206265 \text{ seconds}$$

where α' = angular value of one division and l = length of one division

(*a*) $\alpha' = 20 \text{ seconds};\ l = 2 \text{ mm} = \frac{2}{1000} \text{ m}$

$\therefore$ $$R = \frac{l}{\alpha'} \times 206265 = \frac{2}{1000} \times \frac{206265}{20} = \mathbf{20.62\ m.}$$

(*b*) $\alpha' = 60 \text{ seconds};\ l = 2 \text{ mm} = \frac{2}{1000} \text{ m}$

$\therefore$ $$R = \frac{l}{\alpha'} \times 206265 = \frac{2}{1000} \times \frac{206265}{60} = \mathbf{6.87\ m.}$$

Example 9.22: *If the bubble tube of a level has a sensitiveness of 35″ per 2 mm division, find the error in staff reading on a vertical staff at a distance of 100 m caused by the bubble bending $1\frac{1}{2}$ divisions out of centre.*

Solution: With previous notations, we have

$$\alpha' = \frac{s}{nD} \times 206265 \text{ seconds}$$

where α' = angular value of one division = 35″

s = staff intercept = error in staff reading due to deviation of the bubble

n = number of divisions through which bubble is out = 1.5

D = distance of the staff = 100 m

Substituting the values in the above equation, we get

$$s = \frac{\alpha' nD}{206265} = \frac{35 \times 1.5 \times 100}{206265} \text{ m} = \mathbf{0.025\ m.}$$

Example 9.23: Find the radius of curvature of the bubble tube and the value of each 2 mm division from the following average reading of the ends of the bubble and of a staff 80 m away.

		I	*II*
Staff readings	...	*1.680*	*1.602*
Eye-piece end of bubble	...	*20*	*10*
Object glass end of bubble	...	*10*	*20*

Solution: In the first set, the centre of the bubble has moved $\frac{20-10}{2}$ = 5 divisions towards eye-piece end of the tube. In the second set, the centre of the bubble has moved $\frac{20-10}{2}$ = 5 divisions towards objective end. The total number of divisions through which the bubble has moved = n = 5 + 5 = 10.

The change in staff readings = s = 1.680 – 1.602 = 0.078 m

The radius of curvature of the tube is given by

$$R = \frac{nlD}{s},$$

where n = 10 divisions l = 2 mm = $\frac{2}{1000}$ m; D = 80 m; s = 0.078 m

$$\therefore \qquad R = \frac{10 \times 2 \times 80}{1000 \times 0.078} = \mathbf{20.5\ m.}$$

Also, the value of 2 mm division is given by

$$\alpha' = \frac{s}{nD} \times 206265 \text{ seconds} = \frac{0.078}{10 \times 8} \times 206265 = \mathbf{20.1\ seconds.}$$

9.21 BAROMETRIC LEVELLING

The barometric levelling is based on the fact that the atmospheric pressure varies inversely with the height. As air is a compressible fluid, strata at low level will have a greater density than those at a higher altitude. The higher the place of observation the lesser will be the atmospheric pressure. A barometer is used for the determination of the difference in pressure between two stations and their relative altitudes can then be approximately deduced. The average reading of the barometer at sea level is 30 inch and the barometer falls about 1 inch for every 900 ft of ascent above the sea level. This method of levelling is, therefore, very rough and is used only for exploratory or reconnaissance surveys.

There are two types of barometers:

1. Mercurial barometer
2. Aneroid barometer.

1. **The Mercurial Barometer:** Mercurial barometer is more accurate than the aneroid barometer but is an inconvenient instrument for everyday work due to the difficulty of

carrying it about, and the ease with which it is broken. The mercurial barometer works on the principle of balancing a column of mercury against the atmospheric pressure at the point of observation. There are two main types of mercurial barometers – *Cistern* and *Siphon.* In the *Fortin* type of cistern barometer, the cistern is made of a leather bag contained in a metal tube terminating into a glass cylinder. The height of the mercury in the tube is measured by a vernier working against a scale and the reading to $\frac{1}{500}''$. The level of the mercury in the reservoir is adjustable by means of a thrust screw at its base. The mercury is completely enclosed, and by turning the thumb screw the volume of the reservoir may be reduced until the mercury completely fills it and the barometer tube. By this means, the instrument is rendered extremely portable.

When the barometrical observations are in progress, temperature should be read on two thermometers.

2. **The Aneroid Barometer:** The aneroid barometer though less accurate than the mercurial barometer is far more portable and convenient and is, therefore, used almost exclusively in surveying. It consists of a thin cylindrical metallic box about 8 to 12 cm in diameter hermetically sealed and from which air has been exhausted. The ends of the box are corrugated in circular corrugation, and as the pressure of the atmosphere increases or decreases, they slightly approach or recede from each other. This small movement is magnified by means of a suitable lever arrangement and is transferred finally to a pointer which moves over a graduated arc. Fig. 9.54 shows the essential parts of an aneroid barometer.

The general external appearance of the aneroid barometer is shown in Fig. 9.55.

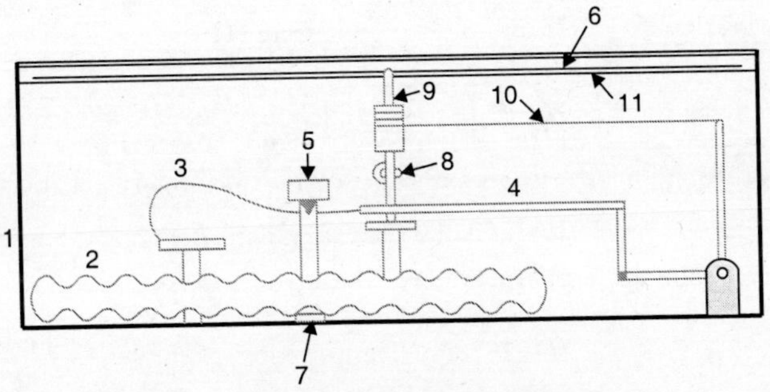

Fig. 9.54 Diagrammatic Section of an Aneroid

1. Outer Casing	2. Corrugated Box	3. Spring
4. Link	5. Knife Edge	6. Pointer
7. Support for Spring	8. Hair Spring	9. Vertical Spindle
10. Chain	11. Scale	

Barometric Formulae

Let it be required to find the difference in elevation H between two points A and B.

Let d_a = density of air at A

d = density of air at any station

Fig. 9.55 Aneroid Barometer.

h = height of mercury column of barometer at any station

L = height of the homogeneous atmosphere on the assumption that its density is constant throughout having a value d_a

p = pressure at A in absolute units

g = acceleration due to gravity

h_1 = barometer reading in cm at the lower station A

h_2 = barometric reading in cm at the higher station B

H = difference in elevation between A and B, in metres.

Then $$p = L.\, d_a g = h.\, d.\, g \qquad ...(1)$$

or $$L = \frac{p}{d_a . g} \qquad ...(2)$$

If g is taken constant, $\frac{p}{d_a}$ is constant by Boyle's law and hence L will be constant.

Let δh = change in barometric reading for a small difference in altitude of δH.

Hence, at a distance δH above A, we have

$$(h - \delta h)\, d.\, g = (L - \delta H)\, d_a .\, g$$

or $$\delta h.\, d = \delta H\, .\, d_a$$

or $$\delta H = \frac{\delta h. d}{d_a} = \frac{\delta h. L}{h}, \text{ from (1)}$$

$\therefore$ $$H = L \int_{h_2}^{h_1} \frac{dh}{h} = L\,(\log_e h_1 - \log_e h_2)$$

Reducing this to common logarithms and substituting the numerical value of L from (2), we get

$$H = 18336.6\,(\log_{10} h_1 - \log_{10} h_2) \text{ (at 32° F and 45° latitude)} \qquad ...(3)$$

Applying a correction for temperature, we get

$$H = 18336.6\,(\log_{10} h_1 - \log_{10} h_2)\left(1 + \frac{t_1 + t_2 - 64°}{900}\right) \qquad ...(4)$$

where t_1 = temperature of air in degree Fahrenheit at A

t_2 = temperature of air in degrees Fahrenheit at B.

Both t_2 and t_1 are measured by detached thermometers. The above expression applies both for mercurial barometer, as well as for the aneroid barometer. However, for mercurial barometer, an additional correction has to be applied for any difference of temperature in the mercury at the two stations. The barometric reading is corrected by the following formula:

$$h_2 = h_2' \,[1 + \alpha\,(t_1' - t_2')] \qquad ...(5)$$

where h_2 = corrected barometric reading at B

h_2' = reading at B

t_1' = mercury temperature at A

t_2' = mercury temperature at B

α = co-efficient of expansion of mercury = 0.00009 per 1° F

The corrected height h_2 is to be substituted in (4).

If the temperatures of the detached thermometers are measured in degrees centigrades, T_1 and T_2, Eq. 4 takes the following form

$$H = 18336.6\,(\log_{10} h_1 - \log_{10} h_2)\left(1 + \frac{T_1 + T_2}{500}\right) \quad ...(6)$$

Another formula given by Laplace is in the following form

$$H = 18393.5\left(1 + \frac{t_1 + t_2 - 64°}{900}\right)(1 + 0.002695 \cos 2\theta) \times \log \frac{h_1}{h_2'\,[1 + \alpha\,(t_1' - t_2')]} \quad ...(7)$$

where θ is the mean latitude of the stations.

Levelling with the Barometer. There are two methods of levelling with a barometer:

1. Method of single oberservations
2. Method of simultaneous observations

1. *Method of Single Observations:* In this method, the barometer is carried from point to point and a single reading is taken at each station; the barometer is brought back to the starting point. The temperature reading is taken at each station. The readings thus obtained involve all atmospheric errors due to the changes in the atmosphere which take place during the interval between the observations.
2. *Method of Simultaneous Observations:* In this method, observations at two stations are taken simultaneously by two barometers previously compared. The aim is to eliminate the errors due to atmospheric changes that take place during the time elapsed between the observations. One barometer is kept at the base or starting point. Another barometer, called the field barometer is taken from station to station and readings of both the barometers taken at predetermined intervals of time. The readings of the field barometer are then compared with those of the barometer at the base. The temperature readings are also taken with each observation.

If temperature is not observed, it alone may introduce an error of as great as 3 m. By simultaneous observations with two barometers and by taking other similar precautions, errors may be reduced to as low as $1\frac{1}{2}$ to 2 m.

Example 9.24: *Find the elevation of the station B from the following data:*

Barometer reading at A:	*78.02 cm at 8 A.M.*
	Temperature of air = 68° F
	78.28 cm at 12 A.M.
	Temperature of air = 72° F
Barometer reading at B:	*75.30 cm at 10 A.M.*
	Temperature of air = 50° F

Elevation of A = 252.5 m

Solution:

The probable reading at A at 10 A.M. $= h_1 = \dfrac{78.02 + 78.28}{2} = 78.15$ cm.

Reading at B at 10 A.M. $= h_2 = 75.30$ cm

$$t_1 \text{ (average), at } A = \frac{68 + 72}{2} = 70° \text{ F at 10 A.M.}$$

$$t_2 \text{ at } B \text{ at 10 A.M.} = 50° \text{ F}$$

Substituting the values in formula (4), we get

$$H = 18336.6 (\log h_1 - \log h_2) \left(1 + \frac{t_1 + t_2 - 64}{900}\right)$$

$$= 18336.6 (\log 78.15 - \log 75.30) \times \left(1 + \frac{70 + 50 - 64}{900}\right) = 315.5 \text{ m}$$

$\therefore$ Elevation of B = 252.5 + 315.5 = **568.0 m.**

9.22 HYPSOMETRY

The working of a *hypsometer* for the determination of altitudes of stations depends on the fact that the temperature at which water boils varies with the atmospheric pressure. A liquid boils when its pressure is equal to the atmoshpheric pressure. The boiling point of vapour water is lowered at higher altitudes since the atmospheric pressure decreases there. A hypsometer essentially consists of a sensitive thermometer graduated to 0.2° F or 0.1° C. The thermometer is held upright in a special vessel in such a way that its bulb is a little above the surface of water contained in a small boiler. A spirit lamp is used to heat the water. Knowing the boiling temperature of water, the atmospheric pressure can be found either from the chart or can be calculated from the following approximate formula:

Pressure in inches of mercury = 29.92 ± 0.586 T_1 ...(1)

where T_1 = the difference of boiling point from 212° F

Having known the atmospheric pressure at the point, elevation can be calculated by using the barometric formula given in the previous article. However, the following formula may also be used to calculate the elevation of the point above datum:

$$E_1 = T_1 (521 + 0.75 T_1) \quad ...(2)$$

Similarly, E_2 at the higher station can also be calculated. The difference in elevation between two points is given by

$$E = (E_1 - E_2) \alpha \quad ...(3)$$

where α = air temperature correction = $\left(1 + \frac{t_1 + t_2 - 64}{900}\right)$

where t_1 = air temperature at lower station

t_2 = air temperature at the higher station.

Water boils at 212°F (100°C) at sea level at atmospheric pressure of 29.921 inches of mercury. A difference of 0.1°F in the reading of the thermometer corresponds to a difference of elevation of about 50 ft. *The method is therefore extremely rough.*

Example 9.25: *Determine the difference in elevation of two stations A and B from the following observations:*

Boiling point at lower station = 210.9° F; Air temperature = 61° F
Boiling point at upper station = 206.5° F; Air temperature = 57° F

Solution: Height of lower point above mean sea level is given by

$$E_1 = T_1\,(521 + 0.75\,T_1);\ \text{where } T_1 = 212° - 210.9° = 1.1°$$

$$\therefore \quad E_1 = 1.1\,(521 + 0.75 \times 1.1) = 574 \text{ feet.}$$

Similarly, height of upper point above mean sea level is given by

$$E_2 = T_2\,(521 + 0.75\,T_2);\ \text{where } T_2 = 212° - 206.5° = 5.5°$$

$$E_2 = 5.5\,(521 + 0.75 \times 5.5) = 2888 \text{ ft.}$$

Air temperature correction

$$= \alpha = 1 + \frac{t_1 + t_2 - 64}{900} = 1 + \frac{61 + 57 - 64}{900} = 1.06$$

$$\therefore \quad \text{Difference in elevation} = H = (E_2 - E_1)\,\alpha = (2888 - 574)\,1.06 = \mathbf{2453\ ft.}$$

PROBLEMS

1. Define the following terms: Benchmark, Parallax, Line of collimation, Level surface, Vertical line, Bubble line, Reduced level, Dip of the horizon, and Backsight.
2. Decribe in brief the essential difference between the following levels:
   Dumpy level, Y-level and Tilting level.
3. What are the different types of levelling staff? State the merits and demerits of each.
4. Describe the 'height of instrument' and 'rise and fall' methods of computing the levels. Discuss the merits and demerits of each.
5. (*a*) Illustrate with neat sketches the construction of a surveying telescope.
   (*b*) Distinguish between the following:
   (*i*) Horizontal plane and level surface
   (*ii*) Line of collimation and line of sight
   (*iii*) Longitudinal section and cross-section.
6. Describe in detail how you would proceed in the field for (*i*) profile levelling, and (*ii*) cross-sectioning.
7. Explain how the procedure of reciprocal levelling eliminates the effect of atmospheric refraction and earth's curvature as well as the effect of inadjustment of the line of collimation.
8. (*a*) R.L. of a factory floor is 100.00′. Staff reading on floor is 4.62 ft and the staff reading when staff is held inverted with bottom touching the tie beam of the roof truss is 12.16 ft. Find the height of the tie beam above the floor.
   (*b*) The following consecutive readings were taken with a dumpy level:
   6.21, 4.92, 6.12, 8.42, 9.81, 6.63, 7.91, 8.26, 9.71, 10.21
   The level was shifted after 4th, 6th and 9th readings. The reduced level at first point was 100 ft. Rule out a page of your answer-book as a level field book and fill all the columns. Use collimation system and apply the usual arithmetical check.
   Indicate the highest and the lowest points. (*A.M.I.E.*)
9. The following staff readings were observed successively with level, the instrument having been moved forward after the second, fourth and eighth readings:
   0.875, 1.235, 2.310, 1.385, 2.930, 3.125, 4.125, 0.120, 1.875, 2.030, 3.765.
   The first reading was taken with the staff held upon a benchmark of elevation 132.135. Enter the readings in level book-form and reduce the levels. Apply the usual checks. Find also the difference in level between the first and the last points.

10. Compare the rise and fall method of reducing levelling notes with the height of collimation method.

It was required to ascertain elevations of *A* and *B*. *A* line of levels was taken from *A* to *B* and then continued to a benchmark of elevation 127.30 ft. The observations are recorded below. Obtain the R.L.'s of *A* and *B*.

B.S.	*I.S.*	*F.S.*	*R.L.*	*Remarks*
3.92				*A*
1.46		7.78		
7.05		3.27		
	2.36			*B*
4.81		0.85		
8.63		2.97		
7.02		3.19		
		4.28	127.30	*B.M.*

(*A.M.I.E.*)

11. The following consecutive readings were taken with a level and 3 metre levelling staff on continuously sloping ground at a common interval of 20 metres:

0.602, 1.234, 1.860, 2.574, 0.238, 0.914, 1.936, 2.872, 0.568, 1.824, 2.722.

The reduced level of the first point was 192.122. Rule out a page of a level field book and enter the above readings. Calculate the reduced levels of the points and also the gradient of the line joining the first and the last points.

12. In running fly-levels from a benchmark of R.L. 384.705, the following readings were obtained.

Backsight 3.215, 1.030, 1.295, 1.855.

Foresight 1.225, 3.290, 2.085.

From the last position of the instrument, six pegs at 25 metres inteval are to be set out on a uniformly falling gradient of 1 in 100, the first peg is to have R.L. of 384.500. Work out the staff readings required for setting the tops of the pegs on the given gradient.

13. The following readings have been taken from the page of an old level book. Reconstruct the page. Fill up the missing quantities and apply the usual checks. Also, calculate the corrected level of the T.B.M. if the instrument is known to have an elevated collimation error of 30″ and backsight foresight distances averaged 40 and 90 metres respectively.

Point	*B.S.*	*I.S.*	*F.S.*	*Rise*	*Fall*	*R.L.*	*Remarks*
1	3.125					×	B.M.
2	×		×	1.325		125.005	T.P.
3		2.320			0.055		
4		×				125.350	
5	×		2.655				T.P.
6	1.620		3.205		2.165		T.P.
7		3.625					
8			×			122.590	T.B.M.

14. (a) Differentiate between 'permanent' and 'temporary' adjustments of level.
    (b) Discuss the effects of curvature and refraction in levelling. Find the correction due to each and the combined correction. Why are these effects ignored in ordinary levelling?
15. In levelling between two points *A* and *B* on opposite sides of a river, the level was set up near *A* and the staff readings on *A* and *B* were 2.642 and 3.228 m respectively. The level was then moved and set up near *B*, the respective staff readings on *A* and *B* were 1.086 and 1.664. Find the true difference in level of *A* and *B*.
16. The following notes refer to reciprocal levels taken with one level:

Instrument Near	*Staff P*	*Reading on Q*	*Remarks*
P	1.824	2.748	Distance *PQ* = 1010 m R.L. of *P* = 126.386
Q	0.928	1.606	

Find (a) the true R.L. of *Q* (b) the combined correction for curvature and refraction, and (c) the angular error in the collimation adjustment.

17. A luminous object on the top of a hill is visible just above the horizon at a certain station at the sea-level. The distance of the top of the hill from the station is 40 km. Find the height of the hill, taking the radius of the earth to be 6370 km.
18. To a person standing on the deck of a ship, a light from the top of a light house is visible just above the horizon. The height of the light in the light-house is known to be 233 yards above M.S.L. If the deck of the ship is 9 yards above M.S.L., work out the distance between the light-house and the ship. Make the necessary assumptions.
19. (a) Explain what is meant by the sensitiveness of a level tube. Describe how you would determine in the field the sensitiveness of a level tube attached to a dumpy level.
    (b) If the bubble tube has a sensitiveness of 23 seconds for 2 mm division, find the error in the staff reading at a distance of 300 ft caused by bubble being one division out of centre.
    (c) Find the error of reading of levelling staff if the observed reading is 12.00′ and at the point sighted the staff is 6" off the vertical through the bottom. *(U.P.)*
20. What are different sources of errors in levelling? How are they eliminated?
21. Describe with the help of a sketch, the working of an aneroid barometer.
22. (a) List out carefully and systematically the field precautions a surveyor should take to ensure good results from levelling field work planned for engineering purposes.
    (b) A 12-mile closed levelling traverse reveals a closing error of 1.56′ on the starting benchmark. Would you consider the work acceptable? Give reasons in support of your answer.
23. (a) Describe briefly the temporary adjustments of a dumpy level.
    (b) Two mile stones *A* and *B* are separated by $6\frac{1}{2}$ miles. A line of levels is run from *A* to *B* and then from *B* to *A*. The differences of levels are found to be

A to *B*	+ 181.34 ft.
B to *A*	– 180.82 ft.

Do you consider the levelling job of an acceptable quality of engineering work? *(A.M.I.E.)*

ANSWERS

8. (*a*) 16.78′
   (*b*) Highest point: Second (R.L. 101.29); Lowest point: Fourth (R.L. 97.79)
9. R.L.'s of change points: 131.775, 132.700, 132.505, 131.505, 137.510, 135.355, 133.620
   Difference in R.L.'s: 1.485 m
10. 116.75; 115.77
11. 192.722, 191.490, 190.864, 190.150 (T.P.), 189.474, 188.452, 187.516 (T.P.), 186.260, 185.362.
    Gradient 1 in 23.82, falling.
12.

Peg No.			Staff reading
1.	...	...	1.000
2.	...	...	1.250
3.	...	...	1.500
4.	...	...	1.750
5.	...	...	2.000
6.	...	...	2.250

13.

Point	*B.S.*	*I.S.*	*F.S.*	*Rise*	*Fall*	*R.L.*	*Remarks*
1.	3.125					123.680	B.M.
2.	2.265		1.800	1.325		125.005	T.P.
3.		2.320			0.055	124.950	
4.		1.920		0.400		125.353	
5.	1.040		2.655		0.735	124.615	T.P.
6.	1.620		3.205		2.165	122.450	T.P.
7.		3.625			2.005	120.445	
8.			1.480	2.145		122.590	T.B.M.

Correct R.L. of T.B.M. = 122.620 metres.

15. 0.582 m, fall.
16. (*a*) 125.585. ; (*b*) 0.069 m. ; (*c*) + 11″
17. 107.76 m.
18. 41.89 miles.
19. (*b*) 0.03 ft. ; (*c*) 0.01 ft.

10 CHAPTER

CONTOURING

10.1 GENERAL

The value of plan or map is highly enhanced if the relative position of the points is represented both horizontally as well as vertically. Such maps are known as *topographic maps*. Thus, in a topographic survey, both horizontal as well as vertical control are required. On a plan, the relative altitudes of the points can be represented by *shading, hachures, form lines* or *contour lines*. Out of these, contour lines are most widely used because they indicate the elevations directly.

Contour

A *contour* is an imaginary line on the ground joining the points of equal elevation. It is a line in which the surface of ground is intersected by a level surface. A *contour line* is a line on the map representing a contour. Fig. 10.1 shows a pond with water at an elevation of 101.00 m as shown in the plan by the water mark. If the water level is now lowered by 1 m, another water mark representing 100.00 m elevation will be obtained. These water marks may be surveyed and represented on the map in the form of contours.

A topographic map presents a clear picture of the surface of the ground. If a map is to a big scale, it shows where the ground is nearly level, where it is sloping, where the slopes are steep and where they are gradual. If a map is to a small scale, it shows the flat country, the hills and valleys, the lakes and water courses and other topographic features.

10.2 CONTOUR INTERVAL

The vertical distance between any two consecutive contours is called *contour interval.* The *contour interval* is kept constant for a contour plan, otherwise the general appearance of the map will be misleading. The horizontal distance between two points on two consecutive contours is known as the *horizontal equivalent* and depends upon the steepness of the ground. The choice of proper contour interval depends upon the following considerations:

(*i*) **The nature of the ground:** The contour interval depends upon whether the country is flat or highly undulated. A contour interval chosen for a flat ground will be highly unsuitable for undulated ground. For every flat ground, a small interval is necessary. If the ground is more broken, greater contour interval should be adopted, otherwise the contours will come too close to each other.

(*ii*) **The scale of the map:** The contour interval should be inversely proportional to the scale. If the scale is small, the contour interval should be large. If the scale is large, the contour interval should be small.

(*iii*) **The purpose and extent of the survey:** The contour interval largely depends upon the purpose and the extent of the survey. For example, if the survey is intended for detailed design work or for accurate earth work calculations, small contour interval is to be used. The extent of survey in such cases will generally be small. In the case of location surveys, for lines of communications and for reservoir and drainage areas, where the extent of survey is large, a large contour interval is to be used.

(*iv*) **Time and expense of field and office work:** If the time available is less, greater contour interval should be used. If the contour interval is small, greater time will be taken in the field survey, in reduction and in plotting the map.

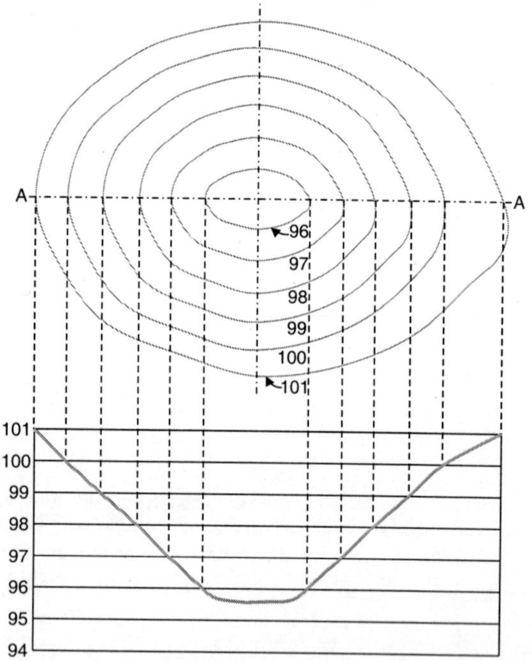

Fig. 10.1

Considering all these aspects, the contour interval for a particular contour plan is selected. This contour interval is kept constant in that plan, otherwise it will mislead the general appearance of the ground. The following table suggests some suitable values of contour interval.

Scale of map	*Type of ground*	*Contour interval (metres)*
Large (1 cm = 10 m or less)	Flat Rolling Hilly	0.2 to 0.5 0.5 to 1 1, 1.5 or 2
Intermediate (1 cm =10 m to 100 m)	Flat Rolling Hilly	0.5, 1 or 1.5 1, 1.5 or 2 2, 2.5 or 3
Small (1 cm = 100 m or more)	Flat Rolling Hilly Mountaineous	1, 2 or 3 2 to 5 5 to 10 10, 25 or 50

The values of contour interval for various purposes are suggested below:

	Purpose of survey	*Scale*	*Interval (metres)*
1.	Building sites	1 cm = 10 m or less	0.2 to 0.5
2.	Town planning schemes, reservoirs, etc.	1 cm = 50 m to 100 m	0.5 to 2
3.	Location surveys	1 cm = 50 m to 200 m	2 to 3

For general topographical work, the general rule that may be followed is as follows:

$$\text{Contour interval} = \frac{25}{\text{No. of cm per km}} \text{ (metres)}$$

or

$$= \frac{50}{\text{No. of inches per mile}} \text{ (feet).}$$

10.3 CHARACTERISTICS OF CONTOURS

The following characteristic features may be used while plotting or reading a contour plan.

1. Two contour lines of different elevations cannot cross each other. If they did, the point of intersection would have two different elevations which is absurd. However, contour lines of different elevations can intersect only in the case of an overhanging *cliff* or a cave (See Fig. 10.2)
2. Contour lines of different elevations can unite to form one line only in the case of a vertical cliff.
3. Contour lines close together indicate steep slope. They indicate a gentle slope if they are far apart. If they are equally spaced, uniform slope is indicated. A series of straight, parallel and equally spaced contours represent a plane surface. Thus, in Fig. 10.3, steep slope in represented at *A-A,* a gentle slope at *B-B,* a uniform slope at *C-C* and a plane surface at *D-D.*

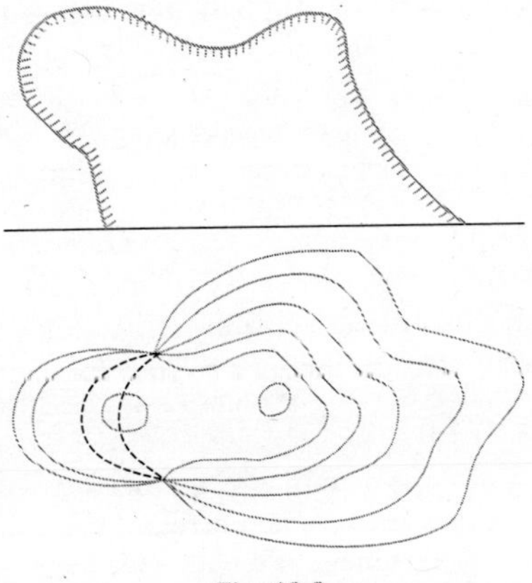

Fig. 10.2

4. A contour passing through any point is perpendicular to the line of steepest slope at that point. This agrees with (3), since the perpendicular distance between contour lines is the shortest distance.
5. A closed contour line with one or more higher ones inside it represents a hill [Fig. 10.4(*a*)]; similarly, a closed contour line with one or more lower ones inside

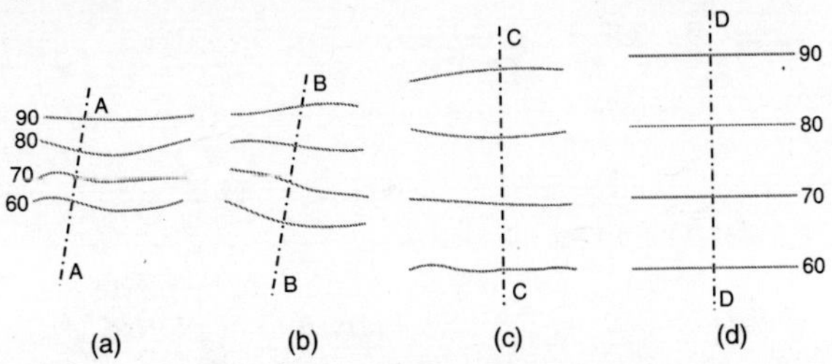

Fig. 10.3

it indicates a depression without an outlet [Fig. 10.4 (*b*)].

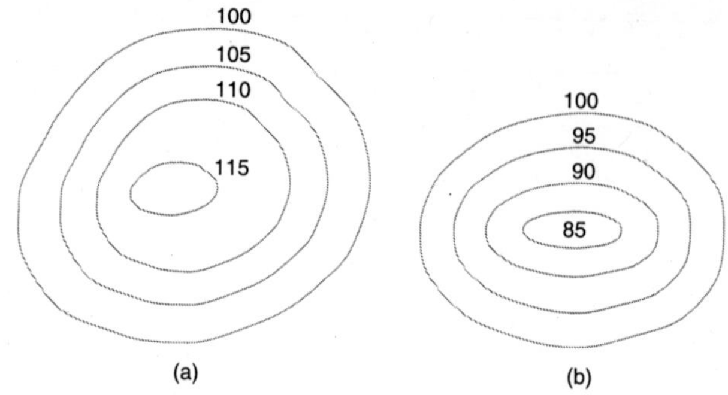

Fig. 10.4

6. To contour lines having the same elevation cannot unite and continue as one line. Similarly, a single contour cannot split into two lines. This is evident because the single line would, otherwise, indicate a knife-edge ridge or depression which does not occur in nature. However, two different contours of the same elevation may approach very near to each other.
7. A contour line must close upon itself, though not necessarily within the limits of the map.
8. Contour lines cross a watershed or ridge line at right angles. They form curves of U-shape round it with the concave side of the curve towards the higher ground (Fig. 10.5).

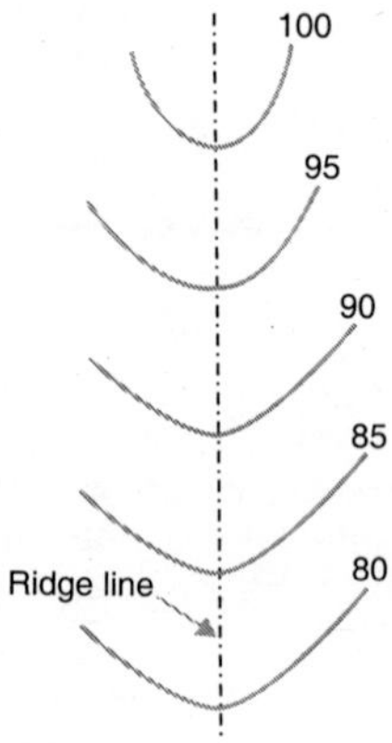

Fig. 10.5

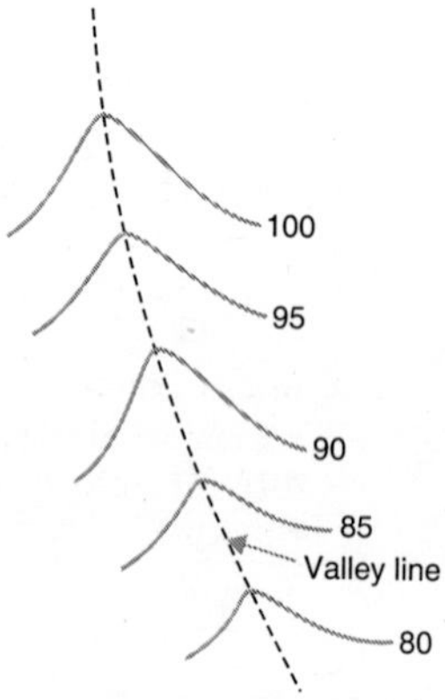

Fig. 10.6

9. Contour lines cross a valley line at right angles. They form sharp curves of V-shape across it with convex side of the curve towards the higher ground (Fig. 10.6). If there is a stream, the contour on either side, turning upstream, may disappear in coincidence with the edge of the stream and cross underneath the water surface.
10. The same contour appears on either sides of a ridge or valley, for the highest horizontal plane that intersects the ridge must cut it on both sides. The same is true of the lower horizontal plane that cuts a valley.

10.4 METHODS OF LOCATING CONTOURS

The location of a point in topographic survey involves both horizontal as well as vertical control. The methods of locating contours, therefore, depend upon the instruments used. In general, however, the field method may be divided into two classes:

(*a*) The direct method. (*b*) The indirect method.

In the *direct method,* the contour to be plotted is actually traced on the ground. *Only those points are surveyed which happen to be plotted.* After having surveyed those points, they are plotted and contours are drawn through them. The method is slow and tedious and is used for small areas and where great accuracy is required.

In the *indirect method,* some suitable guide points are selected and surveyed; the guide points need not necessarily be on the contours. These guide points, having been plotted, serve as basis for the interpolation of contours. This is the method most commonly used in engineering surveys.

Direct Method

As stated earlier, in the indirect method, each contour is located by determining the positions of a series of points *through which* the contour passes. The operation is also sometimes called tracing out contours. The field work is two-fold:

(*i*) Vertical control: Location of points on the contour, and

(*ii*) Horizontal control: Survey of those points.

(*i*) *Vertical Control:* The points on the contours are traced either with the help of a *level and staff* or with the help of a hand level. In the former case, the level is set at a point to command as much area as is possible and is levelled. The staff is kept on the B.M. and the height of the instrument is determined. If the B.M. is not nearby, fly-levelling may be performed to establish a temporary benchmark (T.B.M.) in that area. Having known the height of the instrument, the staff reading is calculated so that the bottom of the staff is at an elevation equal to the value of the contour. For example, if the height of the instrument is 101.80 metres, the staff reading to get a point on the contour of 100.00 metres will be 1.80 metres. Taking one contour at a time (say 100.0 m contour), the staff man is directed to keep the staff on the points on contour so that reading of 1.80 m is obtained every time. Thus, in Fig. 10.7, the dots represent the points determined by this method explained above.

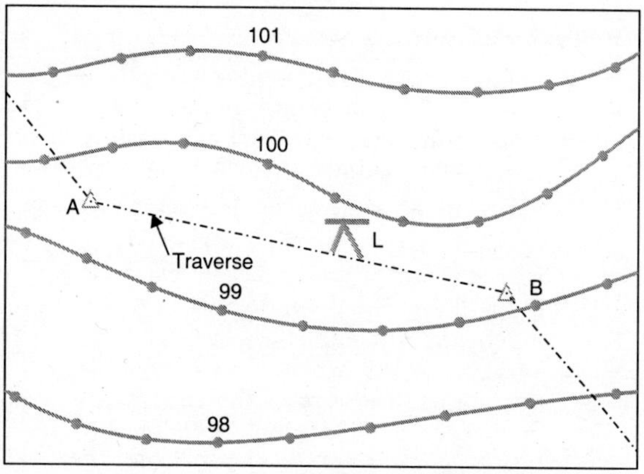

Fig. 10.7

If a hand level is used, slightly different procedure is adopted in locating the points on the contour. A ranging pole having marks at every decimetre interval may be used in conjunction with any type of hand level, preferably an Abney Clinometer. To start with, a point is located on one of the contours, by levelling from a B.M. The starting point must be located on the contour which is a mean of those to be commanded from that position. The surveyor then holds the hand level at that point and directs the rod man till the point on the rod corresponding to the height of the instrument above the ground is bisected. To do this conveniently, the level should be held against a pole at some convenient height, say, 1.50 metres. If the instrument (*i.e.*, the hand level) is at 100 m contour, the reading of the rod to be bisected at each point of 100.5 m, with the same instrument position, will be (1.50 – 0.5) = 1.0 metre. The work can thus be continued.

The staff man should be instructed to insert a lath or twig at the point thus located. The twig must be split to receive a piece of paper on which R.L. of the contour should be written.

(*ii*) *Horizontal Control:* After having located the points on various contours, they are to be surveyed with a suitable control system. The system to be adopted depends mainly on the type and extent of areas. For small area, chain surveying may be used and the points may be located by offsets from the survey lines. In a work of larger nature, a traverse may be used. The traverse may be a theodolite or a compass or a plane table traverse.

In the direct method, two survey parties generally work simultaneously—one locating the points on the contours and the other surveying those points. However, if the work is of a small nature, the points may be located first and then surveyed by the same party.

In Fig. 10.7, the points shown by dots have been surveyed with respect to points *A* and *B* which may be tied by a traverse shown by chain dotted lines.

Indirect Methods

In this method, some guide points are selected along a system of straight lines and their elevations are found. The points are then plotted and contours are then drawn by interpolation. These guide points are not, except by coincidence, points on the contours to be located. While interpolating, it is assumed that the slope between any two adjacent guide points is uniform.

The following are some of the indirect methods of locating the ground points:

(*i*) *By Squares* (Fig. 10.8): The method is used when the area to be surveyed is small and the ground is not very much undulating. The area to be surveyed is divided into a number of squares. The size of the square may vary from 5 to 20 m depending upon the nature of the contour and contour interval. The elevations of the corners of the square are then determined by means of a level and a staff. The contour lines may then be drawn by interpolation. It is not necessary that the squares may be of the same size. Sometimes, rectangles are also used in place of squares. When there are appreciable breaks in the surface between corners, guide points in addition to those at corners may also be used. The squares should be as long as practicable, yet small enough to conform to the inequalities of the ground and to the accuracy required. The method is also known *as spot levelling*.

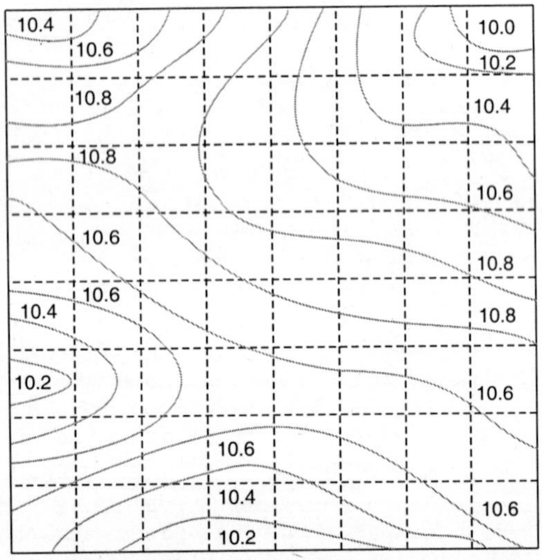

Fig. 10.8 Spot levelling.

(*ii*) *By Cross-sections:* In this method, cross-sections are run transverse to the centre line of a road, railway or canal etc. The method is most suitable for railway route surveys. The spacing of the cross-section depends upon the character of the terrain, the contour interval and the purpose of the survey. The cross-sections should be more closely spaced where the contours curve abruptly, as in ravines or on spurs. The cross-section and the points can then be plotted and the elevation of each point is marked. The contour lines are interpolated on the assumption that there is uniform slope between two points on two adjacent contours. Thus, in Fig. 10.9, the points marked with dots are the points actually surveyed in the field while the points marked with × on the first cross-section are the points interpolated on contours.

The same method may also be used in the *direct method* of contouring with a slight modification. In the method described above, points are taken *almost* at regular intervals on a cross-section. However, the contour points can be located directly on the cross-section as in the direct method. For example, if the height of the instrument is 101.80 and if it is required to trace a contour of 100 m on the ground, the levelling staff is placed on several guide points on the cross-section so that the staff readings on all such points are 1.80 m, and all these points will be on 100 m contour. The guide points of different contours are determined first on one cross-line and then on another instead of first on one contour and then on another, as in the direct method.

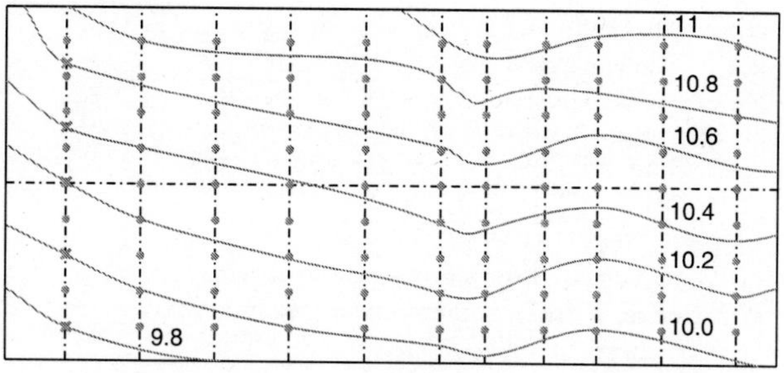

Fig. 10.9

If there are irregularities in the surface between two cross-lines, additional guide points may be located on intermediate cross-lines. If required, some of the cross-lines may also be chosen at any inclination other than 90° to the main line.

(*iii*) *By Tacheometric Method:* In the case of hilly terrain, the tacheometric method may be used with advantages. A tacheometer is a theodolite fitted with stadia diaphragm so that staff readings against all the three hairs may be taken. The staff intercept s is then obtained by taking the difference between the readings against the top and bottom wires. The line of sight can make any inclination with the horizontal, thus increasing the range of instrument observations. The horizontal distances need not be measured, since the tacheometer provides both horizontal as well as vertical control. Thus, if θ is the inclination of the line of sight with horizontal (Fig. 10.10), the horizontal distance (D), between the instrument and the staff, and the vertical difference in elevation (V) between the instrument axis and the point in which the line of sight against the central wire intersects the staff are given by

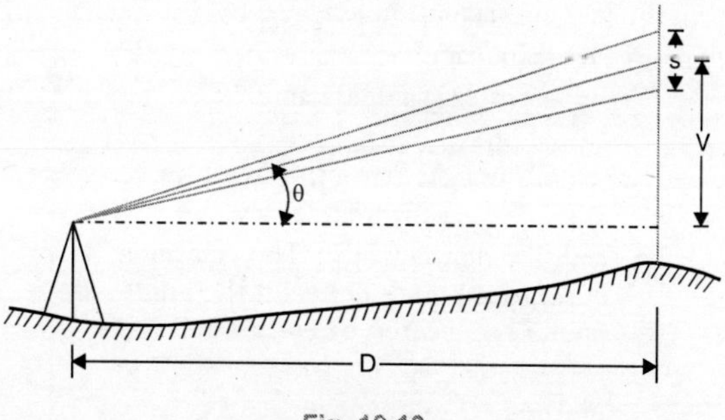

Fig. 10.10

$$D = K_1 s \cos^2 \theta + K_2 \cos \theta \qquad \text{and} \qquad V = D \tan \theta$$

where K_1 and K_2 are instrumental constants.

The tacheometer may be set on a point from which greater control can be obtained. Radial lines can then be set making different angles with either the magnetic meridian or with the first radial line (Fig. 10.11). On each radial line, readings may be taken on levelling staff kept at different points. The point must be so chosen that approximate vertical difference in elevation between two consecutive points is less than the contour interval. Thus, on the same radial line, the horizontal equivalent will be smaller for those two points the vertical difference in elevation of which is greater and *vice versa*.

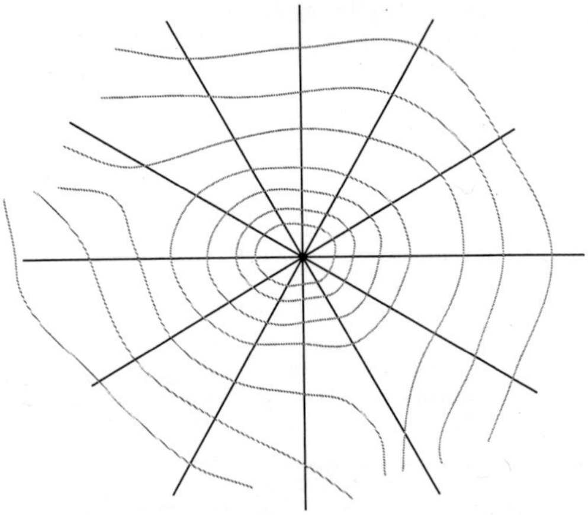

Fig. 10.11

To survey an area connected by series of hillocks, a tacheometric traverse may be run, the tacheometric traverse stations being chosen at some commanding positions. At each traverse station, several radial lines may be run in various directions as required, the horizontal control being entirely obtained by the tacheometer. The traverse, the radial lines and the points can then be plotted. The elevation of each point is calculated by tacheometric formulae and entered, and the contours can be interpolated as usual.

10.5 INTERPOLATION OF CONTOURS

Interpolation of the contours is the process of spacing the contours proportionately between the plotted ground points established by indirect methods. The methods of interpolation are based on the assumption that the slope of ground between the two points is uniform. The chief methods of interpolation are:

(*i*) By estimation

(*ii*) By arithmetic calculations

(*iii*) By graphical method.

(*i*) **By Estimation:** This method is extremely rough and is used for small scale work only. The positions of contour points between the guide points are located by estimation.

(*ii*) **By Arithmetic Calculations:** The method, though accurate, is time consuming. The positions of contour points between the guide points are located by arithmetic calculation. For example, let *A, B, D* and *C* be the guide points plotted on the map, having elevations of 607.4, 617.3, 612.5 and 604.3 feet respectively (Fig. 10.12). Let $AB = BD = CD = CA = 1$ inch on the plan and let it

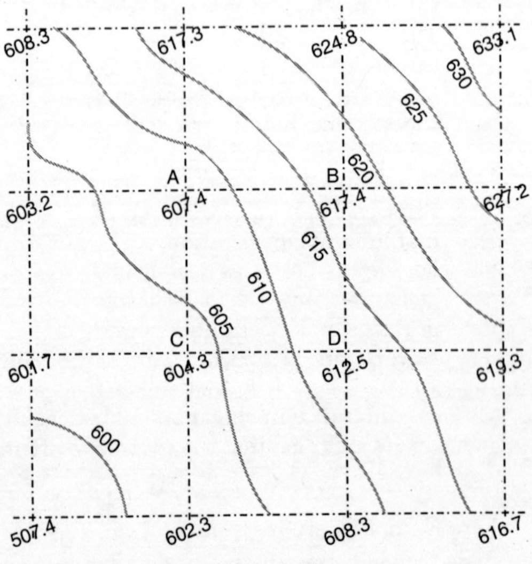

Fig. 10.12

be required to locate the position of 605, 610 and 615 feet contours on these lines. The vertical difference in elevation between A and B is (617.3 – 607.4) = 9.9 ft. Hence, the distances of the contour points from A will be:

$$\text{Distance of 610 ft contour point} = \frac{1}{9.9} \times 2.6 = 0.26'' \text{ (approx.)}$$

$$\text{Distance of 615 ft contour point} = \frac{1}{9.9} \times 7.6 = 0.76'' \text{ (approx.)}$$

These two contour points may be located on AB. Similarly, the position of the contour points on the lines AC, CD, BD and also on AD and BC may be located. Contour lines may then be drawn through appropriate contour points, as shown in Fig. 10.12.

(iii) **By Graphical Method:** In the graphical method, the interpolation is done with the help of a tracing paper or a tracing cloth. There are two methods:

First Method: The first method is illustrated in Fig. 10.13. On a piece of tracing cloth, several lines are drawn parallel to each other, say at an interval representing 0.2 metre. If required, each fifth line may be made heavier to represent each metre interval. Let the bottom line of the diagram, so prepared on the tracing cloth, represent an elevation of 99 m and let it be required to interpolate contours of 99.5, 100 and 100.5 m values between two points A and B having elevations of 99.2 and 100.7 m respectively. Keep the tracing cloth on the line in such a way that point A may lie on a *parallel* representing an elevation of 99.2 metres. Now rotate the tracing cloth on drawing in such a way that point B may lie on a parallel representing 100.7 metres. The points at which the parallels representing 99.5 (point x), 100.0 (point y) and 100.5 (point z) may now be pricked through the respective positions of the contour point on the line AB.

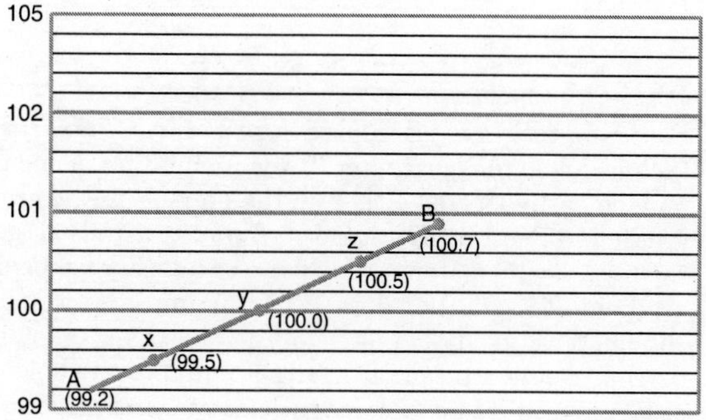

Fig. 10.13

Second Method: The second method is illustrated in Fig. 10.14. A line XY of any convenient length is taken on a tracing cloth and divided into several parts, each representing any particular interval, say 0.2 m. On a line perpendicular to XY at its mid-point, a pole O is chosen and radial lines are drawn joining the pole O and the division on the line XY. Let the bottom radial line represent an elevation of 97.0. If required, each fifth

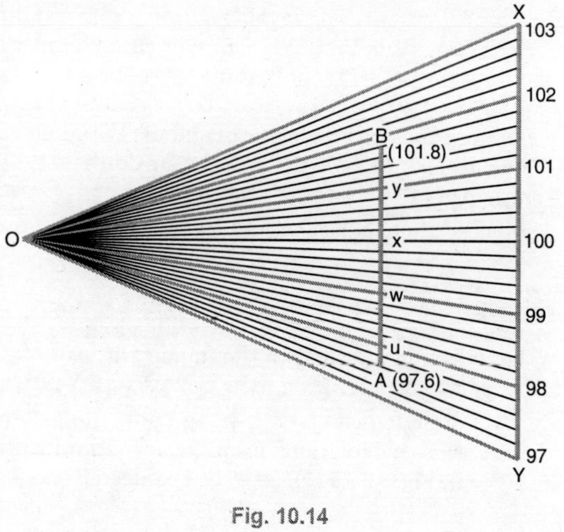

Fig. 10.14

radial line representing one metre interval may be made dark. Let it be required to interpolate contours of 98, 99, 100 and 101 metres elevations between two points *A* and *B* having elevations of 97.6 and 101.8 metres. Arrange the tracing cloth on the line *AB* in such a way that the point *A* and *B* lie simultaneously on radial lines representing 97.6 and 101.8 metres respectively. The points at which radial lines of 98, 99, 100 and 101 metres intersect *AB* may then be pricked through.

Contour Drawing: After having interpolated the contour points between a network of guide points, smooth curves of the contour lines may be drawn through their corresponding contour points. While drawing the contour lines, the fundamental properties of contour lines must be borne in mind. The contour lines should be inked in either black or brown. If the contour plan also shows the features like roads, etc., it is preferable to use brown ink for the contour so as to distinguish it clearly from rest of the features. The value of the contours should be written in a systematic and uniform manner.

10.6 CONTOUR GRADIENT

Contour gradient is a line lying throughout on the surface of the ground and preserving a constant inclination to the horizontal. If the inclination of such a line is given, its direction *from a point* may be easily located either on the map or on the ground. The method of locating the contour gradient *on map* is discussed in the next article (Fig. 10.7). To locate the contour gradient in the field, a clinometer, a theodolite or a level may be used. Let it be required to trace a contour gradient of inclination 1 in 100, starting from a point *A*, with the help of a clinometer. The clinometer is held at *A* and its line of sight is clamped at an inclination of 1 in 100. Another person having a target at a height equal to the height of the observer's eye is directed by the observer to move up or down the slope till the target is bisected by the line of sight. The point is then pegged on the ground. The clinometer is then moved to the point so obtained and another point is obtained in a similar manner. The line between any two pegs will be parallel to the line of sight.

If a level is used to locate the contour gradient, it is not necessary to set the level on the contour gradient. The level is set at a commanding position and reading on the staff kept at the first point is taken. For numerical example, let the reading be 1.21 metres. The reading on another peg *B* (say) distant 20 metres from *A*, with a contour gradient of 1 in 100, will be 1.21 + 0.20 = 1.41 metres. To locate the point *B*, the staff man holds the 20 metres end of chain or tape (with zero metre end at *A*) and moves till the reading on the staff is 1.41 metres. Thus, from one single instrument station several points at a given gradient can be located. The method of calculating the staff readings for several pegs has been explained through numerical examples in Chapter 9 on Levelling.

10.7 USES OF CONTOUR MAPS

The following are some of the important uses of contour maps:

1. **Drawing of Sections:** From a given contour plan, the section along any given direction can be drawn to know the general shape of the ground or to use it for earth work calculations for a given communication line in the direction of the section. Thus, in Fig. 10.15 (*b*), let it be required to draw the section along the line *AB*.

The points in which the line *AB* intersect with various contours are projected on the axis *OX* and their corresponding heights are plotted along the axis *OY* to some scale to get the corresponding contour points which may be joined to get the configuration of ground surface [Fig. 10.15 (*a*)].

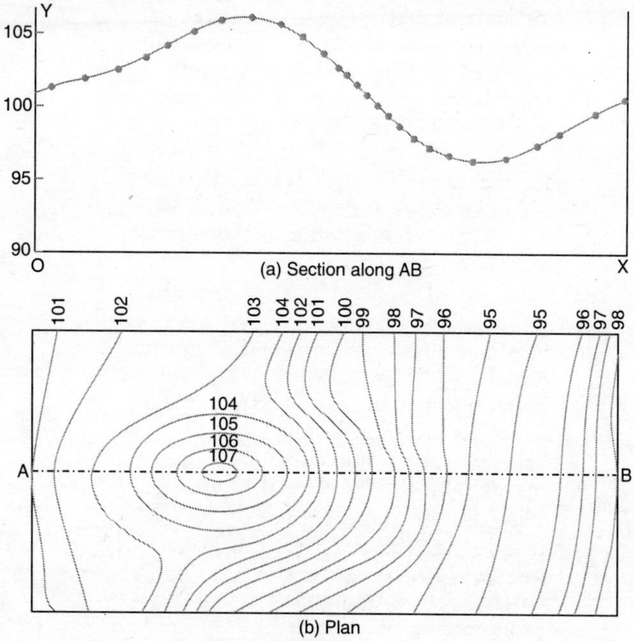

Fig. 10.15

2. **Determination of Intervisibility between two points:** The distances between the triangulation stations are generally several kilometres and before selecting their position it is necessary to determine their intervisibility. A contour map may be used to determine the intervisibility of the triangulation stations. For example, to let it be required to determine the intervisibility of the points *A* and *B* in Fig. 10.16, their elevation being 70 and 102 metres respectively.

Draw line *AB* on the plan. The difference in elevation of *A* and *B* is 102.0 – 70 = 32.0 m. The line of sight between *A* and *B* will have an inclination 32.0 metres in a distance *AB*. Mark on the line of *AB*, the points of elevation of 75, 80, 85, 90, 95 and 100 metres, by calculation. Compare these points with the corresponding points in which the contours cut the line *AB*. Thus, at the point *E* the line of the sight will have an elevation less than

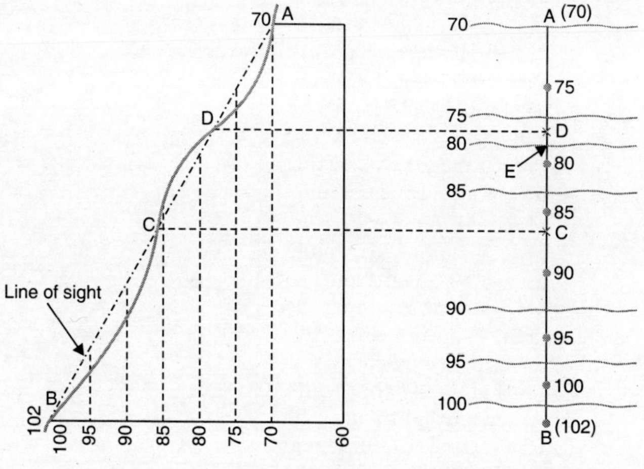

Fig. 10.16

80 metres while the ground has an elevation of 80 metres. Thus, there will be an obstruction and points A and B will not be intervisible. The points C and D at which the line of sight and the ground are at the same elevation can be located. It will be seen by inspection that all other points are clear and there will be no obstruction at other points except for the range CD.

3. **Tracing of Contour Gradients and Location of Route:** A contour plan is very much useful in locating the route of a highway, railway, canal or any other communication line. Let it be required to locate a route, from A to B at an upward gradient of 1 in 25 (Fig. 10.17).

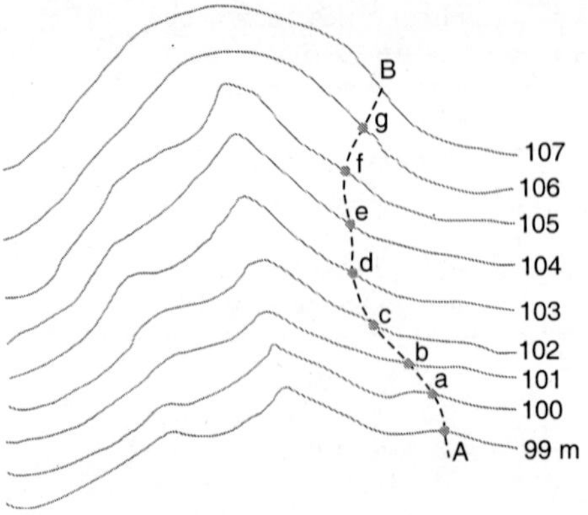

Fig. 10.17

The contours are at an interval of 1 metre. The *horizontal equivalent* will therefore be equal to 25 metres. With A as centre and with a radius representing 25 metres (to the same scale as that of the contour plan) draw an arc to cut the 100 m contour in a. Similarly, with a as centre, cut the 101 m contour in b. Similarly, other points such as c, d, e...., B may be obtained and joined by a line (shown dotted). The route is made to follow this line as closely as possible.

4. **Measurement of Drainage Areas:** A drainage area for a given point in a stream or river can be defined as the area that forms the source of all water that passes that point. A contour plan may be used to trace that line separating the basin from the

Fig. 10.18

rest of the area. The line that marks the limits of drainage area has the following characteristics:

1. It passes through every ridge or saddle that divides the drainage area from other areas.
2. It often follows the ridges.
3. It is always perpendicular to the contour lines.

Such a line is also known as the water-shed line. Fig. 10.18 shows the drainage area enclosed by a line shown by dot and dash. The area contained in a drainage basin can be measured with a planimeter (see Chapter 12). The area shown by hatched lines gives an idea about the extent of the reservoir having a water level of 100 metres.

5. **Calculation of Reservoir Capacity:** The contour plan may be used to calculate the capacity of a reservoir. For example, let it be required to calculate the capacity of reservoir shown in Fig. 10.18, having water elevation of 100.00 metres. The area enclosed in 100, 90, 80 ...... contours may be measured by a planimeter. The volume of water between 100 m and 90 m contour will then be equal to the average areas of the two contours multiplied by the contour interval. Similarly, the volume between the other two successive contours can be calculated. The total volume will then be equal to the sum of the volumes between the successive contours.

Thus, if $A_1, A_2, ..., A_n$ are the areas enclosed by various contours and h is contour interval, the reservoir capacity will be given by

$$V = \Sigma \frac{h}{2}(A_1 + A_2) \text{ by trapezoidal formula}$$

and

$$V = \Sigma \frac{h}{3}(A_1 + 4A_2 + A_3) \text{ by primoidal formula}$$

For detailed study, reference may be made to Chapter 13 on calculation of Volumes.

6. **Intersection of Surfaces and Measurement of Earth Work:** See Chapter 13.

PROBLEMS

1. Describe various methods of contouring. Discuss the merits and demerits of each.
2. Describe with the help of sketches the characteristics of contours.
3. What is grade contour? How will you locate it (*a*) on the ground, (*b*) on the map?
4. Explain, with sketches, the uses of contour maps.
5. Discuss various methods of interpolating the contours.

CHAPTER 11

PLANE TABLE SURVEYING

11.1 GENERAL: ACCESSORIES

Plane tabling is a graphical method of survey in which the field observations and plotting proceed simultaneously. It is means of making a manuscript map in the field while the ground can be seen by the topographer and without intermediate steps of recording and transcribing field notes. It can be used to tie topography by existing control and to carry its own control systems by triangulation or traverse and by lines of levels.

Instruments used

The following instuments are used in plane table survey:

1. The plane table with levelling head having arrangements for (*a*) levelling, (*b*) rotation about vertical axis, and (*c*) clamping in any required position.
2. Alidade for sighting
3. Plumbing fork and plumb bob.
4. Spirit level.
5. Compass.
6. Drawing paper with a rainproof cover.

1. **The Plane Table (Fig. 11.1):** Three distinct types of tables (board and tripod) having devices for levelling the plane table and controlling its orientation are in common use: (*i*) the *Traverse* Table, (*ii*) the *Johnson* Table and (*iii*) the *Coast Survey* Table.

   The Traverse Table: The traverse table consists of a small drawing board mounted on a light tripod in such a way that the board can be rotated about the vertical axis and can be clamped in any position. The table is levelled by adjusting tripod legs, usually by eye-estimation.

   Johnson Table (Fig. 11.2): This consists of a drawing board usually 45 × 60 cm or 60 × 75 cm. The head consists of a ball-and-socket joint and a vertical spindle with two thumb screws on the underside. The ball-and-socket joint is operated by the upper thumb screw. When the upper screw is free, the table may be tilted about the ball-and-socket for levelling. The clamp is then tightened to fix the board in a horizontal position. When the lower screw is loosened, the table may be rotated about the vertical axis and can thus be oriented.

   The Coast Survey Table: This table is superior to the above, two types and is generally used for work of high precision. The levelling of the table is done very accurately with the help of the three foot screws. The table can be turned about the vertical axis and can be fixed in any direction very accurately with the help of a clamp and tangent screw.

2. **Alidade:** A plane table alidade is a straight edge with some form of sighting device. Two types are used: (*i*) plain alidade and (*ii*) telescopic alidade.

   Plain Alidade: Fig. 11.3 shows the simple form and used for ordinary work. It generally consist of a metal or wooden rule with two vanes at the ends. The two vanes or sights are hinged to fold down on the rule when the alidade is not in use. One of the vanes is provided with a narrow slit while the other is open and carries a hair or thin wire. Both the slits thus provide a definite line of sight which can be made to pass through the object to be sighted. The alidade can be rotated about the point representing the instrument station on the sheet so that the line of sight passes through the object to be sighted. A line is then drawn against the working edge (known as the *fiducial edge)* of the alidade. It is essential to have the vanes perpendicular be the surface of the sheet. The alidade is not very much suitable on hilly area since the inclination of the line of sight is limited. A string joining the tops of the two vanes is sometimes provided to use it when sights of considerable inclination have to be taken.

   Telescopic Alidade: (Fig. 11.4) The telescopic alidade is used when it is required to take inlined sights. Also the accuracy and range of sights are increased by its use. It essentially consists of a small telescope with a level tube and graduated arc mounted on horizontal axis. The horizontal axis rests on a A-frame fitted with verniers fixed in position in the same manner as that in a transit. All the parts are finally supported on a heavy rule, one side of which is used as the working edge along which line may be drawn. The inclination of the line of sight can be read on the vertical circle. The horizontal distance between the instrument and the point sighted can be computed by taking stadia readings on the staff kept at the point. The elevation of the point can also be computed by using usual tacheometric relations. Sometimes, to facilitate calculation work, a *Beaman stadia arc* may be provided as an extra. Thus, the observer can very quickly and easily obtain the true horizontal distance from the plane table to a levelling staff placed at the point and the difference in elevation between them. The same geometric principle apply to the alidade as to the transit, but the adjustments are somewhat modified in accordance with the lower degree of accuracy required.

3. **Plumbing Fork:** The plumbing fork (Fig. 11.5), used in large scale work, is meant for centring the table over the point or station occupied by the plane table when the plotted position of that point is already known on the sheet. Also, in the beginning of the work, it is meant for transferring the ground point on to the sheet so that the plotted point and the ground station are in the same vertical line.

   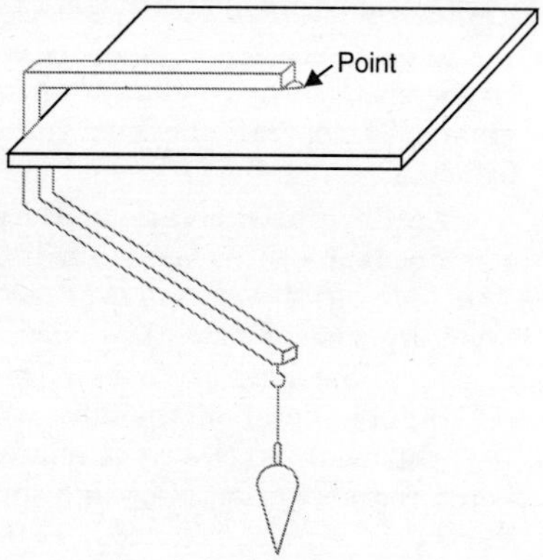

   Fig. 11.5

   The fork consists of a hair pin-shaped light metal frame having arms of equal length, in which a plumb-bob is suspended from the end of the lower arm. The fitting can be placed with the upper arm lying on the top of the table and the lower arm below it, the table being centred when the plumb-bob hangs freely over the ground mark and the pointed end of the upper arm coincides with the equivalent point on the plan.

4. **Spirit Level:** A small spirit level may be used for ascertaining if the table is properly level. The level may be either of the tubular variety or of the circular type, essentially with a flat base so that it can be laid on the table and is truly level when the bubble is central. The table is levelled by placing the level on the board in two positions at right angles and getting the bubble central in both positions.
5. **Compass:** The compass is used for orienting the plane table to magnetic north. The compass used with a plane table is a trough compass in which the longer sides of the trough are parallel and flat so that either side can be used as a ruler or laid down to coincide with a straight line drawn on the paper.
6. **Drawing Paper:** The drawing paper used for plane tabling must be of superior quality so that it may have minimum effect of changes in the humidity of the atmosphere. The changes in the humidity of the atmosphere produces expansion and contraction in different directions and thus alter the scale and distort the map. To overcome this difficulty, sometimes two sheets are mounted with their grains at right angles and with a sheet of muslin between them. Single sheet must be seasoned previous of the use by exposing it alternatively to a damp and a dry atmosphere. For work of high precision, fibre glass sheets or paper backed with sheet aluminium are often used.

11.2 WORKING OPERATIONS

Three *operations* are needed

(*a*) ***Fixing:*** Fixing the table to the tripod.

(*b*) ***Setting:*** (*i*) Levelling the table, (*ii*) Centring, (*iii*) Orientation.

(*c*) ***Sighting the points.***

Setting

Levelling: For small-scale work, levelling is done by estimation. For work of accuracy, an ordinary spirit level may be used. The table is levelled by placing the level on the board in two positions at right angles and getting the bubble central in both directions. For more precise work, a Johnson Table or Coast Survey Table may be used.

Centring: The table should be so placed over the station on the ground that the point plotted on the sheet corresponding to the station occupied should be exactly over the station on the ground. The operation is known as *centring* the plane table. As already described this is done by using a plumbing fork.

Orientation: Orientation is the process of putting the plane-table into some fixed direction so that line representing a certain direction on the plan is parallel to that direction on the ground. *This is essential condition to be fulfilled when more than one instrument station is to be used.* If orientation is not done, the table will not be parallel to itself at different positions resulting in an overall distortion of the map. The processes of centring and orientation are dependent on each other. For orientation, the table will have to be rotated about its vertical axis, thus disturbing the centring. If precise work requires that the plotted point should be exactly over the ground point, repeated orientation and shifting of the whole table are necessary. It has been shown in §11.9 that centring is a needless refinement for small-scale work.

There are two main methods of orienting the plane table:

(*i*) Orientation by means of trough compass.

(*ii*) Orientation by means of backsighting.

(*i*) ***Orientation by trough compass:*** The compass, though less accurate, often proves a valuable adjunct in enabling the rapid approximate orientation to be made prior to the final adjustment. The plane table can be oriented by compass under the following conditions:

(*a*) When speed is more important that accuracy.

(*b*) When there is no second point available for orientation.

(*c*) When the traverse is so long that accumulated errors in carrying the azimuth forward might be greater than orientation by compass.

(*d*) For approximate orientation prior to final adjustment.

(*e*) In certain resection problems.

For orientation, the compass is so placed on the plane table that the needle floats centrally, and a fine pencil line is ruled against the long side of the box. At any other station, where the table is to be oriented, the compass is placed against this line and the table is oriented by turning it until the needle floats centrally. The table is then clamped in position.

(*ii*) ***Orientation by backsighting:*** Orientation can be done precisely by sighting the points already plotted on the sheet. Two cases may arise:

(*a*) When it is possible to set the plane table on the point already plotted on the sheet by way of observation from previous station.

(*b*) When it is not possible to set the plane table on the point.

Case (*b*) presents a problem of *Resection* and has been dealt in § 11.6. When conditions are as indicated in (*a*), the orientation is said to be done by *backsighting*.

To orient the table at the next station, say *B*, represented on the paper by a point *b* plotted by means of line *ab* drawn from a previous station *A*, the alidade is kept on the line *ba* and the table is turned about its vertical axis in such a way that the line of sight passes through the ground station *A*. When this is achieved, the plotted line *ab* will be coinciding with the ground line *AB* (provided the centring is perfect) and the table will be oriented. The table is then clamped in position.

The method is equivalent to that employed in azimuth traversing with the transit. Greater precision is obtainable than with the compass, but an error in direction of a line is transferred to succeeding lines.

Sighting the points: When once the table has been set, *i.e.*, when levelling, centring and orientation has been done, the points to be located are sighted through the alidade. The alidade is kept pivoted about the plotted location of the instrument station and is turned so that the line of sight passes or bisects the signal at the point to be plotted. A ray is then drawn from the instrument station along the edge of the alidade. Similarly, the rays to other points to be sighted are drawn. The points are finally plotted on the corresponding rays either by way of intersection or by radiation as described in the following articles.

11.3 PRECISE PLANE TABLE EQUIPMENT

Modified versions of plane table equipment are now available, having (*i*) precise levelling head, (*ii*) clamp and tangent screw for exact orientation, and (*iii*) telescopic alidade with auto-reduction stadia system. Fig. 11.6 shows the photograph of such a table, with telescopic alidade by Fennel-Kassel. The telescopic alidade is equipped with diagram of Hammer-Fennel auto reduction system (see Chapter 22) giving reduced horizontal distances and difference of elevation directly, using Hammer-Fennel Stadia rod. In addition, the instrument is provided with a vertical circle of glass which can be read by a screw focusing eye-piece along with the telescope. Fig. 11.7 demonstrates the field of view of the telescope as it appears in the case of a horizontal, ascending or descending line of sight.

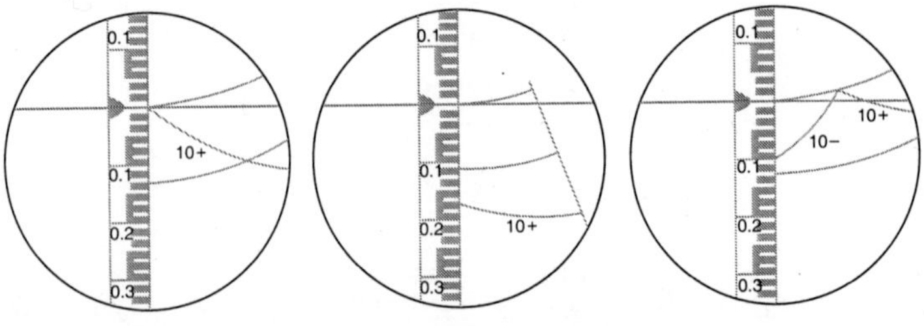

Fig. 11.7 Field of View (Fennel's Auto-Reduction System)

The blade with parallel rule attachment is provided with a circular spirit level and a trough compass for magnetic orientation. A tubular spirit level as well as the tangent screws are built into the pillar, assuring easy operation.

The plane table is locked to the levelling head by means of three screws. The levelling head is provided with clamp and tangent screw for exact orientation. A plumbing fork serves precise centring.

11.4 METHODS (SYSTEMS) OF PLANE TABLING

Methods of plane tabling can be divided into four distinct heads:

1. Radiation. 2. Intersection. 3. Traversing. 4. Resection.

The first two methods are generally employed for locating the details while the other two methods are used for locating the plane table stations.

Radiation

In this method, a ray is drawn from the instrument station towards the point, the distance is measured between the instrument station and that point, and the point is located by plotting to some scale the distance so measured. Evidently, the method is more suitable when the distances are small (within a tape length) and one single instrument can control the points to be detailed. The method has a wider scope if the distances are obtained *tacheometrically* with the help of telescopic alidade (See chapter 22).

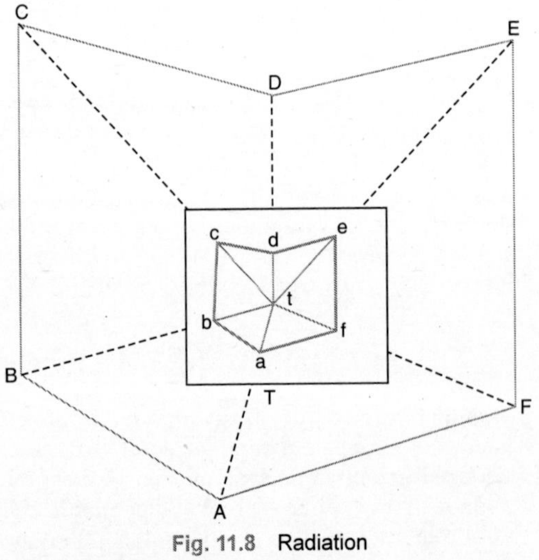

Fig. 11.8 Radiation

The following *steps* are necessary to locate the points from an instrument station T (Fig. 11.8):

1. Set the table at T, level it and transfer the point on to the sheet by means of plumbing fork, thus getting point t representing T. Clamp the table.

2. Keep the alidade touching *t* and sight to *A*. Draw the ray along the fiducial edge of the alidade. Similarly, sight different points *B, C, D, E* etc., and draw the corresponding rays. A pin may be inserted at *t*, and the alidade may be kept touching the pin while sighting the points.
3. Measure *TA, TB, TC, TD, TE* etc., in the field and plot their distances to some scale along the corresponding rays, thus getting *a, b, c, d, e* etc. Join these if needed.

11.5 INTERSECTION (GRAPHIC TRIANGULATION)

Intersection is resorted to when the distance between the point and the instrument station is either too large or cannot be measured accurately due to some field conditions. The location of an object is determined by sighting at the object from two plane table stations (previously plotted) and drawing the rays. The intersection of these rays will give the position of the object. It is therefore very essential to have at least two instrument stations to locate any point. The distance between the two instrument stations is measured and plotted on the sheet to some scale. The line joining the two instrument stations is known as the *base line*. No linear measurement other than that of the base line is made. The point of intersection of the two rays forms the *vertex* of a triangle having the two rays as two sides and the base line as the third line of the triangle. Due to this reason, intersection is also sometimes known as *graphic triangulation.*

Procedure (Fig. 11.9): The following is the *procedure* to locate the points by the method of intersection:

1. Set the table at *A*, level it and transfer the point *A* on to the sheet by way of plumbing fork. Clamp the table.
2. With the help of the trough compass, mark the north direction on the sheet.
3. Pivoting the alidade about *a*, sight it to *B*. Measure *AB* and plot it along the ray to get *b*. The base line *ab* is thus drawn.
4. Pivoting the alidade about *a*, sight the details *C, D, E* etc., and draw corresponding rays.
5. Shift the table at *B* and set it there. Orient the table roughly by compass and finally by backsighting *A*.
6. Pivoting the alidade about *b*, sight the details *C, D, E* etc. and draw the corresponding rays along the edge of the alidade to intersect with the previously drawn rays in *c, d, e* etc. The positions of the points are thus mapped by way of intersection.

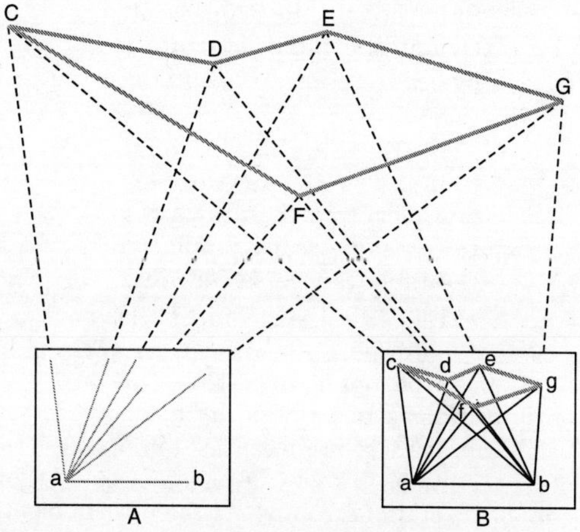

Fig. 11.9 Intersection

The method of intersection is mainly used for mapping details. If this is to be used for locating a point which will be used as subsequent plane table station, the point should be got by way of intersection of at least three or more rays. Triangles should be well conditioned and the angle of intersection of the rays should not be less than 45° in such cases. Graphic triangulation can also proceed without preliminary measurement of the base, as the length of the base line influences only the scale of plotting.

11.6 TRAVERSING

Plane table traverse involves the same principles as a transit traverse. At each successive station the table is set, a foresight is taken to the following station and its location is plotted by measuring the distance between the two stations as in the radiation method described earlier. Hence, traversing is not much different from radiation as far as working principles are concerned—the only difference is that in the case of radiation the observations are taken to those points which are to be detailed or mapped while in the case of traversing the observations are made to those points which will subsequently be used as instrument stations. The method is widely used to lay down survey lines between the instrument stations of a closed or unclosed traverse.

Procedure: (Fig. 11.10)

1. Set the table at *A*. Use plumbing fork for transferring *A* on to the sheet. Draw the direction of magnetic meridian with the help of trough compass.
2. With the alidade pivoted about *a*, sight it to *B* and draw the ray. Measure *AB* and scale off *ab* to some scale. Similarly, draw a ray towards *E*, measure *AE* and plot *e*.
3. Shift the table to *B* and set it. Orient the table accurately by backsighting *A*. Clamp the table.
4. Pivoting the alidade about *b*, sight to *C*. Measure *BC* and plot it on the drawn ray to the same scale. Similarly, the table can be set at other stations and the traverse is completed.

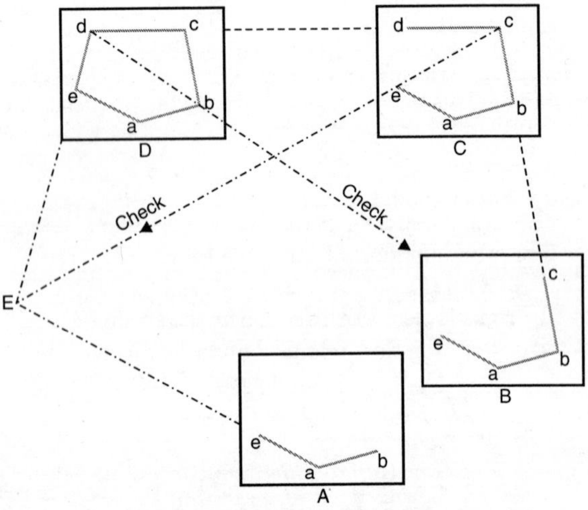

Fig. 11.10 Traversing

It is to be noted here that the orientation is to be done by back-sighting. If there are *n* stations in a closed traverse, the table will have to be set on at least $(n - 1)$ stations to know the error of closure though the traverse may be closed even by setting it on $(n - 2)$ stations. At any station a portion of the traverse may be checked if two or more of the preceding stations are visible and are not in the same straight line with the station occupied.

11.7 RESECTION

Resection is the process of determining the plotted position of the station occupied by the plane table, by means of sights taken towards known points, locations of which have been plotted.

The method consists in drawing two rays to the two points of known location on the plan after the table has been *oriented.* The rays drawn from the unplotted location of the station to the points of known location are called resectors, the intersection of which gives the required location of the instrument station. If the table is not correctly oriented at the station to be located on the map, the intersection of the two resectors will not give the correct location of the station. The problem, therefore, lies in orienting table at the stations and can be solved by the following four methods of orientation.

(*i*) Resection after orientation by compass.

(*ii*) Resection after orientation by backsighting.

(*iii*) Resection after orientation by three-point problem.

(*iv*) Resection after orientation by two-point problem.

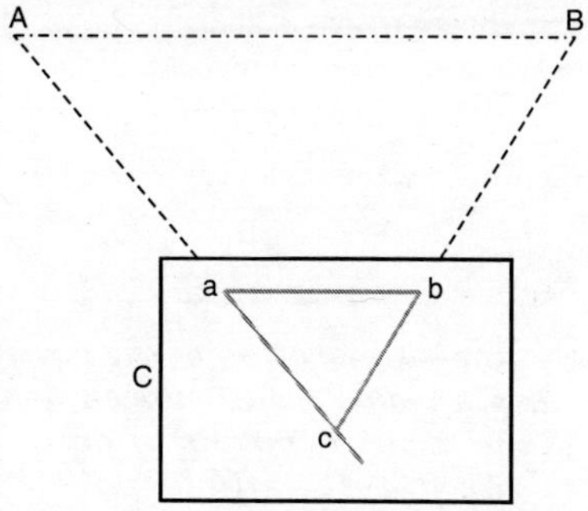

Fig. 11.11 Resection after Orientation by Compass.

(*i*) Resection after orientation by compass

The method is utilised only for small-scale or rough mapping for which the relatively large errors due to orienting with the compass needle would not impair the usefulness of the map.

The method is as follows (Fig. 11.11).

1. Let *C* be the instrument station to be located on the plan. Let *A* and *B* be two visible stations which have been plotted on the sheet as *a* and *b*. Set the table at *C* and orient it with compass. Clamp the table.
2. Pivoting the alidade about *a*, draw a resector (ray) towards *A* ; similarly, sight *B* from *b* and draw a resector. The intersection of the two resectors will give *c*, the required point.

(*ii*) Resection after orientation by backsighting

If the table can be oriented by backsighting along a previously plotted backsight line, the station can be located by the intersection of the backsight line and the resector drawn through another known point. The method is as follows (Fig. 11.12):

1. Let *C* be the station to be located on the plan and *A* and *B* be two visible points which have been plotted on the sheet as *a* and *b*. Set the table at *A* and orient it by backsighting *B* along *ab*.
2. Pivoting the alidade at *a*, sight *C* and draw a ray. Estimate roughly the position of *C* on this ray as c_1.
3. Shift the table to *C* and centre it approximately with respect to c_1. Keep the alidade on the line $c_1\ a$ and orient the table by backsight to *A*. Clamp the table which has been oriented.
4. Pivoting the alidade about *b*, sight *B* and draw the resector *bB* to intersect the ray $c_1\ a$ in *c*. Thus, *c* is the location of the instrument station.

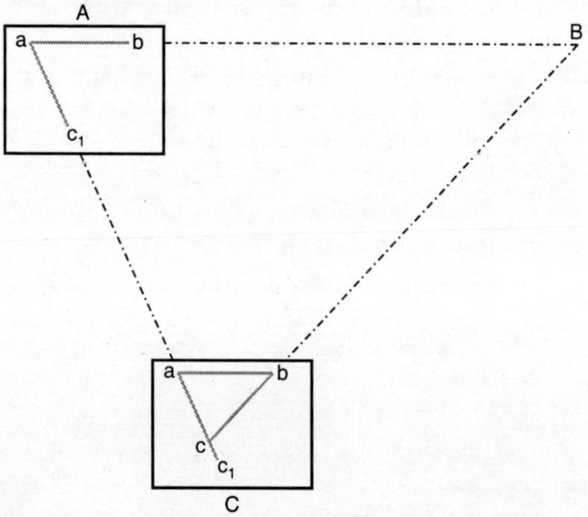

Fig. 11.12 Resection after Orientation by Backsighting.

Resection by Three-point Problem and Two-point Problem

Of the two methods described above, the first method is rarely used as the errors due to local attraction etc., are inevitable. In the second method, it is necessary to set the table on one of the known points and draw the ray towards the station to be located. In the more usual case in which no such *ray* has been drawn, the data must consist of either:

(*a*) Three visible points and their plotted positions (The three-point problem).

(*b*) Two visible points and their plotted positions (The two-point problem).

11.8 THE THREE-POINT PROBLEM

Statement. ***Location of the position, on the plan, of the station occupied by the plane table by means of observations to three well-defined points whose positions have been previously plotted on the plan.***

In other words, it is required to orient the table at the station with respect to three visible points already located on the plan. Let *P* (Fig. 11.13) be the instrument station and *A, B, C* be the points which are located as *a, b, c* respectively on the plan. The table is said to be correctly oriented at *P* when the three resectors through *a, b* and *c* meet at a point and not in a triangle. The intersection of the three resectors in a *point* gives the location of the instrument station. *Thus, in three-point problem, orientation and resection are accomplished in the same operation.*

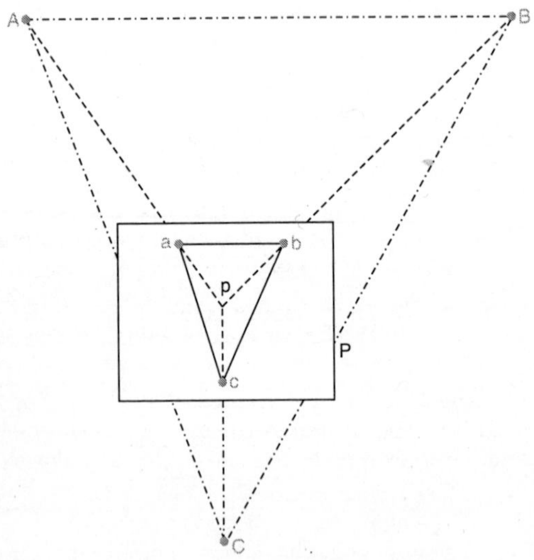

Fig. 11.13 Condition of Correct Orientation

The following are some of the important methods available for the solution of the problem:

1. Mechanical Method (Tracing Paper Method)
2. Graphical Method
3. Lehmann's Method (Trial and Error Method)

1. Mechanical Method (Tracing Paper Method)

The method involves the use of a tracing paper and is, therefore, also known as *tracing paper method.*

Procedure (Fig. 11.14)

Let *A, B, C* be the known points and *a, b, c* be their plotted positions. Let *P* be the position of the instrument station to be located on the map.

1. Set the table on *P*. Orient the table approximately with eye so that *ab* is parallel to *AB*.

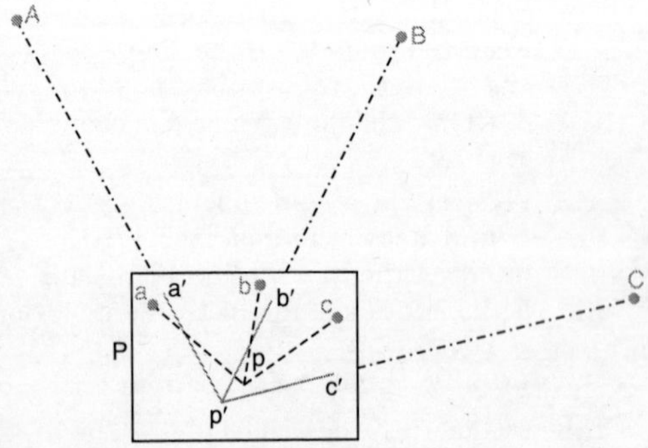

Fig. 11.14

2. Fix a tracing paper on the sheet and mark on it p' as the approximate location of P with the help of plumbing fork.
3. Pivoting the alidade at p', sight A, B, C in turn and draw the corresponding lines $p'a'$, $p'b'$ and $p'c'$ on the tracing paper. These lines will not pass through a, b, and c as the orientation is approximate.
4. Loose the tracing paper and rotate it on the drawing paper in such a way that the lines $p'a'$, $p'b'$ and $p'c'$ pass through a, b and c respectively. Transfer p' on to the sheet and represent it as p. Remove the tracing paper and join pa, pb and pc.
5. Keep the alidade on pa. The line of sight will not pass through A as the orientation has not yet been corrected. To correct the orientation, loose the clamp and rotate the plane table so that the line of sight passes through A. Clamp the table. The table is thus oriented.
6. To test the orientation, keep the alidade along pb. If the orientation is correct, the line of sight will pass through B. Similarly, the line of sight will pass through C when the alidade is kept on pc.

2. Graphical Methods

There are several graphical methods available, but the method given by Bessel is more suitable and is described first.

Bessel's Graphical Solution (Fig. 11.15)

1. After having set the table at station P, keep the alidade on $b\ a$ and rotate the table so that A is bisected. Clamp the table.
2. Pivoting the alidade about b, sight to C and draw the ray xy along the edge of the alidade [Fig. 11.15 (*a*)].
3. Keep the alidade along ab and rotate the table till B is bisected. Clamp the table.
4. Pivoting the alidade about a, sight to C. Draw the ray along the edge of the alidade to intersect the ray xy in c' [Fig. 11.15 (*b*)]. Join cc'.
5. Keep the alidade along $c'c$ and rotate the table till C is bisected. Clamp the table. *The table is correctly oriented* [Fig. 11.15 (*c*)].
6. Pivoting the alidade about b, sight to B. Draw the ray to intersect cc' *in* p. Similarly, if alidade is pivoted about a and A is sighted, the ray will pass through p if the work is accurate.

The points a, b, c' and p form a quadrilateral and all the four points lie along the circumference of a circle. Hence, this method is known as *"Bessel's Method of Inscribed Quadrilateral"*.

In the first four steps, the sighting for orientation was done through a and b, and rays were drawn through c. However, any two points may be used for sighting and the rays drawn towards the third point, which is then sighted in steps 5 and 6.

Alternative Graphical Solution. (Fig. 11.16)

1. Draw a line ae perpendicular to ab at a. Keep the alidade along ea and rotate the plane table till A is bisected. Clamp the table. With b as centre, direct the alidade to sight B and draw the ray be to cut ae in e [Fig. 11.16 (*a*)].
2. Similarly, draw cf perpendicular to bc at c. Keep the alidade along fc and rotate the plane table till C is bisected. Clamp the table.

   With b as centre, direct the alidade to sight B and draw the ray bf to cut cf in f [Fig. 11.16 (*b*)].
3. Join e and f. Using a set square, draw bp perpendicular to ef. Then p represents on the plan the position P of the table on the ground.

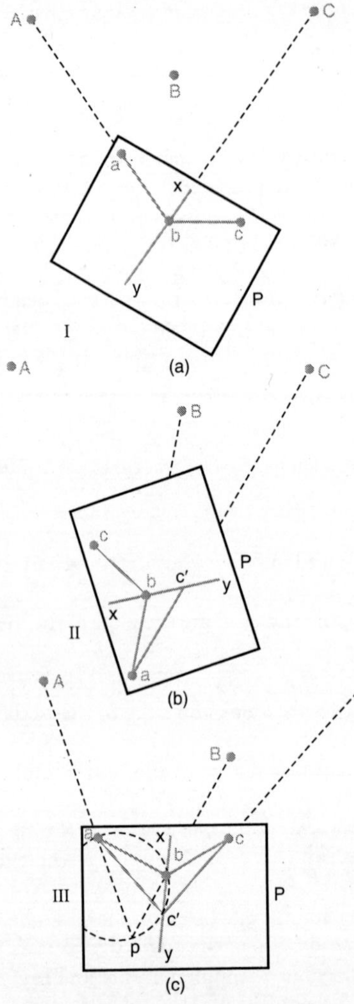

Fig. 11.15 Three-Point Problem: Bessel's Method.

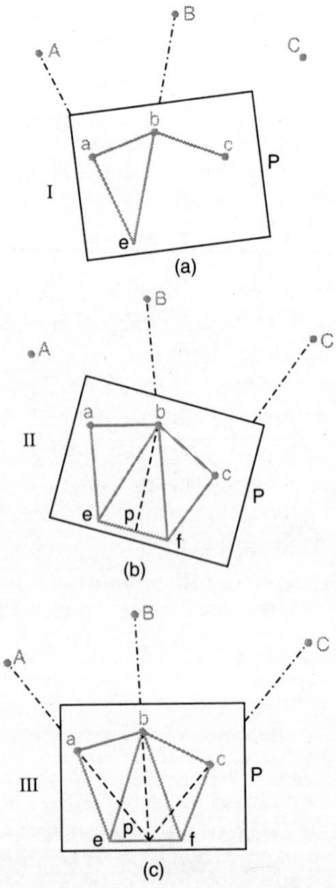

Fig. 11.16

4. To orient the table, keep the alidade along *pb* and rotate the plane table till *B* is bisected. To check the orientation, draw rays *aA*, *cC*, both of which should pass through *p*, as shown in Fig. 11.16 (*c*).

3. Lehmann's Method

We have already seen that the three-point problem lies in orienting the table at the point occupied by the table. In this method, the orientation is done by trial and error and is, therefore, also known as *the trial and error method.*

Procedure. (Refer Fig. 11.17)

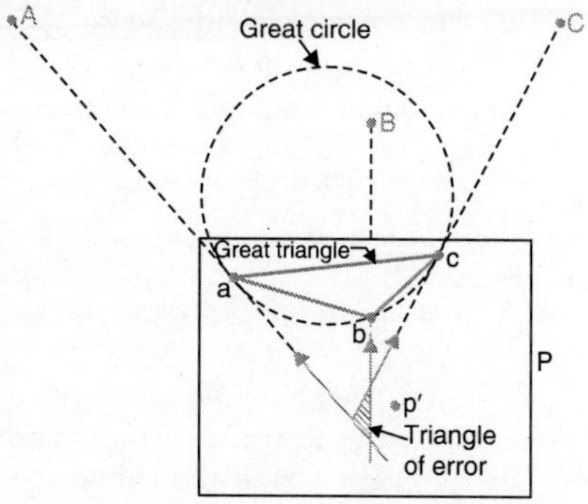

Fig. 11.17 Triangle of Error Method.

1. Set the table at *P* and orient the table approximately so that *ab* is parallel to *AB*. Clamp the table.
2. Keep the alidade pivoted about *a* and sight *A*. Draw the ray. Similarly, draw rays from *b* and *c* towards *B* and *C* respectively. If the orientation is correct, the three rays will meet at one point. If not, they will meet in three points forming one small *triangle of error*.
3. The *triangle of error* so formed will give the idea for the further orientation.

   The orientation will be correct only when the triangle of error is reduced to one point. To do this, choose the point *p′* as shown. The approximate choice of the position may be done with the help of Lehmann's Rules described later.
4. Keep the alidade along *p′a* and rotate the table to sight *A*. Clamp the table. This will give next approximate orientation (but more accurate than the previous one).
5. Keep the alidade at *b* to sight *B* and draw the ray. Similarly, keep the alidade at *c* and sight *C*. Draw the ray. These rays will again meet in one triangle, the size of which will be smaller than the previous triangle of error, if *p′* has been chosen judiciously keeping in the view the Lehmann's Rules.
6. Thus, by successive trial and error, the triangle of error can be reduced to a point.

The final and correct position of the table will be such that the rays *Aa, Bb* and *Cc* meet in one single point, giving the point *p*.

The whole problem, thus, involves a fair knowledge of Lehmann's Rules for the approximate fixation of *p′* so that the triangle of error may be reduced to a minimum.

The lines joining *A, B, C* (or *a, b, c*) form a triangle known as the *Great Triangle*. Similarly, the circle passing through *A, B, C* or (*a, b, c*) is known as the *Great Circle*.

Lehmann's Rules

1. If the station *P* is outside the great triangle *ABC*, the triangle of error will also fall outside the great triangle and the point *p′* should be chosen outside the triangle of error. Similarly, if the station *P* is inside the great triangle, the triangle of error will also be inside the great triangle and the point *p′* should be chosen inside the triangle of error (Fig. 11.18).

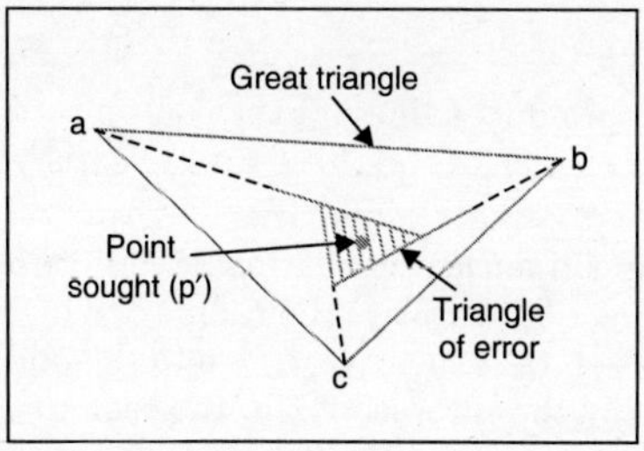

Fig. 11.18

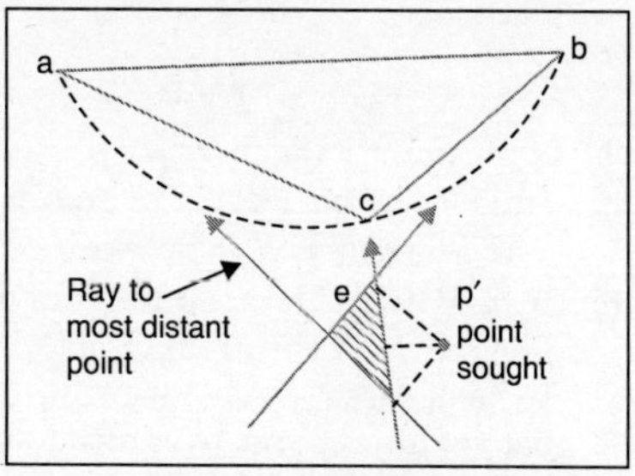

Fig. 11.19

2. The point p' should be so chosen that its distance from the rays Aa, Bb, and Cc is proportional to the distance of P from A, B and C respectively.
3. The point p' should be so chosen that it is to the same side of all the three rays Aa, Bb, and Cc. That is, if point p' is chosen to the right of the ray Aa, it should also be to the right of Bb and Cc (Fig. 11.19).

Though the above rules are sufficient for the location of p', the following sub-rules may also be useful:

3(*a*). If the point P is outside the great circle, the position of p' should be so chosen that the point e (got by the intersection of the two rays drawn to nearer points) is midway between the point p' and the ray to the most distant point (Fig. 11.19).

3(*b*). When P is outside the great triangle but inside the great circle (say in one of the segments of great circle), the point p' must be so chosen that the ray to middle point may lie between p' and the point e which is the intersection of the rays to the other two extreme points (Fig. 11.20).

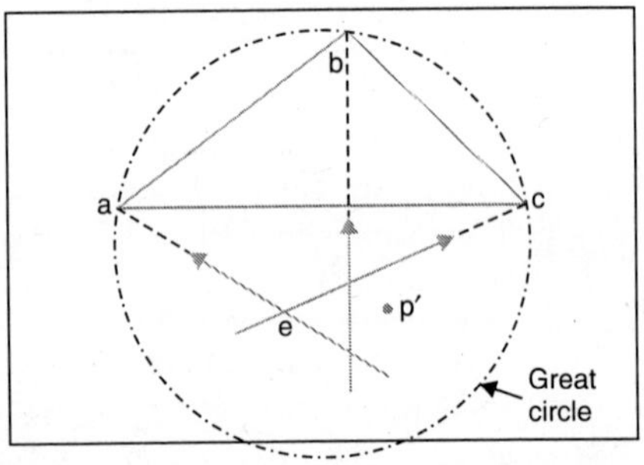

Fig. 11.20

Special Cases: The following are few rules for special cases:

4(*a*). If the positions of A, B, C and P are such that P lies *on* or *near* the side of AC of the great triangle, the point p' must be so chosen that it is in between the two parallel rays drawn to A and C and to the right (or to the same side of each of the rays) of each of the three rays to satisfy Rule 3 (Fig. 11.21).

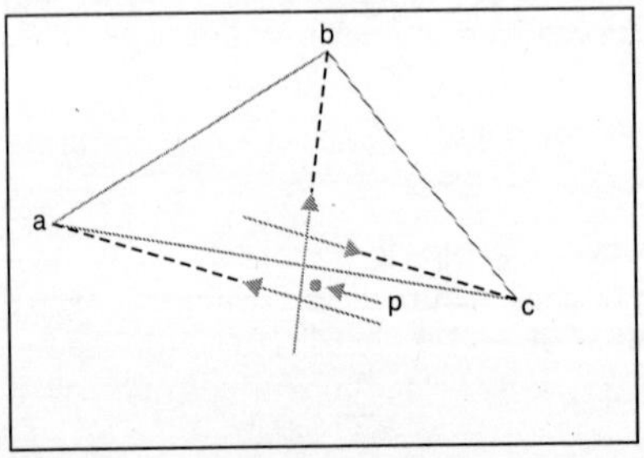

Fig. 11.21

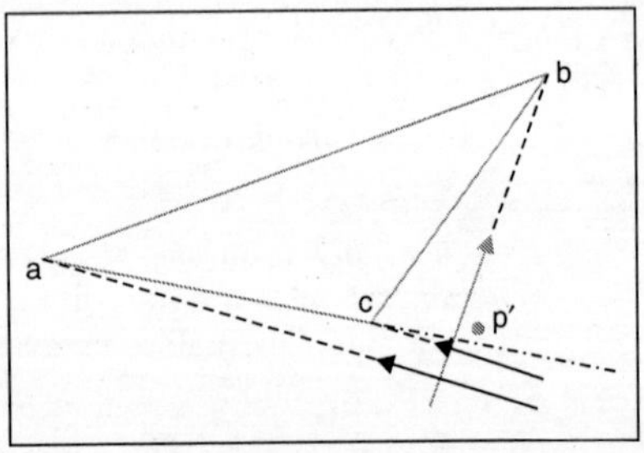

Fig. 11.22

4(*b*). If the point P (as in 4 *a*) lies on or near the prolonged line AC, the point p' must be chosen outside the parallel rays and to the right of each of the three rays to satisfy both Rules 2 and 3 (Fig. 11.22).

4(*c*). If A, B and C happen to be in one straight line the great triangle will be one straight line only and the great circle will be having abc as its arc the radius of which is infinite. In such cases, the point p' must be so chosen that the rays drawn to the middle point is between the point p' and the point e got by the intersection of the rays to the extreme point (Fig. 11.23).

4(*d*). If the positions *A, B, C* and *P* are such that *P* lies on the great circle, the point *p′* cannot be determined by three-point problem because three rays will intersect in one point even when the table is not at all oriented (Fig. 11.24).

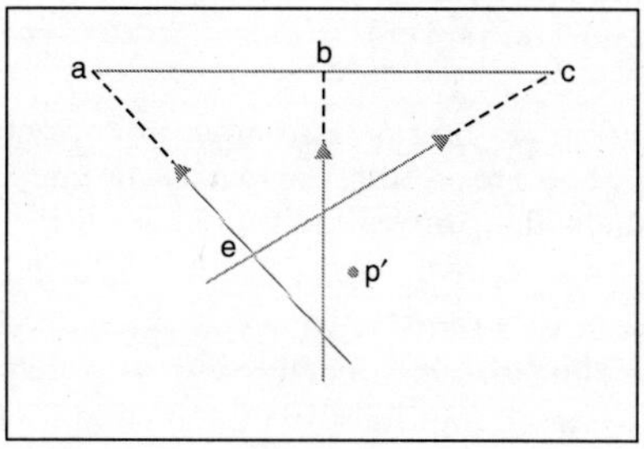

Fig. 11.23

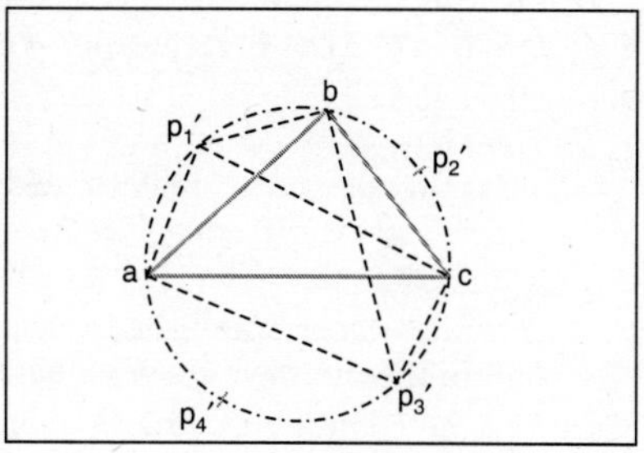

Fig. 11.24

11.9 TWO-POINT PROBLEM

Statement. ***Location of the position on the plan, of the station occupied by the plane table by means of observations to two well defined points whose positions have been previously plotted on the plan.***

Let us take two points *A* and *B*, the plotted positions of which are known. Let *C* be the point to be plotted. The whole problem is to orient the table at *C*.

Procedure. Refer Fig. 11.25

1. Choose an auxiliary point *D* near *C*, to assist the orientation at *C*. Set the table at *D* in such a way that *ab* is approximately parallel to *AB* (either by compass or by eye judgment). Clamp the table.
2. Keep the alidade at *a* and sight *A*. Draw the resector. Similarly, draw a resector from *b* and *B* to intersect the previous one in *d*. The position of *d* is thus got, the degree of accuracy of which depends upon the approximation that has been made in keeping *ab* parallel to *AB*. Transfer the point *d* to the ground and drive a peg.

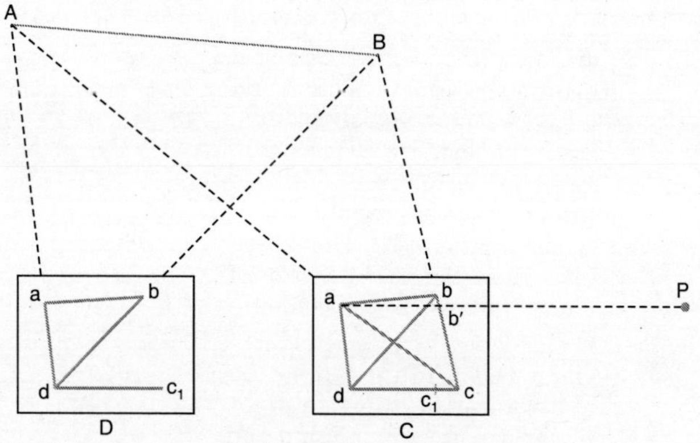

Fig. 11.25 Two-Point Problem

3. Keep the alidade at *d* and sight *C*. Draw the ray. Mark a point c_1 on the ray by estimation to represent the distance *DC*.
4. Shift the table to *C*, orient it (tentatively) by taking backsight to *D* and centre it with reference to c_1. The orientation is, thus, the same as it was at *D*.

5. Keep the alidade pivoted at *a* and sight it to *A*. Draw the ray to intersect with the previously drawn ray from *D* in *c*. Thus, *c* is the point representing the station *C*, with reference to the approximate orientation made at *D*.
6. Pivoting the alidade about *c*, sight *B*. Draw the ray to intersect with the ray drawn from *D* to *B* in *b′*. Thus *b′* is the approximate representation of *B* with respect to the orientation made at *D*.
7. The angle between *ab* and *ab′* is the error in orientation and must be corrected for. In order that *ab* and *ab′* may coincide (or may become parallel) keep a pole *P* in line with *ab′* and at a great distance. Keeping the alidade along *ab*, rotate the table till *P* is bisected. Clamp the table. The table is thus correctly oriented.
8. After having oriented the table as above, draw a resector from *a* to *A* and another from *b* to *B*, the intersection of which will give the position *C* occupied by the table.

It is to be noted here that unless the point *P* is chosen infinitely distant, *ab* and *ab′* cannot be made parallel. Since the distance of *P* from *C* is limited due to other considerations, two-point problem does not give much accurate results. At the same time, more labour is involved because the table is also to be set on one more station to assist the orientation.

Alternative Solution of Two-point Problem (Fig. 11.26)

1. Select an auxiliary point *D very near* to *B* and orient the table there by estimation (making *ba* approximately parallel to *BA*). If *D* is chosen in the line *BA*, orientation can be done accurately.
2. With *b* as centre, sight *B* and draw a ray *Bb*. Measure the distance *BD* and plot the point *d* to the same scale to which *a* and *b* have been previously plotted. Since the distance *BD* is small, any small error in orientation will not have appreciable effect on the location of *d*. The dotted lines show the first position of plane table with approximate orientation.
3. Keep the alidade along *da* and rotate the table to sight *A*, for orientation. Clamp the table. The firm lines show the second position with correct orientation.

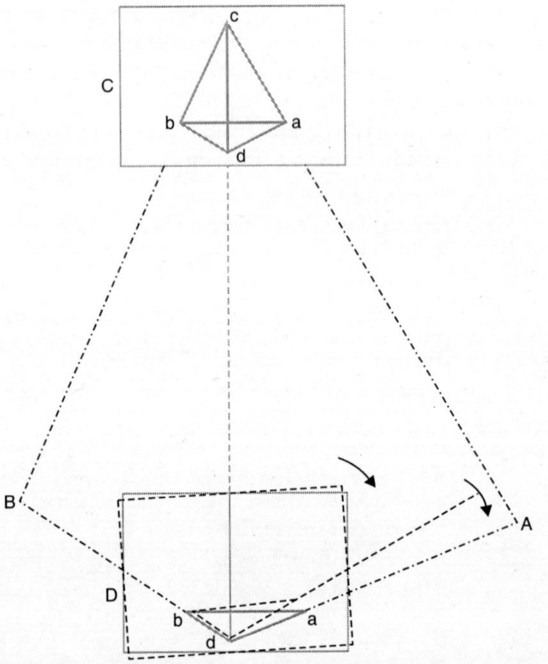

Fig. 11.26 Two-Point Problem

4. With *d* as centre, draw a ray towards *C*, the point to be actually occupied by the plane table.
5. Shift the table to *C* and orient it by backsighting to *D*.

6. Draw a ray to A through a, intersecting the ray dc in c. Check the orientation by sighting B through b. The ray Bb should pass through c if the orientation is correct.

It should be noted that the two-point resection and three point resection give both an orientation as well as fixing.

11.10 ERRORS IN PLANE TABLING

The degree of precision to be attained in plane tabling depends upon the character of the survey, the quality of the instrument, the system adopted and upon the degree to which accuracy is deliberately sacrificed for speed. The various sources of errors may be classified as:

1. **Instrumental Errors:** Errors due to bad quality of the instrument. This includes all errors described for theodolite, if telescopic alidade is used.
2. **Errors of plotting:**
3. **Error due to manipulation and sighting:** These include:
   (*a*) Non-horizontality of board.
   (*b*) Defective sighting.
   (*c*) Defective orientation.
   (*d*) Movement of board between sights.
   (*e*) Defective or inaccurate centring.

(a) Non-horizontality of board: The effect of non-horizontality of board is more severe when the difference in elevation between the points sighted is more.

(b) Defective sighting: The accuracy of plane table mapping depends largely upon the precision with which points are sighted. The plain alidade with open sight is much inferior to the telescopic alidade in the definition of the line of sight.

(c) Defective orientation: Orientation done with compass is unreliable, as there is every possibility of local attraction. Erroneous orientation contribute towards distortion of the survey. This orientation should be checked at as many stations as possible by sighting distant prominent objects already plotted.

(d) Movement of board between sights: Due to carelessness of the observer, the table may be disturbed between any two sights resulting in the disturbance of orientation. To reduce the possibility of such movement, the clamp should be firmly applied. It is always advisable to check the orientation at the end of the observation from a station.

(e) Inaccurate centring: It is very essential to have a proper conception of the extent of error introduced by inaccurate centring, as it avoids unnecessary waste of time in setting up the table by repeated trials.

Let p be the plotted position of P (Fig. 11.27), while the position of exact centring should have been p', so that linear error in centring is $-e = pp'$ and the angular error in centring is $APB - apb = (\alpha + \beta)$.

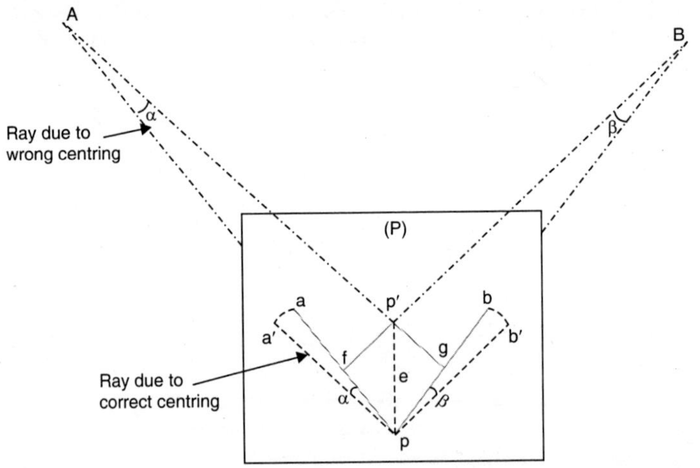

Fig. 11.27 Error due to Wrong Centring.

Drop perpendiculars from p' to ap and bp at f and g respectively.

Then $\quad p'f = AP \sin \alpha \qquad p'g = BP \sin \beta \qquad$ ...(1)

Now $\quad \sin \alpha = \dfrac{p'f}{AP} \approx \alpha$, as α is small and $\sin \beta = \dfrac{p'g}{BP} \approx \beta$, as β is small.

Let us find out the error in the plotting of a and b. Let us say that a' and b' are the positions of A and B for correct centring. Then the error in the position of A and B will be aa' and bb' respectively.

$$aa' = pa \,.\, \alpha = pa.\frac{p'f}{AP} \text{ and } bb' = pb \,.\, \beta = pb.\frac{p'g}{BP} \qquad \text{...(2)}$$

Let $\quad p'f = p'g = e$ metres and s = fractional scale (R.F.)

Then $\quad pa = PA \,.\, s$ and $pb = PB \,.\, s$

Hence from (2),

$$aa' = s\,.\,PA\,.\frac{e}{PA} = s\,.\,e \text{ metres and } bb' = s\,.\,PB\,.\frac{e}{PB} = s\,.\,e \text{ metres.}$$

Hence, we find that the displacement of the points is es metres. If we take 0.25 mm as the unit of precision in plotting,

$$e\,.\,s = aa' = bb' = 0.00025 \text{ metre}$$

$$\therefore \qquad e = \frac{0.00025}{s} \text{ metre} \qquad \text{...(3)}$$

Thus, we have got an expression that the value of e should be less than $\dfrac{0.00025}{s}$ metre.

Centring must be performed with care in large scale work. For a scale of 1 cm = 1 metre, $s = 1/100$.

$$\therefore \qquad e = \frac{0.00025}{1/100} = 0.025 \text{ m} = 2.5 \text{ cm}$$

which shows that for large scale work (such as 1 cm to 1 metre), the maximum value of $e = 2.5$ cm only and centring should be done carefully.

Let us take the case of small scale work also, such as 1 cm = 20 metres.

$$s = \frac{1}{2000}$$

$$\therefore \qquad e = \frac{0.00025}{1/2000} = 0.5 \text{ m} = 50 \text{ cm.}$$

This shows that for such small scales, we can have the position of the ground points within the limits of the board.

Example 11.1: *In setting up the plane table at a station P the corresponding point on the plan was not accurately centred above P. If the displacement of P was 30 cm in a direction at right angles to the ray, how much on the plan would be the consequent displacement of a point from its true position, if,*

(i) scale is: 1 cm = 100 m *(ii) scale is: 1 cm = 2 metres.*

Solution:

Case (i)

$$aa' = e \,.\, s \text{ metres}$$

$$\text{Scale: } 1 \text{ cm} = 100 \text{ m} \quad \therefore \quad s = \frac{1}{10{,}000}$$

$$\therefore \qquad aa' = e \,.\, s = 30 \times \frac{1}{10000} \text{ cm} = \mathbf{0.03 \text{ mm}} \text{ (negligible)}$$

Case (ii)

$$\text{Scale: } 1 \text{ cm} = 2 \text{ m}\,; \quad \therefore \quad s = \frac{1}{200}$$

$$\therefore \qquad aa' = e\, s = \frac{30}{200} = \mathbf{1.5 \text{ mm}} \text{ (large).}$$

11.11 ADVANTAGES AND DISADVANTAGES OF PLANE TABLING

Advantages

1. The plan is drawn by the out-door surveyor himself while the country is before his eyes, and therefore, there is no possibility of omitting the necessary measurements.
2. The surveyor can compare plotted work with the actual features of the area.
3. Since the area is in view, contour and irregular objects may be represented accurately.
4. Direct measurements may be almost entirely dispensed with, as the linear and angular dimensions are both to be obtained by graphial means.
5. Notes of measurements are seldom required and the possibility of mistakes in booking is eliminated.
6. It is particularly useful in magnetic areas where compass may not be used.
7. It is simple and hence cheaper than the theodolite or any other type of survey.

8. It is most suitable for small scale maps.
9. No great skill is required to produce a satisfactory map and the work may be entrusted to a subordinate.

Disadvantages

1. Since notes of measurements are not recorded, it is a great inconvenience if the map is required to be reproduced to some different scale.
2. The plane tabling is not intended for very accurate work.
3. It is essentially a tropical instrument.
4. It is most inconvenient in rainy season and in wet climate.
5. Due to heavyness, it is inconvenient to transport.
6. Since there are so many accessories, there is every likelihood of these being lost.

PROBLEMS

1. (*a*) Discuss the advantages and disadvantages of plane table surveying over other methods.
   (*b*) Explain with sketches, the following methods of locating a point by plane table survey. Also discuss the relative merits and application of the following methods:
   (*i*) Radiation (*ii*) Intersection (*iii*) Resection. (*A.M.I.E.*)
2. Describe briefly the use of various accessories of a plane table.
3. Discuss with sketches, the various methods of orienting the plane table.
4. (*a*) A plane table survey is to be carried out at a scale of 1: 5000. Show that at this scale, accurate centring of the plane table over the survey station is not necessary. What error would be caused in position on a map if the point is 45 cm out of the vertical through the station?
   (*b*) Define three-point problem and show how it may be solved by tracing paper method.
5. Describe, with the help of sketches, Lehmann's Rules.
6. What is two-point problem? How is it solved?
7. What is three-point problem? How is it solved by (*i*) Bessel's method (*ii*) Triangle of error method.
8. What are the different sources of errors in plane tabling? How are they eliminated?
9. (*a*) Describe the method of orienting plane table by backsighting.
   (*b*) Distinguish between 'resection' and 'intersection' methods as applied to plane table surveying.
   (*c*) How does plane table survey compare with chain surveying in point of accuracy and expediency? (*A.M.I.E.*)
10. (*a*) Compare the advantages and disadvantages of plane table surveying with those of chain surveying.
    (*b*) State three-point problem in plane tabling and describe its solution by trial method giving the rules which you will follow in estimating the position of the point sought. (*A.M.I.E.*)

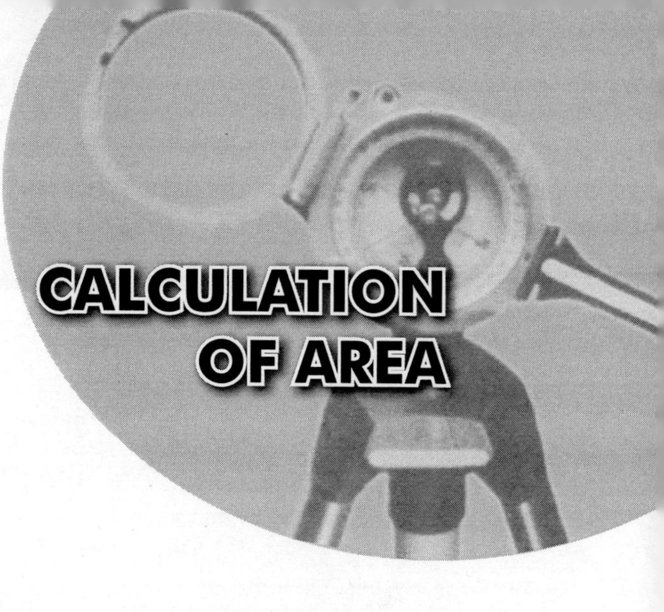

CHAPTER 12

CALCULATION OF AREA

12.1 GENERAL

One of the primary objects of land surveying is to determine the area of the tract surveyed and to determine the quantities of earthwork. The area of land in plane surveying means the area as projected on a horizontal plane. The units of measurements of area in English units are sq. ft or acres, while in metric units, the units are sq. metres or hectares. The following table gives the relation between the two systems.

TABLE 12.1 British Units of Square Measure with Metric Equivalents

Sq. mile	*Acres*	*Square chains*	*Sq. poles or Perches*	*Square yards*	*Square feet*	*Square links*	*Metric Equivalents*
1	640	6,400	102,400	3,097,000	–	–	258.99 ha
	1	10	160	4,840	43,560	100,000	0.40467 ha
		1	16	484	4,356	10,000	404.67 m^2
			1	30.25	272.25	625	25.29 m^2
				1	9	20.7	0.836 m^2
					1	2.3	929 m^2
						1	404.67 cm^2

Note: The standard of square measure is the Acre.

TABLE 12.1(a) Metric Units of Square Measure with British Equivalents

Square kilometre (km^2)	*Hectares (ha)*	*Are (a)*	*Sq. metres (m^2)*	*Sq. centimetres (cm^2)*	*British Equivalents*
1	100	10,000	1,000,000	–	0.3861 sq. mile
	1	100	10,000	–	2.4710 acres
		1	100	1,000,000	1076.4 sq. ft.
			1	10,000	10.764 sq. ft.
				1	0.155 sq. ft.

Note: The standard of square measure is the Are.

12.2 GENERAL METHODS OF DETERMINING AREAS

The following are the general methods of calculating areas:

1. *By computations based directly on field measurements*

These include:

(*a*) By dividing the area into a number of triangles
(*b*) By offsets to base line
(*c*) By latitudes and departures:
  (*i*) By double meridian distance (D.M.D. method)
  (*ii*) By double parallel distance (D.P.D. method)
(*d*) By co-ordinates.

2. *By computation based on measurements scaled from a map.*
3. *By mechanical method:* Usually by means of a planimeter.

12.3 AREAS COMPUTED BY SUB-DIVISION INTO TRIANGLES

In this method, the area is divided into a number of triangles, and the area of each triangle is calculated. The total area of the tract will then be equal to the sum of areas of individual triangles. Fig. 12.1 shows an area divided into several triangles. For field work, a transit may be set up at *O*, and the lengths and directions of each of the lines *OA*, *OB*..... etc. may be measured. The area of each triangle can then be computed. In addition, the sides *AB*, *BC*..... etc. can also be measured and a check may be applied by calculating the area from the three known sides of a triangle. Thus, if two sides and one included angle of a triangle is measured, the area of the triangle is given by

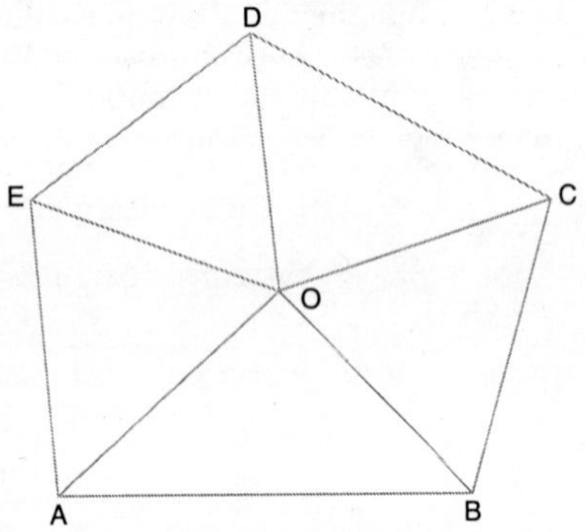

Fig. 12.1

$$\text{Area} = \tfrac{1}{2}\, ab \sin C \qquad ...(12.1)$$

When the lengths of the three sides of a triangle are measured, its area is computed by the equation.

$$\text{Area} = \sqrt{s(s-a)(s-b)(s-c)} \qquad ...(12.2)$$

where $s = \text{half perimeter} = \tfrac{1}{2}(a+b+c)$.

The method is suitable only for work of small nature where the determination of the closing error of the figure is not important, and hence, the computation of latitudes and departure is unnecessary. The accuracy of the field work, in such cases, may be determined by measuring the diagonal in the field and comparing its length to the computed length.

12.4 AREAS FROM OFFSETS TO A BASE LINE: OFFSETS AT REGULAR INTERVALS

This method is suitable for long narrow strips of land. The offsets are measured from the boundary to the base line or a survey line at regular intervals. The method can also be applied to a plotted plan from which the offsets to a line can be scaled off. The area may be calculated by the following rules:

1. Mid-ordinate rule;
2. Average ordinate rule
3. Trapezoidal rule;
4. Simpson's one-third rule.

1. Mid-Ordinate Rule (Fig. 12.1)

The method is used with the assumption that the boundaries between the extremities of the ordinates (or offsets) are straight lines. The base line is divided into a number of divisions and the ordinates are measured at the mid-points of each division, as illustrated in Fig. 12.2.

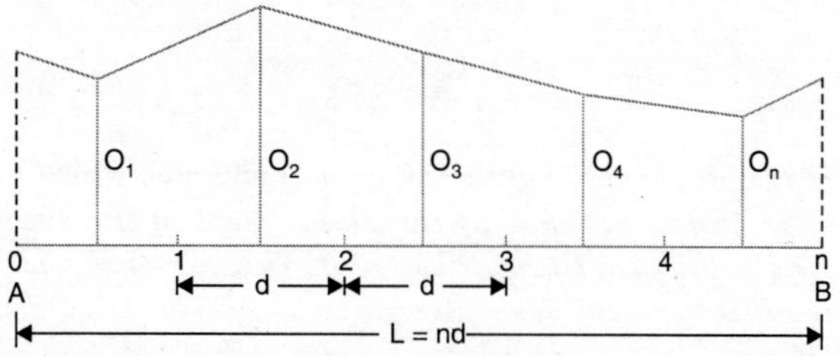

Fig. 12.2

The area is calculated by the formula

$$\text{Area} = \Delta = \text{Average ordinate} \times \text{Length of base}$$

$$= \frac{O_1 + O_2 + O_3 + \ldots\ldots + O_n}{n} L = (O_1 + O_2 + O_3 + \ldots\ldots + O_n)\, d = d\Sigma O \qquad \ldots(12.3)$$

where $O_1, O_2 \ldots$ = the ordinates at the mid-points of each division

ΣO = sum of the mid-ordinates; n = number of divisions

L = length of base line = nd; d = distance of each division

2. Average Ordinate Rule (Fig. 12.3)

This rule also assumes that the boundaries between the extremities of the ordinates are straight lines. The offsets are measured to each of the points of the divisions of the base line.

The area is given by Δ = Average ordinate × Length of the base

$$= \left(\frac{O_0 + O_1 \ldots + O_n}{n+1}\right) L = \frac{L}{(n+1)} \Sigma O \qquad \ldots(12.4)$$

where O_0 = ordinate at one end of the base.

O_n = ordinate at the other end of the base divided into n equal divisions.

$O_1, O_2 \ldots$ = ordinates at the end of each division

3. Trapezoidal Rule (Fig. 12.3)

This rule is based on the assumption that the figures are trapezoids. The rule is more accurate than the previous two rules which are approximate versions of the trapezoidal rule.

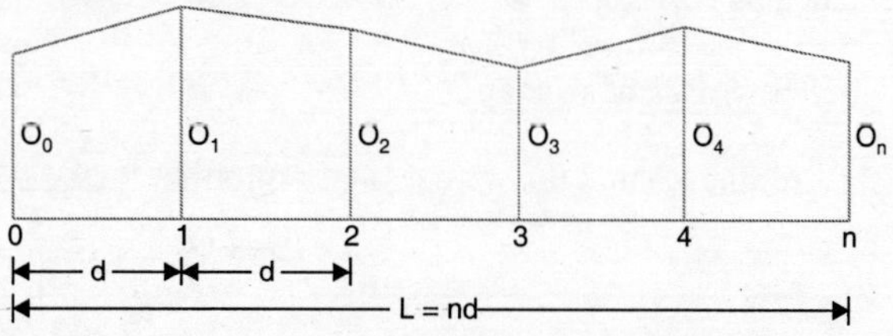

Fig. 12.3

Referring to Fig. 12.3, the area of the first trapezoid is given by

$$\Delta_1 = \frac{O_0 + O_1}{2} d$$

Similarly, the area of the second trapezoid is given by $\Delta_2 = \frac{O_1 + O_2}{2} d$

Area of the last trapezoid (*n*th) is given by

$$\Delta_n = \frac{O_{n-1} + O_n}{2} d$$

Hence the total area of the figure is given by

$$\Delta = \Delta_1 + \Delta_2 + \ldots\ldots \Delta_n = \frac{O_0 + O_1}{2} d + \frac{O_1 + O_2}{2} d + \ldots\ldots + \frac{O_{n-1} + O_n}{2} d$$

or

$$\Delta = \left(\frac{O_0 + O_n}{2} + O_1 + O_2 + \ldots\ldots + O_{n-1} \right) d \qquad \ldots(12.5)$$

Equation (12.5) gives the trapezoidal rule which may be expressed as below:

Add the average of the end offsets to the sum of the intermediate offsets. Multiply the total sum thus obtained by the common distance between the ordinates to get the required area.

4. Simpson's One-Third Rule

This rule assumes that the short lengths of boundary between the ordinates are parabolic arcs. This method is more useful when the boundary line departs considerably from the straight line.

Thus, in Fig. 12.4, the area between the line *AB* and the curve *DFC* may be considered to be equal to the area of the trapezoid *ABCD* plus the area of the segment between the parabolic arc *DFC* and the corresponding chord *DC*.

Let O_0, O_1, O_2 = any three consecutive ordinates taken at regular interval of *d*.

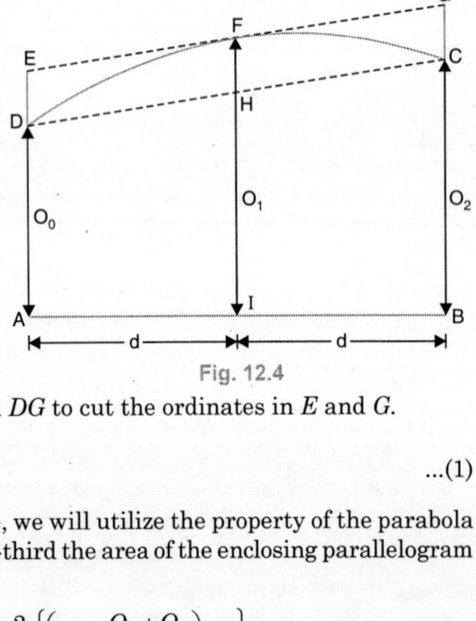

Fig. 12.4

Through *F*, draw a line *EG* parallel to the chord *DG* to cut the ordinates in *E* and *G*.

$$\text{Area of trapezoid} \qquad ABCD = \frac{O_0 + O_2}{2} \,.\, 2d \qquad \ldots(1)$$

To calculate the area of the segment of the curve, we will utilize the property of the parabola that area of a segment (such as *DFC*) is equal to two-third the area of the enclosing parallelogram (such as *CDEG*).

$$\text{Thus, area of segment} \qquad DFC = \frac{2}{3} (FH \times AB) = \frac{2}{3} \left\{ \left(O_1 - \frac{O_0 + O_2}{2} \right) 2d \right\} \qquad \ldots(2)$$

Adding (1) and (2), we get the required area ($\Delta_{1,\,2}$) of first two intervals. Thus,

$$\Delta_{1,\,2} = \frac{O_0 + O_2}{2} . 2d + \frac{2}{3} \left\{ \left(O_1 - \frac{O_0 + O_2}{2} \right) 2d \right\} = \frac{d}{3} (O_0 + 4O_1 + O_2) \ldots(3)$$

Similarly, the area of next two intervals ($\Delta_{3, 4}$) is given by

$$\Delta_{3, 4} = \frac{d}{3}(O_2 + 4O_3 + O_4) \quad ...(4)$$

Area of the last two intervals (Δ_{n-1}, Δ_n) is given by

$$O_{n-1, n} = \frac{d}{3}(O_{n-2} + 4O_{n-1} + O_n) \quad ...(5)$$

Adding all these to get the total area (Δ), we get

$$\Delta = \frac{d}{3}[O_0 + 4O_1 + 2O_2 + 4O_3 \ ...... + 2O_{n-2} + 4O_{n-1} + O_n]$$

or

$$\Delta = \frac{d}{3}[(O_0 + O_n) + 4(O_1 + O_3 + ... + O_{n-1}) + 2(O_2 + O_4 + ... O_{n-2})] \quad ...(12.6)$$

It is clear that the rule is applicable only when the number of divisions of the area is *even i.e.,* the total number of ordinates is *odd.* If there is an odd number of divisions (resulting in even number of ordinates), the area of the last division must be calculated separately, and added to equation 12.6.

Simpson's one-third rule may be stated as follows: ***The area is equal to the sum of the two end ordinates plus four times the sum of the even intermediate ordinates + twice the sum of the odd intermediate ordinates, the whole multiplied by one-third the common interval between them.***

Comparison of Rules: The results obtained by the use of Simpson's rule are in all cases the more accurate. The results obtained by using 'Simpson's rule are greater or smaller than those obtained by using the trapezoidal rule according as the curve of the boundary is concave or convex towards the base line. In dealing with irregularly shaped figures, the degree of precision of either method can be increased by increasing the number of ordinates.

Example 12.1: *The following perpendicular offsets were taken at 10 metres intervals from a survey line to an irregular boundary line:*

3.25, 5.60, 4.20, 6.65, 8.75, 6.20, 3.25, 4.20, 5.65

Calculate the area enclosed between the survey line, the irregular boundary line, and the first and last offsets, by the application of (a) average ordinate rule, (b) trapezoidal rule, and (c) Simpson's rule.

Solution:

(*a*) ***By average ordinate rule***

From equation 12.4 (*a*), we have $\Delta = \frac{L}{n+1}\Sigma O$

Here n = number of divisions = 8; $n + 1$ = number of ordinates = 8 + 1 = 9

L = Length of base = 10 × 8 = 80 m

ΣO = 3.25 + 5.60 + 4.20 + 6.65 + 8.75 + 6.20 + 3.25 + 4.20 + 5.65 = 47.75 m

$\therefore$ $\Delta = \frac{80}{9} \times 47.75$ = **424.44** sq. metres = **4.2444** ares.

(*b*) *By trapezoidal rule*

From Eq. 12.5, $\Delta = \left(\frac{O_0 + O_n}{2} + O_1 + O_2 + ... + O_{n-1}\right)d$

Here $d = 10$ m; $\frac{O_0 + O_n}{2} = \frac{3.25 + 5.65}{2} = 4.45$ m

$O_1 + O_2 ...... O_{n-1} = 5.60 + 4.20 + 6.65 + 8.75 + 6.20 + 3.25 + 4.20 = 38.85$ m

$\Delta = (4.45 + 38.85)\ 10 =$ **433** sq. metres = **4.33** ares

(*c*) *By Simpson's rule*

From Eq. 12.6, $\Delta = \frac{d}{3}\ [(O_0 + O_n) + 4(O_1 + O_2 + ... + O_{n-1}) + 2(O_2 + O_4 ...... + O_{n-2})]$

Here $d = 10$ m; $O_0 + O_n = 3.25 + 5.65 = 8.9$ m

$4\ (O_1 + O_3 + ... O_{n-1}) = 4\ (5.60 + 6.65 + 6.20 + 4.20) = 90.60$

$2\ (O_2 + O_4 + ... O_{n-2}) = 2\ (4.20 + 8.75 + 3.25) = 32.40$

$\therefore$ $\Delta = \frac{10}{3}\ (8.9 + 90.60 + 32.40) =$ **439.67** sq. metres = **4.3967** ares.

Example 12.2: *A series of offsets were taken from a chain line to a curved boundary line at intervals of 15 metres in the following order.*

0, 2.65, 3.80, 3.75, 4.65, 3.60, 4.95, 5.85 m.

Compute the area between the chain line, the curved boundary and the end offsets by (a) average ordinate rule, (b) trapezoidal rule, and (c) Simpson's rule.

Solution:

(*a*) *By average ordinate rule*

From Eq. 12.4 (*a*), we have $\Delta = \frac{L}{n+1}\Sigma O$

Hence $n = 7$; $n + 1 = 8$

$L = nd = 7 \times 15 = 105$ m

$\Sigma O = 0 + 2.65 + 3.80 + 3.75 + 4.65 + 3.60 + 4.95 + 5.85 = 29.25$ m

$\therefore$ $\Delta = \frac{105}{8} \times 29.25 =$ **383.91** sq. m = 3.8391 ares.

(*b*) *By trapezoidal rule*

From equation 12.5 $\Delta = \left(\frac{O_0 + O_n}{2} + O_1 + O_2 + ... O_{n-1}\right)d$

Here $d = 15$ m; $\frac{O_0 + O_n}{2} = \frac{0 + 5.85}{2} = 2.925$ m

$O_1 + O_2 + ... O_{n-1} = 2.65 + 3.80 + 3.75 + 4.65 + 3.60 + 4.95 = 23.40$

$\therefore$ $\Delta = (2.925 + 23.40)\ 15 =$ **394.87** sq. m = 3.9487 ares.

(*c*) *By Simpson's rule*

From equation 12.6, $\Delta = \frac{d}{3}\ [(O_0 + O_n) + 4\ (O_1 + O_3 + ... O_{n-1}) + 2(O_2 + O_4 + ... O_{n-2})]$

Here, $\frac{d}{3} = \frac{15}{3} = 5$ m.

It will be seen that the Simpson's rule is not directly applicable here since the number of ordinates (n) is even. However, the area between the first and seventh offsets may be calculated by Simpson's rule, and the area enclosed between the seventh and last offsets may be found by the trapezoidal rule.

Thus, $(O_0 + O_n) = 0 + 4.95 = 4.95$

$$4(O_1 + O_3 + ...O_{n-1}) = 4(2.65 + 3.75 + 3.60) = 40$$

$$2(O_2 + O_4 + ...O_{n-2}) = 2(3.80 + 4.65) = 16.90$$

$\therefore$ $\Delta' = 5(4.95 + 40 + 16.90) = 309.25$ sq. m.

Area of the last trapezoid $= (4.95 + 5.85)\ \frac{15}{2} = 81.0$ sq. m.

$\therefore$ Total area = 309.25 + 81.0 = **390.25** sq. m = 3.9025 ares.

12.5 OFFSETS AT IRREGULAR INTERVALS

(a) First Method **(Fig. 12.5):** In this method, the area of each trapezoid is calculated separately and then added together to calculate the total area. Thus, from Fig. 12.5,

$$\Delta = \frac{d_1}{2}\ (O_1 + O_2) + \frac{d_2}{2}\ (O_2 + O_3) + \frac{d_3}{2}\ (O_3 + O_4) \qquad ...(12.7)$$

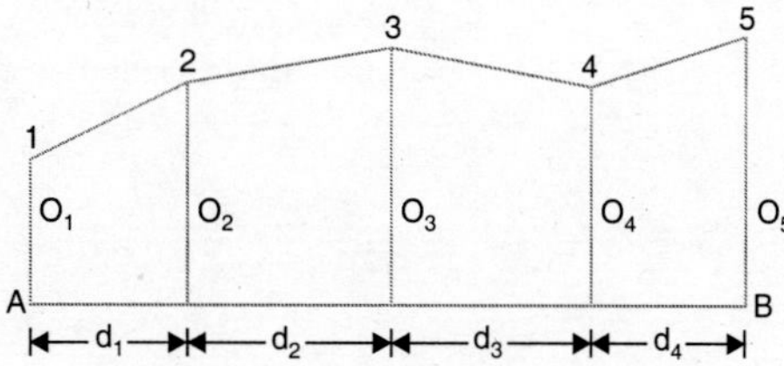

Fig. 12.5

(b) Second Method: By method of co-ordinates: See § 12.7

Example 12.3: The following perpendicular offsets were taken from a chain line to an irregular boundary:

Chainage	*0*	*10*	*25*	*42*	*60*	*75 m*
Offset	*15.5*	*26.2*	*31.8*	*25.6*	*29.0*	*31.5*

Calculate the area between the chain line, the boundary and the end offsets.

Solution:

$$\text{Area of first trapezoid} = \Delta_1 = \frac{10-0}{9}\ (15.5 + 26.2) = 208.5\ \text{m}^2$$

$$\text{Area of second trapezoid} = \Delta_2 = \frac{25-10}{2}\ (26.2 + 31.8) = 435\ \text{m}^2$$

$$\text{Area of third trapezoid} = \Delta_3 = \frac{42-25}{2}\ (31.8 + 25.6) = 487.9\ \text{m}^2$$

$$\text{Area of fourth trapezoid} = \Delta_4 = \frac{60-42}{2}\ (25.6 + 29.0) = 491.4\ \text{m}^2$$

$$\text{Area of fifth trapezoid} = \Delta_5 = \frac{75-60}{2}\ (29.0 + 31.5) = 453.7\ \text{m}^2$$

$\therefore$ Total area $= \Delta = \Delta_1 + \Delta_2 + \Delta_3 + \Delta_4 + \Delta_5$

$$= 208.5 + 435 + 487.9 + 491.4 + 453.7$$
$$= \mathbf{2076.5\ m^2 = 20.765\ ares.}$$

Example 12.4: *The following perpendicular offsets were taken from a chain line to a hedge:*

Chainage (m)	*0*	*15*	*30*	*45*	*60*	*70*	*80*	*100*	*120*	*140*
Offsets (m)	*7.60*	*8.5*	*10.7*	*12.8*	*10.6*	*9.5*	*8.3*	*7.9*	*6.4*	*4.4*

Calculate the area between the survey line, the hedge and the end offsets by (a) Trapezoidal rule (b) Simpson's rule.

Solution:

(*a*) **By Trapezoidal rule:** The interval is constant from first offset to 5th offset. There is another interval between the 5th and 7th offset and a third interval between 7th offset and 10th offset. The total area Δ can, therefore, be divided into three sections.

$$\therefore \qquad \Delta = \Delta_1 + \Delta_2 + \Delta_3$$

where Δ_1 = area of first section; Δ_2 = area of second section
Δ_3 = area of third section; d_1 = interval for first section = 15 m
d_2 = interval for second section = 10 m; d_3 = interval for third section = 20 m

Now
$$\Delta_1 = \left(\frac{7.60+10.6}{2}+8.5+10.7+12.8\right) 15 = 616.3\ \text{m}^2$$
$$\Delta_2 = \left(\frac{10.6+8.3}{2}+9.5\right) 10 = 189.5\ \text{m}^2$$
$$\Delta_3 = \left(\frac{8.3+4.4}{2}+7.9+6.4\right) 20 = 413\ \text{m}^2$$
$$\Delta = 616.5 + 189.5 + 413 = \mathbf{1219}\ \text{m}^2 = 12.19\ \text{ares.}$$

(*b*) **By Simpson's Rule:** The first section and the second section have odd number of ordinates, and therefore, Simpson's rule is directly applicable. The third section has 4 ordinates (even number); the rule is applicable for the first three ordinates only:

$$\Delta_1 = \frac{15}{3}[(7.60 + 10.6) + 4\,(8.5 + 12.8) + 2\,(10.7)] = 624\ \text{m}^2$$
$$\Delta_2 = \frac{10}{3}[(10.6 + 8.3) + 4\,(9.5)] = 189.7\ \text{m}^2$$
$$\Delta_3 = \frac{20}{3}[(8.3 + 6.4) + 4\,(7.9)] + \frac{20}{2}(6.4 + 4.4)$$
$$= 308.6 + 108 = 416.6\ \text{m}^2$$
$$\therefore \qquad \Delta = 624 + 189.7 + 416.6 = \mathbf{1230.3\ m^2} = 12.303\ \text{ares.}$$

12.6 AREA BY DOUBLE MERIDIAN DISTANCES

This method is the one most often used for computing the area of a closed traverse. This method is known as D.M.D. method. To calculate the area by this method, the latitudes and departures of each line of the traverse are calculated. The traverse is then balanced. A reference meridian is then assumed to pass through the *most westerly station* of the traverse and the double meridian distances of the lines are computed.

Meridian Distances

The meridian distance of any point in a traverse is the distance of that point to the reference meridian, measured at right angles to the meridian. The meridian distance of a survey line is defined as the meridian distance of its mid-point. The meridian distance (abbreviated as M.D.) is also sometimes called as the *longitude.*

Thus, in Fig. 12.6, if the reference meridian is chosen through the most westerly station *A,* the meridian distance (represented by symbol *m*) of the line *AB* will be equal to half its departure. The meridian distance of the second line *BC* will be given by

$$m_2 = m_1 + \frac{D_1}{2} + \frac{D_2}{2}$$

Similarly, the meridian distance of the third line *CD* is given by

$$m_3 = m_2 + \frac{D_2}{2} + \left(-\frac{D_3}{2}\right) = m_2 + \frac{D_2}{2} - \frac{D_3}{2}$$

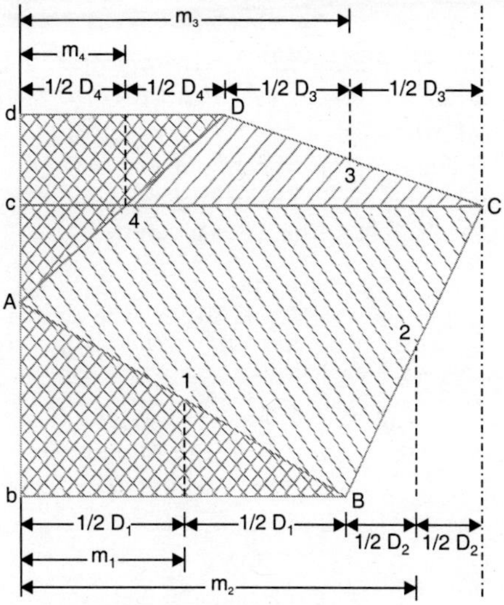

Fig. 12.6

And, the meridian distance of the fourth (last) line *DA* is given by

$$m_4 = m_3 + \left(-\frac{D_3}{2}\right) + \left(-\frac{D_4}{2}\right) = m_3 - \frac{D_3}{2} - \frac{D_4}{2} = \frac{D_4}{2}$$

Hence, the rule for the meridian distance may be stated as follows: ***The meridian distance of any line is equal to the meridian distance of the preceding line plus half the departure of the preceding line plus half the departure of the line itself.***

According to the above, the meridian distance of the first line will be equal to half its departure. In applying the rule, proper attention should be paid to the signs of the departures i.e., positive sign for eastern departure and negative sign for western departure.

Area by Latitudes and Meridian Distances

In Fig. 12.6, east-west lines are drawn from each station to the reference meridian, thus getting triangles and trapeziums. One side of each triangle or trapezium (so formed) will be one of the lines, the *base* of the triangle or trapezium will be the *latitude* of the line, and the *height* of the triangle or trapezium will be the *meridian distance* of that line. Thus,

area of each triangle or trapezium = latitude of the line × meridian distance of the line.

or $$A_1 = L_1 \times m_1$$

The latitude (*L*) will be taken positive if it is a northing, and negative if it is a southing.

In Fig. 12.6, the area of the traverse *ABCD* is equal to the algebraic sum of the areas of *dDCc, CcbB, dDA* and *ABb*. Thus,

$$A = \text{Area of } dDCc + \text{area of } cCBb - \text{area of } dDA - \text{area of } ABb$$

or $$A = L_3 m_3 + L_2 m_2 - L_4 m_4 - L_1 m_1 = \Sigma Lm.$$

(It is to be noted that the quantities $L_4 m_4$ and $L_1 m_1$ bear negative sign since L_4 and L_1 of *DA* and *AB* are negative).

Double Meridian Distance

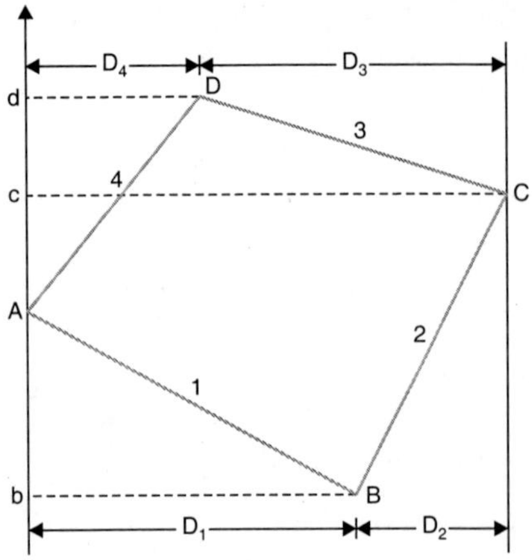

Fig. 12.7

The double meridian distance of a line is equal to the sum of the meridian distances of the two extremities.

Thus, in Fig. 12.7, we have:

Double meridian distance (represented by symbol M) of the first line AB is given by

$$M_1 = m \text{ of } A + m \text{ of } B = 0 + D_1 = D_1$$

Similarly, if M_2, M_3, M_4 are the double meridian distances of the lines BC, CD and DA respectively, we have

$$M_2 = m \text{ of } B + m \text{ of } C$$
$$= D_1 + (D_1 + D_2) = M_1 + D_1 + D_2 = \text{D.M.D. of } AB + \text{Departure of } AB + \text{Departure of } BC$$

$$M_3 = m \text{ of } C + m \text{ of } D$$
$$= (D_1 + D_2) + (D_1 + D_2 - D_3) = M_2 + D_2 - D_3$$
$$= \text{D.M.D. of } BC + \text{Departure of } BC + \text{Departure of } CD$$

and $$M_4 = m \text{ of } D + m \text{ of } A = (D_1 + D_2 - D_3) + (D_1 + D_2 - D_3 - D_4) = M_3 - D_3 - D_4$$
$$= \text{D.M.D. of } CD + \text{Departure of } CD + \text{Departure of } DA$$

Hence, the rule for finding D.M.D. of any line may be stated as follows: ***"The D.M.D. of any line is equal to the D.M.D. of the preceding line plus the departure of the preceding line plus the departure of the line itself."***

Due attention should be paid to the sign of the departure. The D.M.D. of the first line will evidently be equal to its departure. The double meridian distance of the last line is also equal to its departure, but this fact should be used simply as a check.

Area by Latitudes and Double Meridian Distances

In Fig. 12.7, the area of the traverse $ABCD$ is given by

$$A = \text{area of } dDCc + \text{area of } CcbB - \text{area of } dDA - \text{area of } ABb$$

Now, area of $dDCc = \frac{1}{2}(dD + cC)\, cd = \frac{1}{2}(M_3) \times L_3$

That is, area of any triangle or trapezium = Half the product of the latitude of the line and its meridian distance.

Hence $$A = \frac{1}{2}\,[M_3L_3 + M_2L_2 - M_4L_4 - M_1\,L_1]$$

Thus, to find the area of the traverse by D.M.D. method, the following steps are necessary:

1. Multiply D.M.D. of each line by its latitude.
2. Find the *algebraic sum* of these products.
3. The required area will be half the sum.

Area from Departures and Total Latitudes

From Fig. 12.8, the area (A) of $ABCD$ is given by

$$A = \text{area of } ABb + \text{area of } BbcC + \text{area of } dcCD + \text{area of } DdA$$

If L_1', L_2', L_3' are the total latitudes of the ends of the lines, we get

$$A = \tfrac{1}{2}\,[(D_1)(0 - L_1') + (D_2)(- L_1' + L_2') + (- D_3)(L_2' + L_3') + (- D_4)(L_3' + 0)]$$
$$= -\tfrac{1}{2}\,[L_1'(D_1 + D_2) + L_2'(- D_2 + D_3) + L_3'(D_3 + D_4)]$$

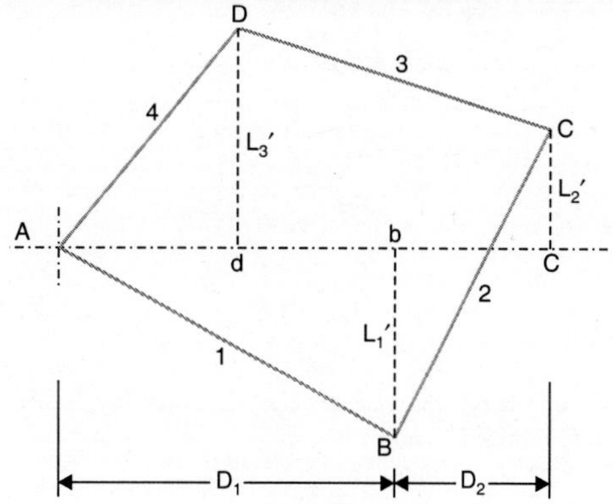

Fig. 12.8

Note: The negative sign to the area has no significance.

Hence, to find the area by this method, the following steps are necessary:

1. Find the *total latitude* (L') of each station of traverse.
2. Find the *algebraic sum* of the departures of the two lines meeting at that station.
3. Multiply the total latitude of each station by the corresponding algebraic sum of the departure (found in 2).
4. Half the algebraic sum of these products gives the required area.

Area by Double Parallel Distances and Departures

A *parallel distance* of any line of a traverse is the perpendicular distance from the middle point of that line to a reference line (chosen to pass through most southerly station) at right angles to the meridian. *The double parallel distance (D.P.D.) of any line is the sum of the parallel distances of its ends.* The principles of finding area by D.M.D. method and D.P.D. method are identical. The rules derived above may be changed to get the corresponding rules for D.P.D. method, by substituting D.P.D. for D.M.D. and 'departure' for 'latitude'. The method is employed as an independent method of checking area computed by D.M.D. method.

12.7 AREA BY CO-ORDINATES

Let (x_1, y_1), (x_2, y_2), (x_3, y_3) and (x_4, y_4) be the co-ordinates of the stations A, B, C, D respectively, of a traverse $ABCD$. If A is the total area of the traverse, we have

A = (Area $aABb$) + (Area $bBCc$) – (Area $cCDd$) – (Area $dDAa$)

$$= \tfrac{1}{2}[(y_1 - y_2)(x_1 + x_2) + (y_2 - y_3)(x_2 + x_3) - (y_4 - y_3)(x_4 + x_3) - (y_1 - y_4)(x_1 + x_4)]$$
$$= \tfrac{1}{2}[y_1(x_2 - x_4) + y_2(x_3 - x_1) + y_3(x_4 - x_2) + y_4(x_1 - x_3)]$$

In general, if we have n stations, we get

$$A = \tfrac{1}{2}[y_1(x_2 + x_n) + y_2(x_3 - x_1) + y_3(x_4 - x_2) + \ldots + y_n(x_1 - x_{n-1})] \qquad \ldots(12.7)$$

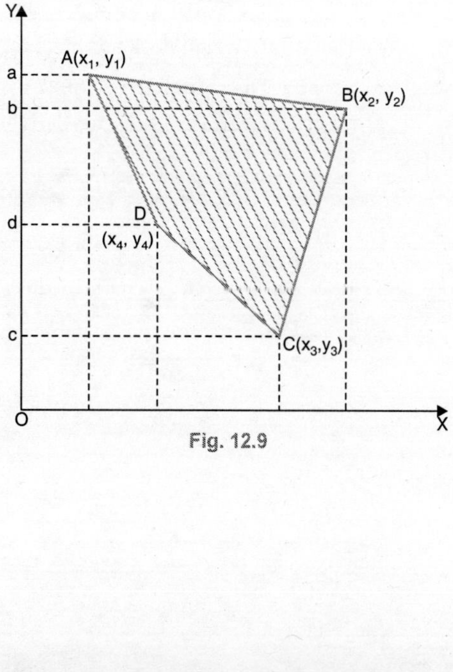

Fig. 12.9

Example 12.5: *The following table gives the corrected latitudes and departures (in metres) of the sides of a closed traverse ABCD:*

Side	Latitude		Departure	
	N	S	E	W
AB	108		4	
BC	15		249	
CD		123	4	
DA	0			257

Compute its area by (i) M.D. method, (ii) D.M.D. method, (iii) Departures and total latitudes, (iv) Co-ordinate method.

Solution:

1. *By meridian distances and latitudes*

$$\text{Area} = \Sigma(mL)$$

Calculate the meridian distance of each line. The calculations are arranged in the tabular form below. By the inspection of the latitudes and departures, point A is the most westerly station. AB is taken as the first line and DA as the last line.

Line	Latitude (L)	Departure (D)	$\frac{1}{2}$ Departure ($\frac{1}{2}$D)	M.D. (m)	Area = mL
AB	+ 108	+ 4	+ 2	2	+ 216
BC	+ 15	+ 249	+ 124.5	128.5	+ 1928
CD	– 123	+ 4	+ 2	255	– 31365
DA	0	– 257	– 128.5	128.5	0
				Sum	– 29221

∴ **Total area** $= \Delta = \Sigma mL = -29221 \text{ m}^2$

Since the negative sign does not have any significance.

∴ The actual area = **29221 m² = 2.9221 hectares.**

2. *By D.M.D. method:* Area $= \frac{1}{2}\Sigma mL$

Line	Latitude (L)	Departure (D)	D.M.D. (m)	Area = mL
AB	+ 108	+ 4	4	+ 432
BC	+ 15	+ 249	257	+ 3855
CD	– 123	+ 4	510	– 62,730
DA	0	– 257	257	0
			Sum	**– 58,443**

∴ **Area** $= \frac{1}{2}\Sigma mL = 29221 \text{ m}^2 =$ **2.9221 hectares.**

3. *By Departure and total latitudes:* Let us first calculate the total latitudes of the point, starting with A as the reference point,

Thus, total latitude of $B = +\,108$

total latitude of $C = +\,108 + 15 = +\,123$

total latitude of $D = +\,123 - 123 = 0$

total latitude of $A = 0 + 0 = 0$

The area $= \frac{1}{2}\Sigma$ (Total latitude × algebraic sum of adjoining departures)

Line	*Latitude (L)*	*Departure (D)*	*Station*	*Total Latitude (L)*	*Algebraic sum of adjoining departures*	*Double area*
AB	+ 108	+ 4				
			B	+ 108	+ 253	+ 27,324
BC	+ 15	+ 249				
			C	+ 123	+ 253	+ 31,119
CD	– 123	+ 4				
			D	0	– 253	0
DA	0	– 257				
			A	0	– 253	0
					Sum	**58,443**

∴ **Area** $= \frac{1}{2}(58{,}443)\ m^2 = 29221\ m^2 =$ **2.9221 hectares.**

4. *By Co-ordinates:* For calculation of area by co-ordinates, it is customary to calculate the independent co-ordinates of all the points. This can be done by taking the co-ordinates of A as (+ 100, + 100). The results are tabulated below:

Line	*Latitude*	*Departure*	*Station*	*Independent co-ordinates*	
				North (y)	*East (x)*
			A	100	100
AB	+ 108	+ 4			
			B	208	104
BC	+ 15	+ 249			
			C	223	353
CD	– 123	+ 4			
			D	100	357
DA	0	– 257			
			A	100	100

Substituting the values of x and y in equation 12.7, we get

$$A = \tfrac{1}{2}[(y_1(x_2 - x_4) + y_3(x_3 - x_1) + y_3(x_4 - x_2) + y_4(x_1 - x_3)]$$
$$= \tfrac{1}{2}[100(208 - 100) + 104\,(223 - 100) + 353\,(100 - 208) + 357\,(100 - 223)]$$
$$= \tfrac{1}{2}(10800 + 12792 - 38124 - 43911) = -\,29221 \text{ m}^2$$

Since the negative sign does not have significance, the area = **2.9221 hectares.**

12.8 AREA COMPUTED FROM MAP MEASUREMENTS

(*a*) By sub-division of the area into geometric figures: The area of the plan is sub-divided into common geometric figures, such as triangles, rectangles, squares, trapezoids etc. The length and latitude of each such figure is scaled off from the map and the area is calculated by using the usual formulae.

(*b*) By sub-division into squares: Fig. 12.10 (*a*): The method consists in drawing squares on a tracing paper each square representing some definite number of square metres. The tracing paper is placed on the drawing and the number of squares enclosed in the figure are calculated. The positions of the fractional squares at the curved boundary are estimated. The total area of the figure will then be equal to the total number of squares multiplied by the factor (*i.e.*, sq. metres) represented by each square.

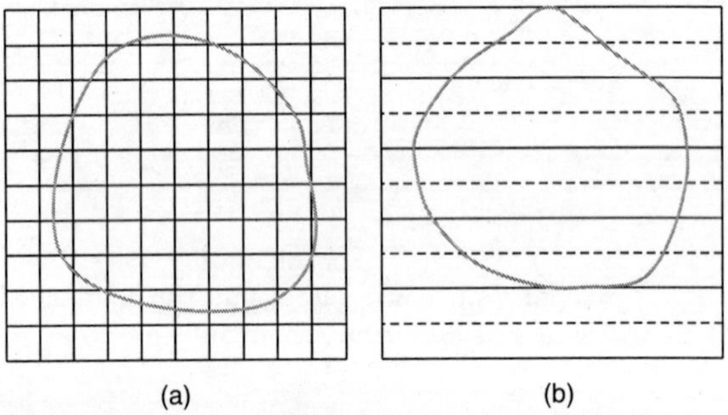

Fig. 12.10

(*c*) By division into trapezoids: Fig. 12.10 (*b*): In this method, a number of parallel lines, at constant distance apart, are drawn on a tracing paper. The constant between the consecutive parallel lines represents some distance in metres or links. Midway between each pair of lines there is drawn another pair of lines in a different colour or dotted. The tracing is then placed on the drawing in such a way that the area is exactly enclosed between two of the parallel lines. The figure is thus divided into a number of strips. Assuming that the strips are either trapezoids or triangles, the area of each is equal to the length of the mid-ordinate multiplied by the constant breadth. The mid-ordinates of the strips are represented by the length of the dotted lines intercepted within the maps. The total sum of these intercepted dotted lines is measured and multiplied by the constant breadth to get the required area. More accuracy will be obtained if the strips are placed nearer.

12.9 AREA BY PLANIMETER

A planimeter is an instrument which measures the area of plan of any shape very accurately. There are two types of planimeters: (1) Amsler Polar Planimeter, and (2) Roller Planimeter. The polar planimeter is most commonly used and is, therefore discussed here.

Fig. 12.11 shows the essential parts of a polar planimeter. It consists of two arms hinged at a point known as the pivot point. One of the two arms carries an anchor at its end, and is known as the anchor arm. The length of anchor arm is generally fixed, but in some of the planimeters a variable length of anchor arm is also provided. The other arm carries a tracing point at its end, and is known as the tracing arm. The length of the tracing arm can be varied by means of a fixed screw and its corresponding slow motion screw. The tracing point is moved along the boundary of the plan the area of which is to be determined. The normal displacement of the tracing arm is measured by means of a wheel whose axis is kept parallel to the tracing arm. The wheel may either be placed between the hinge and the tracing point or is placed beyond the pivot point away from the tracing point. The wheel carries a concentric drum which is divided into 100 divisions. A small vernier attached near the drum reads one-tenth of the drum division.

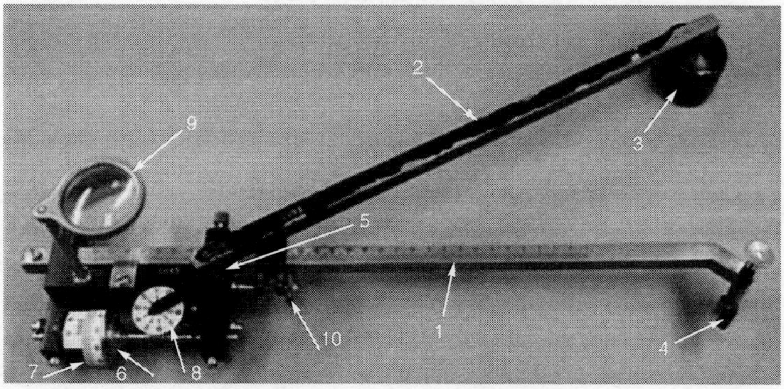

Fig. 12.11 Amsler polar planimeter.

1. Tracing Arm
2. Anchor Arm
3. Anchor
4. Tracing Point
5. Hinge
6. Wheel
7. Graduated Drum
8. Disc
9. Magnifier
10. Adjusting Screw for 1

The complete revolutions of the wheel are read on a disc actuated by a suitable gearing to the wheel. Thus, each reading is of four digits—the units being read on the disc, the tenths and hundredths on the drum, and the thousandths on the vernier. In addition to this, a fixed index near the disc can be utilised to know the number of the times the zero of the disc has crossed the index.

It is clear from Fig. 12.11 that the planimeter rests on three points—the wheel, the anchor point and the tracing point. Out of these three, the anchor point remains fixed in position while the wheel partly rolls and partly slides as the tracing point is moved along the boundary. Since the plane of the wheel is perpendicular to the plane of the centre line of the tracing arm, the wheel measures only normal displacement—when it actually rolls.

To find the area of the plan, the anchor point is either placed outside the area (if the area is small) or it is placed inside the area (if the area is large). A point is then marked on the boundary of area and the tracing point kept exactly over it. The initial reading of the wheel is then taken. The tracing point is now moved *clockwise* along the boundary till it comes to the starting point. The final reading of the drum is taken. The area of the figure is then calculated from the following formula:

$$\text{Area} \quad (\Delta) = M(F - I \pm 10N + C) \qquad \text{...(12.7)}$$

where F = Final reading; I = Initial reading

N = The number of times the zero mark of the dial passes the fixed index mark. Use *plus* sign if the zero mark of the dial passes the index mark in a clockwise direction and *minus* sign when it passes in the anticlockwise direction.

M = A multiplying constant, also sometimes known as the planimeter constant. It is equal to the area per revolution of the roller.

C = Constant of the instrument which when multiplied by M, gives the area of zero circle. The constant C is to be added only when the anchor point is inside the area.

It is to be noted that the tracing point is to be moved in the clockwise direction only. Proper sign must be given to N. The proof of the above formula is given below.

Theory of Planimeter

Fig. 12.12 (*a*) shows the schematic diagram of polar planimeter, where

A_0 = Area to be measured, the anchor point being outside the area.

L = Length of the tracing arm
= Distance between the tracing point and the hinge.

R = Length of anchor arm
= Distance between the pivot and the anchor point.

a = Distance between the wheel and the pivot, the wheel being placed between the tracing point and pivot.

w = Distance rolled by the roller in tracing the area.

A_s = Area swept by the tracing arm.

Fig. 12.12 (*b*) shows the section of the perimeter of the area. Any such movement of the arm is equivalent to two simultaneous motions: (*i*) translation of the tracing arm *TP* in parallel motion and (*ii*) rotation of the tracing arm about the pivot. Fig. 12.12 (*c*) shows the components of the two motions separately. Thus, if the tracing arm sweeps a very small area dA_s, such that dh is the movement in parallel direction and $d\theta$ is the rotation, we have

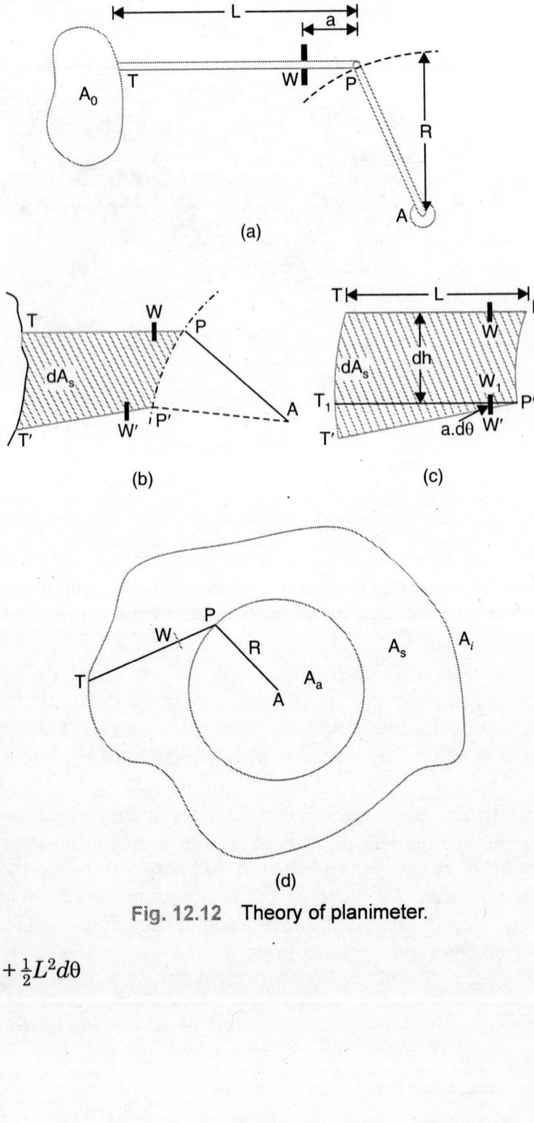

Fig. 12.12 Theory of planimeter.

$$dA_s = Ldh + \tfrac{1}{2}L^2 d\theta$$

Since the recording wheel (W) is placed in plane perpendicular to that of the tracing arm, the wheel records only the movement perpendicular to its axis. If dw is the distance rolled out by the wheel in sweeping the area dA_s, we get

$$dw = dh + ad\theta \quad \text{or} \quad dh = dw - ad\theta \qquad ...(1)$$

Substituting the value of dh in (1), we get

$$dA_s = L(dw - ad\theta) + \tfrac{1}{2} L^2 d\theta \qquad ...(2)$$

When the tracing point is moved along the boundary, the arm moves downwards along one side of area and upwards along the other side. Hence, the net area A_s swept by the tracing arm is equal to the area of the plan (A_0),

Thus $$A_0 = \int dA_s = L\int dw - aL\int d\theta + \tfrac{1}{2}L^2 \int d\theta \qquad ...(3)$$

But $\int dw =$ total distance moved by the wheel $= w$

$\therefore$ $$A_0 = Lw - aL\int d\theta + \tfrac{1}{2}L^2 \int d\theta \qquad ...(4)$$

Now when the anchor point is kept outside the area, the motion of the pivot is *constrained* along the arc of a circle *i.e.*, it never completes one revolution about the anchor point but simply moves along the arc in upward and downward directions so that

$$\int d\theta = 0.$$

Hence from (4), $$A_0 = Lw \qquad ...(12.8)$$

However, if the anchor point is kept inside the area, the pivot point moves along the circle of radius R and completes one revolution when the tracing point is brought back to its original position after tracing the area. Hence the quantity $\int d\theta = 2\pi$

Let A_i = Area of the plan when the anchor point is kept inside the area.

A_a = Area swept by the pivot.

Then, the area $$A_i = A_s + A_a = \int dA_s + \pi R^2 = [L\int dw - aL\int d\theta + \tfrac{1}{2}L^2\int d\theta] + \pi R^2$$

$$= Lw - aL(2\pi) + \frac{1}{2}L^2(2\pi) + \pi R^2 = Lw + \pi(L^2 - 2aL + R^2) \qquad ...(12.9)$$

Thus, equation 12.8 is to be used when the anchor point is outside the area while equation 12.9 is to be used when the anchor point is kept inside the area.

Now $$w = \text{Total distance rolled by the wheel} = \pi D\, n \qquad ...(5)$$

where D = Diameter of the wheel

n = Total change in the reading, due to the movement of the tracing point along the periphery of the area

$= F - I \pm 10N.$

Substituting the value of w in equation 12.9, we get the area A_i.

or $$\Delta = L\pi Dn + \pi(L^2 - 2aL + R^2) = Mn + \pi(L^2 - 2aL + R^2) \qquad ...(12.10\ a)$$

$$= Mn + MC = M(n + C) = M(F - I \pm 10N + C) \qquad ...(12.10\ b)$$

where M = The multiplier $= L\pi D$ = Length of tracing arm × Circumference of the wheel

$$C = \text{Constant} = \frac{\pi(L^2 - 2aL + R^2)}{M}$$

Thus, we get equation 12.10, which was given in the earlier stage. *In the above equation C is to be added only if the anchor point is inside the area.*

Zero Circle

The quantity $MC = n\,(L^2 - 2aL + R^2)$ is known as the area of the zero circle or correction circle. *The zero circle or the circle of correction is defined as the circle round the circumference of which if the tracing point is moved, the wheel will simply slide (without rotation) on the paper without any change in the reading.* This is possible when the tracing arm is held in such a position relative to the anchor arm that the plane of the roller passes through the anchor point *i.e.*, the line joining the anchor point and the wheel is at right angles to the line joining the tracing point and the wheel.

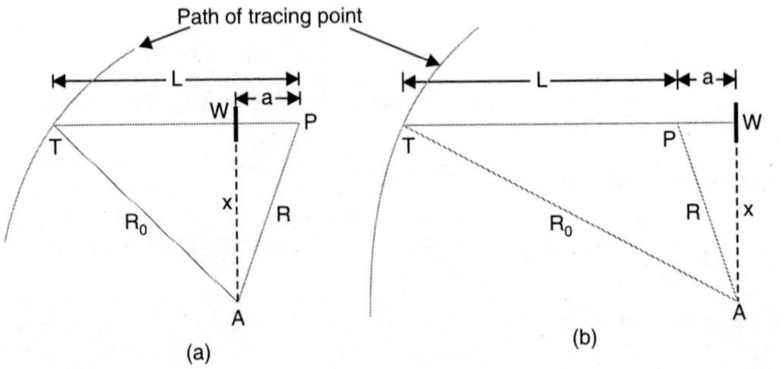

Fig. 12.13

In Fig. 12.13 (*a*), the wheel has been placed between the tracing point (*T*) and the pivot (*P*). Let R_0 be the radius of the zero circle. If x is the perpendicular distance of the wheel W from anchor A, we get

$$R_0 = (L - a)^2 + x^2 = (L - a)^2 + (R^2 - a^2)$$

$$= (L^2 + a^2 - 2La + R^2 - a^2) = (L^2 - 2aL + R^2) \quad \text{...(12.11 } a\text{)}$$

And area of the zero circle $= \pi R_0^{\,2} = \pi(L^2 - 2aL + R^2)$ ...(12.11 *b*)

In Fig. 12.13 (*b*), the wheel has been placed beyond the pivot.

Hence, $R_0^{\,2} = (L + a)^2 + (R^2 - a^2) = L^2 + a^2 + 2aL + R^2 - a^2 = L^2 + 2aL + R^2$ ...(12.11 *c*)

Area of the zero circle $= \pi(L^2 + 2aL + R^2)$ ...(12.11 *d*)

Thus, the general expression for the area of the zero circle can be written as:

Area of the zero circle $= \pi(L^2 \pm 2aL + R^2)$ ...(12.11)

Use + sign if the wheel is beyond the pivot and – sign if the wheel is between the tracing point and the pivot.

To find the area of the zero circle practically, the tracing point is traversed along the perimeter of a figure, one with the anchor point outside the figure, and then with the anchor point inside it.

Since the area swept is the same in both the cases, we get, from equation 12.10 *a*

$$\Delta = [Mn + \pi\,(L^2 \pm 2aL + R^2)] = Mn'$$

$$\therefore \quad \pi(L^2 \pm 2aL + R^2) = M(n' - n) \quad \text{...(12.12)}$$

where n and n' are the two corresponding readings of the wheel. *It is to be noted that n will be positive, if the area of the figure is greater than the area of the zero circle, while it will be negative if the area of the figure is smaller than the area of the zero circle.*

Multiplier Constant (M)

The multiplier constant or the *planimeter constant* is equal to the number of units of area per revolution of the roller. Numerically, it is equal to $L\pi D$. Since the diameter of the roller or wheel is a fixed quantity, the value of M depends on L. Thus, the length of the tracing arm is set to such a length that one revolution of the wheel corresponds to a whole number and convenient value of area. When the figure is drawn to a natural scale, and the area is desired in sq. inches, the value of M is generally kept as equal to 10 sq. in. of area.

For any other setting of the tracing arm, the value of M can be determined by traversing the perimeter of a figure of known area (A), with anchor point outside the figure. Then

$$M = \frac{\text{Known area}}{n'} = \frac{A}{n'} \quad \text{where } n' = \text{Change in the wheel readings}$$

It is to be noted that the value of M and C depends upon the length L which is adjustable. The manufacturers, therefore, supply a table which gives the values of L and C for different convenient values of M.

The manufacturers always supply the values of the vernier setting on the tracing arm with the corresponding values of M and C. The following table is an extract from the values for a typical planimeter.

Scale	*Vernier position on tracing bar*	*Area of one revolution of the measurement wheel (M)*		*Constant (C)*
		Scale	*Actual*	
1 : 1	33.44	100 sq. cm	100 sq. cm	23.521
1 : 1	21.58	10 sq. in.	10 sq. in.	26.430
1 : 48	26.97	200 sq. ft.	12.5 sq. in.	24.569
1 : 24	26.97	50 sq. ft.	12.5 sq. in.	24.569
1 : 50	21.66	0.4 acres	10.04 sq. in.	26.676

Thus, for full scale, value of M = 10 sq. in. in F.P.S. units, and for another setting of tracing bar, the value of M = 100 sq. cm.

Example 12.6: *Calculate the area of a figure from the following readings by a planimeter with anchor point outside the figure:*

Initial reading = 7.875, final reading = 3.086; M = 10 sq. in.

The zero mark on the dial passed the fixed index mark twice in the clockwise direction.

Solution:

$$A = M(F - I \pm 10N + C)$$

Since anchor point is outside, C is not to be used in the formula, M = 10; F = 3.086; I = 7.875; N = + 2

$$A = 10(3.086 - 7.875 + 20) = \mathbf{152.11 \text{ sq. in.}}$$

Example 12.7: *Calculate the area of a figure from the following readings recorded by the planimeter with the anchor point inside the figure.*

Initial reading = 9.918; Final reading = 4.254; M = 100 sq. cm; C = 23.521

It was observed that the zero mark on the dial passed the index once in the anti-clockwise direction.

Solution: The area is given by $A = M(F - I \pm 10\,N + C)$

Here $M = 100$ sq. cm; $I = 9.918$; $F = 4.254$; $C = 23.521$ and $N = -1$

$\therefore$ $A = 100(4.254 - 9.918 - 10 + 23.521) =$ **785.7 sq. cm.**

Example 12.8: *The following readings were obtained when an area was measured by a planimeter the tracing arm being set to the natural scale. The initial and final readings were 2.268 and 4.582. The zero of disc passed the index mark once in the clockwise direction. The anchor point was inside the figure with the value of the constant C of the instrument = 26.430.*

(a) *Calculate the area of the figure.*

(b) *If the area of the figure drawn be to a scale of 1 inch = 64 feet, find the area of the figure.*

Solution: Since the tracing arm was set to the natural scale, the value of $M = 10$ sq. inches.

$$A = M(F - I \pm 10\,N + C)$$

Here $F = 4.582;\quad I = 2.268;\quad N = +1;\quad C = 26.430$

$\therefore$ $A = 10\,(4.582 - 2.268 + 10 + 26.430) =$ **387.44** sq. inches.

The scale being $1'' = 64$ ft. Hence 1 sq. in. = 64×64 sq. ft.

$\therefore$ Area of field $= \dfrac{64 \times 64 \times 387.44}{43560}$ acres = **36.39** acres.

Example 12.9: *The perimeter of a figure is traversed clockwise with the anchor point inside and with the tracing arm so set that one revolution of the roller measured 100 sq. cm on the paper. The initial and final readings are 2.828 and 9.836. The zero mark of the disc passed the fixed index mark twice in the reverse direction. The area of the zero circle is found to be 2352 sq. cm. Find the area of the figure.*

Solution: The area of the figure is given by

$$A = Mn + \pi\,(L^2 - 2aL + A^2) \qquad \text{...(12.10 } a)$$

where $\pi(L^2 - 2aL + A^2)$ = Area of the zero circle = 2352 sq. cm.

$M = 100$ sq. cm.

$n = F - I \pm 10N = 9.836 - 2.828 - 10 \times 2 = -12.992$

$\therefore$ Substituting the values in Eq. 12.10 a, we get

$A = 100\,(-12.992) + 2352 = -1299.2 + 2352 =$ **1052.8** sq. cm.

Example 12.10: *The following observations were made with a planimeter.*

Area	*I.R.*	*F.R.*	*N*
(1) Known area of 60 sq. inches	*2.326*	*8.286*	*0*
(2) Unknown area	*8.286*	*5.220*	*+ 1*

The anchor point was placed outside the figure in both the cases with the same setting of the tracing arm. Calculate:

(1) The multiplier constant and (2) The unknown area.

Solution:

1. *The multiplier constant (M)*

$$A = M(F - I \pm 10N)$$

Substituting the values, we get

$60 = M\,(8.286 - 2.326 + 0)$, from which $M = \dfrac{60}{5.960} =$ **10.027** sq. in.

2. *The unknown area*

$$A = M(F - I + 10\ N) = 10.027\ (5.220 - 8.286 + 10) = \mathbf{69.80}\ \text{sq. in.}$$

Example 12.11: *The following readings were obtained when an area was measured by a planimeter, the tracing arm being so set that one revolution of the wheel measures 10 sq. inches on paper.*

When the anchor point was outside, the initial and final readings were 5.286 and 1.086. The zero mark of the dial passed the index mark once in the clockwise direction. When the anchor point was placed inside the same figure, the initial and final readings were 5.282 and 3.842. The zero mark of the dial passed the index mark twice in the counter clockwise direction.

Find the area of the zero circle.

Solution: ***With the anchor point outside***

$$A = M\ (F - I + 10\ N) = 10\ (1.086 - 5.286 + 10) = \mathbf{58}\ \text{sq. inches.}$$

With the anchor point inside

$$A = M(F - I \pm 10N + C);$$

Here $A = 58$ and $N = -2$

$\therefore$ $58 = 10\ (3.842 - 5.282 - 20 + C)$

or $5.8 = (-21.440 + C)$ from which $C = 5.8 + 21.440 = 27.240$

Area of zero circle $= MC = 27.240 \times 10 = \mathbf{272.40\ sq.\ in.}$

Example 12.12: *The length of the tracing arm between the tracing point and the hinge is 16.6 cm. The distance of the anchor point from the hinge is 22.6 cm. The diameter of the rim of the wheel is 1.92 cm, the wheel being placed between the hinge and the tracing point. The distance of the wheel from the hinge is 1.68 cm. Find the area of one revolution of the measuring wheel and area of the zero circle.*

Solution:

(*i*) Area of one revolution of the measuring wheel $= M$

= Length of tracing arm × Circumference of the wheel

$= L\pi d = 16.6 \times \pi \times 1.92 = \mathbf{100\ cm^2}.$

(*ii*) Area of zero circle $= \pi\ (L^2 \pm 2La + R^2)$

Since the wheel is placed between the hinge and the tracing point, minus sign will be used with $2La$. Hence,

Area of zero circle $= \pi\ (L^2 - 2a + R^2) = \pi(16.6^2 - 2 \times 16.6 \times 1.68 + 22.6^2) = \mathbf{2290\ cm^2}.$

Example 12.13: *Calculate the area of a figure from the following readings recorded by the planimeter with the anchor point inside the figure:*

Initial reading = 2.286, final reading = 8.215

The zero of the counting disc passed the index mark twice in the counter-clockwise direction. Since the constants of the instrument were not available, the following observations were also made:

The distance of the hinge from the tracing point = 4.09″

The distance of the hinge from the anchor point = 6.28″

The perimeter of the wheel = 2.5″

The wheel was placed beyond the hinge at a distance of = 1.22″.

Solution: *Calculation of instrumental constants*

M = Length of the tracing arm × its circumference

$= 4.09 \times 2.5 = 10.225 \text{ in}^2.$

Area of zero circle $= \pi\,(L^2 + 2La + R^2) = \pi\,(4.09^2 + 2 \times 4.09 \times 1.22 + 6.28^2) = 208$ sq. in.

Now $A = M(F - I \pm 10\,N + C) = M\,(F - I \pm 10\,N) + MC$

$= 10.225\,(8.215 - 2.286 - 20) + 208 = -\,143.88 + 208 =$ **64.12 sq. inches.**

PROBLEMS

1. What is Simpson's rule? Derive an expression for it.
2. The following give the values in feet of the offsets taken from a chain line to an irregular boundary:

Distance	0	50	100	150	200	250	300	350	400
Offset	10.6	15.4	20.2	18.7	16.4	20.8	22.4	19.3	17.6

Calculate the area in sq. yards included between the chain line, the irregular boundary and the first and the last offset by Simpson's rule. *(U.P.)*

3. The area of a figure was measured by a planimeter with the anchor point outside the figure and the tracing arm set to the natural scale ($M = 100$ sq. cm). The initial reading was 8.628 and final reading was 1.238. The zero mark of the disc passed the index mark once in the clockwise direction. Calculate the area of the figure.
4. The roller of a planimeter recorded a reading of 1.260 revolutions in the clockwise direction while the measuring area of a rectangular plot 21 cm × 6 cm with the anchor point outside. With the same setting of the tracing arm and the anchor point outside, another figure was traversed and the reading recorded was 2.986 revolutions in the clockwise direction. Find the area of the figure if it is drawn to a scale of 1 cm = 20 metres.
5. A figure is traversed clockwise with the anchor point inside and with the tracing arm so set that one revolution of the roller measures 10 sq. inches on the paper.

I.R. = 3.009; F.R. = 8.547 respectively.

The zero mark of the disc has passed the index mark once in the reverse direction. The area of the zero circle is found to be 164.31 sq. inches. What is the area of the figure? *(U.P.)*

6. What is meant by zero circle? Describe the various methods of determining its area.
The tracing arm of a planimeter is so set that one revolution of the roller corresponds 10 sq. in. A figure is traversed clockwise, first with the anchor point outside and then with the anchor point inside. The observed differences in planimeter readings are 2.342 and – 9.319 respectively. Find the area of the zero circle. *(U.P.)*
7. Describe the polar planimeter and explain its principle. The perimeter of a figure is traversed clockwise, with the anchor point inside and with tracing arm so set that one revolution of the roller measures 10 sq. inches on the paper. The initial and final readings are 3.009 and 8.547 respectively. The zero mark of the disc has passed the fixed index mark once in the reverse direction. The area of the zero circle is found to be 164.31 sq. inches. What is the area of the figure?

ANSWERS

2. 820.38 sq. yds.
3. 261 sq. cm.
4. 11.944 hectares
5. 119.69 sq. inches.
6. 116.6 sq. inches.
7. 119.7 sq. inches.

CHAPTER 13

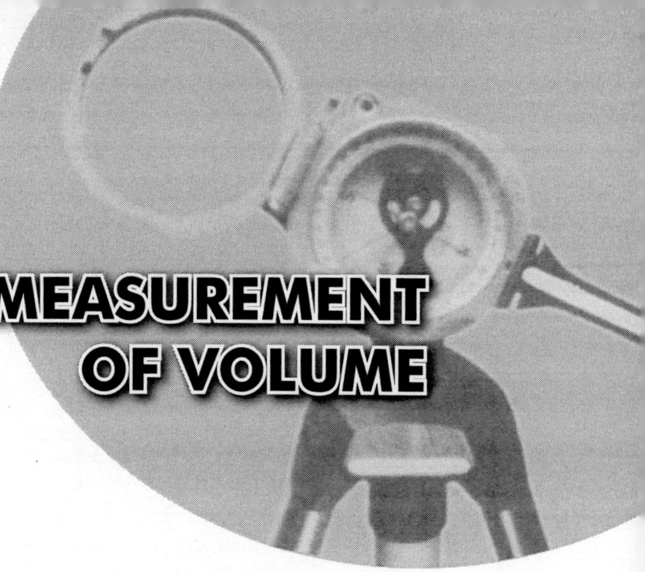

MEASUREMENT OF VOLUME

13.1 GENERAL

There are three methods generally adopted for measuring the volume. They are:

(*i*) From cross-sections

(*ii*) From spot levels

(*iii*) From contours

The first two methods are commonly used for the calculation of earth work while the third method is generally adopted for the calculation of reservoir capacities.

13.2 MEASUREMENT FROM CROSS-SECTIONS

This is the most widely used method. The total volume is divided into a series of solids by the planes of cross-sections. The fundamental solids on which measurement is based are the prism, wedge and prismoid. The spacing of the sections depends upon the character of the ground and the accuracy required in the measurement. The area of the cross-section taken along the line are first calculated by standard formulae developed below, and the volumes of the prismoids between successive cross-sections are then calculated by either trapezoidal formula or by prismoidal formula.

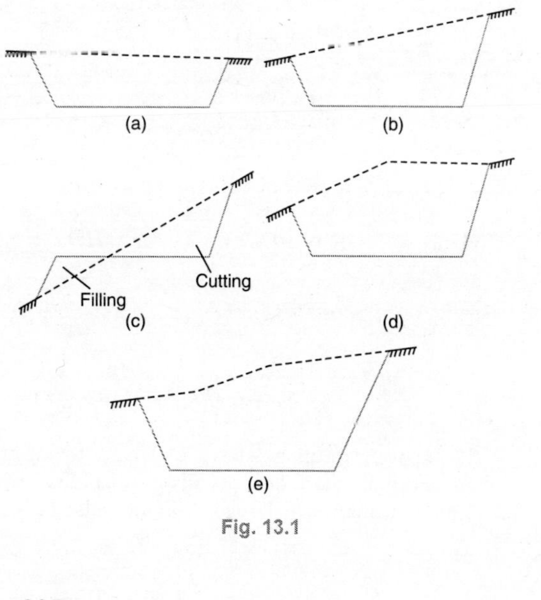

Fig. 13.1

The various cross-sections may be classed as

1. Level section,
   (Figs. 13.1 *a* and 13.2)
2. Two-level section,
   (Figs. 13.1 *b* and 13.3)

3. Side hill two-level section,
   (Figs. 13.1 *c* and 13.4)
4. Three-level section,
   (Figs. 13.1 *d* and 13.5)

and 5. Multi-level section
   (Figs. 13.1 *e* and 13.6)

General notations: Let

b = the constant formation (or sub-grade) width.

h = the depth of cutting on the centre line.

w_1 and w_2 = the side widths, or half breadths, *i.e.*, the horizontal distances from the centre to the intersection of the side slopes with original ground level.

h_1 and h_2 = the side heights, *i.e.*, the vertical distances from formation level to the intersections of the slope with the original surface.

n horizontal to 1 vertical = inclination of the side slopes.

m horizontal to 1 vertical = the transverse slope of the original ground.

A = the area of the cross-section

1. Level Section (Fig. 13.2)

In this case the ground is level transversely.

$$\therefore \quad h_1 = h_2 = h$$

$$w_1 = w_2 = w = \frac{b}{2} + nh$$

$$A = \left\{\frac{b}{2} + \left(\frac{b}{2} + nh\right)\right\} h$$

$$= (b + nh)\, h \qquad ...(13.1)$$

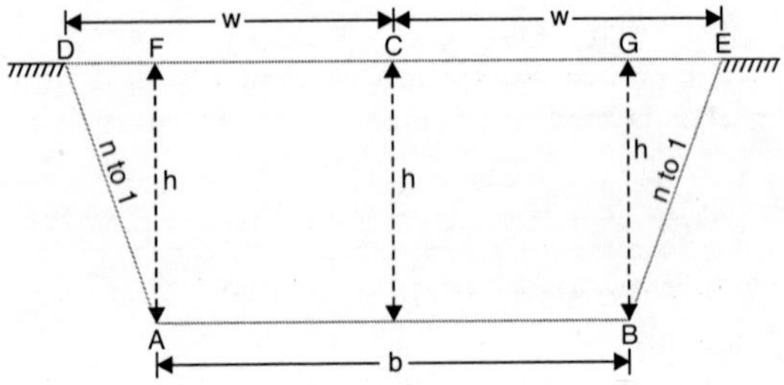

Fig. 13.2

2. Two-Level Section (Fig. 13.3)

Let O be the point on the centre line at which the two side slopes intersect.

Hence $\quad BH : HO :: n : 1 \quad$ or $\quad HO = \dfrac{b}{2n}$

Then area $\quad DCEBA = \Delta\, DCO + \Delta\, ECO - \Delta\, ABO = \dfrac{1}{2}\left\{\left(h + \dfrac{b}{2n}\right) w_1 + \left(h + \dfrac{b}{2n}\right) w_2 - \dfrac{b^2}{2n}\right\}$

$$= \frac{1}{2}\left\{(w_1 + w_2)\left(h + \frac{b}{2n}\right) - \frac{b^2}{2n}\right\} \qquad ...(13.2)$$

The above formula has been derived in terms of w_1 and w_2, and does not contain the term m. The formula is, therefore, equally applicable even if DC and CE have different slopes, provided w_1 and w_2 are known. The formula can also be expressed in terms of h_1 and h_2. Thus,

Area $\quad DCEBA = \Delta\, DAH + \Delta\, EBH + \Delta\, DCH + \Delta\, ECH$

$$= \frac{1}{2}\left\{\frac{b}{2}h_2 + \frac{b}{2}h_1 + hw_2 + hw_1\right\} = \frac{1}{2}\left\{\frac{b}{2}(h_1 + h_2) + h(w_1 + w_2)\right\} \quad ...(13.3)$$

The above expression is independent of m and n. Let us now find the expression for w_1, w_2, h_1 and h_2 in terms of b, h, m and n.

$$BJ = nh_1 \quad ...(1)$$

Also $$BJ = HJ - HB = w_1 - \frac{b}{2} \quad ...(2)$$

$\therefore$ $$nh_1 = w_1 - \frac{b}{2} \quad ...(i)$$

Also, $$w_1 = (h_1 - h)m \quad ...(ii)$$

Substituting the value of w_1 in (*i*), we get

$$nh_1 = (h_1 - h)m - \frac{b}{2}$$

or $$h_1(m - n) = mh + \frac{b}{2}$$

or $$h_1 = \frac{m}{m-n}\left(h + \frac{b}{2m}\right)$$

Substituting the value of h_1 in (*i*), we get

$$w_1 = \frac{b}{2} + nh_1 = \frac{b}{2} + \frac{mn}{m-n}\left(h + \frac{b}{2m}\right) \quad ...(13.4)$$

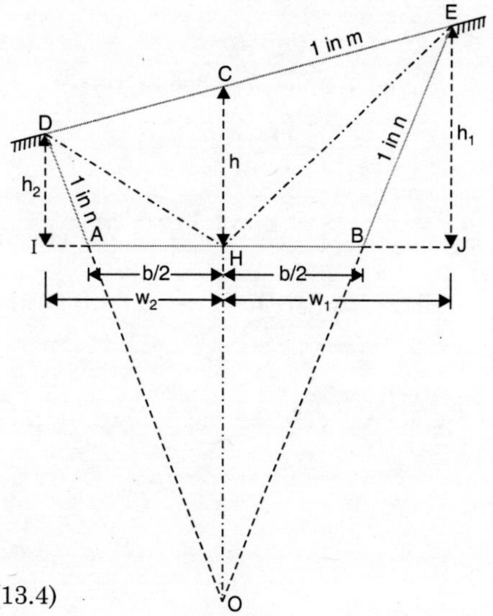

Fig. 13.3

Proceeding in similar manner, it can be shown that

$$h_2 = \frac{m}{m+n}\left(h - \frac{b}{2m}\right) \quad ...(13.5)$$

and $$w_2 = \frac{b}{2} + \frac{mn}{m+n}\left(h - \frac{b}{2m}\right) \quad ...(13.6)$$

Substituting the values of w_1 and w_2 in equation 13.2 and simplifying, we get

$$\text{Area} = \frac{m^2 n}{m^2 - n^2}\left(h + \frac{b}{2n}\right)^2 - \frac{b^2}{4n} \quad ...(13.7)$$

Similarly, substituting the values of w_1, w_2, h_1 and h_2 in equation 13.3, we get

$$\text{Area} = \frac{n\left(\frac{b}{2}\right)^2 + m^2(bh + nh^2)}{(m^2 - n^2)} \quad ...(13.8)$$

3. Side Hill Two-Level Section (Fig. 13.4)

In this case, the ground slope crosses the formation level so that one portion of the area is in cutting and the other in filling.

Now, $BJ = nh_1$

Also, $BJ = HJ - HB = w_1 - \dfrac{b}{2}$

$\therefore \quad nh_1 = w_1 - \dfrac{b}{2}$ ...(*i*)

But $\quad w_1 = (h_1 - h)m$ ...(*ii*)

Solving (*i*) and (*ii*) as before, we get

$$h_1 = \frac{mn}{m-n}\left(\frac{b}{2m}+h\right) \quad ...(13.9)$$

and
$$w_1 = \frac{b}{2}+\frac{mn}{m-n}\left(\frac{b}{2m}+h\right) \quad ...(13.10)$$

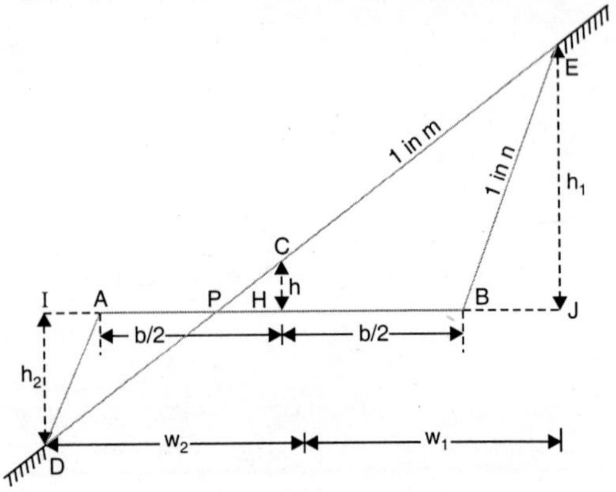

Fig. 13.4

Let us now derive expressions for w_2 and h_2:

$$IA = nh_2$$

Also $\quad IA = IH - AH = w_2 - \dfrac{b}{2}$

$\therefore \quad nh_2 = w_2 - \dfrac{b}{2}$ ...(*iii*)

Also $\quad w_2 = (h + h_2)m$ ...(*iv*)

$\therefore \quad nh_2 = (h + h_2)m - \dfrac{b}{2} \quad$ or $\quad h_2(m - n) = \dfrac{b}{2} - mh$

or
$$h_2 = \frac{mn}{m-n}\left(\frac{b}{2m}-h\right) \quad ...(13.11)$$

Hence
$$w_2 = \frac{b}{2} + nh_2 = \frac{b}{2} + \frac{mn}{m-n}\left(\frac{b}{2m}-h\right) \quad ...(13.12)$$

By inspection, it is clear that the expressions for w_1 and w_2 are similar; also expression of h_1 and h_2 are similar, except for $-h$ in place of $+h$.

Now area of filling = $\Delta PBE = A_1$ (say), And, area of cutting = $\Delta PAD = A_2$ (say).

$$\therefore \quad A_1 = \frac{1}{2}(PB)(EJ) = \frac{1}{2}\left(\frac{b}{2}+mh\right)\left\{\frac{m}{m-n}\left(\frac{b}{2m}+h\right)\right\} = \frac{\left(\frac{b}{2}+mh\right)^2}{2(m-n)} \quad ...(13.13)$$

and
$$A_2 = \frac{1}{2}(AP)(ID) = \frac{1}{2}\left(\frac{b}{2}-mh\right)\left\{\frac{m}{m-n}\left(\frac{b}{2m}-h\right)\right\} = \frac{\left(\frac{b}{2}-mh\right)^2}{2(m-n)} \quad ...(13.14)$$

4. Three-Level Section (Fig. 13.5)

Let 1 in m_1 be the transverse slope of the ground to one side and 1 in m_2 be the slope to the other side of the centre line of the cross-section (Fig. 13.5).

The expressions for w_1, w_2, h_1, and h_2 can be derived in the similar way as for case (2). Thus,

$$w_1 = \frac{m_1 n}{m_1 - n}\left(h + \frac{b}{2n}\right) \quad ...(13.15)$$

$$w_2 = \frac{m_2 n}{m_2 - n}\left(h + \frac{b}{2n}\right) \quad ...(13.16)$$

$$h_1 = \left(h + \frac{w_1}{m_1}\right) = \frac{m_1}{m_1 - n}\left(h + \frac{b}{2m_1}\right) \quad ...(13.17)$$

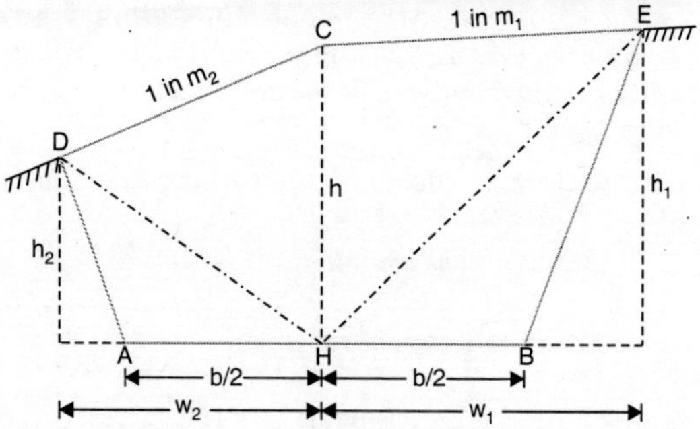

Fig. 13.5

$$h_2 = \left(h - \frac{w_2}{m_2}\right) = \frac{m_2}{m_2 + n}\left(h - \frac{b}{2m_2}\right) \quad ...(13.18)$$

The area $ABECD = \Delta\, AHD + \Delta\, BHE + \Delta\, CDH + \Delta\, CEH$

$$= \frac{1}{2}\left[\left(h_2 \times \frac{b}{2}\right) + \left(h_1 \times \frac{b}{2}\right) + hw_2 + hw_1\right] = \left[\frac{b}{4}(h_1 + h_2) + \frac{h}{2}(w_1 + w_2)\right] \quad ...(13.19)$$

5. Multi-Level Section (Fig. 13.6)

In the multi-level section the co-ordinate system provides the most general method of calculating the area. The cross-section notes provide with x and y co-ordinates for each vertex of the section, the origin being at the central point (H). The x co-ordinates are measured positive to the right and negative to the left of H. Similarly, the y co-ordinates (*i.e.*, the heights) are measured positive for cuts and negative for fills. In usual form, the notes are recorded as below:

$$\frac{h_2}{w_2}\ \frac{h_1}{w_1}\ \frac{h}{0}\ \frac{H_1}{W_1}\ \frac{H_2}{W_2}$$

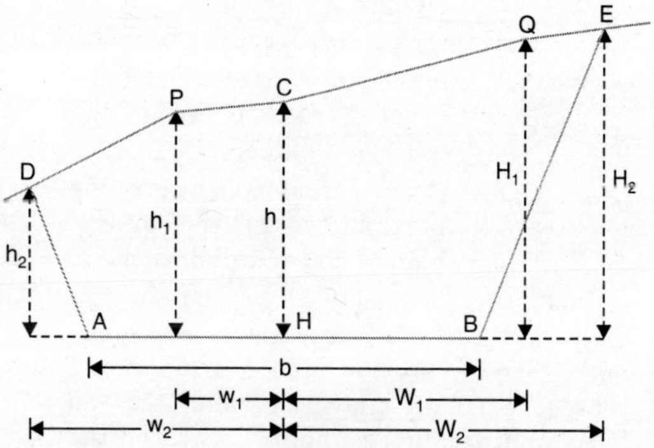

Fig. 13.6

If the co-ordinates are given proper sign and if the co-ordinates of formation points A and B are also included (one at extreme left and other at extreme right), they appear as follows:

$$\frac{0}{-b/2}\ \frac{h_2}{-w_2}\ \frac{h_1}{-w_1}\ \frac{h}{0}\ \frac{H_1}{+W_1}\ \frac{H_2}{+W_2}\ \frac{0}{+b/2}$$

There are several methods to calculate the area. In one of the methods, the opposite algebraic sign is placed on the opposite side of each lower term. The co-ordinates then appear as:

$$\frac{0}{-b/2+}\ \frac{h_2}{-w_2+}\ \frac{h_1}{-w_1+}\ \frac{h}{0}\ \frac{H_1}{+W_1-}\ \frac{H_2}{+W_2-}\ \frac{0}{+b/2-}$$

The area can now be computed by multiplying each upper term by the algebraic sum of the two adjacent lower terms, using the signs facing the upper term. The algebraic sum of these products will be double the area of the cross-section.

Thus, we get

$$A = \tfrac{1}{2}[h_2 (+ b/2 - w_1) + h_1 (+ w_2 + 0) + h (+ w_1 + W_1) + H_1 (0 + W_2) + H_2(- W_1 + b/2)] \quad ...(13.20)$$

For a numerical example, see Example 13.6.

13.3 THE PRISMOIDAL FORMULA

The volumes of the prismoids between successive cross-sections are obtained either by trapezoidal formula or by prismoidal formula. We shall first derive an expression for prismoidal formula.

A prismoid is defined as a solid whose end faces lie in parallel planes and consist of any two polygons, not necessarily of the same number of sides, the longitudinal faces being surface extended between the end planes.

The longitudinal faces take the form of triangles, parallelograms, or trapezium.

Let d = length of the prismoid measured perpendicular to the two end parallel planes.

A_1 = area of cross-section of one end plane.

A_2 = area of cross-section of the other end plane.

M = the mid-area = the area of the plane midway between the end planes and parallel to them.

In Fig. 13.7, let $A_1\ B_1\ C_1\ D_1$ be one end plane and $A_2\ B_2\ C_2\ D_2$ be another end plane parallel to the previous one. Let $P\ Q\ R\ S\ T$ represent a plane midway between the end faces and parallel to them. Let A_m be the area of mid-section. Select any point O in the plane of the mid-section and join it to the vertices of both the end planes. The prismoid is thus divided into a number of pyramids, having the apex at A and bases on end and side faces. The total volume of the prismoid will therefore be equal to the sum of the volume of the pyramids.

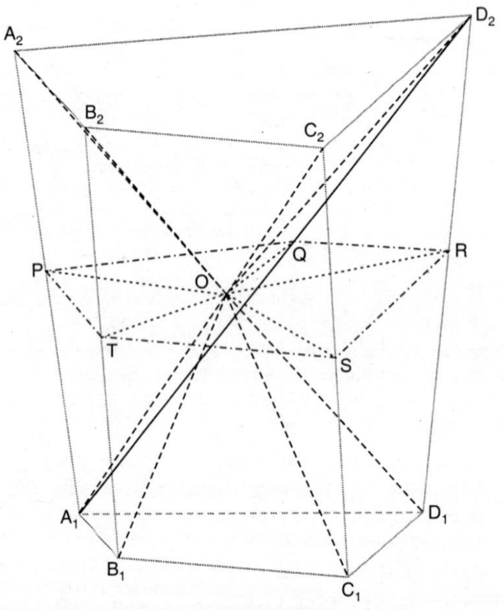

Fig. 13.7

Volume of pyramid $OA_1B_1C_1D_1 = \tfrac{1}{3}\left(\dfrac{d}{2}\right) A_1 = \tfrac{1}{6} A_1 d$

Volume of pyramid $OA_2B_2C_2D_2 = \tfrac{1}{6} A_2 d.$

To find the volumes of pyramids on side faces, consider any pyramid such as $OA_1B_1B_2A_2$.

Its volume $= \tfrac{1}{3}(A_1B_1B_2A_2) \times h$, where h = perpendicular distance of PT from O.

$= \frac{1}{3}(d \times PT)h = \frac{1}{3} d\,(2\Delta OPT) = \frac{2}{3}d\,(\Delta OPT)$

Similarly, volume of another pyramid $OC_1D_1D_2$ on the side face $= \frac{2}{3}d\,(\Delta OSR)$.

$\therefore$ Total volume of lateral (side) pyramids $= \frac{2}{3}d\,(PQRST) = \frac{2}{3}(A_m)$

Hence, total volume of the pyramid $= \frac{1}{6} A_1d + \frac{1}{6} A_2d + \frac{2}{3}A_md$

or
$$V = \frac{d}{6}(A_1 + A_2 + 4A_m) \qquad \text{...(13.21)}$$

Let us now calculate the volume of earth work between a number of sections having area A_1, A_2, A_3, ... A_n spaced at a constant distance d apart. Considering the prismoid between first three sections, its volume will be, from equation 13.21,

$$= \frac{(2d)}{6}(A_1 + 4A_2 + A_3),\ 2d \text{ being the length of the prismoid.}$$

Similarly, volume of the second prismoid of length $2d$ will be

$$= \frac{2d}{6}(A_3 + 4A_4 + A_5),$$

and volume of last prismoid of length $2d$ will be

$$= \frac{2d}{6}(A_{n-2} + 4A_{n-1} + A_n)$$

Summing up, we get the total volume,

$$V = \frac{d}{3}[A_1 + 4A_2 + 2A_3 + 4A_4 ... 2A_{n-2} + 4A_{n-1} + A_n] \qquad \text{...(13.22)}$$

or
$$V = \frac{d}{3}[(A_1 + A_n) + 4(A_2 + A_4 .... A_{n-1}) + 2(A_3 + A_5 .... A_{n-2})]$$

This is also known as Simpson's rule for volume. Here also, it is necessary to have an odd number of cross-sections. If there are even number of sections, the end strip must be treated separately, and the volume between the remaining sections may be calculated by primoidal formula.

13.4 THE TRAPEZOIDAL FORMULA (AVERAGE END AREA METHOD)

This method is based on the assumption that the mid-area is the mean of the end areas. In that case, the volume of the prismoid of Fig. 13.7 is given by

$$V = \frac{d}{2}(A_1 + A_2) \qquad \text{...(1)}$$

This is true only if the prismoid is composed of prisms and wedges only and not of pyramids. The mid area of a pyramid is half the average area of the ends; hence the volume of the prismoid (having pyramids also) is over estimated. However, the method of end area may be acceped with sufficient accuracy since actual earth solid may not be exactly a prismoid. In some cases, the volume is calculated and then a correction is applied, the correction being equal to the difference between the volume as calculated and that which could be obtained by the use of the prismoidal formula. The correction is known as the *prismoidal correction.*

Let us now calculate the volume of earth work between a number of sections having areas A_1, A_2, ... A_n, spaced at a constant distance d.

$$\text{Volume between first two sections} = \frac{d}{2}\,(A_1 + A_2)$$

$$\text{Volume between next two sections} = \frac{d}{2}\,(A_3 + A_4)$$

$$\text{Volume between last two sections} = \frac{d}{2}\,(A_{n-1} + A_n)$$

$$\therefore \quad \text{Total volume} = V = d\left\{\frac{A_1 + A_2}{2} + A_2 + A_3 + \ldots A_{n-1}\right\} \qquad \ldots(13.23)$$

13.5 THE PRISMOIDAL CORRECTION (C_p)

As stated earlier, the prismoidal correction is equal to the difference between the volumes as calculated by the end-area formula and the prismoidal formula. The correction is always *subtractive, i.e.*, it should be subtracted from the volume calculated by the end area formula.

Let us calculate the prismoidal correction for the case when the end sections are level sections. Let A, w_1, w_2, h_1, h_2, etc., refer to the cross-section at one end and A', w_1', w_2', h_1', h_2' etc., to the cross-section at the other end.

Now $\qquad A = h\,(b + nh) \quad$ and $\quad A' = h'\,(b + nh')$

Volume by end area rule is given by

$$V = \frac{d}{2}\,[h(b + nh) + h'(b + nh')] = d\left[\frac{bh}{2} + \frac{bh'}{2} + \frac{nh^2}{2} + \frac{nh'^2}{2}\right] \qquad \ldots(i)$$

Again, the mid-area centre height $= \dfrac{h + h'}{2}$

$$\therefore \quad \text{Mid-area} = \left(\frac{h+h'}{2}\right)\left[b + n\left(\frac{h+h'}{2}\right)\right]$$

Volume by prismoidal formula is given by

$$V = \frac{d}{6}\left[h(b+nh) + h'(b+nh') + 4\left(\frac{h+h'}{2}\right)\times\left(b + \frac{n(h+h')}{2}\right)\right]$$

or

$$V = \frac{d}{6}\,[3bh + 3bh' + 2nh^2 + 2nh'^2 + 2nhh']$$

$$= d\left[\frac{bh}{2} + \frac{bh'}{2} + \frac{nh^2}{3} + \frac{nh'^2}{3} + \frac{nhh'}{3}\right] \qquad \ldots(ii)$$

Subtracting (*ii*) from (*i*), we get the prismoidal correction,

$$C_p = \frac{dn}{6}\,(h - h')^2 \qquad \ldots(13.24)$$

Similarly, the prismoidal correction for other sections can also be derived. The standard expression for C_p are given below.

For two-level section:

$$C_p = \frac{d}{6n}\,(w_1 - w_1')\,(w_2 - w_2') \qquad \ldots(13.25)$$

For side hill two-level section:

$$C_p\ (\text{cutting}) = \frac{d}{12n}\ (w_1 - w_1')\ \left\{\left(\frac{b}{2} + mh\right) - \left(\frac{b}{2} + m'h'\right)\right\} \qquad ...(13.26)$$

$$C_p\ (\text{filling}) = \frac{d}{12n}\ (w_2 - w_2')\ \left\{\left(\frac{b}{2} - mh\right) - \left(\frac{b}{2} - m'h'\right)\right\} \qquad ...(13.27)$$

For three-level section:

$$C_p = \frac{d}{12}\ (h - h')[(w_1 + w_2) - (w_1' + w_2')] \qquad ...(13.28)$$

13.6 THE CURVATURE CORRECTION

The prismoidal and the trapezoidal formulae were derived on the assumption that the end sections are in parallel planes. When the centre line of cutting or an embankment is curved in plan, it is common practice to calculate the volume as if the end sections were in parallel planes, and then apply the correction for curvature. The standard expression for various sections are given below. In some cases, the correction for curvature is applied to the areas of cross-sections thus getting *equivalent areas* and then to use the prismoidal formula.

(*i*) *Level section:* No correction is necessary since the area is symmetrical about the centre line.

(*ii*) *Two-level section and three-level section:*

$$C_c = \frac{d}{6R}(w_1^2 - w_2^2)\left(h + \frac{b}{2n}\right) \qquad ...(13.29)$$

where R is the radius of the curve.

(*iii*) *For a two-level section:* the curvature correction to the area

$$= \frac{Ae}{A} \text{ per unit length} \qquad ...(13.30)$$

where e = the eccentricity, *i.e.*, horizontal distance from the centre line to the

$$\text{centroid of the area} = \frac{w_1\, w_2\ (w_1 + w_2)}{3An} \qquad ...(13.31)$$

The correction is positive if the centroid and the centre of the curvature are to the opposite side of the centre line while it is negative if the centroid and the centre of the curvature are to the same side of the centre line.

(*iv*) *For side hill two-level section:*

$$\text{Correction to area} = \frac{Ae}{R} \text{ per unit length} \qquad ...(13.32)$$

where

$$e = \frac{1}{3}\left(w_1 + \frac{b}{2} - nh\right) \text{ for the larger area} \qquad ...(13.33)$$

and

$$e = \frac{1}{3}\left(w_2 + \frac{b}{2} + nh\right) \text{ for the smaller area} \qquad ...(13.34)$$

Example 13.1: A railway embankment is 10 m wide with side slopes $1\frac{1}{2}$ to 1. Assuming the ground to be level in a direction transverse to the centre line, calculate the volume contained in a length of 120 metres, the centre heights at 20 m intervals being in metres 2.2, 3.7, 3.8, 4.0, 3.8, 2.8, 2.5.

Solution: For a level section, the area is given by $A = (b + nh)h$

Slope is $1\frac{1}{2}$: 1. Hence $n = 1.5$

The areas at different sections will be as under:

$$A_1 = (10 + 1.5 \times 2.2)\ 2.2 = 29.26 \text{ m}^2; \quad A_2 = (10 + 1.5 \times 3.7)\ 3.7 = 57.54 \text{ m}^2$$
$$A_3 = (10 + 1.5 \times 3.8)\ 3.8 = 59.66 \text{ m}^2; \quad A_4 = (10 + 1.5 \times 4.0)\ 4.0 = 64.00 \text{ m}^2$$
$$A_5 = (10 + 1.5 \times 3.8)\ 3.8 = 59.66 \text{ m}^2; \quad A_6 = (10 + 1.5 \times 2.8)\ 2.8 = 39.76 \text{ m}^2$$

and $$A_7 = (10 + 1.5 \times 2.5)\ 2.5 = 34.37 \text{ m}^2$$

Volume by trapezoidal rule is given by

$$V = d\left[\frac{A_1 + A_n}{2} + A_2 + A_3 + \ldots A_{n-1}\right]$$

$$= 20\left[\frac{29.26 + 34.37}{2} + 57.54 + 59.66 + 64.00 + 59.66 + 39.76\right] = \mathbf{6258.9\ m^3}$$

Volume by prismoidal rule is given by

$$V = \frac{d}{3}[A_1 + 4(A_2 + A_4 + A_6 + \ldots) + 2(A_3 + A_5 + \ldots + A_{n-2}) + A_n]$$

$$= \frac{20}{3}[29.26 + 4\,(57.54 + 64.00 + 39.76) + 2\,(59.66 + 59.66) + 34.37] = \mathbf{6316.5\ m^3}.$$

Example 13.2: A railway embankment 400 m long is 12 m wide at the formation level and has the side slope 2 to 1. The ground levels at every 100 m along the centre line are as under:

Distance	*0*	*100*	*200*	*300*	*400*
R.L.	*204.8*	*206.2*	*207.5*	*207.2*	*208.3*

The formation level at zero chainage is 207.00 and the embankment has a rising gradient of 1 in 100. The ground is level across the centre line. Calculate the volume of earthwork.

Solution: Since the embankment level is to have a rising gradient of 1 in 100 the formation level at every section can be easily calculated as tabulated below:

Distance	*Ground*	*Formation level*	*Depth of filling*
0	204.8	207.0	2.2
100	206.2	208.0	1.8
200	207.5	209.0	1.5
300	207.2	210.0	2.8
400	208.3	211.0	2.7

The area of section is given by $A = (b + nh)\, h = (12 + 2h)\, h$

$$A_1 = (12 + 2 \times 2.2)\, 2.2 = 36.08 \text{ m}^2;\; A_2 = (12 + 2 \times 1.8)\, 1.8 = 28.06 \text{ m}^2$$

$$A_3 = (12 + 2 \times 1.5)\, 1.5 = 22.50 \text{ m}^2;\; A_4 = (12 + 2 \times 2.8)\, 2.8 = 49.28 \text{ m}^2$$

and $$A_5 = (12 + 2 \times 2.7)\, 2.7 = 46.98 \text{ m}^2$$

Volume by trapezoidal rule is given by

$$V = d\left[\frac{A_1 + A_n}{2} + A_2 + A_3 + \ldots A_{n-1}\right]$$

$$= 100\left[\frac{36.08 + 46.98}{2} + 28.06 + 22.50 + 49.28\right] = \mathbf{14{,}137\ m^3}$$

Volume by prismoidal rule is given by

$$V = \frac{d}{3}\,[(A_1 + A_5) + 4(A_2 + A_4) + 2(A_3)]$$

$$V = \frac{100}{3}\,[(36.08 + 49.98) + 4(28.06 + 49.28) + (2 \times 22.50)] = \mathbf{14{,}581\ m^3}.$$

Example 13.3: *Find out the volume of earth work in a road cutting 120 metres long from the following data:*

The formation width 10 metres; side slopes 1 to 1; average depth of cutting along the centre of line 5 m; slopes of ground in cross-section 10 to 1.

Solution: The cross-sectional area, in terms of m and n, is given by equation 13.8

Thus, $$A = \frac{n\left(\frac{b}{2}\right)^2 + m^2(bh + nh^2)}{m^2 - n^2}$$

Here $n = 1$; $m = 10$; $h = 5$; and $b = 10$.

$\therefore$ $$A = \frac{1\left(\frac{10}{2}\right)^2 + 10^2(10 \times 5 + 1 \times 5^2)}{10^2 - 1^2} = 76 \text{ m}^2$$

$\therefore$ $V = A \times L = 76 \times 120 =$ **9120** cubic metres.

Example 13.4: *A road embankment 10 m wide at the formation level, with side slopes of 2 to 1 and with an average height of 5 m is constructed with an average gradient 1 in 40 from contour 220 metres to 280 metres. Find the volume of earth work.*

Solution: Difference in level between both the ends of the road

= contour 280 – contour 220 = 60 m

$\therefore$ Length of the road = 60 × 40 = 2400 metres.

Area of the cross-section = $(b + nh)\, h$

Here, $b = 10$ m; $n = 2$ m; $h = 5$ m.

$\therefore$ $A = (10 + 2 \times 5)\, 5 = 100$ sq. m.

$\therefore$ Volume of embankment = Length × Area = 2400 × 100 = **2,40,000** cubic metres.

Example 13.5: *The following notes refer to three level cross-sections at two sections 50 metres apart.*

Station	Cross-section		
1.	$\frac{1.7}{7.7}$	$\frac{2.8}{0}$	$\frac{4.6}{10.6}$
2.	$\frac{2.9}{8.9}$	$\frac{3.7}{0}$	$\frac{6.9}{12.9}$

The width of cutting at the formation level is 12 m. Calculate the volume of cutting between the two stations.

Solution: The area of a three-level cross-section, from Eq. 13.19, is given by

$$A = \left[\frac{h}{2}(w_1 + w_2) + \frac{b}{4}(h_1 + h_2)\right]$$

At station 1,

$$b = 12 \text{ m};\ h = 2.8 \text{ m}$$

$$w_1 = 10.6 \text{ m};\ h_1 = 4.6 \text{ m};\ w_2 = 7.7 \text{ m};\ h_2 = 1.7 \text{ m}$$

$$A_1 = \frac{2.8}{2}(10.6 + 7.7) + \frac{12}{4}(4.6 + 1.7) = 44.52 \text{ m}^2$$

At station 2,

$$b = 12 \text{ m};\ h = 3.7 \text{ m}$$

$$w_1 = 12.9 \text{ m};\ h_1 = 6.9 \text{ m};\ w_2 = 8.9 \text{ m};\ h_2 = 2.9 \text{ m}$$

$$\therefore \quad A_2 = \frac{3.7}{2}(12.9 + 8.9) + \frac{12}{4}(6.9 + 2.9) = 69.73 \text{ m}^2$$

Volume by trapezoidal formula is given by

$$V = \frac{1}{2}(44.52 + 69.73) \times 50 = 2856 \text{ cubic metres.}$$

To calculate the volume by prismoidal rule, let us first calculate the mid-area by assuming the quantities w, h etc., as the average of those at the ends. Thus, for the mid-area, we have

$$b = 12 \text{ m};\ h = \frac{2.8 + 3.7}{2} = 3.25 \text{ m}$$

$$w_1 = \frac{10.6 + 12.9}{2} = 11.75;\ h_1 = \frac{4.6 + 6.9}{2} = 5.75$$

$$w_2 = \frac{7.7 + 8.9}{2} = 8.3;\ h_2 = \frac{1.7 + 2.9}{2} = 2.3$$

$$\therefore \quad A_m = \frac{3.25}{2}(11.75 + 8.3) + \frac{12}{4}(5.75 + 2.30) = 56.73 \text{ m}^3$$

$$V = \frac{L}{6}(A_1 + 4A_m + A_2) = \frac{50}{6}(44.52 + 56.73 \times 4 + 69.73) = \mathbf{2843\ m^3}.$$

Example 13.6: *The following are the notes for a multi-level cross-section for a road. The width of the road bed is 10 m and the side slopes are 1 to 1. Calculate the cross-sectional area.*

$$\frac{2.5}{7.5} \quad \frac{2.9}{5} \quad \frac{3.8}{0} \quad \frac{5.2}{7} \quad \frac{5.8}{10.8}$$

Solution: If the co-ordinates are given proper sign and if the co-ordinates of the formation points are also included (one at extreme left and other at extreme right), they appear as follows:

$$\frac{0}{-5} \quad \frac{2.5}{-7.5} \quad \frac{2.9}{-5} \quad \frac{3.8}{0} \quad \frac{5.7}{+7} \quad \frac{5.8}{+10.8} \quad \frac{0}{+5}$$

Following the method explained in § 13.6 and placing opposite algebraic sign on the opposite side of each lower term, the notes appear as follows:

$$\frac{0}{-5+} \quad \frac{2.5}{-7.5+} \quad \frac{2.9}{-5+} \quad \frac{3.8}{0} \quad \frac{5.2}{+7-} \quad \frac{5.8}{+10.8-} \quad \frac{0}{+5-}$$

The area can be computed by multiplying each upper term by the algebraic sum of two adjacent lower terms, using the signs facing the upper term. The algebraic sum of these products will be double the area of the cross-section.

Thus, we get

$$A = \tfrac{1}{2}[2.5\,(+5-5) + 2.9\,(+7.5+0) + 3.8\,(+5+7) + 5.2\,(+0+10.8) + 5.8(-7+5)]$$
$$= \tfrac{1}{2}[0 + 21.75 + 45.6 + 56.16 - 11.6] = \mathbf{55.96\ m^2}.$$

13.7 VOLUME FROM SPOT LEVELS

In this method, the field work consists in dividing the area into a number of squares, rectangles or triangles and measuring the levels of their corners before and after the construction. Thus, the depth of excavation or height of filling at every corner is known. Let us assume that the four corners of any one square or rectangle are at different elevations but lie in the same *inclined* plane. Assume that it is desired to grade down to a level surface a certain distance below the lowest corner. The earth to be moved will be a right truncated prism, with *vertical edges* at *a, b, c* and *d* [Fig. 13.8 (*a*)]. The rectangle *abcd* represents the horizontal projection of the upper inclined base of the prism and also the lower horizontal base.

Let us consider the rectangle *abcd* of Fig. 13.8 (*a*). If h_a, h_b, h_c and h_d represent the depth of excavation of the four corners, the volume of the right truncated prism will be given by

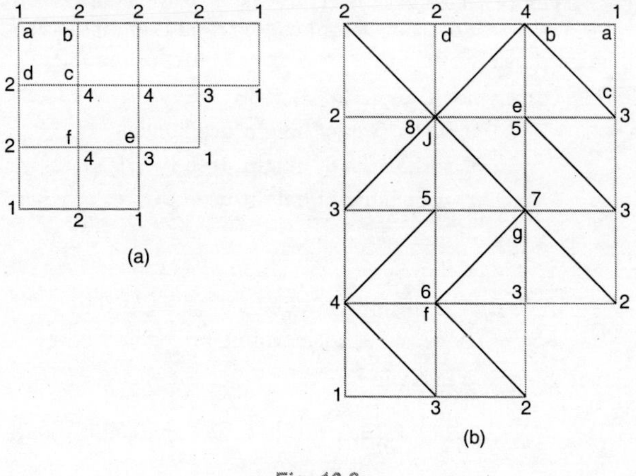

Fig. 13.8

$$V = \left(\frac{h_a + h_b + h_c + h_d}{4}\right) \times A \qquad ...(13.35)$$

= averate height × the horizontal area of the rectangle.

Similarly, let us consider the triangle *abc* of Fig. 13.8 (*b*). If h_a, h_b and h_c are the depths of excavation of the three corners, the volume of the truncated triangular prism is given by

$$V = \left(\frac{h_a + h_b + h_c}{3}\right) \times A \qquad ...(13.36)$$

= (average depth) × horizontal area of the triangle

Volume of a group of rectangles or squares having the same area: Let us now consider a group of rectangles of the same area, arranged as shown in Fig. 13.8 (*a*). It will be seen by inspection that some of the heights are used once only, some heights are common to two rectangles (such as at *b*), some heights are common to three rectangles (such as at *e*), and some heights are common to four rectangles (such as at *f*). Thus, in Fig. 13.8 (*a*), each corner height will be used as many times as there are rectangles joining at the corner (indicated on the figure by numbers).

Let Σh_1 = the sum of the heights used once.

Σh_2 = the sum of the heights used twice.

Σh_3 = the sum of the heights used thrice.

Σh_4 = the sum of the heights used four times.

A = horizontal area of the cross-section of one prism.

Then, the total volume is given by

$$V = \frac{A(1\Sigma h_1 + 2\Sigma h_2 + 3\Sigma h_3 + 4\Sigma h_4)}{4} \qquad ...(13.37)$$

Volume of a group of triangles having equal area [Fig. 13.8 (*b*)]

If the ground is very much undulating, the area may be divided into a number of triangles having equal area. In this case, some corner heights will be used once [such as point *a* of Fig. 13.8 (*b*)], some twice (such as at *d*), some thrice (such as at *c*), some four times (such as at *b*), some five times (such as at *e*), some six times (such as at *f*), and some seven times (such as at *j*). The maximum number of times a corner height can be used is eight. Thus, in Fig. 13.8 (*b*), each corner height will be used as many times as there are triangles joining at the corner (indicated in the figure by numbers).

Let Σh_1 = the sum of height used once.

Σh_2 = the sum of height used twice.

Σh_3 = the sum of height used thrice.

....

....

Σh_8 = the sum of height used eight times.

A = area of each triangle.

The total volume of the group is given by

$$V = \frac{A}{3}(1\Sigma h_1 + 2\Sigma h_2 + 3\Sigma h_3 + 4\Sigma h_4 + 5\Sigma h_5 + 6\Sigma h_6 + 7\Sigma h_7 + 8\Sigma h_8) \qquad ...(13.38)$$

Example 13.7: A rectangular plot ABCD forms the plane of a pit excavated for road work. E is point of intersection of the diagonals. Calculate the volume of the excavation in cubic metres from the following data:

Point	*A*	*B*	*C*	*D*	*E*
Original level	*45.2*	*49.8*	*51.2*	*47.2*	*52.0*
Final level	*38.6*	*39.8*	*42.6*	*40.8*	*42.5*

Length of AB = 50 m and BC = 80 m.

Solution: (Fig. 13.9)

Area of each triangle = $\frac{1}{2} \times 50 \times 40 = 1000$ sq. m.

Take the vertices of each triangle and find the mean depth at each triangle. Thus,

Depth of cutting at $A = 45.2 - 38.6 = 6.6$ m

Depth of cutting at $B = 49.8 - 39.8 = 10.0$ m

Depth of cutting at $C = 51.2 - 42.6 = 8.6$ m

Depth of cutting at $D = 47.2 - 40.8 = 6.4$ m

Depth of cutting at $E = 52.0 - 42.5 = 9.5$ m

Now volume of any truncated triangular prism is given by

$$V = (\text{average height}) \times A = hA$$

For the triangular prism *ABE*

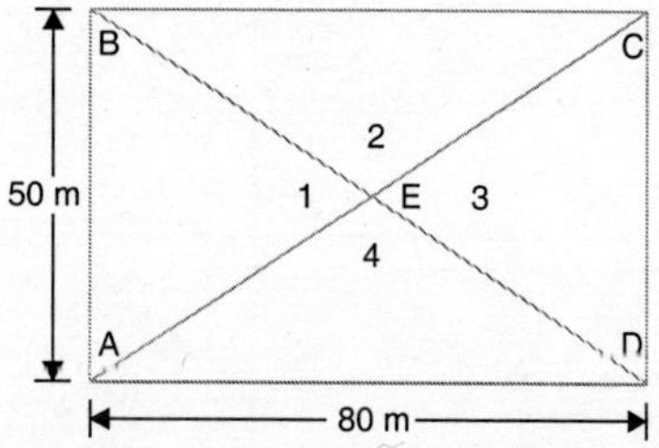

Fig. 13.9

$$h = \frac{6.6 + 10 + 9.5}{3} = 8.7 \text{ m}$$

$$\therefore \quad V_1 = 8.7 \times 1000 = 8700 \text{ m}^3$$

From the prism *BCE*,

$$h = \frac{10 + 8.6 + 9.5}{3} = 9.367 \text{ m}$$

$$\therefore \quad V_2 = 9.367 \times 1000 = 9367 \text{ m}^3$$

For the prism *CDE*,

$$h = \frac{8.6 + 6.4 + 9.5}{3} = 8.167 \text{ m}$$

$$\therefore \quad V_3 = 8.167 \times 1000 = 8167 \text{ m}^3$$

For the prism *DAE*,

$$h = \frac{6.4 + 6.6 + 9.5}{3} = 7.5 \text{ m}$$

$$\therefore \quad V_4 = 7.5 \times 1000 = 7500 \text{ m}^3$$

$$\therefore \quad \text{Total volume} = V_1 + V_2 + V_3 + V_4 = 8700 + 9367 + 8167 + 7500 = \mathbf{33{,}734\ m^3}$$

Alternatively, the total volume may be obtained from equation 13.38. Thus,

$$V = \frac{A}{3}(1\Sigma h_1 + 2\Sigma h_2 + 3\Sigma h_3 + \ldots\ldots\ 8\Sigma h_8) \qquad \ldots(13.38)$$

Here $1\Sigma h_1 = 0$

$$2\Sigma h_2 = 2\,(6.6 + 10 + 8.6 + 6.4) = 63.2$$

(Since height of every outer corner is utilised in two triangles)

$3\Sigma h_3$, $5\Sigma h_5$, $6\Sigma h_6$, $7\Sigma h_7$ and $8\Sigma h_8$ are each zero.

$$4\Sigma h_4 = 4(9.5) = 38$$

Substituting the values in equation 13.38, we get

$$V = \frac{1000}{3} \times (63.2 + 38) = \mathbf{33733\ m^3}$$

Example 13.8: *An excavation is to be made for a reservoir 20 m long 12 m wide at the bottom, having the side of the excavation slope at 2 horizontal to 1 vertical. Calculate the volume of excavation if the depth is 4 metres. The ground surface is level before excavation.*

Solution:

Length of the reservoir at the top $= L + 2nh = 20 + (2 \times 2 \times 4) = 36$ m

Width of the reservoir at the top $= B + 2nh = 12 + (2 \times 2 \times 4) = 28$ m

Length of the reservoir at mid-height $= \dfrac{20+36}{2} = 28$ m

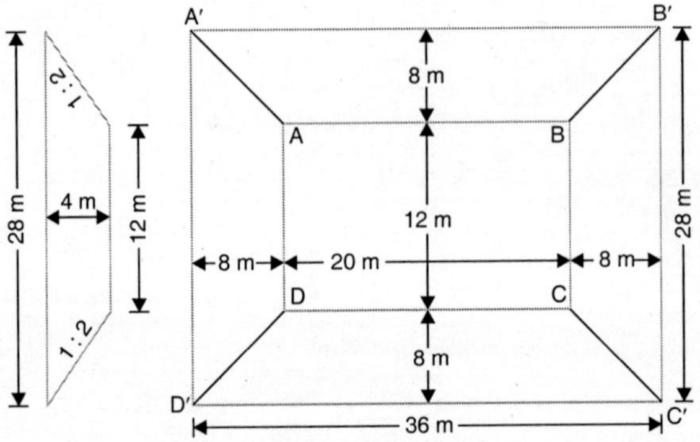

Fig. 13.10

Width of the reservoir at mid-height $= \dfrac{12+28}{2} = 20$ m

Area of the bottom of the reservoir $= 20 \times 12 = 240\ m^2$

Area of the top of the reservoir $= 36 \times 28 = 1008\ m^2$

Area of the reservoir at mid height $= 28 \times 20 = 560\ m^2$

Since the areas *ABCD* and *A'B'C'D'* are in parallel planes spaced 5 m part, prismoidal formula can be used.

$$\therefore \quad V = \frac{h}{6}(A_1 + 4\,A_m + A_2) = \frac{4}{6}(240 + 4 \times 560 + 1008) = \mathbf{2325\ m^2.}$$

Example 13.9: *Calculate the volume of the excavation shown in Fig. 13.11, the side slopes being $1\frac{1}{2}$ horizontal to 1 vertical, and the original ground surface sloping at 1 in 10 in the direction of the centre line of the excavation.*

Solution: Since no two faces are parallel, the solid is not a prismoid and hence prismoidal formula will not be applicable. The total volume will be the sum of the vertical truncated prisms appearing in plan as *ABCD, ABFE, DCGH, ADHE* and *BCGF.*

The depth h at the centre $\dfrac{5+8}{2} = 6.5$ m

The side widths w_1 and w_2 can be calculated from the formulae

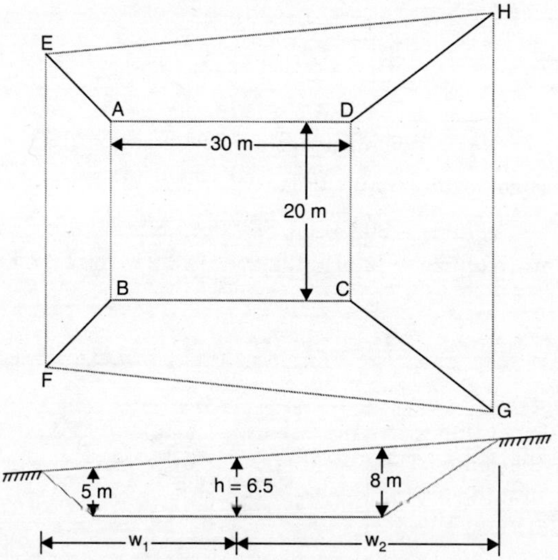

Fig. 13.11

$$w_1 = \frac{b}{2} + \frac{mn}{m-n}\left(h + \frac{b}{2m}\right) = \frac{30}{2} + \frac{10 \times 1.5}{10-1.5}\left(6.5 + \frac{30}{2 \times 10}\right) = 15 + 14.1$$

Horizontal breadth of the slope to the right of DC = 14.1 m

Similarly, $$w_2 = \frac{b}{2} + \frac{mn}{m+n}\left(h - \frac{b}{2m}\right) = \frac{30}{2} + \frac{10 \times 1.5}{10+1.5}\left(6.5 - \frac{30}{2 \times 10}\right) = 15 + 6.52.$$

∴ Horizontal breadth of the slope to the left of BA = 6.52 m

Prism ABCD:

$$\text{Area} = 30 \times 20 = 600 \text{ m}^2$$

$$\text{Average height} = \tfrac{1}{4}(5 + 5 + 8 + 8) = 6.5 \text{ m}$$

$$\therefore \quad \text{Volume} = 600 \times 6.5 = 3900 \text{ m}^3$$

Prism ABFE:

$$\text{Area} = (20 + 6.52)\ 6.52 = 172.9 \text{ m}^2$$

$$\text{Average height} = \tfrac{1}{4}(0 + 0 + 5 + 5) = 2.5 \text{ m}$$

$$\therefore \quad \text{Volume} = 172.9 \times 2.5 = 432.2 \text{ m}^3$$

Prism CDHG:

$$\text{Area} = (20 + 14.1)\ 14.1 = 480.8 \text{ m}$$

$$\text{Average height} = \tfrac{1}{4}(0 + 0 + 8 + 8) = 4 \text{ m}$$

$$\therefore \quad \text{Volume} = 480.8 \times 4 = 1923.2 \text{ m}^3$$

Prisms ADHE and BCGH:

$$\text{Area} = 2\left[(30 + 14.1 + 6.52)\left(\frac{14.1 + 6.52}{2}\right) - \frac{14.1^2}{2} - \frac{6.52^2}{2}\right]$$

$$= 2(521.9 - 99.5 - 21.2) = 802.4 \text{ m}^2$$

$$\text{Average height} = \tfrac{1}{4}(0 + 0 + 5 + 8) = 3.25 \text{ m}$$

$$\therefore \quad \text{Volume} = 802.4 \times 3.25 = 2607.8 \text{ m}^3$$

$$\therefore \quad \text{Total Volume} = 3900 + 432.2 + 1923.2 + 2607.8 = \mathbf{8863.2 \text{ m}^3}.$$

13.8 VOLUME FROM CONTOUR PLAN

As indicated in chapter 10, the amount of earth work or volume can be calculated by the contour plan area. There are four distinct methods, depending upon the type of the work.

(1) By Cross-Sections: It was indicated in chapter 10, that with the help of the contour plan, cross-section of the existing ground surface can be drawn. On the same cross-section, the grade line of the proposed work can be drawn and the area of the section can be estimated either by ordinary methods or with the help of a planimeter.

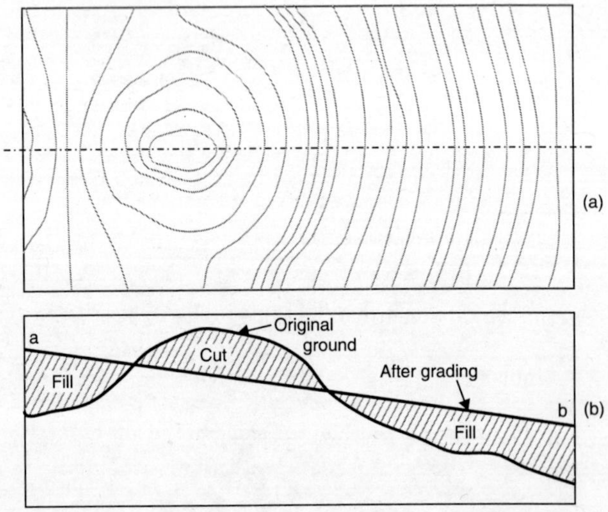

Fig. 13.12

Thus, in Fig. 13.12(*b*), the irregular line represents the original ground while the straight line *ab* is obtained after grading. The area of cut and of fill can be found from the cross-section. The volumes of earth work between adjacent cross-sections may be calculated by the use of average end areas.

(2) By Equal Depth Contours: In this method, the contours of the finished or graded surface are drawn on the contour map, at the same interval as that of the contours. At every point, where the contours of the finished surface intersect a contour of the existing surface, the cut or fill can be found by simply subtracting the difference in elevation between the two contours. By joining the points of equal cut or fill, a set of lines is obtained (represented by thick lines in Fig. 13.13). These lines are the horizontal projections of lines cut from the existing surface by planes parallel to the finished surface. The irregular area bounded by each of these lines can be determined by the use of the planimeter. The volume between any two successive areas is determined by multiplying the average of the two areas by the depth between them, or by prismoidal formula. The sum of the volume of all the layers is the total volume required.

Thus, in Fig. 13.13, the ground contours (shown by thin continuous lines) are at the interval of 1.0 metre. On this a series of straight, parallel and equidistant lines (shown by broken lines) representing a finished plane surface are drawn at the interval of 1.0 metre. At each point in which these two sets of lines meet, the amount of cutting is written. The thick continuous lines are then drawn through the points of equal cut thus getting the lines of 1, 2, 3 and 4 metres cutting. The same procedure may be adopted if the contours of the proposed finished surface are curved in plan.

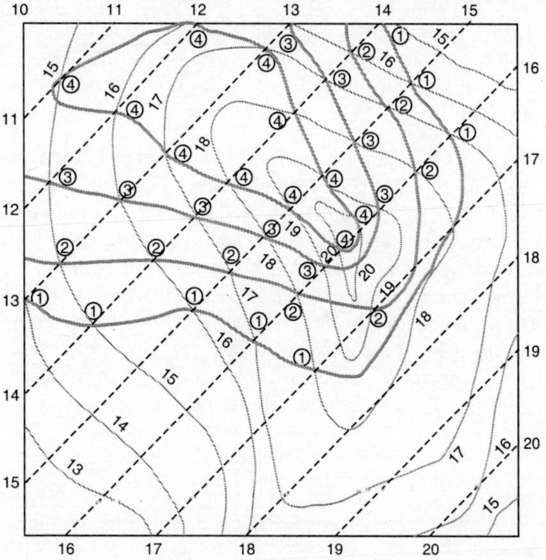

Fig. 13.13

Let A_1, A_2, A_3 ...... etc. be the areas enclosed in each of the thick lines (known as the equal depth contours). This will be the whole area lying within an equal-depth contour line and not that of the strip between the adjacent contour lines.

h = contour interval; V = total volume

Then

$$V = \Sigma \frac{h}{2} (A_1 + A_2) \text{ by trapezoidal formula}$$

or

$$= \Sigma \frac{h}{3} (A_1 + 4A_2 + A_3) \text{ by prismoidal formula.}$$

(3) By Horizontal Planes: The method consists in determining the volumes of earth to be moved between the horizontal planes marked by successive contours.

Thus, in Fig. 13.14, the thin continuous lines represent the ground contours at 1 m interval. The straight, parallel and equidistant lines (shown by broken lines) are drawn to represent the finished plane surface at the same interval. The point p represent the points in which the ground contours and the grade contours of *equal value* intersect. By joining the p-points the line in which the proposed surface cuts the ground is obtained. These lines have been shown by thick lines. *Along thin line no excavation or fill is necessary, but within this line, excavation is necessary and outside this line filling is necessary*. Thus, the extent of cutting between 17 m ground contour and the corresponding 17 m grade contour is also shown by hatched lines. Similary, the extent of cutting between the 16 m ground contour and the corresponding 16 m grade contour is also shown by hatched lines. Proceeding like this, we can mark the extent of earthwork between any two *corresponding* ground and grade contours and the *areas* enclosed in these *extents* can be measured by planimeter. The volume can then be calculated by using end area rule.

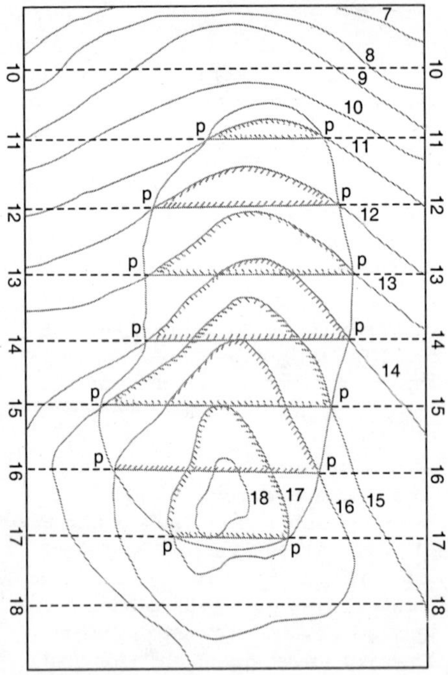

Fig. 13.14

(4) Capacity of Reservoir: This is a typical case of volume in which the finished surface (*i.e.*, surface of water) is level surface. The volume is calculated by assuming it as being divided up into a number of horizontal slices by contour planes. The ground contours and the grade contour, in this case, coincide. The whole area lying within a contour line (and not that of the strip between two adjacent contour lines) is measured by planimeter and the volume can be calculated.

Let $A_1, A_2, A_3 \ldots\ldots, A_n$ = the area of successive contours

h = contour interval

V = capacity of reservoir

Then by *trapezoidal formula,*

$$Vh = \left[\frac{A_1 + A_n}{2} + A_2 + A_3 + \ldots + A_{n-1}\right]$$

By the *prismoidal rule,*

$$V = \frac{h}{3}[A_1 + 4\,A_2 + 2\,A_3 + 4\,A_4 \ldots\ldots 2\,A_{n-2} + 4\,A_{n-1} + A_n]$$

where n is an odd number.

Example 13.10: *The areas within the contour line at the site of reservoir and the face of the proposed dam are as follows:*

Contour	*Area (m²)*	*Contour*	*Area (m²)*
101	1,000	106	1350,000
102	12,800	107	1985,000
103	95,200	108	2286,000
104	147,600	109	2512,000
105	872,500		

Taking 101 as the bottom level of the reservoir and 109 as the top level, calculate the capacity of the reservoir.

Solution: *By trapezoidal formula,*

$$V = h\left(\frac{A_1 + A_n}{2} + A_2 + A_3 \ldots\ldots A_{n-1}\right) = 1\left(\frac{1000 + 2512{,}000}{2} + 12{,}800 + 95{,}200\right.$$

$$\left. + 147{,}600 + 872{,}500 + 1350{,}000 + 1985{,}000 + 2286{,}000\right)$$

$$= 8005{,}600 \text{ m}^2$$

By prismoidal formula

$$V = \frac{h}{3}\,[A_1 + 4(A_2 + A_4 + \ldots) + 2(A_3 + A_5 + \ldots) + A_n]$$

$$= \frac{1}{3}[1000 + 4(12{,}800 + 147{,}600 + 1350{,}000 + 2286{,}000)$$
$$+ 2(95{,}200 + 872{,}500 + 1985{,}000) + 2512{,}000]$$

$\therefore \quad V =$ **7,868,000 m³**.

PROBLEMS

1. What is a prismoid? Derive the prismoidal formula.
2. Derive an expression for trapezoidal formula for volume. Compare it with the prismoidal formula.
3. Explain, with the help of sketches, the use of a contour map for calculation of earth work.
4. How do you determine (*a*) the capacity of a reservoir (*b*) the earth work for a borrow pit?
5. (*a*) Calculate the volume of earth work by Prismoidal formula in a road embankment with the following data:

Chainage along the centre line	0	100	200	300	400
Ground levels	201.70	202.90	202.40	204.70	206.90

Formation level at chainage 0 is 202.30, top width is 2.00 ft side slopes are 2 to 1. The longitudinal gradient of the embankment is 1 in 100 rising. The ground is assumed to be level all across the longitudinal section.

(*b*) If the transverse slope of the ground at chainage 200 is assumed to be 1 in 10, find the area of embankment section at this point.

6. At every 100 ft along a piece of ground, level were taken. They were as follows:

Feet		*G.L.*
0	...	210.00
100	...	220.22
200	...	231.49
300	...	237.90
400	...	240.53
500	...	235.00

A cutting is to be made for a line of uniform gradient passing through the first and last points. What is the gradient? Calculate the volume of cutting on the assumption that the ground at right angles to the centre line is levelled.

Given: Breadth of formation 30 feet; slope of the cutting in each side $1\frac{1}{2}$ to 1. Use prismoidal formula. (*U.P.*)

7. A cutting is to be made through the ground where the cross-slope varies considerably. At *A*, the depth of the cut is 12 ft at the centre line, and cross slope is 8 to 1. At *B* the corresponding figures are 10 ft and 12 to 1 and at *C* 14 ft and 10 to 1. *AB* and *BC* are each 100 feet. The formation width is 30 feet and the side slopes $1\frac{1}{2}$ to 1. Calculate by the prismoidal method the volume of the cutting in cubic yards between *A* and *C*. (*U.P.*)

ANSWERS

5. (*a*) 4013 cubic yds. (*b*) 352.52 sq. ft.
6. 6953 cubic yds.
7. 3919 cubic yds.

CHAPTER 14

MINOR INSTRUMENTS

14.1 HAND LEVEL

A hand level is a simple, compact instrument used for reconnaissance and preliminary survey, for locating contours on the ground and for taking short cross-sections. It consists of a rectangular or circular tube, 10 to 15 cm long, provided with a small bubble tube at the top. A line of sight, parallel to the axis of the bubble tube, is defined by a line joining a pin-hole at the eye end and a horizontal wire at the object end. In order to view the bubble tube at the instant the object is sighted, a small opening, immediately below the bubble, is provided in the tube. The bubble is reflected through this opening on to a mirror, which is inside the tube inclined at 45° to the axis, and immediately under the bubble tube. The mirror occupies half the width of the tube and the objects are sighted through the other half. The line of sight is horizontal when the centre of the bubble appears opposite the cross-wire, or lies on a line ruled on the reflector.

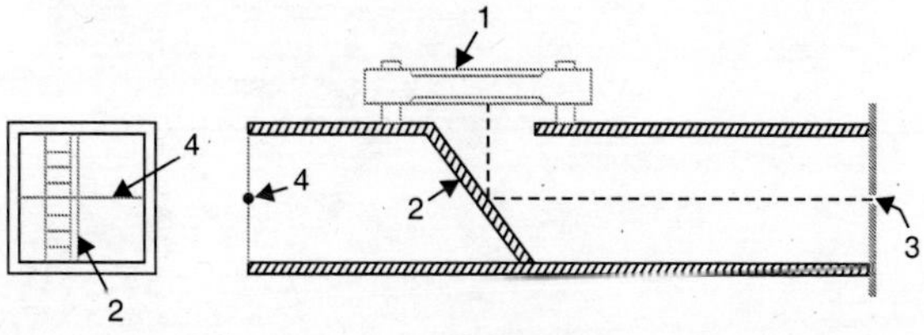

Fig. 14.1 Hand Level.

1. Bubble tube
2. Reflecting mirror
3. Eye slit or hole
4. Cross-wire.

To use the instrument

(*i*) Hold the instrument in hand (preferably against a rod or staff) at the eye level and sight the staff kept at the point to be observed.

(*ii*) Raise or lower the object end of the tube till the image of the bubble seen in the reflector is bisected by the cross-wire.

(*iii*) Take the staff reading against the cross-wire.

In some of the hand levels, telescopic line of the sight may also be provided.

Adjustment of the hand level (Fig. 14.2)

To make the line of sight horizontal when the bubble is centred.

1. Select two rigid supports P and Q at about 20 to 30 metres apart.

2. Hold the level at a point A on the support at P and mark a point D on the other support Q, when the bubble is central.
3. Shift the instrument to Q, hold it at the point D, centre the bubble, and mark the point B where the line of sight strikes the first support. If A and B do not coincide, the instrument requires adjustment.

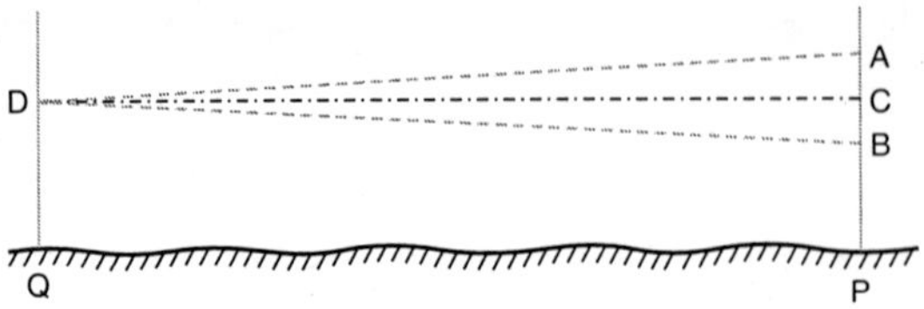

Fig. 14.2

4. Select a point C midway between A and B. With the adjustment screws, raise or lower the cross-wire till the line of sight bisects C.

14.2 ABNEY CLINOMETER (*ABNEY LEVEL*)

Abney level is one of the various forms of clinometers used for the measurement of slopes, taking cross-sections, tracing contours, setting grades and all other rough levelling operations. It is a light, compact and hand instrument with low precision as compared to engineer's level. The abney level consists of the following (Fig. 14.3):

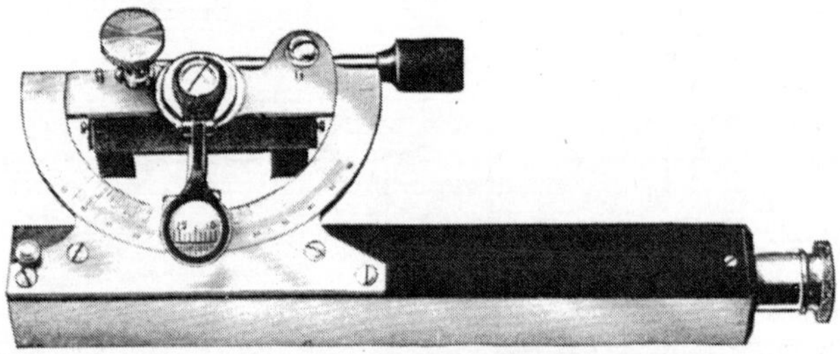

Fig. 14.3 Abney level.
(By courtesy of M/s Vickers Instruments Ltd.)

1. A square sighting tube having peep hole or eye-piece at one end and a cross-wire at the other end. Near the objective end, a mirror is placed at an angle of 45° inside the tube and occupying half the width, as in the hand level. Immediately above the mirror, an opening is provided to receive rays from the bubble tube placed above it. The line of sight is defined by the line joining the peep hole and the cross-wire.
2. A small bubble tube, placed immediately above the openings attached to a vernier arm, which can be rotated either by means of a milled headed screw or by rack and pinion arrangement. The image of the bubble is visible in the mirror.

   When the line of sight is at any inclination, the milled-screw is operated till the bubble is bisected by the cross-wire. The vernier is thus moved from its zero position, the amount of movement being equal to the inclination of the line of sight.
3. A semi-circular graduated arc is fixed in position. The zero mark of the graduations coincides with the zero of the vernier. The reading increases from 0° to 60° (or 90°) in both the directions, one giving the angles of elevation and the other angles of depression.

In some instruments, the values of the slopes, corresponding to the angles, are also marked. The vernier is of extended type having least count of 5′ or 10′.

If the instrument is to be used as a hand level, the vernier is set to read zero on the graduated arc and the level is then used as an ordinary hand level.

The Abney level can be used for (*i*) measuring vertical angles, (*ii*) measuring slope of the ground, and (*iii*) tracing grade contour.

(*i*) *Measurement of vertical angle*

1. Keep the instrument at eye level and direct it to the object till the line of sight passes through it.
2. Since the line of sight is inclined, the bubble will go out of centre. Bring the bubble to the centre of its run by the milled-screw. When the bubble is central, the line of sight *must pass* through the object.
3. Read the angle on the arc by means of the vernier.

(*ii*) *Measurement of slope of the ground*

1. Take a target, having cross-marks, at observer's eye height and keep it at the other end of the line.
2. Hold the instrument at one end and direct the instrument towards the target till the horizontal wire coincides with the horizontal line of the target.
3. Bring the bubble in the centre of its run.
4. Read the angle on the arc by means of the vernier.

(*iii*) *Tracing grade contour:* See § 10.6.

Testing and Adjustment of Abney Level:

1. Fix two rods, having marks at equal heights h (preferably at the height of observer's eye), at two points P and Q, about 20 to 50 metres apart.
2. Keep the Abney level at the point A against the rod at P and measure the angle of elevation α_1 towards the point B of the rod Q.
3. Shift the instrument to Q, hold it against B and sight A. Measure the angle of depression α_2.
4. If α_1 and α_2 are equal, the instrument is in adjustment *i.e.*, the line of sight is parallel to the axis of the bubble tube when it is central and when vernier reads zero.
5. If not, turn the screw so that the vernier reads the mean reading $\frac{\alpha_1 + \alpha_2}{2}$. The bubble will no longer be central.

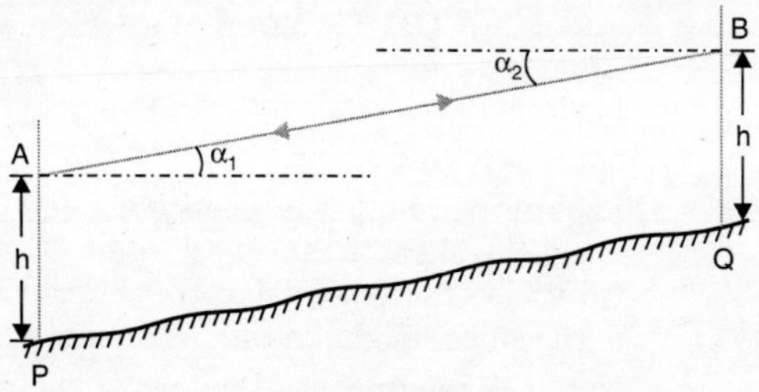

Fig. 14.4

Bring the bubble to the centre of its run by means of its adjusting screws. Repeat the test till correct.

Note: If the adjustment is not done, the index error, equal $\frac{\alpha_1 - \alpha_2}{2}$, may be noted and the correction may be applied to all the observed readings.

14.3 INDIAN PATTERN CLINOMETER (*TANGENT CLINOMETER*)

Indian pattern clinometer is used for determining difference in elevation between points and is specially adopted to plane tabling. The clinometer is placed on the plane table which is levelled by estimation. The clinometer consists of the following:

1. A base plate carrying a small bubble tube and a levelling screw. Thus, the clinometer can be accurately levelled.
2. The eye vane carrying a peep hole. The eye vane is hinged at its lower end to the base plate.
3. The object vane having graduations in degrees at one side and tangent of the angles to the other side of the central opening. The object vane is also hinged at its lower end to the base plate. A slide, provided with a small window and horizontal wire in its middle, can be moved up and down the object vane by a rack and pinion fitted with a milled head. The line of sight is defined by the line joining the peep hole and the horizontal wire of the slide.

When the instrument is not in use, the vanes fold down over the base.

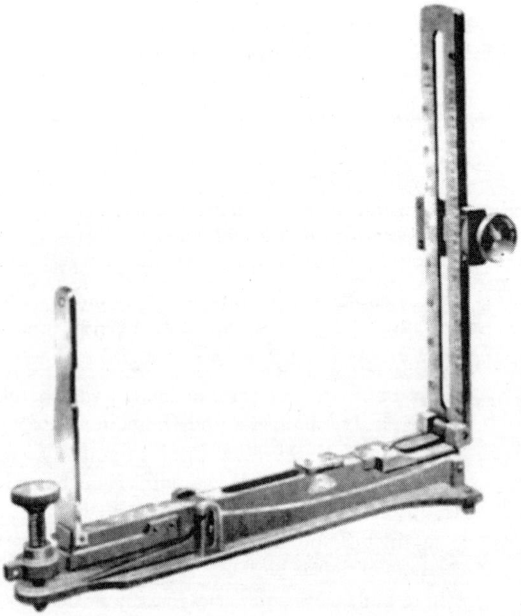

Fig. 14.5 Indian Pattern Clinometer.

Use of Indian Pattern Clinometer with Plane Table

1. Set the plane table over the station and keep the Indian Pattern Clinometer on it.
2. Level the clinometer with the help of the levelling screw.
3. Looking through the peep hole, move the slide of the object vane till it bisects the signal at the other point to be sighted. It is preferable to use a signal of the same height as that of the peep hole above the level of the plane table station.
4. Note the reading, *i.e.* tangent of the angle, against the wire. Thus, the difference in elevation between the eye and the object = distance × tangent of vertical angle = $d \tan \alpha$.

The distance d between the plane table station and the object can be found from the plan. The reduced level of the object can thus be calculated if the reduced level of the plane table station is known.

14.4 BUREL HAND LEVEL (*Fig. 14.6*)

This consists of a simple frame carrying a mirror and a plain glass. The mirror extends half-way across the frame. The plain glass extends to the other half. The frame can be suspended vertically in gimbles. The edge of the mirror forms vertical reference line. The instrument is based on the principle that a ray of light after being reflected back from a vertical mirror along the path of incidence, is horizontal. When the instrument is suspended at eye level, the image of the eye is visible at the edge of the mirror, while the objects appearing through the plain glass opposite the image of the eye are at the level of observer's eye.

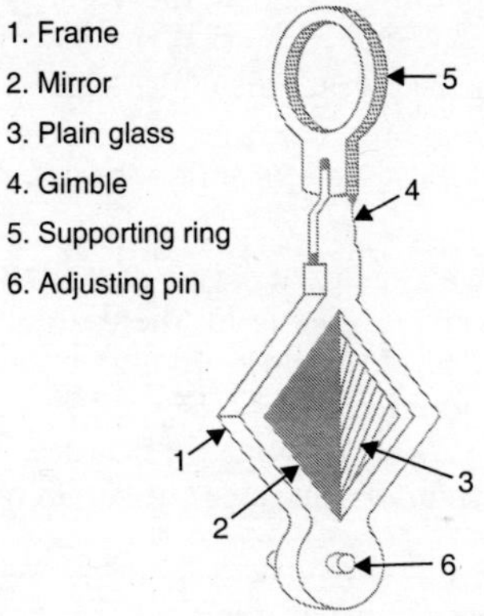

Fig. 14.6 Burel hand level.

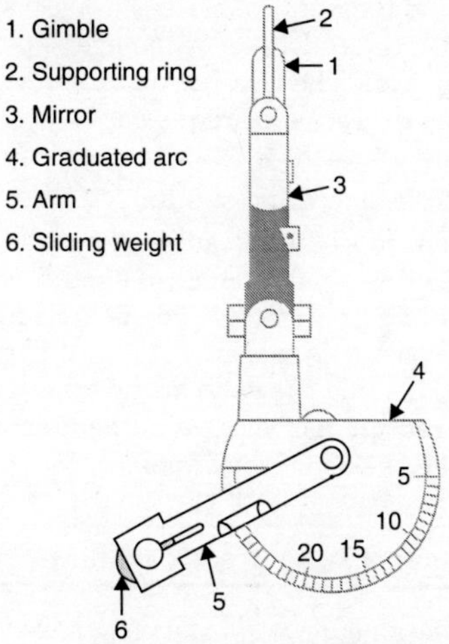

Fig. 14.7 De Lisle's Clinometer.

14.5 DE LISLE'S CLINOMETER (*Fig. 14.7*)

This is another form of clinometer, similar to that of Burel hand level, used for measuring the vertical angles, determining the slope of the ground, and for setting out gradients. This consists of the following:

1. A simple frame, similar to that of a Burel level, carrying a mirror extending half-way across the frame, the objects being sighted through the other half which is open. The frame can be suspended in gimbles.

   The edge of the mirror forms a vertical reference line.
2. A heavy semi-circular arc is attached to the lower end of the frame. The arc is graduated in gradients or slopes from 1 in 5 to 1 in 50. The arc is attached to the vertical axis so that it may be revolved to bring the arc towards the observer (*i.e.* forward) to measure the rising gradients or away from the observer to measure the falling gradients.
3. A radial arm is fitted to the centre of the arc. The arm consists of a bevelled edge which acts as index. By moving the arm along the arc, the mirror can be inclined to the vertical. The inclination to the horizontal of the line from the eye to the point at which it appears in the mirror equals the inclination of the mirror to the vertical. The arm also carries a sliding weight. When the weight is moved to the outer stop (at the end of the arm), it counter balances the weight of the arc in horizontal position and makes the mirror vertical. To make the line of the sight horizontal, the weight is slided to the outer slope and the radial arm is turned back to its fullest extent.

To measure a gradient

1. Slide the weight to the inner stop of the arm. The arc should be turned forward for rising gradients and backward for falling gradients.
2. Suspend the instrument from the thumb and hold it at arm's length in such a position that the observer sees the reflected image of his eye at the edge of the mirror.
3. Move the radial arm till the object sighted through the open half of the frame is coincident with the reflection of the eye. Note the reading on the arc against the bevelled edge of the arm. The reading obtained will be in the form of gradient which can be converted into degrees if so required.

For better results, a vane or target of height equal to the height of observer's eye must be placed at the object and sighted.

A similar procedure is adopted to set a point on a given gradient, say 1 in n. The arm is set on the reading 1 in n. The arc should be turned forward for rising gradients and backward for the falling gradients. A peg is driven at the other end of the line and a vane, equal to the height of observer's eye, is kept there. The instrument man then sights the vane and signals the assistant, holding the vane at the other end, to raise or lower the vane till it is seen coincident with the reflection of the eye in the mirror. The peg is then driven in or out till its top is at the level of the bottom of the vane.

14.6 FOOT-RULE CLINOMETER (*Fig. 14.8*)

A foot-rule clinometer consists of a box wood rule having two arms hinged to each other at one end, with a small bubble tube on each arm. The upper arm or part also carries a pair of sights through which the object can be sighted. A graduated arc is also attached to the hinge, and angles of elevations and depressions can be measured on it. A small compass is also recessed in the lower arm for taking bearings.

To sight an object, the instrument is held firmly against a rod, with the bubble central in the lower arm. The upper arm is then raised till the line of sight passes through the object. The reading is then taken on the arc.

Another common method of using the clinometer is to keep the lower arm on a straight edge laid on the slope to be measured. The rule is then opened until the bubble of the upper arm is central. The reading is then noted.

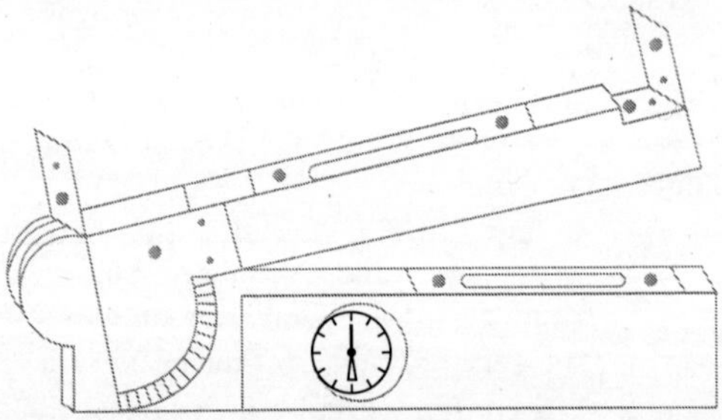

Fig. 14.8 Foot-Rule Clinometer.

14.7 CEYLON GHAT TRACER (*Fig. 14.9*)

It is a very useful instrument for setting out gradients. It essentially consists of a long circular tube having a peep hole at one end and cross-wires at the other ends. The tube is supported by a A-frame having a hole at its top to fix the instrument to a straight rod or stand. The tube is also engraved to give readings of gradients. A heavy weight slides along the tube by a suitable rack and pinion arrangement. The weight, at its top, contains one bevelled edge which slides along the graduations of the bar, and serves as an index. The line of sight is defined by the line joining the hole to the intersection of the cross-wires and its prolongation. When the bevelled edge of the weight is against the zero reading, the line of sight is horizontal. For the elevated gradients, the weight is slided towards the observer. For falling gradients, the weight is slided away from the observer.

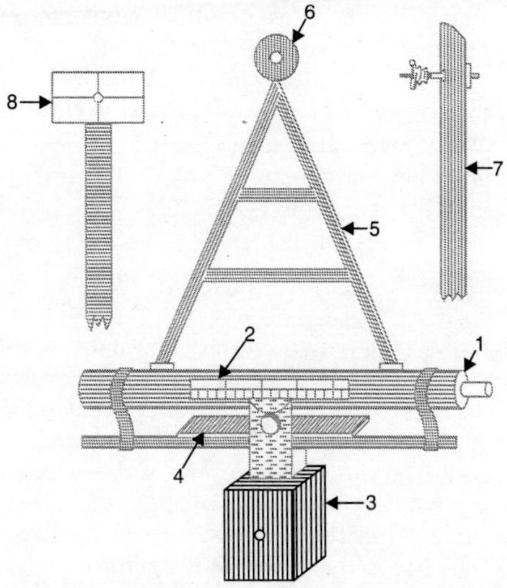

Fig. 14.9 Ceylon Ghat Tracer.

1. Tube 2. Graduations 3. Sliding weight
4. Rack 5. A-Frame 6. Supporting Hole
7. Stand 8. Vane or Target

(a) To measure a slope

1. Fix the instrument on to the stand and hold it to one end of the line. Keep the target at the other end.
2. Looking through the eye hole, move the sliding weight till the line of sight passes through the cross mark of the sight vane.
3. The reading against the bevelled edge of the weight will give the gradient of the line.

(b) To set out a gradient

1. Hold the instrument at one end.
2. Send the assistant at the other end with the target.
3. Slide the weight to set it to the given gradient, say 1 in n.
4. Direct the assistant to raise or lower the target till it is bisected.

Drive a peg at the other end so that the top of the peg is at the same level as that the bottom of the target.

14.8 FENNEL'S CLINOMETER

It is a precise clinometer for the measurement of slopes. It consists of the following parts (Figs. 14.10, 14.11):

1. A telescope for providing line of sight.
2. Two plate levels for checking horizontality of the holding staff.
3. A vertical arc which rotates or tilts along with the tilting of the telescope.
4. A holding staff, and
5. A target mounted on a holding staff of the same height.

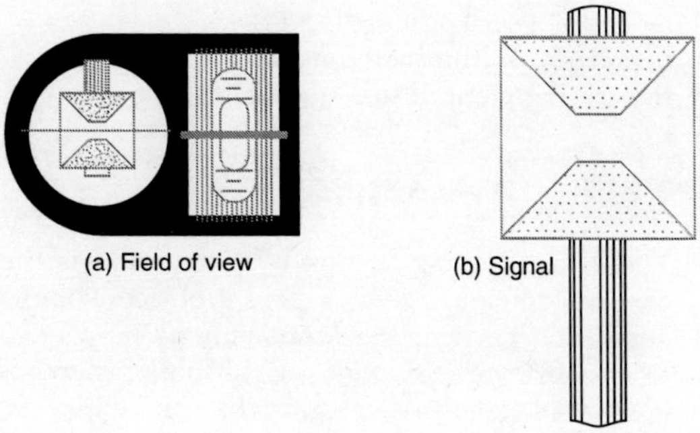

Fig. 14.10

This instrument is specially designed for finding the lines of highways with a predetermined percentage inclination (*i.e.* percentage slope) and for determination of the percentage amount of inclination of existing highways. It has a vertical arc allowing to read slopes upto ± 40% with graduation to 0.5% thus making sure estimation to 0.1%.

The design of the telescope, when inclined, admits the sighted object, the diaphragm with stadia lines and the first spirit level running parallel to the vertical arc can be simultaneously seen in the telescope [Fig. 14.10 (*a*)]. A second spirit level likewise is parallel to the tilting axis.

14.9 THE PANTAGRAPH (*Fig. 14.12*)

A pantagraph is an instrument used for reproducing, enlarging or reducing the maps. It is based on the principle of similar triangles. It consists of two long bars *AB* and *AD* hinged together at *A* and supported on castors or rollers at *B* and *D*. Two short arms *EF* and *GF* are hinged together at *F* and are connected to *AD* and *AB* at *E* and *G* respectively. Thus *AGFE* is a parallelogram of equal sides for all positions of the instrument. The long bar *AD* carries a movable tubular frame which can be slided along it. The sliding frame carries an index and also a heavy weight *Q* which forms the vertical axis of the instrument; the whole instrument moves about the point *Q*. The bar *EF* carries a pencil point *P* attached to a carrier which can also be set to a desired reading on the bar *EF*. The longer arm *AB* carries tracing point at the end *B*. For any setting of the instrument, the point *B*, *P* and *Q* are in a straight line. The original map is kept at *B* and is traced. Correspondingly, the pencil point *P* also moves, but the point *Q* remains fixed in position. Thus, if *B* is moved straight by an amount *BB'*, the point *P* moves to *P'* the ratio between *BB'* and *PP'* being equal to the ratio of reduction. For any position of the tracing point, the points *B'*, *P'* and *Q* are always in a straight line.

If it is desired to enlarge the map, the pencil point is kept at *B*, the tracing point at *P* and the map under the point *P*. The moving frames at *Q* and *P* are set to the same reading equal to the ratio of enlargement. The pencil can be raised off the paper, by means of a cord passing from the pencil round the instrument to the tracing point, if so required.

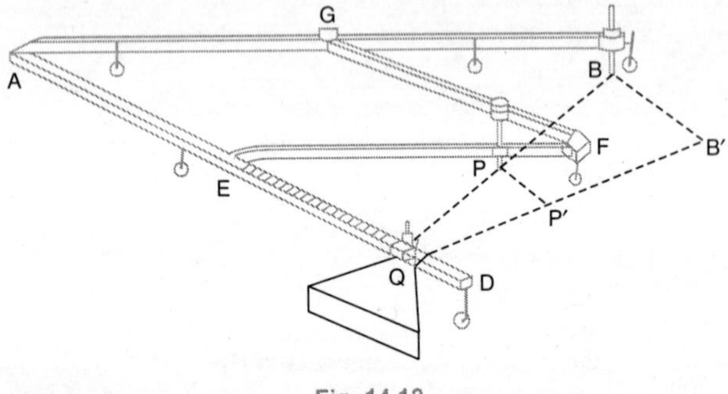

Fig. 14.12

14.10 THE SEXTANT

The distinguishing feature of the sextant is the arrangement of mirrors which enables the observer to sight at two different objects simultaneously, and thus to measure an angle in a single observation. A sextant may be used to measure horizontal angle. It can also be used to measure vertical angles. Essentially, therefore, a sextant consists of fixed glass (*H*) which is silvered to half the height while the upper half is plain. Another glass (*P*) is attached to a movable arm which can be operated by means of a milled head. The movable arm also carries a vernier at the other end. The operation of the sextant depends on bringing the image of one point (R), after suitable reflection in two mirrors, into contact with the image of a second point (L) which is viewed direct, by moving the movable mirror (P). Since the vernier and the movable mirror are attached to the same arm, the movement of the vernier from the zero position gives the required angle subtended by the two objects at the instrument station.

The sextant is based on the principle that when a ray of light is reflected successively from two mirrors, the angle between the first and last directions of ray is twice the angle between the planes of the two mirrors.

Thus, in Fig. 14.13, H is the fixed glass (also known as the horizon glass) and P is the index glass or the movable glass. Let the angle between the planes of two glasses be α when the image of the object R has been, after double reflection, brought in the same vertical line as that of the object L viewed directly through the unsilvered portion of the glass H. Let the rays from both the objects subtend an angle β. PI is the index arm pivoted at P.

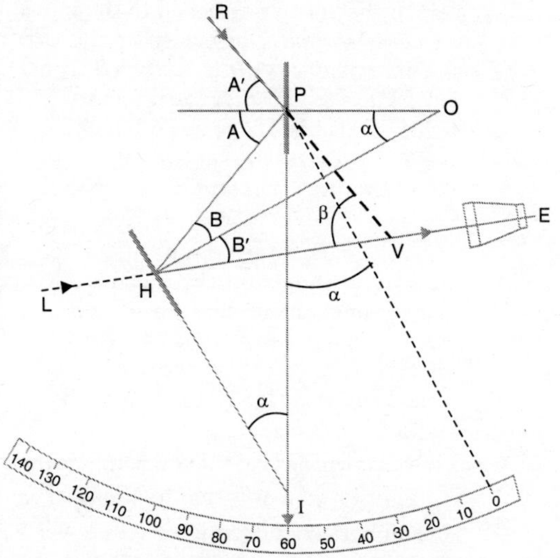

Fig. 14.13 Optical Diagram of a sextant.

Since the angle of incidence is equal to the angle of reflection, we have

$$\angle A = A' \,;\ \angle B = \angle B'$$

or

$$\alpha = \angle A - \angle B \text{ (exterior angle)}$$

$$\beta = \angle A + \angle A' - (\angle B + \angle B') = 2\angle A - 2\angle B = 2(\angle A - \angle B) = 2\alpha$$

or

$$\alpha = \frac{\beta}{2}.$$

Hence the angle between the mirrors is equal to half the actual angle between two objects. While constructing the sextant, the plane of mirror P is so adjusted that it is parallel to the mirror H when the index reads zero. The movement of the mirror P is equal to the movement of the vernier. The scale is numbered in values equal to twice the actual angle so that actual angle between the objects is read directly.

Optical Requirements of the Sextant

1. The two mirrors should be perpendicular to the plane of the graduated arc.
2. When the two mirrors are parallel, the reading on the index should be zero.
3. The optical axis should be parallel to the plane of the graduated arc and pass through the top of the horizon mirror. If only a peep sight is provided in place of telescope, the peep sight should be at the same distance above the arc as the top of the mirror.

There are mainly three types of sextants:

1. Box Sextant
2. Nautical Sextant
3. Sounding Sextant.

(*a*) **Nautical Sextant:** A *nautical sextant* is specially designed for navigation and astronomical purposes and is fairly large instrument with a graduated silver arc of about 15 to 20 cm radius let into a gun metal casting carrying the main parts. With the help of the vernier attached to the index mirror, readings can be taken to 20″ or 10″. A *sounding sextant* is also very similar to the nautical sextant, with a large index glass to allow for the difficulty of sighting an object from a small rocking boat in hydrographic survey. Fig. 14.14 shows a nautical sextant by U.S. Navy.

(*b*) **Box Sextant:** The *box sextant* is small pocket instrument used for measuring horizontal and vertical angles, measuring chain angles and locating inaccessible points. By setting the vernier to 90°, it may be used as an optical square. Fig. 14.15 shows a box sextant.

Fig. 14.15 Box Sextant.

A box sextant consists of the following parts:

1. A circular box about 8 cm in diameter and 4 cm high.
2. A fixed horizon glass, silvered at lower half and plain at upper half.
3. A movable index glass fully silvered.
4. An index arm pivoted at the index glass and carrying a vernier at the other end.
5. An adjustable magnifying glass, to read the angle.
6. A milled-headed screw to rotate the index glass and the index arm.
7. An eye hole or peep hole or a telescope for long distance sighting.
8. A pair of coloured glasses for use in bright sun.
9. A slot in the side of the box for the object to be sighted.

Measurement of Horizontal Angle with Box Sextant

1. Hold the instrument in the right hand and bring the plane of the graduated arc into the plane of the eye and the two points to be observed.
2. Look through the eye hole at the *left hand object* through the lower unsilvered portion of the horizon glass.
3. Turn the milled-headed screw slowly so that the image of the right-hand object, after double reflection, is coincident with the left-hand object ; view directly through the upper half of the horizon glass. Clamp the vernier. If a slow motion screw is provided, bring the images of object into exact coincidence.

The reading on the vernier gives directly the angle.

Note: The vertex (*V*) of an angle measured is not exactly at the eye but at the intersection of the two lines of sight which, for small angles, is considerably behind the eye. For this reason, there may be an appreciable error in the measurement of the angles less than, say, 15°.

Measurement of Vertical Angle with Sextant

Vertical angles may be measured by holding the sextant so that its arc lies in a vertical plane. If it is required to measure the vertical angle between two points, view the lower object directly, and turn the milled headed screw until the image of the higher object appears coincident with the lower one.

Permanent Adjustments of a Sextant

A sextant requires the following four adjustments:

1. To make the index glass perpendicular to the plane of the graduated arc.
2. To make the horizon glass perpendicular to the plane of graduated arc.
3. To make the line of sight parallel to the plane of the graduated arc.
4. To make the horizon mirror parallel to the index mirror when the vernier is set at zero (*i.e.* to eliminate any index correction).

In a box sextant, the index glass is permanently fixed at right angles to the plane of the instrument by the maker. Also, no provision is made for adjustment 3. Hence, only adjustments 2 and 4 are made for a box sextant.

Adjustment 2: Adjustment of horizon glass

(*i*) Set the vernier at approximately zero and aim at some well-defined distant point like a star, with the arc vertical.

(*ii*) Move the index arm back and forth slightly. The image of the star will move up and down.

(*iii*) Adjust the horizon mirror by tilting it forward or backward until, when the index arm is moved, the image of the star, in passing will coincide with the star itself.

Adjustment 4: Elimination of index error

(*i*) Bring the direct and reflected image of a distant point into coincidence. If the vernier does not read zero, the error is called the index error.

(*ii*) Correct the error by turning the horizon glass around an axis perpendicular to the plane of the graduated arc.

If the index error is not large, it is customary not to correct the error, but to apply the correction to the observed readings. An index error should, however, be determined from time to time.

CHAPTER 15

TRIGONOMETRICAL LEVELLING

15.1 INTRODUCTION

Trigonometrical levelling is the process of determining the differences of elevations of stations from observed vertical angles and known distances, which are assumed to be either horizontal or geodetic lengths at mean sea level. The vertical angles may be measured by means of an accurate theodolite and the horizontal distances may either be measured (in the case of plane surveying) or computed (in the case of geodetic observations).

We shall discuss the trigonometrical levelling under two heads:

1. Observations for heights and distances, and
2. Geodetical observations

In the first case, the principles of plane surveying will be used. It is assumed that the distances between the points observed are not large so that either the effect of curvature and refraction may be neglected or proper corrections may be applied *linearly* to the calculated difference in elevation. Under this head fall the various methods of angular levelling for determining the elevations of particular points such as top of chimney, or church spire etc.

In the geodetical observations of trigonometrical levelling, the distance between the points measured is geodetic and is large. The ordinary principles of plane surveying are not applicable. The corrections for curvature and refraction are applied in *angular measure* directly to the observed angles. *The geodetical observations of trigonometrical levelling have been dealt with in the second volume.*

HEIGHTS AND DISTANCES

In order to get the difference in elevation between the instrument station and the object under observation, we shall consider the following cases:

Case 1: Base of the object accessible.

Case 2: Base of the object inaccessible: Instrument stations in the same vertical plane as the elevated object.

Case 3: Base of the object inaccessible: Instrument stations not in the same vertical plane as the elevated object.

15.2 BASE OF THE OBJECT ACCESSIBLE

Let it be assumed that the horizontal distance between the instrument and the object can be measured accurately. In Fig. 15.1, let

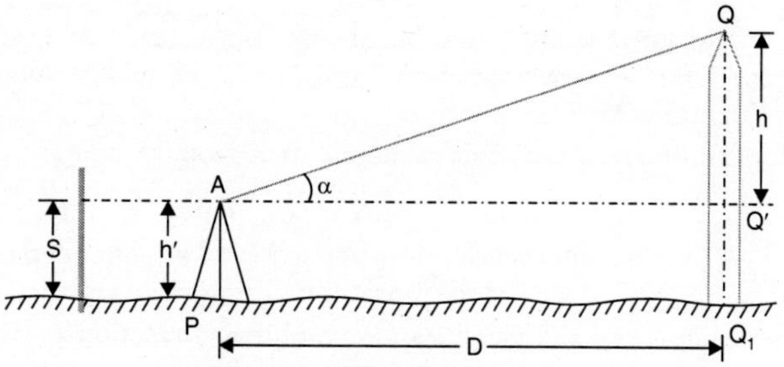

Fig. 15.1 Base accessible.

P = instrument station
Q = point to be observed
A = centre of the instrument
Q' = projection of Q on horizontal plane through A
$D = AQ'$ = horizontal distance between P and Q
h' = height of the instrument at P
$h = QQ'$
S = reading of staff kept at B.M., with line of sight horizontal
α = angle of elevation from A to Q.

From triangle AQQ'; $h = D \tan \alpha$ ...(15.1)

$$\text{R.L. of } Q = \text{R.L. of instrument axis} + D \tan \alpha$$

If the R.L. of P is known,

$$\text{R.L. of } Q = \text{R.L. of } P + h' + D \tan \alpha$$

If the reading on the staff kept at the B.M. is S with the line of sight horizontal,

$$\text{R.L. of } Q = \text{R.L. of B.M.} + S + D \tan \alpha$$

The method is usually employed when the distance A is small. However, if D is large, the combined correction for curvature and refraction can be applied.

In order to get the sign of the combined correction due to curvature and refraction, consider Fig. 15.2. $PP''P'$ is the vertical (or plumb) line through P and $QQ'Q''$ is the vertical line through Q. P' is the projection of P on the horizontal line through Q, while P'' is the projection of P on the level line through Q. Similarly, Q' and Q'' are the projections of Q on horizontal and level lines respectively through P.

If the distance between P and Q is not very large, we can take $PQ' = PQ'' = D = QP'' = QP'$.

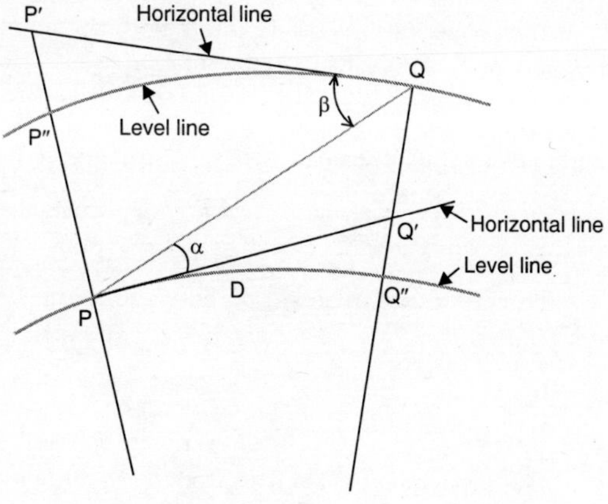

Fig. 15.2

and $\angle QQ'P = \angle QP'P = 90°$ (approximately)

Then $QQ' = D \tan \alpha$

But the true difference in elevation between P and Q is QQ''. Hence the combined correction for curvature and refraction = $Q'Q''$ which should be added to QQ' to get the true difference in elevation QQ''.

Similarly, if the observation is made from Q, we get

$$PP' = D \tan \beta$$

The true difference in elevation is PP''. The combined correction for curvature and refraction = $P'P''$ which should be subtracted from PP' to get the true difference in elevation PP''

Hence we conclude that if the combined correction for curvature and refraction is to be applied *linearly,* its sign is positive for angles of elevation and negative for angles of depression. As in levelling, the combined correction for curvature and refraction in linear measure is given by

$$C = 0.06728\ D^2 \text{ metres, when } D \text{ is in kilometres.}$$

Thus, in Fig. 15.1, R.L. of Q = R.L. of B.M. + S + $D \tan \alpha + C$.

Indirect Levelling. The above principle can be applied for running a line of indirect levels between two points P and Q, whose difference of level is required (Fig. 15.3).

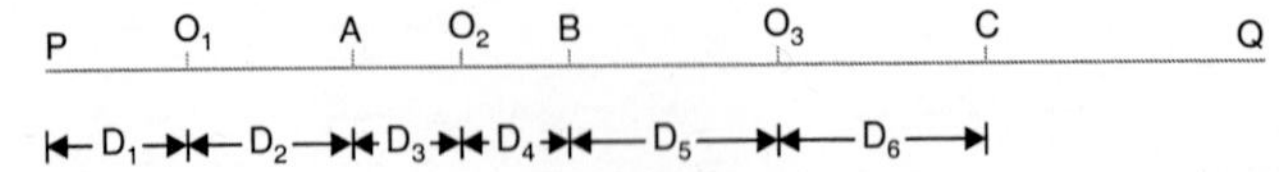

Fig. 15.3

In order to find the difference in elevation between P and Q, the instrument is set at a number of places O_1, O_2, O_3 etc., with points A, B, C etc., as the turning points as shown in Fig. 15.3. From each instrument station, observations are taken to both the points on either side of it, the instrument being set midway between them. Thus, in Fig. 15.4, let O_1 be the first position of the instrument midway P and A. If α_1 and β_1 are the angles observed from O_1 to P and A, we get

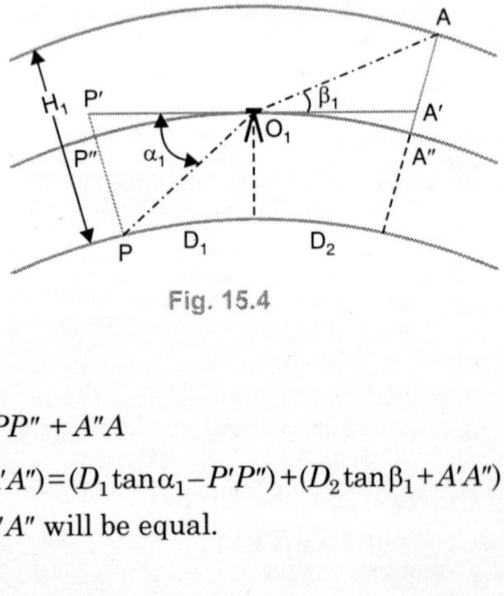

Fig. 15.4

$$PP' = D_1 \tan \alpha_1$$

and $AA' = D_1 \tan \beta_1$

The difference in elevation between A and $P = H_1 = PP'' + A''A$

$$= (PP' - P'P'') + (AA' + A'A'') = (D_1 \tan \alpha_1 - P'P'') + (D_2 \tan \beta_1 + A'A'')$$

If $D_1 = D_2 = D$, $P'P''$ and $A'A''$ will be equal.

Hence $H_1 = D (\tan \alpha_1 + \tan \beta_1)$

The instrument is then shifted to O_2, midway between A and B, and the angles α_2 and β_2 are observed. Then the difference in elevation between B and A is

$$H_2 = D' (\tan \alpha_2 + \tan \beta_2) \quad \text{where} \quad D' = D_3 = D_4$$

The process is continued till Q is reached.

15.3 BASE OF THE OBJECT INACCESSIBLE: INSTRUMENT STATIONS IN THE SAME VERTICAL PLANE AS THE ELEVATED OBJECT

If the horizontal distance between the instrument and the object can be measured due to obstacles etc., two instrument stations are used so that they are in the same vertical plane as the elevated object (Fig. 15.5).

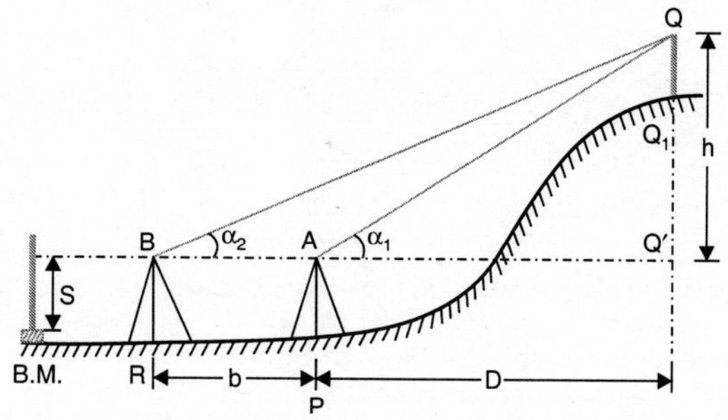

Fig. 15.5 Instrument axes at the same level.

Procedure

1. Set up the theodolite at P and level it accurately with respect to the altitude bubble.
2. Direct the telescope towards Q and bisect it accurately. Clamp both the plates. Read the vertical angle α_1.
3. Transit the telescope so that the line of sight is reversed. Mark the second instrument station R on the ground. Measure the distance RP accurately.
   Repeat steps (2) and (3) for both face observations. The mean values should be adopted.
4. With the vertical vernier set to zero reading, and the altitude bubble in the centre of its run, take the reading on the staff kept at the nearby B.M.
5. Shift the instrument to R and set up the theodolite there. Measure the vertical angle α_2 to Q with both face observations.
6. With the vertical vernier set to zero reading, and the altitude bubble in the centre of its run, take the reading on the staff kept at the nearby B.M.
   In order to calculate the R.L. of Q, we will consider three cases:
   (*a*) when the instrument axes at A and B are at the same level.
   (*b*) when they are at different levels but the difference is small, and
   (*c*) when they are at very different levels.
   (*a*) *Instrument axes at the same level (Fig. 15.5)*

Let $h = QQ'$

α_1 = angle of elevation from A to Q

α_2 = angle of elevation from B to Q

S = staff reading on B.M., taken from both A and B, the reading being the same in both the cases

b = horizontal distance between the instrument stations

D = horizontal distance between P and Q

From triangle AQQ', $h = D \tan \alpha_1$ ...(1)

From triangle BQQ', $h = (b + D) \tan \alpha_2$ ...(2)

Equating (1) and (2), we get

$$D \tan \alpha_1 = (b + D) \tan \alpha_2 \quad \text{or} \quad D(\tan \alpha_1 - \tan \alpha_2) = b \tan \alpha_2$$

or $$D = \frac{b \tan \alpha_2}{\tan \alpha_1 - \tan \alpha_2} \qquad ...(15.2)$$

$\therefore$ $$h = D \tan \alpha_1 = \frac{b \tan \alpha_1 \tan \alpha_2}{\tan \alpha_1 - \tan \alpha_2} = \frac{b \sin \alpha_1 \sin \alpha_2}{\sin (\alpha_1 - \alpha_2)} \qquad ...(15.3)$$

$$\text{R.L. of } Q = \text{R.L. of B.M.} + S + h .$$

(b) Instrument axes at different levels (Figs. 15.6 and 15.7)

Figs. 15.6 and 15.7 illustrate the cases, when the instrument axes are at different levels. If S_1 and S_2 are the corresponding staff readings on the staff kept at B.M., the difference in levels of the instrument axes will be either $(S_2 - S_1)$ if the axis at B is higher or $(S_1 - S_2)$ if the axis at A is higher. Let Q' be the projection of Q on horizontal line through A and Q'' be the projection on horizontal line through B. Let us derive the expressions for Fig. 15.6 when S_2 is greater than S_1.

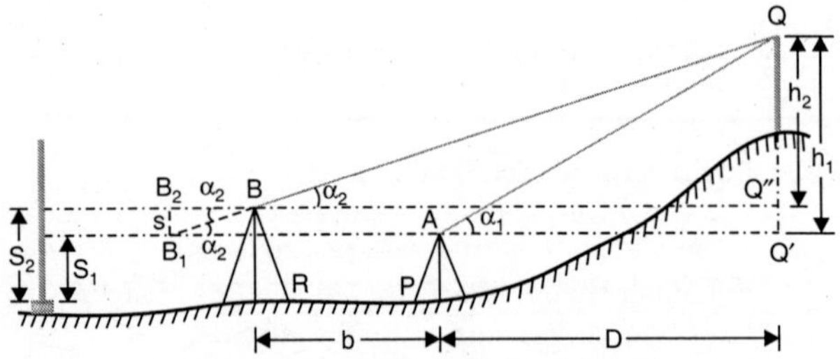

Fig. 15.6 Instrument at Different Levels.

From triangle QAQ', $h_1 = D \tan \alpha_1$ ...(1)

From triangle BQQ'', $h_2 = (b + D) \tan \alpha_2$ ...(2)

Subtracting (2) from (1), we get

$$(h_1 - h_2) = D \tan \alpha_1 - (b + D) \tan \alpha_2$$

But $h_1 - h_2$ = difference in level of instrument axes $= S_2 - S_1 = s$ (say)

$\therefore$ $$s = D \tan \alpha_1 - b \tan \alpha_2 - D \tan \alpha_2$$

or $$D (\tan \alpha_1 - \tan \alpha_2) = s + b \tan \alpha_2$$

or $$D = \frac{s + b \tan \alpha_2}{\tan \alpha_1 - \tan \alpha_2} = \frac{(b + s \cot \alpha_2) \tan \alpha_2}{\tan \alpha_1 - \tan \alpha_2} \qquad ...[15.4\ (a)]$$

Now $$h_1 = D \tan \alpha_1$$

$\therefore$ $$h_1 = \frac{(b + s \cot \alpha_2) \tan \alpha_1 \tan \alpha_2}{\tan \alpha_1 - \tan \alpha_2} = \frac{(b + s \cot \alpha_2) \sin \alpha_1 \sin \alpha_2}{\sin (\alpha_1 - \alpha_2)} \qquad ...[15.5\ (a)]$$

Expression 15.4 (*a*) could also be obtained by producing the lines of sight BQ backwards to meet the line $Q'A$ in B_1. Drawing B_1B_2 as vertical to meet the horizontal line $Q''B$ in B_2, it is clear that with the same angle of elevation if the instrument axis were at B_1, the instrument axes in both the cases would have been at the same elevation. *Hence the distance at which the axes are at the same level is* $AB_1 = b + BB_2 = b + s \cot \alpha_2$. Substituting this value of the distance between the instrument stations in equation 15.2, we get

$$D = \frac{(b + s \cot \alpha_2) \tan \alpha_2}{\tan \alpha_1 - \tan \alpha_2}$$ which is the same as equation [15.4 (*a*)].

Proceeding on the same lines for the case of Fig. 15.7, where the instrument axis at D is higher, it can be proved that

$$D = \frac{(b - s \cot \alpha_2) \tan \alpha_2}{\tan \alpha_1 - \tan \alpha_2} \quad ...[15.4\ (b)]$$

and $$h_1 = \frac{(b - s \cot \alpha_2) \sin \alpha_1 \sin \alpha_2}{\sin (\alpha_1 - \alpha_2)} \quad ...[15.5\ (b)]$$

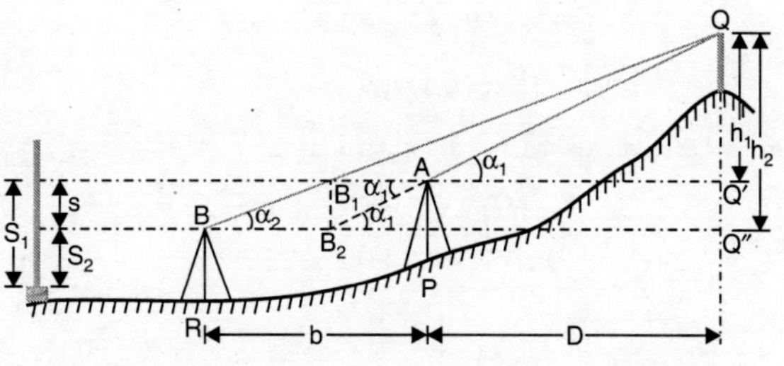

Fig. 15.7 Instrument axes at different levels.

Thus, the general expressions for D and h_1 can be written as

$$D = \frac{(b \pm s \cot \alpha_2) \tan \alpha_2}{\tan \alpha_1 - \tan \alpha_2} \quad ...(15.4)$$

and $$h_1 = \frac{(b \pm s \cot \alpha_2) \sin \alpha_1 \sin \alpha_2}{\sin (\alpha_1 - \alpha_2)} \quad ...(15.5)$$

Use + sign with $s \cot \alpha_2$ when the instrument axis at A is lower and – sign when it is higher than at B.

$$\text{R.L. of } Q = \text{R.L. of B.M.} + S_1 + h_1$$

(c) Instrument axes at very different levels: If $S_2 - S_1$ or s is too great to be measured on a staff kept at the B.M., the following procedure is adopted (Figs. 15.8 and 15.9):

1. Set the instrument at P (Fig. 15.8), level it accurately with respect to the altitude bubble and measure the angle α_1 to the point Q.
2. Transit the telescope and establish a point R at a distance b from P.
3. Shift the instrument to R. Set the instrument and level it with respect to the altitude bubble, and measure the angle α_2 to Q.
4. Keep a vane of height r at P (or a staff) and measure the angle to the top of the vane [or to the reading r if a staff is used (Fig. 15.9)].

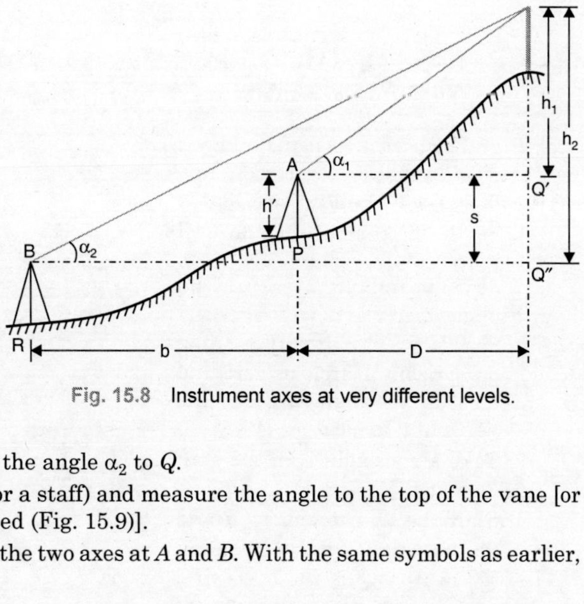

Fig. 15.8 Instrument axes at very different levels.

Let s = Difference in level between the two axes at A and B. With the same symbols as earlier, we have

$$h_1 = D \tan \alpha_1 \qquad ...(1)$$

and $$h_2 = (b + D) \tan \alpha_2 \qquad ...(2)$$

Subtracting (1) from (2), we get

$$(h_2 - h_1) = s = (b + D) \tan \alpha_2 - D \tan \alpha_1$$

or $D (\tan \alpha_1 - \tan \alpha_2) = b \tan \alpha_2 - s$

$$\therefore \quad D = \frac{b \tan \alpha_2 - s}{\tan \alpha_1 - \tan \alpha_2} \qquad ...(3)$$

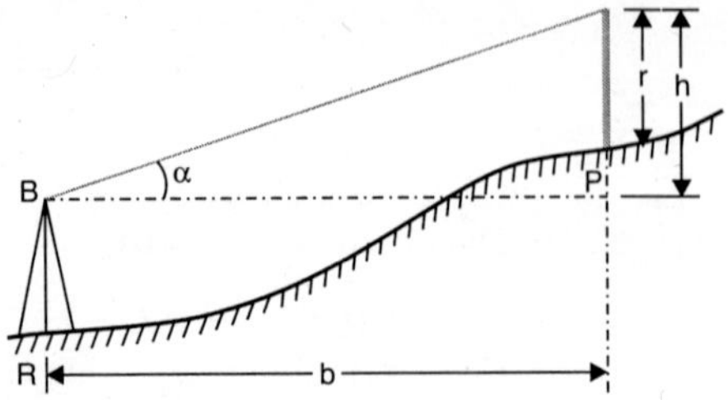

Fig. 15.9

and $$h_1 = D \tan \alpha_1 = \frac{(b \tan \alpha_2 - s) \tan \alpha_1}{\tan \alpha_1 - \tan \alpha_2} = \frac{(b - s \cot \alpha_2) \sin \alpha_1 \sin \alpha_2}{\sin(\alpha_1 - \alpha_2)} \qquad ...[15.5\ (b)]$$

From Fig. 15.9, we have

Height of station P above the axis at $B = h - r = b \tan \alpha - r$.

Height of axis at A above the axis at $B = s = b \tan \alpha - r + h'$

where h' is the height of the instrument at P.

Substituting this value of s in (3) and equation [15.5 (*b*)], we can get D and h_1

Now R.L. of Q = R.L. of $A + h_1$ = R.L. of $B + s + h_1$

= (R.L. of B.M. + backsight taken from B) + $s + h_1$

where $s = b \tan \alpha - r + h'$.

15.4 BASE OF THE OBJECT INACCESSIBLE: INSTRUMENT STATIONS NOT IN THE SAME VERTICAL PLANE AS THE ELEVATED OBJECT

Let P and R be the two instrument stations *not* in the same vertical plane as that of Q. The procedure is as follows:

1. Set the instrument at P and level it accurately with respect to the altitude bubble. Measure the angle of elevation α_1 to Q.
2. Sight the point R with reading on horizontal circle as zero and measure the angle RPQ_1, *i.e.* the horizontal angle θ_1 at P.
3. Take a backsight s on the staff kept at B.M.
4. Shift the instrument to R and measure α_2 and θ_2 there.

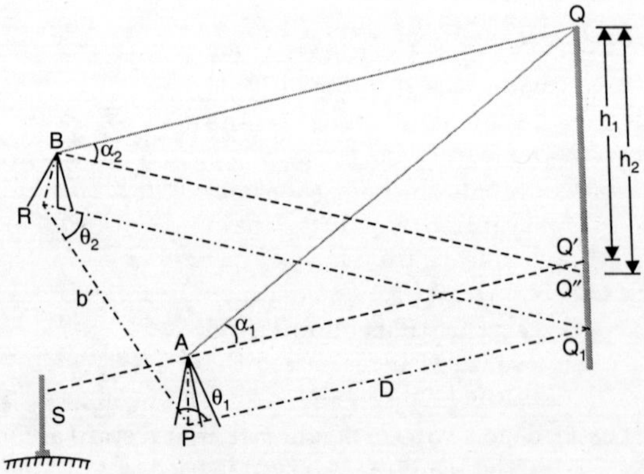

Fig. 15.10 Instrument and the object not in the same vertical plane.

In Fig. 15.10, AQ' is the horizontal line through A, Q' being the vertical projection of Q. Thus, AQQ' is a vertical plane. Similarly, BQQ'' is a vertical plane, Q'' being the vertical projection of Q on a horizontal line through B. PRQ_1 is a horizontal plane, Q_1 being the vertical projection of Q, and R vertical projection of B on a horizontal plane passing through P. θ_1 and θ_2 are the horizontal angles, and α_1 and α_2 are the vertical angles measured at A and B respectively.

From triangle AQQ', $\quad QQ' = h_1 = D \tan \alpha_1 \quad$ ...(1)

From triangle PRQ_1, $\quad \angle PQ_1R = 180° - (\theta_1 + \theta_2) = \pi - (\theta_1 + \theta_2)$

From the sine rule, $$\frac{PQ_1}{\sin \theta_2} = \frac{RQ_1}{\sin \theta_1} = \frac{RP}{\sin [\pi - (\theta_1 + \theta_2)]} = \frac{b}{\sin (\theta_1 + \theta_2)}$$

$$\therefore \quad PQ_1 = D = \frac{b \sin \theta_2}{\sin (\theta_1 + \theta_2)} \quad \text{...(2)}$$

and $$RQ_1 = \frac{b \sin \theta_1}{\sin (\theta_1 + \theta_2)} \quad \text{...(3)}$$

Substituting the value of D in (1), we get

$$h_1 = D \tan \alpha_1 = \frac{b \sin \theta_2 \tan \alpha_1}{\sin (\theta_1 + \theta_2)} \quad \text{...(15.6)}$$

$$\therefore \quad \text{R.L. of } Q = \text{R.L. of B.M.} + s + h_1$$

As a check, $$h_2 = RQ_1 \tan \alpha_2 = \frac{b \sin \theta_1 \tan \alpha_2}{\sin (\theta_1 + \theta_2)}$$

If a reading on B.M. is taken from B, the R.L. of Q can be known by adding h_2 to R.L. of B.

Example 15.1: *An instrument was set up at P and the angle of elevation to a vane 4 m above the foot of the staff held at Q was 9° 30′. The horizontal distance between P and Q was known to be 2000 metres. Determine the R.L. of the staff station Q, given that the R.L. of the instrument axis was 2650.38 m.*

Solution: Height of vane above the instrument axis

$$= D \tan \alpha = 2000 \tan 9° \, 30' = 334.68 \text{ metres.}$$

Correction for curvature and refraction $= \dfrac{6}{7} \dfrac{D^2}{2R}$

or $$C = 0.06728 \, D^2 \text{ metres, } D \text{ is in km}$$

$$= 0.06728 \left(\frac{2000}{1000}\right)^2 = 0.2691 \approx 0.27 \text{ m (+ve)}$$

Ht. of vane above inst. axis = 334.68 + 0.27 = 334.95 m

R.L. of vane = 334.95 + 2650.38 = 2985.33 m

R.L. of Q = 2985.33 – 4 = **2981.33 m.**

Example 15.2: *An instrument was set up at P and the angle of depression to a vane 2 m above the foot of the staff held at Q was 5° 36′. The horizontal distance between P and Q was known to be 3000 metres. Determine the R.L. of the staff station Q, given that staff reading on a B.M. of elevation 436.050 was 2.865 metres.*

Solution: The difference in elevation between the vane and the instrument axis

$$= D \tan \alpha = 3000 \tan 5° \, 36' = 294.152 \text{ m}$$

Combined correction due to curvature and refraction $= \dfrac{6}{7}\dfrac{D^2}{2R}$

or $\quad C = 0.06728\, D^2$ metres, when D is in km $= 0.06728\left(\dfrac{3000}{1000}\right)^2 = 0.606$ m

Since the observed angle is negative, combined correction due to curvature and refraction is subtractive.

Difference in elevation between the vane and the instrument axis

$$= 294.152 - 0.606 = 293.546 = h.$$

R.L. of instrument axis = 436.050 + 2.865 = 438.915 m

∴ R.L. of the vane = R.L. of instrument axis – h = 438.915 – 293.546 = **145.369**

∴ R.L. of Q = 145.369 – 2 = **143.369 m.**

Example 15.3: *In order to ascertain the elevation of the top (Q) of the signal on a hill, observations were made from two instrument stations P and R at a horizontal distance 100 metres apart, the stations P and R being in line with Q. The angles of elevation of Q at P and R were 28° 42′ and 18° 6′ respectively. The staff readings upon the bench mark of elevation 287.28 were respectively 2.870 and 3.750 when the instrument was at P and at R, the telescope being horizontal. Determine the elevation of the foot of the signal if the height of the signal above its base is 3 metres.*

Solution: (Fig. 15.6)

Elevation of instrument axis at P = R.L. of B.M. + staff reading

= 287.28 + 2.870 = 290.15 m

Elevation of instrument axis at R = R.L. of B.M. + staff reading

= 287.28 + 3.750 = 291.03 m

Difference in level of the instrument axis at the two stations

$= s = 291.03 - 290.15 = 0.88$ m

$$\alpha_1 = 28° \, 42' \text{ and } \alpha_2 = 18° \, 6'$$

$$s \cot \alpha_2 = 0.88 \cot 18° \, 6' = 2.69 \text{ m}$$

From equation [15.4 (a)], we have

$$D = \frac{(b + s \cot \alpha_2) \tan \alpha_2}{\tan \alpha_1 - \tan \alpha_2} = \frac{(100 + 2.69) \tan 18°6'}{\tan 28°42' - \tan 18°6'} = \mathbf{152.1 \text{ m.}}$$

∴ $h_1 = D \tan \alpha_1 = 152.1 \tan 28° \, 42' = 83.264$ m

∴ R.L. of foot of signal = R.L. of inst. axis at $P + h_1$ – height of signal

= 290.15 + 83.264 – 3 = **370.414 m**

Check : $(b + D) = 100 + 152.1 = 252.1$ m

∴ $h_2 = (b + D) \tan \alpha_2 = 252.1 \times \tan 18° \, 6' = 82.396$ m

∴ R.L. of foot of signal = R.L. of instrument axis at $R + h_2$ – height of signal

= 291.03 + 82.396 – 3 = **370.426 m.**

Example 15.4: *The top (Q) of a chimney was sighted from two stations P and R at very different levels, the stations P and R being in line with the top of the chimney. The angle of elevation from P to the top of the chimney was 38° 21′ and that from R to the top of the chimney was 21° 18′. The angle of the elevation from R to a vane 2 m above the foot of the staff held at P was 15° 11′. The heights of instrument at P and R were 1.87 m and 1.64 m respectively. The horizontal distance between P and R was 127 m and the reduced level of R was 112.78 m. Find the R.L. of the top of the chimney and the horizontal distance from P to the chimney.*

Solution: (Figs. 15.8 and 15.9)

When the observations were taken from R to P,

$$h = b \tan \alpha = 127 \tan 15° \; 11' = 34.47 \text{ m}$$

$$\text{R.L. of } P = \text{R.L. of } R + \text{height of instrument at } R + h - r$$

$$= 112.78 + 1.64 + 34.47 - 2 = 146.89 \text{ m}$$

$$\text{R.L. of instrument axis at } P = \text{R.L. of } P + \text{height of instrument at } P$$

$$= 146.89 + 1.87 = 148.76 \text{ m} \qquad \text{...}(i)$$

Difference in elevation between the instrument axes = s

$$= 148.76 - (112.78 + 1.64) = 34.34 \text{ m}$$

$$\therefore \quad D = \frac{(b \tan \alpha_2 - s)}{\tan \alpha_1 - \tan \alpha_2} = \frac{127 \tan 21°18' - 34.34}{\tan 38°21' - \tan 21°18'} = \frac{49.52 - 34.34}{0.79117 - 0.38988}$$

$$= 37.8 \text{ m}$$

$$h_1 = D \tan \alpha_1 = 37.8 \tan 38° \; 21' = 29.92 \text{ m}$$

$$\therefore \quad \text{R.L. of } Q = \text{R.L. of instrument axis at } P + h_1 = 148.76 + 29.92 = \mathbf{178.68 \text{ m}}$$

Check:

$$\text{R.L. of } Q = \text{R.L. of instrument axis at } R + h_2$$

$$= (112.78 + 1.64) + (b + D) \tan \alpha_2$$

$$= 114.42 + (127 + 37.8) \tan 21° \; 18' = 114.42 + 64.26 = \mathbf{178.68 \text{ m.}}$$

Example 15.5: *To find the elevation of the top (Q) of a hill, a flag-staff of 2 m height was erected and observations were made from two stations P and R, 60 metres apart. The horizontal angle measured at P between R and the top of the flag-staff was 60° 30′ and that measured at R between the top of the flag-staff and P was 68° 18′. The angle of elevation to the top of the flag-staff P was measured to be 10° 12′ at P. The angle of elevation to the top of the flag-staff was measured to be 10° 48′ at R. Staff readings on B.M. when the instrument was at P = 1.965 m and that with the instrument at R = 2.055 m. Calculate the elevation of the top of the hill if that of B.M. was 435.065 metres.*

Solution: (Fig. 15.10)

Given

$$b = 60 \text{ m}; \; \theta_1 = 60° \; 30'; \; \theta_2 = 68° \; 18'; \; \alpha_1 = 10° \; 12'; \; \alpha_2 = 10° \; 18'$$

$$\therefore \quad PQ_1 = D = \frac{b \sin \theta_2}{\sin (\theta_1 + \theta_2)}$$

and

$$h_1 = D \tan \alpha_1 = \frac{b \sin \theta_2 \tan \alpha_1}{\sin (\theta_1 + \theta_2)} = \frac{60 \sin 68°18' \tan 10°12'}{\sin (60°30' + 68°18')} = 12.87 \text{ m}$$

$\therefore$ R.L. of Q = (R.L. of instrument axis at P) + h_1 = (435.065 + 1.965) + 12.87

= **449.900 m.**

Check :
$$h_2 = \frac{b \sin \theta_1 \tan \alpha_2}{\sin (\theta_1 + \theta_2)} = \frac{60 \sin 60° \, 30' \tan 10° \, 48'}{\sin (60° \, 30' + 68° \, 18')} = 12.78 \text{ m.}$$

$\therefore$ R.L. of Q = R.L. of instrument axis at $R + h_2$ = (435.065 + 2.055) +12.78

= **449.9 m.**

15.5 DETERMINATION OF HEIGHT OF AN ELEVATED OBJECT ABOVE THE GROUND WHEN ITS BASE AND TOP ARE VISIBLE BUT NOT ACCESSIBLE

(a) Base line horizontal and in line with the object: Let A and B be the two instrument stations, b apart. The vertical angles measured at A are α_1 and α_2, and those at B are β_1 and β_2, corresponding to the top (E) and bottom (D) of the elevated object. Let us take a general case of instruments at different heights, the difference being equal to s.

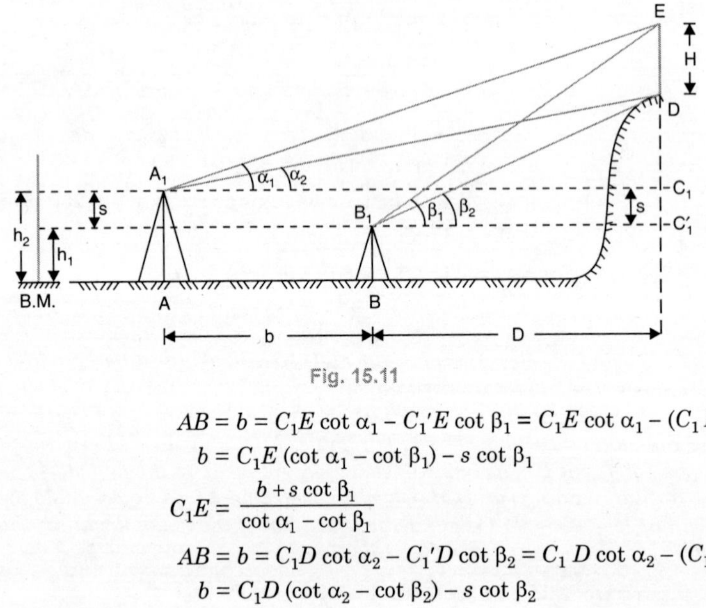

Fig. 15.11

Now
$$AB = b = C_1E \cot \alpha_1 - C_1'E \cot \beta_1 = C_1E \cot \alpha_1 - (C_1E + s) \cot \beta_1$$

$\therefore$
$$b = C_1E (\cot \alpha_1 - \cot \beta_1) - s \cot \beta_1$$

or
$$C_1E = \frac{b + s \cot \beta_1}{\cot \alpha_1 - \cot \beta_1} \quad ...(1)$$

Also,
$$AB = b = C_1D \cot \alpha_2 - C_1'D \cot \beta_2 = C_1 D \cot \alpha_2 - (C_1D + s) \cot \beta_2$$

or
$$b = C_1D (\cot \alpha_2 - \cot \beta_2) - s \cot \beta_2$$

or
$$C_1D = \frac{b + s \cot \beta_2}{\cot \alpha_2 - \cot \beta_2}$$

$\therefore$
$$H = C_1E - C_1D = \frac{b + s \cot \beta_1}{\cot \alpha_1 - \cot \beta_1} - \frac{b + s \cot \beta_2}{\cot \alpha_2 - \cot \beta_2} \quad ...(15.7)$$

If heights of the instruments at A and B are equal, $s = 0$

$\therefore$
$$H = b\left[\frac{1}{\cot \alpha_1 - \cot \beta_1} - \frac{1}{\cot \alpha_2 - \cot \beta_2}\right] \quad ...[15.7\ (a)]$$

Horizontal distance of the object from B

$$EC_1' = D \tan \beta_1 \quad \text{and} \quad DC_1' = D \tan \beta_2$$

$\therefore$

$$EC_1' - DC_1' = H = D (\tan \beta_1 - \tan \beta_2)$$

or

$$D = \frac{H}{\tan \beta_1 - \tan \beta_2} \qquad \text{...[15.7 (b)]}$$

where H is given by Eq. 15.7.

(*b*) **Base line horizontal but not in line with the object:** Let A and B be two instrument stations, distant b. Let α_1 and α_2 be the vertical angles measured at A, and β_1 and β_2 be the vertical angle measured at B, to the top (*E*) and bottom (*D*) of the elevated object. Let θ and φ be the horizontal angles measured at A and B respectively.

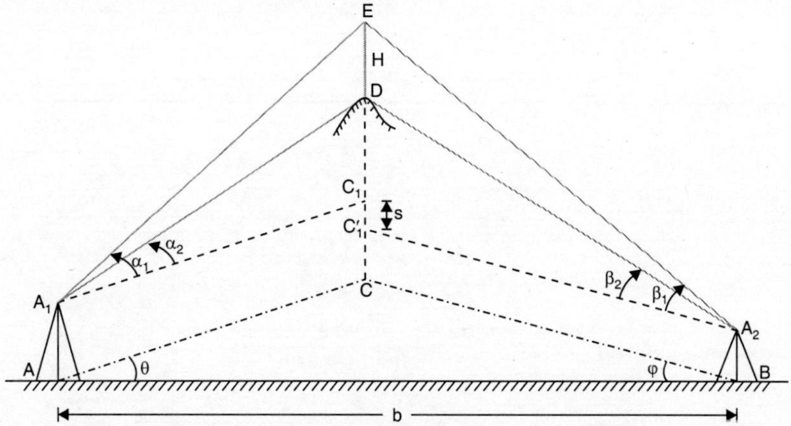

Fig. 15.12

From triangle ACB,

$$\frac{AC}{\sin \varphi} = \frac{BC}{\sin \theta} = \frac{AB}{\sin (180° - \theta - \varphi)}$$

$\therefore$

$$AC = b \sin \varphi \operatorname{cosec} (\theta + \varphi)$$

and

$$BC = b \sin \theta \operatorname{cosec} (\theta + \varphi)$$

Now

$$H = ED = A_1C_1 (\tan \alpha_1 - \tan \alpha_2) = AC (\tan \alpha_1 - \tan \alpha_2)$$

or

$$H = b \sin \varphi \operatorname{cosec} (\theta + \varphi) (\tan \alpha_1 - \tan \alpha_2) \qquad \text{...[15.8 (a)]}$$

Similarly,

$$H = ED = BC_1' (\tan \beta_1 - \tan \beta_2) = BC (\tan \beta_1 - \tan \beta_2)$$

or

$$H = b \sin \theta \operatorname{cosec} (\theta + \varphi) (\tan \beta_1 - \tan \beta_2) \qquad \text{...[15.8 (b)]}$$

15.6 DETERMINATION OF ELEVATION OF AN OBJECT FROM ANGLES OF ELEVATION FROM THREE INSTRUMENT STATIONS IN ONE LINE

Let A, B, C be three instrument stations in one horizontal line, with instrument axes at the same height. Let E' be the projection of E on the horizontal plane through ABC, and let

$EE' = h$. Let α, β and γ be the angles of elevation of the object E, measured from instruments at A, B and C respectively. Also let $AB = b_1$ and $BC = b_2$, be the measured horizontal distances.

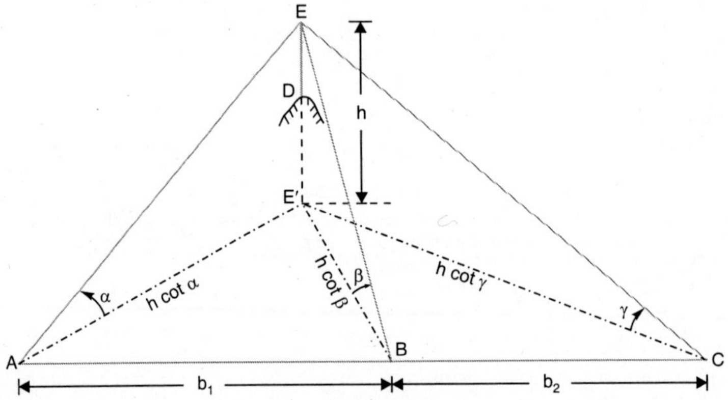

Fig. 15.13

From triangle $AE'B$, we have from *cosine rule*

$$\cos\varphi = \frac{h^2\cot^2\alpha + b_1^2 - h^2\cot^2\beta}{2b_1 h\cot\alpha} \quad ...(1) \qquad ...(15.9)$$

Also, from triangle $AE'C$, $\cos\varphi = \dfrac{h^2\cot^2\alpha + (b_1+b_2)^2 - h^2\cot^2\gamma}{2(b_1+b_2)h\cot\alpha}$ ...(2)

Equating (1) and (2), $\dfrac{h^2\cot^2\alpha + b_1^2 - h^2\cot^2\beta}{2b_1 h\cot\alpha} = \dfrac{h^2\cot^2\alpha + (b_1+b_2)^2 - h^2\cot^2\gamma}{2(b_1+b_2)h\cot\alpha}$

or $\quad (b_1 + b_2)\,[h^2(\cot^2\alpha - \cot^2\beta) + b_1^2] = b_1[h^2(\cot^2\alpha - \cot^2\gamma) + (b_1 + b_2)^2]$

or $\quad h^2\,[(b_1 + b_2)(\cot^2\alpha - \cot^2\beta) - b_1(\cot^2\alpha - \cot^2\gamma)] = b_1(b_1 + b_2)^2 - b_1^2(b_1 + b_2)$

or
$$h^2 = \frac{(b_1+b)[b_1(b_1+b_2) - b_1^2]}{(b_1+b_2)(\cot^2\alpha - \cot^2\beta) - b_1(\cot^2\alpha - \cot^2\gamma)}$$

$$= \frac{(b_1+b_2)b_1 b_2}{(b_1+b_2)(\cot^2\alpha - \cot^2\beta) - b_1(\cot^2\alpha - \cot^2\gamma)}$$

or
$$h = \left[\frac{b_1 b_2(b_1+b_2)}{b_1(\cot^2\gamma - \cot^2\beta) + b_2(\cot^2\alpha - \cot^2\beta)}\right]^{1/2} \qquad ...(15.10)$$

If $\quad b_1 = b_2 = b$

$$h = \frac{\sqrt{2}\,b}{(\cot^2\gamma - 2\cot^2\beta + \cot^2\alpha)^{1/2}} \qquad ...[15.10\ (a)]$$

Example 15.6: *Determine the height of a pole above the ground on the basis of following angles of elevation from two instrument stations A and B, in line with the pole.*

Angles of elevation from A to the top and bottom of pole: 30° and 25° Angles of elevation from B to the top and bottom of pole: 35° and 29°

Horizontal distance AB = 30 m.

The readings obtained on the staff at the B.M. with the two instrument settings are 1.48 and 1.32 m respectively.

What is the horizontal distance of the pole from A?

Solution: (Refer Fig. 15.11)

$$s = 1.48 - 1.32 = 0.16 \text{ m}$$

$$b = 30 \text{ m}; \alpha_1 = 30°; \alpha_2 = 25°; \beta_1 = 35°; \beta_2 = 29°$$

Substituting the values in Eq. 15.7.

$$H = \frac{b + s\cot\beta_1}{\cot\alpha_1 - \cot\beta_1} - \frac{b + s\cot\beta_2}{\cot\alpha_2 - \cot\beta_2}$$

$$= \frac{30 + 0.16\cot 35°}{\cot 30° - \cot 35°} - \frac{30 + 0.16\cot 29°}{\cot 25° - \cot 29°}$$

$$= 99.47 - 88.96 = \mathbf{10.51\ m}$$

Also, $$D = \frac{H}{\tan\beta_1 - \tan\beta_2} = \frac{10.51}{\tan 35° - \tan 29°} = 72.04 \text{ m}$$

∴ Distance of pole from $A = b + D = 30 + 72.04 =$ **102.04 m.**

Example 15.7: *A, B and C are stations on a straight level line of bearing 110° 16′48″. The distance AB is 314.12 m and BC is 252.58 m. With instrument of constant height of 1.40 m, vertical angles were successively measured to an inaccessible up station E as follows:*

At A: 7° 13′40″

At B: 10° 15′00″

At C: 13° 12′10″

Calculate (a) the height of station E above the line ABC

(b) the bearing of the line AE

(c) the horizontal distance between A and E.

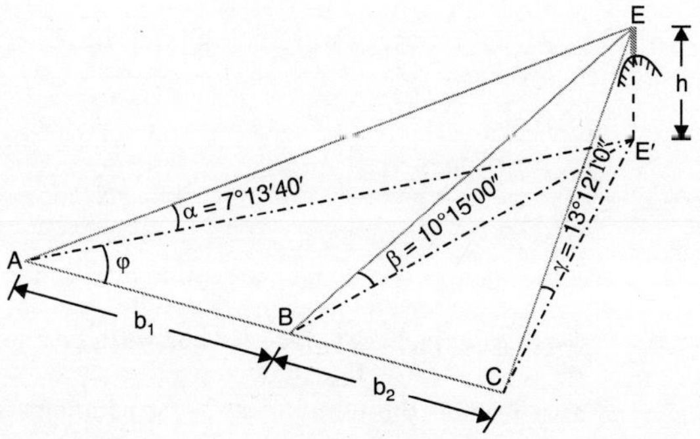

Fig. 15.14

Solution: Refer Fig. 15.14.

Given: $\alpha = 7° 13'40''$;

$\beta = 10° 15' 00''$;

$\gamma = 13° 12' 10''$;

$b_1 = 314.12$ m;

and $b_2 = 252.58$ m

Substituting the values in Eq. 15.10, we get

$$EE' = h = \left[\frac{b_1 b_2 (b_1 + b_2)}{b_1(\cot^2\gamma - \cot^2\beta) + b_2(\cot^2\alpha - \cot^2\beta)}\right]^{1/2}$$

$$= \left[\frac{314.12 \times 252.58(314.12 + 252.58)}{314.12\,(\cot^2 13^\circ 12'10'' - \cot^2 10^\circ 15'00'') + 252.58(\cot^2 7^\circ 13'40'' - \cot^2 10^\circ 15'00'')}\right]^{1/2}$$

$= 104.97$ m

$\therefore$ Height of E above ABC = 104.97 + 1.4 = **106.37 m**

Also, From Eq. 15.9.

$$\cos\varphi = \frac{h^2(\cot^2\alpha - \cot^2\beta) + b_1^2}{2b_1 h \cot\alpha}$$

$$= \frac{(104.97)^2(\cot^2 7^\circ 13'40'' - \cot^2 10^\circ 15'00'') + (314.12)^2}{2 \times 314.12 \times 104.97 \cot 7^\circ 13'40''}$$

$= 0.859205$

or $\varphi = 30^\circ\ 46'21''$

Hence bearing of AE = 110° 16′48″ – 30° 46′21″

= **79° 30′27″**

Length $AE' = h \cot\alpha$ = 104.97 cot 7° 13′40″

= **827.70 m.**

PROBLEMS

1. A theodolite was set up at a distance of 200 m from a tower. The angle of elevations to the top of the parapet was 8° 18′ while the angle of depression to the foot of the wall was 2° 24′. The staff reading on the B.M. of R.L. 248.362 with the telescope horizontal was 1.286 m. Find the height of the tower and the R.L. of the top of the parapet.
2. To determine the elevation of the top of a flag-staff, the following observations were made:

Inst. station	*Reading on B.M.*	*Angle of elevation*	*Remarks*
A	1.266	10° 48′	R.L. of B.M. = 248.362
B	1.086	7° 12′	

Stations A and B and the top of the aerial pole are in the same vertical plane.
Find the elevation of the top of the flag-staff, if the distance between A and B is 50 m.

3. Find the elevation of the top of a chimney from the following data:

Inst. station	*Reading on B.M.*	*Angle of elevation*	*Remarks*
A	0.862	18° 36′	R.L. of B.M. = 421.380 m
B	1.222	10° 12′	Distance AB = 50 m

Stations A and B and the top of the chimney are in the same vertical plane.

4. The top (Q) of a chimney was sighted from two stations P and R at very different levels, the stations P and R being in line with the top of the chimney. The angle of elevation from P to the top of chimney was 36° 12′ and that from R to the top of the chimney was 16° 48′. The angle of elevation from R to a vane 1 m above the foot of the staff held at P was 8° 24′. The heights of instrument at P and R were 1.85 m and 1.65 m respectively. The horizontal distance between P and R was 100 m and the R.L. of R was 248.260 m. Find the R.L. of the top of the chimney and the horizontal distance from P to the chimney.

ANSWERS

1. 37.558 m; 278.824 m
2. 267.796 m
3. 442.347 m
4. 290.336; 33.9 m

16 CHAPTER

PERMANENT ADJUSTMENTS OF LEVELS

16.1 INTRODUCTION

Permanent adjustments consist in setting essential parts into their true positions relatively to each other. Accurate work can often be done with an instrument out of adjustment, provided certain special methods eliminating the errors are followed. Such special methods involve more time and extra labour. Almost all surveying instruments, therefore, require certain field adjustments from time to time.

Method of Reversion

The principle of reversion is very much used in all adjustments. By reversing the instrument or part of it, the error becomes apparent. The magnitude of apparent error is double the true error because reversion simply places the error as much to one side as it was to the opposite side before reversion. Example may be taken of a set square, the two sides of which have an error e in perpendicularity (Fig. 16.1). By reversing the set square, the apparent error becomes $2e$.

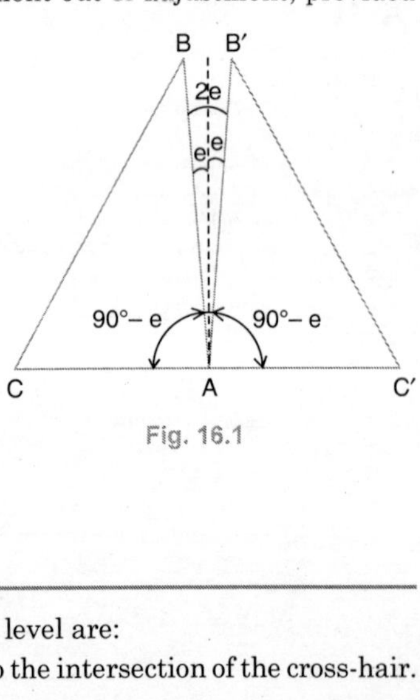

Fig. 16.1

16.2 ADJUSTMENTS OF DUMPY LEVEL

(*a*) **The Principal lines:** The principal lines in a dumpy level are:

(*i*) The line of sight joining the centre of the objective to the intersection of the cross-hair.

(*ii*) Axis of the level tube.

(*iii*) The vertical axis.

(*b*) **Conditions of Adjustments:** The requirements that are to be established are:

(*i*) The axis of the bubble tube should be perpendicular to the vertical axis (Adjustment of the level tube).

(*ii*) The horizontal cross-hair should lie in a plane perpendicular to the vertical axis (Adjustment of cross-hair ring).

(*iii*) The line of collimation of the telescope should be parallel to the axis of the bubble tube (Adjustment of line of sight).

(*c*) **Adjustments**

(I) *Adjustment of Level Tube*

1. ***Desired Relation:*** The axis of the bubble tube should be perpendicular to the vertical axis when the bubble is central.
2. ***Object:*** The object of the adjustment is to make the vertical axis truly vertical so as to ensure that once the instrument is levelled up (see temporary adjustments), the bubble will remain central in all directions of sighting.
3. ***Necessity:*** Once the requirement is accomplished the bubble will remain central for all directions of pointing of telescope. This is necessary merely for convenience in using the level.
4. ***Test:***
   (*i*) Set the instrument on firm ground. Level the instrument in the two positions at right angles to each other as the temporary adjustment.
   (*ii*) When the telescope is on third foot screw, turn the telescope through 180°.
   (*iii*) If the bubble remains central, the adjustment is correct. If not, it requires adjustment.
5. ***Adjustment:***
   (*i*) Bring the bubble half way back by the third foot screw.
   (*ii*) Bring the bubble through the remaining distance to centre by turning the capston nuts at the end of the level tube.
   (*iii*) Repeat the test and adjustment until correct.
6. ***Principle involved:*** This is the case of single reversion in which the apparent error is double the true error. Referring to Fig. 16.2, (90° – e) is the angle between the vertical axis and the axis of the bubble tube. When the bubble is centred, the vertical axis makes an angle e with the true vertical. When the bubble is reversed, axis of the bubble tube is displaced by an angle $2e$.

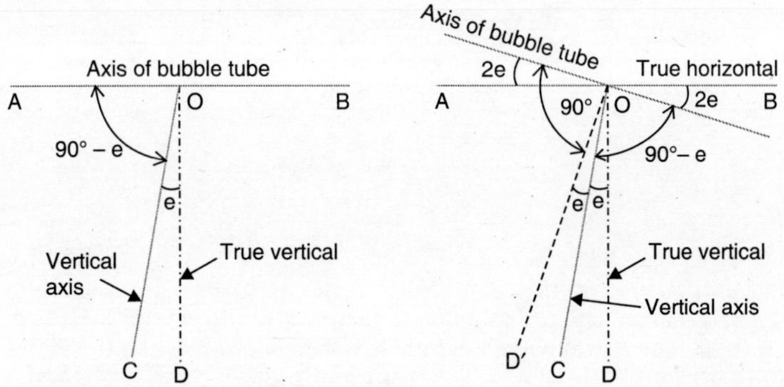

(a) First position of bubble tube

(b) Position after reversal

Fig. 16.2

Fig. 16.3 explains clearly how the principle of reversion has been applied to the adjustment. In Fig. 16.3 (*a*), the bubble tube is attached to the plate *AB* with unequal supports *x* and *y* so that the bubble is in the centre even when the plate *AB* is inclined and, therefore, the vertical axis of the instrument is also inclined. *uw* represents the axis of the bubble tube which coincides with the horizontal *wh*. *uv* represents a line parallel to *AB*, making an angle *e* with the axis of the bubble tube. If the plate *AB* is kept stationary and the bubble tube is lifted off and turned end for end, as shown in Fig. 16.3 (*b*), the bubble will go to the left hand end of the tube. In this position, the axis of the bubble tube *uw* still makes an angle *e* with the line *uv*, but in the downward direction. The axis of the bubble tube has, therefore, been turned through an angle $(e + e) = 2e$ from *uh*. In order to coincide the axis *uw* of the bubble tube with line *uv*, bring the bubble half way towards the centre by making the supports *y* and *x* equal (by capstan screws). The axis of the bubble has thus been made parallel to the plate *AB*, but the bubble is not yet in the centre and the line *AB* is still inclined to the horizontal [Fig. 16.3 (*c*)]. In order to make *AB* horizontal (and to make the vertical axis truly vertical), use the foot screw till the bubble comes in the centre. Fig. 16.3 (*d*) shows this condition in which *x* and *y* are of equal lengths, the bubble is central and the vertical axis is truly vertical.

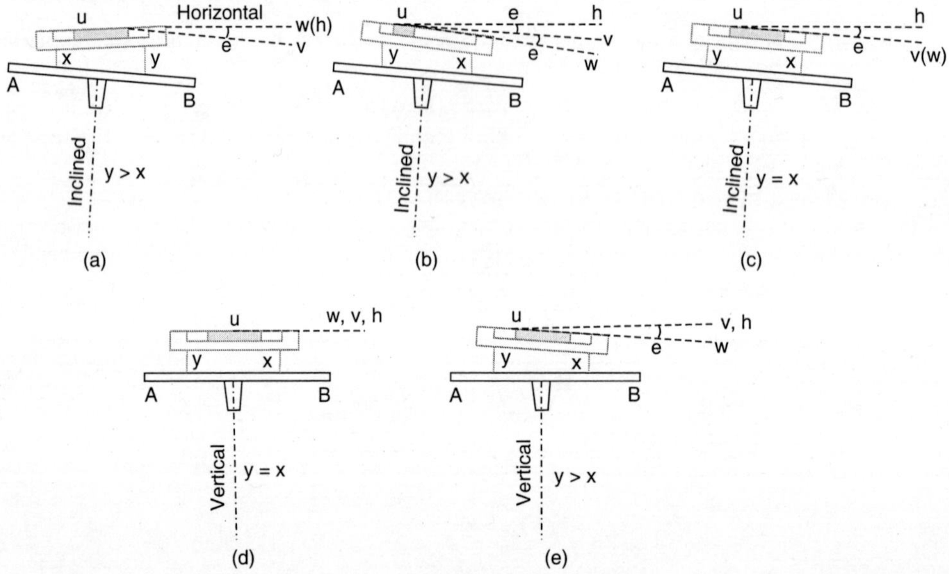

Fig. 16.3

Note 1. For ordinary work, this adjustment is not an essential requirement, but is made merely for the sake of convenience in using the level. *If adjustment No. III is perfect,* the line of sight will be truly horizontal when the bubble is central, even when the plate *AB* is inclined as shown in Fig. 16.3 (*a*). Now when the line of sight is directed towards the staff in any other direction, the bubble will go out of centre, which may be brought to centre by foot screws and the line of sight will be truly horizontal. The change in elevation of the line of sight so produced will be negligible for ordinary work. For subsequent pointings also, the bubble may be brought to centre similarly, at the expense of time and labour. *Thus the adjustment is not at all essential, but is desirable for speedy work and convenience.*

2. In Fig. 16.3 (*e*), it has been shown that if the bubble is brought half-way towards the centre by foot screws, the plate *AB* will be horizontal, but the axis of the bubble tube will be inclined and the line of sight will also be inclined if the instrument is otherwise correct. The vertical axis will, of course, be truly vertical.

(II) *Adjustment of Cross-Hair Ring*

(*i*) *Desired Relation*: The horizontal cross-hair should lie in a plane perpendicular to the vertical axis.

(*ii*) *Object*: The object of the adjustment is to ensure that the horizontal cross-hair lie in a plane perpendicular to the vertical axis.

(*iii*) *Necessity*: Once the desired relation is accomplished, the horizontal cross-hair will lie in horizontal plane, the bubble being in the centre.

(*iv*) *Test*: (1) Sight a well defined object *A* (about 60 m away) at one of the horizontal hair. (2) Rotate the end level slowly about its spindle until the point *A* is traced from one end of the hair to the other hair. (3) If the point does not deviate from the hair, the adjustment is correct. If it deviates, the adjustment is out.

(*v*) *Adjustment*: Loose the capstan screws of the diaphragm and turn it slightly until by further trial the point appears to travel along the horizontal hair. Refer Fig. 16.4.

Note: It is not necessary to level the instrument when the test is carried out.

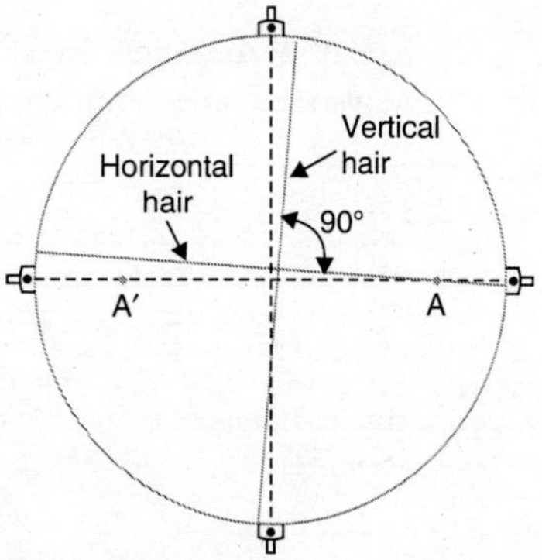

Fig. 16.4

(III) *Adjustment of Line of Collimation: (Two-peg Test)*

(*i*) *Desired Relation:* The line of collimation of the telescope should be parallel to the axis of the bubble tube.

(*ii*) *Necessity:* Once the desired relation is accomplished, line of sight will be horizontal when the bubble is in the centre, regardless of the direction in which the telescope is pointed. This adjustment is very necessary, and is of prime importance, since the whole function of the level is to provide horizontal line of sight.

(*iii*) *Test and Adjustment:*

Two-peg Test: Method A (Refer Fig. 16.5.)

1. Choose two points *A* and *B* on fairly level ground at a distance of about 100 or 120 metres. Set the instrument at a point *C*, very near to *A*, in such a way that the eye-piece almost touches the staff kept at *A*.
2. With the staff kept at *A*, take the reading through the objective. The cross-hairs will not be visible, but the field of view will be very small so that its centre may be determined keeping a pencil point on the staff. The reading so obtained may be called true rod reading with sufficient precision. Sight the rod kept at the point *B* and take the staff reading. Take the difference of the two staff readings, which will give the apparent difference in elevation.

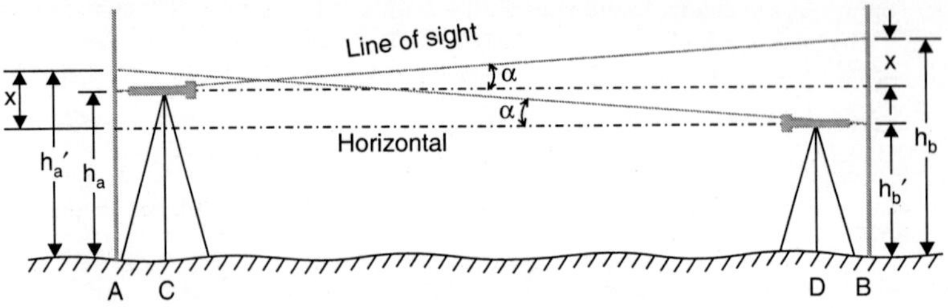

Fig. 16.5 Two peg test (Method *A*).

Apparent difference in elevation = $h = h_a - h_b$. ...(1)

3. Move the instrument to a point *D*, very near to *B* and set it so that the eye-piece almost touches the staff kept at *B*.
4. Sighting through the objective, take the reading on the staff kept at *B*. Read the staff kept at the point *A*. Find the difference of the two readings, thus getting another apparent difference in elevation.

h' = Apparent differences in elevation = $h_a' - h_b'$. ...(2)

5. If the two apparent differences in elevation, calculated in steps (3) and (4) are the same, the instrument is in adjustment. If not, it requires adjustment.
6. Calculate the correct difference in elevation, as in the case of reciprocal levelling.

$$H = \text{Correct difference in elevation} = \frac{(h_a - h_b) + (h_a' - h_b')}{2} \quad ...(3)$$

(If *H* comes out positive, point *B* is higher than point *A* and if *H* comes negative, point *B* is lower than point *A*).

7. Knowing the correct difference in elevation between the points, calculate the correct staff readings at the points when the instrument is at point *D* *if it were in adjustment.*

Correct staff reading at $A = (H + h_b')$

[Use proper algebraic sign for *H* from Eq. (3)]

8. Keep the staff at *A* and sight it through the instrument set up at *D*. Loose the capstan screws of diaphragm and raise or lower diaphragm so as to get the same staff reading as calculated in (7). The test is repeated for checking.

The line of sight will thus be perfectly horizontal.

Two-peg Test: Method B (Ref. Fig. 16.6)

1. Choose two points *A* and *B* on fairly level ground at a distance of about 90 or 100 metres. Set the instrument at a point *C*, exactly midway between *A* and *B*.
2. Keep the staff, in turn at *A* and *B*, and take the staff readings when the bubble is exactly centred.
3. Calculate the difference in elevation between the two points. It is to be noted that since point *C* is midway, the difference in the two staff readings will give the correct difference in elevation even if the line of sight is inclined.

Correct difference in elevation $H = h_a - h_b$.

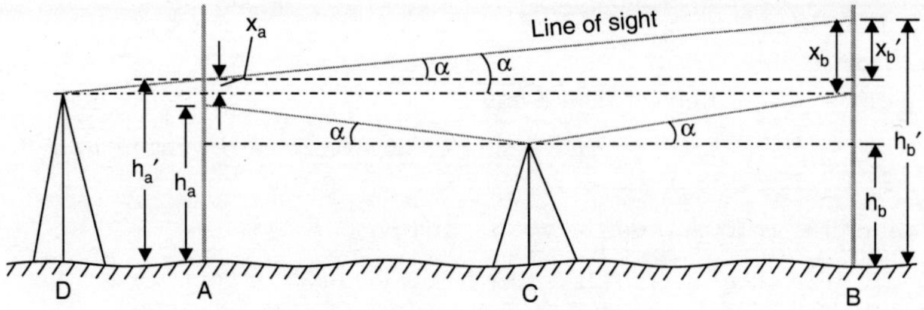

Fig. 16.6 Two-peg test (Method *B*).

4. Move the level and set it on a point *D*, about 20 to 25 metres from *A*, *preferably in line with the pegs.* Take the readings on the staff kept at *A* and *B*. Let the readings be h_a' and h_b' respectively.
5. Calculate difference in elevation between *A* and *B*, by the above staff readings. Thus $H' = h_a' - h_b'$. If the difference comes to be the same as found in (3), the instrument is in adjustment. If not, it requires adjustment.
6. The inclination of the line of sight in the net distance *AB* will be given by

$$\tan \alpha = \frac{H - H'}{AB} = \frac{(h_a - h_b) - (h_a' - h_b')}{AB}$$

The errors in the rod reading at *A* and *B* will be given numerically, by

$$x_a = \frac{(H - H')}{AB} DA \quad \text{and} \quad x_b = \frac{(H - H')}{AB} (DA + AB).$$

It is to be noted that, for positive values of H and H', the line of sight will be inclined upwards or downwards according as H' is lesser or greater than H. Similarly, for negative values of H and H', the line of sight will be inclined upwards or downwards according as H' is greater or lesser than H.

7. Calculate the correct reading at *B*, by the relation

$$h = h_b' \mp x_b.$$

Use + sign with the arithmetic value x_b if the line of sight is inclined downwards and use – sign with the arithmetic values of x_b if the line of sight is inclined upwards.

Loose the capstan screws of the diaphragm to raise or lower it (as the case may be) to get the correct reading *h* on the rod kept at *B*.

For the purpose of check, the correct reading at *A* can be calculated equal to $h_a' \mp x_a$ and seen whether the same staff reading is obtained after the adjustment.

Example 16.1: *A Dumpy level was set up at C exactly midway between two pegs A and B 100 metres apart. The readings on the staff when held on the pegs A and B were 2.250 and 2.025 respectively. The instrument was then moved and set up at a point D on the line BA produced, and 20 metres from A. The respective staff reading on A and B were 1.875 and 1.670. Calculate the staff readings on A and B to give a horizontal line of sight.*

Solution: (Fig. 16.6)

When the instrument is at C.

The difference in elevation between A and B

$$= H = 2.250 - 2.025 = 0.225 \text{ m}, B \text{ being higher.}$$

When the instrument is at D

Apparent difference in elevation between A and B

$$= H' = 1.875 - 1.670 = 0.205, B \text{ being higher.}$$

Since the apparent difference of level is not equal to the true difference, the line of collimation is out of adjustment.

∴ The inclination of the line of sight in the net distance AB will be

$$\tan \alpha = \frac{H - H'}{AB} = \frac{0.225 - 0.205}{100} = \frac{0.020}{100}$$

Since H' is lesser than H, the line of sight is inclined upwards.

$$\therefore \text{ Correct staff reading at } A = 1.875 - AD \tan \alpha = 1.875 - \frac{20 \times 0.020}{100} = \mathbf{1.871}$$

$$\text{and correct staff reading at } B = 1.670 - DB \tan \alpha = 1.670 - \frac{120 \times 0.020}{100} = \mathbf{1.646.}$$

Check: True difference in elevation = 1.871 – 1.646 = 0.225 m.

Example 16.2: *The following observations were made during the testing of a dumpy level:*

Instrument at	*Staff reading on*	
	A	*B*
A	*1.702*	*2.244*
B	*2.146*	*3.044*

Distance AB = 150 metres.

Is the instrument in adjustment? To what reading should the line of collimation be adjusted when the instrument was at B? If R.L. of A = 432.052 m, what should be the R.L. of B?

Solution: (Fig. 16.5)

When the instrument was at A:

Apparent difference in elevation between A and B

$$= 2.244 - 1.702 = 0.542, \quad B \text{ being lower.}$$

When the instrument was at B:

Apparent difference in elevation between A and B

$$= 3.044 - 2.146 = 0.898, \quad B \text{ being lower.}$$

∴ True difference in elevation between A and B

$$= \frac{0.542 + 0.898}{2} = 0.720 \text{ m}, B \text{ being lower.}$$

When the instrument is at B, the apparent difference in elevation is 0.898 and is more than the true difference. Hence in this case, the reading obtained at A is *lesser* than the true reading.

The line of sight is therefore inclined *downards* by an amount 0.898 – 0.720 = 0.178 m in a distance of 150 m.

Staff reading at A for collimation adjustment = 2.146 + 0.178 = **2.324 m**

Check: True difference in elevation = 3.044 – 2.324 = 0.720

∴ R.L. of B = 432.052 – 0.720 = **431.332 m.**

Example 16.3: *In a two peg test of a dumpy level, the following readings were taken:*

(*i*) *The instrument at C midway between A and B AB = 100 m*	*The staff reading on A = 1.682* *The staff reading on B = 1.320*
(*ii*) *The instrument near A*	*The staff reading on A = 1.528* *The staff reading on B = 1.178*

Is the line of collimation inclined upwards or downwards and how much? With the instrument of A, what should be the staff reading on B in order to place the line of collimation truly horizontal?

Solution: *When the instrument is at C*

True difference in level A and B = 1.682 – 1.320 = 0.362 m, A being higher.

When the instrument is near A:

Apparent difference in elevation = 1.528 – 1.178 = 0.350, B being higher.

Since the apparent difference in level is lesser than the true one, the staff reading at B is greater than the true one for this instrument position. The line of sight is, therefore, inclined *upwards.*

The amount of inclination = 0.362 – 0.350 = **0.012 m in 100 m**

Correct staff reading at B for collimation to be truly horizontal

= 1.178 – 0.012 = **1.166 m**

Check: True difference in level = 1.528 – 1.166 = 0.362 m

16.3 ADJUSTMENT OF TILTING LEVEL

(*a*) **Principal Lines:** The principal lines in a tilting level are:

(*i*) The line of sight and (*ii*) The axis of the level tube.

(*b*) **The Conditions of Adjustments:** The tilting level has greatest advantage over other levels as far as adjustments are concerned. Since it is provided with a tilting screw below the objective end of the telescope, it is not necessary to bring the bubble exactly in the centre of its run with the foot screw; the tilting screw may be used to bring the bubble in the centre for each sight. *Therefore, it is not essential for tilting level that the vertical axis should be truly vertical.* The only condition of the adjustment is that the line of collimation of the telescope should be parallel to the axis of bubble tube (adjustment of line of sight).

(*c*) **Adjustment of line of Sight**

(*i*) ***Desired Relation:*** The line of collimation of the telescope should be parallel to the axis of the bubble tube.

(*ii*) *Object:* The object of the test is to ensure that the line of sight rotates in horizontal plane when the bubble is central.

(*iii*) *Necessity:* The same as for dumpy level.

(*iv*) *Test and Adjustment:* (See Figs. 16.5 and 16.6)

The same methods are applied as for Dumpy level. In either of the methods, the correct staff reading is calculated and the line of sight is raised or lowered with the help of the tilting screw to read the calculated reading. By doing so, the bubble will go out of centre. The adjustable end of the bubble is, then, lifted or lowered till the bubble comes in the centre of the run.

The test is repeated till correct.

16.4 ADJUSTMENTS OF WYE LEVEL

(*a*) **Principle Lines:** The principal lines to be considered are:

(*i*) The line of sight. (*ii*) The axis of the collars.

(*iii*) The axis of the level tube. (*iv*) The vertical axis through the spindle of the level.

(*b*) **Conditions of Adjustment**

Case A: ***When the level tube is attached to the telescope, the following are the conditions of adjustment:***

(*i*) The line of sight should coincide with axis of the collars (adjustment of line of sight).

(*ii*) The axis of the level tube should be parallel to the line of sight and both of these should be in the same vertical plane (Adjustment of level tube).

(*iii*) The axis of the level tube should be perpendicular to the vertical axis.

Case B: ***When the level tube is on the stage under telescope***

(*i*) The line of sight should coincide with the axis of the collars (adjustment of line of sight).

(*ii*) The axis of the level tube should be perpendicular to the vertical axis.

(*iii*) The line of sight should be parallel to the axis of the level tube.

(*c*) **Adjustments of Wye Level**

Case A

1. **Adjustment of line of Sight**

(*i*) *Desired Relation:* The line of sight should coincide with the axis of the collars.

(*ii*) *Necessity:* The fulfilment of this condition of the adjustment is of prime importance. If the line of collimation does not coincide with the axis of the collars (or axis of wyes), when the telescope is rotated about its longitudinal axis, the line of sight will generate a cone and, therefore, the line of sight will be parallel to the axis of the bubble tube only in one particular position of the telescope in the wye.

(*iii*) *Test:*

1. Set the instrument and carefully focus a well-defined point at a distance of 50 to 100 metres.
2. Loose the clips and rotate the telescope through 180° about its longitudinal axis. Fasten the clips.
3. Sight the point again. If the line of sight strikes the same point, the instrument is in adjustment. If not, it requires adjustment.

(*iv*) *Adjustment:*

1. If both the hairs are off the point, adjust each by bringing it *halfway* back by the diaphragm screws.
2. Repeat the test on a different point till in the final test the intersection of the cross-hairs remains on the point *throughout* a complete revolution of the telescope.

(*v*) *Principle Involved:* The principle of single reversion has been used. Refer to Fig. 16.7(*a*). The line of sight is inclined by *e* upwards to the axis of the collars before the reversion. After the reversion, it is inclined by the same amount *e* downwards to the axis of the collars. The apparent error is, therefore, twice the actual error. Similar discussion will also hold good when the vertical hair is also either to the left or to the right of the true position [Fig. 16.7(*b*)].

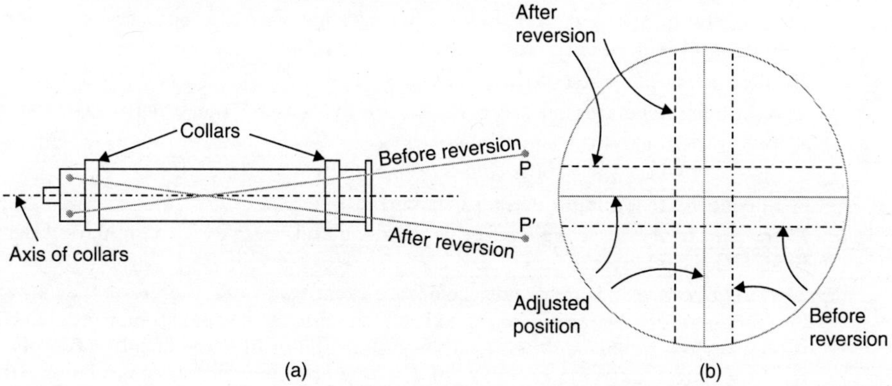

Fig. 16.7

(*vi*) *Notes*

1. It is not necessary to level the instrument so long as the wyes remain in the fixed position.
2. In a well made instrument, the optical axis of the instrument coincides with the axis of the collars. If it is not coincident, the defect can be remedied only by the makers.
3. Since both the hairs are to be adjusted in one single operation, the adjustment is to be done by trial-and-error so that error in both ways is adjusted by half the amount.
4. In order to test the accuracy of the objective focusing slide, the test should be repeated on a point very near the instrument, say 5 to 6 metres away. If the instrument is out of adjustment for this second point, either (*a*) the objective slide does not move parallel to the axis of the collars or (*b*) the optical axis does not coincide with the axis of collars. The objective slide should be adjusted if it is adjustable.

2. Adjustment of Level Tube

(*i*) *Desired Relation:* The axis of the level tube should be parallel to the line of sight and both of these should be in the same vertical plane.

(*ii*) *Necessity:* Once the desired relation is accomplished, the line of sight will be horizontal when the bubble is in the centre, regardless of the direction in which the telescope is pointed. This adjustment is very necessary, and of prime importance, since the whole function of the level is to provide horizontal line of sight.

(*iii*) *Test and Adjustment:* For both the axes to be in the same vertical plane:

1. Level the instrument carefully keeping the telescope parallel to two foot screws.
2. Turn the telescope slightly in the wyes about its longitudinal axis. If the bubble remains central, the instrument is in adjustment. If not, bring the bubble central by means of a small horizontal screw which controls the level tube laterally. Repeat the test till correct. It is to be noted that since no reversion is made, the whole error is to be adjusted by the horizontal screw.

(*iv*) *Test and Adjustment:* For both the axis to be parallel:

1. Level the instrument by keeping the telescope over two foot screws. Clamp the horizontal motion of the telescope.
2. Loose the clips, take out the telescope gently and replace it end for end.
3. If the bubble remains in the centre, it is in adjustment. If not, it requires adjustment.
4. To adjust it, loose the capstan screws of the level tube to raise or lower it, as the case may be till the bubble *comes halfway* towards the centre.
5. Repeat the test and adjustment till correct.

(*v*) *Principle involved:* Single reversion is done and, therefore, the apparent error is twice the actual error.

(*vi*) *Note:* The reversion is made with reference to the wyes and, therefore, the axis of the bubble tube is made parallel to the axis which joins the bottom of the wyes. However, the axis of the bubble tube may not be set parallel to the line of collimation by this test due to the following reasons: (*a*) The line of collimation may not be parallel to the axis of wyes if adjustment (I) is not correct, (*b*) Even if adjustment (I) is made first, the collars may not be true circles of equal diameter. This test is, therefore, not suitable in such cases. The test and adjustment can then be made by two-peg test method as in the case of dumpy level and the correction, if necessary, is made by the level tube adjusting screws.

3. Adjustment for Perpendicularity of Vertical Axis and Axis of Level Tube

(*i*) *Desired Relation:* The axis of the level tube should be perpendicular to the vertical axis.

(*ii*) *Necessity:* Once the requirement is accomplished, the bubble will remain central for all directions of pointing of the telescope.

(*iii*) *Test*

1. Centre the bubble in the usual manner.
2. Turn the telescope through 180° in horizontal plane. If the bubble does not remain central, the instrument requires adjustment.

(*iv*) *Adjustment*

1. Bring the bubble halfway back by foot screws and half by raising or lowering one wye relative to the other by means of screws which join the base of the wye to the stage.
2. Repeat the test and adjustment till correct.

Case B

1. **Adjustment of Line of Sight:** Same as for case A.
2. **Adjustment for the Perpendicularity of the Vertical Axis and Level Tube:** The test is done in the same way as adjustment (*iii*) for *case A,* but the error is adjusted half by means of foot screws and half by means of capstan screws of the bubble tube.
3. **Adjustment for Parallelism of Line of Sight to the Axis of the Level Tube**

   (*i*) *Test*

   1. Level the instrument carefully by keeping the telescope parallel to two foot screws. Clamp the motion about vertical axis.
   2. Keep a level rod in the line of sight and take the reading.
   3. Reverse the telescope end for end in the wyes and again sight the staff.
   4. If the reading is the same, the instrument is in adjustment. If not, it requires adjustment.

   (*ii*) *Adjustment*

   Bring the line of sight to the mean reading on the staff by means of adjusting screws under one wye.

17 CHAPTER

PRECISE LEVELLING

17.1 INTRODUCTION

Precise levelling is used for establishing bench marks with great accuracy at widely distributed points. The precise levelling differs from the ordinary levelling in the following points:

(*i*) High grade levels and stadia rods are used in precise levelling.

(*ii*) Length of sight is limited to 100 m in length.

(*iii*) Rod readings are taken against the three horizontal hairs of the diaphragm.

(*iv*) Backsight and foresight distances are precisely kept equal, the distances being calculated from stadia hair readings.

(*v*) Two rodmen are employed and backsight and foresight are taken in quick succession.

(*vi*) The adjustments of the *precise level* are tested *daily* and the correction applied to the rod readings. The rod is standardized frequently.

The precise levelling can be classified under the following three heads, depending upon the permissible errors:

First order : permissible error = 4 mm $\sqrt{K}$ or 0.017 ft $\sqrt{M}$

Second order : permissible error = 8.4 mm $\sqrt{K}$ or 0.035 ft $\sqrt{M}$

Third order : permissible error = 12 mm $\sqrt{K}$ or 0.05 ft $\sqrt{M}$.

For most of the engineering surveys, permissible error of closure of a level circuit is 0.1 $\sqrt{M}$ or 24 mm $\sqrt{K}$. The construction engineer, therefore, is accustomed to refer to any of the three higher orders as precise levelling.

17.2 THE PRECISE LEVEL

The precise levelling instrument has, generally, a telescope of greater magnifying power (40 to 50 D). It is provided with three parallel plate screws and a very sensitive bubble which is brought to the centre for each reading by a fine tilting drum placed under the eyepiece. Thus, the line of sight can be made horizontal even when the instrument as a whole is not exactly level.

The bubble can be seen from the eyepiece end of the telescope by reflection in the small prism above the bubble tube. Coincidence system is used for centering the bubble, as shown in

Fig. 17.1. An adjustable mirror placed immediately below the bubble tube illuminates the bubble. One-half of each extremity of the bubble is reflected by the prism in the long rectangular casing immediately above the bubble tube into the small prism box. When the bubble is not perfectly central, the reflections of the two halves appear as shown in Fig. 17.1 (*a*). When the bubble is central, the reflection of the two halves makes one curve as shown in Fig. 17.1 (*b*). The bubble tube generally has sensitiveness of 10 seconds of arc per 2 mm graduation.

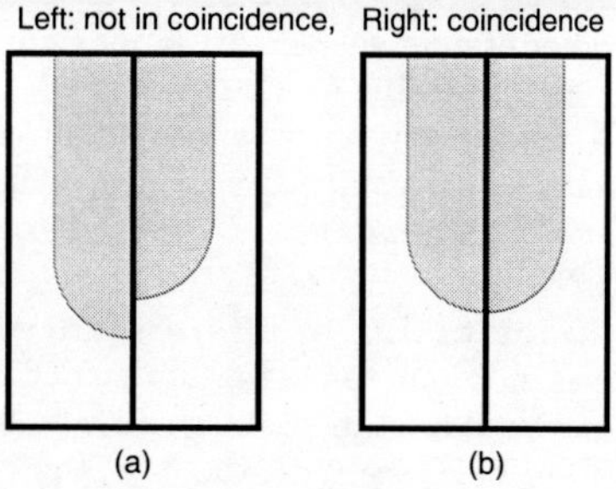

Fig. 17.1

17.3 WILD N-3 PRECISION LEVEL

Fig. 17.2 shows the photograph of Wild N-3 precision level for geodetic levelling of highest precision, construction of bridges, measurements of deformation and deflection, determination of the sinking of dams, mounting of large machinery etc. Apart from the main telescope, the level contains two optical micrometers placed to the left of the eyepiece—one is meant for viewing the coincidence level and the other is for taking the micrometer reading (Both the auxiliary telescopes are not visible in the photograph since right-hand view has been shown). The tilting screw (2) has fine pitch and is placed below the eyepiece and for fine movement in azimuth, it also contains a horizontal tangent screw (4). The micrometer knob (6) is used for bringing the image of the particular staff division line accurately between the V-line of the graticule plate.

The centring of the bubble is done by means of prism-system in which the bubble-ends are brought to coincidence (Fig. 17.1). The optical micrometer is used for reading the staff. Fig. 17.3 shows the field of view through all three eyepieces. The graticule has a horizontal hair to the right half and has two inclined hairs, forming V-shape, to the left hair. After having focused the objective, the approximate reading of the staff may be seen. The optical micrometer is used for fine reading of staff. By turning the knob (6) for micrometer, the plane parallel glass plate mounted in front of the objective is tilted and the image of the particular staff division line is thus brought accurately between the V-lines of the graticule plate. This displacement of the line of sight, to a maximum of 10 mm, is read on a bright scale in the measuring eyepiece to $\frac{1}{100}$ mm. Thus, the staff reading (Fig. 17.3) is 148 + 0.653 = 148.653 cm. An invar rod (Fig. 17.6) is used with this level. The manufacturers claim an accuracy of ± 0.001 inch in a mile of single levelling.

17.4 THE COOKE S-550 PRECISE LEVEL

Fig. 17.4 shows the photograph of the Cooke S-550 precise level manufactured by M/s Vickers Instruments Ltd. used for geodetic levelling, determination of dam settlement and ground subsidence, machinery installation, and large scale meteorology. The telescope spirit vial is illuminated by a light diffusing window. The vial is read through the telescope eyepiece by an optical coincidence system. The telescope is fitted with a calibrated fine levelling screw, one revolution tilting the telescope through a vertical angle corresponding to 1 : 1000. The micrometer head is sub-divided into fifty parts, one division, therefore, being equal to 1 in 50,000. The extent of calibration is twenty revolutions, corresponding to an angle of 1 in 50.

The reticule has vertical line, stadia lines, horizontal line and micrometer setting V. The level vial has a sensitivity of 12″ per 2 mm. The manufacturers claim an accuracy of ± 0.02 inch/mile or ± 0.3 mm/km of single levelling.

For taking accurate staff reading, the micrometer screw is turned till the particular staff division line is brought in coincidence with the V of the reticule. This is accomplished by a parallel plate micrometer (Fig. 17.5) which measures the interval between the reticule line and the nearest division on the staff to an accuracy of 0.001 ft.

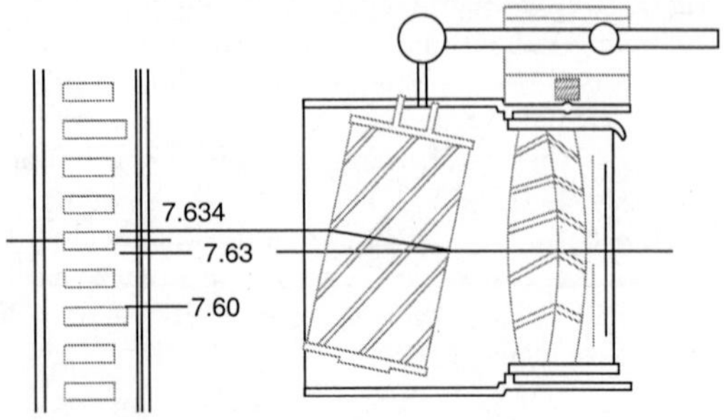

Fig. 17.5

The device consists of parallel plate of glass which may be fitted to displace the rays of light entering the objective. The displacement is controlled by a micrometer screw (6) calibrated to give directly the amount of the interval.

17.5 ENGINEER'S PRECISE LEVEL (*FENNEL*)

Fig. 17.6 shows the photograph of Fennel's A 0026 precise Engineer's level with optical micrometer. It is equipped with a tilting screw and a horizontal glass circle. The coincidence of the bubble ends can be directly seen in the field of view of telescope. This assures exact centering of the bubble, when the rod is read. Fig. 17.7 (*a*) shows the telescope field of view when spirit level is not horizontal. Fig. 17.7 (*b*) shows the telescope field of view when the spirit level is horizontal. The sensitivity of tubular spirit level is 2″ per 2 mm. The optical micrometer is used for fine reading of staff. Fig. 17.7 (*c*) shows the field of view of optical micrometer for fine reading of the staff. The telescope has magnification of 32 dia. The horizontal glass circle—reading 10 minutes, estimation 1 minute—renders the instrument excellent for levelling tacheometry when used in conjuction with the Reichenback stadia hairs.

17.6 FENNEL'S FINE PRECISION LEVEL

Fig. 17.8 shows Fennel's 0036 fine precision level with optical micrometer. The length of the telescope, including optical micrometer is 15 inches, with $2\frac{1}{4}$ inch aperture of object glass and a magnifying power of 50 ×. The sensitivity of circular spirit level is 6′ while that of the tubular spirit level 10″ per 2 mm.

The bubble ends of the main spirit level are kinematically supported in the field of view, where they are read in coincidence (Fig. 17.8). A scale, arranged in the field of view, provides the reading of differences of variation of the bubble. The instrument is provided with wooden precision rod as well as invar tape rod, 3 m long with half centimetre graduated. Centimetre reading is directly read in the field of view of the telescope. Fine reading of the staff is read through separate microscope mounted adjacent the eyepiece. A scale permits direct readings of 1/10 of the rod interval and estimations of 1/100. Thus, in Fig. 17.9, the rod reading is = 244 + 0.395 = 244.395. A mean error of ± 0.3 to ± 0.5 mm per kilometre of double levelling is well obtainable with this instrument, if all precautions of precise levelling work are compiled with.

17.7 PRECISE LEVELLING STAFF

For levelling of the highest precision, an invar rod is used. Fig. 17.10 shows invar rod by M/s Wild Heerbrugg Ltd. An invar band bearing the graduation is fitted to a wooden staff, tightly fastened at the lower end and by a spring at the upper end. Thus any extension of the staff has no influence on the invar band. The thermal expansion of the invar is practically nil. The graduations are of 1 cm. Two graduations mutually are displaced against each other to afford a check against gross errors. The length of graduations is 3 m. For measuring, the rod is always set up on an iron base plate. Detachable stays are provided for accurately and securely mounting the invar levelling staff. Once the rod is approximately vertical, the ends of stays are clamped tight. By means of the slow motion screw, the spherical level of the rod can be centred accurately.

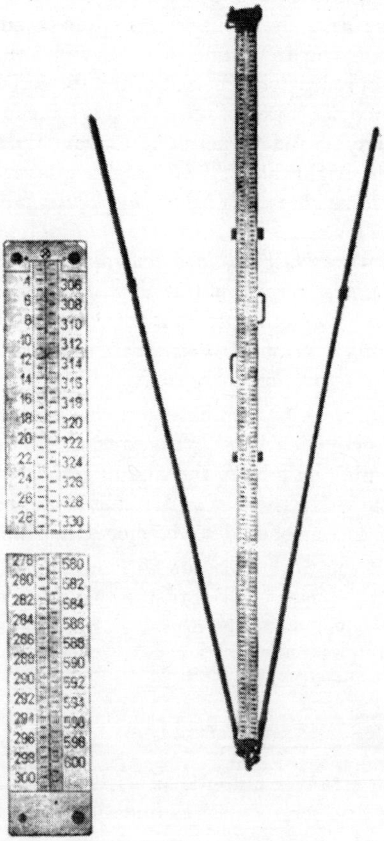

Fig. 17.10 Invar precision levelling rod (By Courtesy of M/s Wild Heerbrugg Ltd.)

17.8 FIELD PROCEDURE FOR PRECISE LEVELLING

Two rod men are used; they may be designated rod man *A* and rod man *B*. The rod *A* is called the B.M. rod. The rod *A* is held on the benchmark and the *B* rod on the turning point. After setting the level, micrometer is set at the *reversing point.* The longitudinal bubble is brought to its centre by micrometer screw before taking any reading. The first reading is taken on *A* rod and the second reading is taken on *B* rod placed at the turning point such that the backsight and foresight distances are approximately equal. For each reading, all the three wires are read. When the instrument is moved, the *B* rod is left at the first turning point and the *A* rod is moved to the second turning point. At the second set up the level man reads rod A (foresight) first and then the rod B (backsight). When the instrument is moved again, the A rod is held where it is and the B rod is moved. At third set up of the level, the level man reads rod A (backsight) first and the rod B (foresight) next. *Thus, at alternate set up the foresight is read before the backsight and at every set up the A rod is read first and B rod next. The procedure neutralizes the effect of changing conditions like sinking of the level or changing refraction.* If it should happen that the B rod normally comes to the B.M. at the end of a section of levels, it is not used. Instead, the A rod is moved to the B.M. Thus, both sights at this instrument position are taken on the A rod. *This procedure eliminates any difference in index correction of the rod.*

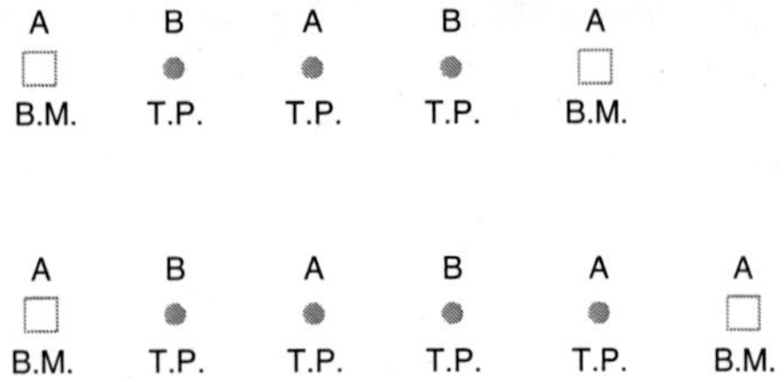

Fig. 17.11

In order to eliminate serious systematic errors due to the variations in temperature and refraction, each *section* is to be checked by a forward and a backward running—the forward running may be in the morning and backward in the afternoon. The difference in elevation obtained by these two runs should be checked within the limits of accuracy desired. The length of a section should not be more than 1200 metres.

If the work proceeds without interruption and no sudden change in temperature occurs, it is sufficient to record the two rod temperatures at the beginning and end of the section. The level should be protected from the sun. A rod level must be used to plumb the rod at all readings.

17.9 FIELD NOTES

The arrangement of level notes is almost similar to that of ordinary levelling, except that all the three cross-wire readings are taken and recorded. A line is drawn after three readings and average is found. This average gives the backsight or foresight reading at the point. The intervals between the top and central-wire and between bottom and central-wire readings are computed. The difference between these two interval readings should not be more than 0.005 ft or another set of readings must be taken. The difference between the top and bottom-wire readings (or the sum of the above calculated intervals) is a *measure* of the distance from the level to the rod and is called the *distance reading.* Starting with the first backsight, the distance reading of each successive backsight is added. Similarly, the distance reading of each successive foresight is also added. Thus, at any turning point, the sum thus formed gives the *total* of distances of backsights or foresights, as the case may be. The sum of total backsight distances must approximately be equal to the sum of total foresight distance at any turning point. The table below shows a page from precise level book.

A Page From Precise Level Book

Station	B.S.	H.I.	F.S.	Elv.	Distance		Remarks
					B.S.	F.S.	
	2.623						
B.M.	3.346				0.723		
	4.070				0.724		
	3.346	528.125′		524.779′	1.447		
	3.825		3.986				
T.P. 1	4.506		4.706		0.681	0.720	
	5.189		5.428		0.683	0.722	
	4.507	527.925′	4.707	523.418′	2.811	1.444	
	4.685		3.628				
T.P. 2	5.610		4.280		0.925	0.652	
	6.534		4.930		0.924	0.650	
	5.610	529.256′	4.279	523.646′	4.660	2.746	
			4.960				
B.M.			5.890			0.930	
			6.822			0.932	
			5.891	523.365′		4.608	
	13.463		14.877	524.779			
Check			13.463	523.365			
		Fall	1.414	1.414	Fall		

17.10 DAILY ADJUSTMENTS OF PRECISE LEVEL

The adjustments of a precise level should be *tested* daily. If the adjustments are out by permissible amount, corrections are applied to the observations of the day's work. If, however, the adjustments are out by appreciable amount, they are adjusted. The following adjustments are made:

(*i*) Adjustment for circular bubble,
(*ii*) Adjustment for prism mirror,
(*iii*) Adjustment for the size of the bubble tube,
(*iv*) Adjustment for the line sight, and
(*v*) Adjustment for the reversing point.

(*i*) **Adjustment for circular bubble:** Centre the circular bubble by means of foot screws. Reverse the telescope. If the bubble moves from the centre, bring it half way back by means of the adjusting screws.

(*ii*) **Adjustment of the prism mirror:** With the right eye in position at the eyepiece, sight the prism mirror with the left eye. Swing the mirror until the bubble appears to be evenly situated to the centre line.

(*iii*) **Adjustment for the size of the bubble tube:** This adjustment can be made only if the level vial has an adjustable air chamber. If it has air chamber, the length of the bubble can be changed by tilting the chamber. Thus, to enlarge the bubble, tilt the eyepiece and upward and to decrease it, turn the eyepiece end downward.

(*iv*) **Adjustment for the line of sight:** The test of the parallelism of the line of sight and the axis of the bubble tube is of prime importance and shall be made daily. It may not be necessary to make the adjustment daily. However, the error is determined and correction is applied to the observed readings.

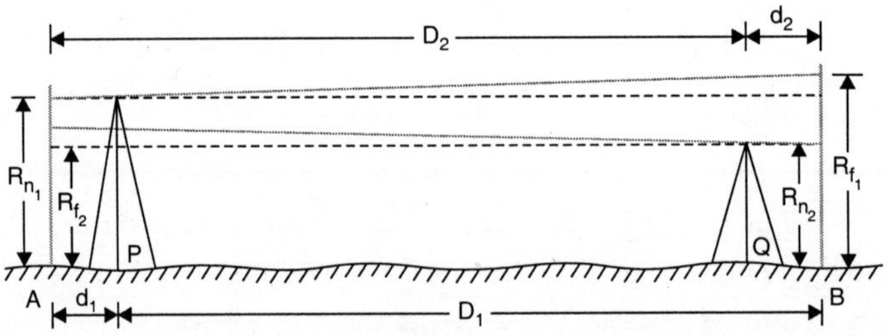

Fig. 17.12

To test the adjustment, two points A and B are selected about 120 m apart. The level is first set at P, near to A, at a distance d_1 from A and D_1 from B. Let the reading obtained at A be R_{n_1} and that at B be R_{f_1}, the suffix n and f being used to denote the readings on near and far points. The instrument is then moved to a point Q, near to B, at distance d_2 from B and D_2 from A. Let the reading obtained at A be R_{f_2} and at B be R_{n_2}. Let c = slope of the line of sight = tan α.

When the instrument is at P

True difference in elevation between A and $B = (R_{f_1} - cD_1) - (R_{n_1} - cd_1)$ ...(1)

When the instrument is at Q

The difference in elevation between A and $B = (R_{n_2} - cd_2) - (R_{f_2} - cD_2)$ ...(2)

Equating these two and solving for c, we get

$$c = \frac{(R_{n_1} + R_{n_2}) - (R_{f_1} + R_{f_2})}{(D_1 + D_2) - (d_1 + d_2)} = \frac{\text{Sum of near rod readings} - \text{Sum of far rod readings}}{\text{Sum of far distances} - \text{Sum of near distances}}$$

Knowing c, the correction to any rod reading can be calculated.

The line of sight will be inclined downwards if c has plus sign and will be inclined upwards if c has minus sign. If the value of c comes out to be more than 0.00005 (*i.e.* 0.005 m in 100 m), adjustments should be made by calculating the correction for a staff kept at 90 m distance from the instrument.

(*v*) **Adjustment of the reversing point:** The reversing point is a particular reading on the micrometer screw at which the bubble will remain central after reversal, when the vertical axis of the level is truly vertical. To find the reversing point, the bubble tube is centred exactly and the micrometer reading is noted. The telescope is then reversed, the bubble again centred and the micrometer reading is noted. The reversing point is then half-way between the two micrometer readings. The adjustment is not *essential* but is merely *necessary* for quick centring of the bubble at all times. Whenever the instrument is being levelled, the micrometer screw should be set at the reversion point.

18

CHAPTER

PERMANENT ADJUSTMENTS OF THEODOLITE

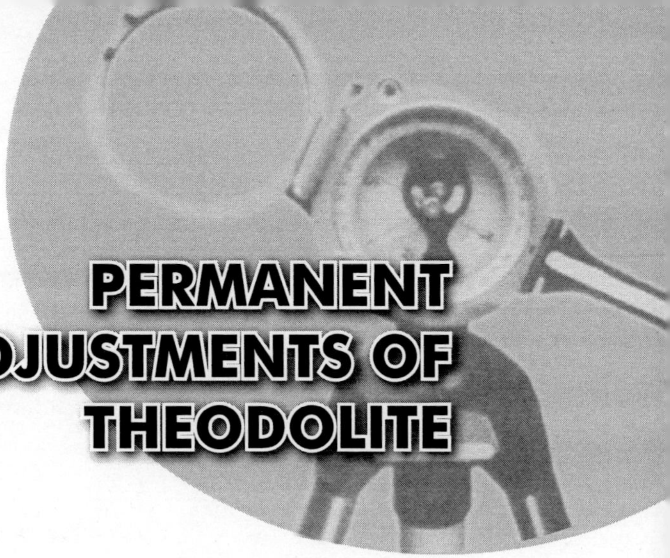

18.1 GENERAL

The *fundamental lines* of a transit are as follows:

1. The vertical axis
2. The horizontal axis
3. The line of collimation (or line of sight)
4. Axis of plate level
5. Axis of altitude level
6. Axis of the striding level, if provided.

The following *desired relations* should exist between these lines:

1. *The axis of the plate level must lie in a plane perpendicular to the vertical axis.*

   If this condition exists, the vertical axis will be truly vertical when the bubble is in the centre of its run.

2. *The line of collimation must be perpendicular to the horizontal axis at its intersection with the vertical axis. Also, if the telescope is external focusing type', the optical axis, the axis of the objective slide and the line of collimation must coincide.*

   If this condition exists, the line of sight will generate a vertical plane when the telescope is rotated about the horizontal axis.

3. *The horizontal axis must be perpendicular to the vertical axis.*

   If this condition exists, the line of sight will generate a vertical plane when the telescope is plunged.

4. *The axis of the altitude level (or telescope level) must be parallel to the line of collimation.*

   If the condition exists, the vertical angles will be free from index error due to lack of parallelism.

5. *The vertical circle vernier must read zero when the line of collimation is horizontal.*

   If this condition exists, the vertical angles will be free from index error due to displacement of the vernier.

6. *The axis of the striding level (if provided) must be parallel to the horizontal axis.*

If this condition exists, the line of sight (if in adjustment) will generate a vertical plane when the telescope is plunged, the bubble of striding level being in the centre of its run.

The *permanent adjustments* of a transit are as follows:

1. Adjustment of plate level
2. Adjustment of line of sight
3. Adjustment of the horizontal axis
4. Adjustment of altitude bubble and vertical index frame.

18.2 ADJUSTMENT OF PLATE LEVEL

(*i*) **Desired Relation:** *The axis of the plate bubble should be perpendicular to the vertical axis when the bubble is central.*

(*ii*) **Object:** The object of the adjustment is to make the vertical axis truly vertical; to ensure that, once the instrument is levelled up, the bubble will remain central for all directions of sighting.

(*iii*) **Necessity:** Once the requirement is accomplished, the horizontal circle and also the horizontal axis of the telescope will be truly horizontal, provided both of these are perpendicular to the vertical axis.

(*iv*) **Test:**

1. Set the instrument on firm ground. Level the instrument in the two positions at right angles to each other as in temporary adjustment.
2. When the telescope is on the third foot screw, swing it through 180°.

If the bubble remains central, adjustment is correct.

(*v*) **Adjustment:**

1. If not, level the instrument with respect to the altitude bubble till it remains central in two positions at right angles to each other.
2. Swing the telescope through 180°. If the bubble moves from its centre, bring it back *halfway* with the levelling screw and half with the clip screw.
3. Repeat till the altitude bubble remains central in all positions. The vertical axis is now truly vertical.
4. Centralize the plate levels(s) of the horizontal plate with capstan headed screw.

It is assumed that the altitude bubble is fixed on the index frame.

(*vi*) **Principle involved:** This is the case of single reversion in which the apparent error is double the true error. See also permanent adjustment (1) of a dumpy level, chapter 16.

18.3 ADJUSTMENT OF LINE OF SIGHT

(*i*) **Desired Relation:** *The line of sight should coincide with the optical axis of the telescope.*

(*ii*) **Object:** The object of the adjustment is to place the intersection of the cross-hair in the optical axis. Thus, both horizontal as well as vertical hair are to be adjusted.

(*iii*) **Necessity:**

(*a*) *Horizontal hair:* This adjustment is of importance only in the case of external focusing telescope in which the direction of line of sight will change while focusing if the horizontal hair does not intersect the vertical hair in the same point in which the optical axis does.

(b) *Vertical hair:* If the adjustment is accomplished, the line of collimation will be perpendicular to the horizontal axis (since the optical axis is placed permanently perpendicular to the horizontal axis by the manufacturers) and hence the line of sight will sweep out a plane when the telescope is plunged.

(iv) **Test for horizontality and verticality of hairs:** Before the adjustment is made, it is necessary to see if the vertical and horizontal hairs are truly vertical and horizontal when the instrument is levelled up. To see this, level the instrument carefully, suspend a plumb bob at some distance and sight it through the telescope by careful focusing. If the image of the plumb bob string is parallel to the vertical hair, the latter is vertical. If not, loose the capstan screws of the diaphragm and rotate it till the vertical hair coincides with the image of the string. The horizontal hair will then be horizontal.

Adjustment of Horizontal Hair (Fig. 18.1)

(v) **Test:**

1. Level the instrument carefully with all clamps fixed. Take a reading on a staff placed some distance apart (say 100 m). Note also the reading on the vertical circle.
2. Unclamp the lower clamp, transit the telescope and swing it through 180°. Set the same reading on the vertical circle and see the staff. If the same reading is obtained, the horizontal hair is in adjustment.

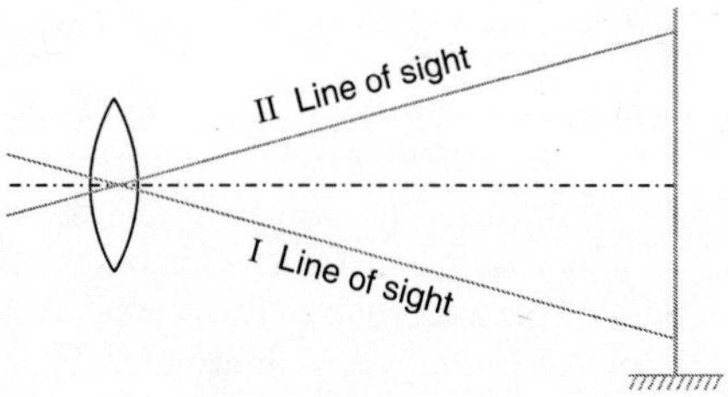

Fig. 18.1

(vi) **Adjustment:**

1. If not, adjust the horizontal hair by top and bottom capstan screws of the diaphragm until the reading on the staff is the mean of the two.
2. Repeat the test till the adjustment is correct.

Adjustment of Vertical Hair (Fig. 18.2)

(vii) **Test:**

1. Set the instrument on a level ground so that a length of about 100 m is available to either side of it. Level it.
2. Sight a point *A* about 100 m away. Clamp the horizontal movement.
3. Transit the telescope and establish a point *B* to the other side at the same level as *A*, such that *OA* = *OB* (approx.).
4. Unclamp the horizontal movement and turn the telescope to sight *A* again.
5. Transit the telescope. If it intersects *B*, the line of sight is perpendicular to the horizontal axis.

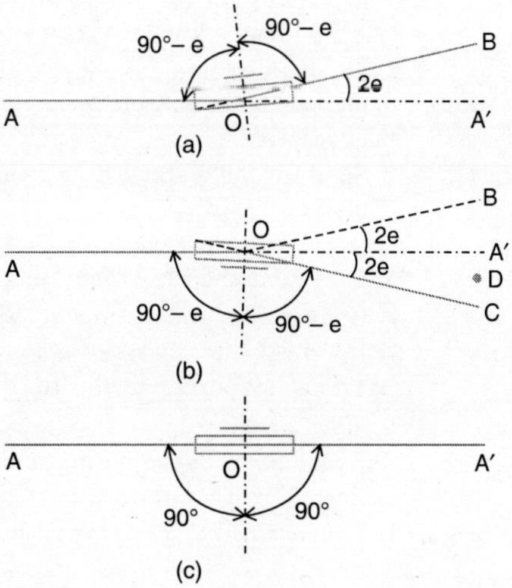

Fig. 18.2

(*viii*) **Adjustment:**

1. If not, mark the point C in the line of sight and at the same distance as that of B.
2. Join C and B and establish a point D towards B such that $CD = \frac{1}{4} CB$. [Fig. 18.2(*b*)].
3. Using the side capstan screws of the diaphragm bring the vertical hair to the image of D.
4. Repeat till there is no error on changing the face, as illustrated in Fig. 18.2(*c*).

(*ix*) **Principle involved:** This is *double* application of the principle of reversion. Transiting the telescope once *doubles* the error; transiting a second time (after changing the face) again doubles the error on the *opposite side,* so that total *apparent* error is four times the *true* error.

18.4 ADJUSTMENT OF THE HORIZONTAL AXIS

(*i*) **Desired Relation:** *The horizontal axis should be perpendicular to the vertical axis.*

(*ii*) **Object:** The object of the adjustment is to make the horizontal axis perpendicular to the vertical axis so that it is perfectly horizontal when the instrument is levelled.

(*iii*) **Necessity:** If adjustment (2) is done the line of sight will move in plane when the telescope is plunged; this adjustment ensures that this plane will be a vertical plane. This is essential when it is necessary to move the telescope in the vertical plane while sighting the objects.

(*iv*) **Test:** The test is known as the *spire test:*

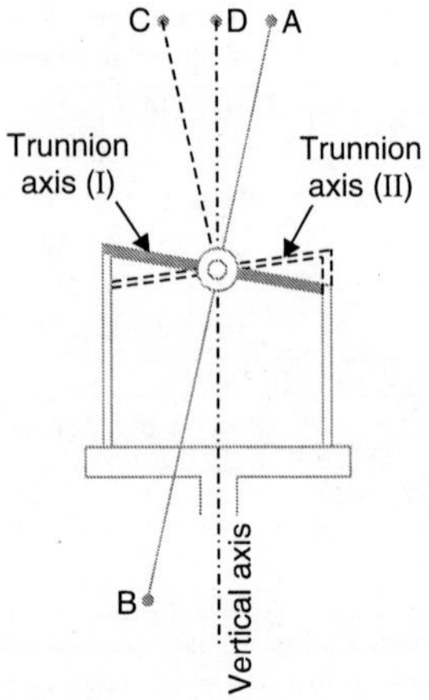

Fig. 18.3 Spire Test.

1. Set up the instrument near a high building or any other high well-defined point such as the final of a steeple etc. Level it.
2. Sight the well-defined high point A. Clamp the horizontal plates.
3. Depress the telescope and sight a point B on the ground as close to the instrument as possible.
4. Change face and again sight B. Clamp the horizontal plates.
5. If, on raising telescope to sight A, an imaginary point C is sighted, the horizontal axis is not perpendicular to the vertical axis.

(*v*) **Adjustment:**

1. By means of the adjusting screws at the trunnion support on one standard, bring the line of sight to an imaginary point D half way between A and C.
2. Repeat until C coincides with A when the telescope is raised after backsighting B.

18.5 ADJUSTMENT OF ALTITUDE LEVEL AND VERTICAL INDEX FRAME

General: The procedure for this adjustment depends upon whether the clip screw and the vertical circle tangent screw are provided on the same arm or on different arms, and also upon whether the altitude bubble is provided on the index frame or on telescope. There are, therefore, the following cases:

(*a*) *Clip and tangent screws on separate arms*

(*i*) altitude level on index arm.

(*ii*) altitude level on telescope.

(*b*) *Clip and tangent screws on the same arm*

(*i*) altitude level on index arm.

(*ii*) altitude level on telescope.

In case *a*(*i*), *a*(*ii*) and *b*(*i*), both the adjustments, *i.e.*, adjustment of altitude level and adjustment of vertical index frame, are done togther. In case *b* (*ii*), the adjustment of altitude level is done first by two-peg test (see § 16.2) and then the vertical index frame is adjusted. However, in most of the modern theodolites, with the object of securing better balance, the vertical circle clamp and tangent screw are placed on one side of the telescope and the clip screw on the other. It is, therefore, intended to discuss case (*a*) only, which is the most usual case.

(a) Clip and Tangent Screws on Separate Arms

Object: *To make the line of sight horizontal when the bubble is central and the vertical circle reading is zero.*

Necessity: If this is not achieved, the vertical circle reading will not be zero when the bubble is central and the line of sight is horizontal. The reading on the vernier, when the line of sight is horizontal, is known as *index error,* which will have to be added to or subtracted from the observed readings if the adjustment is not made.

(i) Altitude Bubble on Index Frame

Test:
1. Level the instrument with respect to plate levels.
2. Bring the altitude bubble in its centre by using the clip screw.
3. Set the vertical circle reading to zero by vertical circle clamp and tangent screw.
4. Observe a levelling staff held 75 or 100 m away and note the reading.
5. Release the vertical circle clamp, transit the telescope and swing by 180°. Re-level the bubble by clip screw, if necessary.
6. Set the vertical circle reading to zero.
7. Again read the staff held on the same point. If the reading is unchanged, the adjustment is correct.

Adjustment:
1. If not, bring the line of collimation on to the mean reading by turning the vertical circle tangent screw.
2. Return the vernier index to zero by means of clip screw.
3. Bring the bubble of the altitude level central by means of its adjusting capstan screw.

(ii) Altitude Bubble on the Telescope

Test:
1. Level the instrument with reference to the plate levels, set the vertical circle to read zero by means of vertical circle clamp and tangent screw.
2. Bring the telescope level central by the foot screws. Observe a levelling staff about 100 m away and note the reading.
3. Loose the vertical circle clamp, transit the telescope and again set the vertical circle

to read zero. Swing through 180°. Re-level if necessary and again read the staff held on the same point. If the reading is unchanged, the adjustment is correct.

Adjustment:

1. If not, bring the line of collimation on to the mean reading by turning the vertical tangent screw.
2. Return the vernier index to zero by means of clip screw.
3. Bring the bubble of the level tube central by means of adjusting screws attaching it to the telescope.
4. Repeat till no error is discovered.

PROBLEMS

1. Give a list of the permanent adjustments of a transit theodolite and state the object of each of the adjustment. Describe how you would make the trunnion axis perpendicular to the vertical axis.
2. What is spire test? How is it carried?
3. Explain the adjustment for making the axis of the spirit level over T-frame of the vertical circle perpendicular to the vertical axis of die theodolite.

CHAPTER 19

PRECISE THEODOLITES

19.1 INTRODUCTION

The instruments for geodetic survey require great degree of refinement. In earlier days of geodetic surveys, the required degree of refinement was obtained by making greater diameter of the horizontal circles. The great theodolite of *Ordinance Survey* had a diameter of 36″. These large diameter theodolites were replaced by the micrometer theodolites (similar in principle to the old 36″ and 24″ instruments) such as the Troughton and Simm's 12″ or the Parkhurst 9″. However, more recently the tendency has been to replace the micrometer theodolites by others of the double reading type (glass arc) such as the Wild, Zeiss and Tavistock having diameters of $5\frac{1}{2}''$ and 5″ respectively. The distinguishing features of the double reading theodolite with optical micrometers are as follows:

(*i*) They are small and light.

(*ii*) The graduations are on glass circle, and are much finer.

(*iii*) The mean of the two readings on opposite sides of the circle is read directly in an auxiliary eye-piece generally besides the telescope. This saves the observing time, and also saves disturbance of the instrument.

(*iv*) No adjustments for micrometer run are necessary.

(*v*) It is completely waterproof and dust proof.

(*vi*) It is electrically illuminated.

There are two types of instruments used in the triangulation of high precision.

1. The repeating theodolite.
2. The direction theodolite.

1. **The Repeating Theodolite:** The characteristic feature of the repeating theodolite is that it has a double vertical axis (two centres and two clamps). It has two or more verniers to read to 20, 10 or 5 seconds. The ordinary transit is the repeating theodolite. The vernier theodolite by M/s Vickers Instruments Ltd. and the Watts Microptic Theodolite No. 1, fall under this category.
2. **The Direction Theodolite:** The direction theodolite has only one vertical axis, and a single horizontal clamp-and-tangent screw which controls the rotation about the vertical axis. Optical micrometers are used to read fractional parts of the smallest divisions of the graduated circle. The direction theodolite is used for very precise work needed in

the first order or second order triangulation survey. Wild T-2, T-3 and T-4 theodolites fall under this category, and will be discussed here.

19.2 WATTS MICROPTIC THEODOLITE NO. I

Messers Hilger and Watts Ltd. manufacture three models of optical microptic theodolites No. 1. No. 2 and No. 5. Out of the three, No. 1 is most precise having a least count of 1″ while in No. 2, the reading can be taken directly to 20″ and by estimation to 5″. Fig. 19.1 shows Watts Microptic Theodolite No. 1.

The Theodolite has horizontal and vertical circles of glass and images of both are brought together with that of the micrometer scale, into the field of view of the reading eyepiece. Both circles are divided directly to 20 minutes and are figured at each degree. Finer sub-divisions are read from the lowest scale (Fig. 19.2) which gives the micrometer reading. It is divided at 20 second intervals and figured at every 5 minutes. The two circle scales are read against patented indexes which facilitate precise setting. They take the form of hollow triangular pointers which indicate light displacements of the lines beneath them. A small lateral displacement of a line result in a relatively large asymmetry of small triangles of light beneath the pointer on either side of it. Estimated readings may easily be made to 5 seconds.

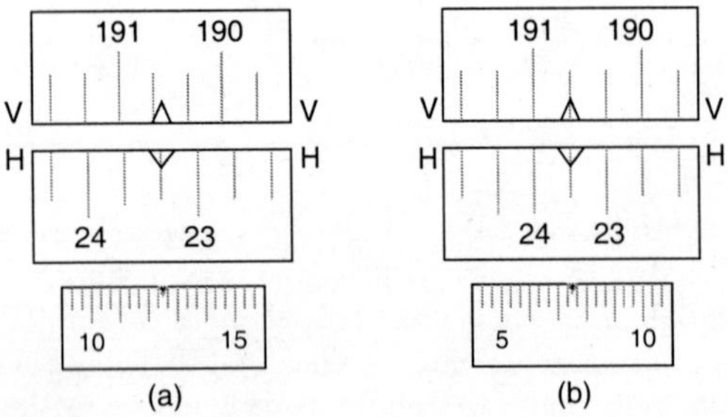

Coincidence for horizontal circle reading 23°32′30″ | Coincidence for vertical circle reading 190°47′30″

Fig. 19.2 View in Micrometer of Watts Microptic Theodolite No. 1.

In use, the micrometer is adjusted until the nearest division of the circle being observed is brought into coincidence with the index. The reading of the micrometer scale is then added to that of the circle to give the instrument reading. Fig. 19.2 (*a*) shows the field of view when *coincidence* has been made for the horizontal circle reading, using the optical micrometer screw. The reading on the horizontal circle is 23° 20′ and that on the micrometer is 12′ 30″. The total reading on the horizontal circle is, therefore, 23° 20′ + 12′ 30″ = 23° 32′ 30″. Fig. 19.2 (*b*) shows the same field of view when coincidence has been made for the vertical circle reading. The reading on the vertical circle is 190° 40′ and that on micrometer is 7′ 30″. The total reading on the vertical circle is, therefore, 190° 40′ + 7′ 30″ = 190° 47′ 30″.

19.3 FENNEL'S PRECISE THEODOLITE

Fig. 19.3 shows the photograph of Fennel's precise theodolite '*Themi*'. The instrument has following specifications:

1. **Horizontal circle**

Diameter : 5 in.

Graduation : 360° to 1/6°.

Reading by micrometer microscopes ensuring easy estimation to ....2″.

2. **Vertical circle**

Diameter	:	4 in.
Graduation	:	360° to 1/12°.

Reading by vernier microscopes to 30″.

3. **Telescope**

Length of telescope	:	$8\frac{11}{16}$ in.
Aperture of object glass	:	$1\frac{7}{16}$ in.
Focusing	:	Internal
Min. Focus	:	$8\frac{1}{5}$ ft.
Magnification	:	26 dia.

The horizontal circle is read with the help of micrometer microscopes. Fig. 19.4

(*a*) shows the image after the target has been aimed at. This position is shown as 'zero position'. In the lower half of the field of view the *graduation* is seen while the *secondary graduation* appears at the upper half. *Double-line index* (lower half of figure) is used for setting of graduation, while *single-line index* is used for setting of secondary graduation. Fig. 19.4

(*b*) marks the field of view as it is seen when the graduation line which may originally appear at the left of the firm double-line index has been placed keenly amidst the double-line index by means of the micrometer screw on the microscope. By this arrangement, the secondary graduation has been posed automatically and mark in the figure 3′ 16^d (^{d} = double seconds). Reading as per figure thus is found to be 38° 23′ 16^d or 38° 23′ 32″.

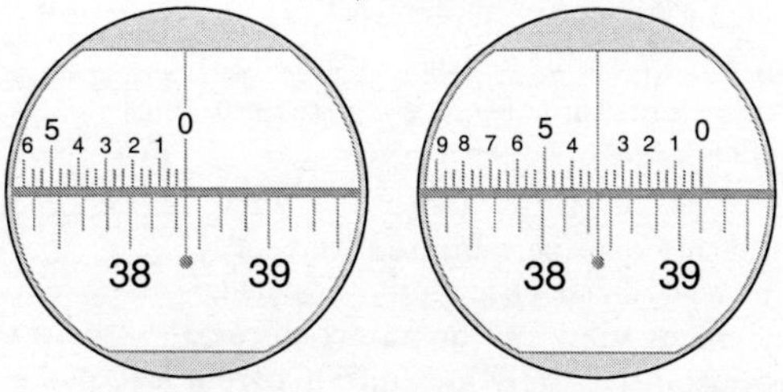

(a) Zero position (b) Reading position (38° 23′32″)

Fig. 19.4 Example of horizontal circle reading.

The vertical circle is read by simple vernier microscope. Fig. 19.5 shows the example of vertical circle reading. The reading after setting to reading position is 129° 34′ 00″.

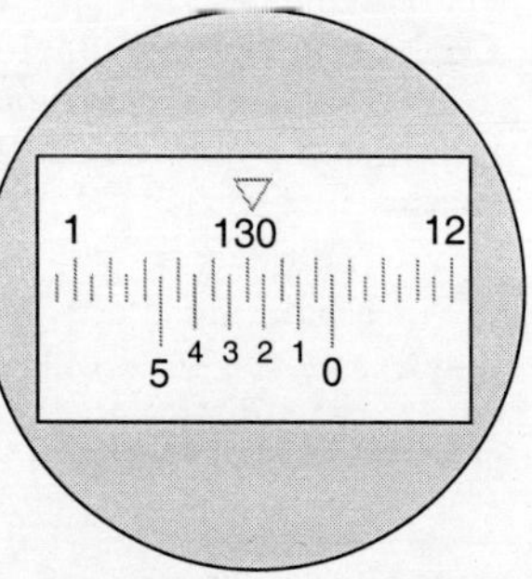

Fig. 19.5 Vertical circle reading.

19.4 WILD T-2 THEODOLITE

Fig. 19.6 shows the photograph of Wild T-2 theodolite. Both circles are made of glass. The diameter of horizontal circle is 90 mm and that of vertical circle is 70 mm and both are illuminated through adjustable mirrors. The artificial illumination required at night or in tunnels

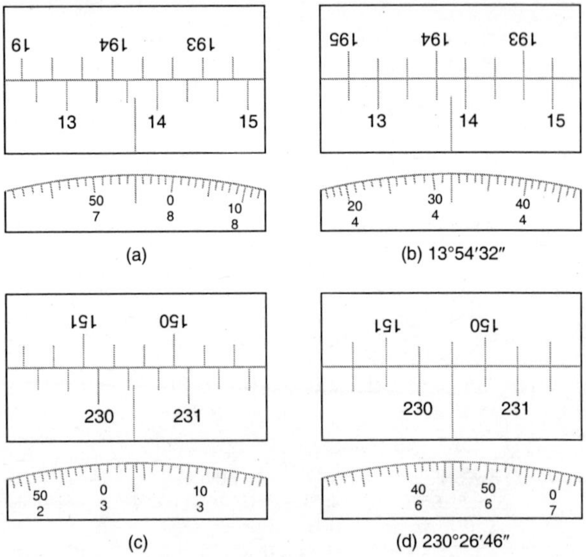

Fig. 19.7 Views in Micrometer of Wild T-2 Theodolite.

is supplied by an electric lamp replacing the mirror. The telescope is of internal focusing type having an over-all length of 148 mm. The vertical axis system consists of the axle bush and the vertical axis turning therein on ball bearings, which is automatically centred by the weight of the instrument. The glass circle is mounted on the outer side of the axle bush and is oriented as desired by drive knob. Since there is only one set of clamp and tangent screw for the motion about the vertical axis, the angles are measured by direction method only. This is, therefore a *direction theodolite.*

The readings are made with the microscope mounted adjacent to the telescope eyepiece. In the field of view of the micrometer appear the circle graduations from two parts of the circle 180° apart. The circle is divided in 20-minute intervals. The appearance of field of view is shown in each of the rectangles of Fig. 19.7.

Coincidence system is used to take the readings. Fig. 19.4(*a*) shows the field of view before co-incidence. The rectangle shows the scale reading, while the micrometer scale, seen underneath the rectangle, shows single second graduation over a range from 0 to 10 minutes. The lower numbers in the micrometer scale is thus rotated, the positions of the circles that appear in the rectangle are moved *optically* simultaneously by equal amounts in opposite directions till coincidence occurs, as shown in Fig. 19.7 (*b*) where the reading is 13° 54′ 32″. Fig. 19.7 (*c*) shows another illustration before coincidence and Fig. 19.7 (*d*) shows the same after coincidence where the final reading is 230° 26′ 46″.

Since both sides of the circle are moved simultaneously, a coincidence occurs every time they are moved 11 minutes. The micrometer scale, therefore, has a range of only 10 minutes to ensure reaching a coincidence. When coincidence occurs, the index line will either be against a 20 minute line or half-way between two 20 minute lines.

19.5 THE TAVISTOCK THEODOLITE

The Tavistock theodolite is a precision theodolite and derives its name from the fact that it was the outcome of a conference held in 1926 at Tavistock in Devon between instrument makers and British Government survey officers. Fig. 19.8 shows the Cooke's Tavistock theodolite manufactured by Messrs Vickers Instruments Ltd., England.

The horizontal and vertical circles are graduated every 20 minutes on the glass annuli. A single optical micrometer is provided for both circles, the circle reading eyepiece being situated parallel to the main telescope. A control on the standard of the instrument enables the observer to select which circle is to be viewed. Both circles are illuminated by a single mirror.

The images of divisions, diametrically opposite each other, are made to coincide when setting the micrometer. The reading can be taken direct to one second and be estimated to 0.25″ or 0.5″.

An optical plummet for centring over a ground mark is incorporated. The horizontal circle is rotated by level pinion, the engagement being controlled in an impersonal manner by cam connected to the cover over the control screw. A single slow motion screw is provided in azimuth.

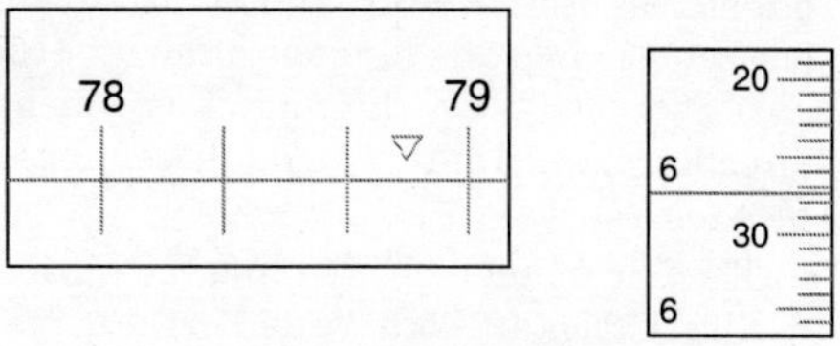

Fig. 19.9 View in the Micrometer of Cooke Tavistock Theodolite.

Fig. 19.9 shows the field of view of the reading micrometer at coincidence. The coincidence is made by the micrometer setting theodolite. Coincidence takes place at intervals of 10 minutes, the coarse and fine readings always being additive, provided the observer notes whether the coincidence takes place opposite the reading mark or symmetrically on either side of the reading mark (as illustrated) in which case 10 minutes must be added to the coarse reading, short of the reading mark, in addition to the micrometer reading. Thus, the reading illustrated is 78° 56′ 27″.5.

In another model of Geodetic Tavistock theodolite manufactured by Messers Vickers Instruments Ltd., the reading can be taken direct to 0.5 second of arc on the horizontal circle and 1 second on the vertical circle.

19.6 THE WILD T-3 PRECISION THEODOLITE

Fig. 19.11 shows the Wild T-3 precision theodolite meant for primary triangulation. Both the horizontal and vertical circles are made of glass. The graduation interval of horizontal circle is 4′ and that of the vertical is 8′. The readings can be taken on the optical micrometer direct to 0.2″ and by estimation to 0.02″. The following is the technical data:

Magnification 24, 30 or 40 ×

Clear diameter 2.36 in. (60 mm)

Shortest focusing distance 15 ft (4.5 m)

Normal range ..... 20 to 60 miles (32 km to 96 km)

Field of view at 1000 ft .... 29 ft (8.84 m)

Length of telescope 10.2 in. (260 mm)

Sensitivity of alidade level, 7″ per 2 mm

Sensitivity of collimation level 12″ per 2 mm

Coincidence adjustment of vertical circle level to 0.2″

Diameter to horizontal circle 5.5 in. (140 mm)

Graduation interval of horizontal circle 4′

Diameter of vertical circle 3.8 in. (97 mm)

Graduation interval of vertical circle 8′

Graduation interval of micrometer drum 0.2″.

The vertical axis system consists of the axle bush and the vertical axis turning therein on ball bearings, which is automatically centred by the weight of the instrument. The glass circle is mounted on the outer side of the axle bush and is oriented as desired by drive knob. Since there is only one set of clamp and tangent screws for the motion about vertical axis, the angles are measured by direction method only. This is, therefore, a *direction theodolite*.

The micrometers for reading the horizontal and vertical circles are both viewed in the same eyepiece which lies at the side of the telescope. In the field of view of the micrometer appear the circle graduations from two parts of circle 180° apart, separated by a horizontal line. The horizontal circle is divided in 4′ interval. The appearance of field of viev, is shown in Fig. 19.10 in which the top window shows the circle readings. A vertical line in the bottom half of the window serves as an index from which the coarse readings are taken. The lower window is graduated to seconds readings and carries a pointer. Coincidence system is used to take the readings. To read the micrometer, micrometer knob is turned so that the two sets of graduations in the upper window appear to coincide one another, and finally coincide. The seconds readings will then be given by the scale and pointer in the lower window. *The reading on the seconds scale in the bottom window is one-half of the proper reading.* Hence, the number of seconds which are read on this scale must either be doubled, or opposite graduations in the upper window should be brought into coincidence twice and the two readings on the seconds scale added together, as illustrated in Fig. 19.10.

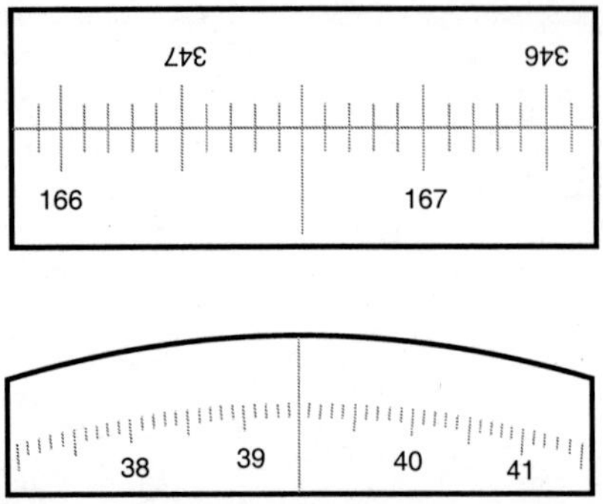

Circle reading	166° 40′
1st drum reading	39″.3
2nd drum reading	39″.4
	166° 41′ 18″.7

Fig. 19.10

To view the horizontal circle reading, an inverted knob is turned in a clockwise direction; to view the vertical circle reading, the knob is turned in the reverse direction. Thus, the same eyepiece can be used for taking the readings of both the circles.

19.7 THE WILD T-4 UNIVERSAL THEODOLITE (FIG. 19.12)

The Wild T-4 is a theodolite of utmost precision for first order triangulation, the determination of geographic positions and taking astronomical observations. The instrument has a horizontal circle of 250 mm (9.84″) which is almost double the diameter of that of T-3 model. The reading can thus be taken with greater accuracy. The theodolite is of the 'broken telescope' type; that is, the image formed in the telescope is viewed through an eyepiece placed at one end of the trunnion axis which is made hollow. The graduation interval on horizontal circle is 2′ with direct reading to 0.1″ on optical micrometer. The other technical data is as follows:

Telescope power: 65 ×

Clear objective glass apertute: 60 mm (2.36″)

Azimuth (horizontal) circle on glass: 360°

Diameter of scale: 250 mm (9.84″)

Interval between divisions: 2′

Direct readings to: 0″.1

Elevation (vertical) circle on glass, 360°

Diameter of scale: 145 mm 5.71″

Interval between divisions: 4″

Direct readings to: 0″.2

Setting circle, for telescope angle of sight

Interval of divisions: 1°

Scale reading microscope interval: 10′

Angles can be estimated to: 1′

Sensitivity of suspension level: 1″ of elevation circle level: 5″ of Horrebow level (both): 1″ – 2″

The vertical and azimuth circles are both equipped with a reading micrometer which gives automatically the arithmetic mean of two diametrically opposed readings. Fig. 19.13 shows the example of circle readings.

The eyepiece is equipped with the so-called longitude micrometer for accurate recording of a star's transit. The reversal of the horizontal axis and telescope is carried out by a special hydraulic arrangement which ensures freedom from vibration. Electrical lighting, to illuminate both circle and field, is built into the body.

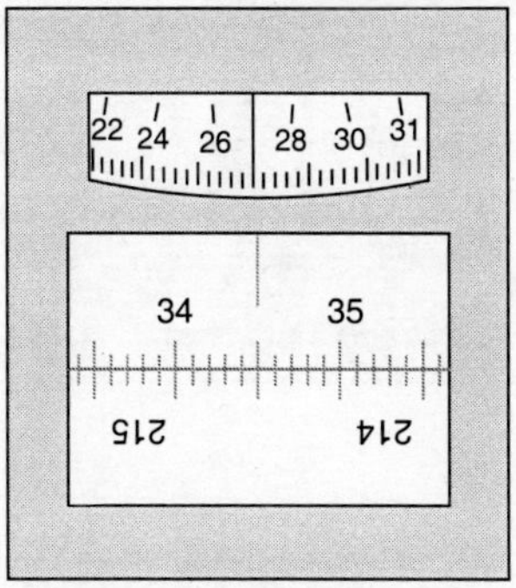

Example of a vertical circle reading
34° 25′ 26.9″
Example of a horizontal circle reading
146° 27′ 19.2″

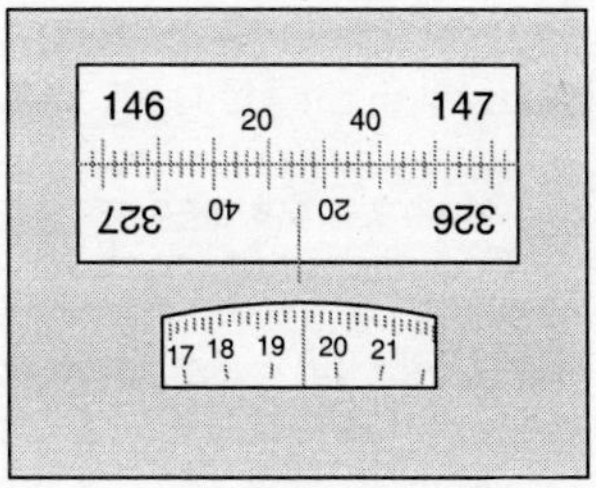

Fig. 19.13

20 CHAPTER

SETTING OUT WORKS

20.1 INTRODUCTION

Whereas surveying is the process of producing a plan or map of a particular area, setting out begins with the plan and ends with some particular engineering structure correctly positioned in the area. Most of the techniques and equipment used in surveying are also used in setting out. It is important to realise that setting out is simply one application of surveying. In many cases, insufficient importance is attached to the process of setting out, and it tends to be rushed to save time. This attitude may result in errors, causing delays which leave construction machinery and plant idle, resulting in additional costs.

There are two *aims* when undertaking setting out operations:

1. The structure to be constructed must be set out correctly in all three dimensions—both relatively and absolutely, so that it is of correct size, in the correct *plan position* and at correct *level*.
2. The setting out process, once begun, must proceed quickly, without causing any delay in construction programme.

20.2 CONTROLS FOR SETTING OUT

The setting out of work requires the following two controls:

(*a*) Horizontal control

(*b*) Vertical control.

20.3 HORIZONTAL CONTROL

Horizontal control points/stations must be established within or near the construction area. The horizontal control consists of reference marks of known plan position, from which salient points of the designated structure may be set out. For big structures of major importance, *primary* and *secondary* control points may be used (Fig. 20.1).

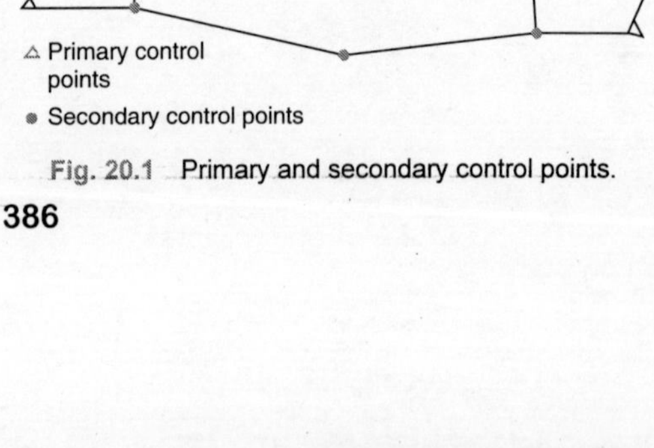

Fig. 20.1 Primary and secondary control points.

The primary control points may be the triangulation stations. The secondary control points are referred to these primary control stations. The co-ordinates of secondary points may be found by traversing methods. These secondary control points provide major control at the site. Hence, it should be located as near to the construction, but sufficiently away so that these points are not disturbed during construction operations. In the process of establishing these control points, the well known principle of *'working from whole to part'* is applied.

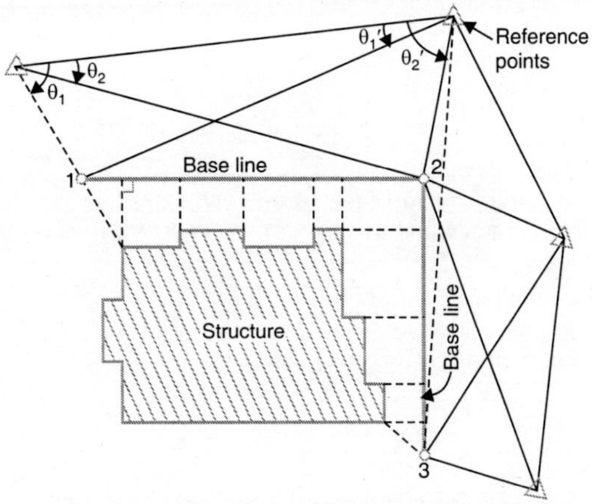

Fig. 20.2 Base line tied to reference points.

Base line: The control points can also be used to establish a *base line* on which the setting out is based, as shown in Fig. 20.2. In order to increase the accuracy at the site, two base lines, mutually perpendicular to each other are some times used.

Reference grids: Reference grids are used for accurate setting out of works of large magnitude. The following types of reference grids are used:

(*i*) Survey grid.

(*ii*) Site grid.

(*iii*) Structural grid.

(*iv*) Secondary grid.

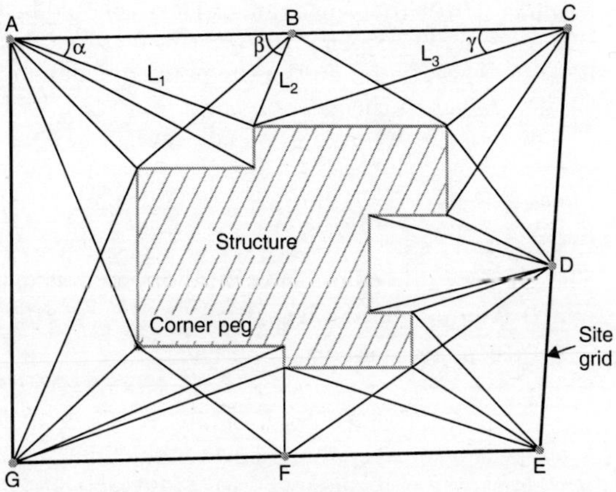

Fig. 20.3 Site grid.

Survey grid is the one which is drawn on the survey plan, from the original traverse. Original traverse stations form the control points of this grid. The *site grid,* used by the designer, is the one with the help of which actual setting out is done. As far as possible, the site grid should be actually the survey grid. All the design positions (points) are related in terms of site grid co-ordinates (Fig. 20.3). The points of the site grid are marked with wooden or steel pegs set in concrete. These grid points may be in sufficient number, so that each design point is set out with reference to at least two, and preferably three, grid points.

The *structural grid* is used when the structural components of the building (such as column etc.) are large in number and are so positioned that these components cannot be set out from the site grid with sufficient accuracy. The structural grid is set out from the site-grid points. The *secondary grid* is established inside the structural, to establish internal details of building, which are otherwise not visible directly from the structural grid.

Construction and Protection of Control Points

The control points of any grid has to be so constructed and protected that they are not disturbed during the course of construction. For non-permanent stations, wooden pegs may be used. However, for longer life, steel bolts, embedded in concrete block 600 mm × 600 mm may be used. The station may be etched on the top of the bolt.

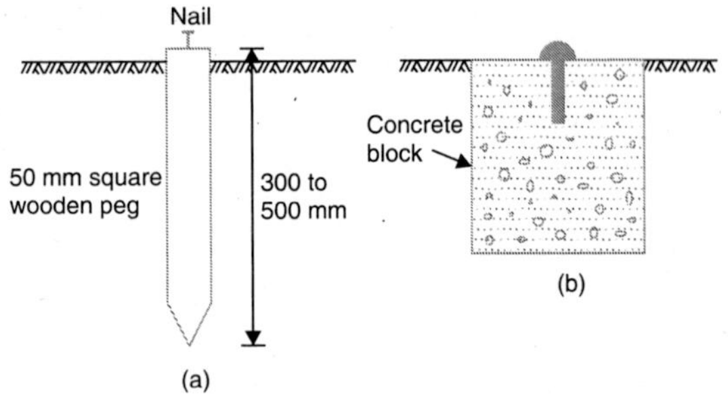

Fig. 20.4 Control points.

20.4 VERTICAL CONTROL

The vertical control consists of establishment of reference marks of known height relative to some specified datum. All levels at the site are normally reduced to a nearby *bench mark,* usually known as *master bench mark* (MBM). This master bench mark is used to establish a number of *transferred bench marks* or *temporary bench marks* (TBM) with an accuracy of levelling within ± 0.010 m. The position of TBM's should be fixed during the initial site reconnaissance. Wherever possible, permanent existing features should be used as TBM. Each TBM is referenced by a number or letter on the site plan, and should be properly related to the agreed MBM. All TBM's should be checked, properly protected, and should be re-checked at regular intervals. The distance between any two adjacent TBM's should not exceed 100 m.

20.5 SETTING OUT IN VERTICAL DIRECTION

The setting out of points in vertical direction is usually done with the help of following rods:

(*i*) Boning rods and travellers
(*ii*) Sight rails
(*iii*) Slope rails or batter boards
(*iv*) Profile boards.

Boning rods: A boning rod consists of an upright pole having a horizontal board at its top, forming a T-shaped rod. Boning rods are made in sets of three, and may consist of three T-shaped rods, each of equal size and shape, or two rods identical to each other and a third one consisting of a longer rod with a detachable or movable T-piece. The third one is called a *travelling rod* or a traveller.

Traveller: A traveller is a special type of boning rod in which the horizontal piece can be moved along a graduated vertical staff, and can be conveniently clamped at any desired height (Fig. 20.6).

Sight rails: A sight rail consists of a horizontal cross-piece nailed to a single upright or pair of uprights driven into the ground. The upper edge of the cross-piece is set to a convenient height above the required plane of the structure, and should be at a height above the ground to enable a man to conveniently align his eye with the upper edge. Various forms of sight rails are shown in Fig. 20.7. The single sight rail shown in Fig. 20.7 (*a*) is used for road works, footings and small diameter pipes, while at corners of buildings, sight rail shown in Fig. 20.7 (*b*) is used. For trenches and large diameter pipes, sight rail shown in Fig. 20.7 (*c*) is used. A *stepped*

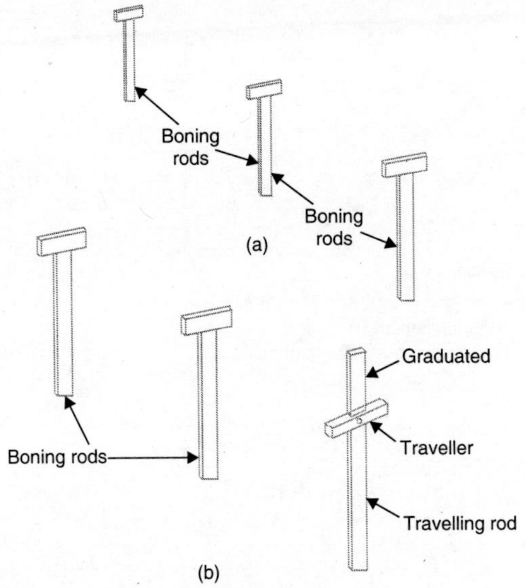

Fig. 20.5 (*a*) Three boning rods, (*b*) Two boning rods with a travelling rod.

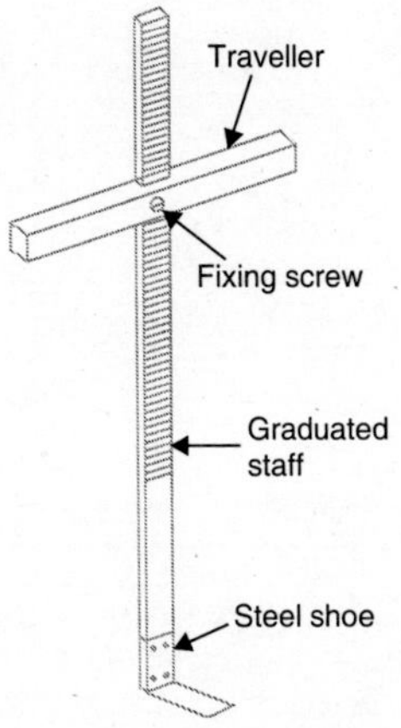

Fig. 20.6 Travelling rod.

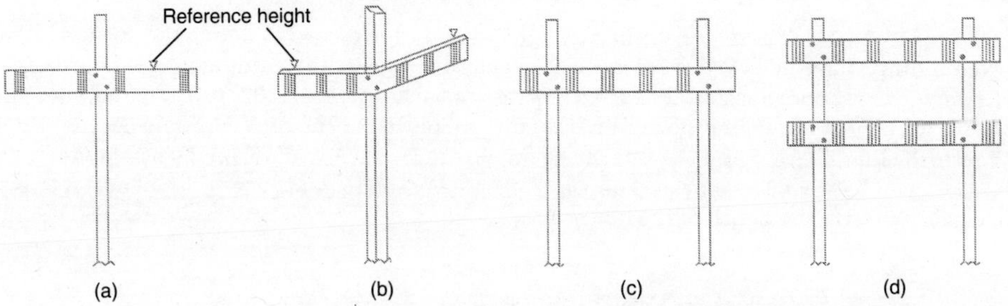

Fig. 20.7 Various forms of sight rails.

sight rail or *double sight rail,* shown in Fig. 20.7 (*d*) is used in highly undulating or falling ground.

Slope rails or batter boards. These are used for controlling the side slopes in embankments and in cuttings. These consists of two vertical poles with a sloping board nailed near their top. Fig. 20.8 (*a*) shows the use of slope rails for construction of an embankment. The slope rails define a plane parallel to the proposed slope of the embankment, but at some suitable vertical distance above it. Travellers are used to control the slope during filling operations. However, the slope rails are set at some distance x from the toe of the slope, to prevent it from disturbance during the earth work operations. Fig. 20.8 (*b*) shows the use of slope rails in cutting.

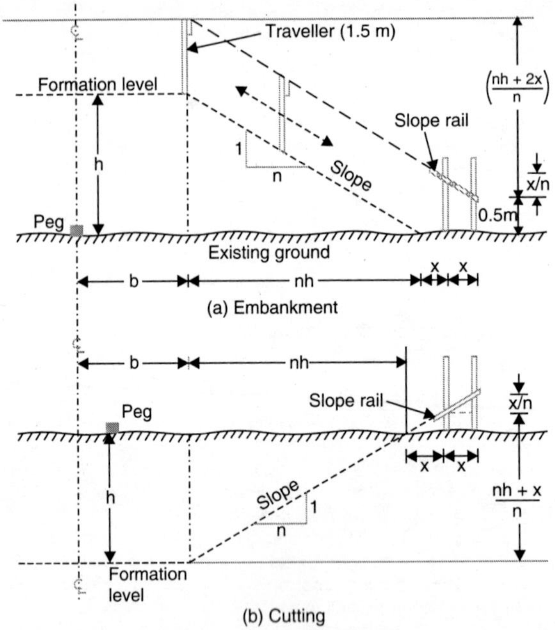

Fig. 20.8 Use of slope rails.

Profile boards: These are similar to sight rails, but are used to define the corners or sides of a building. A profile board is erected near each corner peg. Each unit of profile board consists of two verticals, one horizontal board and two cross-boards Fig. [20.9 (*a*)]. Fig. 20.9 (*b*) shows the alternative arrangement. Nails or saw cuts are placed at the tops of profile boards to define the width of foundation and the line of the outside face of the wall. A spring or piano-wire may be stretched between the marks of opposite profile boards to guide the width of cut. A traveller is used to control the depth of the cut.

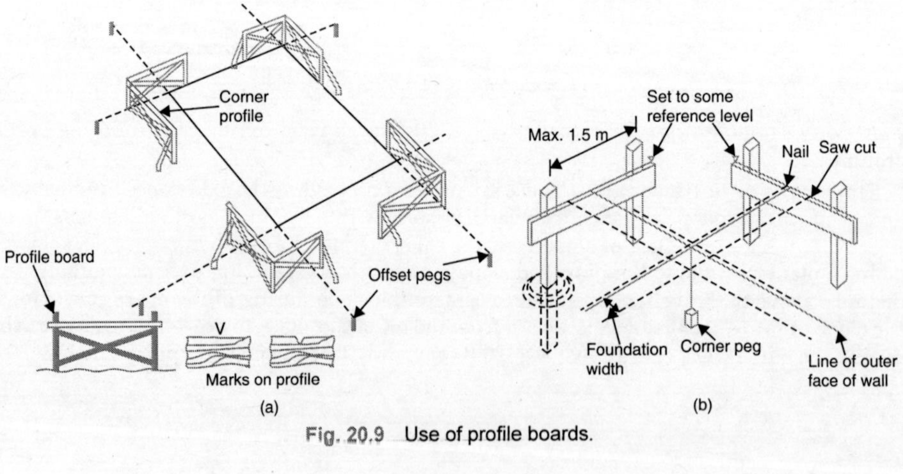

Fig. 20.9 Use of profile boards.

20.6 POSITIONING OF STRUCTURE

After having established the horizontal and vertical control points, the next operation is to locate the design points of the structure to be constructed. Following are some of the commonly used methods:

1. From existing detail
2. From co-ordinates.

When a single building is to be constructed, its corners (or salient design points) may be fixed by running a line between corners of existing building and offsetting from this. However, where an existing building or features are not available, the design points are co-ordinated in terms of site grid or base line. This can be achieved by the following:

(*i*) *Setting out by polar co-ordinates:* In this, the *distance* and bearing of each design point is calculated from at least three site grid points, as illustrated in Fig. 20.10 (*a*).

(*ii*) *By intersection* with two theodolites stationed at two stations of site grid, using bearings and checking the intersection from a third station.

(*iii*) *By offsetting* from the base line.

Offset pegs: It has been illustrated in Fig. 20.3 that the corners of a building can be set out by polar measurements from the stations of site grid. Corner pegs can then be driven in the ground. However, during the excavation of the foundations, these corner pegs get dislocated. To avoid the labour of relocation of these corner points, extra pegs, known as *offset pegs* are located on the lines of the sides of the building but offset back from true corner points, as shown in Fig. 20.11. The offset distance should be sufficient so that offset pegs are not disturbed during earthwork operations.

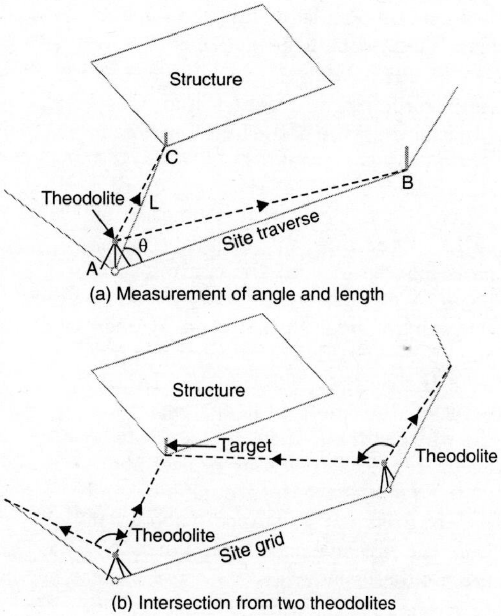

Fig. 20.10 Positioning of design points.

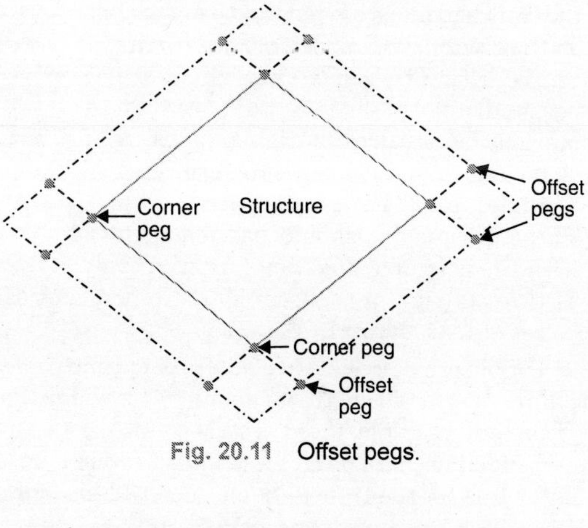

Fig. 20.11 Offset pegs.

20.7 SETTING OUT FOUNDATION TRENCHES OF BUILDINGS

The *setting out* or *ground tracing* is the process of laying down the excavation lines and centre lines etc., on the ground, before excavation is started. After the foundation design is done, a *setting out plan,* sometimes also known as *foundation layout plan,* is prepared to some suitable scale (usually 1 : 50). The plan is fully dimensioned.

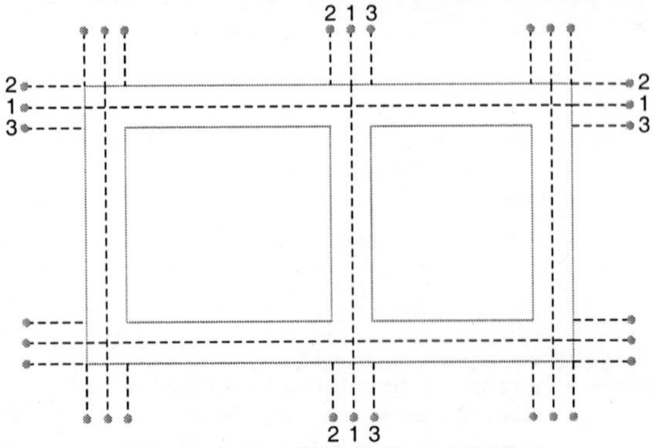

Fig. 20.12 Setting out with the help of pegs.

For setting out the foundations of small buildings, the centre line of the longest outer wall of the building is first marked on the ground by stretching a string between wooden or mild steel pegs driven at the ends. This line serves as reference line. For accurate work, nails may be fixed at the centre of the pegs. Two pegs, one on either side of the central pegs, are driven at each end of the line. Each peg is equidistant from the central peg, and the distance between the outer pegs correspond to the width of the foundation trench to be excavated. Each peg may project about 25 to 50 mm above the ground level and may be driven at a distance of about 2 m from the edge of excavation so that they are not disturbed.

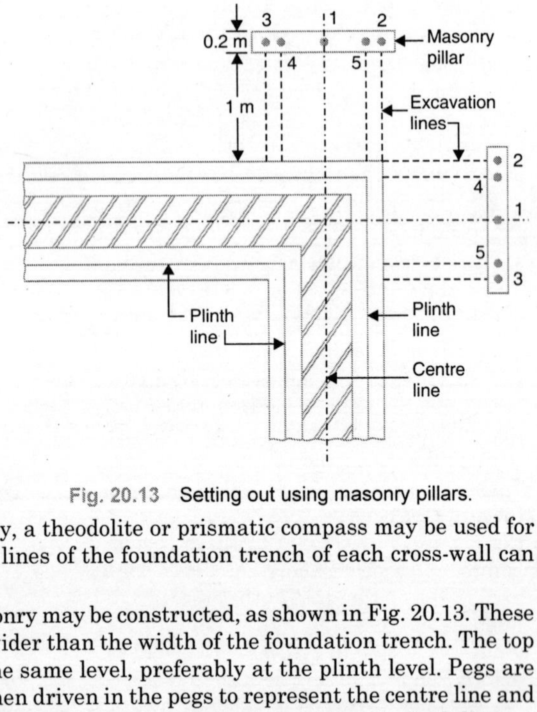

Fig. 20.13 Setting out using masonry pillars.

When string is stretched joining the corresponding pegs (say 2-2) at the extremities of the line, the boundary of the trench to be excavated can be marked on the ground with dry lime powder. The centre lines of other walls, which are perpendicular to the long wall, are then marked by setting out right angles. A right angle can be set out by forming a triangle with 3, 4 and 5 units long sides. These dimensions should be measured with the help of a steel tape. Alternatively, a theodolite or prismatic compass may be used for setting out right angles. Similarly, other lines of the foundation trench of each cross-wall can be set out, as shown in Fig. 20.12.

For big project, reference pillars of masonry may be constructed, as shown in Fig. 20.13. These pillars may be 20 cm thick, about 15 cm wider than the width of the foundation trench. The top of the pillars is plastered, and is set at the same level, preferably at the plinth level. Pegs are embedded in these pillars and nails are then driven in the pegs to represent the centre line and outer lines of the trench. Sometimes, additional nails are provided to represent plinth lines.

21 CHAPTER

SPECIAL INSTRUMENTS

21.1 INTRODUCTION

In the earlier chapters, we have studied some routine instruments which serve normal surveying operations. However, some special instruments are now available to conduct surveys for some special purpose or special operations. In this chapter, we shall study the following special instruments:

1. Site square
2. Automatic level
3. Convertible transit level
4. Special Compasses
5. Brunton universal pocket transit
6. Mountain compass-transit.

21.2 THE SITE SQUARE

As indicated in chapter 4, a site square can be used to set two lines at right angles to each other. Fig. 21.1 (*a*) shows the sketch of a site square while Fig. 21.1 (*b*) shows its photographic view. Basically, it consists of a cylindrical metal case containing two telescopes the lines of sight of which are mutually set at right angles to each other by the manufacturer. The site square is

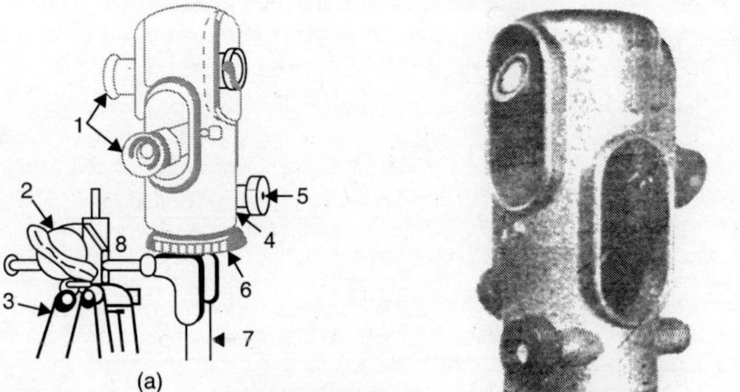

1. Telescopes
2. Clamp
3. Tripod
4. Cylindrical Metal Case
5. Fine Setting Screw
6. Knurled Ring
7. Datum Rod
8. Clamp Arm

Fig. 21.1 The Site Square.

fixed to steel pin set on the top of a metal tripod by means of a *clamp arm* and a clamp screw. The instrument is levelled with reference to a circular bubble, using this clamp screw. By this arrangement, the instrument is so mounted that it is some distance away from the tripod. A datum rod is screwed into the base of the instrument. This datum rod contains a spiked extension at the bottom.

Setting up the Site Square: Let datum rod of the site square be set on a datum peg (known as the instrument station), by placing the clamp of the datum arm over the tripod pin kept at upper most position. The arm is so positioned that it is nearly horizontal, with the clamp about mid-distance along the arm. The tripod is placed in such a position that the datum rod is approximately over the datum peg. The site square is then placed on the top of the datum rod. The instrument is secured by turning the knurled screw in clockwise direction. Thus, the instrument has been set on the datum rod. However, it is capable of being rotated. The instrument is now levelled with reference to the circular bubble, by holding the instrument with one hand, releasing the clamp screw, and moving the instrument slightly till the bubble is in centre. The clamp screw is then secured.

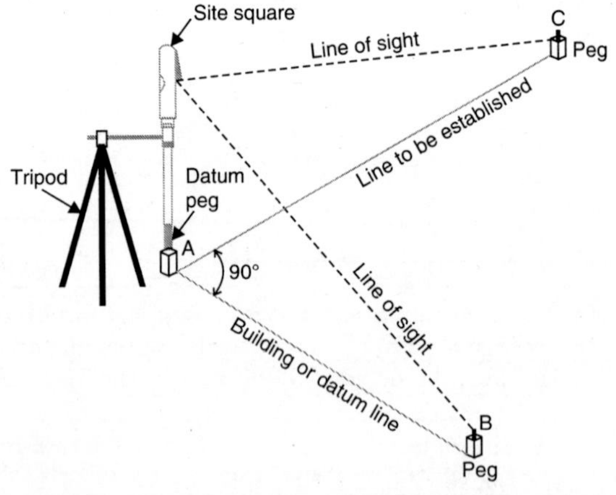

Fig. 21.2 Setting out a Right Angle.

Setting out a Right Angle: Let it be required to set a line AC *at* right angles to a given datum line AB, at the datum station A (Fig. 21.2)

The tripod is so set near peg A that the datum rod is exactly over the datum peg A. Line AB is the building line or datum line. The instrument is so rotated that one telescope is on the datum line AB. The instrument is locked in position, and the fine setting screw is rotated so that the line of sight bisects the station mark on peg B. A sight is now made through the other telescope and a ranging rod is held as near as possible at the right angles to line AB. The observer now signals the person holding the ranging rod so that the line of sight exactly bisects the ranging rod. A peg is now inserted at the base C of the ranging rod.

21.3 AUTOMATIC OR AUTOSET LEVEL

An *automatic level* or *autoset level* contains an optical compensator which maintains a level line of collimation even though the instrument may be tilted as much as 15 minutes of arc. In conventional levelling instrument, the line of collimation is made horizontal by means of long bubble tube. This is a time consuming job. In such a conventional instrument if the bubble is not

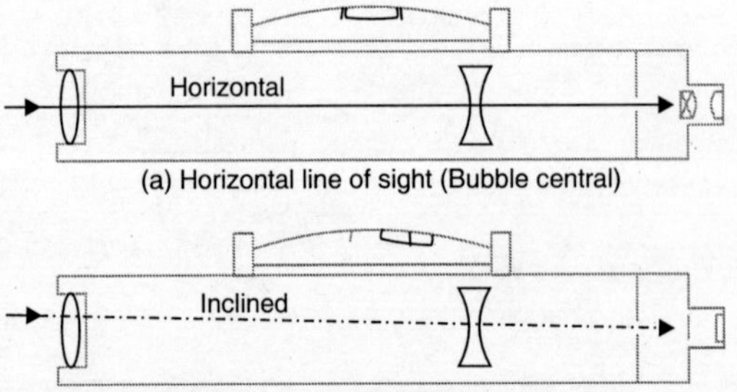

(a) Horizontal line of sight (Bubble central)

(b) Inclined line of sight (Bubble out of centre)

Fig. 21.3 Conventional Level.

in the centre of its run, the vertical axis will not be truly vertical, and the line of collimation will be tilted instead of being horizontal. Fig. 21.3 shows a conventional levelling instrument showing both (*a*) horizontal line of sight as well as (*b*) inclined line of sight.

In an autoset level, spirit bubble is no longer required to set a horizontal line of collimation. In such a level, the line of collimation is directed through a system of *compensators* which ensure that the line of sight viewed through the telescope is horizontal even if the optical axis of the telescope tube itself is not horizontal. A circular bubble is used to level the instrument approximately 15′ of the vertical, either with the help of footscrew arrangements or a quickest device. An automatic or autoset level is also sometimes known as a *self-levelling level* or *pendulum level.*

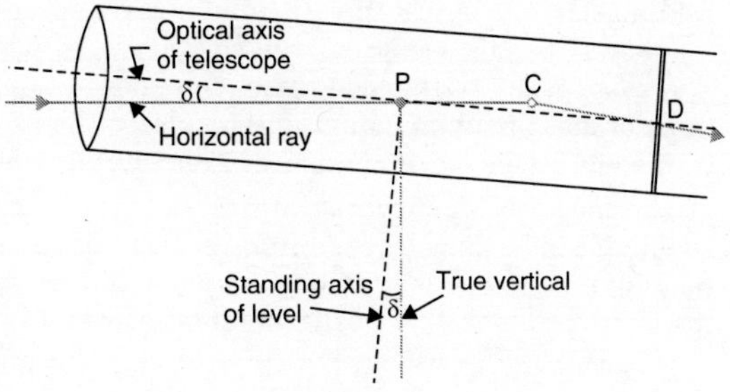

Fig. 21.4 Principle of Compensator.

Fig. 21.4. shows the principle of the compensators. The small angle δ(< 15′) between the standing axis and vertical axis tilts the telescope by the same amount δ. Point *P* is the point of rotation of the telescope. The compensator located at *C* deviate all horizontal rays of light entering the telescope tube (at the same height as *P*) through the centre of cross-hairs *D*.

The compensating systems may be of two types (*a*) *free suspension compensators,* and (*b*) *Mechanical compensators*. The former type consists of two prisms on a suspended mount within the telescope tube. If the autoset level is tilted, the compensating system hangs like a plumb bob and keeps the horizontal ray on the cross-hairs automatically. The mechanical compensators consist of a fixed roof prism above two swinging prisms supported on four metallic tapes forming a cross spring flexure pivot. The ingenuity of design ensures a frictionless suspension having a repetition of setting better than 1 second of arc. Both the systems use air damping system, in which the compensator is attached to prism moving in a closed cylinder. This will reduce the oscillation of the light-weight compensator.

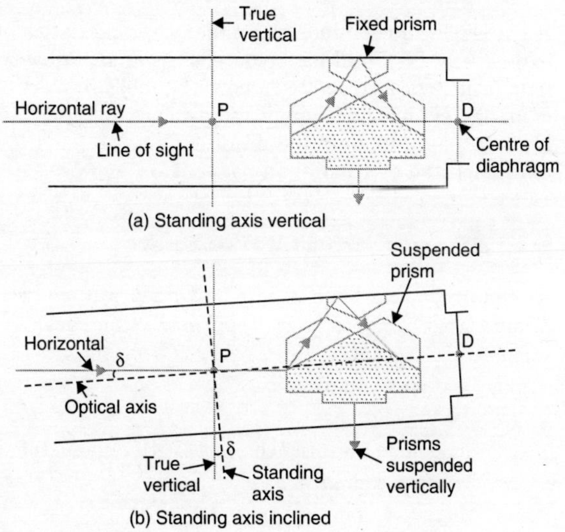

Fig. 21.5 The Compensator System.

The commercially available autoset levels use different forms of compensators. Fig. 21.6 (*a*) shows an autoset level Ni 2 by Zeiss. Fig. 21.6 (*b*) shows the photograph of the compensator used in Zeiss Ni 2 level.

Wild NA2 and NAK2 Automatic Levels: Wild NA2 automatic level is useful for levelling of all types and all orders of accuracy. The advantages of the automatic level is that as soon as the circular bubble is central, the line of sight is horizontal for all pointings of the telescope. Elimination of the traditional tubular level speed up work and improves accuracy. Fig. 21.7 (*a*) shows the cut section of the compensator, which is essentially a pendulum with prism. The suspension system comprises four flexed tapes made of special alloy to ensure faultless functioning. Pneumatic damping insulates the pendulum from the influence of strong winds, traffic vibrations etc. and ensures great stability of the line of sight.

The model NAK2 incorporates a horizontal circle for angle measurements. Tacheometric levelling can be done by combining stadia and angular measurements with height readings. A reading microscope is provided near the main eyepiece, for circle reading. Fig. 21.7 (*b*) shows the photograph of Wild NAK2 automatic level.

In both NA2 as well as NAK2 levels, a press button is provided just below the eyepiece for compensator control. Pressing this button gives the compensator a gental tap, so that the observer sees the staff image swing smoothly away and then float gently back to give the horizontal line of sight. This check, which takes less than a second, is technically perfect, as the pendulum itself is activated and swings through its full range. It is also immediately apparent if the circular bubble is not centred.

21.4 TRANSIT LEVEL

A transit level combines the major characteristics of both a level as well as a transit, used extensively for building layouts, road and highway works, excavation measurements and foundation works. Fig. 21.8 shows Fennel's convertible transit level. For measuring or layout of angles, clamps and tangent screws provide exact sighting. Horizontal circle and vertical arc have verniers reading to 5 minutes. Fig. 21.9 shows the photograph of builder's transit level manufactured by Keuffel and Esser Co.

21.5 SPECIAL COMPASSES

In chapter 5, we have studied two types of compasses for the measurement of bearings: (*i*) surveyors compass, and (*ii*) prismatic compass. We shall now consider the following special purpose compasses:

(*i*) Geologist compass
(*ii*) Mining compass
(*iii*) Suspension mining compass with clinometer
(*iv*) Brunton compass
(*v*) Mountain compass transit.

1. **Geologist compass:** Fig. 21.10 shows the photograph of geologist's compass. It is very suitable for the determination of magnetic bearings and slopes of layers. The angle between, the layer direction and magnetic north is measured by means of the compass dial, the knife-edged ground plate being set horizontally and rectangular to the gradient line. For measuring the inclination angle, the edge of the ground plate is set upon the gradient line of the layer. The compass taken upright, the clinometer hinge indicates the slope angle.

2. **Mining compass:** For works below ground level, the mining compass is more suitable, instead of the geologists compass. Fig. 21.11 shows the photograph of a mining compass. String hooks are provided as finder sights and for bearing measurements with the aid of the string.
3. **Suspension mining compass with clinometer:** It basically consists of a compass box connected with a suspension frame. The string of the suspension frame is set along the dip of the strata, and its slope is measured with the help of a large diameter clinometer with plumb bob. Fennel Kassel manufactures two variations : (*i*) Kassel type, and (*ii*) Freiberg type.

Fig. 21.12 shows the photograph of the Kassel type mining compass. The compass is connected by *hinges* with suspension frame which has the advantage of easy packing and taking less space in the container. The clamping screw of the knife-edged magnetic needle is placed on the brim of the compass ring. The horizontal circle is divided at intervals of 1 degree and figured every 10 degrees. The clinometer, made from light metal, has a diameter of 9.4 inch and is graduated to 1/3 degree.

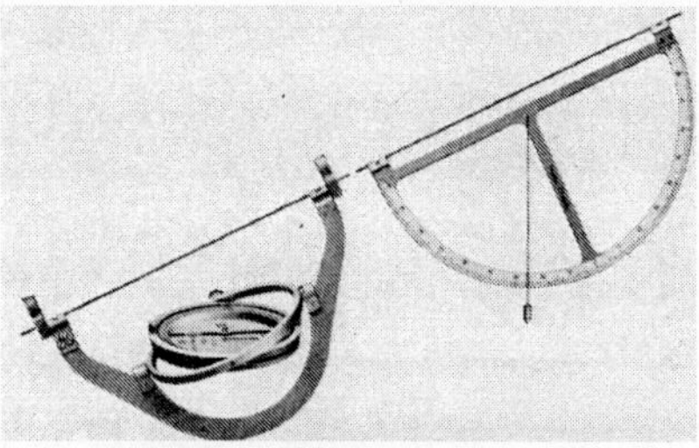

Fig. 21.13 Freiberg Type Mining Suspension Compass with Clinometer.

Freiberg type compass with clinometer is shown in Fig. 21.13. The functions of the mining compass of Freiberg type are exactly the same as with Kassel type. Its mechanical features depart in two things from the Kassel type, viz, the rigid connection of the compass suspension with the frame and the clamping screw to be placed centrically, under the compass box.

21.6 BRUNTON UNIVERSAL POCKET TRANSIT

Brunton's Universal pocket transit is one of the most convenient and versatile instrument for preliminary surveying on the surface or underground. It is suitable for forestry, geological and mining purposes, and for simple contour and tracing work. The main part of Brunton Pocket Transit is the magnetic compass with a 5 cm long magnetic needle pivoting on an agate cap. Special pinion arrangement provides for the adjustment of the local variation of the declination with a range of ± 30°. For accurate centring purposes a circular spirit bubble is built in. A clinometer connected with a tubular spirit bubble covers measurement of vertical angles within a range of ± 90°. Fig. 21.14 shows the photograph of Brunton Universal pocket transit along with box containing various accessories.

The Brunton pocket transit comprises a wide field of application for which it is equipped with the following special accessories:

1. *Camera tripod* for measurement of horizontal and vertical angles.
2. *Plane table* for using the compass as on alidade.
3. *Protractor base plate* for protracting work in the field or in the office.
4. *Suspension plate* for use of the instrument as a mining compass.
5. *Brackets* for suspension plate

Measurement of Horizontal Angles: Horizontal and vertical angles can be measured by using the camera tripod with the ball joint. For measuring horizontal angles, the compass box has to be screwed on the ball joint until the locking pin will fit into the socket which is imbedded in the compass case. For more precise centring, a plumb bob can be fastened at the plumb hook of the tripod. Accurate setting of the instrument is accomplished with a circular spirit bubble. The north end of the needle indicates magnetic bearing on the compass graduation.

Measurement of vertical angle: For measuring vertical angles, the compass has to be fitted in the ball joint. The observations have to be carried out with completely opened mirror by sighting through the hole of the diopter ring and the pointer. Before readings can be taken, the tubular bubble which is connected with the clinometer arm has to be centered by turning the small handle mounted at the back of the compass. Using the instrument in this vertical position, it is necessary to lock the needle to prevent the agate cap and the pivot from being damaged.

Use as a mining compass: Brunton compass can be fitted on the suspension plate and be used as mining compass. The compass is correctly positioned on the plate when the locking pin fits into the socket. Then, the North-South line of the compass is parallel to the longitudinal axis of the suspension plate.

For vertical angle measurements, the hook hinges have be fitted. The brackets prevent the suspension outfit from sliding along the rope. Before readings of vertical circle can be taken, accurate centring of the clinometer arm bubble is necessary.

Use with plane table: The compass in connection with the protector base plate can be used for protecting work in the field or in the office. The parallelism of the base plate edges and the line of sight of the compass is secured when the locking pin on the plate fits accurately into the socket. This combination gives the possibility to employ the compass as an alidade for minor plane table surveys.

21.7 MOUNTAIN COMPASS-TRANSIT

A mountain compass-transit (also known as compass theodolite) basically consists of a compass with a telescope. Both these are mounted on a levelling head which can be mounted on a tripod. For movement of the instrument about vertical axis, a clamp and tangent screw is used. For measurement of vertical angles, the telescope can rotate about the trunnion axis, provided with a clamp and slow motion screw. The instrument is levelled with respect to a circular bubble mounted on the upper plate, and a longitudinal bubble tube mounted on the telescope. Fig. 21.15 shows the photograph of a compass transit by Breithaupt Kassel. The instrument is suitable for compass traversing, reconnaissance, contour works, and for the purposes of forest departments. The eccentric telescope admits steep sights (in mountainous area), being provided with stadia hairs for optical distance measurings (tacheometric surveying). A telescope reversion spirit level suits the determination of the station-height as well as auxiliary levelling. The vertical circle is graduated to 1° and reading with vernier can be taken upto 6′. The compass ring is graduated to 1° and reading can be estimated to 6′.

22

CHAPTER

TACHEOMETRIC SURVEYING

22.1 GENERAL

Tacheometry (or Tachemetry or Telemetry) is a branch of angular surveying in which the horizontal and vertical distances of points are obtained by optical means as opposed to the ordinary slower process of measurements by tape or chain. The method is very rapid and convenient. Although the accuracy of Tacheometry in general compares unfavourably with that of chaining, it is best adapted in obstacles such as steep and broken ground, deep ravines, stretches of water or swamp and so on, which make chaining difficult or impossible. The accuracy attained is such that under favourable conditions the error will not exceed 1/1000 , and if the purpose of a survey does not require greater accuracy, the method is unexcelled. The primary object of tacheometry is the preparation of contoured maps or plans requiring both the horizontal as well as vertical control. Also, on surveys of higher accuracy, it provides a check on distances measured with the tape.

22.2 INSTRUMENTS

An ordinary transit theodolite fitted with a stadia diaphragm is generally used for tacheometric survey. The stadia diaphragm essentially consists of one stadia hair above and the other an equal distance below the horizontal cross-hair, the stadia hairs being mounted in the same ring and in the same vertical plane as the horizontal and vertical cross-hairs. Fig. 22.1 (*a* to *f*) shows the different forms of stadia diaphragm commonly used.

The telescope used in stadia surveying are of three kinds:

1. the simple external–focusing telescope.
2. the external-focusing anallactic telescope (Porro's telescope).
3. the internal–focusing telescope.

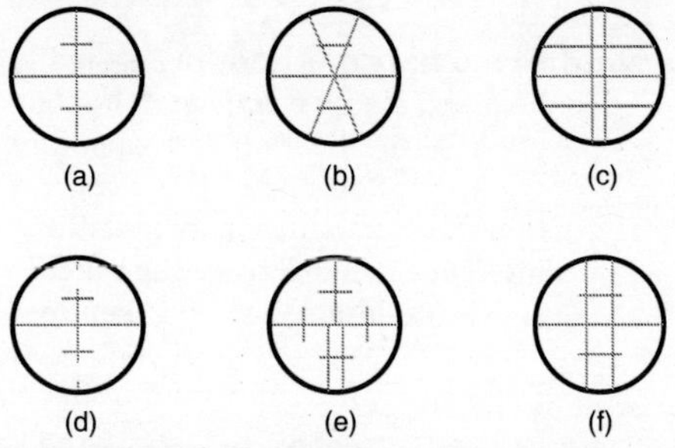

Fig. 22.1 Various Patterns of Stadia Diaphragm.

The first type is known as *stadia theodolite*, while the second type is known as *'tacheometer'*. The *'tacheometer'* (as such) has the advantage over the first and the third type due to the fact that the additive constant of the instrument is zero. However, the internal focusing telescope is becoming more popular, though it has a very small additive constant. Some of the latest patterns of internal focusing telescope may be regarded as strictly anallactic (see § 22.7).

A *tacheometer* must essentially incorporate the following features:

(*i*) The multiplying constant should have a nominal value of 100 and the error contained in this value should not exceed 1 in 1000.

(*ii*) The axial horizontal line should be exactly midway between the other two lines.

(*iii*) The telescope should be truly anallactic.

(*iv*) The telescope should be powerful having a magnification of 20 to 30 diameters.

The aperture of the objective should be 35 to 45 mm in diameter in order to have a sufficiently bright image.

For small distances (say upto 100 metres), ordinary levelling staff may be used. For greater distances a stadia rod may be used. A stadia rod is usually of one piece, having 3 to 5 metres length. The pattern of graduations should be bold and simple. Fig. 22.2 shows two typical patterns of graduations. For smaller distances, a stadia rod graduated in 5 mm (*i.e.* 0.005 m) may be used, while for longer distances, the rod may be graduated in 1 cm (*i.e.* 0.01 m).

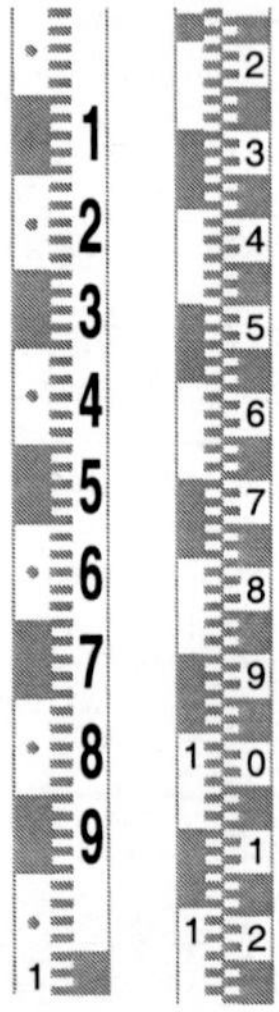

Fig. 22.2 Stadia Rods.

22.3 DIFFERENT SYSTEMS OF TACHEOMETRIC MEASUREMENT

The various systems of tacheometric survey may be classified as follows:

1. The stadia system
   (*a*) Fixed hair method or stadia method
   (*b*) Movable hair method, or subtense method.
2. The tangential system.
3. Measurements by means of special instruments.

The principle common to all the systems is to calculate the horizontal distance between two points *A* and *B* and their difference in elevation, by observing (*i*) the angle at the instrument at *A* subtended by a known short distance along a staff kept at *B*, and (*ii*) the vertical angle to *B* from *A*.

(*a*) **Fixed hair method**. In this method, observation (*i*) mentioned above is made with the help of a stadia diaphragm having stadia wires at fixed or constant distance apart. The readings on the staff corresponding to all the three wires are taken. The staff intercept, *i.e.*, the difference of the readings corresponding to top and bottom stadia wires will therefore, depend on the distance of the staff from the instrument. When the staff intercept is more than the length of the staff, only half intercept is read. For inclined sights, readings may be taken by keeping the staff either vertical or normal to the line of sight. This is the most common method in tacheometry and the name 'stadia method' generally bears reference to this method.

(*b*) **Subtense method.** This method is similar to the fixed hair method except that the stadia interval is variable. Suitable arrangement is made to vary the distance between the stadia hair so as to set them against the two targets on the staff kept at the point under observation. Thus, in this case, the staff intercept, *i.e.*, the distance between the two targets is kept fixed while the stadia interval, *i.e.*, the distance between the stadia hairs is variable. As in the case of fixed hair method, inclined sights may also be taken.

The tangential method. In this method, the stadia hairs are not used, the readings being taken against the horizontal cross-hair. To measure the staff intercept, two pointings of the instruments are, therefore, necessary. This necessitates measurement of vertical angles twice for one single observation.

THE STADIA METHOD

22.4 PRINCIPLE OF STADIA METHOD

The stadia method is based on the principle that the ratio of the perpendicular to the base is constant in similar isosceles triangles.

In Fig. 22.3 (*a*), let two rays OA and OB be equally inclined to the central ray OC. Let A_2B_2, A_1B_1 and AB be the staff intercepts. Evidently,

$$\frac{OC_2}{A_2B_2} = \frac{OC_1}{A_1B_1} = \frac{OC}{AB}$$

$$= \text{constant } k = \frac{1}{2}\cot\frac{\beta}{2}.$$

This constant k entirely depends upon the magnitude of the angle β. If β is made equal to 34′ 22″.64, the constant $k = \frac{1}{2}\cot 17' \, 11''.32 = 100$.

In this case, the distance between the staff and the point O will be 100 times the staff intercept. In actual practice, observations may be made with either horizontal line of sight or with inclined line of sight. In the latter case, the staff may be kept either vertically or normal to the line of sight. We shall first derive the *distance-elevation formulae* for the horizontal sights.

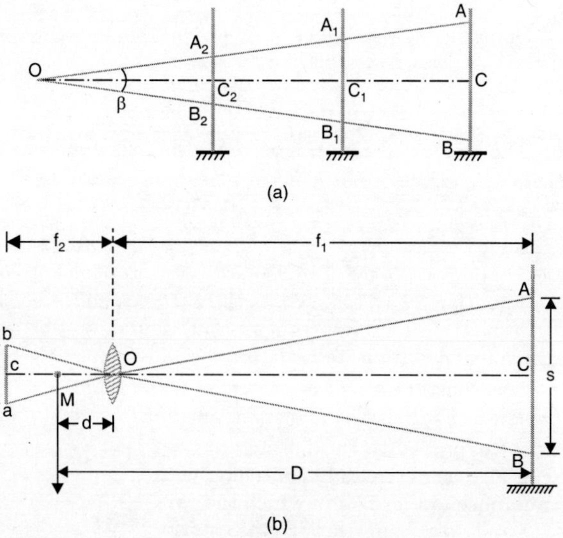

Fig. 22.3 Principle of Stadia Method.

Horizontal sight consider Fig. 22.3 (*b*) in which O is the optical centre of the objective of an *external focusing telesope.*

Let A, C and B = The points cut by the three lines of sight corresponding to the three wires.

b, c and a = Top, axial and bottom hairs of the diaphragm.

$ab = i$ = interval between the stadia hairs (stadia interval)

$AB = s$ = staff intercept.

f = focal length of the objective

f_1 = Horizontal distance of the staff from the optical centre of the objective.

f_2 = Horizontal distance of the cross-wires from O.

d = Distance of the vertical axis of the instrument from O.

D = Horizontal distance of the staff from the vertical axis of the instrument.

M = Centre of the instrument, corresponding to the vertical axis.

Since the rays BOb and AOa pass through the optical centre, they are straight so that, Δs AOB and aOb are similar. Hence,

$$\frac{f_1}{f_2} = \frac{s}{i} \qquad \text{...(i)}$$

Again, since f_1 and f_2 are conjugate focal distances, we have from lens formula,

$$\frac{1}{f} = \frac{1}{f_2} + \frac{1}{f_1} \qquad \text{...(ii)}$$

Multiplying throughout by ff_1, we get $f_1 = \frac{f_1}{f_2} f + f$

Substituting the values of $\frac{f_1}{f_2} = \frac{s}{i}$ in the above, we get

$$f_1 = \frac{s}{i} f + f \qquad \text{...(iii)}$$

The horizontal distance between the axis and the staff is

$$D = f_1 + d$$

or

$$D = \frac{f}{i} s + (f + d) = ks + C \qquad \text{...(iv)}$$

Equation 22.1 is known as the *distance equation*. In order to get the horizontal distance, therefore, the staff intercept s is to be found by subtracting the staff readings corresponding to the top and bottom stadia hairs.

The constant $k = f/i$ is known as the *multiplying constant* or *stadia interval factor* and the constant $(f + d) = C$ is known as the *additive constant* of the instrument.

Alternative Method. Equation 22.1 can also be derived alternatively, with reference to Fig. 22.4 in which the rays Bb' and Aa' passing through the exterior principal focus F, become parallel to the optical axis. The rays Aa and Bb pass through O and remain undeviated.

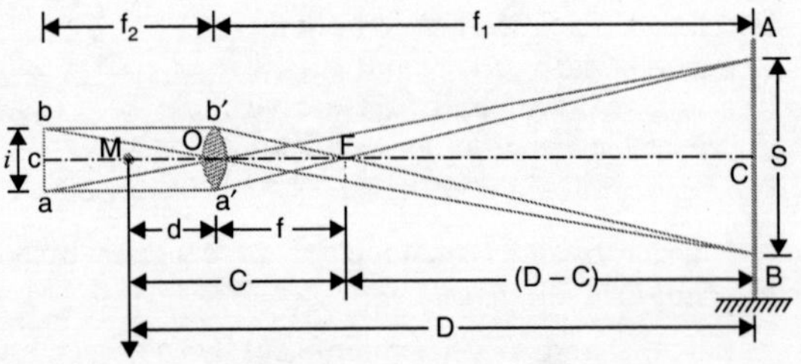

Fig. 22.4 Principle of Stadia Method.

Since the stadia interval ab is fixed in magnitude, the points a' and b' are fixed. Again, since F is also fixed, being the exterior principal focus of the objective, the angle AFB is fixed in magnitude.

From similar triangles AFB and $a'Fb'$, we have

$$\frac{FC}{AB} = \frac{OF}{a'b'} = \frac{f}{i} \text{ or } FC = \frac{f}{i} AB = \frac{f}{i} s$$

Distance from the axis to the staff is given by

$$D = FC + (f + d) = \frac{f}{i} s + (f + d) = k\, s + C \qquad \text{...(22.1)}$$

Note: *Since point F is the vertex of the measuring triangle, it is also sometimes called the anallactic point.*

Elevation of the Staff Station. Since the line of sight is horizontal, the elevation of the staff station can be found out exactly in the same manner as the levelling. Thus,

Elevation of staff station = Elevation of instrument axis – Central hair reading.

Determination of constants ***k* and *C***

The values of the multiplying constant k and the additive constant C can be computed by the following methods:

1st Method. In this method, the additive constant $C = (f + d)$ is measured from the instrument while the multiplying constant k is computed from field observations:

1. Focus the instrument to a distant object and measure along the telescope the distance between the objective and cross-hairs.

$$\frac{1}{f} = \frac{1}{f_1} + \frac{1}{f_2}$$

Since f_1 is very large in this case, f is approximately equal to f_2, *i.e.*, equal to the distance of the diaphragm from the objective.

2. The distance d between the instrument axis and the objective is variable in the case of external focusing telescope, being greater for short sights and smaller for long sights. It should, therefore be measured for average sight. Thus, the additive constant $(f + d)$ is known.
3. To calculate the multiplying constant k, measure a known distance D_1, and take the intercept s_1, on the staff kept at that point, the line of sight being horizontal. Using equation 22.1,

$$D_1 = ks_1 + C \qquad \text{or} \qquad k = \frac{D_1 - C}{s_1}$$

For average value, staff intercepts, s_2, s_3, etc., can be measured corresponding to distance D_2, D_3 etc., and mean value can be calculated.

Note: *In the case of some external focusing instruments, the eye-piece diaphragm unit moves during focusing. For such instruments d is constant and does not vary while focusing.*

2nd Method. In this method, both the constants are determined by field observations as under:

1. Measure a line, about 200 m long, on fairly level ground and drive pegs at some interval, say 50 metres.
2. Keep the staff on the pegs and observe the corresponding staff intercepts with horizontal sight.
3. Knowing the values of D and s for different points, a number of simultaneous equations can be formed by substituting the values of D and s in equation 22.1. The simultaneous solution of successive pairs of equations will give the values of k and C, and the average of these can be found.

For example, if s_1 is the staff intercept corresponding to distance D_1 and s_2 corresponding to D_2 we have

$$D_1 = ks_1 + C \quad ...(i) \quad \text{and } D_2 = ks_2 + C \qquad ...(ii)$$

Subtracting (*i*) from (*ii*), we get $k = \dfrac{D_2 - D_1}{s_2 - s_1}$ ...(22.2)

Substituting the values of k in (*i*), we get

$$C = D_1 - \frac{D_2 - D_1}{s_2 - s_1} s_1 = \frac{D_1 s_2 - D_1 s_1 - D_2 s_1 + D_1 s_1}{s_2 - s_1}$$

or

$$C = \frac{D_1 s_2 - D_2 s_1}{s_2 - s_1} \qquad ...(22.3)$$

Thus, equations 22.2 and 22.3 give the values of k and C.

22.5 DISTANCE AND ELEVATION FORMULAE FOR STAFF VERTICAL: INCLINED SIGHT

Let P = Instrument station; Q = Staff station

M = Position of instruments axis; O = Optical centre of the objective

A, C, B = Points corresponding to the readings of the three hairs

$s = AB$ = Staff intercept; i = Stadia interval

θ = Inclination of the line of sight from the horizontal

L = Length MC measured along the line of sight

$D = MQ'$ = Horizontal distance between the instrument and the staff

V = Vertical intercept, at Q, between the line of sight and the horizontal line.

h = Height of the instrument

r = Central hair reading

β = Angle between the two extreme rays corresponding to stadia hairs.

Draw a line $A'CB'$ (Fig. 22.5) normal to the line of sight OC.

$\angle AA'C = 90° + \dfrac{\beta}{2}$, being the exterior angle of the $\Delta COA'$.

Similarly, from $\Delta COB'$, $\angle OB'C = \angle BB'C = 90° - \dfrac{\beta}{2}$.

Since $\dfrac{\beta}{2}$ is very small (its value being equal to 17′11″ for $k = 100$), $\angle AA'C$ and $\angle BB'C$ may be approximately taken equal to 90°.

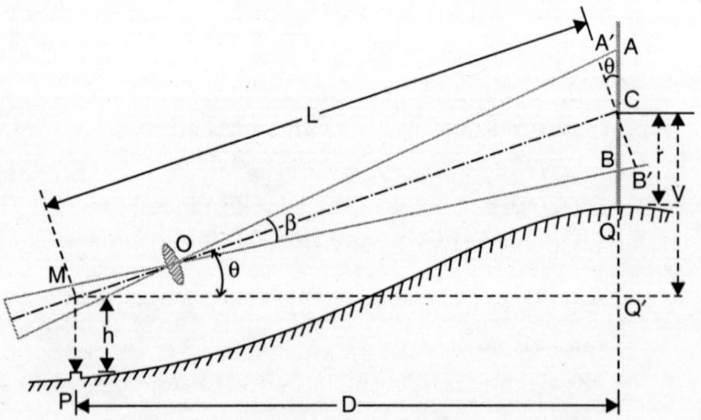

Fig. 22.5 Elevated Sight: Vertical Holding.

$$\therefore \quad \angle AA'C = \angle BB'C \approx 90°$$

From $\Delta ACA'$, $A'C = AC \cos\theta$ or $A'B' = AB \cos\theta = s \cos\theta$ ...(*i*)

Since the line $A'B'$ is perpendicular to the line of sight OC, equation 22.1 is directly applicable. Hence, we have

$$MC = L = k \,.\, A'B' + C = ks \cos\theta + C \qquad ...(ii)$$

The horizontal distance

$$D = L \cos\theta = (ks \cos\theta + C) \cos\theta.$$

or

$$D = ks \cos^2\theta + C \cos\theta \qquad ...(22.4)$$

Similarly,

$$V = L \sin\theta = (ks \cos\theta + C) \sin\theta = ks \cos\theta.\ \sin\theta + C \sin\theta$$

or

$$V = ks\, \frac{\sin 2\theta}{2} + C \sin\theta \qquad ...(22.5)$$

Thus, equations, 22.4 and 22.5 are the distance and elevation formulae for inclined line of sight.

(*a*) **Elevation of the staff station for angle of elevation:** If the line of sight has an angle of elevation θ, as shown in Fig. 22.5, we have

Elev. of staff station = Elev. of instrument station + h + V − r.

(*b*) **Elevation of the staff station for the angle of depression:** Fig. 22.6,

Elevation of Q = Elevation of $P + h - V - r$.

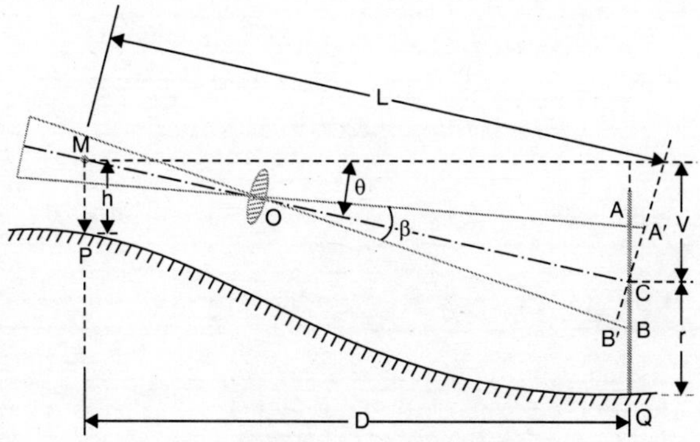

Fig. 22.6 Depressed Sight: Vertical Holding.

22.6 DISTANCE AND ELEVATION FORMULAE FOR STAFF NORMAL

Fig. 22.7 shows the case when the staff is held normal to the line of sight.

Case (*a*) **Line of sight at an angle of elevation θ** (Fig. 22.7)

Let $AB = s$ = Staff intercept; $CQ = r$ = Axial hair reading

With the same notations as in the previous case, we have

$$MC = L = ks + C$$

The horizontal distance between P and Q is given by

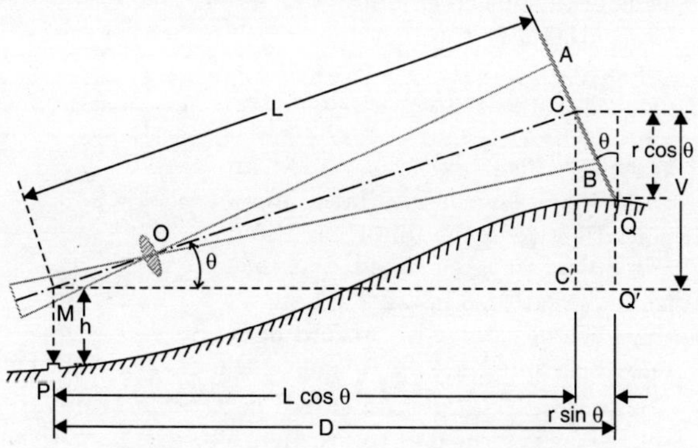

Fig. 22.7 Elevated Sight: Normal Holding.

$$D = MC' + C'Q' = L \cos \theta + r \sin \theta = (ks + C) \cos \theta + r \sin \theta \quad ...(22.6)$$

Similarly, $V = L \sin \theta = (ks + C) \sin \theta$ ...(22.7)

Elev. of Q = Elev. of $P + h + V - r \cos \theta$.

Case (*b*) Line of sight at an angle of depression θ

When the line of sight is depressed downwards (Fig. 22.8)

$$MC = L = ks + C$$

$$D = MQ' = MC' - Q'C'$$

$$= L \cos \theta - r \sin \theta$$

$$= (ks + C) \cos \theta - r \sin \theta \quad ...(22.8)$$

$$V = L \sin \theta = (ks + C) \sin \theta \quad ...(22.9)$$

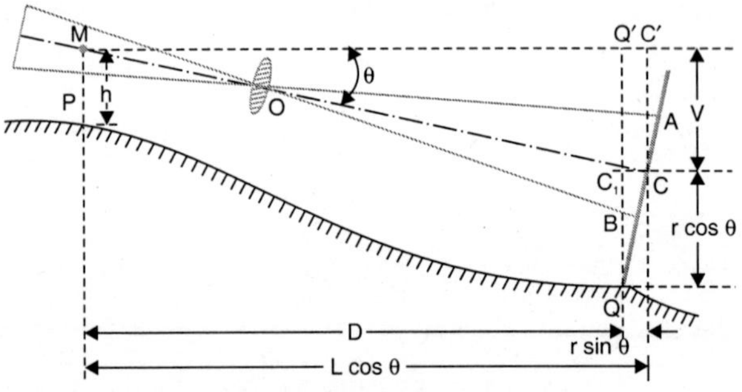

Fig. 22.8 Depressed Sight: Normal Holding.

Elev. of Q = Elev. of $P + h - V - r \cos \theta$

22.7 THE ANALLACTIC LENS

In the distance formula $D = ks + C$, the staff intercept s is proportional to $(D - C)$ which is the distance between the staff and the exterior principal focus of the objective (see Fig. 22.4). This is because the vertex of the measuring triangle (or anallactic point) falls at the exterior principal focus of the objective and *not* at the vertical axis of the instrument. In 1840, Porro devised the *external focusing anallactic telescope*, the special feature of which is an additional (convex) lens, called an *anallactic lens* (or anallatic lens), placed between the diaphragm and the objective at a fixed distance from the latter. Fig. 22.9 (*a*) shows the lines of sight with an ordinary telescope, and Fig. 22.9 (*b*) shows the lines of sight with an anallactic lens.

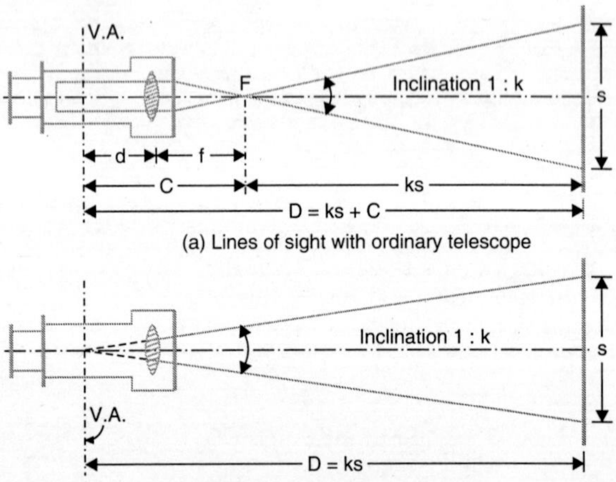

Fig. 22.9

The word *'anallactic'* means 'unalterable' or 'invariable'; by the provision of anallactic lens, the vertex is formed at the vertical axis and its position is always fixed irrespective of the staff position. The anallactic lens is generally provided in external focusing telescope only and not in internal focusing telescope since the latter is virtually anallactic due to very small additive constant.

Theory of Anallactic Lens : Horizontal Sights

Fig. 22.10 shows the optical diagram of an external focusing anallactic telescope.

Let O = Optical centre of the objective

N = Optical centre of the anallactic lens

M = Position of the vertical axis of the instrument

F' = Exterior principal focus of the anallactic lens

A, B = Points on the staff corresponding to the stadia wires

a_1, b_1 = Corresponding points on objective

a_3, b_3 = Corresponding points on anallatic lens

a, b = Position of stadia wires

a_2, b_2 = Corresponding points if there were no anallactic lens

f_1 and f_2 = The conjugate focal length of the objective

D = Distance of the staff from the vertical axis

d = Distance of the vertical axis from the objective

m = Distance of the diaphragm from the objective

n = Distance of the anallactic lens from the objective

f = Focal length of object glass

f' = Focal length of the anallactic lens

i = Stadia interval

$s = AB$ = staff intercept.

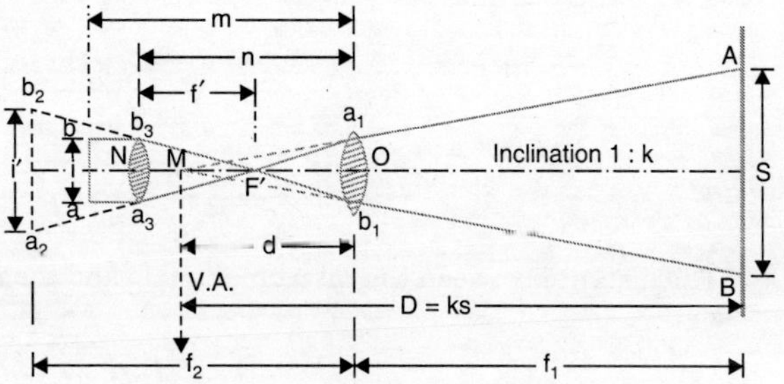

Fig. 22.10 Theory of Anallactic Lens.

The rays emanating from A and B (corresponding to stadia wires) along AM and BM are refracted by the object glass and meet at a point F'. The distance between the anallactic lens and the objective glass is so fixed that the point F' happens to be the exterior principal focus of the anallactic lens. Hence, the rays passing through F' will emerge in a direction parallel to the axis of the telescope after being refracted by the anallactic lens. Then ab is the inverted image of the length AB of the staff; the points a and b correspond to the stadia wires. If the anallactic lens was not interposed, the rays would have formed a virtual image $b_2\, a_2$ at a distance f_2 from the object glass.

From the conjugate relationship for the objective

$$\frac{1}{f} = \frac{1}{f_1} + \frac{1}{f_2} \qquad \text{...}(i)$$

Since the length of AB and $a_2\, b_2$ are proportional to their distance from O,

$$\frac{s}{i'} = \frac{f_1}{f_2} \qquad \text{...}(ii)$$

For the anallactic lens, ab and $a_2\, b_2$ are conjugate, and their distances $(f_2 - n)$ and $(m - n)$ from N are connected by the conjugate relationship:

$$\frac{1}{f'} = -\frac{1}{(f_2 - n)} + \frac{1}{m - n} \qquad \text{...}(iii)$$

The minus sign with $(f_2 - n)$ has been used since both ab and $a_2\, b_2$ are to the same side of N. Since the length of ab and $a_2\, b_2$ are proportional to their distances from N, we get

$$\frac{i'}{i} = \frac{f_2 - n}{m - n} \qquad \text{...}(iv)$$

In order to obtain an expression for D, let us eliminate f_2, m and i' from the above equations. Multiplying (ii) and (iv), we get

$$\frac{s}{i} = \frac{f_1}{f_2} \cdot \frac{f_2 - n}{m - n}$$

But $$\frac{f_1}{f_2} = \frac{f_1 - f}{f} \text{ and } f_2 = \frac{ff_1}{f_1 - f}, \text{ from } (i)$$

and $$\frac{f_2 - n}{m - n} = \frac{f_2 - n + f'}{f'}, \text{ from } (iii)$$

Hence $$\frac{s}{i} = \frac{f_1 - f}{f} \cdot \frac{f_2 - n + f'}{f'} = \frac{f_1 - f}{f} \left\{ \frac{\left(\frac{ff_1}{f_1 - f} \right) - n + f'}{f'} \right\}$$

$$= \frac{ff_1 + (f_1 - f)(f' - n)}{ff'} = \frac{f_1 (f + f' - n)}{ff'} + \frac{f(n - f')}{ff'}$$

$\therefore$ $$f_1 = \frac{s}{i} \cdot \frac{ff'}{f + f' - n} - \frac{f(n - f')}{f + f' - n}$$

The distance between the instrument axis and the staff is given by

$$D = (f_1 + d) = \frac{ff'}{(f + f' - n)\, i} s - \frac{f(n - f')}{f + f' - n} + d = ks + C \qquad \text{...(22.10)}$$

where $$k = \frac{ff'}{(f + f' - n)\, i} \qquad \text{...(22.10 } a)$$

and $$C = d - \frac{f(n - f')}{f + f' - n} \qquad \text{...(22.10 } b)$$

In order that D should be proportional to s, the additive constant C should vanish.

Hence $$\frac{f(n - f')}{f + f' - n} = d$$

which is secured by placing the anallactic lens, such that

$$n = f' + \frac{fd}{(f + d)} \qquad \text{...(22.11)}$$

Thus, if equation 22.11 is satisfied, the apex of the tacheometric angle will be situated at the centre of the trunnion axis.

The value of f' and i must be so arranged that the multiplier $\frac{ff'}{(f+f'-n)\,i}$ is a suitable number, say 100. If all these conditions are fulfilled, equation 22.10 reduces to

$$D = ks = 100 \text{ s.} \quad \text{...(22.12)}$$

Anallactic Telescope: Inclined Sight

It has been shown in Fig. 22.10 that if the conditions of equation 22.11 are satisfied, the vertex of the anallactic angle will be formed at the centre of instrument (M). Fig. 22.11 shows the case of an inclined sight, from which the distance-elevation formulae can be directly derived.

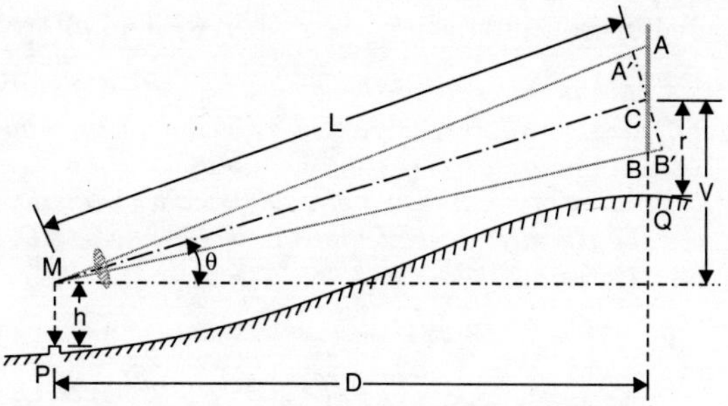

Fig. 22.11 Anallactic Lens: Inclined Sight.

With the same notations as that of Fig. 22.5, we have

$$MC = L = k.A'B' = ks \cos\theta \quad \text{...(i)}$$

$$D = L\cos\theta = ks\cos^2\theta \quad \text{...(ii)}$$

and

$$V = L\sin\theta = ks\cos\theta\sin\theta = \frac{ks}{2}\sin 2\theta \quad \text{...(iii)}$$

$$\text{R.L. of } Q = \text{R.L. of } P + h + V - r$$

Comparative merits of anallactic telescope and the simple external focusing telescope.

The following are the merits of a telescope fitted with an anallactic lens:

1. In the ordinary external focusing telescope, the additive constant is a nuisance since it increases the labour of reduction and necessitates the use of special computation tables or charts. Due to anallactic lens, the additive constant vanishes and the computations are made quicker.
2. As a rule, the anallactic lens is sealed against moisture or dust.
3. The loss of light may be compensated by the use of slightly larger object glass.

The following are the arguments in favour of simple external focusing telescope:

1. It is simple and reliable.
2. The anallactic lens absorbs much of the incident light.
3. The anallactic lens cannot be easily cleaned.
4. If the anallactic lens is adjustable, it is a potential source of error unless proper field check is made from time to time.

Internal Focusing Telescope

We have seen in the principle of stadia method that the staff intercept s is not directly proportional to D, because the additive constant C comes in picture. By the introduction of an anallactic lens in an external focusing telescope, however, this additive constant can be reduced to zero. It should be remembered that an anallactic lens is fitted to external focusing telescope only and *not* in internal focusing telescope since the additive constant in the latter is extremely small (varying between 5 cm to 15 cm only). In some of the modern theodolites, the internal focusing telescopes have zero additive constant. *Thus, an internal focusing telescope is virtually anallactic.*

Since the focal length of the 'objective system' (*i.e.*, object lens and sliding lens) varies with the distance of the object focused, the theory of internal focusing stadia telescope is rather complicated. In general, the standard formulae developed for an anallactic telescope may be used in reducing the readings taken with an internal focusing telescope.

Example 22.1: *A tacheometer was set up at a station A and the readings on a vertically held staff at B were 2.255, 2.605 and 2.955, the line of sight being at an inclination of + 8° 24′. Another observation on the vertically held staff at B.M. gave the readings 1.640, 1.920 and 2.200, the inclination of the line of sight being +1°6′. Calculate the horizontal distance between A and B, and the elevation of B if the R.L. of B.M. is 418.685 metres. The constants of the instruments were 100 and 0.3.*

Solution:

(*a*) *Observation to B.M.:* $V = ks\dfrac{\sin 2\theta}{2} + C\sin\theta$

Here, $k = 100;\ s = 2.200 - 1.640 = 0.560\text{ m};\ C = 0.3\text{ m}$

$\therefore \quad V = \frac{1}{2} \times 100 \times 0.56 \sin 2°\,12' + 0.3 \sin 1°\,6' = 1.075 + 0.006 = 1.081\text{ m}$

Elevation of collimation at the instrument = 418.685 + 1.920 – 1.081 = **419.524 m**

(*b*) *Observation to B:*

$$s = 2.955 - 2.255 = 0.700\text{ m};\ \theta = 8°\,24'$$

$$D = ks\cos^2\theta + C\cos\theta = 100 \times 0.7\cos^2 8°\,24' + 0.3 \times \cos 8°\,24'$$

$$= 68.506 + 0.2968 \approx \mathbf{68.80\ m}$$

$$V = ks\,\tfrac{1}{2}\sin 2\theta + C\sin\theta = \tfrac{1}{2} \times 100 \times 0.7 \sin 16°\,48' + 0.3\sin 8°\,24'$$

$$= 10.116 + 0.044 = 10.160$$

R.L. of B = 419.524 + 10.160 – 2.605 = **427.079 m.**

Example 22.2: *The elevation of a point P is to be determined by observations from two adjacent stations of a tacheometric survey. The staff was held vertically upon the point, and the instrument is fitted within an anallactic lens, the constant of the instrument being 100. Compute the elevation of the point P from the following data, taking both the observations as equally trustworthy:*

Inst. station	*Height of axis*	*Staff point*	*Vertical angle*	*Staff readings*	*Elevation of station*
A	1.42	P	+2° 24′	1.230, 2.055, 2.880	77.750 m
B	1.40	P	–3° 36′	0.785, 1.800, 2.815	97.135 m

Also, calculate the distance of A and B from P.

Solution: (*a*) *Observation from A to P:*

$$s = 2.880 - 1.230 = 1.65\text{ m}$$

$$D = ks \cos^2 \theta = 100 \times 1.65 \cos^2 2° 24' = \mathbf{164.7\ m}$$

$$V = ks \frac{\sin 2\theta}{2} = \tfrac{1}{2} \times 100 \times 1.65 \sin 4° 48' = 6.903$$

$$\text{R.L. of } P = 77.750 + 1.420 + 6.903 - 2.055 = 84.018 \text{ m}$$

(*b*) *Observation from B to P:*

$$s = 2.815 - 0.785 = 2.03 \text{ m}$$

$$D = ks \cos^2 \theta = 100 \times 2.03 \cos^2 3° 36' = \mathbf{202.2\ m.}$$

$$V = ks \tfrac{1}{2} \sin 2\theta = \tfrac{1}{2} \times 100 \times 2.03 \sin 7° 12' = 12.721 \text{ m}$$

R.L. of $P = 97.135 + 1.40 - 12.721 - 1.800 = 84.014$

Average elevation of $P = \tfrac{1}{2}(84.018 + 84.014) = \mathbf{84.016\ m.}$

Example 22.3: *Determine the gradient from a point A to a point B from the following observations made with a tacheometer fitted with an anallactic lens. The constant of the instrument was 100 and the staff was held vertically:*

Inst. station	*Staff point*	*Bearing*	*Vertical angle*	*Staff readings*
P	*A*	*134°*	*+ 10° 32′*	*1.360, 1.915, 2.470*
	B	*224°*	*+ 5° 6′*	*1.065, 1.885, 2.705*

Solution:

(*a*) *Observation from P to A:*

$$s = 2.470 - 1.360 = 1.11 \text{ m}$$

$$D = ks \cos^2 \theta = 100 \times 1.11 \cos^2 10° 32' = 107.3 \text{ m}$$

$$V = ks \tfrac{1}{2} \sin 2\theta = \tfrac{1}{2} \times 100 \times 1.11 \sin 21° 4' = 19.95 \text{ m}$$

Difference in elevation between A and instrument axis

$= 19.95 - 1.915 = 18.035$ m (*A* being higher)

(*b*) *Observation from P to B:*

$$s = 2.705 - 1.065 = 1.64 \text{ m}$$

$$D = ks \cos^2 \theta = 100 \times 1.64 \times \cos^2 5° 6' = 162.7 \text{ m}$$

$$V = ks \tfrac{1}{2} \sin 2\theta = \tfrac{1}{2} \times 100 \times 1.64 \sin 10° 12' = 14.521 \text{ m}$$

Difference in elevation between B and instrument axis

$= 14.521 - 1.885 = 12.636$ m (*B* being higher)

(*c*) *Gradient from A to B:*

Distance $AP = 107.3$ m; Distance $BP = 162.7$ m

$$\angle APB = 224° - 134° = 90°$$

$$AB = \sqrt{AP^2 + BP^2} = \sqrt{(107.3)^2 + (162.7)^2} = 194.9 \text{ m}$$

Difference in elevation between A and B

$$= 18.035 - 12.636 = 5.399 \qquad (A \text{ being higher})$$

$$\text{Gradient from } A \text{ to } B = \frac{5.399}{194.9} = \mathbf{1 \text{ in } 36.1} \text{ (falling)}.$$

Example 22.4: *Following observations were taken from two traverse stations by means of a tacheometer fitted with an anallactic lens. The constant of the instruments is 100.*

Inst. station	*Staff station*	*Height of Inst.*	*Bearing*	*Vertical angle*	*Staff readings*
A	C	*1.38*	*226° 30′*	*+10° 12′*	*0.765, 1.595, 2.425*
B	D	*1.42*	*84° 45′*	*–12° 30′*	*0.820, 1.840, 2.860*

Co-ordinates of station A 212.3 N 186.8 W
Co-ordinates of station B 102.8 N, 96.4 W
Compute the length and gradient of the line CD, if B is **6.50** *m higher than A.*

Solution:

(a) ***Observation from A to C:***

$$s = 2.425 - 0.765 = 1.66 \text{ m}$$

Distance $AC = k.s \cos^2\theta = 100 \times 1.66 \cos^2 10° 12' = 160.8$ m

$$V = \frac{k.s \sin 2\theta}{2} = \frac{100 \times 1.66}{2} \sin 20° 24' = 28.931 \text{ m}$$

Let the elevation of $A = 100.00$ m

$\therefore$ R.L. of $C = 100 + 1.38 + 28.931 - 1.595 = 128.716$ m.

(b) ***Observation from B to D:***

$$s = 2.860 - 0.820 = 2.040 \text{ m}$$

$$\text{Distance } BD = k.s \cos^2\theta = 100 \times 2.04 \cos^2 12° 30' = 194.4 \text{ m}$$

$$V = k.s \frac{\sin 2\theta}{2} = \frac{100 \times 2.040}{2} \sin 25° = 43.107 \text{ m}$$

R.L. of $B = 100 + 6.50 = 106.50$ m

R.L. of $D = 106.50 + 1.42 - 43.107 - 1.84 = 62.973$ m

(c) ***Length and gradient of CD:***

Length of $AC = 160.8$ m; R.B. of AC = S 46° 30′ W

Hence AC is in the third quadrant.

Latitude of $AC = -\ 160.8 \cos 46° 30' = -\ 110.7$

Departure of $AC = -\ 160.8 \sin 46° 30' = -\ 116.6$

Length of $BD = 194.4$ m; R.B. of BD = N 84° 45′ E

Hence BD is in the first quadrant.

$$\text{Latitude of } BD = 194.4 \cos 84° 45' = +17.8$$

$$\text{Departure of } BD = 194.4 \sin 84° 45' = +193.6$$

Now, total latitude of $A = +212.3$	Total departure of $A = -186.8$
Add latitude of $AC = -110.7$	Add departure of $AC = -116.6$
Total latitude of $C = +101.6$	Total departure of $C = -303.4$
Similarly, Total latitude of $B = +102.8$	Total departure of $B = -96.4$
Add latitude of $BD = +17.8$	Add departure of $BD = +193.6$
Total latitude of $D = +120.6$	Total departure of $D = +97.2$

Thus, the total co-ordinates of the points C and D are known.

$$\text{Latitude of line } CD = \text{Total latitude of } D - \text{Total latitude of } C$$
$$= 120.6 - 101.6 = +19.0$$

and

$$\text{Departure of line } CD = \text{Total departure of } D - \text{Total departure of } C$$
$$= 97.2 - (-303.4) = +400.6$$

The line CD is, therefore, in the fourth quadrant.

Length $CD = \sqrt{(19.0)^2 + (400.6)^2}$ = **401.1 m**

∴ Gradient of $CD = (128.716 - 62.973) \div 401.1$ = **1 in 6.1** (falling).

Example 22.5: *A tacheometer is set up at an intermediate point on a traverse course PQ and the following observations are made on a vertically held staff:*

Staff station	*Vertical angle*	*Staff intercept*	*Axial hair readings*
P	*+ 8° 36′*	*2.350*	*2.105*
Q	*+ 6° 6′*	*2.055*	*1.895*

The instrument is fitted with an anallactic lens and the constant is 100. Compute the length of PQ and reduced level of Q, that of P being 321.50 metres.

Solution:

(*a*) *Observation from the instrument to P:*

$$s = 2.350;\ \theta = 8° 36'$$

Distance to $P = k \,.\, s \cos^2 \theta = 100 \times 2.350 \times \cos^2 8° 36' = 229.75$ m

$$V = k \,.\, s \frac{\sin 2\theta}{2} = \frac{100 \times 2.350}{2} \sin 17° 12' = 34.745$$

Difference in elevation between P and the instrument axis

$$= 34.745 - 2.105 = 32.640 \text{ m} \quad (P \text{ being higher}).$$

(*b*) ***Observation from the instrument to Q:***

$$s = 2.055;\ \theta = 6°\ 6'$$

$$\text{Distance to } Q = k\ .\ s\cos^2\theta = 100 \times 2.055\cos^2 6°\ 6' = 203.18 \text{ m}$$

$$V = k\ .\ s\frac{\sin 2\theta}{2} = \frac{100 \times 2.055}{2}\sin 12°\ 12' = 21.713 \text{ m}$$

Difference in elevation between Q and the instrument axis

$$= 21.713 - 1.895 = 19.818 \ (Q \text{ being higher})$$

Since the tacheometer is set up at an intermediate point on the line PQ, the distance

$$PQ = 229.75 + 203.18 = \mathbf{432.93\ m.}$$

Difference in elevation of P and Q

$$= 32.640 - 19.818 = 12.822 \ (P \text{ being higher})$$

$$\text{R.L. of } Q = \text{R.L. of } P - 12.822$$

$$= 321.50 - 12.822 = \mathbf{308.678\ m.}$$

Example 22.6: *To determine the multiplying constant of a tacheometer, the following observations were taken on a staff held vertically at distance, measured from the instrument:*

Observation	*Horizontal distance in metres*	*Vertical angle*	*Staff intercept*
1	*50*	*+ 3° 48′*	*0.500 m*
2	*100*	*+ 1° 06′*	*1.000 m*
3	*150*	*+ 0° 36′*	*1.500 m*

The focal length of the object glass is 20 cm and the distance from the object glass to trunnion axis is 10 cm. The staff is held vertically at all these points. Find the multiplying constant.

Solution:

$$C = (f + d) = 0.20 + 0.10 = 0.30 \text{ m}$$

(*i*) ***First observation:***

$$D = ks\cos^2\theta + C\cos\theta$$

$$\therefore \quad 50 = k \times 0.500\cos^2 3°\ 48' + 0.30\cos 3°\ 48'; \quad \text{or} \quad k = 99.84.$$

(*ii*) ***Second observation:***

$$100 = k \times 1\cos^2 1°\ 6' + 0.3\cos 1°\ 6'; \quad \text{or} \quad k = 99.74.$$

(*iii*) ***Third observation:***

$$150 = k \times 1.5\cos^2 0°\ 36' + 0.3\cos 0°\ 36'; \quad \text{or} \quad k = 99.81.$$

$$\text{Average value of } k = \tfrac{1}{3}(99.84 + 99.74 + 99.81) = \mathbf{99.8}$$

Example 22.7: *Two distances of 20 and 100 metres were accurately measured out and the intercepts on the staff between the outer stadia webs were 0.196 m at the former distance and 0.996 at the latter. Calculate the tacheometric constants.*

Solution: Let the constants be k and C

For the first observation $20 = ks + C = k \times 0.196 + C$ ...(*i*)

For the second observation $100 = k \times 0.996 + C$

Subtracting (*i*) from (*ii*), we get $k(0.996 - 0.196) = 100 - 20$

From which $k = \mathbf{100}$

Substituting in (*i*), we get $C = 20 - 0.196 \times 100 = \mathbf{0.4\ m.}$

Example 22.8: *Two sets of tacheometric readings were taken from an instrument station A, the reduced level of which was 100.06 m to a staff station B.*

(*a*) *Instrument P—multiplying constant 100, additive constant 0.06 m, staff held vertical.*

(*b*) *Instrument Q—multiplying constant 90, additive constant 0.06 m, staff held normal to the line of sight.*

Instrument	*At*	*To*	*Ht of Inst.*	*Vertical angle*	*Stadia readings (m)*
P	*A*	*B*	*1.5 m*	*26°*	*0.755, 1.005, 1.255*
Q	*A*	*B*	*1.45 m*	*26°*	*?*

What should be the stadia readings with instrument Q ?

Solution:

(*i*) ***Observations with instrument P: Staff vertical***

$$AB = ks \cos^2 \theta + C \cos \theta$$

$$s = 1.255 - 0.755 = 0.5;\ k = 100;\ C = 0.06 \text{ m}$$

$$\therefore \quad AB = 100 \times 0.5 \cos^2 26° + 0.06 \cos 26° = 40.45 \text{ m}$$

$$V = AB \tan \theta = 40.45 \tan 26° = 19.73$$

$$\therefore \quad \text{R.L. of } B = 100.06 + 1.5 + 19.73 - 1.005 = 120.285.$$

(*ii*) ***Observations with instrument Q: Staff normal***

Let the stadia readings be r_1, r and r_2

$$s = r_1 - r_2 = 2(r_1 - r)$$

$$AB = ks \cos \theta + C \cos \theta + r \sin \theta$$

or $$40.45 = 90\, s \cos 26° + 0.06 \cos 26° + r \sin 26°$$

or $$80.89\, s + 0.4384\, r = 40.4 \quad \text{...(1)}$$

Also $$V = ks \sin \theta + C \sin \theta = 90\, s \sin 26° + 0.06 \sin 26° = (39.46\, s + 0.03)$$

$$\therefore \quad \text{R.L. of } B = 100.06 + 1.45 + V - r \cos 26° = 101.51 + (39.46\, s + 0.03) - 0.8988\, r$$

$$= 101.54 + 39.46\, s - 0.8988\, r$$

But $$\text{R.L. of } B = 120.285$$

$$\therefore \quad 120.285 = 101.54 + 39.46\, s + 0.8988\, r$$

or $$39.46\, s - 0.8988\, r = 18.745 \quad \text{...(2)}$$

Solving equations (1) and (2), we get

$$s = 0.49 \text{ m}; \quad r = 0.63 \text{ m}$$

$\therefore$ $$s = 2(r_1 - r)$$

or $$r_1 = \frac{0.49}{2} + r = 0.245 + 0.63 = 0.875$$

and $$r_2 = r_1 - s = 0.875 - 0.49 = 0.385$$

Hence the readings are **0.385, 0.63, 0.875.**

Example 22.9: *With a tacheometer stationed at P, sights were taken on three points A, B and C as follows:*

Ins. at	*To*	*Vertical angle*	*Stadia readings*	*Remarks*
P	*A*	*– 4° 30′*	*2.405, 2.705, 3.005*	*R.L. of A = 107.08 m*
				Staff held normal
	B	*0° 00′*	*0.765, 1.070, 1.375*	*R.L. of B = 113.41 m*
				Staff held vertical
	C	*+ 2° 30′*	*0.720, 1.700, 2.680*	*Staff held vertical*

The telescope was of the draw tube type and the focal length of the object glass was 25 cm. For the sights to A and B, which were of equal length, the distance of the object glass from the vertical axis was 12 cm. For sight to C, the distance of object glass from the vertical axis was 11 cm.

Calculate (a) the spacing of the cross-hairs in the diaphragm and (b) the reduced level to C.

Solution:

(i) Observation from P to A: (Ref. Fig. 22.8)

$$s = 3.005 - 2.405 = 0.6 \text{ m}$$

$$L = \frac{f}{i} \cdot s + (f + d) = \frac{25 \times 0.6}{i} + \frac{25 + 12}{100} = \frac{15}{i} + 0.37$$

$$V = L \sin \theta = L \sin 4° \, 30' = 0.0785 \, L = 0.0785 \left(\frac{15}{i} + 0.37\right) = \frac{1.175}{i} + 0.029$$

$$r \cos \theta = 2.705 \cos 4° \, 30' = 2.7$$

R.L. of instrument collimation

$$= 107.08 + r \cos \theta + V = 107.08 + 2.7 + \left(\frac{1.175}{i} + 0.029\right) = 109.78 + \frac{1.175}{i} \quad ...(1)$$

(ii) Observation from P to B:

Since the line of collimation is horizontal, its level = 113.41 + 1.07 = 114.48 ...(2)

Equating (1) and (2), we get

$$109.78 + \frac{1.175}{i} = 114.48 \quad \text{or} \quad i = 0.25 \text{ cm} = \mathbf{2.5\ mm}$$

$$\therefore \quad k = \frac{f}{i} = \frac{25}{0.25} = 100$$

(*iii*) ***Observation from P to C:***

$$C = \frac{25+11}{100} = 0.36; \; s = 2.68 - 0.72 = 1.96 \text{ m}$$

$$V = ks\,\tfrac{1}{2}\sin 2\theta + C\sin\theta = 100 \times 1.96 \times \tfrac{1}{2}\sin 5° + 0.36\sin 2.5° \approx 8.555$$

$$\therefore \quad \text{R.L. of } C = 114.48 + 8.555 - 1.70 = \mathbf{121.335\ m.}$$

Example 22.10: *A theodolite is fitted with an ordinary telescope in which the eye-piece end moves in focusing, the general description being as follows:*

Focal length of objective f = 23 cm. Fixed distance d between the objective and vertical axis 11.5 cm; diaphragm: lines on glass in cell which may be withdrawn.

It is desired to convert the instrument into an anallactic tacheometer by inserting an additional positive lens in a tube and ruling a new diaphragm so as to give a multiplier of 100 for intercepts on a vertical staff; and in this connection it is found that 19 cm will be a convenient value for the fixed distance between the objective and the anallactic lens.

Determine: (a) a suitable value of the focal length f' of the anallactic lens, and

(b) the exact spacing of the lines on the diaphragm.

Solution: With our notations, we have $d = 11.5$ cm; $f = 23$ cm; $n = 19$ cm; $k = 100$.

It is required to determine f' and i.

From equation 22.11, we have $n = f' + \dfrac{fd}{f+d}$ ...(22.11)

$$\therefore \quad f' = n - \frac{fd}{f+d} = 19 - \frac{23 \times 11.5}{23 + 11.5} = 19 - 7.65 = \mathbf{11.35\ cm.}$$

From equation 22.10 (*a*), $k = \dfrac{ff'}{(f + f' - n)i}$

or

$$i = \frac{ff'}{k(f + f' - n)} = \frac{23 \times 11.35}{100(23 + 11.35 - 19)} = \mathbf{0.17\ cm.}$$

Example 22.11: *An anallactic tacheometer in use on a remote survey was damaged and it was decided to use a glass diaphragm not originally designed for the instrument. The spacing to the outer lines of the new diaphragm was 1.25 mm, focal lengths of the object glass and anallactic lens 75 mm, fixed distance between object glass and trunnion axis 75 mm, and the anallactic lens could be moved by an adjusting screw between its limiting positions 75 mm and 100 mm from the object glass. In order to make the multiplier 100, it was decided to adjust the position of the anallactic lens, or if this proved inadequate, to graduate a special staff for use with the instrument. Make calculation to determine which course was necessary and if a special staff is required, determine the correct calibration and the additive constant (if any).*

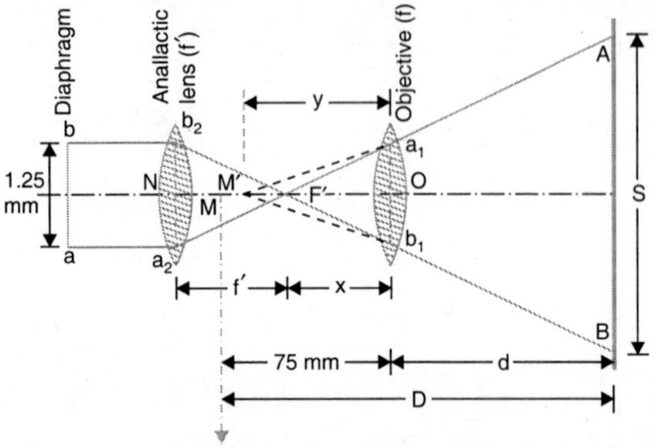

Fig. 22.12 Design of Anallactic Telescope.

Solution: The optical diagram is shown in Fig. 22.12. Since the telescope is no longer anallactic, the apex (M') of the tacheometric angle does not form at the trunnion axis (M) of the instrument, but slightly away from it.

Considering one ray (Aa) through the object glass, we have

$$\frac{1}{f} = \frac{1}{f_1} + \frac{1}{f_2};$$

or $$f_2 = \frac{ff'}{f_1 - f}$$

where f = Focal length of objective = 7.5 cm

f_1 = Objective distance = $M'O = -y$; f_2 = Image distance = $F'O = x$

Substituting the values, we get

$$x = \frac{7.5(-y)}{(-y) - 7.5} = \frac{7.5y}{7.5 + y} \quad ...(1)$$

From similar triangles $M'a_1b_1$ and $M'AB$

$$\frac{a_1b_1}{y} = \frac{AB}{d+y} = \frac{s}{d+y} \quad ...(1a)$$

From similar triangles $F'a_2b_2$ and $F'a_1b_1$

$$\frac{a_1b_1}{x} = \frac{a_2b_2}{f'} = \frac{0.125}{7.5} = \frac{1}{60} \quad ...(1b)$$

Eliminating a_1 b_1 from (1a) and (1b), we get

$$\frac{sy}{d+y} = \frac{x}{60} \quad ...(2)$$

Substituting the value of x from (1), we get

$$\frac{sy}{d+y} = \frac{1}{60}\left\{\frac{7.5y}{7.5+y}\right\} = \frac{y}{8(7.5+y)}$$

from which $d = 8(7.5 + y)\, s - y$

Hence $D = d + 7.5 = 8(7.5 + y)\, s - y + 7.5 \quad ...(3)$

For the multiplier to be 100, we have $8(7.5 + y) = 100$; or $y = \frac{100}{8} - 7.5 = 5$ cm.

Substituting the value of y in (1), we get

$$x = \frac{7.5 \times y}{7.5 + y} = \frac{7.5 \times 5}{7.5 + 5} = 3 \text{ cm.}$$

Hence the anallactic lens should be placed at a distance of 3 + 7.5 = 10.5 cm from the object glass. But since the maximum distance through which the lens can be moved is 10.0 cm only, this is not possible. Placing it at a distance of 10 cm from the object glass, we have

$$x = 10 - 7.5 = 2.5 \text{ cm} = \frac{7.5y}{7.5+y}$$

$$\therefore \quad y = 3.75 \text{ cm}$$

Substituting the values of x and y in (3), we get the tacheometric formula,

$$D = 8(7.5 + 3.75)\, s - 3.75 + 7.5 = 90\, s + 3.75 \text{ (cm)}$$
$$= 90\, s + 0.0375 \text{ (metres).}$$

If it is desired to have the multiplier constant as 100, a specially graduated staff having its graduations longer in the ratio of $\frac{10}{9}$ *will have to be used.*

Example 22.12: *Find upto what vertical angle, sloping distances may be taken as horizontal distance in stadia work, so that the error may not exceed 1 in 400. Assume that the instrument is fitted with an anallactic lens and that the staff is held vertically.*

Solution: Let the angle be θ.

True horizontal distance = $D = ks \cos^2 \theta$; Sloping distance = $L = ks$

$$\frac{\text{Sloping distance}}{\text{Horizonal distance}} = \frac{L}{D} = \frac{ks}{ks \cos^2 \theta} = \sec^2 \theta \quad \text{...(1)}$$

If the error is 1 in 400, we have

$$\frac{L}{D} = \frac{400+1}{400} = \frac{401}{400} \quad \text{...(2)}$$

In the limiting case, equating (1) and (2), we get

$$\sec^2 \theta = \frac{401}{400} \text{ or } \theta = \sec^{-1}\left(\sqrt{\frac{401}{400}}\right) = \mathbf{2° 51'45'' \approx 2° 52'}.$$

Example 22.13: *State the error that would occur in horizontal distance with an ordinary stadia telescope in which the focal length is 25 cm, the multiplier constant 100, and the additive constant 35 cm, when an error of 0.0025 cm exists in the interval between the stadia lines.*

Solution: The horizontal distance is given by $D = \frac{f}{i} s + C$

If δD = error in distance and δi = error in the stadia interval, we get

$$\delta D = -s \frac{f}{i^2} . \delta i \quad \text{...(1)}$$

Now, $\frac{f}{i} = 100$ or $i = \frac{f}{100} = \frac{25}{100} = 0.25$ cm.

Substituting the values of $\frac{f}{i}$, i and δi in (1), we get

$$\delta D = -s. \frac{f}{i} . \frac{1}{i} . \delta i = -s\,(100)\left(\frac{1}{0.25}\right)(0.0025) = -s.$$

Thus, the error in the distance is numerically equal to the staff intercept.

THE SUBTENSE METHOD

22.8 PRINCIPLE OF SUBTENSE (OR MOVABLE HAIR) METHOD: VERTICAL BASE OBSERVATIONS

In the stadia principle, we have seen that whatever may be the distance between the staff and the tacheometer, the tacheometric angle is always a constant for a given telescope. The staff intercept, which forms the *base* of stadia measurement, varies with the distance of the staff from the instrument. The principle of subtense method is just the reverse of it. In this case, as illustrated in Fig. 22.13, the staff intercept *s* forms the fixed base while the tacheometric angle β changes with the staff position. This can be attained by sighting a graduated staff having the targets at some fixed distance apart (say 3 metres or 10 ft) and changing the interval *i* between the stadia wires till the lines of sights corresponding to the stadia wires bisect the targets. If the staff position is now changed, the value of *i* is changed. In subtense measurement, the base may be kept either horizontal or vertical. If the base is vertical, the method is known as *'vertical base subtense method'* and the angle at *F* can be measured with the help of special diaphragm. If the base is horizontal, the method is known as *'horizontal base subtense method'* and the angle at *F* can be measured with the horizontal circle of the theodolite by the method of repetition.

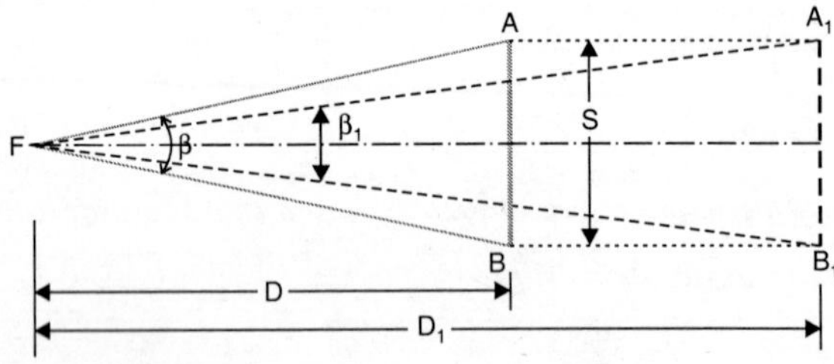

Fig. 22.13 Principle of Substense Method.

Fig. 22.14 shows the optical diagram for observations through a substense theodolite. For the staff at point *P*, the rays *Aa′* and *Bb′*, passing through the exterior focus *F* of the objective, become parallel to the principal axis after refraction. The points *a* and *b* correspond to the positions of stadia wires for this observation so that the lines of sight intersect the targets at *A* and *B*. Similarly, the dashed lines show the corresponding optical diagram for another staff position at *Q*, the staff intercept being the same.

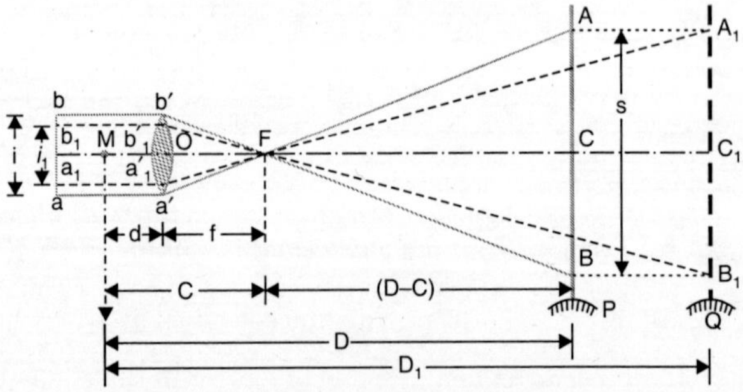

Fig. 22.14 Vertical Base Subtense Method.

Let $AB = s$ = Staff intercept = distance between the targets

$ab = i$ = Stadia interval measured at the diaphragm

F = Exterior principal focus of the objective

M = Centre of the instrument

a', b' = Points on the objective corresponding to *A* and *B*.

Other notations are same as earlier.

From Δs $a'b'F$ and FAB

$$\frac{FC}{s} = \frac{FO}{a'b'} = \frac{f}{i} \quad \text{or} \quad FC = \frac{f}{i}\, s$$

$$\therefore \qquad D = FC + MF = \frac{f}{i}\, s + (f + d)$$

Thus, the expression for the subtense measurement is the same as for the stadia method. The only difference is that in this expression, s is fixed quantity while i is variable. Due to this reason, the multiplying factor $\frac{f}{i}$ varies with the staff position and is no longer constant. The stadia interval i is measured with the help of micrometer screws (Fig. 22.15) having a pitch p.

Let m be the total number of the revolutions of the micrometer screws for the staff intercept s. Then $i = mp$. Substituting the value of i, we get

$$D = \frac{f}{mp}\, s + (f + d) \text{ or } D = \frac{Ks}{m} + C \qquad \text{...(22.13)}$$

where $K = \frac{f}{p}$ = constant for an instrument and C = additive constant.

If, however, e is the index error, expression 22.13 reduces to

$$D = \frac{Ks}{m - e} + C \qquad \text{...(22.13a)}$$

Expression 22.13 can be extended for inclined sights also exactly in the same manner as for stadia method. Thus, for inclination θ and staff vertical, we have

$$D = \frac{K.\, s}{m - e} \cos^2\theta + C\cos\theta \qquad \text{...(22.13b)}$$

$$V = \frac{K.\, s}{m - e} \cdot \frac{\sin 2\theta}{2} + C\sin\theta = D\tan\theta \qquad \text{...(22.13c)}$$

The Subtense Diaphragm

Since the accuracy of subtense method mainly depends upon the measurement of the stadia interval i, the theodolite must have arragements for measuring it with accuracy and speed. Fig. 22.15 (a) shows diagrammatically the form of a stadia diaphragm for this purpose.

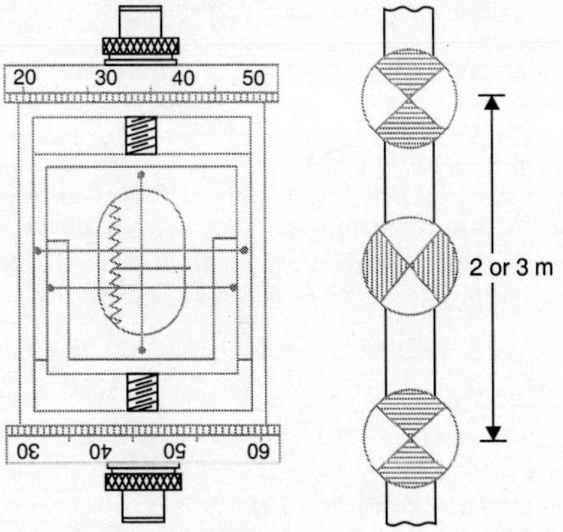

(a) Subtense diaphragm (b) Rod with targets

Fig. 22.15

Each hair of the stadia diaphragm can be moved by a separate sliding frame actuated by a micrometer screw with a large graduated head. The number of complete turns on the screw are directly visible in the field of view, and the fractions are read on the graduated head. When both the hairs coincide with the central mark of the comb they are in the plane of the line of sight and the reading on each graduated head should be zero. When an observation is made, both the heads are rotated till each hair bisects its target. Fig. 22.15 (b) shows one form of the rod fitted with three targets.

Determination of Constants *K* and *C*

The instrumental constants K and C can best be determined by measuring the additive constant C along the telescope (as in the case of stadia method) and observing the micrometer readings corresponding to staff kept at some measured distance. Let D_1 and D_2 = Measured distances from the instrument

s = Fixed distance between the two targets

m_1 = Sum of the two micrometer readings when the staff is kept at distance D_1

m_2 = Sum of two micrometer readings when the staff is kept at distance D_2

e = Index error.

Substituting the corresponding values in equation 22.13 (a), we get

$$D_1 = \frac{K.s}{m_1 - e} + C \quad \text{or} \quad K.s = (D_1 - C)(m_1 - e) \qquad ...(i)$$

and
$$D_2 = \frac{K.s}{m_2 - e} + C \quad \text{or} \quad K.s = (D_2 - C)(m_2 - e) \qquad ...(ii)$$

Equating (i) and (ii), we get $(D_1 - C)(m_1 - e) = (D_2 - C)(m_2 - e)$

From which
$$e = \frac{(D_2 - C)m_2 - (D_1 - C)m_1}{(D_2 - D_1)} \qquad ...(22.14)$$

Substituting the value of e in (i), we get

$$K = \frac{D_1 - C}{s}\left\{m_1 - \frac{(D_2 - C)m_2 - (D_1 - C)m_1}{(D_2 - D_1)}\right\}$$

or
$$K = \frac{D_1 - C}{s(D_2 - D_1)}(m_1 D_2 - m_1 D_1 - m_2 D_2 + m_2 C + m_1 D_1 - m_1 C)$$

or
$$K = \frac{(D_1 - C)(D_2 - C)(m_1 - m_2)}{s(D_2 - D_1)} \qquad ...(22.15)$$

Merits and Demerits of Movable Hair Method

The 'movable hair method' or the vertical base subtense method, though more accurate, has become almost obsolete due to lack in the speed in the field. Moreover, since the variable m comes in the denominator (equation 22.13), the computations are not quicker. Long sights can be taken with greater accuracy than in stadia method, since only targets are to be bisected, but this advantage may be neutralised unless i is measured very accurately. *The term 'subtense method' is now more or less exclusively applied to horizontal base subtense method.*

22.9 HORIZONTAL BASE SUB-TENSE MEASUREMENTS

In this method, the base AB is kept in a horizontal plane and the angle AOB is measured with the help of the horizontal circle of the theodolite.

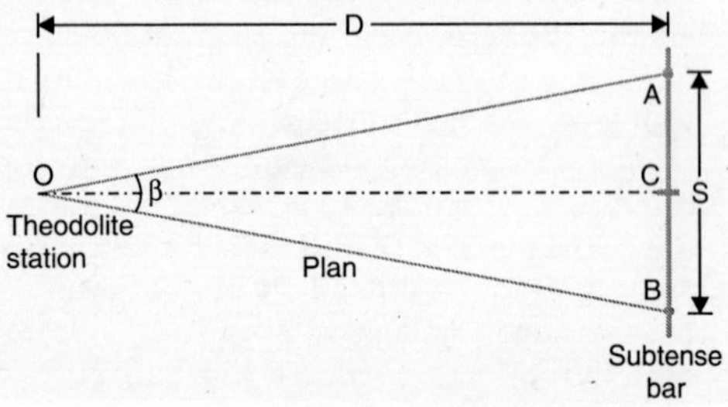

Fig. 22.16 Horizontal Base Subtense Method.

Thus, in Fig. 22.16, let AB be the horizontal base of a length s and let O be the position of the instrument meant for measuring the horizontal angle *AOB*. If the line *AB* is perpendicular to the line *OC*, where *C* is midway between *A* and *B*, we have from ΔOAC,

$$D = \tfrac{1}{2} s \cot \frac{\beta}{2} = \frac{s}{2 \tan \beta / 2} \qquad \text{...(22.16)}$$

Equation 22.16 is the standard expression for the horizontal distance between *O* and *C*. If β is small, we get

$$\tan \frac{\beta}{2} = \tfrac{1}{2} \beta, \text{ where } \beta \text{ is in radians}$$

$$= \frac{1}{2} \frac{\beta}{206265}, \text{ if } \beta \text{ is in seconds (since 1 radian} = 206265 \text{ seconds)}$$

Substituting in Eq. 22.16, we get

$$D = \frac{s \times 206265}{\beta} \text{ where } \beta \text{ is in seconds} \qquad \text{...(22.17)}$$

The accuracy of the expression 22.17 depends upon the size of angle. For similar angles (say upto 1°), expression 22.17 may be taken as exact enough for all practical purposes. The angle at *O* is generally measured with the help of a theodolite, while a *subtense bar* is used to provide the base *AB*.

The Subtense Bar

For measurements of comparatively short lines in a traverse, a subtense bar may be used as the subtense base. Fig. 22.7 shows a subtense bar mounted on a tripod. The length of the base is generally 2 metres (6 ft) or 3 metres (10 ft). The distance between the two targets is exactly equal to the length of the base. In order that the length of base may not vary due to temperature and other variations, the vanes are attached to the invar rod. The invar rod is supported at a number of points in a duralumin tube provided with a spirit level.

Fig. 22.17 Subtense Bar (Kern Instruments).

The bar is centrally supported on a levelling head for accurate centring and levelling. A clamp and slow motion screw is also provided to rotate the bar about its vertical axis. Either a pair of sights or a small telescope is provided at the centre of the bar to align it perpendicular to the line *OC* joining the theodolite station and the centre of the bar. It should be noted that in order that equation 22.17 is valid, the longitudinal axis of the subtense bar must be perpendicular to the line *OC*.

The angle *AOB* at *O* is usually measured with a theodolite, preferably by method of repetition. It should be noted here that the difference in elevation between theodolite station *O* and subtense bar station *C* does not affect the magnitude of the angle *AOB*, since the angle *AOB* is always measured on the horizontal circle of the theodolite. If, however, the angle *AOB* is measured with the help of a sextent, it will have to be reduced to *horizontal*.

Effect of Angular Error on Horizontal Distance: It is evident from equation 22.17 that for a given length (s) of the subtense base, D is inversely proportional to the angle β. Hence the negative error in the measurement of the angle will produce a positive error in the distance D and *vice versa.*

Let the angular error be $\delta\beta$ (negative), and the resulting linear error be δD (positive). Then, we have

$$s = D\beta = (D + \delta D)(\beta - \delta\beta)$$

$$\therefore \quad \frac{D+\delta D}{D} = \frac{\beta}{(\beta-\delta\beta)} \quad \text{or} \quad \frac{(D+\delta D)-D}{D} = \frac{\beta-(\beta-\delta\beta)}{\beta-\delta\beta}$$

or $$\frac{\delta D}{D} = \frac{\delta\beta}{\beta-\delta\beta} \quad \text{From which, } \delta D = \frac{D\delta\beta}{(\beta-\delta\beta)} \qquad \text{...(22.18 } a\text{)}$$

Similarly, if $\delta\beta$ is positive, it can be shown that the resulting error δD (negative) is given by

$$\delta D = \frac{D\delta\beta}{\beta+\delta\beta} \qquad \text{...(22.18 } b\text{)}$$

If, however, $\delta\beta$ is very small in comparison to β, we have

$$\delta D = \frac{D\delta\beta}{\beta} \qquad \text{...(22.18)}$$

Long Non-rigid Subtense Bases: For the measurement of comparatively long lines, the subtense base must also be correspondingly long to preserve a suitable D/s ratio. The non-rigid base may vary in length from 20 metres to 150 metres. For traversing on large scale, a compact well-designed subtense base outfit is very much useful. *'Hunter's Short Base'* designed by Dr. de Graaf-Hunter is a typical type of outfit used by Survey of India. If consists of four lengths of steel tape each of 66′ length, connected by swivel joints. The base is supported on two low-end tripods and three intermediate bipods, one at the end of each tape. Targets are ingeniously mounted on the terminal tripods where correct amount of tension is applied by attaching a weight to a lever arm connected to one of the end targets. The whole base outfit weighs only 20 lb and can be set up in a few minutes. If a shorter length is required, the intermediate supports and tape lengths may be dispensed with. An effective base apparatus like this goes a long way towards solving the main difficulty of subtense measurements on a large scale.

Example 22.14: *The stadia intercept read by means of a fixed hair instrument on a vertically held staff is 1.05 metres, the angle of elevation being 5° 36′. The instrument constants are 100 and 0.3. What would be the total number of turns registered on a movable hair instrument at the same station for a 1.75 metres intercept on a staff held on the same point, the vertical angle in this case being 5° 24′ and the constants 1000 and 0.5?*

Solution: (*a*) ***Observations by means of fixed hair instrument:***

$$D = ks\cos^2\theta + C\cos\theta = 100 \times 1.05\cos^2 5°\,36' + 0.3\cos 5°\,36'$$

$$= 104.29 \text{ m.}$$

(*b*) ***Observations by means of movable hair instrument:***

$$D = \frac{K}{n} s\cos^2\theta + C\cos\theta$$

$$\therefore \quad 104.29 = \frac{1000}{n} 1.75 \cos^2 5° 24' + 0.5 \cos 5° 24'$$

$$\therefore \quad \frac{1734.5}{n} = 103.8 \quad \text{or} \quad n = \frac{1734.5}{103.8} = \mathbf{16.71}$$

Example 22.15: *The constant for an instrument is 850, the value of C = 0.5 m, and intercept s = 3 m. Calculate the distance from the instrument to the staff when the micrometer reading are 4.628 and 4.626 and the line of sight is inclined at + 10° 36 '. The staff was held vertical.*

Solution: Sum of micrometer readings = n = 4.628 + 4.626 = 9.254

$$D = \frac{K \cdot s}{n} \cos^2 \theta + C \cos \theta = \frac{850 \times 3}{9.254} \cos^2 10° 36' + 0.5 \cos 10° 36'$$

$$= \mathbf{226.7\ m.}$$

Example 22.16: *The horizontal angle subtended at a theodolite by a subtense bar with vanes 3 m apart is 12' 33''. Calculate the horizontal distance between the instrument and the bar. Also find (a) the error of horizontal distance if the bar was 3° from being normal to the line joining the instrument and bar stations; (b) the error of the horizontal distance if there is an error of 1'' in the measurement of the horizontal angle at the instrument station.*

Solution:

$$\beta = 12' 33'' = 753''$$

From equation 22.17, $$D = \frac{206265}{\beta} s = \frac{206265}{753} \times 3 = \mathbf{821.77\ m.}$$

(*a*) The above distance was calculated on the assumption that the bar was normal to the line joining the instrument and bar station. If, however, the bar is not normal, the correct horizontal distance is given by

$$D' = D \cos \beta = 821.77 \cos 3° = 820.64 \text{ m}$$

$$\therefore \quad \text{Error} = D' - D = 821.77 - 820.64 = 1.13 \text{ m}$$

$$\text{Ratio of error} = \frac{e}{D'} = \frac{1.13}{820.64} = \mathbf{1\ in\ 726.}$$

(*b*) If there is an error of 1'' in the measurement of the angle at the instrument.

we have $$\delta D = \frac{D\delta\beta}{\beta} = \frac{821.77 \times 1}{753} = \mathbf{1.09\ m.}$$

22.10 HOLDING THE STAFF

There are two methods of holding the staff rod in the stadia method:

(*i*) Vertical holding.

(*ii*) Normal holding.

(*i*) **Vertical Holding:** In order to keep the errors of verticality within very narrow limits, the staff should be held strictly vertical. Since the margin of allowable error is very narrow, some sort of device must be used to ascertain the verticality of the rod. The plummets and pendulums, if used for this purpose, are clumsy and too much at the mercy of the wind. A neater method is to fit a small circular spirit level or a single level tube with its axis perpendicular to the face of the staff.

Fig. 22.18 shows two patterns of circular levels. The folding pattern (Fig. 22.18*a*) is attached to the rear side of the staff and perpendicular to it so that the staff is vertical when the bubble is cetral. It must be screwed on very firmly and adequately guarded so that it does not catch-in things or get broken at the hinges. Fig. 22.18 (*b*) shows a circular level mounted on a strong bracket. Circular levels are useful in indicating whether the staff is out of plumb in *any* direction. However, since slight deviation of the staff in lateral directions is not much important, a single level tube rigidly attached to the staff may be used with advantage.

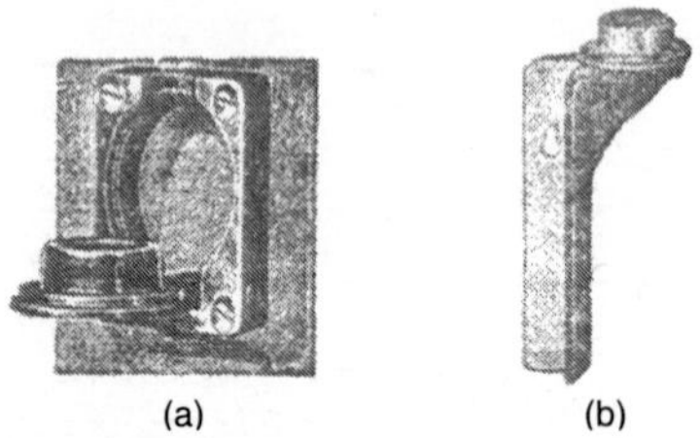

(a) (b)

Fig. 22.18 Levels for holding the staff.

The method of vertical holding of the staff is most commonly adopted for the following reasons: (*a*) The staff can be held plumb easily, and (*b*) The reduction of stadia notes are less laborious and greatly simplified by the use of stadia tables or charts.

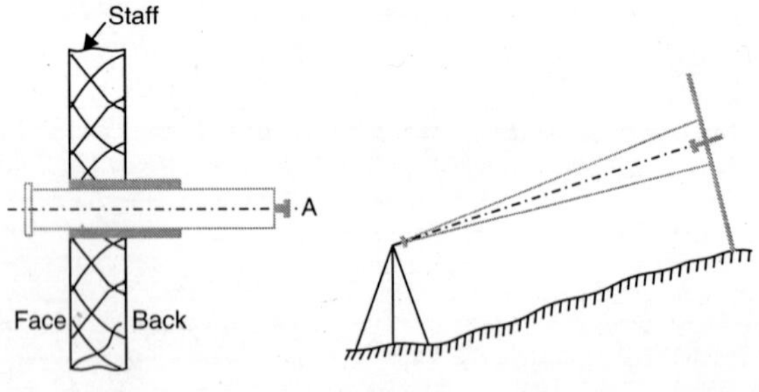

(a) (b)

Fig. 22.19 The Peep Sight.

(*ii*) **Normal Holding:** The staff can be held normal to the line of sight either with the help of a *peep sight* or with the help of a *detector*. A peep sight enables the staffman to ascertain the correct position himself, and may be in the form of either a pair of open sights on a metal bar for short sights or a telescope for very long sights. The line of sights provided by a deep sight must be perpendicular to the face of the staff.

Fig. 22.19 (*a*) shows an ordinary peep sight consisting of a metal tube fitting in a metal socket machined for this purpose. At *A*, a small hole is provided while a pair of cross-hairs is provided at *B*. The staff is inclined slowly, either towards the instrument or away from it as the case may be, till the line of sight bisects the telescope. The reading is then taken. Fig. 22.19 (*b*) illustrates how a peep sight is used. The tube may also be fitted with lenses forming a small telescope to assist the staffman in setting the rod for long sights. Strictly, the peep sight should be attached to the rod at the reading of the central hair, but it is sufficient to place it at the height of eye of the staffman. The advantages of normal holding are:

(*i*) For a given amount of error in the direction, the errors caused in the distances and elevations are less serious in the normal holding than in the vertical holding. In cases where accuracy is essential, angles are large, and the staff has no reliable plumbing device, the only way out of the difficulty is to observe the normal staff.

(*ii*) The accuracy in the direction of the staff can also be judged by transit man.

22.11 METHODS OF READING THE STAFF

There are three methods of observing the staff for distance and altitude : (*i*) the conventional three-hair method; (*ii*) the height of instrument method; and (*iii*) the even-angle method. The observations consist of the staff intercept (s), the middle hair reading (r), and the vertical angle (θ).

(*a*) The Conventional Three-Hair Method:

Steps:

(*i*) Sight the staff and using the vertical circle tangent screw, bring the apparent lower hair to bear exactly on some convenient reading (say 0.5 m or 1 m).

(*ii*) Read the apparent upper hair.

(*iii*) Read the middle (or axial) hair.

(*iv*) Read the vertical angle to the nearest minute or closer in important observations.

The advantages of this method are that staff is easier to be read (since only two readings are uneven values) and the subtractions for finding s and checking its accuracy are easier.

(*b*) The Height of Instrument Method:

Steps:

(*i*) Sight the staff and bring the middle hair to the reading equal to the height of the instrument, thus making r equal to h.

(*ii*) Read the two stadia hairs.

(*iii*) Read the vertical angle.

The main purpose of using this method is to facilitate in calculating the elevation of the staff since r is equal to h. However, the disadvantages of this method are:

(*i*) all the three readings are uneven;

(*ii*) in some cases r cannot be made equal to h;

(*iii*) it adds to the difficulty of the field work and has nothing to offer in return.

(*c*) The Even-angle Method:

Steps:

(*i*) Sight the staff and with the help of the vertical circle tangent screw, bring the zero of the vernier into exact coincidence with the nearest division on the vertical circle. The even angles generally employed are multiples of 20′.

(*ii*) Read the stadia hairs.

(*iii*) Read the middle hair.

The main advantages of this method are: (*i*) since the even angles are multiples of 20′, the trouble of measuring a vertical angle is saved; (*ii*) the computations are simpler.

22.12 STADIA FIELD WORK

General Arrangement of Field Work: The tacheometric survey can be put to a great variety of uses, the principal being the following:

1. Plane surveying involving location of points in plan, but no elevations.
2. Rapid sectioning on steep ground, involving elevations of points and their location along a line.

3. Topography, involving elevations of points as well as their location in plan.
4. Contouring, involving the location or setting out and surveying of level contour lines.

When stadia methods are to be used for filling in detail, adequate control are highly desirable. It is advisable to carry out the following preliminary operations:

1. To establish a sufficient number of well-selected stations for exercising horizontal control.
2. To determine the reduced level of these stations.
3. To determine the position of at least one control point with respect to some well established station (*e.g.* a nearby trig-station) whose co-ordinates are-known.

For vast surveys, horizontal control points are as a rule fixed by a triangulation, but occasionally, a combination of triangulation and traversing may be employed with advantage. When the tract to be surveyed is sufficiently narrow that half of its breadth is within the sighting range of the instrument, the survey can be controlled by an open traverse approximately along the centre line of the strip. For moderate areas, the arrangement may consist of a single main traverse from which numerous circuits are projected. When the survey is too broad on a single traverse, the control may be furnished either by a triangulation or by a series of traverse.

Triangulation: If triangulation is used to fix the horizontal control points (or tacheometer stations) the first step is the establishment of a suitable base. This may be accomplished:

1. By making use of major control points such as trig-stations.
2. By measurement with a steel tape.
3. By subtense measurement.

The first method is the most suitable and accurate if a pair of convenient trig-stations within early reach of the area to be surveyed, since the length of the line joining them and its bearing are known precisely. Second method may be used if such stations are not situated nearby. The third method of establishing the base by subtense measurement can be employed in any sort of difficult country.

Traversing: The lengths of the traverse courses may be measured either by tape or tacheometrically. Similarly, the elevations of the instrument stations can be determined, either by spirit levelling or by tacheometrical levelling, depending upon the degree of accuracy required. The tacheometric methods for determining the lengths of traverse line and the elevations of stations can be used only in small-scale work.

Tacheometer Stations: It is desirable that main stations should be fixed and surveyed before the tacheometric detail work is pursued. The best tacheometer station is one which commands a clear view of the area to be surveyed within the range of observations. With regard to elevation, it should be so suited that the use of large vertical angles is avoided. The great majority of tacheometer stations are generally the stadia traverse station. Skill in selecting the best stations is largely the result of observations and experience.

Field Party: For surveys of small extents, a surveyor and a staffman are sufficient; but for surveys of large extent in a rough country, the field party may consist of:

1. The Surveyor or Chief of the party for the over-all control of the survey.
2. The instrument man to take the actual observations.
3. The recorder to record the readings taken by the instrument man.
4. Two or four staffmen, depending upon the expertness of the instrument man.
5. Labourers for clearing and transport.

Tacheometric Observations: The following are the usual operations:

1. *Setting up the instrument:* This consists of:
   (*a*) Setting the instrument exactly over the station mark, and
   (*b*) Levelling it carefully.

The instrument should first be levelled up with respect to the plate levels and then with respect to the altitude bubble. In general if the altitude bubble deviates only by one division during a complete revolution of the instrument about its vertical axis, the instrument may be regarded as level. However, for all important observations, the bubble should be central when the middle hair is read.

2. *Measuring the height of the instrument:* The height of the instrument (H.I.) is the vertical distance from the top of the peg to the centre of the object glass and should be measured with the vertical vernier set to zero and the altitude bubble central. This observation is very important since all observations for altitude are practically worthless unless the height of axis is recorded.
3. *Orienting the instrument:* Since a number of rays or directions of sight may emerge from one station, the instrument should be properly oriented when zero is clamped on the horizontal circle. The reference line passing through the instrument may be a true meridian or magnetic meridian or arbitrary meridian. If the reference line is a trye meridian or magnetic meridian, reading on the horizontal circle should be zero when the line of sight along that meridian and the angles to different rays or directions will directly be their whole circle bearings. If, however, the instrument is oriented with reference to another station of the survey, the circle should read the bearing of this station when the line of sight is directed to it. Once an instrument has been correctly oriented, the position of the circle should not be disturbed until all the readings at the station are completed.
4. *Observing staff held on bench mark:* In order to know the elevation of the centre of the instrument, the staff should be kept on the nearest B.M. and tacheometric observations should be taken to the staff. If the B.M. is not nearby, the staff should be observed on a point of known elevation, or flying levels may be run from the B.M. to establish one near the area.
5. *Observations of distance and altitude:* In order to know the horizontal distance and elevation of the representative points, the following observations are made on the staff:
   (*i*) Stadia hair readings
   (*ii*) Axial hair readings
   (*iii*) Angle of elevation or depression of line of sight.

The staff may be held either vertical or normal to the line of sight. The three methods of observing the staff have already been discussed.

The observations to various points are known as *side shots*. Observations can be taken more quickly and systematically if all the stations are along the radial lines through the station at some constant angular interval. For general work, the bearings should be observed to 5′ and the vertical angles read to the nearest 1′.

Field Book: The Table below is the usual form of booking the field notes.

STADIA FIELD BOOK

Ins. Station	*Ht. of Instrument*	*Staff Station*	*Bearing*	*Vertical angle*	*Stadia Hair Reading Top / Bottom*	*Axial Hair Reading*	*Stadia Intercept*	*D*	*V*	*R.L.* *Intst. Station*	*R.L.* *Staff Station*	*Remarks*
1	*2*	*3*	*4*	*5*	*6*	*7*	*8*	*9*	*10*	*11*		*12*
P	1.42 m	*A*	30°	+ 2° 24′	2.880 / 1.230	2.055	1.65	164.7	6.903	77.750	84.018	
		B	94° 30′	– 3° 36′	2.815 / 0.785	1.800	2.03	202.1	12.73	77.750	64.640	

22.13 THE TANGENTIAL METHOD

In the tangential method, the horizontal and vertical distances from the instrument to the staff station are computed from the observed vertical angles to the vanes fixed at a constant distance apart upon the staff. The stadia hairs are, therefore, not used and the vane is bisected every time with the axial hair. Thus, two vertical angles are to be measured—one corresponding to each vane. There may be three cases of the vertical angles:

(*i*) Both angles are angles of elevation.

(*ii*) Both angles are angles of depression.

(*iii*) One angle of elevation and the other of depression.

Case I. Both Angles are Angles of Elevation:

Let P = Position of the instrument

Q = Staff station

M = Position of instrument axis

A, B = Position of vanes

s = Distance between the vanes—staff intercept

α_1 = Angle of elevation corresponding to A

α_2 = Angle of elevation corresponding to B

D = Horizontal distance between P and $Q = MQ'$

V = Vertical intercept between the lower vane and the horizontal line of sight

h = Height of the instrument = MP

r = Height of the lower vane above the foot of the staff

= Staff reading at lower vane = BQ

From $\Delta MBQ'$, $\quad V = D \tan \alpha_2 \quad$ ...(*i*)

From $\Delta AMQ'$, $\quad V + s = D \tan \alpha_1 \quad$ ...(*ii*)

Subtracting (*i*) from (*ii*), we get

$$s = D \tan \alpha_1 - D \tan \alpha_2$$

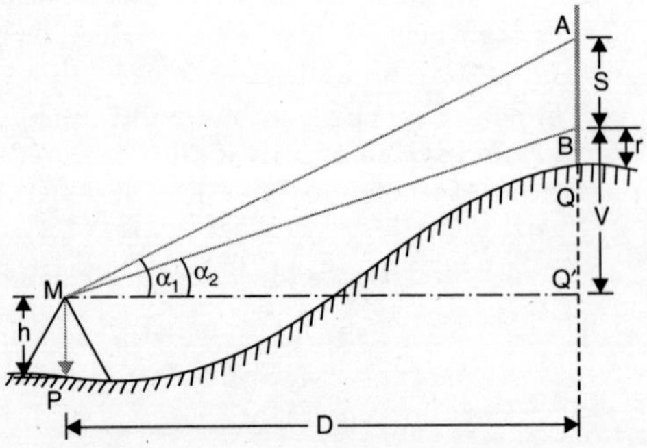

Fig. 22.20 Tangential Method: Angles of Elevation

$$\therefore \quad D = \frac{s}{\tan \alpha_1 - \tan \alpha_2} \qquad ...(22.19)$$

$$= \frac{s \cos \alpha_1 \cos \alpha_2}{\sin (\alpha_1 - \alpha_2)} \qquad ...(22.19\ a)$$

$$V = D \tan \alpha_2 = \frac{s \tan \alpha_2}{\tan \alpha_1 - \tan \alpha_2} \qquad ...(22.20)$$

$$= \frac{s \cos \alpha_1 \sin \alpha_2}{\sin (\alpha_1 - \alpha_2)} \qquad ...(22.20\ a)$$

Elevation of Q = (Elevation of station + h) + $V - r$.

Case II. Both Angles are Angles of Depression:

With the same notations as earlier

$$V = D \tan \alpha_2 \quad ...(i) \quad \text{and} \quad V - s = D \tan \alpha_1 \qquad ...(ii)$$

Subtracting (*ii*) from (*i*), we get

$$s = D \tan \alpha_2 - D \tan \alpha_1$$

$$\therefore \quad D = \frac{s}{\tan \alpha_2 - \tan \alpha_1}$$

$$= \frac{s \cos \alpha_1 \cos \alpha_2}{\sin (\alpha_2 - \alpha_1)} \qquad ...(22.21)$$

$$\therefore \quad V = D \tan \alpha_2 = \frac{s \tan \alpha_2}{\tan \alpha_2 - \tan \alpha_1}$$

$$= \frac{s \cos \alpha_1 \sin \alpha_2}{\sin (\alpha_2 - \alpha_1)}. \qquad ...(22.22)$$

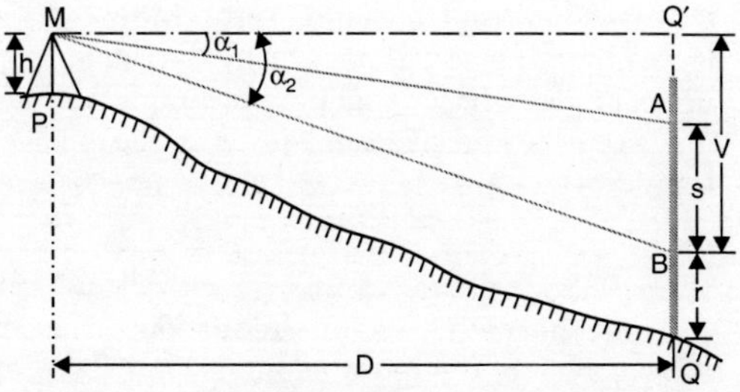

Fig. 22.21 Tangential Method: Angles of Depression.

Elevation of Q = (Elevation of $P + h$) – $V - r$.

Case III. One Angle of Elevation and other of Depression:

$$V = D \tan \alpha_2 \qquad ...(i)$$

$$\text{and} \quad s - V = D \tan \alpha_1 \qquad ...(ii)$$

Adding (*i*) and (*ii*), we get

$$s = D \tan \alpha_1 + D \tan \alpha_2$$

$$\therefore \quad D = \frac{s}{\tan \alpha_1 + \tan \alpha_2}$$

$$= \frac{s \cos \alpha_1 \cos \alpha_2}{\sin (\alpha_1 + \alpha_2)} \qquad ...(22.23)$$

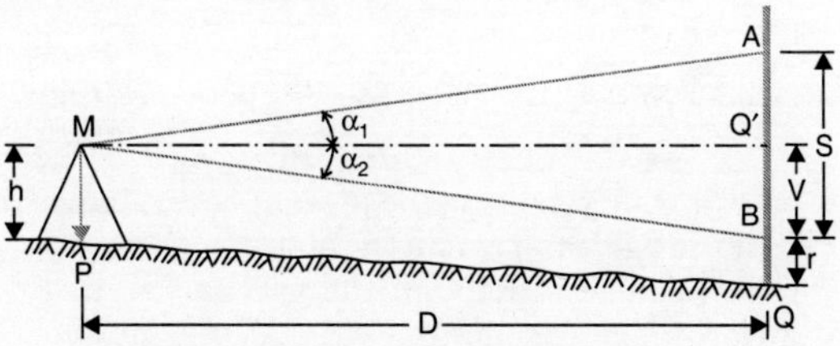

Fig. 22.22 One Angle of Elevation.

$$\text{and} \quad V = D \tan \alpha_2 = \frac{s \tan \alpha_2}{\tan \alpha_1 + \tan \alpha_2} = \frac{s \cos \alpha_1 \sin \alpha_2}{\sin (\alpha_1 + \alpha_2)} \qquad ...(22.24)$$

Elevation of Q = Elevation of $P + h - V - r$.

Methods of Application: The principle of tangential measurement can be applied in practice by measuring the angles α_1 and α_2 subtended with the horizontal by the two rays of the measuring triangle *MAB*. The tangential measurement can be applied in two ways:

(*i*) The base *AB* may be of constant length *s* and the angles α_1 and α_2 may be measured for each position of the staff. The method is sometimes known as the *constant base tangential measurement.*

(*ii*) The angles α_1 and α_2 may be special 'pre-selected' angles and the base *s* may be of variable length depending upon the position of the staff. The method is sometimes known as the *variable base tangential measurement.*

Constant-Base Tangential Measurement: Airy's Method

In this method, a staff having two targets at contant distance *s* apart is used at every station and the angle α_1 and α_2 measured. The method is sometimes known as *Airy's Method.* Equations 22.19 to 22.24 are used, depending upon the signs of α_1 and α_2. Though the observations in this case are simpler than the variable base method, the computations are more tedious.

Variable Base Tangential Measurements: System of Percentage Angles

In the above method, the angles α_1 and α_2 are to be measured accurately and the reduction is rather tedious. A better method is to use selected values α_1 of α_2 and measure the variable base (*i.e.,* staff intercept) on a uniformly graduated staff. The variable base method using the system of percentage angles was devised by Barcenas, a Spanish surveyor. The method consists in making use of angles whose tangents are simple fractions of 100, like 0.03 or 3%, 0.12 or 12% etc. These angles can be laid off accurately with the aid of an appropriate scale on the vertical circle and the computations are easier. If α_1 and α_2 are consecutive angles whose tangents differ by 1%, we get

$$D = \frac{s}{\tan \alpha_1 - \tan \alpha_2} = \frac{s}{0.01} = 100\ s.$$

Thus, the method enables reductions to be performed mentally. By reference to trigonometric tables, a list of the required angles may be prepared as follows:

Tangent	*Angle to nearest second*			*Tangent*	*Angle to nearest second*		
	°	′	″		°	′	″
0.01	0	34	24	0.06	3	26	01
0.02	1	08	45	0.07	4	00	15
0.03	1	43	06	0.08	4	34	26
0.04	2	17	26	0.09	5	08	34
0.05	2	51	45	0.10	5	42	38

Fergusson's Percentage Unit System: The only difficulty in using the percentage system is that the angles shown in the above table cannot be set out accurately on the vertical circle of an ordinary theodolite. Mr. J.D. Fergusson, however, has devised a system for the division of the circle so as to get the percentage angles directly.

Fig. 22.23 illustrates the method of division devised by Fergusson. A circle, inscribed in a square is divided into eight octants. Each of the eight octants is of length equal to the radius of the circle and is divided into 100 *equal* parts. Lines are then drawn from the centre to these points, thus dividing each octant into 100

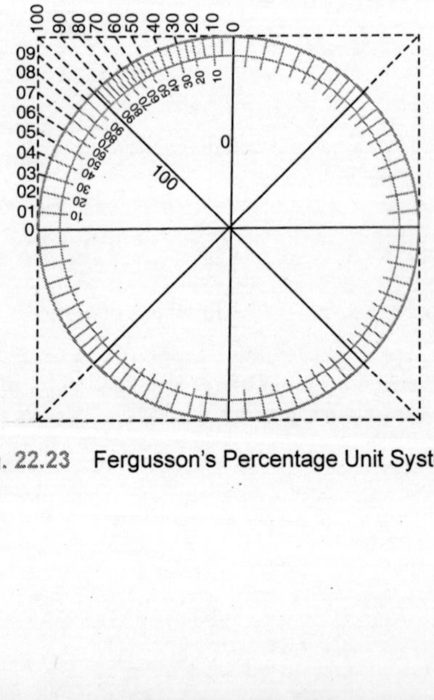

Fig. 22.23 Fergusson's Percentage Unit System.

unequal parts. The points of division on circle are then marked from 0 to 100 as shown. Since vernier cannot be used to subdivide these unequal parts, a spiral drum micrometer is used to take the readings to 0.01 of a unit.

Effect of Angular Error in Tangential Measurement

In order to find the resulting error in the measured horizontal distance due to error in the measurement of α_1 and α_2 let us assume that the probable error in measuring each of these angles is 20″. Let $\delta\alpha_1 = +20''$ and $\delta\alpha_2 = -20''$ giving a combined angular error of 40″.

Let D_1 = corresponding horizontal distance; s = staff intercept = 3 m (say)

D = correct horizontal distance.

Thus,
$$D_1 = \frac{s}{\tan(\alpha_1 + 20'') - \tan(\alpha_2 - 20'')}$$

where α_1 and α_2 are the correct angles.

Now $\tan(\alpha_1 + 20'') = \tan\alpha_1 + a_1$ and $\tan(\alpha_2 - 20'') = \tan\alpha_2 - a_2$

where a_1 and a_2 are the tangent difference corresponding to a difference of 20″.

$$\therefore \quad D_1 = \frac{s}{(\tan\alpha_1 - \tan\alpha_2) + (a_1 + a_2)} \qquad \text{...(22.25)}$$

If will be noticed from the trigonometric tables that the difference between a_1 and a_2 is slight. Let $a_1 = a_2 = a$ and $\tan\alpha_1 - \tan\alpha_2 = q$

Then
$$s = D(\tan\alpha_1 - \tan\alpha_2) = D_1\{(\tan\alpha_1 - \tan\alpha_2) + 2a\}$$

or
$$s = D_q = D_1(q + 2a)$$

$$\frac{D}{D_1} = \frac{q + 2a}{q} \quad \text{or} \quad \frac{D - D_1}{D_1} = \frac{e}{D_1} = \frac{2a}{q}$$

But
$$\frac{e}{D_1} = \frac{e}{D} \quad \text{(very closely)}$$

$$\therefore \quad r = \frac{e}{D} = \frac{2a}{q} = \frac{2aD}{s}; \text{ where } \quad r = \text{ratio of error.} \qquad \text{...(22.26)}$$

As an example, let $D = 60$ m; $\alpha_1 = 5°$ and $s = 3$ m

Then $\tan 5° = 0.0874887$; $a_1 = 0.0000978$

Taking $a_1 = a = 0.0000978$, we get from Eq. 22.26

$$r = \frac{e}{D} = \frac{2aD}{s} = \frac{2 \times 0.0000978 \times 60}{3} = \frac{1}{256}$$

It is evident, therefore, that in this system of tacheometry, the angles must be measured very accurately. It can be shown that if r is not to exceed 1/500, the permissible angular error is about ± 5″ for rays of about 125 metres and for base of 3 metres.

Example 22.17: *The vertical angles to vanes fixed at 1 m and 3 m above the foot of the staff held vertically at a station A were +2° 30′ and + 5° 48′ respectively. Find the horizontal distance and the reduced level of A if the height of the instrument, determined from observation on to a bench mark is 438.556 metres above datum.*

Solution: (Fig. 22.20). From equation 22.19 (*a*), we have

$$D = \frac{s \cos \alpha_1 \cos \alpha_2}{\sin (\alpha_1 - \alpha_2)} = \frac{2 \cos 5°48' \cos 2°30'}{\sin (5°48' - 2°30')} = \mathbf{34.53\ m.}$$

$$V = D \tan \alpha_2 = 34.53 \tan 2° 30' = \mathbf{1.508\ m}$$

$$\therefore \quad \text{R.L. of } A = 438.556 + 1.508 - 1 = \mathbf{439.064\ m.}$$

Example 22.18: *An observation with a percentage theodolite gave staff readings of 1.052 and 2.502 for angles of elevation of 5% and 6% respectively. On sighting the graduation corresponding to the height of the instrument axis above the ground, the vertical angle was 5.25%. Compute the horizontal distance and the elevation of the staff station if the instrument station has an elevation of 942.552 metres.*

Solution: $\tan \alpha_1 = 0.06$ and $\tan \alpha_2 = 0.05$

$$\therefore \quad D = \frac{s}{\tan\alpha_1 - \tan\alpha_2} = \frac{2.502 - 1.052}{0.06 - 0.05} = \mathbf{145\ m.}$$

$$V = D \tan \alpha_2 = 145 \times 0.05 = \mathbf{7.25\ m}$$

Let the angles to the graduation corresponding to the height of the instrument be α_3 so that $\tan \alpha_3 = 0.0525$. If s' is the corresponding staff intercept, we have

$$D = \frac{s'}{\tan\alpha_1 - \tan\alpha_3} = \frac{s'}{0.06 - 0.0525}$$

or $$s' = D\,(0.06 - 0.0525) = 145 \times 0.0075 = 1.088 \text{ m}$$

If r is the staff reading corresponding to the height of the instrument, we have

$$r = 2.502 - 1.088 = 1.414 \text{ m}$$

R.L. of staff = R.L. of I.A. + V – 1.052 = (942.552 + 1.414) + 7.250 – 1.052 = **950.164 m.**

22.14 REDUCTION OF STADIA NOTES

After having taken the field observations, the distance and elevations of the points can be calculated by the use of various tacheometric formulae developed earlier. If the number of points observed is less, a log table may be used to solve the tacheometric formulae. However, for surveys of large extent where the number of points observed is much more, calculation or reduction of stadia notes is done quickly either with the help of tacheometric tables, charts, diagrams or by mechanical means.

Tacheometric Tables: If distances are required only to the nearest quarter of metre, the value $C \cos \theta$ may be taken either as C or simply as $\frac{1}{4}$ m. It can be shown that for distances not exceeding 100 m, the distance reading may be taken as the horizontal distance for vertical angles upto 3°. If the additive constant C is ignored alltogether in conjuction with the above approximation, the two errors tend to compensate. Various forms of tacheometric tables are available, a simple form being given on next page. A complete set of stadia tables (in slightly different form) for angle of elevation upto 30°, is given in the Appendix.

Example: *Let $s = 1.5$ m; $\theta = 3° 36'$; $C = 0.3$ m.*

From the table, for $\theta = 3° 36'$, we get

Hor. correction	Diff. Elev.
0.39	6.27
$C = 0.00$	0.02

Distance reading $= 100 \times 1.5 = 150$ m

Horizontal distance $= 150 - (0.39 \times 1.5) + (0.3 - 0.00) =$ **149.71 m**

and $V = (6.27 \times 1.5) + 0.02 =$ **9.43 m.**

STADIA REDUCTION TABLE

Minutes	0°		1°		2°		3°	
	Hor. Corr.	Diff. Elev.	Hor. Corr.	Diff Elev.	Hor. Corr.	Diff Elev.	Hor. Corr.	Diff Elev.
0	0.00	0.00	0.03	1.74	0.12	3.49	0.27	5.23
2	0.00	0.06	0.03	1.80	0.13	3.55	0.28	5.28
4	0.00	0.12	0.03	1.86	0.13	3.60	0.29	5.34
6	0.00	0.17	0.04	1.92	0.13	3.66	0.29	5.40
8	0.00	0.23	0.04	1.98	0.14	3.72	0.30	5.46
10	0.00	0.29	0.04	2.04	0.14	3.78	0.31	5.52
12	0.00	0.35	0.04	2.09	0.15	3.84	0.31	5.57
14	0.00	0.41	0.05	2.15	0.15	3.89	0.32	5.63
16	0.00	0.47	0.05	2.21	0.16	3.95	0.32	5.69
18	0.00	0.52	0.05	2.27	0.16	4.01	0.33	5.75
20	0.00	0.58	0.05	2.33	0.17	4.07	0.34	5.80
22	0.00	0.64	0.06	2.38	0.17	4.13	0.34	5.86
24	0.00	0.70	0.06	2.44	0.18	4.18	0.35	5.92
26	0.01	0.76	0.06	2.50	0.18	4.24	0.36	5.98
28	0.01	0.81	0.07	2.56	0.19	4.30	0.37	6.04
30	0.01	0.87	0.07	2.62	0.19	4.36	0.37	6.09
32	0.01	0.93	0.07	2.67	0.20	4.42	0.38	6.15
34	0.01	0.99	0.07	2.73	0.20	4.47	0.38	6.21
36	0.01	1.05	0.08	2.79	0.21	4.53	0.39	6.27
38	0.01	1.11	0.08	2.85	0.21	4.59	0.40	6.32
40	0.01	1.16	0.08	2.91	0.22	4.65	0.41	6.38
42	0.01	1.22	0.09	2.97	0.22	4.71	0.41	6.44
44	0.02	1.28	0.09	3.02	0.23	4.76	0.42	6.50
46	0.02	1.34	0.10	3.08	0.23	4.82	0.43	6.56
48	0.02	1.40	0.10	3.14	0.24	4.88	0.44	6.61
50	0.02	1.45	0.10	3.20	0.24	4.94	0.44	6.67

(*Contd.*)

(Contd.)

Minutes	0°		1°		2°		3°	
	Hor. Corr.	*Diff. Elev.*	*Hor. Corr.*	*Diff Elev.*	*Hor. Corr.*	*Diff Elev.*	*Hor. Corr.*	*Diff Elev.*
52	0.02	1.51	0.11	3.26	0.25	4.99	0.45	6.73
54	0.02	1.57	0.11	3.31	0.26	5.05	0.46	6.79
56	0.03	1.63	0.11	3.37	0.26	5.11	0.47	6.84
58	0.03	1.69	0.12	3.43	0.27	5.17	0.48	6.90
60	0.03	1.74	0.12	3.49	0.27	5.23	0.49	6.96
$C = 0.2$ m	0.00	0.00	0.00	0.01	0.00	0.01	0.00	0.01
$C = 0.3$ m	0.00	0.00	0.00	0.01	0.00	0.01	0.00	0.02
$C = 0.4$ m	0.00	0.00	0.00	0.01	0.00	0.02	0.00	0.02

Reduction Diagrams: Various forms of reduction diagrams are available, a simple form being suggested below:

Reduction diagram for horizontal correction: For inclined sights, the horizontal distance is given by $D = ks \cos^2 \theta$.

If the line of sight is assumed to be horizontal, the horizontal distance is given by $D' = ks$.

Horizontal correction $= D' - D = ks - ks \cos^2 \theta = ks \sin^2 \theta$

The value $k.s$ is known as *distance reading*. Fig. 22.24 shows the reduction diagram for horizontal correction for a tacheometer having $k = 100$.

To prepare the diagram (Fig. 22.24), the scale of distance reading upto 300 metres is set out along the vertical line. On the horizontal line at 300 m reading, the values of horizontal correction ($= ks \sin^2 \theta = 300 \sin^2 \theta$) is marked off for vertical angle increasing by a suitable interval (say by 10′ or 5′). These points are joined to the origin to get various radial lines. Since the horizontal correction is directly proportional to the distance reading for a given angle, these radial lines give horizontal correction for other distance readings on the scale.

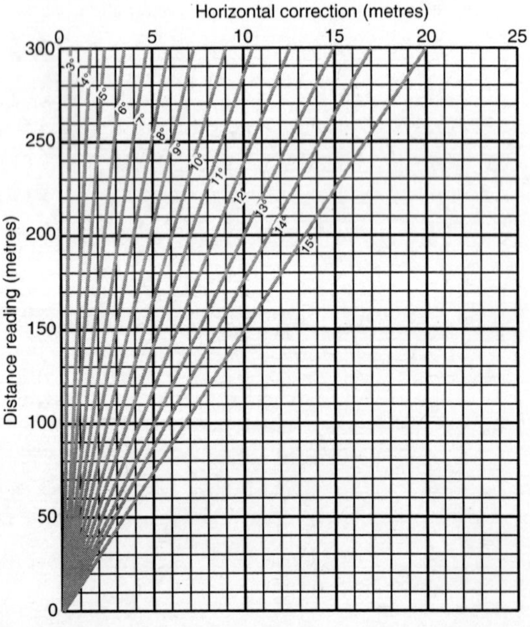

Fig. 22.24 Reduction Diagram for Horizontal Correction.

Example: If $s = 1.5$ m, the distance reading $= 100 \times 1.5 = 150$ m. From the diagram (Fig. 22.24), the horizontal correction (for $\theta = 13°$) $= 7.6$ m.

Hence the correct horizontal distance $= 150 - 7.6 = 142.4$ m.

Reduction diagram of vertical component: To construct the reduction diagram for the vertical component ($V = ks \frac{1}{2} \sin 2\theta$), the distance reading (= ks) is set off on the horizontal scale and the vertical component upto a maximum value of 30 metres on the vertical scale as shown in Fig. 22.25. Upto $\theta = 5°46'$ the values of V are calculated when the distance reading is 300 m. These calculated values of V are marked off on the vertical scale and joined to the origin by straight lines. Beyond $\theta = 5° 46'$, the values of horizontal correction to give $V = 30$ m are calculated for various angles. The calculated values of distance reading are marked off on the top horizontal line and joined to the origin by straight lines. The radial lines may be drawn for angles at interval of every 5′ or 10′ depending upon the size of diagram.

To use the diagram, let $s = 1.5$ m, distance reading = $ks = 100 \times 1.5 = 150$. If $\theta = 4°$, we get $V = 10.5$ m from the diagram. Thus, the observations may be reduced still more rapidly by the use of the reduction diagram.

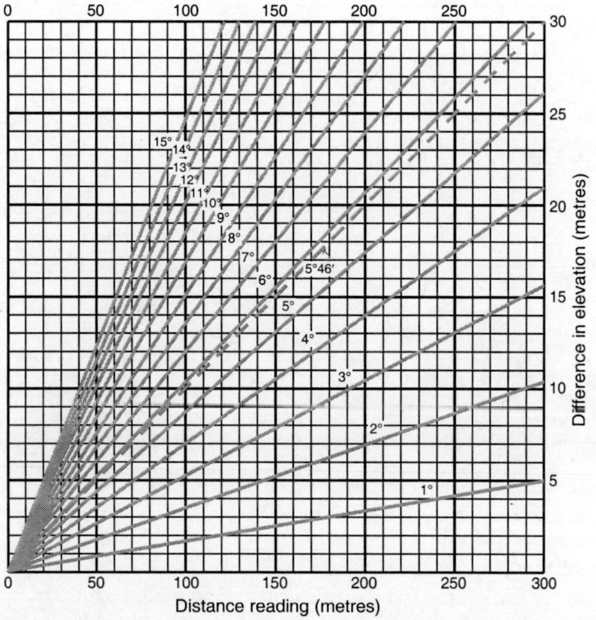

Fig. 22.25 Reduction Diagram for Vertical Component.

Reduction by Mechanical Means: Various forms of slide rules are available with the help of which the observation may be reduced mechanically.

22.15 SPECIAL INSTRUMENTS

1. Beaman Stadia Arc

Beaman stadia arc is a special device fitted to tacheometer and plane table alidades. Its use facilitates the determination of differences of elevation and horizontal distance without the use of stadia tables or stadia slide rule. The arc carries two scales H and V having their central points marked 0 and 50 respectively. A common index is used to read both the scales.

The Beaman stadia arc is designed on the fact that reductions are simplified if the only values of θ used were those for which $\frac{1}{2}\sin 2\theta$ is a convenient figure. The following is the list of angles.

$\frac{1}{2}$ *sin 2θ*	*θ to nearest second*			$\frac{1}{2}$ *sin 2θ*	*θ to nearest second*		
	°	′	″		°	′	″
0.01	0	34	23	0.06	3	26	46
0.02	1	08	46	0.07	4	01	26
0.03	1	43	12	0.08	4	36	12
0.04	2	17	39	0.09	5	11	06
0.05	2	52	11	0.10	5	46	07

The divisions of the V scale are of such magnitude that $\frac{1}{2}\sin 2\theta$ for each graduation is a magnitude of 0.01. When the index reads 51 (or 49), the line of sight is inclined by an angle corresponding to the first division on the arc, or $\frac{1}{2}\sin 2\theta = 0.01$, which gives $\theta = 34'\,23''$.

Hence

$$V = ks\,\tfrac{1}{2}\sin 2\theta = 100\,s \times 0.01$$

or $$V = s,$$

When $k = 100$ and $C = 0$.

The second division (numbered 52 or 48) is positioned at an angular value of $\theta = 1°\,8'46''$ so that $\frac{1}{2}\sin 2\theta = 0.02$ and hence

$V = 100\,s(0.02) = 2\,s$.

Similarly, the third division (numbered 53 or 47) is positioned at an angular value of $\theta = 1°43'12''$ and $V = 3\,s$ and so on. Since the central graduation of V-scale is marked 50, a reading of less than 50 indicates that the telescope is inclined downward, while reading greater than 50 shows it is inclined upward. The value of V is then given by

$$V = s \times (\textit{Reading on V scale} - 50)$$

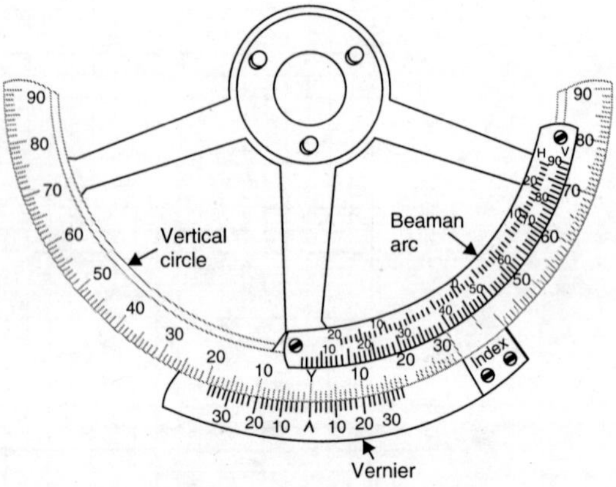

Fig. 22.26 Beaman Stadia Arc.

When the staff is sighted, the staff intercept s is noted. If the index is not against the whole number reading of the V-scale the tangent screw is used to bring the nearest Beaman arc graduation exactly coincident with the stadia index. The middle wire reading is now taken and the reading on V-scale is noted. *It should be remembered that by tilting the line of sight slightly for this operation, there is no appreciable change in the value of s.*

On the horizontal or H-scale, the divisions are of such values as to represent the percentage by which the observed stadia reading is to be reduced to obtain the corresponding distance. In other words, the H-scale reading multiplied by the staff intercept gives the horizontal correction to be subtracted from the distance reading.

Example:

Central wire reading = *1.425 m*

Reading on V-scale = *58*

Reading on H-scale = *4*

Staff intercept = *1.280 m*

Elevation of I.A. = *100.00*

$$V = 1.280 \times (58 - 50) = +\ 10.24 \text{ m}$$

$$\text{Elevation of staff} = 100 + 10.24 - 1.425 = 108.815 \text{ m}$$

$$\text{Horizontal correction} = 1.28 \times 4 = 5.12$$

$$\text{Horizontal distance} = (1.28 \times 100) - 5.12 = 122.88 \text{ m.}$$

(assuming $k = 100$ and $C = 0$).

2. The Jeffcott Direct Reading Tacheometer

This instrument, invented by the late Dr. H.H. Jeffcott, enables the horizontal and vertical components (more or less) directly on the staff, thus saving the labour of calculation. The diaphragm of the instrument carries three pointers, the middle one being fixed and the other two movable (Fig. 22.27).

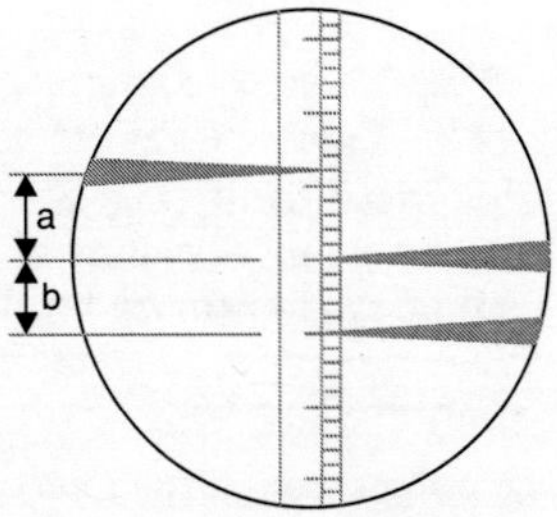

Fig. 22.27 The Jeffcott Direct Reading Tacheometer.

The intercept between the fixed pointer and the distance pointer (right hand movable pointer) multiplied by 100 gives the horizontal distance D. Similarly, the intercept between the fixed pointer and the elevation pointer (left hand movable pointer) multiplied by 10, gives the vertical component V. The telescope of the instrument is anallactic. The staff readings are taken by first setting the fixed pointer at a whole foot mark and then reading the other two pointers.

Each movable pointer is mounted on one end of a lever. The other ends of these levers ride on respective cams (*i.e.* distance cam and elevation cam). The cams are fixed in altitude and so shaped that the interval between the pointer is adjusted automatically to correspond to the angle of the telescope.

The Jeffcott direct reading tacheometer could not be entirely a success due to the following defects:

1. Pointers are inconvenient to read with.
2. Half intercepts cannot be measured.
3. Effect of parallax is unavoidable.

3. The Szepessy Direct Reading Tacheometer

The instrument, invented by a Hungarian has the distinction of being the most successful of those in the tangential group, and use the percentage angles. A scale of tangents of vertical angles is engraved on a glass arc which is fixed to the vertical circle cover. The scale is divided to 0.005 and numbered at every 0.01. Thus the graduation 10 corresponds to the angle whose tangent is 0.10 or 10%. By means of prisms, this scale is reflected in the view of the eye-piece, and when the staff is sighted, the image of the staff is seen along side that of the scale (Fig. 22.28).

Procedure for reading the staff:

1. Sight the staff and clamp the vertical circle at some convenient position.
2. Using the vertical circle tangent screw, bring a whole number division, say 14, opposite the horizontal cross-hair. Note the axial reading.
3. Read the staff intercept between 14 and 13 (or 14 and 15) numbered divisions. The staff intercept multiplied by 100 gives the horizontal distance D. Alternatively, the intercept between 13 and 15 may be measured and multiplied by 50 to get D.
4. The vertical component V is obtained by multiplying the intercept by the numbered division brought opposite the axial hair.

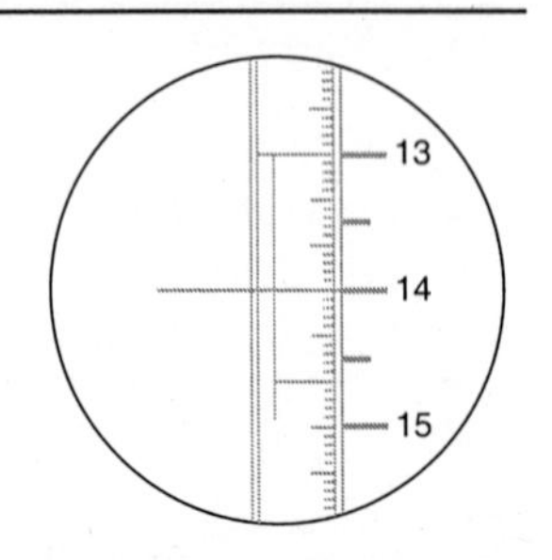

Fig. 22.28 The Szepessy Direct Reading Tacheometer

For example, if s = 1.48 m and the number against the axial hair =14. Then,

$$D = 1.48 \times 100 = 148 \text{ m} \quad \text{and} \quad V = 1.48 \times 14 = 20.70 \text{ m}$$

22.16 THE AUTO-REDUCTION TACHEOMETER (*HAMMER-FENNEL*)

This instrument (Fig. 22.33) permits both the distance and the difference of altitude to be read by a single reading of a vertically held staff—thus reducing tacheometric operation to the simplicity of ordinary levelling.

Special Auto-reduction Device

Looking through the telescope the field of view is found to be divided into 2 halves one of which is designed for the vision of the staff while the second half shows the very diagram of a special type shown in Fig. 22.29.

In Fig. 22.29, there are four curves marked by the letters N, E, D and d. N is the zero curve. E means the curve for reading distances. D illustrates the double curve to be applied for elevation angles upto ± 14°. d is the double curve for greater elevation angles upto ± 47°. The curve lines for elevation angles are marked +, and the curve lines to depth angles are marked –.

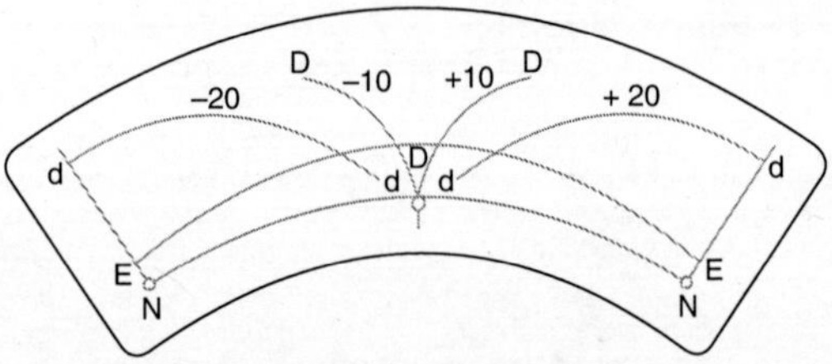

Fig. 22.29 Special Auto-Reduction Device (Hammer-Fennel)

By tilting the telescope up and down, the diagram appears to pass across its field of view. The multiplications to be applied are:

100 for reading the distance (curve E)

10 for reading the difference of altitude (curve D)

20 for reading the difference of altitude (curve d)

The zero-curve appears to touch the zero-line continuously at point of intersection with the vertical edge of the prisms. In taking a reading of the staff, the perpendicular edge of the prism should be brought into line with the staff in such a way that the zero curve bisects the specially marked zero-point of the rod, the zero point being 1.40 m above the ground. Then reading is effected with the distance curve and the respective height curve.

The reading, now taken on the staff with the distance-curve multiplied by 100 gives the distance between instrument and staff, while the reading taken on the staff with the height curve multiplied by 20 or 10 respectively gives the diference in height between the staff position and the instrument station. No other observations or calculations are necessary.

Figs. 22.30 to 22.32 illustrate how readings are taken.

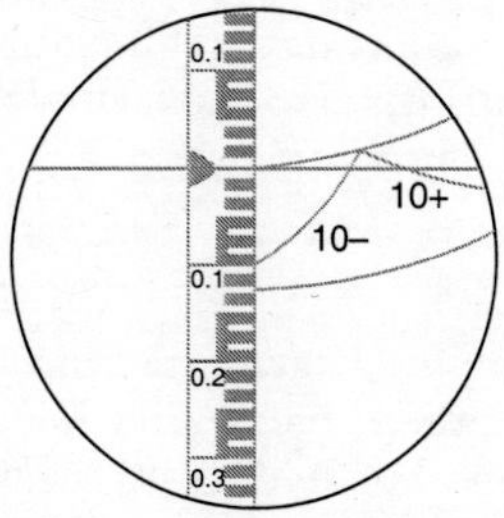

Fig. 22.30 Telescope Depressed.

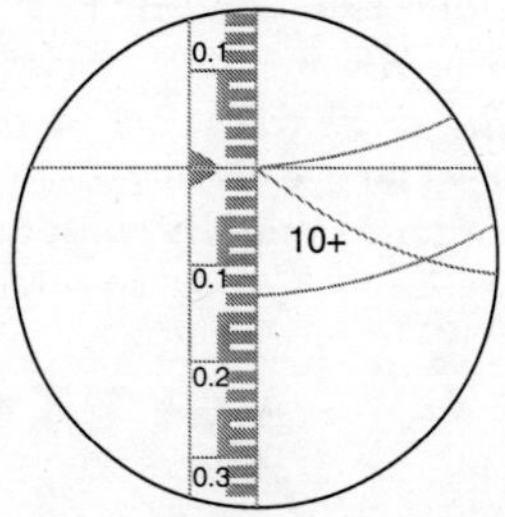

Fig. 22.31 Telescope Horizontal.

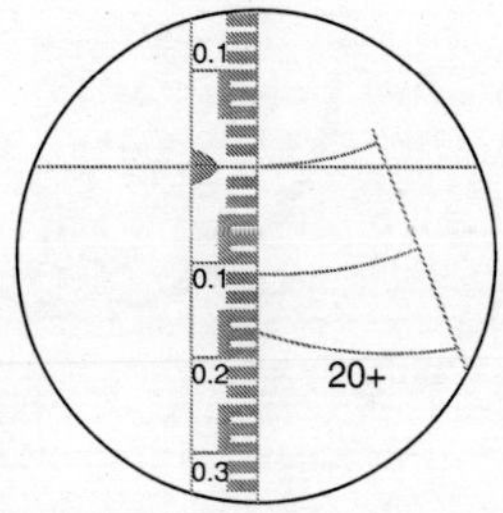

Fig. 22.32 Telescope Elevated.

1. ***Telescope depressed (Fig. 22.30)***
   Reading of distance curve : 0.126
   Reading of height curve : – 0.095
   (with – 10 mark)
   ∴ Horizontal distance = 0.126 × 100 = 12.6 m
   and difference in height = – 0.095 × 10 = – 0.95 m
2. ***Telescope horizontal (Fig. 22.31)***
   Reading of distance curve : 0.134
   Reading of height curve : ± 0.0
   (with + 10 mark)
   ∴ Horizontal distance = 0.134 × 100 = 13.4 m
   and difference in height = ± 0.0 × 10 = ± 0.0 m
3. ***Telescope elevated (Fig. 22.32)***
   Reading of distance curve : 0.113
   Reading of height curve : + 0.175
   (with + 20 mark)
   ∴ Horizontal distance = 0.113 × 100 = 11.3 m
   Difference in height = + 0.175 × 20 = + 3.50 m

22.17 WILD'S RDS REDUCTION TACHEOMETER (Figs. 22.34 and 22.35)

This is also an auto-reduction instrument with a set of curves designed for use with a vertical staff. The credit for the principle of the reducing device goes to Hammer.

In the first telescope position, which is the standard position for distance and height measurements, the vertical circle is on the left hand side and the curve plate on the right hand side of the telescope. The focusing knob is mounted on the right in the telescope trunnion axis. The curves are etched on the glass circle which revolves about the trunnion axis and is located to the right of the telescope. A prism, inside the telescope, projects the image on the plane of the diagram circle and at the same time, rotates it by 90°. Other prism and lenses transfer this image into the reticule plate mounted ahead of the telescope. This plate has a vertical centre line and a horizontal line. Thus, the diagram lines falling in the field of vision appear free of parallax in the plane of hair lines and the image is erect again, although the path of light rays has been broken.

For distance finding, the constant 100 can be used throughout; *thus one centimetre on the rod is equal to one metre of horizontal range.* For difference in height the following constants are chosen:

10 from 0° to 5° ;	20 from 4° to 10°
50 from 9° to 23° ;	100 from 22° to 44°

By this device, the lines used for measuring height always remain between the zero line and the range reading line, which practically rules out any confusion. In order to simplify the mental computations, the height lines have not been marked with the multiplication constants 10, 20 etc. but with the figures + 0.1, + 0.2, $\frac{1}{2}$, + 1, when the telescope is aimed up, and – 0.1, – 0.2, – $\frac{1}{2}$ and – 1 when aimed down. The observer reads heights in the same way as he does distances and multiplies the readings by factors given.

Heights are always referred to the point on the rod which coincides with the zero line. The one metre mark can conveniently be taken as zero. The stadia rod, equipped with a telescope leg, allows for the setting of the metre mark at the instrument height as read on the centering rod, in order to simplify subsequent height computation.

Figs. 22.36 to 22.39 illustrate how the readings are taken.

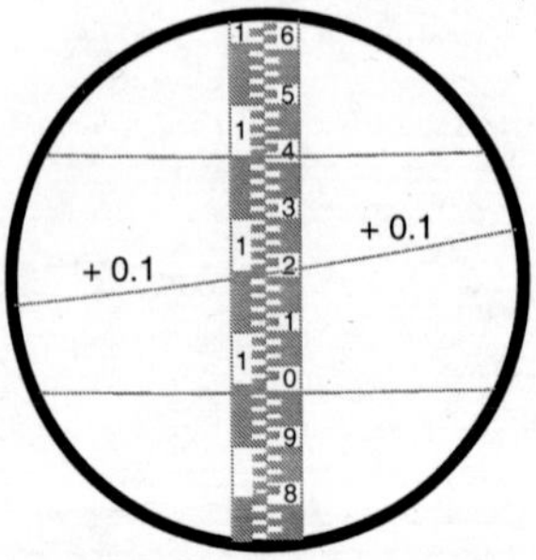

Fig. 22.36 Distance = 41.3 m
Height = + 0.1 × 21.7 = + 21.7 m

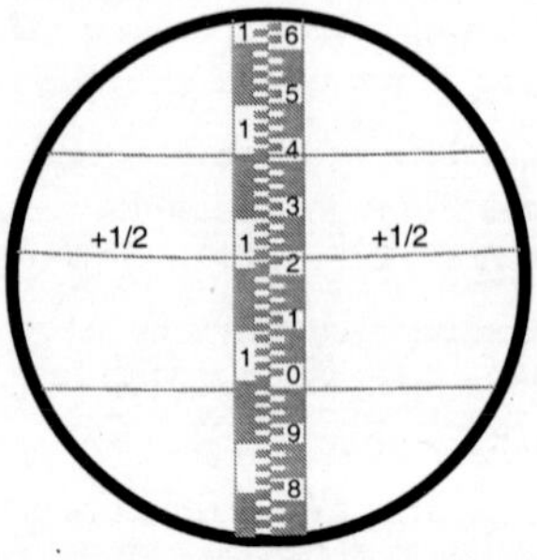

Fig. 22.37 Distance = 35.5 m
Height = + $\frac{1}{2}$ × 21.8 = + 10.9 m

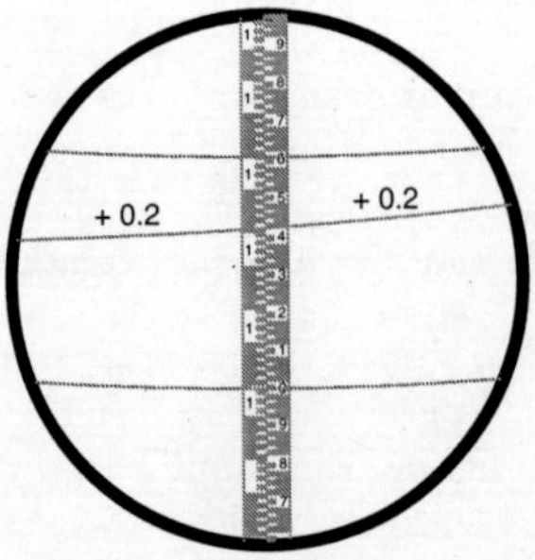

Fig. 22.38 Distance = 57.2 m
Height = + 0.2 × 40.1 = + 8.02 m

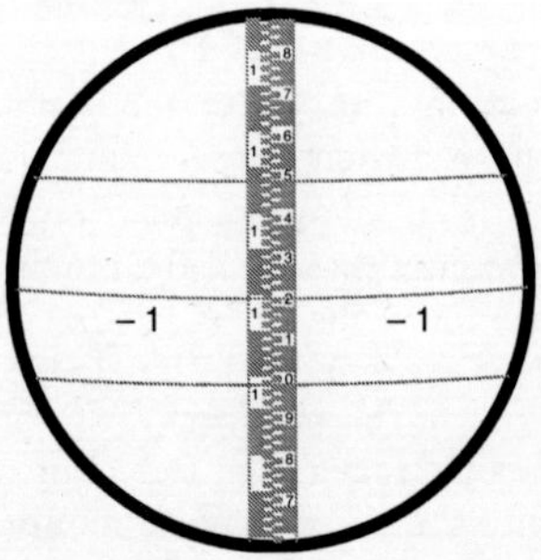

Fig. 22.39 Distance = 48.5 m
Height = – 1 × 21.7 = – 21.7 m

The vertical circle image appears on top and the horizontal circle image at the bottom of the field of vision, in both telescope positions. The minute graduations of the micrometer scales increase from left to right, in the same manner as when reading. The smallest graduation interval is one minute. Fig. 22.40 shows the examples of reading, as appearing in the field of view.

The vertical circle reading is 86° 32′.5 while the horizontal circle reading (A_z) is 265° 28′.5.

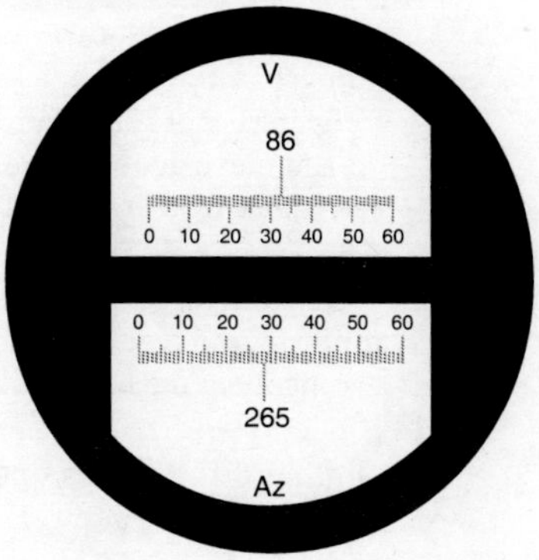

Fig. 22.40 Vertical and Horizontal Circle Readings in Wild RDS Tacheometer.

22.18 THE EWING STADI-ALTIMETER (WATTS) (Fig. 22.41)

This ingenious device, designed by Mr. Alistair Ewing, an experienced Australian surveyor, converts a normal theodolite easily and quickly to a direct reading tacheometer, without interfering with its normal function in any way.

The construction of the altimeter is in two parts—the cylindrical scale unit, which is mounted on one of the theodolite uprights and the *optical reader,* mounted on the telescope or transit axis (Figs. 22.41 and 22.42). The index of the reader is bright pinpoint of light which appears superimposed on the scale of the drum. The scale comprises two sets of curves, reproduced upon the surface of the cylinders. The two sets of curves, called intercept lines are formed at sufficiently frequent intervals for accurate reading and are distinguished by a difference in colour. They represent the reduction equations:

$$\text{Difference in level} = 100\, s\, \tfrac{1}{2} \sin 2\theta$$

$$\text{Horizontal distance correction} = 100\, s \sin^2 \theta.$$

Methods of use: After the usual adjustment of the theodolite, the stadi-altimeter is set to zero, the telescope is directed on to the staff, and the stadia intercept *s* is read. The cylinder is rotated until the curve equal to 100 *s* is in coincidence with the reader index. The difference of level may then be read directly from the external circular scale.

To obtain the reduced level of the staff base, the stadi-altimeter is set in the first instance to the reduced height of the theodolite, instead of zero. The telescope is directed on to the staff, and the intercept is read; it is then pointed so that the centre web cuts the staff reading equal to the height of the theodolite. The height scale reading then gives the reduced level of the staff base.

22.19 ERRORS IN STADIA SURVEYING

The various sources of errors which arise in tacheometry may be divided into three heads:

(*i*) Instrumental errors.

(*ii*) Errors due to manipulation and sighting.

(*iii*) Errors due to natural causes.

(*i*) **Instrumental Errors:** They consists of:

1. *Errors due to imperfect adjustment of the tacheometer:* The effects of inadjustments of various parts on the accuracy have already been discussed in the chapter on thodolite. However, with reference to tacheometric observations, the

accuracy in the determination of distances and elevations are dependent upon: (*a*) the adjustment of altitude level, (*b*) the elimination or determination of index error, and (*c*) accuracy of reading to the vertical circle. Since all these three have serious effects on the elevations, proper care should be taken to adjust the altitude bubble and to see that the altitude bubble is in centre of its run when observations are taken.

2. *Errors due to erroneous divisions on the stadia rod:* Since the accuracy in the determination of staff intercept depends on the graduations, the latter should be bold, uniform and free of errors. The stadia rod should be standardised and corrections for erroneous length, should be applied if necessary.
3. *Errors due to incorrect value of multiplying and additive constants:* To eliminate the errors due to this, the constants should be determined from time to time, under the same conditions that occur in the field.

(*ii*) **Errors due to manipulation and sighting:** They consist of errors due to:

1. Inaccurate centering and bisection.
2. Inaccurate levelling of the instrument.
3. Inaccurate reading to the horizontal and vertical circles.
4. Focusing (or parallax).
5. Inaccurate estimation of the staff intercept.
6. Incorrect position of the staff.

(*iii*) **Errors due to natural causes:** They comprise errors due to:

1. Wind.
2. Unequal refraction.
3. Unequal expansion.
4. Bad visibility.

22.20 EFFECT OF ERRORS IN STADIA TACHEOMETRY, DUE TO MANIPULATION AND SIGHTING

1. **Error due to staff tilted from normal:** In Fig. 22.43, AB is the correct normal holding while A_2B_2 is the *incorrect normal holding,* the angle of tilt being α. Line A_1B_1 is parallel to AB. If the angle of tilt α is small, we have

$$A_1B_1 \approx AB = s$$

Let s_1 (= A_2B_2) be the *observed* staff intercept, because of incorrect holding, while actual staff intercept would be s (= AB) if there is no angle of tilt.

Now $$A_1B_1 = A_2B_2 \cos \alpha$$

or $$s = s_1 \cos \alpha \qquad ...(i)$$

Error in distance $OC = k\, s_1 - k\, s$

$$\text{Ratio of error, } e = \frac{k\, s_1 - k\, s}{k s_1} = 1 - \frac{s}{s_1}$$

or $$e = 1 - \cos \alpha \quad ...(ii) \quad ...(22.27)$$

This shows that the error is independent of the inclination (θ) of line of sight.

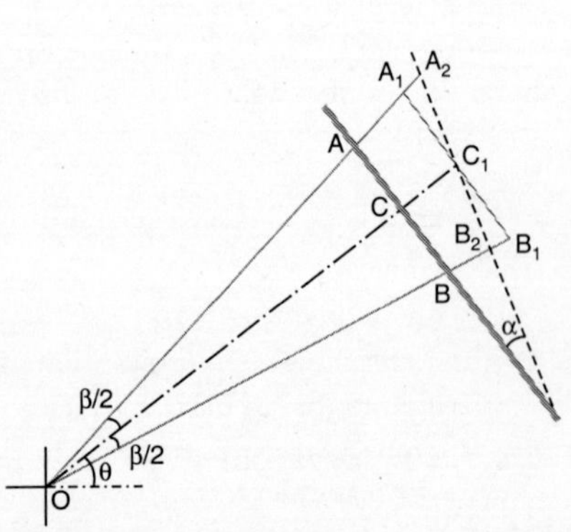

Fig. 22.43

2. **Error due to angle of elevation θ: normal holding of staff:** Let there be an error $\delta\theta$ in the measurement of angle of elevation θ. From Eq. 22.6, we have

$$D = L \cos \theta + r \sin \theta.$$

Differentiating this, we get

$$\frac{\delta D}{\delta \theta} = - L \sin \theta + r \cos \theta$$

$$\therefore \quad \delta D = (- L \sin \theta + r \cos \theta)\, \delta\theta \qquad ...(22.28)$$

3. **Error due to staff tilted from vertical:** In Fig. 22.44, *A, C, B* show the stadia readings when the staff is *truly vertical,* while line *A′CB′* is the corresponding line normal to the line of sight *OC*.

However, let the staff be inclined by an angle α from vertical, *away from the observer,* so that A_1, C_1 and B_1, are the points corresponding to the readings of the three hairs, and $A_1'B_1'$ is the corresponding line normal to the line of sight $O\,C_1$. Then $\angle A_1C_1A_1' = \theta + \alpha$. Also, since angle β/2 is very small, lines $A'B'$ and $A_1'B_1'$ may be taken perpendicular to OAA_1 and OBB_1.

Also, $\quad A'B' = AB \cos \theta = s \cos \theta \qquad ...(i)$

$$A_1'B_1' = A_1B_1 \cos (\theta + \alpha) = s_1 \cos (\theta + \alpha) \qquad ...(ii)$$

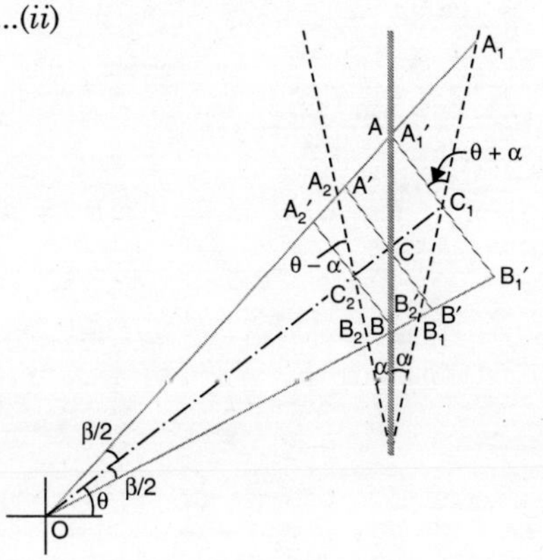

Fig. 22.44

Assuming $A'B' \approx A_1'B_1'$, we have

$$s \cos \theta \approx s_1 \cos (\theta + \alpha)$$

$$\therefore \quad s = \frac{s_1 \cos (\theta + \alpha)}{\cos \theta} \qquad ...(22.28\ a)$$

Similarly, if the staff is inclined by α from vertical towards the observer, $\angle A_2'C_2A_2 = \theta - \alpha$ and

$$s = \frac{s_1 \cos (\theta - \alpha)}{\cos \theta} \qquad ...(22.28\ b)$$

Eqs. 22.28 (*a*) and 22.28 (*b*) are for the angle of elevation θ. Similarly, for the angle of depression θ, the corresponding expressions will be

$$s = \frac{s_1 \cos (\theta - \alpha)}{\cos \theta} \qquad ...(22.28\ c)$$

and

$$s = \frac{s_1 \cos (\theta + \alpha)}{\cos \theta} \qquad ..(22.28\ d)$$

for the tilt away from the observer and towards the observer respectively.

In general, therefore, we have $s = \dfrac{s_1 \cos (\theta \pm \alpha)}{\cos \theta}$ ...(22.29)

where s_1 is the *observed intercept* while s is the *true intercept* for staff truly vertical.

For an anallactic telescope,

True distance $\quad D = ks \cos^2 \theta = \dfrac{ks_1 \cos (\theta \pm \alpha)}{\cos \theta} . \cos^2 \theta \qquad ...(i)$

Incorrect distance $\quad D_1 = ks_1 \cos^2 \theta \qquad ...(ii)$

$\therefore$ Error $\quad e = D - D_1 = k\, s_1 \cos^2 \theta \left[\dfrac{\cos (\theta \pm \alpha)}{\cos \theta} - 1 \right] \qquad ...(22.29\ a)$

Error e expressed as a ratio $= \dfrac{D - D_1}{D_1} = \dfrac{k s_1 \cos^2\theta \left[\dfrac{\cos(\theta \pm \alpha)}{\cos\theta} - 1\right]}{k s_1 \cos^2\theta}$

or $$e = \frac{\cos(\theta \pm \alpha)}{\cos\theta} - 1 \qquad \text{...(22.29 } b\text{)}$$

or $$e = \frac{\cos\theta\cos\alpha \pm \sin\theta\sin\alpha - \cos\theta}{\cos\theta} = \cos\alpha \pm \tan\theta\sin\alpha - 1$$

...(22.29 c)

If α is small (usually < 5°) $e = \pm\,\alpha\tan\theta$ ...(22.30)

4. **Error due to stadia intercept assumption:** In Fig. 22.5, we have assumed that for $\beta/2$ to be small, angles $AA'C$ and $BB'C$ will be each equal to 90°, and consequently, $A'B' = AB\cos\theta = s\cos\theta$.

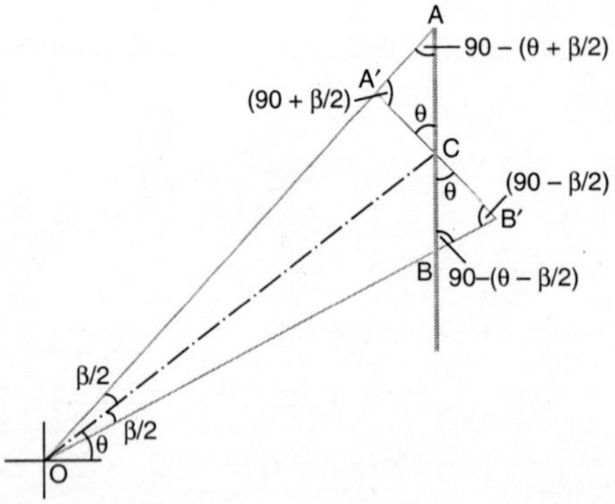

Fig. 22.45

Actually, $\angle AA'C = 90° + \beta/2$ and

$\angle BB'C = 90° - \beta/2$,

as shown in Fig. 22.45

Also, $\angle A'AC = 90° - (\theta + \beta/2)$

and $\angle B'BC = 90° - (\theta - \beta/2)$

For $k = 100$,

$$\frac{\beta}{2} = \tan^{-1}\left(\frac{1}{200}\right)$$

$$= \frac{206265}{200}\text{ sec.} = \mathbf{0°\ 17'\ 11''.35}$$

Let $AC = s_1$ and $CB = s_2$

Now from $\Delta CA'A$,

$$A'C = s_1 \frac{\sin[90° - (\theta + \beta/2)]}{\sin(90° + \beta/2)} = s_1 \frac{\cos(\theta + \beta/2)}{\cos\beta/2} = s_1 \frac{\cos\theta\cos\beta/2 - \sin\theta\sin\beta/2}{\cos\beta/2}$$

From $\Delta CB'B$,

$$CB' = s_2 \frac{\sin[90° - (\theta - \beta/2)]}{\sin(90° - \beta/2)} = s_2 \frac{\cos(\theta - \beta/2)}{\cos\beta/2} = s_2 \frac{\cos\theta\cos\beta/2 + \sin\theta\sin\beta/2}{\cos\beta/2}$$

$$\therefore \quad A'C + CB' = s_1 \frac{\cos\theta\cos\beta/2 - \sin\theta\sin\beta/2}{\cos\beta/2} + s_2 \frac{\cos\theta\cos\beta/2 + \sin\theta\sin\beta/2}{\cos\beta/2}$$

$$= s_1(\cos\theta - \sin\theta\tan\beta/2) + s_2(\cos\theta + \sin\theta\tan\beta/2)$$

$$\therefore \quad A'B' = (s_1 + s_2)\cos\theta + (s_2 - s_1)\sin\theta\tan\beta/2 \qquad \text{...(22.30 } a\text{)}$$

or $$A'B' = AB\cos\theta + (s_2 - s_1)\sin\theta\tan\beta/2 \qquad \text{...(22.31)}$$

Hence the error in assuming $A'B' = AB\cos\theta$ is equal to the magnitude of the second term $(s_2 - s_1)\sin\theta\tan\beta/2$.

5. **Error due to vertical angle measurement:** For vertical holding of staff, the horizontal distance, using an allactic telescope, is given by

$$D = k\, s \cos^2 \theta$$

$$\therefore \qquad \delta D = -\, 2k\, s \cos \theta \sin \theta\, \delta\theta$$

where $\delta\theta$ is the error in the measurement of vertical angle θ.

Now ratio
$$\frac{\delta D}{D} = \frac{2k\, s \cos \theta \sin \theta}{k\, s \cos^2 \theta} \,.\, \delta\theta = 2 \,.\, \tan \theta\, \delta\theta \qquad \text{...(22.32)}$$

Normally, the staff is graduated to 10 mm, capable of estimation to ± 1 mm. Since the multiplying factor *(k)* is usually 100, this would represent ± 100 mm.

Let us assume an *overall accuracy* of 1 in 1000 (representing 100 mm in 100 m).

$$\therefore \qquad \frac{\delta D}{D} = \frac{1}{1000}$$

Substituting in Eq. 22.32 we have

$$\frac{1}{1000} = 2 \tan \theta \,.\, \delta\theta \quad \text{or} \quad \delta\theta = \frac{1}{2000} \cot \theta \qquad \text{...(22.33)}$$

For
$$\theta = 30°, \quad \delta\theta = \frac{206265}{2000} \cot 30° \text{ seconds} = 178 \text{ seconds} \approx 3'.$$

Hence in order to conform to an overall accuracy of 1 in 1000, the angle θ need be measured to an accuracy of 3′.

6. **Error due to reading the staff:** We have seen above that for a staff graduated to 10 mm, estimation can be made to ± 1 mm. As both stadia lines need be read, the error in the stadia intercept would be $\sqrt{2}$ mm, *i.e.* 1.4 mm.

Thus $\qquad \delta s = 1.4$ mm.

Now $\qquad D = k\, s \cos^2 \theta$

$$\therefore \qquad \delta D = k \cos^2 \theta \,.\, \delta s \qquad \text{...(22.34)}$$

Similarly
$$V = \frac{1}{2} k\, s \sin 2\theta$$

$$\therefore \qquad \delta V = \frac{1}{2} k \sin 2\theta \,.\, \delta s \qquad \text{...(22.35)}$$

Taking $\qquad k = 100$ and $\delta s = 1.4$ mm, we have

$$k \,.\, \delta s = 100 \times 1.4 \text{ mm} = 0.14 \text{ m},$$

$$\therefore \qquad \delta D = 0.14 \cos^2 \theta$$

and $\qquad \delta V = 0.07 \sin 2\theta$

The value of δD and δV, for various values of inclination θ are as under

θ	δD	δV
0°	0.140 m	0.000 m
1°	0.140 m	0.002 m
2°	0.140 m	0.005 m

(Contd.)

(Contd.)

3°	0.140 m	0.007 m
4°	0.139 m	0.010 m
5°	0.139 m	0.012 m
7.5°	0.138 m	0.018 m
10°	0.136 m	0.024 m
15°	0.131 m	0.035 m
20°	0.124 m	0.045 m
25°	0.115 m	0.054 m
30°	0.105 m	0.061 m

From the above table, we conclude that unless the angles are less than 4°, the horizontal distance should not be quoted better than 0.1 m while the levels should not be quoted better than 0.01 m.

Example 22.19: *Observations were taken with a tacheometer having additive constant equal to zero and multiplying constant equal to 100, and an intercept of 0.685 m with a vertical angle of 12° was recorded on a staff believed to be vertical. Actually, the staff which was 3.5 m long, was 100 mm out of plumb leaning backwards away from the instrument. Compute the error in the horizontal distance.*

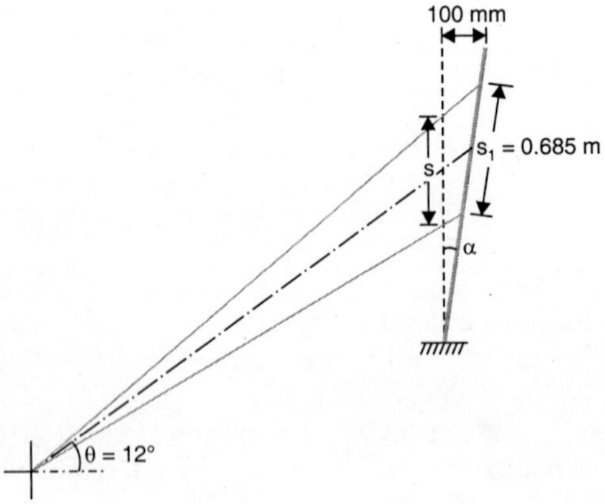

Fig. 22.46

Solution: Angle of tilt,

$$\alpha = \tan^{-1}\frac{0.100}{3.5} = 1°\ 38'\ 12''$$

From Eq. 22.28 (*a*)

$$s = s_1\frac{\cos(\theta+\alpha)}{\cos\theta} = 0.685 \times \frac{\cos(12°\ 00'\ 00'' + 1°\ 38'\ 12'')}{\cos(12°\ 00'\ 00'')} = 0.681$$

Now, $\quad D = k\, s \cos^2\theta$

$\therefore \quad \delta D = k\cos^2\theta\,.\,\delta s = 100\cos^2 12° \times (0.685 - 0.681) = \mathbf{0.425\ m}$

Alternatively, from Eq. 22.29 (*a*)

$$\delta D = k\, s\cos^2\theta\left[\frac{\cos(\theta+\alpha)}{\cos\theta} - 1\right]$$

$$= 100 \times 0.685\cos^2 12°\left[\frac{\cos(12° + 1°\ 38'\ 12'')}{\cos 12°} - 1\right] = \mathbf{0.425\ m}$$

Example 22.20: *A theodolite has a tacheometric multiplying constant of 100 and an additive constant of zero. The centre reading on a vertical staff held at point B was 2.292 m when sighted* *from A. If the vertical angle was + 25° and the horizontal distance AB 190.326 m, calculate the other staff readings and snow that the two intercept intervals are not equal.*

Using these values, calculate the level of B if A is 37.950 m A.O.D. and the height of the instrument 1.35 m. (U.L.)

Solution: From Eq. 22.4,

$$D = k\, s \cos^2 \theta$$

$\therefore$

$$s = \frac{D}{k \cos^2 \theta} = \frac{190.326}{100 \cos^2 25^\circ} = 2.317 \text{ m}$$

Refer Fig. 22.45.

Inclined distance $MC = L = D \sec \theta = 190.326 \sec 25^\circ = 210.002$ m

Now

$$2\, s_0 = \frac{L}{100} \qquad \therefore \; s_0 = \frac{L}{200} = \frac{210.002}{200} = 1.050 \text{ m}$$

By sine rule,

$$s_1 = \frac{s_0 \cos \beta/2}{\cos(\theta + \beta/2)}, \text{ where } \frac{\beta}{2} = 0^\circ\, 17'\, 11''.35.$$

$$= \frac{1.050 \cos (0^\circ 17' 11''.35)}{\cos (25^\circ 17' 11''.35)} = 1.161 \text{ m}$$

Similarly, by sine rule,

$$s_2 = \frac{s_0 \cos \beta/2}{\cos(\theta - \beta/2)} = \frac{1.050 \cos (0^\circ 17' 11''.35)}{\cos (24^\circ\, 42'\, 48''.65)} = 1.156 \text{ m}$$

Alternatively

$$s_1 = D\,[\tan(\theta + \beta/2) - \tan \theta]$$

$$= 190.326\,[\tan 25^\circ\, 17'\, 11''.35 - \tan 25^\circ] = 1.161 \text{ m, as above}$$

and

$$s_2 = D\,[\tan \theta - \tan(\theta - \beta/2)]$$

$$= 190.326\,(\tan 25^\circ - \tan 24^\circ\, 42'\, 48''.65] = 1.156 \text{ m, as above.}$$

(We note that s_1 and s_2 are not equal)

Check: $s_1 + s_2 = 1.161 + 1.156 = 2.317$ m

Hence the staff readings are :

Upper : 2.292 + 1.161 = 3.453

Lower : 2.292 – 1.156 = 1.136

Check: $s = 2.317$

Now $V = D \tan \theta = 190.326 \tan 25^\circ = 88.750$ m

$\therefore$ R.L. of $B = 37.950 + 88.750 + 1.350 - 2.292 =$ **125.758 m.**

PROBLEMS

1. Describe the conditions under which tacheometric surveying is advantageous. Explain how you would obtain in the field the constants of a tacheometer. Upto what vertical angle may sloping

distance by taken as horizontal distance without the error exceeding 1 in 200, the staff being held vertically and the instrument having an anallactic lens ? (*U.L.*)

2. Sighted horizontally, a tacheometer reads 1.645 and 2.840 corresponding to the stadia wires, on a vertical staff 120 m away. The focal length of the object glass is 20 cm and the distance from the object glass to the trunnion axis is 15 cm. Calculate the stadia interval.
3. Two distances of 50 and 80 metres were accurately measured out, and the intercepts on the staff between the outer stadia webs were 0.496 at the former distance and 0.796 at the latter. Calculate the tacheometric constants.
4. An external focusing theodolite with stadia hairs 2 mm apart and object glass of 15 cm focal length is used as a tacheometer. If the object glass is fixed at a distance of 25 cm from the trunnion axis, determine the tacheometric formula for distance in terms of staff intercept.
5. A tacheometer was set up at station *A* and the following readings were obtained on a vertically held staff:

Station	*Staff Station*	*Vertical Angle*	*Hair Readings*	*Remarks*
A	B.M.	– 2° 18′	3.225, 3.550, 3.875	R.L. of B.M. = 437.655 m
	B	+ 8° 36′	1.650, 2.515, 3.380	

Calculate the horizontal distance from *A* to *B* and the R.L. of *B*, if the constants of the instrument were 100 and 0.4.

6. To determine the distance between two points *C* and *D*, and their elevations, the following observations were taken upon a vertically held staff from two traverse stations *A* and *B*. The tacheometer was fitted with an anallactic lens, the constant of the instrument being 100:

Traverse Station	*Ht. of Inst.*	*Co-ordinates*		*Staff Station*	*Bearing*	*Vertical Angle*	*Staff Readings*
		N	*E*				
A	1.58	218.3	164.7	*C*	330° 20′	+ 12° 12′	1.255, 1.860, 2.465
B	1.50	518.2	207.6	*D*	20° 36′	+ 10° 36′	1.300, 1.885, 2.470

Calculate : (*i*) The distance *CD*;

(*ii*) The R.L.'s of *C* and *D*, if those of *A* and *B* were 432.550 m and 436.865 m respectively;

(*iii*) The gradient from *C* to *D*.

7. A line was levelled tacheometrically with a tacheometer fitted with an anallactic lens, the value of the constant being 100. The following observations were made, the staff having been held vertically:

Instrument Stations	*Height of Axis*	*Staff at*	*Vertical Angle*	*Staff Reading*	*Remarks*
A	1.38	*B.M.*	– 1° 54′	1.02, 1.720, 2.420	R.L. 638.55 m
A	1.38	*B*	+ 2° 36′	1.220, 1.825, 2.430	
B	1.40	*C*	+ 3° 6′	0.785, 1.610, 2.435	

Compute the elevations of *A*, *B* and *C*.

8. Two sets of tacheometric readings were taken from an instrument station *A*, the reduced level of which was 15.05 ft to a staff station *B*.

(*a*) Instrument *P*—multiplying constant 100, additive constant 14.4 in., staff held vertical.

(*b*) Instrument *Q*—multiplying constant 95, additive constant 15.5 in., staff held normal to the line of sight.

Instrument	*At*	*To*	*Ht. of Instrument*	*Vertical Angle*	*Stadia Readings*
P	*A*	*B*	4.52	30°	2.35/3.31/4.27
Q	*A*	*B*	4.47	30°	

What should be the stadia readings with instrument *Q*? (*U.L.*)

9. An ordinary theodolite is to be converted into an anallactic tacheometer with a multiplier of 100 by an insertion of a new glass stadia diaphragm and an additional convex lens. Focal length of object glass is 15 cm, fixed at a distance of 10 cm from the trunnion axis. A focusing slide carries the eye-piece. If a suitable lens of 10 cm focal length is available for the anallactic lens, calculate the fixed distance at which this must be placed from the objective and the spacing of the stadia hairs on the diaphragm.

10. The stadia intercept read by means of a fixed hair instrument on a vertically held staff is 2.250 metres, the angle of elevation being 3° 42′. The instrument constants are 100 and 0.4 m. What would be the total number of turns registered on a movable hair instrument at the same station for a 2.0 metres intercept on a staff held on the same point? The vertical angle in this case is 5° 30′ and the constants 1000 and 0.4 m?

11. The constant for an instrument is 1200 and the value of additive constant is 0.4 metres. Calculate the distance from the instrument to the staff when the micrometer readings are 6.262 and 6.258, the staff intercept is 2.5 m and the line of sight is inclined at + 6° 30′, the staff being held vertically.

12. The vertical angles to vanes fixed at 0.5 m and 3.5 m above the foot of the staff held vertically at a point were – 0° 30′ and +1° 12′ respectively. Find the horizontal distance and the reduced level of the point, if the level of the instrument axis is 125.380 metres above datum.

13. Explain how a subtense bar is used with a theodolite to determine the horizontal distance between two points.

    The horizontal angle subtended at a theodolite by a subtense bar with vanes 3 m apart is 15′ 40″. Compute the horizontal distance between the instrument and the bar.

    Deduce the error of horizontal distance if the bar were 2° from being normal to the line joining the instrument and bar station.

14. What are the different methods employed in tacheometric survey? Describe the method most commonly used. (*A.M.I.E.*)

15. Explain how you would determine the contants of a tacheometer. What are the advantages of an anallactic lens used in a tacheometer?

16. Describe anyone form of subtense micrometer, and show clearly how you would determine the value of the additive constant in the case of a subtense micrometer in which there may be an initial reading of micrometer head when the fixed and the moving lines coincide, the focal length of the objective and the pitch of the mircometer screw being known. (*U.L.*)

17. Show the arrangement of the lenses in an ordinary anallactic telescope.

    In a telescope of this type the focal lengths of the objective and anallactic lenses are 24 cm and 12 cm respectively and the constant distance between this is 19.5 cm for a multiplier of 100.

    Determine the error that would occur in horizontal distance *D* when the reading intercepts 2 metres, with an error of one hundredth of a mm in the 1.75 mm interval between the subtense lines.

18. In the event of a broken diaphragm in an anallactic telescope with a multiplier of 100, it is required to determine the exact spacing of the lines on glass for a new diaphragm, the focal lengths of the objective and anallactic lenses being 30 cm and 15 cm respectively and the distance between the

objective and the trunnion axis 12 cm. Also determine the distance between the anallactic lens and the objective.

19. An anallactic telescope has a multiplying constant of 100. The focal lengths of the object glass and anallactic lens are 11 cm and 9 cm respectively. If the stadia interval i is 1.5 mm, calculate the distance between the two lenses and the distances between the vertical axis and the object glass.

ANSWERS

1. 4° 3′
2. 2 mm
3. $k = 100$; $C = 0.4$ m
4. $D = 75\ s + 0.4$ metres
5. $D = 169.5$; R.L. of $B = 466.95$
6. (*i*) 335.8 m
   (*ii*) R.L. of $C = 457.27$; R.L. of $D = 457.62$
   (*iii*) 1 in 959.2
7. 643.528, 648.567, 657.267
8. 1.95; 2.82; 3.68
9. 16 cm; $\frac{1}{6}$ cm
10. 8.844
11. 236.9 m
12. 101.1 m; 1 in 123.998 m
13. 658.29 m; 1 in 1688
17. 1.14 m
18. 23.57 cm; 2.1 mm
19. 13.4 cm; 7.33 cm.

CHAPTER 23

ELECTRONIC THEODOLITES

23.1 INTRODUCTION

Theodolites, used for angular measurements, can be classified under three categories:

(*i*) Vernier theodolites

(*ii*) Microptic theodolites (optical theodolites) and

(*iii*) Electronic theodolites.

Vernier theodolites (such as Vicker's theodolite) use verniers which have a least count of 10″ to 20″. However, microptic theodolites use optical micrometers, which may have least count of as small as 0.1″. Wild T-1 T-16, T-2, T-3 and T-4 and other forms of Tavistock theodolites fall under this category. Thus the optical theodolites are the most accurate instruments where in the readings are taken with the help of optical micrometers. However in electronic theodolites, absolute angle measurement is provided by a dynamic system with *opto-electronic scanning*. The electronic theodolites are provided with control panels with key boards and liquid crystal displays. The LCDs with points and symbols present the measured data clearly and unambiguously. The key-board contains multi-function keys. The main operations require only a single key-stroke. The electronic theodolites work with electronic speed and efficiency. They measure electronically and open the way to electronic data acquisition and data processing. We shall consider here the following two models of electronic theodolites manufactured by M/s Wild Heerbrugg Ltd.

(*i*) Wild T-1000 electronic theodolite

(*ii*) Wild T-2000 and T-2000 S electronic theodolite.

23.2 WILD T-1000 'THEOMAT'

Wild electronic theodolites are known as 'Theomat'. Fig. 23.1 shows the photograph of Wild T-1000 electronic theodolite. Although it resembles a conventional theodolite (*i.e.*, optical theodolite) in size and weight, the T-1000 works with electronic speed and efficiency. It measures electronically and opens the way to electronic data aquisition and data processing. It has 30 × telescope which gives a bright, high-contrast, erect image. The coarse and fine focusing ensures that the target is seen sharp and clear. Pointing is fast and precise, even in poor observing conditions. The displays and reticle plate can be illuminated for works in mines and tunnels and at night.

The theodolite has two control panels, each with key-board and two liquid-crystal displays. It can be used easily and quickly in both positions. Fig. 23.2 shows the control panel of T-1000.

The LCDs with points and symbols present the measured data clearly and unambiguously. The key-board has just six multifunction keys. The main operations require only a single keystroke. Accepted keystrokes are acknowledged by a beep. Colour-coding and easy-to-follow key sequences and commands make the instrument remarkably easy to use.

The theodolite has an absolute electronic-reading system with position-coded circles. There is no initialization procedure. Simply switch on and read the results. Circle reading is instantaneous. The readings update continuously as the instrument is turned. Readings are displayed to 1″. The standard deviation of a direction measured in face left and face right is 3″.

The theodolite has practice-tested *automatic index*. A well-damped pendulium compensator with 1″ setting accuracy provides the reference for T-1000 vertical circle readings. The compensator is built on the same principles as the compensator used in Wild automatic levels and optical theodolites. Thus with T-1000, one need not rely on a plate level alone. Integrated circuits and microprocessors ensure a high level to performance and operating comfort. Automatic self-checks and diagonostic routines makes the instrument easy to use.

DIST			Distance measurement
REC			Recording
ALL			Measurement and recording
DSP	Hz		Display Hz-circle and Hz-distance
SET	TRK		Tracking
SET	SET	Hz	Set horizontal-circle reading to zero

Fig. 23.3 Typical Commands in T-1000 Electronic-Theodolite (Wild Heerbrugg).

T-1000 theodolite has electronic clamp for circle setting and repetition measurements. Using simple commands, one can set the horizontal circle reading to zero or to any value. The theodolite can be operated like a conventional theodolite using any observing procedure, including the repetition method. In addition to the conventional clockwise measurements, horizontal circle readings can be taken counter-clockwise. Horizontal-collimation and vertical index errors can be determined and stored permanently. The displayed circle readings are corrected automatically. Displayed heights are corrected for earth curvature and mean refraction.

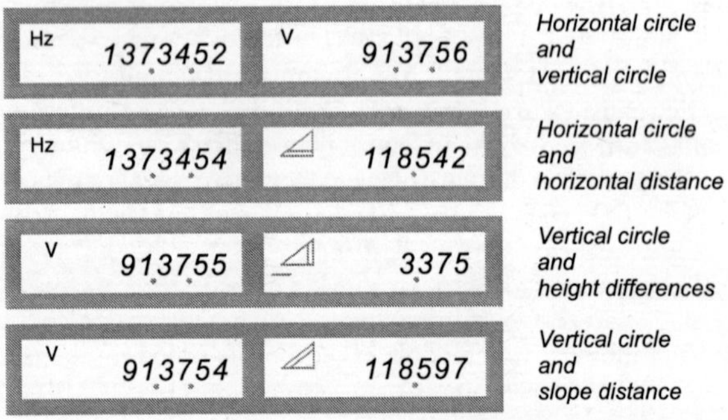

Fig. 23.4 Typical-Displays on the Panels of T-1000.

As stated earlier, the whole instrument is controlled from the key-board. Fig. 23.3 gives details of typical command obtained by pressing corresponding keys. Fig. 23.4 gives typical display values obtained by pressing different keys. The power for T-1000 theodolite is obtained from a small, rechargeable 0.45 Ah Ni Cd battery which plugs into the theodolite standards.

Wild T-1000 theodolite is fully compatible. It is perfectly modular, having the following uses:

(*i*) It can be used alone for angle measurement only.

(*ii*) It combines with Wild *Distomat* for angle and distance measurement.

(*iii*) It connects to GRE 3 data terminal for automatic data aquisition.

(*iv*) It is compatible with Wild theodolite accessories.

(*v*) It connects to computers with RS 232 interface.

Fig. 23.5 depicts diagrammatically, all these functions.

'Distomat' is a registered trade name used by Wild for their *electromagnetic distance measurement* (EDM) instruments (see Chapter 24). Various models of distomats, such as DI-1000, DI-5, DI-5S, DI-4/4L etc. are available, which can be fitted on the top of the telescope of T-1000 theodolite. The telescope can transit for angle measurements in both the positions. No special interface is required. With a distomat fitted to it, the theodolite takes both angle and distance measurements. Wild DI-1000 distomat is a miniaturized EDM, specially designed for T-1000. It integrates perfectly with the theodolite to form the ideal combination for all day-to-day work. Its range is 500 m on to 1 prism and 800 m to 3 prisms, with a standard deviation of 5 mm + 5 ppm. For larger distances, DI-5S distomat can be fitted, which has a range of 2.5 km to 1 prism and 5 km to 11 prisms. For very long distances, latest long-range DI-3000 distomat, having a range of 6 km to 1 prism and maximum range of 14 km in favourable conditions can be fitted. Thus, with a distomat, T-1000 becomes *electronic total station.*

The T-1000 theodolite attains its full potential with the GRE 3 data terminal. This versatile unit connects directly to the T-1000. Circle readings and slope distances are transferred from the theodolite. Point numbering, codes and information are controlled from the GRE 3.

23.3 WILD T-2000 THEOMAT

Wild T-2000 Theomat (Fig. 23.6 *a*) is a high precision electronic angle measuring instrument. It has microprocessor controlled angle measurement system of highest accuracy. Absolute angle measurement is provided by a dynamic system with opto-electronic scanning (Fig. 23.7). As the graduations around the full circle are scanned for every reading, circle graduation error cannot occur. Scanning at diametrically opposite positions eliminates the effect of eccentricity. Circle readings are corrected automatically for index error and horizontal collimation error. Thus angle measurements can be taken in one position to a far higher accuracy than with conventional theodolites. For many applications, operator will set the displays for circle reading to 1″, but for the highest precision the display can be set to read to 0.01″. For less precise work, circle readings can be displayed to 10″. Distances are displayed to 1 mm and 0.01 ft. With good targets, the standard deviation of the mean of a face-left and a face-right observations is better than 0.5″ for both the horizontal and vertical circles.

The theodolite has self-indexing maintenance free liquid compensator. The compensator provides the reference for vertical angle measurements. It combines excellent damping with high precision and allows accurate measurements unaffected by strong winds, vibrations etc.

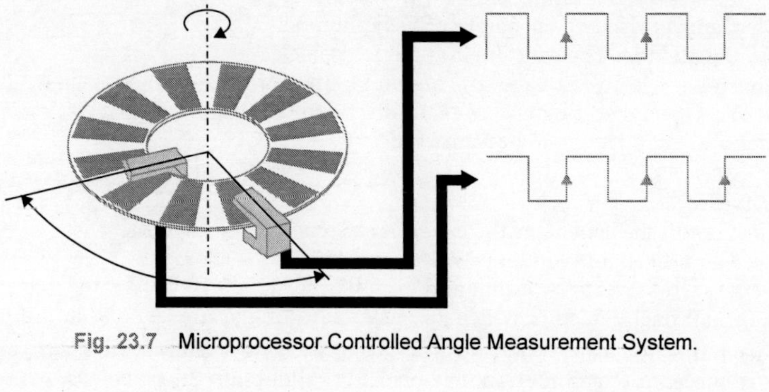

Fig. 23.7 Microprocessor Controlled Angle Measurement System.

The instrument has two angle measuring modes: *single* and *tracking. Single mode* is used for angle measurements of highest accuracy. Hz and/or circle readings are displayed at the touch of a key. *Tracking* provides continuous single measurement with displays updated as the theodolite is turned. Tracking is used for rapid measurements, turning the theodolite to set a bearing or following a moving target. The horizontal circle reading can be set to zero or any value by means of the key-board.

The whole instrument is operated from a central panel comprising a waterproof key-board and three liquid-crystal displays, shown in Fig. 23.8. The key need only the slightest touch. One display guides the operator, the other two contain data. The displays and telescope reticle can be illuminated for work in the dark. Fig. 23.9 illustrates typical commands along with corresponding key to be used.

To measure angles, touch	HZV
To measure and record angles, touch	REC
To measure angles, distances heights and co-ordinates, touch	DIST
To measure and record angles, distances, heights and co-ordinates, touch	ALL

That's all there is to its a single keystroke for the main operations.

Fig. 23.9 Typical Commands.

Various parameters such as a circle orientation station co-ordinates and height scale correction and additive constant can be entered and stored. All are retained until over-written by new values. They cannot be lost even when the instrument is switched off. As circle readings are corrected for index error and horizontal collimation error, one control panel is in position. It is perfectly sufficient for many operations. However, for maximum convenience, particularly when measurements in both positions are required, the instrument is available with a control panel on each side.

The instrument uses rechargeable plug-in internal battery (NiCd, 2 Ah, 12 V DC) which is sufficient for about 1500 angle measurements or about 550 angle and distance measurements. The instrument switches off automatically after commands and key sequences. The user can select a switch off time of 20 seconds or three minutes. This important power saving feature is made possible by the non-volatile memory. There is no loss of stored information when the instrument switches off.

Clamps and drives are coaxial. The drive screws have two speeds: fast for quick aiming, slow for fine pointing. Telescope focusing is also two-speed. An optical plummet is built into the alidade. The carrying handle folds back to allow the telescope to transit with distomat fitted. Horizontal and vertical setting circles facilitate turning into a target and simplify setting-out work.

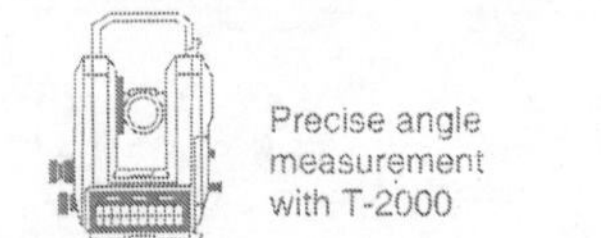

Fig. 23.10 T-2000: Modular Approach.

Modular Approach

The T-2000 offers all the benefits of the modular approach. It can be used as a theodolite combined with any distomat and connects to GRE 3 data terminal and computers. Fig. 23.10 illustrates diagrammetrically this modular approach which provides for easy upgrading at any time at minimum cost.

Wild theodolite accessories fit the T-2000: optional eye-pieces, filters, eye-piece prism, diagonal eye-piece, auto-collimation, eye-piece, parallel-plate micrometers, pentaprism, solar prism, auxiliary lenses etc. Wild tribachs, targets, distomat reflectors, target lamps, subtence bar, optical plummets and equipment for deformation measurements are fully compatible with the T-2000.

Two Way Data Communication

Often, in industry and construction, one or more instruments have to be connected on line to a computer. Computation is in real time. Results are available immediately. To facilitate connection, interface parameters of the T-2000 instruments can be set to match those of the computer. Communication is two-way. The instrument can be controlled from the computer. Prompt messages and information can be transferred to the T-2000 displays. Of particular interest is the possibility of measuring objects by intersection from two theodolites (Fig. 23.11).

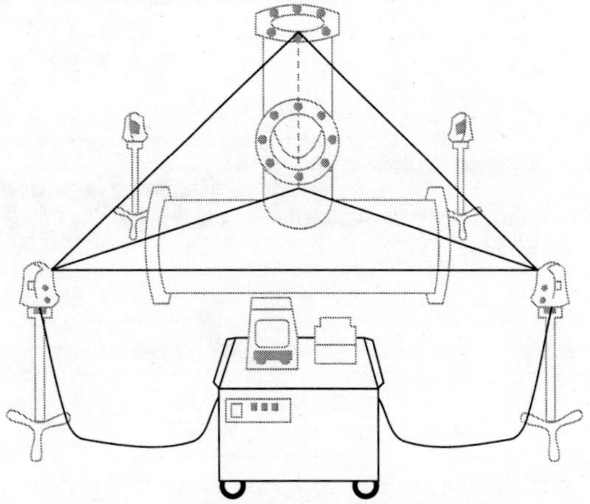

Fig. 23.11 RMS (Remote Measuring System) Intersection Method.

Two T-2000 type instruments can be connected to the Wild GRE 3 Data Terminal. Using the Mini-RMS program, co-ordinates of intersected points are computed and recorded. The distance between any pair of object points can be calculated and displayed. For complex applications and special computations, two or more T-2000 or T-2000 S can be used with the *Wild-Leitz RMS 2000 Remote Measuring System*.

23.4 WILD T-2000 S 'THEOMAT'

Wild T-2000 S [Fig. 23.6 (*b*)] combines the pointing accuracy of a *special telescope* with the precision of T-2000 dynamic circle measuring system. This results in angle measurement of the highest accuracy. The telescope is *panfocal* with a 52 mm objective for an exceptionally bright, high contrast image. It focuses to object 0.5 m from the telescope. The focusing drive has coarse and fine movements.

Magnification and field of view vary with focusing distance. For observations to distant targets, the field is reduced and magnification increased. At close range, the field of view widens and magnification is reduced. This unique system provides ideal conditions for observation at every distance. With the standard eye-piece, magnification is 43 × with telescope focused to infinity. Optional eye-pieces for higher and lower magnification can also be fitted.

Stability of the line of sight with change in focusing is a special feature of the T-2000 S telescope. It is a true alignment telescope for metrology, industry and optical tooling industry. T-2000 S can also be fitted with a special target designed for pointing to small targets.

A special target can also be built into the telescope at the intersection of the horizontal and vertical axes. The target is invaluable for bringing the lines of sight of two T-2000 S exactly into coincidence. This is the usual preliminary procedure prior to measuring objects by the RMS intersection method.

For fatigue-free, maximum-precision auto-collimation measurements, the telescope is available with an auto-collimation eye-piece with negative reticle (green cross).

Like T-2000, the T-2000 S takes all Wild Distomats. It can also be connected the GRE 3 Data Terminal.

CHAPTER 24

ELECTROMAGNETIC DISTANCE MEASUREMENT (EDM)

24.1 INTRODUCTION

There are three methods of measuring distance between any two given points:

1. Direct distance measurement (DDM), such as the one by chaining or taping.
2. Optical distance measurement (ODM), such as the one by tacheometry, horizontal subtense method or telemetric method using optical wedge attachments.
3. Electromagnetic distance measurement (EDM) such as the one by geodimeter, tellurometer or distomat etc.

The method of direct distance measurement is unsuitable in difficult terrain, and sometimes impossible when obstructions occur. The problem was overcome after the development of optical distance measuring methods. But in ODM method also, the range is limited to 150 to 150 m and the accuracy obtained is 1 in 1000 to 1 in 10000. Electromagnetic distance measurement (EDM) enables the accuracies upto 1 in 10^5, over ranges upto 100 km.

EDM is a general term embracing the measurement of distance using electronic methods. In electromagnetic (or electronic) method, distances are measured with instruments that rely on propagation, reflection and subsequent reception of either radio, visible light or infrared waves. There are in excess of fifty different EDM systems available. However, we will discuss here the following instruments:

(*i*) Geodimeter
(*ii*) Tellurometer
(*iii*) Distomats.

24.2 ELECTROMAGNETIC WAVES

The EDM method is based on generation, propagation, reflection and subsequent reception of electromagnetic waves. The type of electromagnetic waves generated depends on many factors but principally, on the nature of the electrical signal used to generate the waves. The evolution and use of radar in the 1939-45 war resulted in the application of radio waves to surveying. However, this was suitable only for defence purposes, since it could not give the requisite accuracy for geodetic surveying. E. Bergestrand of the Swedish Geographical Survey, in collaboration with the manufacturers, Messrs AGA of Sweden, developed a method based on the propagation of *modulated light waves* using instrument called *geodimeter*. Another instrument, called *tellurometer* was developed, using radio waves. Modern short and medium range EDM instruments (such as Distomats) commonly used in surveying use *modulated infra-red waves.*

Properties of Electromagnetic Waves

Electromagnetic waves, though extremely complex in nature, can be represented in the form of periodic sinusoidal waves shown in Fig. 24.1. It has the following properties:

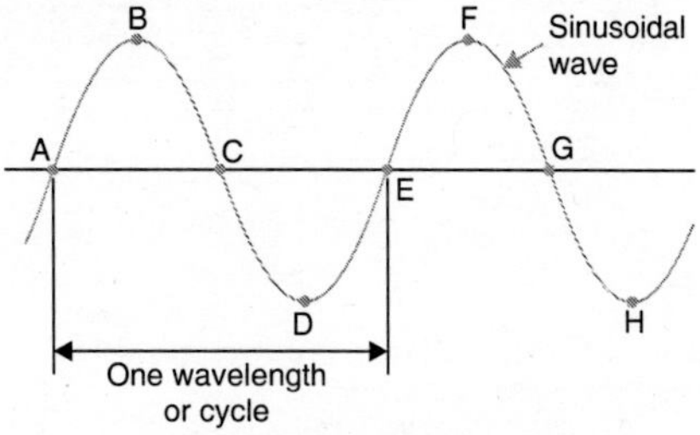

Fig. 24.1 Periodic Sinusoidal Waves.

1. The waves completes a *cycle* in moving from identical points *A* to *E* or *B* to *F* or *D* to *H*.
2. The number of times the wave completes a cycle in one second is termed as *frequency* of the wave. The frequency is represented by *f* hertz (Hz) where 1 hertz (Hz) is one cycle per second. Thus, if the frequency *f* is equal to 10^3 Hz, it means that the waves completes 10^3 cycles per second.
3. The length traversed in one cycle by the wave is termed as *wavelength* and is denoted by λ (metres). Thus, the *wavelength* of a wave is the distance between two identical points (such as *A* and *E* or *B* and *F*) on the wave.
4. The *period* is the time taken by the wave to travel through one cycle or one wavelength. It is represented by *T* seconds.
5. The *velocity* (*v*) of the wave is the distance travelled by in one second.

The frequency, wavelength and period can all vary according to the wave producing source. However, the velocity *v* of an electromagnetic wave depends upon the medium through which it is travelling. The velocity of wave in a vacuum is termed as *speed of light,* denoted by symbol *c*, the value of which is presently known to be 299792.5 km/s. For simple calculations, it may be assumed to be 3×10^8 m/s.

The above properties of an electromagnetic wave can be represented by the relation,

$$f = \frac{c}{\lambda} = \frac{1}{T} \qquad \text{...(24.1)}$$

Another property of the wave, known as *phase* of the wave, and denoted by symbol φ, is a very convenient method of identifying fraction of a wavelength or cycle, in FDM. One cycle or wavelength has a phase ranging from 0° to 360°. Various points *A, B, C* etc. of Fig. 24.1 has the following phase values:

Point →	*A*	*B*	*C*	*D*	*E*	*F*	*G*	*H*
Phase φ°	0	90	180	270	360 (or 0)	90	180	270

Fig. 24.2 gives the electromagnetic spectrum. The type of electromagnetic wave is known by its wavelength or its frequency. However, all these travel with a velocity *approximately* equal to 3×10^8 m/s. *This velocity forms the basis of all electromagnetic measurements.*

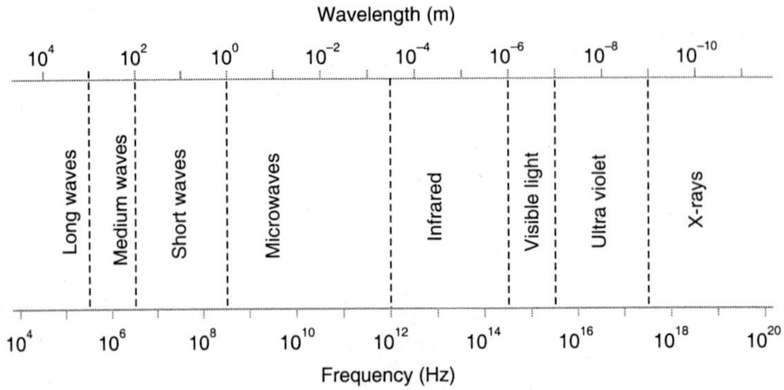

Fig. 24.2 Electromagnetic Spectrum.

Measurement of Transit Times

Fig. 24.3 (*a*) shows a survey line *AB*, the length *D* of which is to be measured using EDM equipment placed at ends *A* and *B*. Let a transmitter be placed at *A* to propagate electromagnetic waves towards *B*, and let a receiver be placed at *B*, along with a timer. If the timer at *B* *starts* at the instant of transmission of wave from *A*, and *stops* at the instant of reception of incoming wave at *B*, the *transit time* for the wave from *A* and *B* in known.

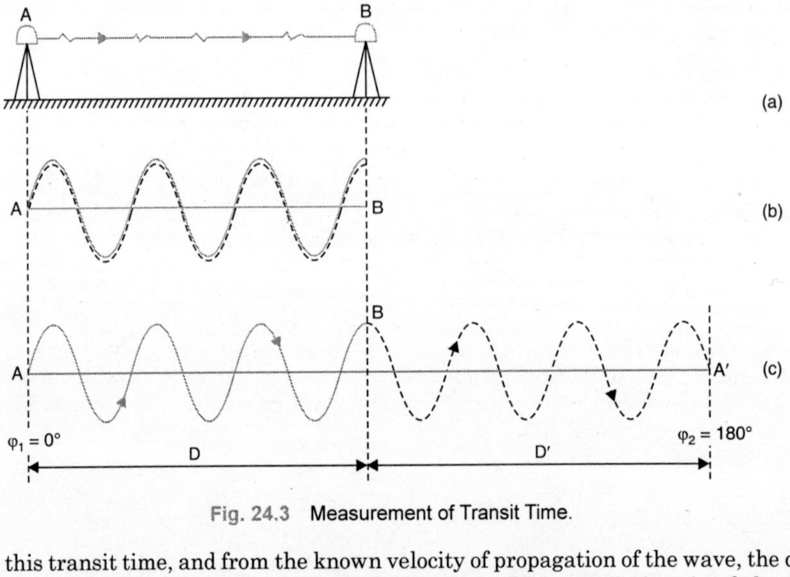

Fig. 24.3 Measurement of Transit Time.

From this transit time, and from the known velocity of propagation of the wave, the distance *D* between *A* and *B* can be easily computed. However, this transit time is of the order of 1×10^{-6} s which requires very advanced electronics. Also it is extremely difficult to start the timer at *B* when the wave is transmitted at *A*. Hence, a reflector is placed at *B* instead of a receiver. This reflector reflects the waves back towards *A*, where they are received Fig. 24.3 (*b*). Thus the equipment at A acts both as a transmitter as well as receiver. The *double transit time* can be easily measured at A. This will require EDM timing devices with an accuracy of $\pm 1 \times 10^{-9}$ s.

Phase Comparison

Generally, the various commercial EDM systems available do not measure the transit time directly. Instead, the distance is determined by measuring the phase difference between the transmitted and reflected signals. This phase difference can be expressed as fraction of a cycle which can be converted into units of time when the frequency of wave is known. Modern techniques can easily measure upto $\frac{1}{1000}$ part of a wavelength.

In Fig. 24.3 (*b*), the wave transmitted from A towards B is instantly reflected from B towards A, and is then received at A, as shown by dotted lines. The same sequence is shown in Fig. 24.3(*c*) by opening out the wave, wherein A and A' are the same. The distance covered by the wave is

$$2D = n\lambda + \Delta\lambda \qquad ...(24.2)$$

where d = distance between A and B

λ = wavelength

n = whole number of wavelengths travelled by the wave

$\Delta\lambda$ = fraction of wavelength travelled by the wave.

The measurement of component $\Delta\lambda$ is known as *phase comparison* which can be achieved by electrical phase detectors.

Let φ_1 = phase of the wave as it is transmitted at A

φ_2 = phase of the wave as it is received at A'

Then
$$\Delta\lambda = \frac{\text{phase difference in degrees}}{360^\circ} \times \lambda \text{ or } \Delta\lambda = \frac{(\varphi_2 - \varphi_1)^\circ}{360^\circ} \times \lambda \qquad ...(24.3)$$

The determination of other component $n\lambda$ of equation 24.2 is referred to as *resolving the ambiguity of the phase comparison*, and this can be achieved by any one of the following methods.

(*i*) by increasing the wavelength manually in multiples of 10, so that a coarse measurement of D is made, enabling n to be deduced.

(*ii*) by measuring the line AB using three different (but closely related) wavelengths, so as to form three simultaneous equations of the form

$$2D = n_1\lambda_1 + \Delta\lambda_1;\ 2D = n_2\lambda_2 + \Delta\lambda_2;\ 2D = n_3\lambda_3 + \Delta\lambda_3$$

The solution of these may give the value of D.

In the latest EDM equipment, this problem is solved automatically, and the distance D is displayed.

For example, let λ for the wave of Fig. 24.3 (*c*) be 20 m. From the diagram, $n = 6$, $\varphi_1 = 0°$ and $\varphi_2 = 180°$.

$$\therefore \qquad 2D = n\lambda + \Delta\lambda = n\lambda + \frac{\varphi_2 - \varphi_1}{360^\circ} \times \lambda$$

or
$$2D = (6 \times 20) + \frac{180 - 0}{360} \times 20$$

$$\therefore \qquad D = 65 \text{ m}.$$

This measurement of distance by EDM is analogous to the measurement of AB by taping, wherein

$$D = ml + \Delta l$$

where l = length of tape = 20 m (say)

m = whole no. of tapes = 3

Δl = remaining length of the tape in the end bay

Hence, the recording in the case of taping will be $D = 3 \text{ m} \times 20 + 5 = 65$ m.

24.3 MODULATION

As stated above, EDM measurements involve the measurement of fraction $\Delta\lambda$ of the cycle. Modern phase comparison techniques are capable of resolving to better than $\frac{1}{1000}$ part of a wavelength. Assume ± 10 mm to be the accuracy requirement for surveying equipment, this must represent $\frac{1}{1000}$ of the measuring wavelength. This means that $\lambda = 10 \times 1000$ mm = 10 m, which is a maximum value. However, by use of modern circuitory, λ can be increased to 40 m, which corresponds to $f = 7.5 \times 10^6$ Hz. Thus, the lowest value of f that can be used is 7.5×10^6 Hz. At present, the range of frequencies that can be used in the measuring process is limited to approximately 7.5×10^6 to 5×10^8 Hz.

In order to increase the accuracy, it is desirable to use an extremely high frequency of propagation. However, the available phase comparison techniques cannot be used at frequencies greater than 5×10^8 Hz which corresponds to a wavelength $\lambda = 0.6$ m. On the other hand, the lower frequency value in the range of 7.5×10^6 to 5×10^8 Hz is not suitable for direct transmission through atmosphere because of the effects of interference, reflection, fading and scatter.

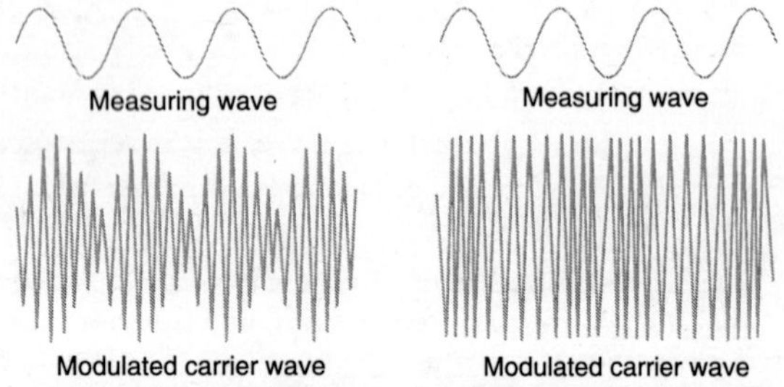

(a) Amplitude modulation (b) Frequency modulation

Fig. 24.4 Modulation.

The problem can be overcome by the technique of *modulation* wherein the measuring wave used for phase comparison is superimposed on a *carrier wave* of much higher frequency. EDM uses two methods of modulating the carrier wave:

(*a*) Amplitude modulation. (*b*) Frequency modulation.

In *amplitude modulation,* the carrier wave has constant frequency and the modulating wave (the measuring wave) information is conveyed by the amplitude of the carrier waves. In the *frequency modulation,* the carrier wave has constant amplitude, while its frequency varies in proportion to the amplitude of the modulating wave. *Frequency modulation* is used in all microwave EDM instruments while *amplitude modulation* is done in *visible light instruments* and infrared instruments using higher carrier frequencies.

24.4 TYPES OF EDM INSTRUMENTS

Depending upon the type of carrier wave employed, EDM instruments can be classified under the following three heads:

(*a*) Microwave instruments
(*b*) Visible light instruments
(*c*) Infrared instruments.

For the corresponding frequencies of carrier waves, reader may refer back to Fig. 24.2. It is seen that all the above three categories of EDM instruments use short wavelengths and hence higher frequencies.

1. Microwave Instruments

These instruments come under the category of long range instruments, where in the carrier frequencies of the range of 3 to 30 GHz (1 GHz = 10^9) enable distance measurements upto 100 km range. *Tellurometer* come under this category.

Phase comparison technique is used for distance measurement. This necessitates the erection of some form of *reflector* at the remote end of the line. If *passive reflector* is placed at the other end, a weak signal would be available for phase comparison. Hence an electronic signal is required to be erected at the reflecting end of the line. This instrument, known as *remote instrument* is identical to the *master instrument* placed at the measuring end. The *remote instrument* receives the transmitted signal, amplifies it and transmits it back to the master in exactly the phase at which it was received. This means that microwave EDM instruments require two instruments and two operators. Frequency modulation is used in most of the microwave instruments. The method of varying the measuring wavelength in multiplies of 10 is used to obtain an unambiguious measurement of distance. The microwave signals are radiated from small aerials (called *dipoles)* mounted in front of each instrument, producing directional signal with a beam of width varying from 2° to 20°. Hence the alignment of master and remote units is not critical. Typical maximum ranges for microwave instruments are from 30 to 80 km, with an accuracy of ± 15 mm to ± 5 mm/km.

2. Visible Light Instruments

These instruments use visible light as carrier wave, with a higher frequency, of the order of 5×10^{14} Hz. Since the transmitting power of carrier wave of such high frequency falls off rapidly with the distance, the range of such EDM instruments is lesser than those of microwave units. A *geodimeter* comes under this category of EDM instruments.

The carrier, transmitted as light beam, is concentrated on a signal using lens or mirror system, so that signal loss does not take place. Since the beam divergence is less than 1°, accurate alignment of the instrument is necessary. *Corner-cube prisms,* shown in Fig. 24.5 are used as reflectors at the remote end. These prisms are constructed from the corners of glass cubes which have been cut away in a plane making an angle of 45° with the faces of the cube.

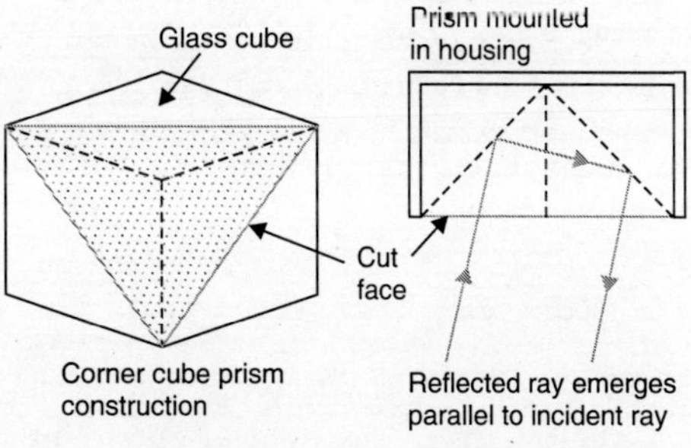

Fig. 24.5 Corner Cube Prism.

The light wave, directed into the cut-face is reflected by highly silvered inner surfaces of the prism, resulting in the reflection of the light beam along a parallel path. This is obtainable over a range of angles of incidence of about 20° to the normal of the front face of the prism. *Hence the alignment of the reflecting prism towards the main EDM instrument at the receiver (or transmitting) end is not critical.*

The advantage of visible light EDM instruments, over the microwave EDM instruments is that only one instrument is required, which work in conjunction with the inexpensive corner cube reflector. *Amplitude modulation* is employed, using a form of electro-optical shutter. The line is measured using three different wavelengths, using the same carrier in each case. The EDM instrument in this category have a range of 25 km, with an accuracy of ± 10 mm to ± 2 mm/km. The recent instruments use pulsed light sources and highly specialised modulation and phase comparison techniques, and produce a very high degree of accuracy of ± 0.2 mm to ± 1 mm/km with a range of 2 to 3 km.

3. Infrared Instruments

The EDM instruments in this group use near infrared radiation band of wavelength about 0.9 μ m as carrier wave which is easily obtained from gallium arsenide (Ga As) infrared emitting diode. These diodes can be very easily directly *amplitude modulated* at high frequencies. Thus, modulated carrier wave is obtained by an inexpensive method. Due to this reason, there is predominance of infrared instruments in EDM. Wild Distomats fall under this category of EDM instruments.

The power output of the diodes is low. Hence the range of these instruments is limited to 2 to 5 km. However, this range is quite sufficient for most of the civil engineering works. The EDM instruments of this category are very light and compact, and these can be theodolite mounted. This enables angles and distances to be measured simultaneously at the site. A typical combination is Wild Dl 1000 infrared EDM with Wild T 1000 electronic theodolite ('Theomat'). The accuracy obtainable is of the order of ± 10 mm, irrespective of the distance in most cases.

The carrier wavelength in this group is close to the visible light spectrum. Hence infrared source can be transmitted in a similar manner to the visible light system using geometric optics, a lens/mirror system being used to radiate a highly collimated beam of angular divergence of less than 15′. Corner cube prisms are used at the remote end, to reflect the signal.

Electronic tacheometer, such as Wild TC 2000 'Tachymat' is a further development of the infrared (and laser) distance measurer, which combines theodolite and EDM units. Microprocessor controlled angle measurement give very high degree of accuracy, enabling horizontal and vertical angles, and the distances (horizontal, vertical, inclined) to be automatically displaced and recorded.

24.5 THE GEODIMETER

The method, based on the propagation of *modulated light waves,* was developed by E. Bergestrand of the Swedish Geographical Survey in collaboration with the manufacturer, M/s AGA of Sweden. Of the several models of the geodimeter manufacured by them, model 2-A can be used only for observations made at night while model-4 can be used for limited day time observations.

Fig. 24.6 shows the schematic diagram of the geodimeter. Fig. 24.7 shows the photograph of the front panel of model-4 geodimeter mounted on the tripod. The main instrument is stationed at one end of the line (to be measured) with its back facing the other end of the line, while a reflector (consisting either of a spherical mirror or a reflex prism system) is placed at the other end of the line.

The light from an incandescent lamp (1) is focused by means of an achromatic condenser and passed through a Kerr cell (2). The Kerr cell consist of two closely spaced conducting plates, the space between which is filled with nitrobenzene. When high voltage is applied to the plates of the cell and a ray of light is focused on it, the ray is split into two parts, each moving with different velocity. Two Nicol's prisms (3) are placed on either side of the Kerr cell. The light leaving the first Nicol's prisms is plane polarised. The light is split into two (having a phase difference) by the Kerr cell. On leaving the Kerr cell, the light is recombined. However, because of phase difference, the resulting beam is elliptically polarised. Diverging light from the second polariser can be focused to a parallel beam by the transmitter objective, and can then be reflected from a mirror lens to a large spherical concave mirror.

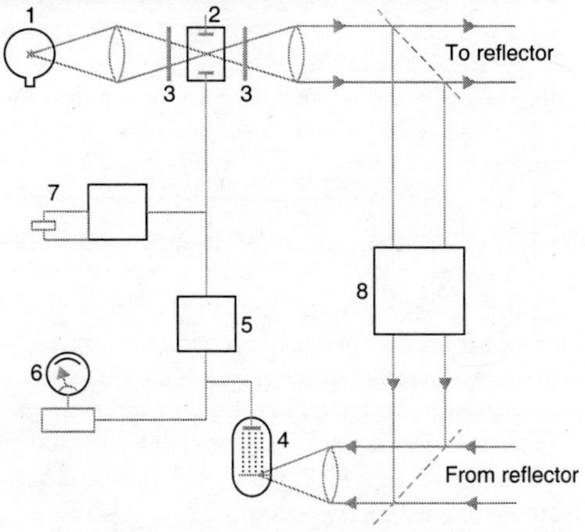

1. Incandescent lamp
2. Kerr cell
3. Nicol's prisms
4. Photo tube
5. Variable electrical delay unit
6. Null indicator
7. Crystal controlled oscillator
8. Variable light delay unit.

Fig. 24.6 Schematic Diagram of the Geodimeter.

On the other end of the line being measured is put a reflex prism system or a spherical mirror, which reflects the beam of light back to the geodimeter. The receiver system of the geodimeter consists of spherical concave mirror, mirror lens and receiver objective. The light of variable intensity after reflection, impinges on the cathode of the photo tube (4). In the photo tube, the light photons impinge on the cathode causing a few primary electrons to leave and travel, accelerated by a high frequency voltage, to the first dynode, where the secondary emission takes place. This is repeated through a further eight dynodes. The final electron current at the anode is some hundreds of thousand times greater than that at the cathode. The sensitivity of the photo tube is varied by applying the high frequency-Kerr cell voltage between the cathode and the first dynode. The low frequency vibrations are eliminated by a series of electrical chokes and condensers. The passages of this modulating voltage through the instrument is delayed by means of an adjustable electrical delay unit (5). The difference between the photo tube currents during the positive and negative bias period is measured on the *null indicator* (6) which is a sensitive D.C. moving coil micro-ammeter. In order to make both the negative and positive current intensities equal (*i.e.* in order to obtain null-point), the phase of the high frequency voltage from the Kerr cell must be adjusted ± 90° with respect to the voltage generated by light at the cathode.

Thus, the light which is focused to a narrow beam from the geodimeter stationed at one end to the reflector stationed at the other end of the line, is reflected back to the photo multiplier. The variation in the intensity of this reflected light causes the current from the photo multiplier to vary where the current is already being varied by the direct signal from the crystal controlled oscillater (7). The phase difference between the two pulses received by the cell are a measure of the distance between geodimeter and the reflector (*i.e.*, length of the line).

The distance can be measured at different frequencies. On Model-2A of the geodimeter, three frequencies are available. Model-4 has four frequencies. Four phase positions are available on the *phase position indicator*. Changing phase indicates that the polarity of the Kerr cell terminals of high and low tension are reversed in turn. The 'fine' and 'coarse' delay switches control the setting of the electrical delay between the Kerr cell and the photo multiplier. The power required is obtained from a mobile gasoline generator. Model-4 has a night range of 15 metres to 15 km, a daylight range of 15 to 800 metres and an average error of ± 10 mm ± five millionth of the distance. It weighs about 36 kg without the generator.

24.6 THE TELLUROMETER

In the Tellurometer, high frequency radio waves (or microwaves) are used intead of light waves. It can be worked with a light weight 12 or 24 volt battery. Hence the instrument is highly portable. Observations can be taken both during day as well as night, while in the geodimeter, observations are normally restricted in the night. However, two such Tellurometers are required, one to be stationed at each end of the line, with two highly skilled persons, to take observations. One instrument is used as the *master set* or *control set* while the other instrument is used as the *remote set* or *slave set*. In Model MRA-2 (manufactured by M/s. Cooke, Troughton and Simms Ltd), each set can either be used as the master set or remote set by switching at 'master' and 'remote' positions respectively. Fig. 24.8 shows the photograph of Tellurometer (Model MRA-2).

Fig. 24.9 shows the block diagram of the Tellurometer, first designed by Mr. T.L. Wadley of the South African Council for Scientific and Industrial Research. Radio waves are emitted by the master instrument at a frequency of 3000 Mc.s. (3×10^9 c.p.s.) from a klystron and have superimposed on them a crystal controlled frequency of 10 Mc.s. The high frequency wave is termed as *carrier wave*. Waves at high frequencies can be propagated in straight line paths other than long distance much more readily. The low frequency wave is known as the *pattern wave* and is used for making accurate measurements. The light frequency pattern wave is thus said to be frequency modulated (F.M.) by low frequency pattern wave. This modulated signal is received at the remote station where a second klystron is generating another carrier wave at 3033 Mc.s. The difference between the two high frequencies, *i.e.* 3033 – 3000 = 33 Mc.s.

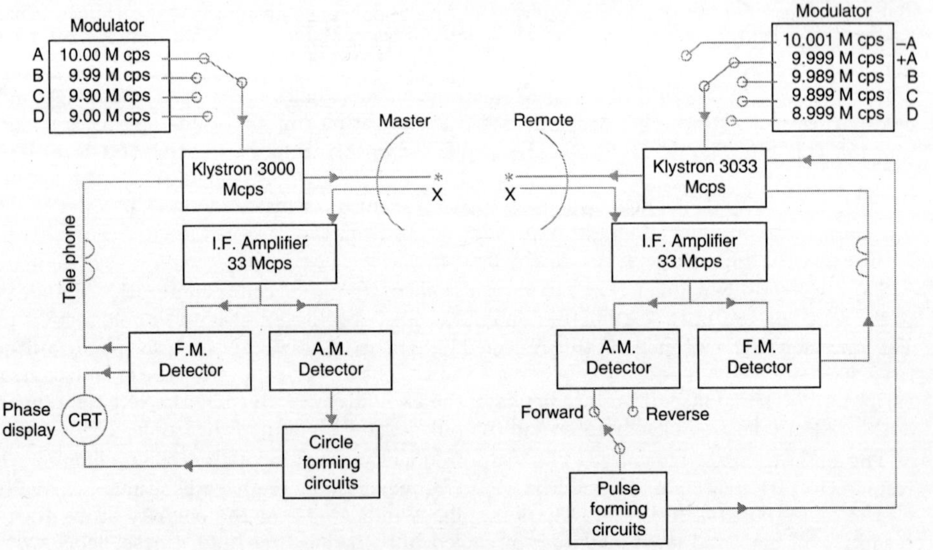

Fig. 24.9 Block Diagram of the Tellurometer System.

(known as *intermediate frequency*) is obtained by an electrical 'mixer', and is used to provide sufficient sensitivity in the internal detector circuits at each instrument. In addition to the carrier wave of 3033 Mc.s., a crystal at the remote station is generating a frequency of 9.999 Mc.s. This is *heterodyned* with the incoming 10 Mc.s. to provide a 1 k c.p.s. signal. The 33 Mc.s. intermediate frequency signal is amplitude modulated by 1 k c.p.s. signal. The amplitude modulated signal passes to the amplitude demodulator, which detects the 1 k c.p.s. frequency. At the pulse forming circuit, a pulse with a repetition frequency of 1 k c.p.s. is obtained. This pulse is then applied to the klystron and frequency modulates the signal emitted, *i.e.*, 3033 Mc.s. modulated by 9.999 Mc.s. and pulse of 1 k c.p.s. This signal is received at the master station. A further compound heterodyne process takes place here also, whereby the two carrier frequencies subtract to give rise to an intermediate frequency of 33 Mc.s. The two *pattern frequencies* of 10 and 9.999 Mc.s. also subtract to provide 1 k c.p.s *reference frequency* as amplitude modulation. The *change in the phase between this and the remote 1 k c.p.s. signal is a measure of the distance.* The value of phase delay is expressed in time units and appear as a *break* in a circular trace on the oscilloscope cathode ray tube.

Four low frequencies (*A*, *B*, *C* and *D*) of values 10.00, 9.99, 9.90 and 9.00 M c.p.s. are employed at the master station, and the values of phase delays corresponding to each of these are measured on the oscilloscope cathode ray tube. The phase delay of *B*, *C* and *D* are subtracted from *A* in turn. The *A* values are termed as 'fine readings' and the *B*, *C*, *D* values as 'coarse readings'. The oscilloscope scale is divided into 100 parts. The wavelength of 10 Mc.s. pattern wave as approximately 100 ft. (30 m) and hence each division of the scale represents 1 foot on the two-way journey of the waves or approximately 0.5 foot on the length of the line. The final readings of A, $A - B$, $A - C$ and $A - D$ readings are recorded in millimicro seconds (10^{-9} seconds) and are converted into distance readings by assuming that the velocity of wave propagations as 299,792.5 km/sec. It should be noted that the success of the system depends on a property of the *heterodyne process,* that the phase difference between two heterodyne signal is maintained in the signal that results from the mixing.

24.7 WILD 'DISTOMATS'

Wild Heerbrugg manufacture EDM equipment under the trade name 'Distomat', having the following popular models:

1. Distomat DI 1000
2. Distomat DI 5S
3. Distomat DI 3000
4. Distomat DIOR 3002
5. Tachymat TC 2000 (Electronic tacheometer)

1. Distomat DI 1000

Wild Distomat DI 1000 is very small, compact EDM, particularly useful in building construction, civil engineering construction, cadastral and detail survey, particularly in populated areas where 99% of distance measurements are less than 500 m. It is an EDM that makes the tape redundant. It has a range of 500 m to a single prism and 800 m to three prisms (1000 m in favourable conditions), with an accuracy of 5 mm + 5 ppm. It can be fitted to all Wild theodolites, such as T 2000, T 2000 S, T 2 etc.

The infrared measuring beam is reflected by a prism at the other end of the line. In the five seconds that it takes, the DI 1000 adjusts the signal strength to optimum level, makes 2048 measurements on two frequencies, carries out a full internal calibration, computes and displays the result. In the tracking mode 0.3 second updates follow the initial 3-second measurement. The whole sequence is automatic. One has to simply point to the reflector, touch a key and read the result.

The Wild modular system ensures full compatibility between theodolites and Distomats. The DI 1000 fits T 1, T 16 and T 2 optical theodolites, as shown in Fig. 24.10(*a*). An optional keyboard can be used. It also combines with Wild T 1000 electronic theodolite and the Wild T 2000 informatics theodolite to form fully electronic *total station* [Fig. 24.10(*b*)]. Measurements, reductions and calculations are carried out automatically. The DI 1000 also connects to the GRE 3 data terminal [Fig. 24.10(*c*)]. If the GRE 3 is connected to an electronic theodolite with DI 1000, all information is transferred and recorded at the touch of a single key. The GRE can be programmed to carry out field checks and computations.

When DI 1000 distomat is used separately, it can be controlled from its own keyboard. There are only three keys on the DI 1000, each with three functions, as shown in Fig. 24.11. Colour coding and a logical operating sequence ensure that the instrument is easy to use. The keys control all the functions. There are no mechanical switches. The liquid-crystal display is unusually large for a miniaturized EDM. Measured distances are presented clearly and unambiguously with appropriate symbols for slope, horizontal distance, height and setting out. In test mode, a full check is provided of the display, battery power and return signal strength. An audible tone can be activated to indicate return of signal. Scale (ppm) and additive constant (mm) settings are displayed at the start of each measurement. Input of ppm takes care of any atmospheric correction, reduction to sea level and projection scale factor. The mm input corrects for the prism type being used. The microprocessor permanently stores ppm and mm values and applies them to every measurement. Displayed heights are corrected for earth curvature and mean refraction.

DI 1000 is designed for use as the standard measuring tool in short range work. A single prism reflector is sufficient for most tasks. For occasional longer distance (upto 800 m), a three prism reflector can be used. The power is fed from NiCd rechargeable batteries.

2. Distomat DI 5S

Wild DI 5S is a medium range infrared EDM controlled by a small powerful microprocessor. It is multipurpose EDM. The 2.5 km range to single prism covers all short-range requirements: detail, cadastral, engineering, topographic survey, setting out, mining, tunnelling etc. With its 5 km range to 11 prisms, it is ideal for medium-range control survey: traversing, trigonometrical heighting, photogrammetric control, breakdown of triangulation and GPS networks etc. Finely, tuned opto-electronics, a stable oscillator, and a microprocessor that continuously evaluates the results, ensure the high measuring accuracy of 3 mm + 2 ppm standard deviation is standard measuring mode and 10 m + 2 ppm standard deviation in tracking measuring mode.

Fig. 24.12 shows the view of DI 5S. It has three control keys, each with three functions. There are no mechanical switches. A powerful microprocessor controls the DI 5S. Simply touch the DIST key to measure. Signal attenuation is fully automatic. Typical measuring time is 4 seconds. In tracking mode, the measurement repeats automatically every second. A break in the measuring beam due to traffic etc., does not affect the accuracy. A large, liquid-crystal display shows the measured distance clearly and unambiguously throughout the entire measuring range of the instrument. Symbols indicate the displayed values. A series of dashes shows the progress of the measuring cycle. A prism constant from – 99 mm to + 99 mm can be input for the prism type being used. Similarly, ppm values from – 150 ppm to + 150 ppm can be input for automatic compensation for atmospheric conditions, height above sea level and projection scale factor. These values are stored until replaced by new values. The microprocessor corrects every measurement automatically.

DI 5S can be also fitted to Wild electronic theodolites T 1000 and T 2000 [Fig. 24.13(*a*)] or to Wild optical theodolites T 1, T 16, T 2. Fig. 24.13(*b*) the infrared measuring beam is parallel to the line of signal. Only a single pointing is needed for both angle and distance measurements. When fitted to an optical theodolite, an optional keyboard [Fig. 24.13(*b*)] covert it to efficient low cost effective *total station*. The following parameters are directly obtained for the corresponding input values (Fig. 24.14):

(*a*) Input the vertical angle for

(*i*) Horizontal distance

(*ii*) Height difference corrected for earth curvature and mean refraction.

(*b*) Input the horizontal angle for

(*i*) Coordinate differences ΔE and ΔN.

(*c*) Input the distance to be set out for

(*i*) ΔD, the amount by which the reflector has to be moved forward or back.

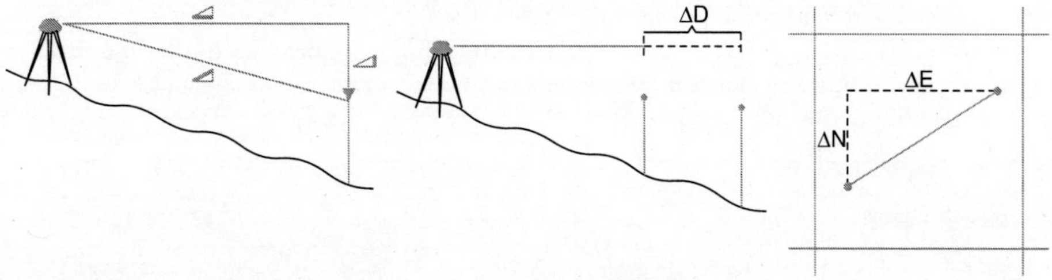

Fig. 24.14

When fitted to an electronic theodolite (T 1000 or T 2000) DI 5S transfers the slope distance to the theodolite. The following reductions (Fig. 24.9) are carried out in the theodolite microprocessor.

The DI 5S can also be connected to GRE 3 data terminal for automatic data acquisition. The EDM is powered from a NiCd rechargeable battery. When used on a Wild electronic theodolite, DI 5S is powered from the theodolites' internal battery.

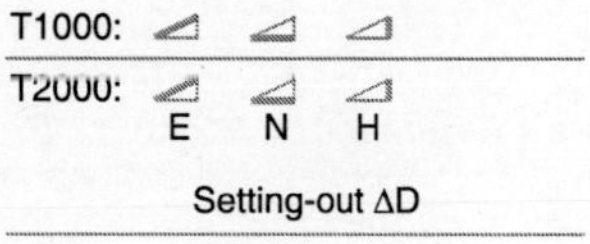

Fig. 24.15

3. Distomats DI 3000 and DI 3002

Wild DI 3000 distomat is a long range infrared EDM in which infrared measuring beam is emitted from a *laser diode*. Class I laser products are inherently safe; maximum permissible exposure cannot be exceeded under any condition, as defined by International Electrotechnical Commission.

The DI 3000 is a *time-pulsed* EDM. The time needed for a pulse of infrared light to travel from the instrument to the reflector and back is measured. The displayed result is the mean of hundreds or even thousands of time-pulsed measurements. The pulse technique has the following important advantages:

(*i*) **Rapid measurement:** It provides 0.8 second rapid measurement for detail surveys, tacheometry, setting out etc. It is advantageous for long range measurements in turbulent atmospheric conditions.

(*ii*) **Long range:** Its range is 6 km to 1 prism in average conditions and 14 km to 11 prisms in excellent conditions.

(*iii*) **High accuracy:** Accuracy is 5 mm + 1 ppm standard deviation. A calibrated quartz crystal ensures 1 ppm frequency stability throughout the temperature range – 20°C to + 60°C. In tracking mode, accuracy is 10 mm + 1 ppm.

(*iv*) **Measurement to moving targets:** For measuring to moving targets, the times-pulse measuring technique is very advantageous. There are practically no limits to the speed at which an object may move. For this purpose, a reflector should be suitably attached to the object or vehicle to which measurements have to be made. The distomat can be (*a*) manually controlled, (*b*) connected to Wild GRE 3 data terminal for automatic recording or (*c*) connected on-line to a computer for remote control and real-time processing results. The following important operations can be achieved on moving objects:

(*a*) ***Offshore surveys:*** DI 3000 can be mounted on electronic theodolite for measuring to ships, dredgers and pipe laying barges, positioning oil rigs, controlling docking manoeuvres etc. (Fig. 24.16).

(*b*) ***Controlling objects on rails:*** DI 3000 can be connected on-line to computer for controlling the position of cranes, gantries, vehicles, machinery on rails, tracked equipments etc. (Fig. 24.17).

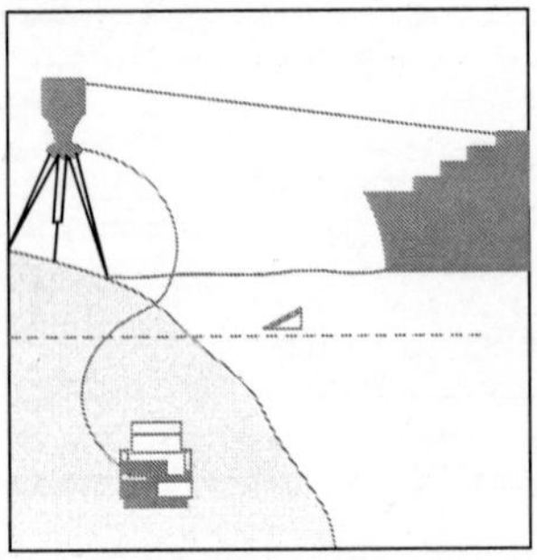

Fig. 24.16

Fig. 24.17

(*c*) ***Monitoring movements in deformation surveys:*** DI 3000 can be connected with GRE 3 or computer for continuous measurement to rapidly deforming structures, such as bridges undergoing load tests (Fig. 24.18).

(*d*) ***Positioning moving machinery:*** DI 3000 can be mounted on a theodolite for continuous determination of the position of mobile equipment (Fig. 24.19).

Fig. 24.18

Fig. 24.19

The DI 3000 is also ideal all-round EDM for conventional measurements in surveying and engineering: control surveys, traversing, trigonometrical heighting, breakdown of GPS networks, cadastral, detail and topographic surveys, setting out etc. It combines with Wild optical and electronic theodolites. It can also fit in a yoke as stand-alone instrument.

Fig. 24.20 shows a view of DI 3000 distomat, with its control panel, mounted on a Wild theodolite. The large easy to read LCD shows measured values with appropriate signs and symbols. An acoustic signal acknowledges key entries and measurement. With the DI 3000 on an optical theodolite, reductions are via the built in keyboard. For cadastral, detail, engineering and topographic surveys, simply key in the vertical circle reading. The DI 3000 displays slope and horizontal distance and height difference. For traversing with long-range measurements, instrument and reflector heights can be input the required horizontal distance. The DI 3000 displays the amount by which the reflector has to be moved forward or back. All correction parameters are stored in the non-volatile memory and applied to every measurements. Displayed heights are corrected for earths curvature and mean refraction.

4. Distomat DIOR 3002

The DIOR 3002 is a special version of the DI 3000. It is designed specifically for distance measurement without reflector. Basically, DIOR 3002 is also time pulsed Infrared EDM. When used without reflectors, its range varies from 100 m to 250 m only, with a standard deviation of 5 mm to 10 mm. The interruptions of beam should be avoided. However, DIOR 3002, when used with reflectors have a range of 4 km to 1 prism, 5 km to 3 prisms and 6 km to 11 prisms.

Although, the DIOR 3002 can fitted on any of the main Wild theodolites, the T 1000 electronic theodolite is the most suitable. When used without reflectors, it can carry the following operation.

(*i*) *Profile and cross-sections* (Fig. 24.21): DIOR 3002 with an electronic theodolite, can be used for measuring tunnel profiles and cross-sections, surveying stopes, caverns, interior of storage tanks, domes etc.

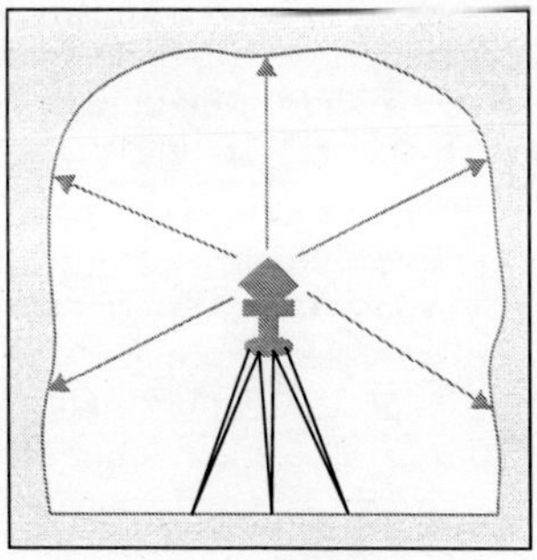

Fig. 24.21

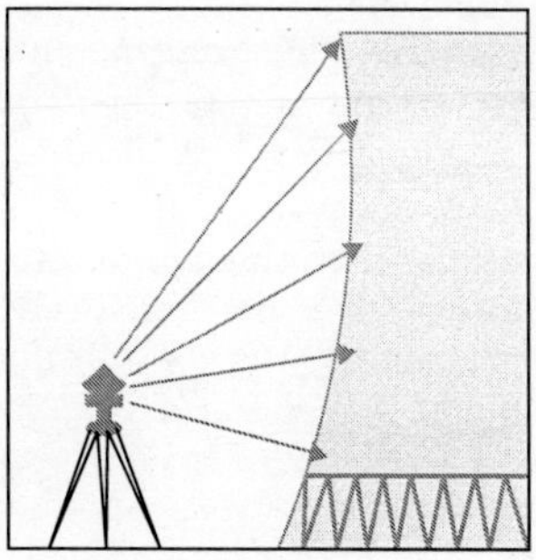

Fig. 24.22

(*ii*) *Surveying and monitoring buildings, large objects quarries, rock faces, stock piles* (Fig. 24.22): DIOR 3002 with a theodolite and data recorder can be used for measuring and monitoring large objects, to which access is difficult, such as bridges, buildings, cooling towers, pylons, roofs, rock faces, towers, stock piles etc.

(*iii*) *Checking liquid levels, measuring to dangerous or touch sensitive surfaces* (Fig. 24.23): DIOR 3002 on line to a computer can be used for controlling the level of liquids in storage tanks, determining water level in docks and harbours, measuring the amplitude of waves around oil rigs etc., also for measuring to dangerous surfaces such as furnace linings, hot tubes, pipes and rods.

(*iv*) *Landing and docking manoeuvres* (Fig. 24.24): It can be used for measuring from helicopters to landing pads and from ships to piers and dock walls.

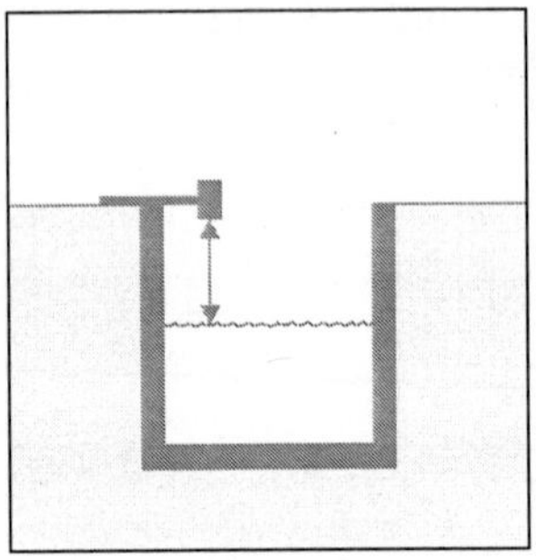

Fig. 24.23

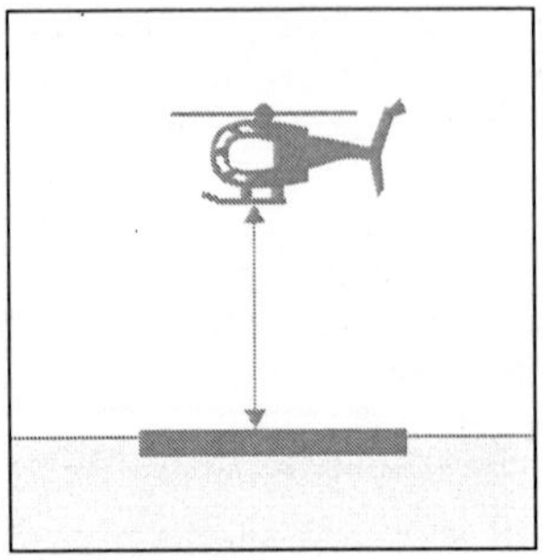

Fig. 24.24

5. Wild 'Tachymat' TC 2000

Wild TC 2000 (Fig. 24.25) is a fully integrated instrument. It combines in one instrument the advantages of the T 2000 informatics theodolite with the distance measuring capabilities of Wild distomats. For applications where distances and angles are always required, and instrument with built-in EDM is particularly useful. Wild TC 2000 having built-in EDM is a single package *total station* which can be connected to Wild GRE 3 data terminal. The same telescope is used for observing and distance measurement. The infrared measuring beam coincides with the telescope line of sight.

The telescope is panfocal, magnification and field of view vary with focusing distance. When focusing to distant targets, the magnification is 30 ×. Over shorter distances, the field widens and the magnification is reduced for easy pointing to the prism, the telescope with coarse and fine focusing is used for both angle and distant measurement.

The whole unit, theodolite and built-in EDM, is operated from the keyboard. Angles and distances can be measured in both telescope positions. Single attenuation and distance measurements are fully automatic. Normal distance measurement takes 6.5 seconds with a standard deviation of 3 mm ± 2 ppm. In tracking mode, the display updates at 2.5 seconds intervals and the standard deviation is 10 mm to 20 mm. The 2 km range to a single prism covers all short range work. Maximum range is about 4 km in average atmospheric conditions.

Keyboard control. The entire equipment—angle and distance measuring and recording—is controlled from the keyboard. The multifunctional capability of the instrument makes it suitable for almost any task.

Pair of displayed values: The panel directly displays angles, distances, heights and co-ordinates of the observed point where the signal (reflector prism) is kept (Fig. 24.26). Height above datum and station co-ordinates can be entered and stored.

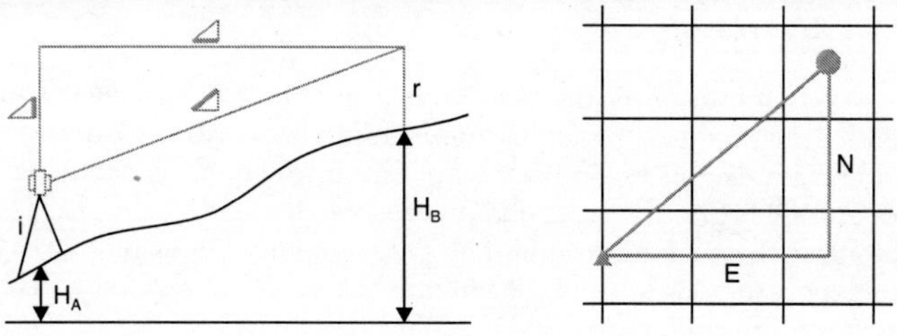

Fig. 24.26

The following pairs are displayed:

(*i*) Hz circle	V circle
(*ii*) Hz circle	Horizontal distance
(*iii*) Height difference	Height above datum
(*iv*) Slope distance	V circle
(*v*) Easting	Northing.

Remote object height (ROH): The direct height readings of inaccessible objects, such as towers and power lines, the height difference and height above datum changes with telescope. However, both the pairs of values are displayed automatically. The microprocessor applies the correction for earth curvature and mean refraction. Corrected heights are displayed (Fig. 24.27).

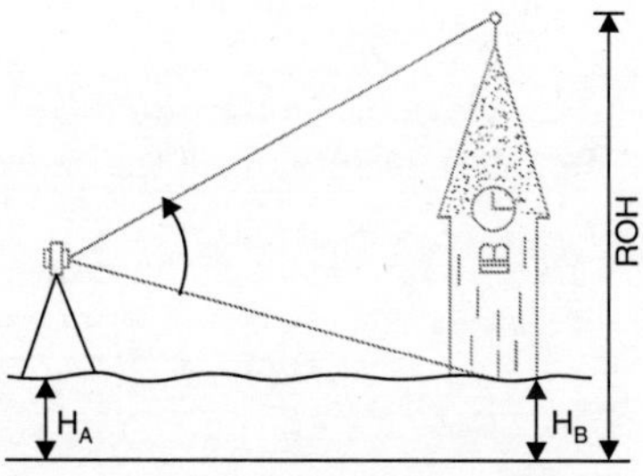

Fig. 24.27 Determination of ROH.

Traversing program: The coordinates of the reflector and the bearing on the reflector can be stored for recall at the next set-up. Thus, traverse point coordinates are available in the field and closures can be verified immediately (Fig. 24.28).

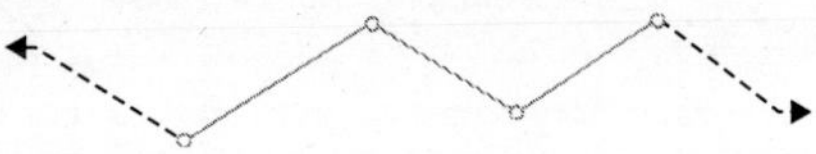

Fig. 24.28 Traversing.

Setting out for direction, distance and height (Fig. 24.29): The required direction and horizontal distance can be entered. The instrument displays:

(*i*) The angle through which the theodolite has to be turned.

(*ii*) The amount by which the reflector has to be moved.

And by means of remote object height (ROH) capability, markers can be placed at the required height above datum.

Setting out can be fully automated with GRE 3 data terminal. The bearings and distances to the points to be set out are computed from the stored coordinates and transferred automatically to the TC 2000 *total station.*

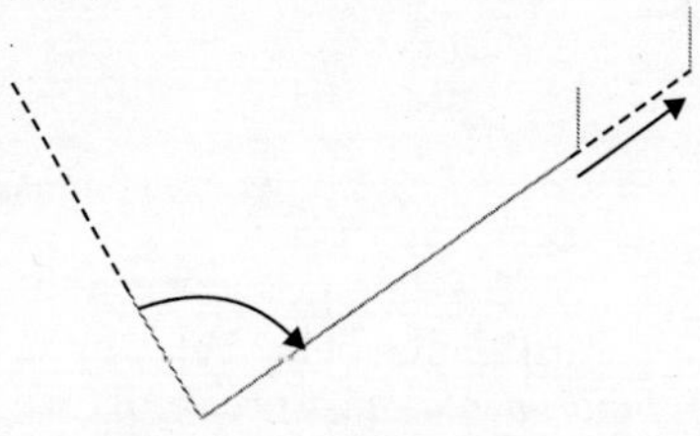

Fig. 24.29 Setting Out.

Differences in Hz and V: For locating targets and for real time comparisons of measurements in deformation and monitoring surveys, it is advantageous to display angular differences in the horizontal and vertical planes between a required direction and the actual telescope pointing.

24.8 TOTAL STATION

A total station is a combination of an electronic theodolite and an electronic distance meter (EDM). This combination makes it possible to determine the coordinates of a reflector by aligning the instruments cross-hairs on the reflector and simultaneously measuring the vertical and horizontal angles and slope distances. A microprocessor in the instrument takes care of recording, readings and the necessary computations. The data is easily transferred to a computer where it can be used to generate a map. Wild, 'Tachymat' TC 2000, described in the previous article is one such *total station* manufactured by M/s Wild Heerbrugg.

As a teaching tool, a total station fulfills several purposes. Learning how to properly use a total station involves the physics of making measurements, the geometry of calculations, and statistics for analysing the results of a traverse. In the field, it requires team work, planning, and careful observations. If the total station is equipped with data-logger it also involves interfacing the data-logger with a computer, transferring the data, and working with the data on a computer. The more the user understands how a total station works, the better they will be able to use it.

Fundamental measurements: When aimed at an appropriate target, a total station measures three parameters (Fig. 24.31)

1. The *rotation* of the instrument's optical axis from the instrument north in a horizontal plane: i.e. *horizontal angle.*
2. The *inclination* of the optical axis from the local vertical i.e. *vertical angle.*

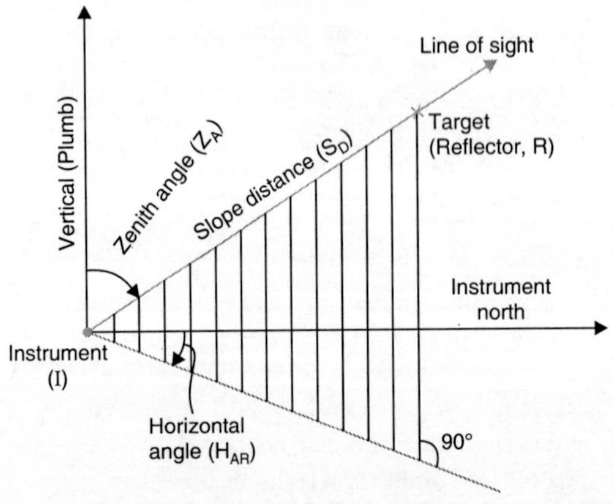

Fig. 24.31 Fundamental Measurements Made by a Total Station.

3. The distance between the instrument and the target i.e. *slope distance.*

All the numbers that may be provided by the total station are derived from these three *fundamental measurements.*

1. **Horizontal Angle:** The horizontal angle is measured from the zero direction on the horizontal scale (or *horizontal circle*). When the user first sets up the instrument the choice of the zero direction is made—this is *Instrument North.* The user may decide to set zero (North) in the direction of the long axis of the map area, or choose to orient the instrument approximately to True, Magnetic or Grid North. The zero direction should

be set so that it can be recovered if the instrument was set up at the same location at some later date. This is usually done by sighting to another benchmark, or to a distance recognizable object. Using a magnetic compass to determine the orientation of the instrument is not recommended and can be very inaccurate. Most total stations can measure angle to at least 5 seconds, or 0.0013888°. The best procedure when using a Total Station is to set a convenient "north" and carry this through the survey by using backsights when the instrument is moved.

2. **Vertical Angle:** The vertical angle is measured relative to the local vertical (plumb) direction. The vertical angle is usually measured as a *zenith angle* (0° is vertically up, 90° is horizontal, and 180° is vertically down), although one is also given the option of making 0° horizontal. The zenith angle is generally easier to work with. The telescope will be pointing downward for zenith angles greater than 90° and upward for angles less than 90°.

Measuring vertical angles requires that the instrument be exactly vertical. It is very difficult to level an instrument to the degree of accuracy of the instrument. Total stations contain an internal sensor (the vertical compensator) that can detect small deviations of the instrument from vertical. Electronics in the instrument then adjust the horizontal and vertical angles accordingly. The compensator can only make small adjustments, so the instrument still has to be well levelled. If it is too far out of level, the instrument will give some kind of "tilt" error message.

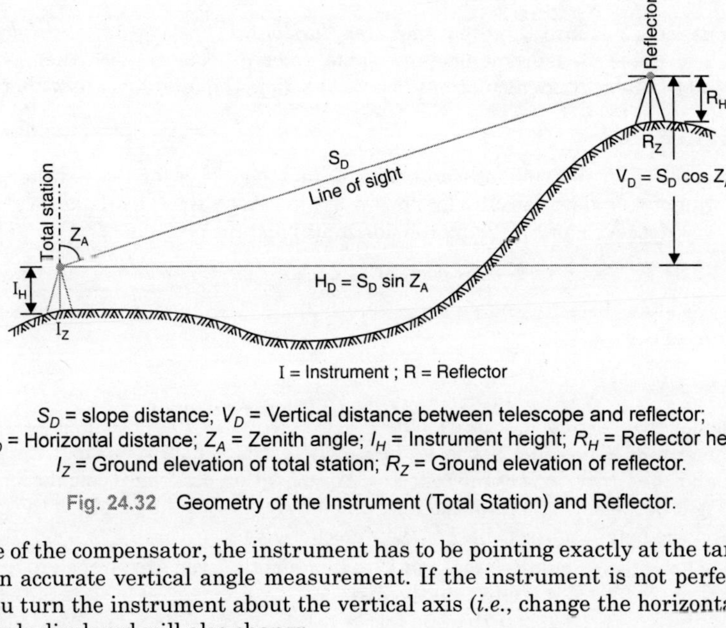

S_D = slope distance; V_D = Vertical distance between telescope and reflector; H_D = Horizontal distance; Z_A = Zenith angle; I_H = Instrument height; R_H = Reflector height; I_Z = Ground elevation of total station; R_Z = Ground elevation of reflector.

Fig. 24.32 Geometry of the Instrument (Total Station) and Reflector.

Because of the compensator, the instrument has to be pointing exactly at the target in order to make an accurate vertical angle measurement. If the instrument is not perfectly levelled then as you turn the instrument about the vertical axis (*i.e.*, change the horizontal angle) the vertical angle displayed will also change.

3. **Slope Distance:** The instrument to reflector distance is measured using an Electronic Distance Meter (EDM). Most EDM's use a Gallium Aresnide Diode to emit an infrared light beam. This beam is usually modulated to two or more different frequencies. The infrared beam is emitted from the total station, reflected by the reflector and received and amplified by the total station. The received signal is then compared with a reference

signal generated by the instrument (the same signal generator that transmits the microwave pulse) and the phase-shift is determined. This phase shift is a measure of the travel time and thus the distance between the total station and the reflector.

This method of distance measurement is not sensitive to phase shifts larger than one wavelength, so it cannot detect instrument-reflector distances greater than 1/2 the wavelength (the instrument measures the two-way travel distance). For example, if the wavelength of the infrared beam was 4000 m then if the reflector was 2500 m away the instrument will return a distance of 500 m.

Since measurement to the nearest millimetre would require very precise measurements of the phase difference, EDM's send out two (or more) wavelengths of light. One wavelength may be 4000 m, and the other 20 m. The longer wavelength can read distances from 1 m to 2000 m to the nearest metre, and then the second wavelength can be used to measure distances of 1 mm to 9.999 m. Combining the two results gives a distance accurate to millimetres. Since there is overlap in the readings, the meter value from each reading can be used as a check.

For example, if the wavelengths are $\lambda_1 = 1000$ m and $\lambda_2 = 10$ m, and a target is placed 151.51 metres away, the distance returned by the λ_1 wavelength would be 151 metres, the λ_2 wavelength would return a distance of 1.51 m. Combining the two results would give a distance of 151.51 m.

Basic Calculations

Total Stations only measure three parameters: *Horizontal Angle*, *Vertical Angle*, and *Slope Distance*. All of these measurements have some error associated with them, however for demonstrating the geometric calculations, we will assume the readings are without error.

Horizontal Distance

Let us use symbol I for instrument (total station) and symbol R for the reflector. In order to calculate coordinates or elevations it is first necessary to convert the slope distance to a horizontal distance. From inspection of Fig. 24.32 the horizontal distance (H_D) is

$$H_D = S_D \cos (90° - Z_A) = S_D \sin Z_A \quad ...(1) \quad ...(24.4)$$

where S_D is the slope distance and Z_A is the zenith angle. The horizontal distance will be used in the coordinate calculations.

Vertical Distance

We can consider two vertical distances. One is the *Elevation Difference* (dZ) between the two points on the ground. The other is the *Vertical Difference* (V_D) between the tilting axis of the instrument and the tilting axis of the reflector. For elevation difference calculation we need to know the height of the tilting-axis of the instrument (I_H), that is the height of the center of the telescope, and the height of the center of the reflector (R_H).

The way to keep the calculation straight is to imagine that you are on the ground under the instrument (Fig. 24.32). If you move up the distance I_H, then travel horizontally to a vertical line passing through the reflector then up (or down) the vertical distance (V_D) to the reflector, and then down to the ground (R_H) you will have the elevation difference dZ between the two points on the ground. This can be written as

$$dZ = V_D + (I_H - R_H) \quad ...(2) \quad ...(24.5)$$

The quantities I_H and R_H are measured and recorded in the field. The vertical difference V_D is calculated from the vertical angle and the slope distance (see Fig. 24.32)

$$V_D = S_D \sin (90° - Z_A) = S_D \cos Z_A \qquad ...(3) \quad ...(24.6)$$

Substituting this result (3) into equation (2) gives

$$dZ = S_D \cos Z_A + (I_H - R_H) \qquad ...(4) \quad ...(24.7)$$

where dZ is the change in elevation with respect to the ground under the total station. We have chosen to group the instrument and reflector heights. Note that if they are the same then this part of the equation drops out. If you have to do calculations by hand it is convenient to set the reflector height the same as the instrument height.

If the instrument is at a known elevation, I_Z, then the elevation of the ground beneath the reflector, R_Z, is

$$R_z = I_Z + S_D \cos Z_A + (I_H - R_H) \qquad ...(5) \quad ...(24.8)$$

Coordinate Calculations

So far we have only used the vertical angle and slope distance to calculate the elevation of the ground under the reflector. This is the Z-coordinate (or elevation) of a point. We now want to

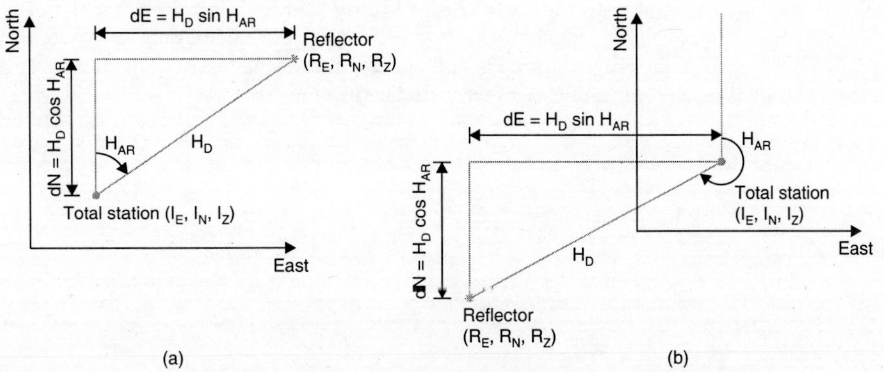

Fig. 24.33 Computation of East and North Coodinates of the Reflector.

calculate the X- (or East) and Y- (or North) coordinates. The zero direction set on the instrument is instrument north. This may not have any relation on the ground to true, magnetic or grid north. The relationship must be determined by the user. Fig. 24.33 shows the geometry for two different cases, one where the horizontal angle is less than 180° and the other where the horizontal angle is greater than 180°. The sign of the coordinate change positive in Fig. 24.33 (*a*) and negative in Fig. 24.33 (*b*) is taken care of by the trigonometric functions, so the same formula can be used in all cases. Let us use symbol E for easting and N for northing, and symbol I for the instrument (*i.e.* total station) and R for the reflector. Let R_E and R_N be the easting and northing of the reflector and I_E and I_N be the easting and northing of the instrument (*i.e.* total station).

From inspection of Fig. 24.33 the coordinates of the reflector relative to the total station are:

$$dE = \text{Change in Easting} = H_D \sin H_{AR}$$

$$dN = \text{Change in Northing} = H_D \cos H_{AR}$$

where H_D is the horizontal distance and H_{AR} is the horizontal angle measured in a clockwise sense from instrument north. In terms of fundamental measurments (*i.e.* equation 1) this is the same as

$$dE = S_D \sin Z_A \sin H_{AR} \quad ...(24.9)$$

$$dN = S_D \cos (90° - Z_A) \cos H_{AR} = S_D \sin Z_A \cos H_{AR} \quad ...(24.10)$$

If the easting and northing coordinates of the instrument station are known (in grid whose north direction is the same as instrument north) then we simply add the instrument coordinates to the change in easting and northing to get the coordinates of the reflector. The coordinates of the ground under the reflector, in terms of fundamental measurements are:

$$R_E = I_E + S_D \sin Z_A \sin H_{AR} \quad ...(24.11)$$

$$R_N = I_N + S_D \sin Z_A \cos H_{AR} \quad ...(24.12)$$

$$R_Z = I_Z + S_D \cos Z_A + (I_H - R_H) \quad ...(24.13)$$

where I_E, I_N, and I_Z are the coordinates of the total station and R_E, R_N, R_Z are the coordinates of the ground under the reflector. These calculations can be easily done in a spreadsheet program.

All of these calculations can be made within a total station, or in an attached electronic notebook. Although it is tempting to let the total station do all the calculations, it is wise to record the three fundamental measurements. This allows calculations to be checked, and provides the basic data that is needed for a more sophisticated error analysis.

CHAPTER A

APPENDIX

Example A-1: *Given: A base line is measured with a steel tape. It is approximately 1000 m long. Calculate the correct length of the base line at M.S.L. when the pull at the standardisation equals 15 kg. The pull applied is 23 kg, cross-sectional area of the tape is 0.0645* cm^2, *E* = *2.11* × 10^6 *kg/*cm^2*, Temperatures* T_m *and* T_0 *be 35°C and 15°C respectively. The difference of level between the two ends of the base line is 2.0 m. The radius of earth R = 6400 km.*

Elevation of base line above M.S.L. is 1000. $\alpha = 12 \times 10^{-6}$.

Required: *The correct length of the line after applying the following corrections.*

(*i*) *Temperature correction* (*ii*) *Pull correction*

(*iii*) *Slope correction* (*iv*) *M.S.L. correction*

(Engg. Services)

Solution:

(*i*) Temperature correction, $C_l = \alpha (T_m - T_0) L = 12 \times 10^{-6} (35 - 15) 1000 = 0.240 \text{ m} (+)$

(*ii*) Pull correction, $C_p = \dfrac{P - P_0}{AE} L = \dfrac{23-15}{0.0645 \times 2.11 \times 10^6} \times 1000 = 0.0588 \text{ m} (+)$

(*iii*) Slope correction, $C_{sl} \simeq \dfrac{h^2}{2L} \simeq \dfrac{2^2}{2 \times 1000} = 0.0020 \text{ m} (-)$

(*iv*) M.S.L. correction, $C_{msl} = \dfrac{Lh}{R} = \dfrac{1000 \times 1000}{6400 \times 1000} = 0.1563 (-)$

$\therefore$ Total correction = 0.2400 + 0.0588 – 0.0020 – 0.1563 = + 0.1405 m

$\therefore$ Corrected length of base line = 1000 + 0.1405 = **1000.1405 m**

Example A-2: *The plan of an old survey plotted to a scale of 10 m to 1 cm carried a note stating that 'the chain was 0.8 links (16 cm) too short'. It was also found that the plan has shrunk so that a line originally 10 cm long was 9.77 cm. The area of a plot on the available plan was found to be 58.2 sq.cm. What is the correct area of the plan in hectares?*

(U.P.S.C. Engg. Services Exam.)

Solution: Present area of the plot on the survey plan = 58.2 sq. cm.

This area is on the shrunk plan.

Now 9.77 cm on shrunk plan =10 cm of original plan

$\therefore \quad (9.77)^2 = (10)^2$

$\therefore$ Correct area of un-shrunk plan $= \dfrac{(10)^2}{(9.77)^2} \times 58.2 = 60.9725 \text{ cm}^2$

Let us now take into account the faulty length of the chain.

Let us assume that the chain used for the survey was of 30 m designated length.

Actual (or erroneous) length (L') of chain = 30 – 0.16 = 29.84 m

Now True area on plan $= \left(\dfrac{L'}{L}\right) \times$ measured area $= \left(\dfrac{29.84}{30}\right)^2 \times 60.9725 = 60.3238 \text{ cm}^2$

Scale of plan: 1 cm = 10 m or $1 \text{ cm}^2 = (10)^2 \text{ m}^2$

$\therefore$ Field area of the plot = $60.3238\,(10)^2$ = **6032.38 m²** = **0.603238** hectares

Example A-3: (a) *The length of an offset is 16 m. The maximum error in its length is 6.5 cm and scale used is 1 cm = 20 m. What is the maximum permissible error in the laying of the direction of the offset so that the maximum displacement dos not exceed 0.5 mm?*

(b) *A road 1557 m long was found, when measured by a defective 30 m chain, to be 1550 m. How much correction does the chain need?*

(c) *To find the width of a river flowing from West to East, two points A and B are fixed along the bank 500 m apart. The bearing on the ranging rod (point C) on the other bank of the river as observed from A and B are 45° and 330° respectively. Determine the width of the river.*

(d) *If the magnetic bearing of a line AB is 312° 45′ and the declination of the place is 2° 32′ W, find the true bearing of the line BA and express it in quadrantal system.* ***(U.P.S.C. Asst. Engg. C.P.W.D. Exam.)***

Solution: Refer Fig. 4.10

(*a*) Let α = maximum permissible angular error.

Length of offset, l = 16 m ; e = 6.5 cm = 0.065 m ; scale : 1 cm = 20 m

Maximum displacement = 0.5 mm = 0.05 cm

Displacement of point due to incorrect direction $= P_2 P_1 = l \sin\alpha = 16 \sin\alpha$

Max. error in the length of the offset $= PP_1 = 0.065$ m

$\therefore$ Max. displacement due to both errors $= PP_2 = \sqrt{(16\sin\alpha)^2 + (0.065)^2}$,

$\therefore$ Max. displacement on paper $= \dfrac{\sqrt{(16\sin\alpha)^2 + (0.065)^2}}{20}$ cm = 0.05 cm (given)

$\therefore \quad (16 \sin\alpha)^2 + (0.065)^2 = (0.05 \times 20)^2$

or $\quad 256 \sin^2\alpha = 0.995775$

which gives $\quad \alpha = 3°.576 \approx$ **3° 35′**

(*b*) Let the correction to the chain length be ΔL

$\therefore$ Incorrect length of chain, $L' = L + \Delta L = 30 + \Delta L$

Now $\quad l = l'\left(\dfrac{L'}{L}\right)$

$$\therefore \qquad L' = \frac{l}{l'} L$$

$$\text{or} \qquad 30 + \Delta L = \frac{1557}{1550} \times 30 = 30.135 \text{ m}$$

$$\therefore \qquad \Delta L = 30.135 - 30 = 0.135 \text{ m} = \mathbf{13.5 \text{ cm}}$$

(*c*) See Fig. A-l.

$$\angle CAB = 45° \,;\, \angle CBA = 60°$$

$$\therefore \qquad \angle ACB = 180° - (45° + 60°) = 75°$$

Applying sine formula for ΔACB,

$$AC = \frac{\sin 60°}{\sin 75°} \times 500 = 448.29 \text{ m}$$

$$BC = \frac{\sin 45°}{\sin 75°} \times 500 = 366.03$$

Now $\qquad CD = AC \sin 45° = 448.29 \sin 45°$

$= \mathbf{316.99 \text{ m}}$

Alternatively, $CD = BC \sin 60° = 366.03 \sin 60° = 316.99$ m

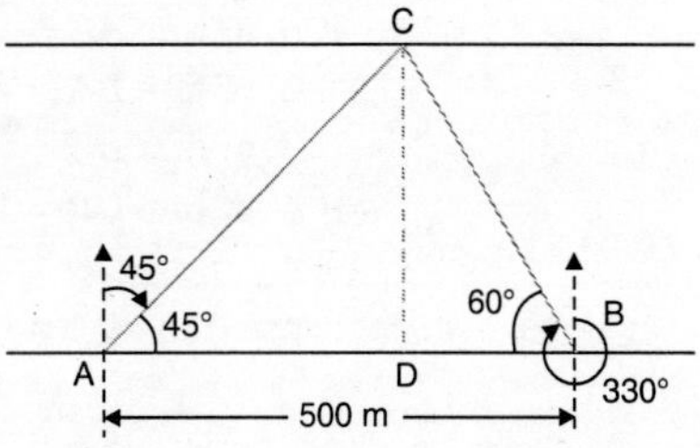

Fig. A-1

(*d*) See Fig. A-2.

True bearing of $\quad AB = 312° 45' - 2°32' = 310° 13'$

$\therefore$ True bearing of $\quad BA = 310° 13' - 180° = \mathbf{130° 13'}$

Quadrantal T.B. of $\quad BA = S\,(180° - 130° 13')\,E$

$= \mathbf{S\ 549°47'\ E}$

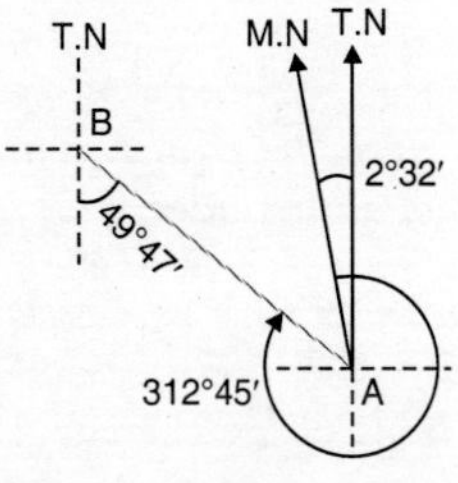

Fig. A-2

Example A-4: *Following is the data regarding a closed compass traverse ABCD taken it a clockwise direction.*

(*i*) *Fore bearing and back bearing at station A = 50° and 130°*

(*ii*) *Fore bearing and back bearing of line CD = 206° and 26° respectively*

(*iii*) *Included angles ∠B = 100° and ∠C = 105°*

(*iv*) *Local attraction at station C = 2° W*

All the observations were free from all the errors except local attraction.

From the above data, calculate (a) local attraction at stations A and D and (b) corrected bearings of all the lines

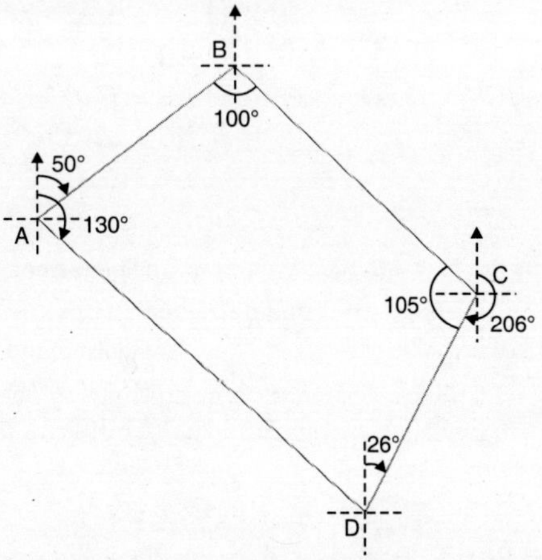

Fig. A-3

Solution: The F.B. and B.B. of line *CD* differ exactly by 180°. Hence either both stations *C* and *D* are free from local attraction, or are *equally affected* by local attraction. Since station *C* has a local attraction of 2° W, station *D* also has a local attraction of 2° W. Due to this, all the recorded bearings at *C* and *D* are 2° *more* than the correct values.

$\therefore$ Corrected F.B. of $CD = 206° - 2° = 204°$

and corrected B.B. of $CD = 26° - 2° = 24°$

Let us first calculate the included angles.

$\angle BAD = 130° - 50° = 80°.$

$\angle ADC = 360° - (80° + 100° + 105°) = 75°$

Now B.B. of line $CD = 24°$

F.B. of line $DA = 24° + (360° - 75°) = 309°$

$\therefore$ B.B. of line $DA = 309° - 180° = 129°$

F.B. of line $AB = 129° - 80° = 49°$

B.B. of line $BA = 49° + 180° = 229°$

F.B. of line $BC = 229° - 100° = 129°$

B.B. of line $CB = 129° + 180° = 390°$

$\therefore$ F.B. of line $CD = 309° - 105° = 204°$

Local attraction at A = Observed F.B. of AB – corrected F.B. of AB

$= 50° - 49° = 1°$ W

Answer: Local attraction at **A** = **1° W**

Local attraction at **D** = **2° W**

Example A-5: *In an anticlockwise traverse ABCA, all the sides were equal. Magnetic fore bearing of side BC was obtained as 20° 30′. The bearing of sun was also observed to be 182° 20′ at the local noon, with a prismatic compass. Calculate the magnetic bearings and true bearings of all the sides of the traverse. Tabulate the results and draw a neat sketch to show the bearings.*

Solution:

(*a*) ***Computation of magnetic bearings***

$\angle ABC = \angle BCA - \angle CAB = 60°$

Now F.B. of $BC = 20° 30'$

$\therefore$ B.B. of $BC = 20° 30' + 180° = 200° 30'$

F.B. of $CA = 200° 30' + 60° = 260° 30'$

B.B. of $CA = 260° 30' - 180° = 80° 30'$

F.B. of $AB = 80° 30' + 60° = 140° 30'$

B.B. of $AB = 140° 30' + 180° = 320° 30'$.

F.B. of $BC = 320° 30' + 60° - 360° = 20° 30'$

Hence OK.

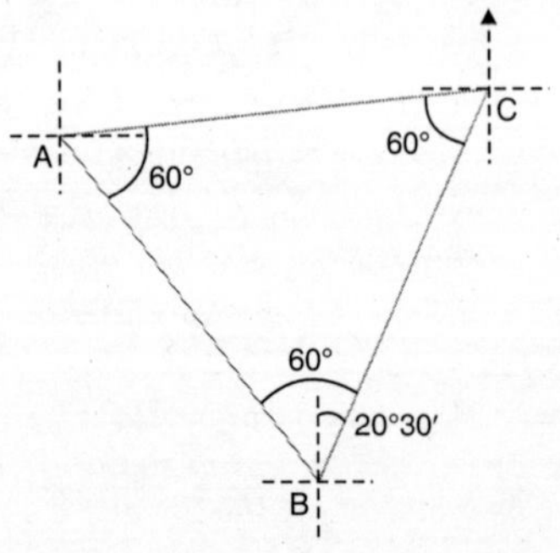

Fig. A-4

(*b*) ***Computation of true bearings***

True bearing of sun at local noon = 180°

Measured magnetic bearing of sun = 182° 20′

$\therefore$ Declination = 182° 20′ – 180° = 2° 20′ W

The true bearings of various lines can be calculated by subtracting 2° 20′ from the corresponding magnetic bearings, and the results can be tabulated as shown below.

Line	Magnetic bearing		True bearing	
	F.B.	B.B.	F.B.	B.B.
AB	140° 30′	320° 30′	138° 10′	318° 10′
BC	20° 30′	200° 30′	18° 10′	198° 10′
CA	260° 30′	80° 30′	258° 10′	78° 10′

Example A-6: *Three ships A, B and C started sailing from a harbour at the same time in three directions. The speed of all the three ships was the same, i.e. 40 km/hour. Their bearings were measured to be N 65° 30′E, S 64° 30′ E and S 14° 30′ E. After an hour, the captain of ship B determined the bearings of the other two ships with respect to his own ship. After that he found out the distances. Calculate the value of bearings and distances which might have been determined by the captain of ship B.*

Solution: (*a*) *Consider triangle OCB*

$\angle COB = 64° 30' - 14° 30' = 50°$

$\angle OCB = \angle OBC = \frac{1}{2}(180° - 50°) = 65°$

Now B.B. of OB = N 64° 30′ W = 295° 30′ (W.C.B.)

∴ F.B. of BC = 295° 30′ – 65° = 230° 30′

= **S 50° 30′ W**

Distance $BC = 2 \times 40 \cos 65°$ = **33.809 km**

(*b*) *Consider triangle OAB*

$\angle AOB = 180° - (65° 30' + 64° 30') = 50°$

$\angle OBA = \angle OAB = \frac{1}{2}(180° - 50) = 65°$

Now B.B. of OB = 295° 30′ (found earlier)

∴ F.B. of BA = 295° 30′ + 65°

= 360° 30′ = 0° 30′ = N 0° 30′ E

Distance $BA = 2 \times 40 \cos 65°$ = **33.809 km**

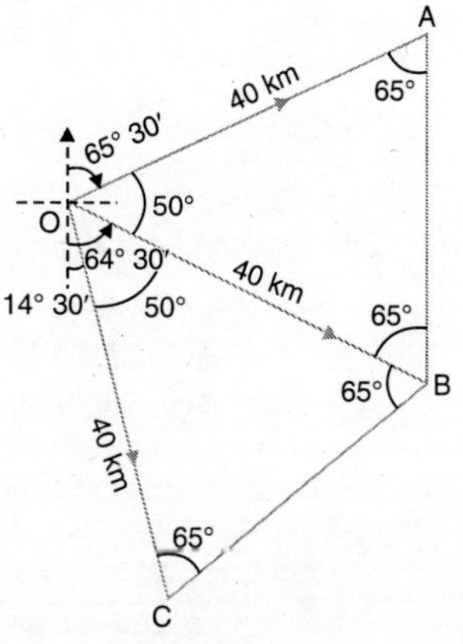

Fig. A-5

Example A-7: *The measured lengths and bearings of the sides of a closed traverse ABCDE, run in the counter-clockwise direction are tabulated below. Calculate the lengths of CD and DE*

Line	Length (m)	Bearings
AB	239	N 00° 00′ E
BC	164	N 25° 12′ W
CD	?	S 75° 06′ E
DE	?	S 56° 24′ E
EA	170	N 35° 30′ E

Solution: The traverse in shown in Fig. A-6. Let the unknown lengths of

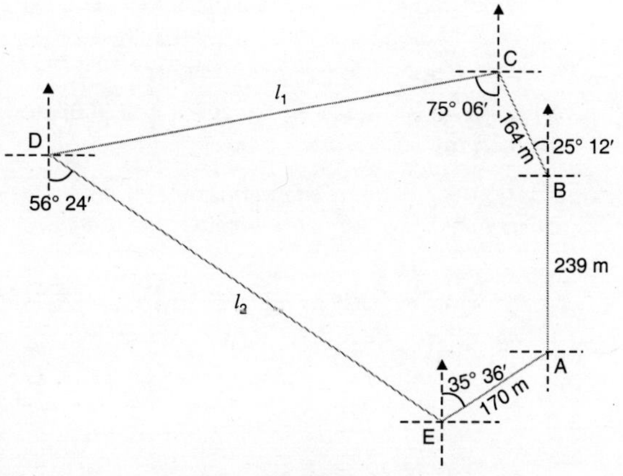

Fig. A-6

CD and DE be l_1 and l_2 respectively. The bearings of all the lines are known. For the whole traverse, $\Sigma L = 0$ and $\Sigma D = 0$.

$\therefore \quad 239 \cos 0° + 164 \cos 25° 12' - l_1 \cos 75° 06' - l_2 \cos 56° 24' + 170 \cos 35° 36' = 0$

or $$0.2571\, l_1 + 0.5534\, l_2 = 525.62 \quad ...(1)$$

Also, $239 \sin 0° - 164 \sin 25° 12' - l_1 \sin 75° 06' + l_2 \sin 56° 24' + 170 \sin 35° 36° = 0$

or $$0.9664\, l_1 - 0.8329\, l_2 = 29.13 \quad ...(2)$$

Solving Eqs. (1) and (2), we get

$$l_1 = \mathbf{606.10} \quad \text{and} \quad l_2 = \mathbf{668.22\ m}$$

Example A-8: *Using the data of a closed traverse given below, calculate the lengths of the lines BC and CD.*

Line	*Length(m)*	*W.C.B.*
AB	*275.2*	*14° 31′*
BC	–	*319° 42′*
CD	–	*347° 15′*
DE	*240.0*	*5° 16′*
EA	*1566.4*	*168° 12′*

Also, sketch the traverse

Solution: The traverse is shown in Fig. A-7. Let the lengths of BC and CD be l_1 and l_2 respectively. Since the traverse is closed, we have

$$\Sigma L = 0$$

and $$\Sigma D = 0$$

$\therefore \quad 275.2 \cos 14° 31' + l_1 \cos 319° 42' + l_2 \cos 347° 15' + 240 \cos 5° 16' + 1566.4 \cos 168° 12' = 0$

or $$0.7627\, l_1 + 0.9753\, l_2 = 1027.90 \quad ...(1)$$

and $275.2 \sin 14° 31' + l_1 \sin 319° 42' + l_2 \sin 347° 15' + 240 \sin 5° 16' + 1566.4 \sin 168° 12' = 0$

or $$0.6468\, l_1 + 0.2207\, l_2 = 411.33 \quad ...(2)$$

Solving Eqs. (1) and (2), we get l_1 = **376.96 m** and l_2 = **759.14 m**

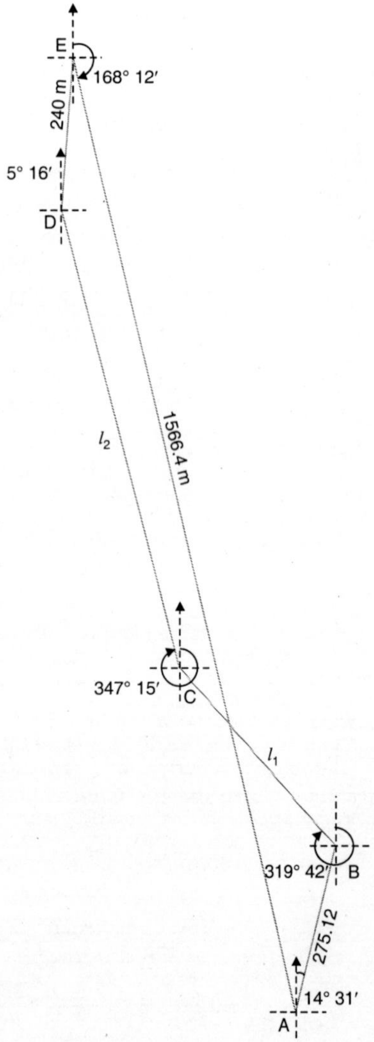

Fig. A-7

Example A-9: *From a common point A, traverses are conducted on either side of a harbour as follows:*

Traverse	*Line*	*Length (m)*	*W.C.B.*
1	*AB*	*240*	*85° 26′*
	BC	*120*	*125° 11′*
2	*AD*	*270*	*175° 50′*
	DE	*600*	*85° 07′*

Calculate (a) distance from C to a point F on DE due south of C, and (b) distance EF.

Solution: The two traverses are shown in Fig. A-8. The combined traverse *ABCFDA* is a closed one, in which the sides *CF* (= l_1) and *FD* (= l_2) are not known. However, these can be determined from the fact that for the composite traverse, $\Sigma L = 0$ and $\Sigma D = 0$.

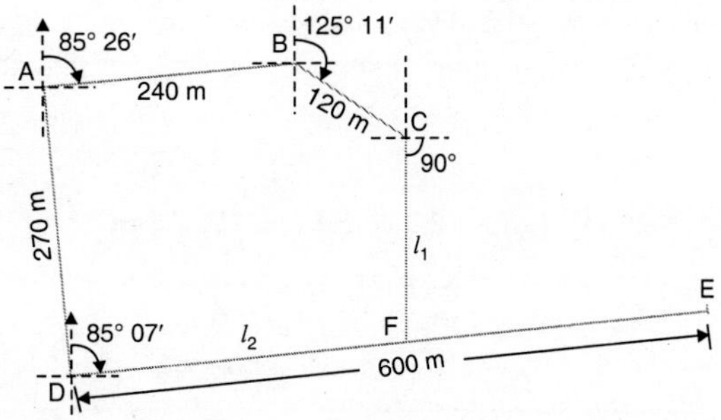

Fig. A-8

$\therefore$ 240 cos 85° 26′ + 120 cos 125° 11′ $- l_1 + l_2$ cos 265° 07′ + 270 cos 355° 50′ = 0

or $l_1 + 0.0851\, l_2 = 219.25$ ...(1)

and 240 sin 85° 26′ + 120 sin 125° 11′ + 0 + l_2 sin 265° 07′ + 270 sin 355° 50′ = 0

or $0.9964\, l_2 = 317.70$ ...(2)

From (2) $l_2 = 318.85$ Substituting the value of l_2 in (1),

we get l_1 = **192.12 m**

Distance $EF = 600 - l_1 = 600 - 318.85$ = **281.15 m**

Example A-10: *A clockwise traverse ABCDEA was surveyed with the following results. AB = 161.62 m; BC = 224.38 m; CD = 158.83 m ∠BAE = 128° 10′ 20″ ; ∠DCB = 84° 18′ 10″ ;*

∠CBA = 102° 04′ 30″ ; ∠EDC = 121° 30′ 30″

The angle AED and the sides DE and EA could not be measured direct.

Assuming no error in the survey, find the missing lengths and their bearings if AB is due north.

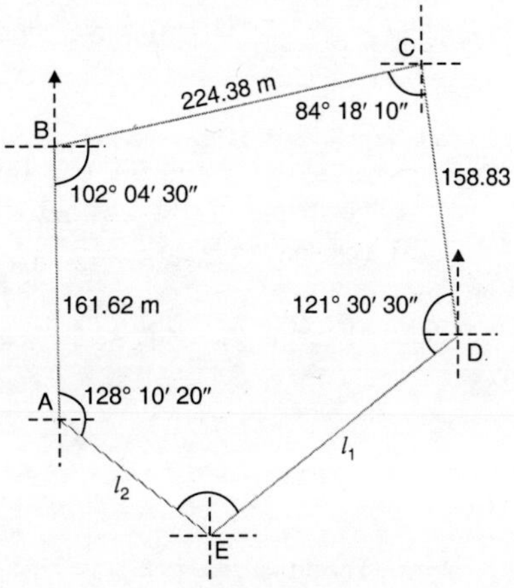

Fig. A-9

Solution: Fig. A-9 shows the sketch of the traverse.

Total interior angles

$= (2N - 4)\, 90° = 540°$

$\therefore$ ∠*AED* = 540° – (128° 10′ 20″ + 102° 04′ 30″ + 84° 18′ 10″ + 121° 30′ 30″) = 103° 56′ 30″

Taking the W.C.B. of *AB* as 0° 0′ 0″, the bearing of other lines are:

BC : 180° 0′ 00″ – 102° 04′ 30″ = 77° 55′ 30″

CD : (77° 55′ 30″ + 180°) – 84° 18′ 10″ = 173° 37′ 20″

DE : (173° 37′ 20″ + 180°) – 121° 30′ 30″ = **232° 06′ 50″**

EA : (233° 06′ 50″ – 180°) – (360° – 103° 56′ 30″) = 308° 10′ 20″

AB : (308° 10′ 20″ – 180°) – 128° 10′ 20″ = 0° 0′ 0″ (check)

Now, let the lengths of *DE* and *EA* be l_1 and l_2.

For the whole traverse, we have $\Sigma L = 0$ and $\Sigma D = 0$

$\therefore$ 161.62 cos 0° 0′ 0″ + 224.38 cos 77° 55′ 30″ + 158.33 cos 173° 37′ 20″ + l_1 sin 232° 06′ 50″ + l_2 cos 308° 10′ 20″ = 0 or $0.6141\, l_1 - 0.618\, l_2 = 51.37$ ...(1)

and $161.62 \sin 0° 00' 00'' + 224.38 \sin 77° 55' 30'' + 158.33 \sin 173° 37' 20''$
$+ l_1 \sin 232° 06' 50'' + l_2 \sin 308° 10' 20'' = 0$

or $0.7892\, l_1 + 0.786\, l_2 = 237.00$ ...(2)

Solving (1) and (2), we get $l_1 =$ **192.55 m** and $l_2 =$ **108.21 m**

Example A-11: *ABCD is a closed traverse in which the bearing of AD has not been observed and the length of BC has been missed to be recorded. The rest of the field record is as follows*

Line	*Bearing*	*Length (m)*
AB	*181° 18'*	*335*
BC	*90° 00'*	*?*
CD	*357° 36'*	*408*
DA	*?*	*828*

Calculate the bearing of AD and the length of BC. ***(Engg. Services)***

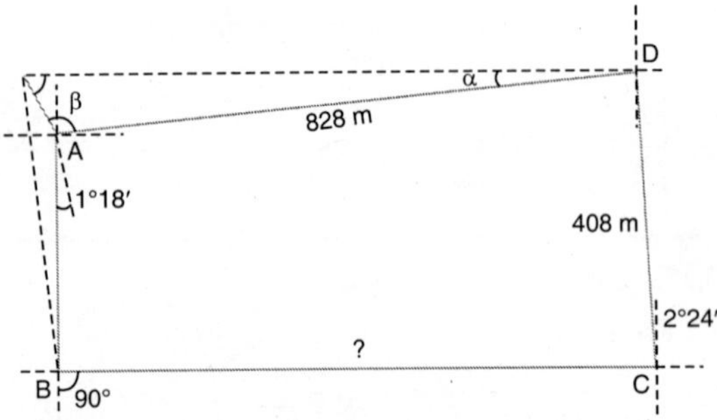

Fig. A-10

Solution:

(*a*) *Semi-Analytical solution.* In order to bring the affected sides adjacent, draw *DA* parallel to *CB* and *BA'* parallel to *CD*, both meeting at *A'*. Let l be the length and θ be the bearings of the closing line *A'A*.

For the closed traverse *ABA'*,

$$\Sigma L = 335 \cos 181° 18' + 408 \cos 357° 36 + l \cos \theta = 0$$

or $-334.91 + 407.64 + l \cos \theta = 0$

or $l \cos \theta = -72.73$ ...(1)

Also, $\Sigma D = 335 \sin 181° 18' + 408 \sin 357° 36' + l \sin \theta = 0$

or $-7.60 - 17.09 + l \sin \theta = 0$

∴ $l \sin \theta = +24.69$ ...(2)

From (1) and (2), we get $l = \sqrt{(72.73)^2 + (24.69)^2} = 76.81$ m

Since the latitude is –ve and departure is positive, *A'A* lies in the second quadrant.

∴ $\theta = \tan^{-1} \dfrac{24.69}{72.73} = 18° 45'$, ∴ W.C.B. of *A'A* = 161° 15'

∴ $\angle A A' D = \gamma =$ Bearing of $A'A$ – bearing of $A'D = 161° 15' - 90° = 71° 15'$

Now $$\frac{AA'}{\sin \alpha} = \frac{A'D}{\sin \beta} = \frac{AD}{\sin \gamma} = \frac{828}{\sin 71°15'}$$

∴ $$\alpha = \sin^{-1}\left[\frac{76.81 \sin 71°15'}{828}\right] = 5° 2'$$

$$\beta = 180° - (71° 15' + 5° 2') = 103° 43'$$

∴ $$BC = A'D = 828 \frac{\sin 103°43'}{\sin 71°15'} = \mathbf{849.49\ m}$$

Bearing of DA = Bearing of $DA' - \alpha = 270° - 5° 2' =$ **264° 58′**

(*b*) ***Analytical method***

Let us use suffixes 1, 2, 3, 4, for lines *AB, BC, CD* and *DA*.

$$\therefore \quad \Sigma L = 0 = 335 \cos 181° 18' + l_2 \cos 90° + 408 \cos 357° 36' + 828 \cos \theta_4$$

or

$$-334.91 + 0 + 407.64 + 824 \cos \theta_4 = 0$$

From which $\theta_4 = \cos^{-1} \dfrac{-72.73}{828} =$ **264° 58′**

Also, $\Sigma D = 0 = 335 \sin 181° 18' + l_2 \sin 90° + 408 \sin 357° 36' + 828 \sin 264° 58'$

From which $l_2 =$ **849.49 m**

Example A-12: *An open traverse was run from A to E in order to obtain the length and bearing of the line AE which could not be measured direct, with the following results:*

Line	*Length*	*W.C.B.*
AB	*82 m*	*261° 41′*
BC	*87 m*	*9° 06′*
CD	*74 m*	*282° 22′*
DE	*100 m*	*71° 30′*

Find by calculation the required information.

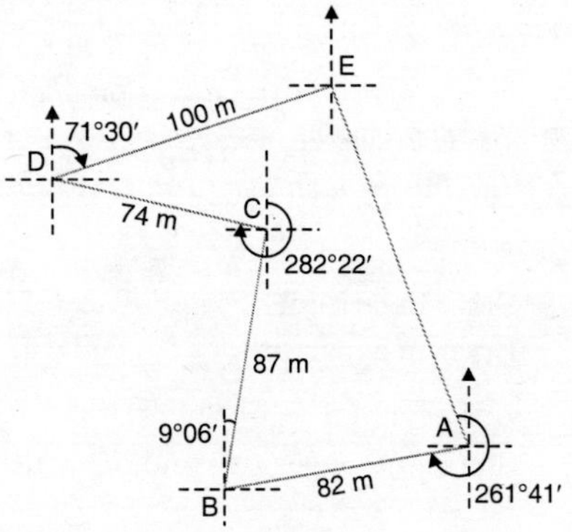

Fig. A-11

Solution: Refer Fig. A-11

The length and bearing of line *AE* is required.

Since *ABCDEA* is a closed traverse,

We have

$$\Sigma L = 0 \quad \text{and} \quad \Sigma D = 0$$

$$\therefore \quad L_{EA} = -\Sigma L' \quad \text{and} \quad D_{EA} = -\Sigma D'$$

or

$$L_{AE} = \Sigma L' \quad \text{and} \quad D_{AE} = \Sigma D'$$

where $\Sigma L'$ and $\Sigma D'$ are for the four lines *AB*, *BC*, *CD* and *DE*.

The computations are done in a tabular form below.

Line	*Length (m)*	*W.C.B.*	*Latitude*	*Departure*
AB	82	261° 41′	– 11.85	– 81.14
BC	87	9° 06′	+ 85.91	+ 13.76
CD	74	282° 22′	+ 15.85	– 72.28
DE	100	71° 30′	+ 33.38	+ 94.26
			$\Sigma L' = +123.28$	$\Sigma D' = -45.40$

$$\therefore \quad L_{AE} = \Sigma L' = +123.28 \quad \text{and} \quad D_{AE} = \Sigma D' = -45.40$$

Since latitude is +ve and departure is negative, line *AB* is in fourth quadrant.

$$\therefore \quad l_{AE} = \sqrt{(123.28)^2 + (45.40)^2} = \mathbf{131.37\ m}$$

$$\theta = \tan^{-1}\frac{45.40}{123.28} = 20°\,13'; \textbf{W.C.B.}\text{ of}AE = 360° - 20°\,13' = \mathbf{339°\,47'}$$

Example A-13: *The following table gives data of consecutive coordinates in respect of a closed theodolite traverse ABCDA*

Station	*N*	*S*	*E*	*W*
A	*240*			*160*
B	*160*		*239*	
C		*239*	*160*	
D		*160*		*240*

From the above data, calculate

(i) Magnitude and direction of closing error

(ii) Corrected consecutive coordinates of station B, using transit rule

(iii) Independent coordinates of station B, if those of A are (80, 80)

Solution:

Error in latitude, $\Delta L = \Sigma L = 240 + 160 - 239 - 160 = 1$

Error in departure, $\Delta D = \Sigma D = -160 + 239 + 160 - 240 = -1$

$\therefore$ Closing error $= \sqrt{(+1)^2 + (-1)^2} = \mathbf{1.414\ m}$

Since ΔL is positive and ΔD is negative, the line of closure is in 4th quadrant

$$\theta = \tan^{-1}\frac{1.0}{1.0} = 45°.\ \textbf{W.C.B.}\text{ of closing error} = 360° - 45° = \mathbf{315°}$$

Arithmetic sum of latitudes = 240 + 160 + 239 + 160 = 799

Arithmetic sum of departures = 100 + 239 + 160 + 240 = 799

$$\therefore \text{Correction to latitude of } AB = -\frac{1.0}{799} \times 160 - 0.20$$

$$\text{Correction to departure of } AB = +\frac{1.0}{799} \times 239 = +0.30$$

Hence corrected consecutive co-ordinates of B are:

N: 160 – 0.2 = **159.80**

E: 239 + 0.3 = **239.30**

Independent coordinates of station B

N: 80 + 159.80 = **239.80**

E: 80 + 239.30 = **319.30**

Example A-14: *In a traverse ABCDEFG, the line BA is taken as the reference meridian. The latitudes and departures of the sides AB, BC, CD, DE and EF are:*

Line	*AB*	*BC*	*CD*	*DE*	*EF*
Latitude	*– 95.20*	*– 45.22*	*+ 47.24*	*+ 48.55*	*+ 87.78*
Departures	*0.00*	*+ 58.91*	*+ 63.74*	*– 37.44*	*+ 29.63*

If the bearing of FG is N 75°47′ W and its length is 71.68 m, find the length and bearing of GA.

Solution: Bearing of FG = N 75° 47′ W

Length of FG = 71.68 m

∴ Latitude of FG = + 71.68 cos 75° 47′ = + 17.60

Departure of FG = – 71.68 sin 75°47′ = – 69.48

The traverse is shown diagrammatically in Fig. A-12. Since traverse $ABCDEFG$ is a closed one, we have

$$L_{GA} = -\Sigma L' \text{ and } D_{GA} = -\Sigma D'$$

The computations are arranged in the tabular form below.

Line	*Latitude*	*Departure*
AB	– 95.20	0.00
BC	– 45.22	+ 58.91
CD	+ 47.24	+ 63.74
DE	+ 48.55	– 37.44
EF	+ 87.78	+ 29.93
FG	+ 17.60	– 69.48
	ΣL′ = + 60.75	ΣD′ = + 45.66

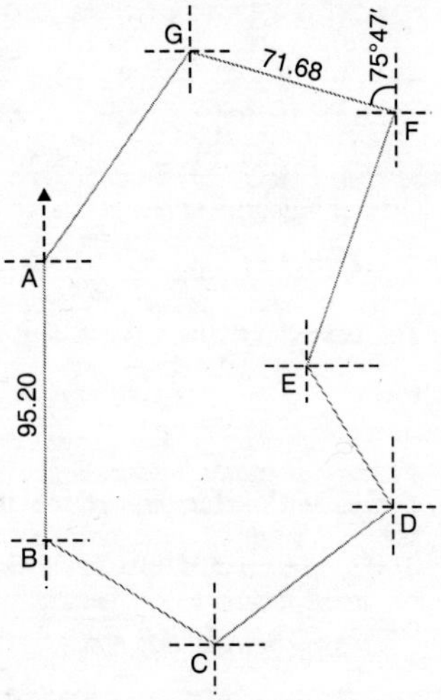

Fig. A-12

∴ $L_{GA} = -\Sigma L' = -60.75$

and $D_{GA} = -\Sigma D' = -45.66$

Hence the line GA is in the third quadrant.

$$GA = \sqrt{(60.75)^2 + (45.66)^2} \approx 76.00 \text{ m}$$

$$\theta = \tan^{-1}\frac{45.66}{60.75} = 36°\ 56'$$

∴ Bearing of GA = **S 36° 56′ W**

Example A-15: *Calculate latitudes, departures and closing error for the following traverse and adjust using Bowditch's rule.*

Line	*Length (m)*	*Whole circle bearing*
AB	*89.31*	*45° 10′*
BC	*219.76*	*72° 05′*
CD	*151.18*	*161° 52′*
DE	*159.10*	*228° 43′*
EA	*232.26*	*300° 42′*

(Engg. Services)

Solution: The computations are arranged in tabular form below:

Line	*Length (m)*	*W.C.B.*	*Latitude (m)*			*Departure (m)*		
			calculated	*correction*	*corrected*	*calculated*	*correction*	*corrected*
AB	89.31	45° 10′	+ 62.97	– 0.06	+ 62.91	+ 63.34	– 0.02	+ 63.32
BC	219.76	72° 05′	+ 67.61	– 0.13	+ 67.48	+ 209.10	– 0.06	+ 209.04
CD	151.18	161° 52′	– 143.67	– 0.09	– 143.76	+ 47.05	– 0.04	+ 47.01
DE	159.10	228°43′	– 104.97	– 0.10	– 105.07	– 119.56	– 0.04	– 119.60
EA	232.26	300° 42′	+ 118.58	– 0.14	+ 118.34	– 199.71	– 0.06	– 199.77
Sum	851.61		+ 0.52	– 0.52	0.00	+ 0.22	– 0.22	0.00

Correction for latitude of any line $= -\dfrac{0.52}{851.61} \times$ Length of that line

Correction for departure of any line $= \dfrac{-0.22}{851.61} \times$ length of that line.

$$\text{closing error} = \sqrt{(0.52)^2 + (0.22)^2}$$

$$= 0.565 \text{ m}$$

Angle of error of closure is given by $\theta = \tan^{-1}\dfrac{0.22}{0.52} = 22°\,56'$

Relative accuracy

$$= \frac{0.565}{851.61} = \mathbf{1 \text{ in } 1507}$$

The traverse is shown in Fig. A-13.

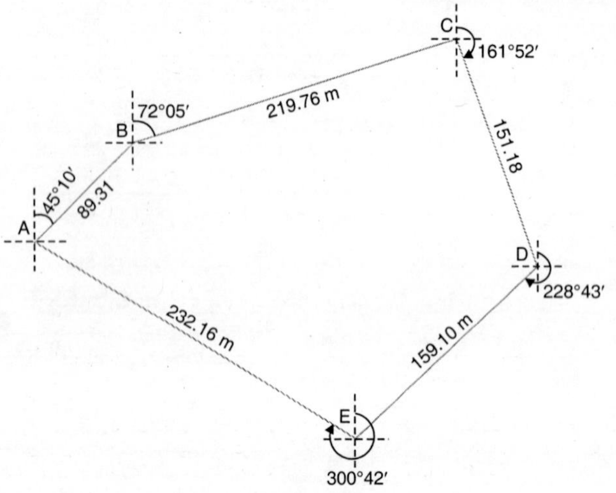

Fig. A-13

Example A-16: *The following measurements were obtained when surveying a closed traverse ABCDEA*

Line	*EA*	*AB*	*BC*
Length (m)	*95.24*	*181.45*	*103.64*
∠DEA = 93° 14′	*∠EAB = 122°36′*	*∠ABC = 131°42′*	*∠BCD = 95°43′*

It is not possible to occupy D, but it could be observed from both C and E. Calculate the angle CDE and the lengths CD and DE, taking DE as the datum and assuming all observations to be correct.

Solution: Refer. Fig. A-14.

Computation of ∠CDE and bearings of all lines

Theoretical sum of interior angles

$= (2N - 4)\ 90° = 540°$

$\therefore \quad \angle CDE = 540° - (93°\ 14' + 122°\ 36' + 131°\ 42' + 95°\ 43')$

$= \mathbf{96°\ 45'}$

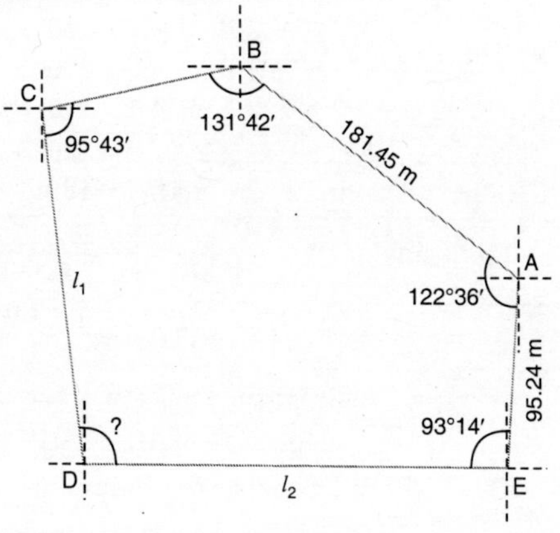

Fig. A-14

Let us take the W.C.B. bearing of *DE* as 90°.

$\therefore$ Bearing of *ED* $= 90° + 180° = 270°$

$\therefore$ Bearing of *EA* $= 270° + 93°\ 14' - 360° = 3°\ 14'$

$\therefore$ Bearing of *AE* $= 3°\ 14' + 180° = 183°\ 14'$

Bearing of *AB* $= 183°\ 14' + 122°\ 36' = 305°\ 50'$

Bearing of *BA* $= 305°\ 50' - 180° = 125°\ 50'$

Bearing of *BC* $= 125°\ 50' + 131°\ 42' = 257°\ 32'$

Bearing of *CB* $= 257°\ 32' - 180° = 77°\ 32'$

Bearing of *CD* $= 77°\ 32' + 95°\ 43' = 173°\ 15'$

Bearing of *DC* $= 173°\ 15' + 180° - 353°\ 15'$

$\therefore$ Bearing of *DB* $= 353°\ 15' + 96°\ 45' - 360°$

$= 90°\ 00'$ (check)

Let the lengths of *CD* and *DE* be l_1 and l_2 respectively.

For the closed traverse *DEABCDA* $\Sigma L = 0$ and $\Sigma D = 0$

$\therefore \quad 95.24 \cos 3°\ 14' + 181.45 \cos 305°\ 50' + 103.64 \cos 257°\ 32' + l_1 \cos 173°\ 15' + l_2 \cos 90° = 0$

or $\quad 95.09 + 106.23 - 22.37 - 0.9931\, l_1 + 0 = 0$

From which $\quad l_1 = CD = \mathbf{180.19\ m}$

Similarly, $\quad 95.24 \sin 3°\ 14' + 181.45 \sin 305°\ 50' + 103.64 \sin 257°\ 32' + 180.19 \sin 173°\ 15' + l_2 \sin 90° = 0$

$5.37 - 147.11 - 101.20 + 21.18 + l_2 = 0$

From which $\quad l_2 = DE = \mathbf{221.76\ m}$

Example A-17: *The bearings of two inaccessible stations A and B, taken from station C, were 220° and 148° 30′ respectively. The coordinates of A and B were as under:*

Station	*Easting*	*Northing*
A	180	120
B	240	90

Calculate the independent coordinates of C

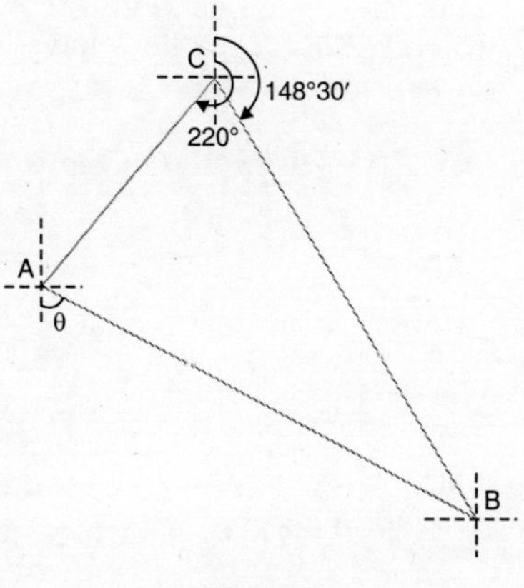

Fig. A-15

Solution: In order to calculate the independent coordinates of *C*, we need either the length *AC* or *BC*.

Length $\quad AB = \sqrt{(240-180)^2 + (90-120)^2}$

$= 67.08$ m

$$\theta = \tan^{-1} \frac{240-180}{120-90} = 63°.435 = 63°\ 26'$$

$\therefore$ W.C.B. of $AB = 180° - 63° 26' = 116° 34'$

W.C.B. of $AC = 220° - 180° = 40°$

$\therefore$ $\angle CAB = 116° 34' - 40° = 76° 34'$

$\angle ACB = 220° - 148° 30' = 71° 30'$

$\angle ABC = 180° - (76° 34' + 71° 30') = 31° 56'$

From sine rule, $\dfrac{AC}{\sin 31° 35'} = \dfrac{BC}{\sin 76° 34'} = \dfrac{AB}{\sin 71° 30'} = \dfrac{67.08}{\sin 17° 30'}$

$\therefore$ $AC = \dfrac{67.08}{\sin 71° 30'} \sin 31° 56' = 37.46$ m

Now latitude of $AC = 37.46 \cos 40° = 28.70$ m

departure of $AC = 37.46 \sin 40° = 24.08$ m

Hence the independent coordinates of C are

Easting = 180 + 24.08 = 204.08

Northing = 120 + 28.70 = 148.70

Example A-18: *It is not possible to measure the length and fix the direction of AB directly on account of an obstruction between the stations A and B. A traverse ACDB was, therefore, run and the following data were obtained.*

Line	*Length (m)*	*Reduced bearing*
AC	*63*	*N 55° E*
CD	*92*	*S 65° E*
DB	*84*	*S 25° E*

Find the length and direction of line BA. It was also required to fix a station E on line BA such that DE will be perpendicular to BA. If there is no obstruction between B and E, calculate the data required for fixing the station as required.

Solution: Fig. A-16 shows the traverse. The computations are done in a tabular form below, where W.C.B. of the lines have been entered for convenience.

Line	*Length*	*W.C.B.*	*Latitude*	*Departure*
AC	63	55°	+ 36.14	+ 51.61
CD	92	115°	– 38.88	+ 83.38
DB	84	155°	– 76.13	+ 35.50
		sum	– 78.87	+ 170.49

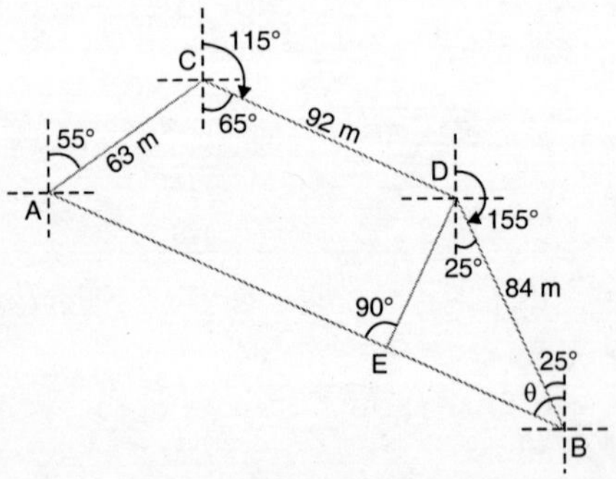

Fig. A-16

$\therefore$ $L_{BA} = -\Sigma L = +78.87$

and $D_{BA} = -\Sigma D = -170.49$

Hence *BA* is in the fourth quadrant.

$\theta = \tan^{-1} \dfrac{170.49}{78.87} = 65° 10'$, and **W.C.B.** ***BA*** **= 294° 50′**

$$\text{Length of } BA = \sqrt{(78.87)^2 + (170.49)^2} = \mathbf{187.85\ m}$$

Now in triangle DEB, $\angle DEB = 90°$ and $\angle EBD = 65°\ 10' - 25° = 40°\ 10'$

Now $DE = DB \sin EBD = 84 \sin 40°\ 10' =$ **54.18 m**

and $BE = DB \cos EBD = 84 \cos 40°\ 10' =$ **64.19 m**

Example A-19: *The magnetic bearing of the sun at noon is 160°. Find the variation.*

(Engg. Services)

Solution: This question is based on Example 5.8.

At noon, the sun is exactly on the geographical meridian.

∴ True bearing of sun = 180°

Magnetic bearing of sun = 160°

Now, True bearing = Magnetic bearing + Declination.

∴ 180° = 160° + Declination

∴ Declination = 180° – 160° = 20°

As the sign is positive, the variation is east.

∴ Variation = **20° E**

Example A-20: *Select the correct answer in each of the following and show the calculations made in arriving at the answer:*

(a) A uniform slope was measured by the method of stepping. If the difference in level between two points is 1.8 m and the slope distance between them is 15 m, the error is approximately equal to

(i) Cumulative, + 0.11 m *(ii) Compensating, ± 0.11 m*

(iii) Cumulative, – 3.11 m *(iv) None of these*

(b) A standard steel tape of length 30 m and cross-section 15 × 1.0 mm was standardised at 25°C and at 30 kg pull. While measuring a base line at the same temperature, the pull applied was 40 kg. If the modules of elasticity of the steel tape is 2.2 × 10^6 kg/cm^2, the correction to be applied is

(i) – 0.000909 m *(ii) + 0.0909 m*

(iii) + 0.000909 m *(iv) None of these*

(c) The bearing of AB is 190° and that of CB is 260° 30′. The included angle ABC is

(i) 80° 30′ *(ii) 99° 30′*

(iii) 70° 30′ *(iv) None of these*

(d) A dumpy level was set up at mid-point between pags A and B, 80 m apart and the staff readings were 1.32 and 1.56. When the level was set up at a point 10 m from A on BA produced, the staff readings obtained at A and B were 1.11 and 1.39. The correct staff reading from this set up, at B should be

(i) 1.435 *(ii) 1.345*

(iii) 1.425 *(iv) None of these*

(e) The desired sensitivity of a bubble tube with 2 mm division is 30″. The radius of the bubble tube should be

(i) 13.75 m *(ii) 3.44 m*

(iii) 1375 m *(iv) None of these*

(U.P.S.C. Asst. Engg. C.P.W.D., Exam.)

Solution:

(*a*) Horizontal distance $D = (l^2 - h^2)^{1/2}$

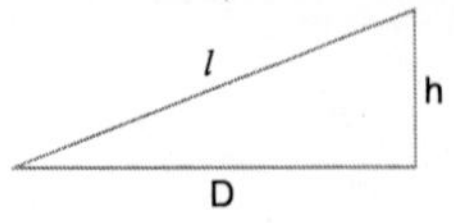

Fig. A-17

$$= L\left(1 - \frac{h^2}{l^2}\right)^{1/2} \simeq l\left(1 - \frac{1}{2}\frac{h^2}{l^2}\right)$$

$$\therefore \quad \text{Error } e = l - D = l - l + \frac{1}{2}\frac{h^2}{l} \simeq \frac{1}{2}\frac{h^2}{l} = \frac{1}{2}\frac{(1.8)^2}{15} = 0.108 \simeq 0.11 \text{ m}$$

Hence error = + **0.11** m (cumulative). Hence correct answer is (*i*)

(*b*) Correction for tension or pull

$$= C_p = \frac{(P - P_0)\,L}{AE}$$

Here, $P = 40$ kg ; $P_0 = 30$ kg,

$A = 1.5 \times 10.1$ cm^2 ; $E = 2.2 \times 10^6$ kg/cm^2 ; $L = 30$ m

$$\therefore \quad C_p = \frac{(40-30)\,30}{1.5 \times 0.1 \times 2.2 \times 10^6} = \mathbf{0.000909\ m.}$$

Hence answer (*iii*) is correct.

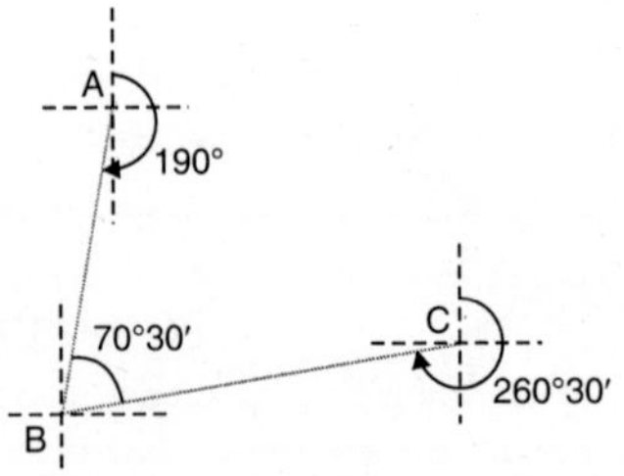

Fig. A-18

(*c*) Bearing of AB = 190°

∴ Bearing of BA = 190° – 180° = 10°

Also, Bearing of CB = 260° 30′

∴ Bearing of BC = 260° 30′ – 180° = 80° 30′

∴ Included angle ABC = Bearing of BC – Bearing of BA

= 80° 30′ – 10° = **70° 30′**. Hence correct answer is (*iii*)

(*d*) *Instrument at mid-point*

The collimation error is balanced.

∴ True difference in level between A and B = 1.56 – 1.32 = 0.24 m (*B* being lower)

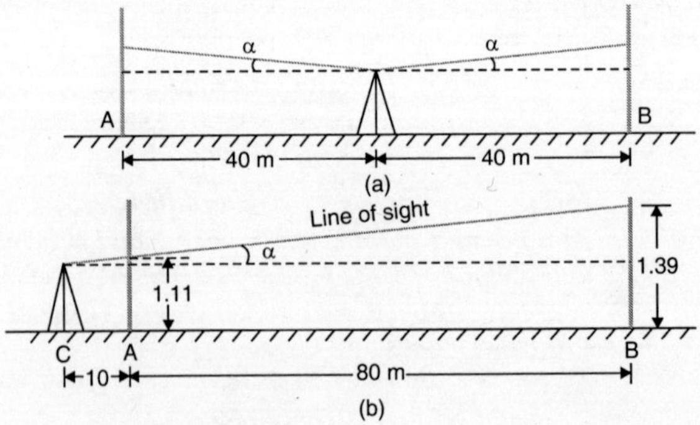

Fig. A-19

Instrument at 10 m from A on BA produced

∴ Apparent difference in elevation = 1.39 – 1.11 = 0.28 m (*B* being lower). If the line of collimation were horizontal, the correct reading at *B* would be

= Reading at A + True difference in level between A and B
= 1.11 + 0.24 = 1.35 m

Since the actual reading at B is 1.39 m, the line of collimation is elevated upwards (Fig. A-19 *b*)

$\therefore$ Collimation error in 80 m = 1.39 – 1.35 = 0.04

Hence collimation error in 90 m = $0.04 \times \frac{90}{80} = 0.045$ m

$\therefore$ Correct staff reading on B = 1.39 – 0.045 = **1.345 m**

Hence correct answer is (*ii*).

(*e*) Sensitivity, $\alpha' = \frac{l}{R} \times 206265$ seconds

Here $\alpha' = 30''$ and $l = 2$ mm

$\therefore$ $R = \frac{l}{\alpha'} \times 206265 = \frac{2}{30} \times 206265 = 13751 \text{ mm} \approx$ **13.75 m**

Hence correct answer is (*i*).

Example A-21: Calculate the latitudes, departures and closing error for the following travers and adjust the using Bowditch's rule.

Line	*Length (m)*	*Whole circle Bearing*
AB	*89.31*	*45°10′*
BC	*219.76*	*72°05′*
CD	*151.18*	*161°52′*
DE	*159.10*	*228°43′*
EA	*232.26*	*300°42′*

(U.P.S.C. Engg. Services Exam.)

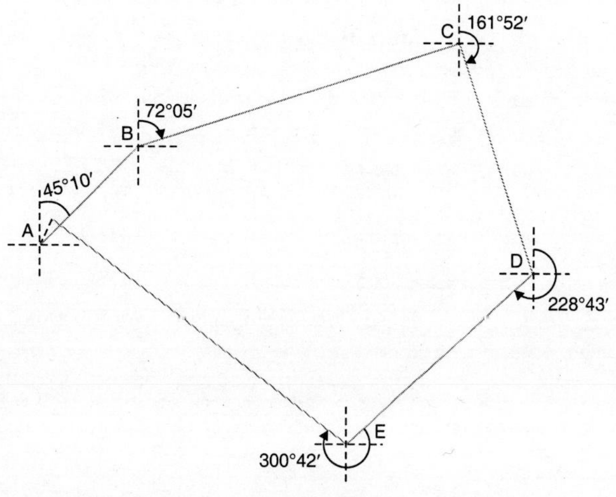

Fig. A-20

Solution: Fig. A-20 shows the traverse *ABCDEA′*, in which *AA′* is the closing error. Table below shows the computations for latitude and departure of various lines of the traverse.

Line	*Length (m)*	*W.C.B.*	*Reduced bearing*	*Latitude*	*Departure*
AB	89.31	45° 10′	N 45° 10′ E	+ 62.97	+ 63.34
BC	219.76	72° 05′	N 72° 05′ E	+ 67.61	+ 209.10
CD	151.18	161° 52′	S 18° 08′ E	– 143.67	+ 47.05
DE	159.10	228° 43′	S 48° 43′ W	– 104.97	– 119.56
EA	232.26	300° 42′	N 59° 18′ W	+ 118.58	– 199.71
			Sum	+ 0.52	+ 0.22

∴ Closing error, $e = \sqrt{(0.52)^2 + (0.22)^2} = \mathbf{0.565\ m}$

$$\theta = \tan^{-1}\frac{0.22}{0.52} = 22.932 = 22°\ 55'\ 56''$$

Total correction for latitude = – 0.52; Total correction for departure = – 0.22

Σl = Perimeter of traverse = 89.31 + 219.76 + 151.18 + 159.10 + 232.26 = 851.61

According to the Bowditch rule:

Correction for latitude, $C_L = \Sigma L \dfrac{1}{\Sigma l} = -0.52 \times \dfrac{l}{851.61} = -6.106 \times 10^{-4}\, l$ ...(1)

Correction for departure, $C_D = \Sigma D \dfrac{1}{\Sigma l} = -0.2 \times \dfrac{l}{851.61} = -2.3485 \times 10^{-4}\, l$ ...(2)

The computations for the corrections for latitude and departure of each line, along with the corrected latitude and departure are arranged in a tabular form below.

Line	*Latitude*			*Departure*		
	Latitude	*Correction*	*Corrected Latitude*	*Departure*	*Correction*	*Corrected Departure*
AB	+ 62.97	– 0.05	+ 62.92	+ 63.34	– 0.02	+ 63.32
BC	+ 67.61	– 0.13	+ 67.48	+ 209.10	– 0.06	+ 209.04
CD	– 143.67	– 0.09	– 143.76	+ 47.05	– 0.04	+ 47.01
DE	– 104.97	– 0.10	– 105.07	– 119.56	– 0.04	– 119.60
EA	+ 118.58	– 0.15	+ 118.43	– 199.71	– 0.06	– 199.77
	Sum	– 0.52	0.00		– 0.22	0.00

Example A-22: *For a railway project, a straight tunnel is to be run between two points P and Q whose co-ordinates are given below:*

Point	*Co-ordinates*	
	N	*E*
P	*0*	*0*
Q	*4020*	*800*
R	*2110*	*1900*

It is desired to sink a shaft at S, the mid-point of PQ. S is to be fixed from R, the third known point.

Calculate (i) the coordinates of S, (ii) Length of RS, (iii) the bearing of RS.

(U.P.S.C. Engg. Services Exam.)

Solution:

(*i*) *Coordinates of S*

$$\text{Northing} = \frac{0 + 4020}{2} = \mathbf{2010\ m}$$

$$\text{Easting} = \frac{0 + 800}{2} = \mathbf{400\ m}$$

(ii) Length RS

ΔN between R and S = 2110 – 2010 = 100 m

ΔE between R and S = 1900 – 400 = 1500

$\therefore$ $RS = \sqrt{100^2 + 1500^2}$ = **1503.33 m**

(iii) Bearing of RS

Let the reduced bearing of RS be θ

$$\therefore \quad \tan\theta = \frac{\Delta E}{\Delta N} = \frac{1500}{100} = 15$$

$$\therefore \quad \theta = \tan^{-1} 15 = \mathbf{S\ 86°\ 11'\ 09''\ W}$$

$$\text{W.C.B. of } RS = 180° + \theta = 180° + 86°\ 11'\ 09''$$

$$= \mathbf{266°\ 11'\ 09''}$$

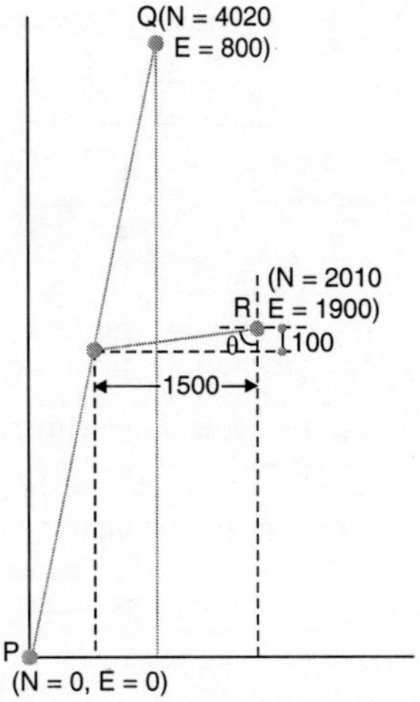

Fig. A-21

Example A-23: (*a*) *The following perpendicular offsets were taken at 30 m intervals from a base line of an irregular boundary line:*

5.8, 12.2, 17.0, 16.2, 18.4, 16.3, 24.6, 22.2, 18.4 and 17.2

Calculate the area enclosed between the base line, the irregular boundary line and the second and the last offsets by average ordinate rule.

(*b*) *The following are the bearings of the sides of a closed traverse PQRSTUVW. Compute the corrected bearings for the local attraction.*

Line	*Foreward Bearing*	*Backward Bearing*
PQ	*39° 00′*	*215° 30′*
QR	*75° 12′*	*255° 42′*
RS	*127° 06′*	*308° 36′*
ST	*145° 18′*	*325° 18′*
TU	*160° 12′*	*339° 12′*
UV	*214° 36′*	*35° 12′*
VW	*287° 24′*	*107° 00′*
WP	*347° 42′*	*170° 00′*

(*c*) *Compute the missing data**

B.S.	*F.S.*	*H.I.*	*R.L.*	*Remarks*
1.605		*	*400.50*	*Change point*
	– 1.015		*	*Bench mark*

(U.P.S.C. Asst Engineers, C.P.W.D. Exam.)

Solution: (*a*) See Fig. A-22. Average ordinate is given by

$$O_{av} = \frac{1}{9}(12.2 + 17.0 + 16.2 + 18.4 + 16.3 + 24.6 + 22.2 + 18.4 + 17.2) = 18.056 \text{ m}$$

Length = 8 × 30 = 240 m

$\therefore$ Area = O_{av} × length = 18.056 × 240 = **4333.33 m²**

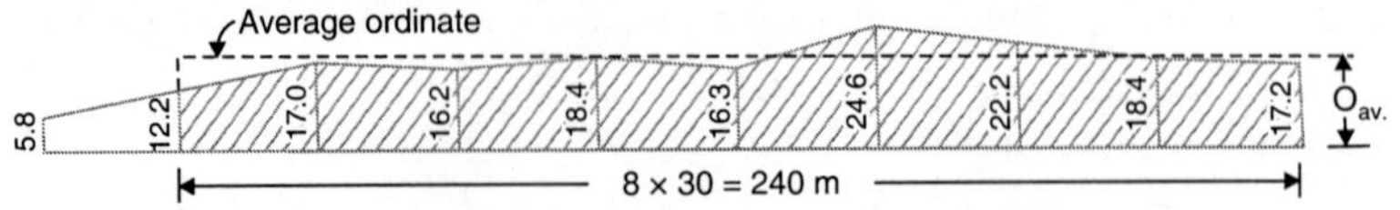

Fig. A-22

(*b*) By inspection, we find that *ST* is the only line whose forebearing and back bearing differ exactly by 180°. *Hence both S and T are free from local attraction.* Hence the bearing of *TU* and *SR* are correct.

Thus, correct bearing of *TU* = 160° 12′

∴ Correct bearing of *UT* = 160° 12′ + 180° = 340° 12′

But observed bearing of *UT* = 339° 12′

∴ Error at *U* = 339° 12′ – 340° 12′ = – 1°

∴ Correction at *U* = + 1°

∴ Corrected bearing of *UV* = 214° 36′ + 1° = 215° 36′

and correct bearing of *VU* = 215° 36′ – 180° = 35° 36′

But observed bearing of *VU* = 35° 12′

∴ Error at *V* = 35° 12′ – 35° 36′ = – 24′ and correction at *V* = + 24′

∴ Corrected bearing of *VW* = 287° 24′ + 24′ = 287° 48′

and correct bearing of *WV* = 287° 48′ – 180° = 107° 48′

But observed bearing of *WV* = 107° 00′

∴ Error at W = 107° 00′ – 107° 48′ = – 48′ and correction at W = + 48′

∴ Corrected bearing of *WP* = 347° 42′ + 48′ = 348° 30′

∴ Correct bearing of *PW* = 348° 30 – 180° = 168° 30′

But observed bearing of *PW* = 170° 00′

∴ Error at *P* = 170° 00′ – 168° 30′ = + 1°30′

and correction at P = – 1° 30′

corrected bearing of *PQ* = 39° 00′ – 1° 30′ = 37° 30′

and correct bearing of *QP* = 37° 30′ + 180° = 217° 30′

But observed bearing of *QP* = 215° 30′

∴ Error at *Q* = 215° 30′ – 217° 30′ = – 2° 0′ and correction at *Q* = + 2° 0′

Corrected bearing of *QR* = 75° 12′ + 2° = 77° 12′

and correct bearing of *RQ* = 77° 12′ + 180° = 257° 12′

But observed bearing of *RQ* = 255° 42′

Error at *R* = 255° 42′ – 257° 12′ = – 1° 30′ and correction at *R* = + 1° 30′

∴ Corrected bearing of *RS* = 127° 06′ + 1° 30′ = 128° 36′

∴ Correct bearing of *SR* = 128° 36′ + 180° = 308° 36′

But observed bearing of *SR* = 308° 36′

Hence error at *S* = 0° 0′ (as expected). This is a check on computations. The corrected bearings of various sides of the traverse are shown in Fig. A-23.

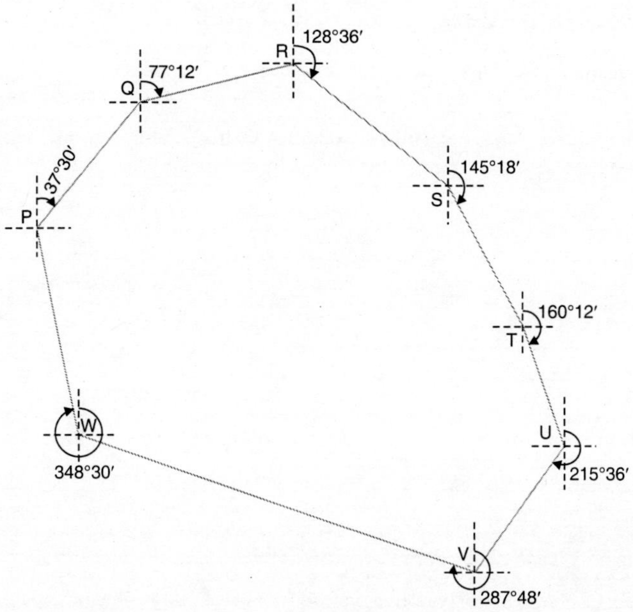

Fig. A-23

(c) ***Computation of missing data:*** In the tabular form below.

B.S.	*F.S.*	*H.I.*	*R.L.*	*Remarks*
1.605		402.105	400.500	Change point
	– 1.015		403.120	Benchmark

H.I. = R.L. of change point + B.S. reading = 400.50 + 1.605 = **402.105 m**

R.L. of B.M. = H.I. – F.S. reading = 402.105 – (– 1.015) = **403.120 m**

Example A-24: *To determine the distance between two points X and Y and their elevations, the following observations were taken upon vertically held staves from two traverse stations R and S. The tacheometer was fitted with an anallactic lens and the instrument constant was 100.*

Traverse station	*R.L.*	*Ht. of Instrument (m)*	*Co-ordinates L*	*D*	*Staff station*	*Bearing*	*Vertical angle*	*Staff Readings*		
R	*1020.60*	*1.50*	*800*	*1800*	*X*	*15°14′*	*+ 8° 9′*	*1.10*	*1.85*	*2.60*
S	*1021.21*	*1.53*	*950*	*2500*	*Y*	*340°18′*	*+ 2° 3′*	*1.32*	*1.91*	*2.50*

Compute the distance XY, the gradient from X to Y and the bearing of XY.

(*U.P.S.C. Engg. Services Exam.*)

Solution: (*a*) ***Observation from R to X***

Horizontal distance $RX = \frac{f}{i} s_1 \cos^2 \theta_1 + 0$

Here $\frac{f}{i} = 100$; $s_1 = 2.60 - 1.10 = 1.50$ m ; $\theta_1 = 8° 9'$; $r_1 = 1.85$

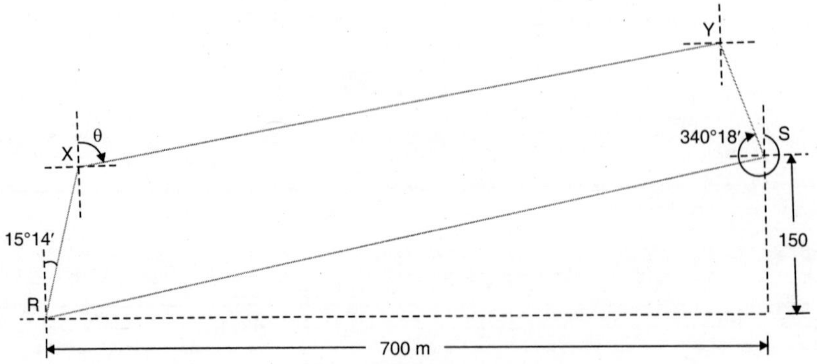

Fig. A-24

$\therefore \quad RX = 100 \times 1.5 \cos^2 8° 9' = 146.99$ m

$$V_1 = \frac{f}{i} s_1 \frac{\sin 2\theta_1}{2} = \frac{100 \times 1.5}{2} \sin 16° 18' = 21.05 \text{ m}$$

$\therefore$ R.L. of X = R.L. of R + H.I. + $V_1 - r_1$

= 1020.60 + 1.5 + 21.05 – 1.85 = 1041.30 m

Latitude of RX = 146.99 cos 15° 14′ = 141.83 m

Department of RX = 146.99 sin 15° 14′ = 38.62 m

$\therefore$ Easting of X = Easting of R + Departure of RX = 1800 + 38.62 = 1838.62 m

Northing of X = Northing of R + Latitude of RX = 800 + 141.83 = 941.83 m

(*b*) ***Observations from S to Y***

Horizontal distance $SY = \frac{f}{i} s_2 \cos^2 \theta_2 = 100\,(2.50 - 1.32) \cos^2 2° 3' = 117.85$ m

Also, $V_2 = \frac{f}{i} s_2 \frac{\sin 2\theta_2}{2} = 100\,(2.50 - 1.32) \frac{\sin 4° 6'}{2} = 4.22$ m

$\therefore$ R.L. of Y = R.L. of S + H.I. + $V_2 - r_2$ = 1021.21 + 1.53 + 4.22 – 1.91 = 1025.05

Latitude of SY = 117.85 cos (360° – 340° 18′) = 110.95 m

Departure of SY = – 117.85 sin (360° – 340° 18′) = – 39.73 m

$\therefore$ Easting of Y = Easting of S – Departure of SY = 2500 – 39.73 = 2460.27 m

Northing of Y = Northing of **S** – Latitude of SY = 950 + 110.95 = 1060.95 m

(*c*) ***Computations of line XY***

ΔN = Northing of Y – Northing of X = 1060.95 – 941.83 = 119.12 m

ΔE = Easting of Y – Easting of X = 2460.27 – 1838.62 = 621.65 m

$\therefore$ Distance $XY = \sqrt{\Delta N^2 + \Delta E^2} = \sqrt{(119.12)^2 + (621.65)^2}$ = **632.96** m

If θ is the R.B. of *XY*, we have $\theta = \tan^{-1}\dfrac{\Delta E}{\Delta N} = \tan^{-1}\dfrac{621.65}{119.12} = 79°.153 = \mathbf{79°9'9''}$

Gradient of $XY = \dfrac{\Delta h}{L} = \dfrac{1041.30 - 1025.05}{632.96} = \dfrac{1}{38.95}$, *i.e.* **1 in 38.95** (Falling)

Example A-25: *A closed traverse has the following lengths and bearings*

Line	*Length*	*Bearing*
AB	*200.0 m*	*Roughly East*
BC	*98.0 m*	*178°*
CD	*Not obtained*	*270°*
DA	*86.4 m*	*1°*

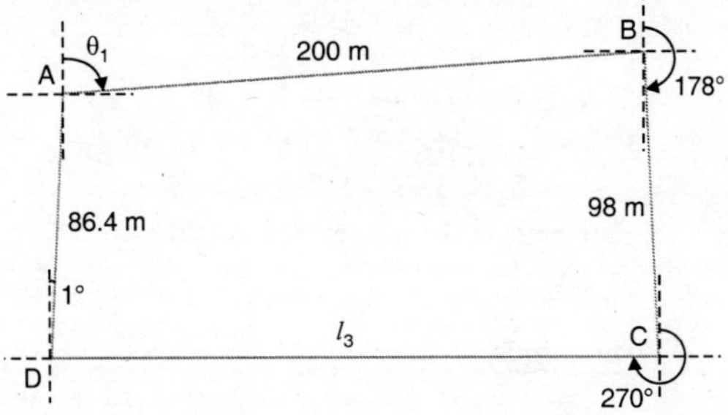

Fig. A-25

The length CD could not be measured due to some obstruction to chaining. The bearing of AB could not be taken as station A is badly affected by local attraction. Find the exact bearing of the side AB and calculate, the length CD. **(Engg. Services)**

Solution: The above question is based on example 8.3 of the book, with change in data.

Let us use suffixes 1, 2, 3 and 4 for lines *AB, BC, CD* and *DA* respectively. Thus the bearing θ, of line *AB* and length l_3 of line *CD* are unknowns. The computations for latitude (*L*) and departure (*D*) of each line are done in the tabular form below.

S.No.	*Line*	*Length (m)*	*Bearing*	*Latitude (L) m*	*Departure (D) (m)*
1	*AB*	200	Roughly east	$200 \cos \theta_1$	$200 \sin \theta_1$
2	*BC*	98	178°	– 97.94	3.420
3	*CD*	*h*	270°	0	$-l_3$
4	*DA*	86.4	1°	86.387	1.508
			Σ	$200 \cos \theta_1 - 11.553$	$200 \sin \theta_1 + 4.928 - l_3$

Since the traverse in closed, we have

$$\Sigma L = 200 \cos \theta_1 - 11.553 = 0 \text{ from which } \theta_1 = 86°.6885 = \mathbf{86°\ 41'}$$

Also, $$\Sigma D = 200 \sin \theta_1 + 4.923 - l_3 = 0$$

∴ $$l_3 = 200 \sin \theta_1 + 4.923 = 200 \sin 86°.6885 + 4.923 = \mathbf{204.59\ m}$$

Example A-26: *In order to determine the elevation of top Q of a signal on a hill, observations were made from two stations P and R. The stations P, R and Q were on the same plane.*

If the angles of elevation of the top Q of the signal measured at P and R were 25° 35′ and 15° 05′ respectively, determine the elevation of the foot of the signal if the height of the signal above its base was 4 m.

The staff readings upon the bench mark (RL 105.42) were respectively 2.755 and 3.855 m when the instrument was at P and at R. The distance between P and R was 120 m.

(Engg. Services)

Solution: Let D be the horizontal distance between the base of the signal and instrument at P.

From geometry,

$$h_1 = D \tan 25°\ 35' \text{ and}$$
$$h_2 = (120 + D) \tan 15°\ 5'$$

$$\therefore \quad h_1 - h_2 = D \tan 25°\ 35' - (120 + D) \tan 15°\ 5'$$

But $h_1 - h_2 = 3.855 - 2.755 = 1.1$ m

$\therefore D (\tan 25°\ 35' - \tan 15°\ 5') - 120 \tan 15°\ 5' = 1.1$

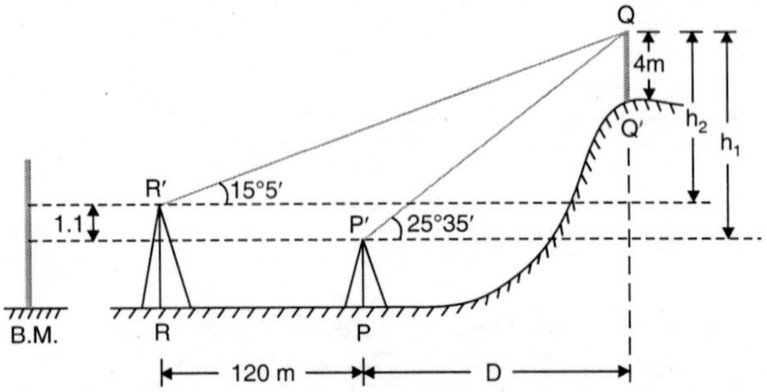

Fig. A-26

From which $\qquad D = \dfrac{1.1 + 120 \tan 15°\ 5'}{\tan 25°\ 35' - \tan 15°\ 5'} = 159.811$ m

Now $\qquad h_1 = D \tan 25°\ 35' = 159.811 \tan 25°\ 35' = 76.512$ m

$\therefore$ Elevation of $\qquad Q$ = Elev. of inst. axis at $P + h_1$

$= (105.42 + 2.755) + 76.512 = 184.687$ m

$\therefore$ Elevation of foot of signal $= 184.687 - 4.0 =$ **180.687 m**

Check: $\qquad h_2 = (b + D) \tan 15°\ 5' = (120 + 159.811) \tan 15°\ 5' = 75.411$

$\therefore$ Elevation of $\qquad Q = 105.42 + 3.855 + 75.411 = 184.686$

and Elevation of $\qquad Q' = 184.686 - 4 = 180.686$

Example A-27: *The following readings were noted in a closed traverse*

Line	*F.B.*	*B.B.*
AB	*32°*	*212°*
BC	*77°*	*262°*
CD	*112°*	*287°*
DE	*122°*	*302°*
EA	*265°*	*85°*

At which station do you suspect local attraction? Find correct bearings of lines. What will be the true fore bearings (as reduced bearings) of lines, if the magnetic declination was 12°W?

(Engg. Services)

Solution: From the given data, we observe that the difference between F.B. and B.B. of lines *AB, DE* and *EA* are 180°. Hence stations *A, B, D* and *E* are free from local attraction. Only station *C* suffer from local attraction.

Let us start with station *B* which is free from local attraction.

Hence bearing of BC = 77° which is correct.

Hence bearing of CB = 77° + 180° = 257°.

But observed bearing of CB = 262°

$\therefore \qquad$ Error at C = 262° – 257° = + 5°

and $\qquad$ Correction at C = – 5°

∴ Corrected bearing of $CD = 112° - 5° = 107°$

and corrected bearing of $DC = 107° + 180° = 287°$ = observed bearing of DC.

Also, True bearing = Magnetic bearing – declination = magnetic bearing – 12°.

The results are presented in the tabular form below.

Line	*F.B.*	*B.B.*	*Difference between F.B. and B.B.*	*Corrected Bearing*		*True Forebearing*
				F.B	*B.B.*	
AB	32°	212°	180°	32°	212°	N 20°E
BC	77°	262°	185°	77°	257°	N 65°E
CD	112°	287°	175°	107°	287°	S 85°E
DE	122°	302°	180°	122°	302°	S 70°E
EA	265°	85°	180°	265°	85°	S 73°W

Example A-28: *The following readings were taken with a level and a 4 m staff. Draw up a level book page and reduce the levels by*

(a) *the rise and fall method*

(b) *the height of collimation method*

0.683, 1.109, 1.838, 3.399, (3.877 and 0.451) C.P., 1.405, 1.896, 2.676 B.M. (31.126 A.O.D.), 3.478, (3.999 and 1.834) C.P., 0.649, 1.706

(c) *Highlight fundamental mistakes in the above levelling operation*

(d) *What error would occur in the final level if the staff has been wrongly extended and a plain gap of 0.012 has occurred at the 2 m section joint?* **(U.L.)**

Solution: (a) ***Booking by Rise and fall method***

Steps (*i*) The first reading is back sight while the next three readings are intermediate sights. The fifth reading a fore sight while the sixth reading is a back sight on a change point. Seventh to tenth readings are intermediate sights, including the one on the B.M. Eleventh readings is a fore sight and 12th reading is a back sight on the change point. 13th reading is an intermediate sight while the last reading is a fore sight. Enter these readings in appropriate columns.

(*ii*) Find rise and fall of each staff station.

(*iii*) Starting with the B.M., reduce levels *below* by normal method and *above* by reversing falls for rises and vice versa.

(*iv*) Apply normal checks.

B.S.	*I.S.*	*F.S.*	*Rise*	*Fall*	*R.L.*	*Remarks*
0.683					36.545	
	1.109			0.426	36.119	
	1.838			0.729	35.390	
	3.399			1.561	33.829	
0.451		3.877		0.478	33.351	Change point C.P. 1
	1.405			0.954	32.397	
	1.896			0.491	31.906	

	2.676			0.780	31.126	B.M. 31.126 A.O.D.
	3.478			0.802	30.324	
1.834		3.999		0.521	29.803	CP. 2
	0.649		1.185		30.988	
		1.706		1.057	29.931	
Sum 2.968		9.582	1.185	7.799	29.931	Checked
(–) 9.582			(–) 7.799		(–) 36.545	
– 6.614			– 6.614		– 6.614	

(*b*) ***Booking by height of collimation method***

Steps (*i*) Book all the readings in appropriate columns, as explained in (*ai*) above (*ii*) Height of collimation for second setting = R.L. of B.M. + I.S. reading on B.M.

$$= 31.126 + 2.676 = 33.802$$

(*iii*) R.L. of C.P.1 = H.I. – B.S. on C.P. 1 = 33.802 – 0.451 = 33.351

(*iv*) Height of collimation of first setting = R.L. of C.P.1 + F.S. on C.P.1

$$= 33.351 + 3.877 = 37.228$$

(*v*) R.L. of C.P.2 = H.I. in second setting – F.S. on C.P.2

$$= 33.802 – 3.999 = 29.830$$

(*vi*) Height of collimation of third setting = R.L. of CP2 + B.S. on C.P.2

$$= 29.830 + 1.834 = 31.637$$

(*vii*) Thus height of collimation of all the three settings of the level are known. The R.L.'s of first point, intermediate sights and last point can be computed as usual.

(*viii*) Apply the normal checks.

B.S.	***I.S.***	***F.S.***	***Height of collimation (or H.I.)***	***R.L.***	***Remarks***
0.683			37.228	36.545	
	1.109			36.119	
	1.838			35.390	
	3.399			33.826	
0.451		3.877	33.802	33.351	C.P. 1
	1.405			32.397	
	1.896			31.906	
	2.676			31.126	B.M. 31.126 m A.O.D.
	3.478			30.324	
1.834		3.999	31.637	29.803	C.P.2
	0.649			30.988	
		1.706		29.931	
Sum: 2.968		9.582		29.931	Checked
(–) 9.582				(–) 36.545	
– 6.614				– 6.614	

(c) ***Fundamental levelling mistakes***

The question high lights three fundamental levelling mistakes

(*i*) The most important sight on the B.M. should not be an intermediate sight, as this can not be checked.

(*ii*) The staff has not been correctly assembled, with the result that all the readings above 2 m are wrong.

(*iii*) Since there is no *circuit closure,* there is no check on field work.

(*d*) ***Error due to wrong extension of staff***

All readings greater than 2 m will be 0.012 mm too small. However, the final level value will be affected only by B.S. and F.S. reading *after* the R.L. of datum, *i.e.* after 31.126, though I.S. on B.M. will be treated as B.S. for booking purposes.

(*i*) B.S. (I.S.) of 2.676 should be 2.664
B.S. of 1.834 will remain as 1.834
sum 4.498

(*ii*) F.S. of 3.999 should be 3.987
F.S. of 1.706 will remain as 1.706
sum 5.693

Difference: B.S. – F.S. = 4.498 – 5.693 = – 1.195

∴ R.L. of last point = 31.126 – 1.195 = 29.931

Existing R.L. of last point, with faulty staff reading = 29.931

Hence the B.S. and F.S. are affected in the same manner and the final value is not altered.

Example A-29: *The following readings were observed with a level*

1.143 (B.M. 34.223), 1.765, 2.566, 3.819 (CP.), 1.390, 2.262, 0.664, 0.433 (CP), 3.722, 2.886, 1.618, 0.616 (T.B.M. value thought to be 35.290 m).

(*a*) *Reduce the levels by rise and fall method*

(*b*) *Calculate the level of the T.B.M. if the line of collimation was tilted upwards at an angle of 6 min. and each back sight length was 90 m and the foresight length 30 m.*

(*c*) *Calculate the level of the T.B.M. if the staff was not held upright but leaning backwards at 5° to the vertical in all cases.* ***(U.L.)***

Solution:

(*a*) ***Reduction of levels by rise and fall method:*** See Table below

B.S.	I.S.	F.S.	Rise	Fall	R.L. Remarks	
1.143					34.223	B.M. 34.223
	1.765			0.622	33.601	
	2.566			0.801	32.800	
1.390		3.819		1.253	31.547	C.P.
	2.262			0.872	30.675	
	0.664		1.598		32.273	
3.722		0.433	0.231		32.504	C.P.
	2.886		0.836		33.340	
	1.618		1.268		34.608	
		0.616	1.002		35.610	T.B.M. 35.290
6.225		4.868	4.935	3.548	35.610	Checked
(–) 4.868			(–) 3.548		(–) 34.223	
1.387			1.387		1.387	

(*b*) ***Effect of tilting of line of collimation*** (See Fig. A-27 (*a*))

$$\text{Error} = e = 30.0 \text{ m} \left(\frac{\pi}{180 \times 60} \times 6 \right) = 0.0524 \text{ m per 30 m}$$

If *b* and *f* are back sight and for sight readings, true difference in level per set-up

$$= (b - 3e) - (f - e) = (b - f) - 2e$$

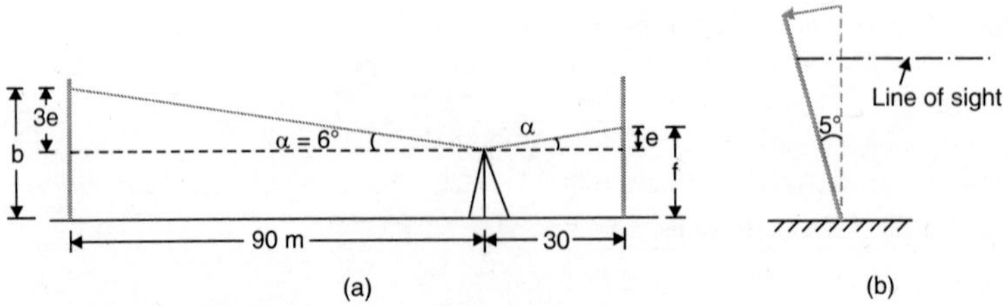

Fig. A-27

$$\text{Total length of B.S.'s} = 3 \times 90 = 270 \text{ m}$$

$$\text{Total length of F.S.'s} = 3 \times 30 = 90 \text{ m}$$

$\therefore$ Effective difference in length = 3 × 60 = 180 m

$$\therefore \quad \text{Error} = \frac{0.0524}{30} \times 180 = 0.314 \text{ m}$$

Hence sum of B.S. is effectively too large by 0.314 m.

$\therefore$ True difference in level = 1.387 – 0.314 = 1.073

$\therefore$ R.L. of T.B.M. = 34.223 + 1.073 = **35.296 m**

[Check: 35.610 – 0.314 = 35.296 m]

(*c*) ***Effect of tilting of staff*** (See Fig. A-27 *b*)

If the staff is tilted, all the readings will be too large.

True reading = observed reading × cos 5°

Apparent difference in level = Σ B.S. – Σ F.S. = 1.387

True difference in level = (Σ B.S.) cos 5° – (Σ F.S.) cos 5°

= (Σ B.S. – Σ F.S.) cos 5° = 1.387 cos 5° = 1.382

$\therefore$ R.L. of T.B.M. = 34.223 + 1.382 = **35.605 m**

Example A-30: *The following observations were taken during the testing of a dumpy level.*

Instrument at	*Staff reading on*	
	A	*B*
A	*1.275*	*2.005*
B	*1.040*	*1.660*

Is the instruments in adjustment? To what reading should the line of collimation be adjusted when the instrument was at B? ***(U.P.S.C. Engg. Services Exam.)***

Solution: When the level is at A, apparent difference is level = 2.005 – 1.275 = 0.73, A being higher. When the level is at B, apparent difference in level = 1.660 – 1.040 = 0.62, A being higher. Since both these values are not equal, *the instrument is not in adjustment.*

$$\text{True difference in level} = \frac{0.73 + 0.62}{2} = 0.675 \text{ m}$$

When the level is at B, the line of collimation should adjust to read on A = 1.660 – 0.675 = **0.985 m.**

Example A-31: *The following readings have been taken from a page of an old level book. It is required to reconstruct the page. Fill up the missing quantities and apply the usual checks. Also, calculate the corrected level of the TBM if the instrument is known to have an elevated collimation error of 30″ and backsight, foresight distances averaged 40 m and 90 m respectively.*

Point	*B.S.*	*I.S.*	*F.S.*	*Rise*	*Fall*	*R.L.*	*Remarks*
1	3.125					?	B.M.
2	?		?	1.325		125.505	T.P.
3		2.320			0.055		
4		?				125.850	
5	?		2.655				T.P.
6	1.620		3.205		2.165		T.P.
7		3.625					
8			?			123.090	T.B.M.

(U.P.S.C. Engg. Services Exam.)

Solution: The solution is done is the following steps.

1. F.S. of point 2 = B.S. of point 1 – Rise of point 2 = 3.125 – 1.325 = 1.800
2. R.L. of point 1 = R.L. of point 2 – Rise of point 2 = 125.505 – 1.325 = 124.180
3. B.S. of station 2 = I.S. of point 3 – Fall of point 3 = 2.320 – 0.055 = 2.265
4. R.L. of point 3 = R.L. of point 2 – Fall of point 3 = 125.505 – 0.055 = 125.450
5. Rise of point 4 = R.L. of point 4 – R.L. of point 3 = 125.850 – 125.450 = 0.400
6. I.S. of point 4 = I.S. of point 3 – Rise of point 3 = 2.320 – 0.400 = 1.920
7. Fall of point 5 = F.S. of point 5 – I.S. of point 4 = 2.655 – 1.920 = 0.735
8. R.L. of point 5 = R.L. of point 4 – Fall of point 5 = 125.850 – 0.735 = 125.115
9. B.S. of point 5 = F.S. of point 6 – Fall of point 6 = 3.205 – 2.165 = 1.040
10. R.L. of point 6 = R.L. of point 5 – Fall of point 6 = 125.115 – 2.165 = 122.950
11. Fall of point 7 = I.S. of point 7 – B.S. of point 6 = 3.625 – 1.620 = 2.005
12. R.L. of point 7 = R.L. of point 6 – Fall of point 7 = 122.950 – 2.005 = 120.945
13. Rise of point 8 = R.L. of point 8 – R.L. of point 7 = 123.090 – 120.945 = 2.145
14. F.S. of point 8 = I.S. of point 7 – Rise of point 8 = 3.625 – 2.145 = 1.480

The computations are arranged in tabular form below along with the missing quantities *underlined.*

Point	B.S.	I.S.	F.S.	Rise	Fall	R.L.	Remarks
1	3.125					124.180	B.M.
2	2.265		1.800	1.325		125.505	T.P.
3		2.320			0.055	125.450	
4		1.920		0.400		125.850	
5	1.040		2.655		0.735	125.115	T.P.
6	1.620		3.205		2.165	122.950	T.P.
7		3.625			2.005	120.945	
8			1.480	2.145		123.090	T.B.M.
Sum	8.050		9.140	3.870	4.960		

Arithmetic checks

$$\Sigma \text{ B.S.} - \Sigma \text{ F.S.} = \Sigma \text{ Rise} - \Sigma \text{ Fall} = \text{Last R.L.} - \text{First R.L.}$$

or $$8.050 - 9.140 = 3.870 - 4.960 = 123.090 - 124.180 = -1.090 \text{ (checked)}$$

Value of corrected T.B.M.

Since the collimation line is elevated, each backsight and foresight reading will be too great.

Error is each backsight reading = 40 tan 30″

Error is each foresight reading= 90 tan 30″

∴ Difference is errors of one set of B.S. and F.S. readings = 50 tan 30″ = 0.00727

Since there are four sets of readings, total error = 4 × 0.00727 = 0.029 m

Treating the B.S. readings to be correct, relative increase in sum of the F.S. readings = 0.029 m

∴ Corrected sum of F.S. readings = 9.140 – 0.029 = 9.111

∴ Corrected difference in the level of B.M. and T.B.M. = 9.111 – 8.050 = 1.061

∴ Corrected R.L. of T.B.M. = 124.180 – 1.061 = **123.119 m**

Example A-32: *The following consecutive readings were taken with a level and 5 metre levelling staff an continuously slopping ground at a common interval of 25 metres: 0.450, 1.120, 1.875, 2.905, 3.685, 4.500, 0.520, 2.150, 3.205 and 4.485*

Given: The reduced level of the change point was 250.000

Rule out a page of level field book and enter the above readings.

Calculate the reduced levels of the points by rise and fall method and also the gradient of the line joining the first and the last point. ***(U.P.S.C. Asst. Engineers C.P.W.D. Exam.)***

Solution:

S.No.	B.S.	I.S.	F.S.	Rise	Fall	R.L.	Remarks
1	0.450					254.050	
2		1.120			0.670	253.380	
3		1.875			0.755	252.625	
4		2.905			1.030	251.595	
5		3.685			0.780	250.815	
6	0.520		4.500		0.815	250.00	Change point
7		2.150			1.630	248.370	
8		3.205			1.055	247.315	
9			4.485		1.280	246.035	
Σ	0.970		8.985	0.000	8.015		

Arithmetic checks

$$\Sigma \text{ B.S.} - \Sigma \text{ F.S.} = \Sigma \text{ Rise} - \Sigma \text{ Fall} = \text{R.L. of last point} - \text{R.L. of first point}$$

or $$0.970 - 8.985 = 0.000 - 8.015 = 246.035 - 254.050$$

or $$-8.015 = -8.015 = -8.015 \quad \textbf{\textit{(checked)}}$$

$$\text{Gradient of line} = \frac{254.050 - 246.035}{25 \times 8} = \frac{1}{\mathbf{24.95}}$$

Example A-33: *In levelling between two points A and B on opposite banks of river, the level was set up near A and the staff readings on A and B were 1.570 and 2.875 respectively. The level was then moved and set up near B and the respective staff readings on B and A were 2.055 and 0.850. Find the difference of level between A and B.*

(U.P.S.C., C.P.W.D. Asst. Engineers Exam.)

Solution:

Instrument near A: Apparent difference in level between A and B = 2.875 – 1.570 = 1.305 m, A being higher.

Instrument near B: Apparent difference in level between A and B = 2.055 – 0.805 = 1.205 m, A being higher.

$$\therefore \quad \text{True difference in level between } A \text{ and } B = \frac{1.305 + 1.905}{2} = \mathbf{1.255 \text{ m},} \text{ A being higher.}$$

Example A-34: *Determine the reduced level of a church spire at C from the following observations taken from two stations A and B, 50 m apart.*

Angle BAC = 60° and angle ABC = 50°

Angle of elevation from A to the top of spire = 30°

Angle of elevation from B to the top of spire = 29°

Staff reading from A on bench mark of reduced level 25.00 = 2.500 m

Staff reading from B on the same bench mark = 0.50 m

(Engg. Services)

Solution: Let C be the church spire (Fig. A-28)

From triangle ACB, $\angle ACB = 180° - (60° + 50°) = 70°$

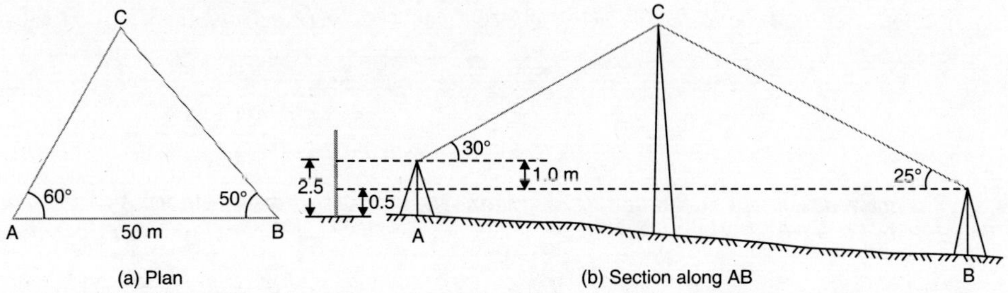

Fig. A-28

$\therefore \quad AC = \dfrac{AB}{\sin 70^\circ} \times \sin 50^\circ = \dfrac{50}{\sin 70^\circ} \sin 50^\circ = 40.76$ m; $BC = \dfrac{50}{\sin 70^\circ} \sin 60^\circ = 46.08$ m

(*a*) ***Observations from A to C***

$$\text{R.L. of } C = \text{R.L. of B.M.} + \text{B.S. reading} + AC \tan 30^\circ$$
$$= 25.00 + 2.50 + 40.76 \tan 30^\circ = 51.033 \text{ m}$$

(*b*) ***Observations from B to C***

$$\text{R.L. of } C = \text{R.L. of B.M.} + \text{B.S. reading} + BC \tan 29^\circ$$
$$= 25.00 + 0.50 + 46.08 \tan 29^\circ = 51.043 \text{ m}$$

$$\therefore \quad \text{Average elevation of } C = \frac{51.033 + 51.043}{2} = \mathbf{51.038 \text{ m}}$$

Example A-35: *A railway embankment is 16 m wide with side slopes 2 to 1. Assume the ground to be level in direction transverse to the centre line. Calculate the volume contained in a length of 100 m, the centre height at 20 m intervals being in m: 2.0, 4.5, 4.0, 3.5, 2.5, 1.5. Use trapezoidal rule.* ***(U.P.S.C. Engg. Services Exam.)***

Solution: Given $b = 16$ m ; $n = 2$

$$A = (b + nh)h$$
$$A_1 = (16 + 2 \times 2)\ 2 = 40 \text{ m}^2$$
$$A_2 = (16 + 2 \times 4.5)\ 4.5 = 112.5 \text{ m}^2$$
$$A_3 = (16 + 2 \times 4)\ 4.0 = 96 \text{ m}^2$$
$$A_4 = (16 + 2 \times 3.5)\ 3.5 = 80.5 \text{ m}^2$$
$$A_5 = (16 + 2 \times 2.5)\ 2.5 = 52.5 \text{ m}^2$$
$$A_6 = (16 + 2 \times 1.5)\ 1.5 = 28.5$$

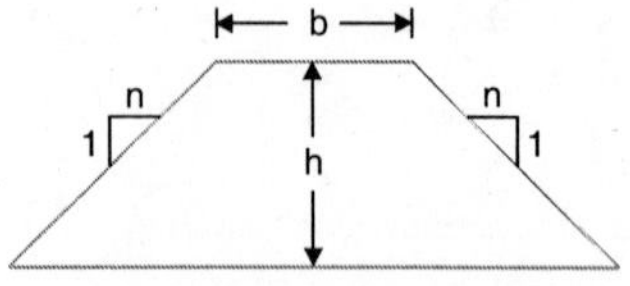

Fig. A-29

Volume, from trapezoidal formula, is given by Eq. 13.23

$$V = d\left[\frac{A_1 + A_n}{2} + A_2 + A_3 + \ldots\ldots A_{n-1}\right] \quad \ldots(13.23)$$

$$= 20\left[\frac{40 + 28.5}{2} + 112.5 + 96 + 80.5 + 52.5\right] = \mathbf{7515 \text{ m}^3}$$

Example A-36: *To determine the gradient between two points A and B, a tacheometer was set up at another station C and the following observations were taken keeping the staff vertical.*

Staff at	*Vertical angle*	*Stadia Readings*
A	+ 4° 20′0″	1.300, 1.610, 1.920
B	+ 0° 10′40″	1.100, 1.410, 1.720

If the horizontal angle ACB is 35° 20′, determine the average gradient between A and B. Take K = 100 and C = 0.0 ***(Engg. Services)***

Solution:

(*a*) ***Observations from C to A:*** $s = 1.920 - 1.300 = 0.620$ m

$$D = Ks \cos^2 \theta + C \cos \theta = 100 \times 0.620 \cos^2 (4^\circ\ 20'\ 00'')$$

≈ 61.65 m

$$V = Ks\,\frac{\sin 2\theta}{2} = 100 \times 0.620\,\frac{\sin 8^\circ\, 40'}{2}$$

$= 4.671$ m

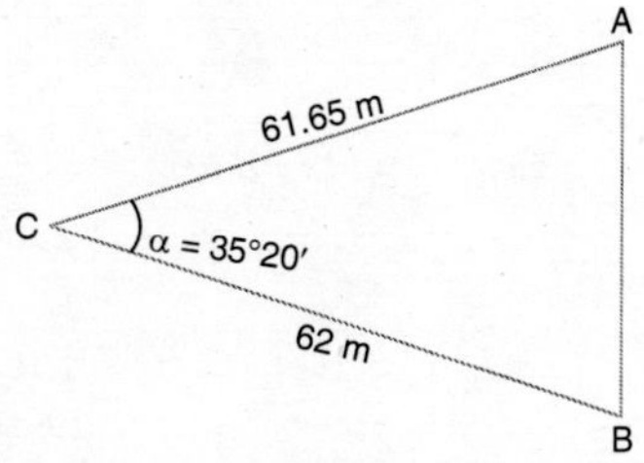

Fig. A-30

∴ Difference in level between A and C

$= 4.671 - 1.610 = 3.061$ (A being higher)

(*b*) ***Observations from C to B:***

$$s = 1.720 - 1.100 = 0.620 \text{ m}$$

$$\therefore \quad D = 100 \times 0.620 \cos^2 (0^\circ\, 10'\, 40'') \approx 62 \text{ m}$$

$$V = 100 \times 0.60\, \frac{\sin 0'\, 21'\, 20''}{2} = 0.186 \text{ m}$$

∴ Difference in level between B and $C = 0.186 - 1.410 = -1.224$ m (B being lower)

(*c*) ***Distance AB and gradient from A to B***

Fig. A-30 shows the plan, in which $\angle ACB = \alpha = 35^\circ\, 20'$, $AC = 61.6$ m and $BC = 62$ m.

By cosine formula, $AB^2 = c^2 = a^2 + b^2 - 2ab \cos C$

or $AB^2 = (62)^2 + (61.65)^2 - 2 \times 62 \times 61.65 \cos 35^\circ\, 20'$

From which $AB = 37.53$ m

Difference in elevation between A and $B = 3.061 - (-1.224) = 4.285$ m

$$\therefore \quad \text{Gradient from } A \text{ to } B = \frac{4.285}{37.53} = \frac{1}{8.758}$$ (*i.e.* **1 in 8.758** falling)

Example A-37: *An observer standing on the deck of a ship just sees a light house. The top of the light house in 49 m above the sea level and the height of observer's eye is 9 on above the sea level. Find the distance of the obscrver from the light house.*

(U.P.S.C. Engg. Services Exam.)

Solution: Refer Example 9.12 and Fig. 9.40.

Let A be the position of the top of light house and B be the position of observer's eye. Let AB be tangential to water surface at O.

The distances d_1 and d_2 are given by

$$d_1 = 3.8553 \sqrt{C_1} \text{ km} = 3.8553 \sqrt{49} = 26.987 \text{ km}$$

and

$$d_2 = 3.8553 \sqrt{C_2} \text{ km} = 3.8553 \sqrt{9} = 11.566 \text{ km}$$

∴ Distance between A and $B = d_1 + d_2 = 26.987 + 11.566 =$ **38.553 km**

Example A-38: *The following observations were made in running fly levels from a bench mark of RL 60.65*

Back sight : 0.964, 1.632, 1.105, 0.850

Fore sight : 0.948, 1.153, 1.984

Five pages at 20 m interval are to be set on falling gradient of 1 in 100 m, from the last position of the instrument. The first peg is to be at RL 60.

Work out the staff readings required for setting the pegs and prepare the page of the level book.

(U.P.S.C. Engg. Services Exam.)

Solution:

S.No.	Distance	B.S.	I.S.	F.S.	H.I.	R.L.	Remarks
1		0.964			61.614	60.650	
2		1.632		0.948	62.298	60.666	
3		1.105		1.153	62.250	61.145	
4		0.850		1.984	61.116	60.266	
5	0		1.116			60.000	Peg 1
6	20		1.316			59.800	Peg 2
7	40		1.516			59.600	Peg 3
8	60		1.716			59.400	Peg 4
9	80			1.916		59.200	Peg 5
Check		Σ	4.551	6.001		60.650	
				– 4.551		– 59.200	
				1.450		1.450	Checked

Example A-39: *The field level book readings from a fly level are as follows:*

Staff station	*R.L.*	*B.S.*	*F.S.*	*Remarks*
B.M.-1	*100.000*	*3.635*	–	
A	×	×	*2.375*	
B	*104.150*	*4.220*	*1.030*	
C	*106.650*	*3.990*	×	
B.M.-2	*108.00*	–	×	

Find out the missing values marked (×) and perform the arithmetic check.

(Engg. Services)

Solution: The solution is done in the tabular form below:

Staff Station	*B.S.*	*F.S.*	*Height of Instrument*	*R.L.*	*Remarks*
B.M.-l	3.635		103.635	100.00	B.M.-1
A	(×) 3.92	2.375	105.180	(×) 101.26	C.P.
B	4.220	1.030	108.370	104.150	C.P.
C	3.990	(×) 1.72	110.64	106.650	C.P.
B.M.-2		(×) 2.64		108.00	B.M.-2
Σ	15.765	7.765			

Steps:

1. Height of instrument in the first setting = 100.000 + 3.635 = 103.635
2. R.L. of *A* – H.I. – F.S. = 103.635 – 2.375 = 101.26 m
3. H.I. for second setting = R.L. of B + F.S. on *B* = 104.150 + 1.030 = 105.18

4. B.S. on A = H.I. – R.L. of A = 105.18 – 101.26 = 3.92
5. H.I. for third setting = R.L. of B. + B.S. on B = 104.150 + 4.220 = 108.370
6. F.S. on C = H.I. in third setting – R.L. of C = 108.370 – 106.650 = 1.72
7. H.I. in the 4th setting = R.L. of C + B.S. on C = 106.650 + 3.990 = 110.64
8. F.S. on BM 2 = H.I. in 4th setting – R.L. of B.M. 2 = 110.64 – 108.00 = 2.64

Check: Σ B.S. – Σ F.S. = 15.765 – 7.765 = 8.0 = Last R.L. – First R.L. = 108.00 – 100.00

Example A-40: *Levelling was done between stations A and F, starting with backsight at A. Various backsights taken were in the following sequence: 2.3, 2.3, – 1.6 and X. The sum of all the fore sights was found to be 3.00. Also, it was known that F is 0.6 m higher than A. Find the value of X. How many foresights do you expect?*

Solution: Since F is 0.6 m higher that A,

We have: R.L. of F – R.L. of A = *0.6* m ...(1)

Also, we have Σ B.S. – Σ F.S. = Last R.L. – First R.L.

$\therefore$ Σ B.S. – Σ F.S. = 0.6 m

$\therefore$ Σ B.S. = 0.6 + Σ F.S. = 0.6 + 3.0 = 3.6 ...(2)

But Σ B.S. = 2.3 + 2.3 + (– 1.6) + X = 3.0 + X

$\therefore$ 3.0 + X = 3.6

or X = 3.6 – 3.0 = **0.6 m**

Since each instrument setting consists of one B.S. and one F.S., the number of foresights are always equal to number of backsights. Hence number of foresights = 4.

Example A-41: *The readings below were obtained from an instrument station B using an anallatic tacheometer having the following constants: focal length of the object glass 203 mm, focal length of anallatic lens 114 mm, distance between object glass and anallatic lens 178 mm, spacings of outer cross-hairs 1.664 mm.*

Instrument at	*Height of instrument*	*To*	*Bearing*	*Vertical angle*	*Stadia readings*
B	1.503 m	A	69° 30′ 00″	+ 5° 00′ 00″	0.658/1.055/1.451
		C	159° 30′ 00″	0° 0′ 00″	2.231/2.847/3.463

The staff was held vertical for both observations.

Bore holes were sunk at A, B and C to expose a plane bed of rock, the ground surface being respectively 11.918 m, 10.266 m and 5.624 m above the rock plane. Given that the reduced level of B was 36.582 m, determine the line of steepest rock slope relative to the direction AB.

(U.L.)

Solution:

(*a*) ***Determination of multiplying constant***

Given: f = 203 mm; f' = 114 mm; n = 178 mm ; i = 1.664 mm

The multiplying constant k is given by:

$$k = \frac{ff'}{(f + f' - n)\,i} = \frac{203 \times 114}{(203 + 114 - 178)\,1.664} = 100.05 \approx \mathbf{100}$$

(b) ***Observations to A:***

$$\theta = 5° \, 00' \, 00''; \; s = 1.451 - 0.658 = 0.793 \text{ m}$$

∴ Horizontal Distance $BA = ks \cos^2 \theta = 100 \times 0.793 \cos^2 5° \, 00' \, 00'' = 78.698$ m

$$V_A = ks \frac{\sin 2\theta}{2} = \frac{100 \times 0.793}{2} \sin 10° \, 00' \, 00'' = 6.885 \text{ m}$$

∴ R.L. of $A = 36.582 + 1.503 + 6.885 - 1.055 = 43.915$ m

(c) ***Observation to C:***

$$\theta = 0° \, 00' \, 00''; \; s = 3.463 - 2.231 = 1.232 \text{ m}$$

∴ Horizontal distance $BC = k \, s \cos^2 \theta = 100 \times 1.232 \cos^2 0° = 123.20$ m

$$V_B = ks \frac{\sin 2\theta}{2} = 0$$

∴ R.L. of $C = 36.582 + 1.503 + 0 - 2.847 = 35.238$ m

(d) ***Determination of line of steepest rock slope:*** Refer Fig. A-31.

Let us first find the levels of rock at A, B and C.

At A, G.L. = 43.915 ; Depth of rock = 11.918 m

∴ Rock level at $A = 43.915 - 11.918 = 31.997$ m

At B: G.L. = 36.582 m; Depth of rock = 10.266 m

∴ Rock level at $B = 36.582 - 10.266 = 26.316$ m

At C: G.L. = 35.238; Rock depth = 5.624 m

∴ Rock level = 35.238 − 5.624 = 29.614

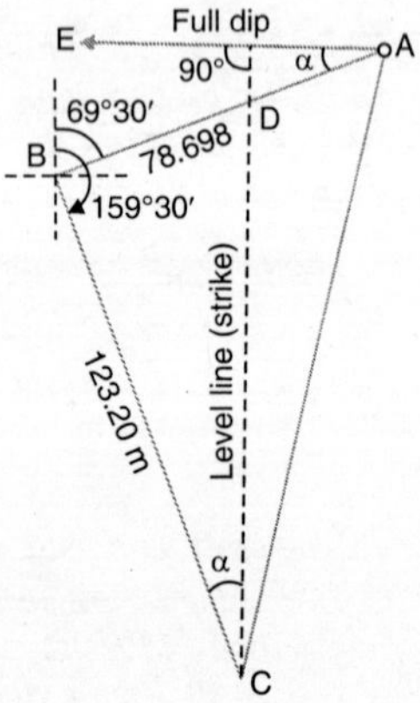

Fig. A-31

$$\text{Gradient of rock along } AB = \frac{31.997 - 26.316}{78.698} = \frac{1}{13.853}$$

Let D the point on AB where the rock level is equal to the rock level at C (*i.e.* 29.614 m)

$$\therefore \text{ Length } AD = (31.997 - 29.614) \times \frac{13.853}{1} = 33.012 \text{ m}$$

∴ Length $BD = 78.698 - 33.012 = 45.686$ m

Line CD is thus a *level line* (or *strike*) *i.e.* a line of zero slope. Hence the line of steepest slope (or *full dip*) *i.e.* line AE, will be at 90° to CD.

Now Let $\angle BCD = \alpha$; $\angle ABC = 159° \, 30' \, 00'' - 69° \, 30' \, 00'' = 90°$

Hence the line AE of the steepest slope is also inclined at angle α to AB.

Now, from triangle ABC,

$$\alpha = \tan^{-1} \frac{BD}{BC} = \tan^{-1} \frac{45.686}{123.20} = 20°21'$$

∴ Bearing of full dip = bearing of AE = bearing of $AB + \alpha$

$$= (69° \, 30' + 180°) + 20° \, 21' = \mathbf{269° \, 51'}$$

INDEX